# Enhancing Crop Genepool Use

## Capturing Wild Relative and Landrace Diversity for Crop Improvement

# Enhancing Crop Genepool Use

## Capturing Wild Relative and Landrace Diversity for Crop Improvement


*Edited by*

**Nigel Maxted**

*School of Biosciences*
*The University of Birmingham*
*Edgbaston*
*Birmingham*
*UK*

**M. Ehsan Dulloo**

*Bioversity International*
*Via dei Tre Denari 472/a*
*00057 Maccarese Rome*
*Italy*

*and*

**Brian V. Ford-Lloyd**

*School of Biosciences*
*The University of Birmingham*
*Edgbaston*
*Birmingham*
*UK*



www.cabi.org

**CABI is a trading name of CAB International**

CABI
Nosworthy Way
Wallingford
Oxfordshire OX10 8DE
UK

Tel: +44 (0)1491 832111
Fax: +44 (0)1491 833508
E-mail: info@cabi.org
Website: www.cabi.org

CABI
745 Atlantic Avenue
8th Floor
Boston, MA 02111
USA

Tel: +1 (0)617 682 9015
E-mail: cabi-nao@cabi.org



A catalogue record for this book is available from the British Library, London, UK.

**Library of Congress Cataloging-in-Publication Data**

Names: Maxted, Nigel, editor. | Dulloo, Mohammad Ehsan, editor. | Ford-Lloyd, Brian, editor.
Title: Enhancing crop genepool use : capturing wild relative and landrace diversity for crop improvement / edited by Nigel Maxted, M. Ehsan Dulloo, Brian V. Ford-Lloyd.
Description: Boston, MA : CABI, [2016] | Includes bibliographical references and index.
Identifiers: LCCN 2015039033 | ISBN 9781780646138 (h : alk. paper)
Subjects: LCSH: Crops--Genetics. | Crop improvement. | Germplasm resources, Plant.
Classification: LCC SB106.G46 E54 2016 | DDC 631.5/233--dc23 LC record available at http://lccn.loc.gov/2015039033

ISBN-13: 978 1 78064 613 8

Commissioning editor: Rachael Russell
Editorial assistant: Emma McCann
Production editor: Tim Kapp

Typeset by SPi, Pondicherry, India
Printed and bound in the UK by CPI Group (UK) Ltd, Croydon, CR0 4YY

# Contents

# Dedication

### J. Trevor Williams (1938–2015): Germplasm conservation pioneer

Trevor Williams, Executive Secretary and first Director of the International Board for Plant Genetic Resources (IBPGR) from 1978 to 1990, passed away at the age of 76 on 30 March 2015 after a long illness.

It is appropriate to recognize that Trevor's contributions to genetic conservation and use can be seen today in the PGR Secure project conference themes covered by this book: characterization techniques; conservation strategies; facilitating crop wild relative and landrace use; and informatics development. The directions that Trevor set for IBPGR almost 40 years ago encouraged, among others, international collaboration, sponsored collecting, conservation, and research for better conservation and use. Furthermore, Trevor served as Executive Secretary of the European Cooperative Programme for Plant Genetic Resources (ECP/GR) from 1985.

Born in June 1938, Trevor developed a lifelong interest in plants from an early age. During the Second World War, he was evacuated from the family home in Cheshire, UK, to live with his grandparents in the Isle of Man, and there he learned to name many of the plants in the local flora. Although he initially went up to Cambridge University in 1956 to study medicine, he transferred to botany after one year, and graduated MA in 1959. He was awarded his PhD in 1962 for a thesis on the weed biology of *Chenopodium* under the supervision of population biologist Professor John Harper at the University of Wales, Bangor (now Bangor University). Joining the Geobotanical Institute at ETH Zurich in 1964, he carried out ecological research on fertilized wet meadows in Switzerland, southern Germany and France with phytogeographer Heinz Ellenberg, and was awarded a Doctor of Natural Sciences degree in 1968.

Between Bangor and ETH, he taught briefly at Goldsmiths College, University of London, then at Lanchester Polytechnic (now Coventry University) on his return to the UK in 1968. In September 1969, Trevor first became directly involved with genetic resources when he joined the Department of Botany at the University of Birmingham. He was appointed Lecturer in Botany and Course Tutor for the recently established MSc course on Conservation and Utilization of Plant Genetic Resources. Trevor worked alongside department head Professor Jack Hawkes, the renowned genetic resources scientist and potato expert, to develop the MSc course that successfully provided training in the theory and practice of genetic conservation and use for more than three decades. An inspirational mentor and teacher, Trevor encouraged many students to follow a career in genetic resources conservation, and indeed many became heads of their own national programmes.

After 7 years, Rome beckoned and he was seconded from Birmingham for 2 years to join the Food and Agriculture Organization (FAO) as Senior Genetic Resources Officer in the Crop Ecology and Genetic Resources Unit, home since 1974 to IBPGR. In 1978, he resigned from Birmingham when he was appointed the Executive Secretary of IBPGR.

In 1974, when IBPGR began operations, there was just a handful of genebanks or *ex situ* collections across the globe where germplasm could be stored safely. Furthermore, few germplasm collecting missions had been made to collect germplasm systematically from farmers' fields before it was lost forever. Under Trevor's dynamic leadership at IBPGR, many countries were helped to set up genetic resources programmes where none had previously existed. In turn, as genetic resources conservation became widely recognized as an important component of enhanced food security, demands for training and research on different conservation approaches increased, as did demand for technical information and assistance (both financial and technical) to collect and conserve crop genetic resources. IBPGR sponsored over 800 collecting trips in 100 or more countries, with the result that almost 211,000 germplasm samples were placed in genebanks around the world. This was a prudent move, as subsequent events have borne out.

Early on, IBPGR promoted international collaboration between crop experts to form crop advisory committees and working groups to develop collecting and conservation guidelines, especially for difficult-to-conserve germplasm. IBPGR published descriptor lists for more than 70 crops, as well as ecogeographic studies of crop genepools, practical manuals and handbooks in several languages. The application of information technology and databases to manage germplasm collections was another initiative that prospered under Trevor's direction. IBPGR regional programmes in Latin America, South and South-east Asia, and in West and East Africa, provided its important outreach capacity to national germplasm systems. IBPGR also established a Seed Handling Unit at Kew-Wakehurst Place in 1988 (followed, in subsequent years, by units at the National University of Singapore and in Turrialba, Costa Rica) to process seeds from collecting missions sponsored by IBPGR and prepare them for long-term conservation in a designated genebank.

Trevor championed the development of a research capability within IBPGR. By the end of the 1980s, staff had been recruited and research networks initiated on genetic diversity, seed physiology, germplasm health and *in vitro* conservation for crops that could not be stored as seeds, in partnership with national scientists around the world.

Leaving IBPGR in 1990, Trevor's legacy was already established. There were hundreds of genebanks or *ex situ* collections around the world, as well as the national capacity and technical expertise to manage these valuable resources. Furthermore, IBPGR was a very different organization from that set up in 1974. Its own research capability had grown significantly, and in 1991 it became the International Plant Genetic Resources Institute (IPGRI), independent of FAO, and from 2006, Bioversity International. As a former colleague comments, 'Trevor enjoyed a risk and it was his ability to take a chance that I'm convinced enabled IBPGR to grow and for many of us to have had the careers or experiences that we did have. There would never have been an IPGRI, Bioversity [International] or [Global Crop Diversity] Trust if it wasn't for Trevor.' One might also say the same for projects like PGR Secure that benefit from the early IBPGR support to genetic resources conservation and use.

After IBPGR, Trevor moved to Washington, DC, to become Director of the International Fund for Agricultural Research (IFAR) and the IBPGR-sponsored Tropical Trees Program (TTP). He became a scientific and policy adviser to the Bamboo and Rattan Research Network in Asia, initially an informal network supported by Canada's International Development Research Centre (IDRC). He co-authored a pivotal report to identify priority species of bamboo and rattan that recognized their value for local communities, and society at large. Trevor's genetic resources interests also grew to encompass underutilized crops. He helped found the International Centre for Underutilized Crops in Southampton, UK, joining its Board, and was Chair from 1998 to 2005.

It can be argued that one of the most important legacies of Trevor's contribution to the conservation of plant genetic resources can be found deep beneath the Arctic permafrost on the island of Spitsbergen, where millions of seeds of crop varieties and wild species – many collected through the efforts of IBPGR in the 1970s and 1980s – are stored safely in the Svalbard Global Seed Vault (SGSV), popularly known as the 'Doomsday Vault'. Opened in 2008, the SGSV is the culmination of the plan set in motion three decades earlier by IBPGR to establish an international network of regional and national genebanks. Even the idea of duplicating genebank collections under the permafrost was first mooted by IBPGR as early as 1988. Without the germplasm safely conserved in genebanks around the world, there might have been less incentive to develop and apply the whole range of innovative techniques that are unlocking the mysteries of genetic diversity, and making it accessible to plant breeders globally. The genetic resources community owes Trevor Williams a great debt.

**M.T. Jackson**
Bromsgrove, Worcestershire, UK

# Contributors

**Abberton, M.**, International Institute of Tropical Agriculture (IITA), P.M.B. 53220, Ibadan, Oyo State, Nigeria.

**Abulaila, K.**, National Center for Agricultural Research and Extension, PO Box 639, Baq'a 19381, Jordan.

**Abu Taleb, T.**, Royal Botanic Garden, Tell Ar-Rumman, Al Balqa Governorate, Jordan.

**Achicanoy, H.A.**, International Center for Tropical Agriculture (CIAT), Km 17, Recta Cali-Palmira, Cali, Colombia.

**Adobor, S.**, University of Nottingham, Loughborough, LE12 5RD, UK.

**Al-Assaf, A.**, Department of Agricultural Economics & Agribusiness, Faculty of Agriculture, University of Jordan, Amman, Jordan.

**Alcázar, C.**, Bioversity International, Regional Office for the Americas, C/O CIAT, Apartado Aereo 6713, Cali, Colombia.

**Alercia, A.**, Bioversity International, Via dei Tre Denari 472/a, 00057 Maccarese, Rome, Italy. E-mail: a.alercia@cgiar.org

**Al-Sheyab, F.**, National Center for Agricultural Research and Extension, PO Box 639, Baq'a 19381, Jordan.

**Alvarez, C.**, Centro Andino de Educacion y Promociona (CADEP), Jirón Atahualpa 482, Tahuantinsuyo, Cusco, Peru.

**Amri, A.**, International Center for Agricultural Research in the Dry Areas (ICARDA), PO Box 6299, Rabat, Morocco.

**Appels, J.**, Wageningen University, Development Economics Group, Hollandseweg 1, 6706KN Wageningen, the Netherlands.

**Baktay, B.**, Research Centre for Agrobiodiversity, Külsömezö 15, H-2766 Tápiószele, Hungary. E-mail: baktay@mail.nodik.hu

**Bani-Hani, R.**, Nature Protection Directorate, Ministry of Environment, PO Box 1408, Amman 11941, Jordan.

**Barata, A.M.**, Banco Português de Germoplasma Vegetal, Instituto Nacional de Investigação Agrária e Veterinária, I.P. (INIAV, I.P.), Quinta S. José, S. Pedro de Merelim, 4700-859 Braga, Portugal. E-mail: Ana.Barata@iniav.pt

**Bari, A.**, International Center for Agricultural Research in the Dry Areas (ICARDA), PO Box 6299, Rabat, Morocco.

**Beltrán, M.**, Bioversity International, Regional Office for the Americas, C/O CIAT, Apartado Aereo 6713, Cali, Colombia.

**Bernau, V.**, International Center for Tropical Agriculture (CIAT), Km 17, Recta Cali-Palmira, Cali, Colombia.

**Bertero, D.**, Faculty of Agronomy, University of Buenos Aires, Buenos Aires, Argentina. E-mail: bertero@agro.uba.ar

**Berthouly-Salazar, C.**, Institut de Recherche pour le Développement (IRD), UMR Diversité, Adaptation et Développement des Plantes (DIADE), 911 Avenue Agropolis, 34394 Montpellier, France Cedex 5, France. E-mail: cecile.berthouly@ird.fr

**Bertin, P.**, Earth and Life Institute, Université catholique de Louvain, Croix du Sud 2 – 11 (L7.05.11), 1348 Louvain-la-Neuve, Belgium.

**bint Ali, B.**, Royal Botanic Garden, Tell Ar-Rumman, Al Balqa Governorate, Jordan.

**Bleeckx, T.**, Earth and Life Institute, Université catholique de Louvain, Croix du Sud 2 – 11 (L7.05.11), 1348 Louvain-la-Neuve, Belgium.

**Bodesmo, M.**, Department of Agricultural, Nutritional and Environmental Sciences, University of Perugia, Borgo XX Giugno 74, 06121 Perugia, Italy.

**Bonnave, M.**, Earth and Life Institute, Université catholique de Louvain, Croix du Sud 2 – 11 (L7.05.11), 1348 Louvain-la-Neuve, Belgium. E-mail: maxime.bonnave@gmail.com

**Boukar, O.**, International Institute of Tropical Agriculture (IITA), PMB 53220, Ibadan, Oyo State, Nigeria.

**Broekgaarden, C.**, Wageningen UR Plant Breeding, Wageningen University & Research Centre, PO Box 16, 6700 AA Wageningen, the Netherlands.

**Bülow, L.**, Julius Kühn-Institut, Institute for Breeding Research on Agricultural Crops, Erwin-Baur-Str. 27, 06484 Quedlinburg, Germany.

**Castañeda-Álvarez, N.P.**, School of Biosciences, University of Birmingham, Birmingham, B15 2TT, UK, and Decision and Policy Analysis Program, International Center for Tropical Agriculture (CIAT), Km 17, Recta Cali-Palmira, Cali, Colombia. E-mail: n.p.castaneda@cgiar.org

**Castellanos-Uribe, M.**, University of Nottingham, Loughborough, LE12 5RD, UK.

**Ccanto, R.**, Grupo Yanapai, Jirón Arequipa 421, Huancayo, Peru.

**Ceccarelli, S.**, Via delle Begonie 2, 63100 Ascoli Piceno, Italy.

**Couderc, M.**, Institut de Recherche pour le Développement (IRD), UMR Diversité, Adaptation et Développement des Plantes (DIADE), 911 Avenue Agropolis, 34394 Montpellier, France Cedex 5, France.

**de Haan, S.**, International Potato Center (CIP), Avenida La Molina 1895, Apartado 1895, Lima 12, Peru. E-mail: s.dehaan@cgiar.org

**de Lorenzo, C.**, IMIDRA – Finca el Encín, A-II km 38.200 Apdo. 127, 28800 Alcalá de Henares, Madrid, Spain.

**De Paola, D.**, Institute of Bioscience and Bioresources, National Research Council (CNR), Via Amendola 122/O, Bari, Italy.

**Dias, S.**, Bioversity International, Via dei Tre Denari 472/a, 00057 Maccarese, Rome, Italy.

**Drucker, A.G.**, Bioversity International, Via dei Tre Denari 472/a, 00057 Maccarese, Rome, Italy. E-mail: a.drucker@cgiar.org

**Dulloo, M.E.**, Bioversity International, Via dei Tre Denari 472/a, 00057 Maccarese, Rome, Italy. E-mail: e.dulloo@cgiar.org

**El-Oqlah, A.**, Department of Biology, Yarmouk University, Shafiq Irshidat St, Irbid 21163, Jordan.

**Endresen, D.T.F.**, UiO Natural History Museum, University of Oslo, PO Box 1172 Blindern, 0318 Oslo, Norway.

**Falk, D.**, Professor Emeritus, Department of Plant Agriculture, Crop Science Building, University of Guelph, Ontario, N1G 2W1, Canada. E-mail: dfalk@uoguelph.ca

**Fatokun, C.**, International Institute of Tropical Agriculture (IITA), PMB 53220, Ibadan, Oyo State, Nigeria.

**Fernández, I.**, IMIDRA – Finca el Encín. A-II km 38.200 Apdo. 127, 28800 Alcalá de Henares, Madrid, Spain.

**Fielder, H.**, School of Biosciences, University of Birmingham, Birmingham, B15 2TT, UK. E-mail: hannahfielder88@gmail.com

**Finkers, R.**, Wageningen UR Plant Breeding, Wageningen University and Research Centre, PO Box 16, 6700 AA Wageningen, the Netherlands. E-mail: richard.finkers@wur.nl

**Fitzgerald, H.**, Botany Unit FI-00014 Finnish Museum of Natural History, University of Helsinki, P.O. Box 7 (Unioninkatu 44), Finland. E-mail: heli.fitzgerald@helsinki.fi

**Ford-Lloyd, B.V.**, School of Biosciences, University of Birmingham, Birmingham, B15 2TT, UK.

**Frese, L.**, Julius Kühn Institute, Federal Research Centre for Cultivated Plants, Institute for Breeding Research on Agricultural Crops, Erwin-Baur-Str. 27, 06484 Quedlinburg, Germany.

**Freudenthaler, P.**, Austrian Agency for Health and Food Safety (AGES), Department for Plant Genetic Resources, Wieningerstraße 8, 4020 Linz, Austria.

**Galluzzi, G.**, Bioversity International, Regional Office for the Americas, C/O CIAT, Apartado Aereo 6713, Cali, Colombia.

**Gasi, F.**, Faculty of Agriculture and Food Sciences, University of Sarajevo, 71 000 Sarajevo, Bosnia and Herzegovina.

**Gelvonauskis, B.**, Plant Gene Bank, Stoties g. 2, LT-58343 Akademija, Kėdainių r., Lithuania.

**Ghazanfar, S.**, Royal Botanic Gardens, Kew, Richmond, Surrey, TW9 3AB, UK.

**Gupta, V.**, Division of Germplasm Conservation, National Bureau of Plant Genetic Resources, Pusa Campus, New Delhi 110012, India. E-mail: vgupta1123@rediffmail.com

**Haddad, N.**, West Asia Regional Program (WARP) ICARDA, PO Box 950764, Amman 11195, Jordan.

**Haghparast, R.**, Dryland Agricultural Research Institute (DARI), PO Box 119, Maragheh, Islamic Republic of Iran.

**Halewood, M.**, Bioversity International, Via dei Tre Denari 472/a, 00057 Maccarese (Fiumicino), Rome, Italy. E-mail: m.halewood@cgiar.org

**Heinonen, M.**, Natural Resources Institute Finland, Genetic Diversity, Alimentum, 31600 Jokioinen, Finland. E-mail: maarit.heinonen@luke.fi

**Ilori, C.**, Department of Crop Protection and Environmental Biology, University of Ibadan, Nigeria.

**Iriondo, J.M.**, Universidad Rey Juan Carlos, Departamento de Biología, Geología, Física y Química Inorgánica, Área de Biodiversidad y Conservación, C/Tulipán, s/n, 28933 Móstoles, Madrid, Spain. E-mail: jose.iriondo@urjc.es

**Janni, M.**, Institute of Bioscience and Bioresources, National Research Council (CNR), Via Amendola 122/O, Bari, Italy.

**Janssen, B.**, ServiceXS, Plesmanlaan 1d, 2333 BZ Leiden, the Netherlands.

**Jarvis, A.**, International Center for Tropical Agriculture (CIAT), Km 17, Recta Cali-Palmira, Cali, Colombia, and CGIAR Research Program on Climate Change, Agriculture and Food Security (CCAFS), Km 17, Recta Cali-Palmira, Cali, Colombia.

**Juarez, H.**, International Potato Center (CIP), Avenida La Molina 1895, Apartado 1895, Lima 12, Peru.

**Kak, A.**, Division of Germplasm Conservation, National Bureau of Plant Genetic Resources, Pusa Campus, New Delhi 110012, India.

**Kalazich, J.**, Instituto de Investigaciones Agropecuarios (INIA), Carretera Panamericana Sur K. 8 Norte, Osorno, Chile.

**Karpavičienė, B.**, Institute of Botany, Nature Research Centre, Žaliųjų Ežerų g. 49, LT-08406 Vilnius, Lithuania.

**Kell, S.P.**, School of Biosciences, University of Birmingham, Birmingham, B15 2TT, UK. E-mail: s.kell@bham.ac.uk

**Khoury, C.K.**, International Center for Tropical Agriculture (CIAT), Km 17, Recta Cali-Palmira, Cali, Colombia, and Centre for Crop Systems Analysis, Wageningen University, PO Box 430, 6700 AK, Wageningen, the Netherlands.

**Kik, C.**, Centre for Genetic Resources, the Netherlands (CGN), Wageningen University and Research Centre, PO Box 16, 6700 AA Wageningen, the Netherlands.

**Kinoshita Kinoshita, K.**, Universidad Rey Juan Carlos, Departamento de Biología, Geología, Física y Química Inorgánica, Área de Biodiversidad y Conservación, C/Tulipán, s/n, 28933 Móstoles, Madrid, Spain.

**Kitner, M.**, Palacký University in Olomouc, Faculty of Science, Department of Botany, Šlechtitelů 11, 783 71 Olomouc-Holice, Czech Republic.

**Korakhashvili, A.**, National Academy of Sciences of Georgia, Agmashenebeli Alley, Georgia Agrarian University, Tbilisi, 0151, Georgia. E-mail: akorakhashvili@yahoo.com

**Korpelainen, H.**, Department of Agricultural Sciences, PO Box 27, Latokartanonkaari 5, FI-00014 University of Helsinki, Finland.

**Křístková, E.**, Palacký University in Olomouc, Faculty of Science, Department of Botany, Šlechtitelů 11, 783 71 Olomouc-Holice, Czech Republic.

**Labokas, J.**, Institute of Botany, Nature Research Centre, Žaliųjų Ežerų g. 49, LT-08406 Vilnius, Lithuania. E-mail: j.labokas@gmail.com

**Lázaro, A.**, IMIDRA – Finca el Encín, A-II km 38.200 Apdo. 127, 28800 Alcalá de Henares, Madrid, Spain. E-mail: almudena.lazaro@madrid.org

**Lebeda, A.**, Palacký University in Olomouc, Faculty of Science, Department of Botany, Šlechtitelů 11, 783 71 Olomouc-Holice, Czech Republic. E-mail: ales.lebeda@upol.cz

**León, B.**, Plant Resources Center, University of Texas at Austin, Austin, Texas, USA.

**Libreros, D.**, Bioversity International, Regional Office for the Americas, C/O CIAT, Apartado Aereo 6713, Cali, Colombia.

**Lima, J.M.**, Direcção Regional de Agricultura e Pescas do Norte (DRAPN), Rua Dr.° Francisco Duarte, 365, 1°, 4715-017 Braga, Portugal.

**Mackay, M.**, Queensland Alliance for Agriculture and Food Innovation (QAAFI), The University of Queensland, Brisbane St Lucia, Queensland, Australia.

**Magos Brehm, J.**, School of Biosciences, University of Birmingham, Birmingham, B15 2TT, UK, and Centro de Biologia Ambiental, Faculdade de Ciências da Universidade de Lisboa, Portugal. E-mail: joanabrehm@gmail.com

**Mariac, C.**, Institut de Recherche pour le Développement (IRD), UMR Diversité, Adaptation et Développement des Plantes (DIADE), 911 Avenue Agropolis, 34394 Montpellier, France Cedex 5, France.

**Maxted, N.**, School of Biosciences, University of Birmingham, Birmingham, B15 2TT, UK. E-mail: nigel.maxted@dial.pipex.com

**May, S.**, University of Nottingham, Loughborough, LE12 5RD, UK.

**Mendes Moreira, P.**, Department of Agronomy, Coimbra College of Agriculture, Polytechnic Institute of Coimbra, Bencanta, 3045-601 Coimbra, Portugal.

**Mieslerová, B.**, Palacký University in Olomouc, Faculty of Science, Department of Botany, Šlechtitelů 11, 783 71 Olomouc-Holice, Czech Republic.

**Negri, V.**, Department of Agricultural, Nutritional and Environmental Sciences, University of Perugia, Borgo XX Giugno 74, 06121 Perugia, Italy. E-mail: valeria.negri@unipg.it

**Neuhaus, G.**, Julius Kühn-Institut, Institute for Breeding Research on Agricultural Crops, Erwin-Baur-Str. 27, 06484 Quedlinburg, Germany.

**Nilsen, L.-B.**, Swiss Federal Institute of Technology Zurich (ETHZ), Center for Development and Cooperation (NADEL), Rämistrasse 101, 8092 Zurich, Switzerland.

**Nobrega, H.**, ISOPlexis Genebank, Biotechnology and Genetic Resources Group, University of Madeira, 9000-390 Funchal, Portugal.

**Oliveira, J.**, Direcção Regional de Agricultura e Pescas do Norte (DRAPN), Rua Dr.° Francisco Duarte, 365, 1°, 4715-017 Braga, Portugal.

**Otondo, A.**, Instituto Nacional de Innovación Agropecuaria y Forestal (INIAF), Calle Batallón Colorados 24, La Paz, Bolivia.

**Ousseini, I.S.**, Institut de Recherche pour le Développement (IRD), UMR Diversité, Adaptation et Développement des Plantes (DIADE), 911 Avenue Agropolis, 34394 Montpellier, France Cedex 5, France.

**Oyatomi, O.**, International Institute of Tropical Agriculture (IITA), PMB 53220, Ibadan, Oyo State, Nigeria. E-mail: o.oyatomi@cgiar.org

**Pacicco, L.**, Department of Agricultural, Nutritional and Environmental Sciences, University of Perugia, Borgo XX Giugno 74, 06121 Perugia, Italy.

**Palmé, A.**, Nordic Genetic Resource Center, Box 41, SE-230 53 Alnarp, Sweden.

**Parra-Quijano, M.**, International Treaty on Plant Genetic Resources for Food and Agriculture, Food and Agriculture Organization of the United Nations, Viale delle Terme di Caracalla 00153, Rome, Italy.

**Pelgrom, K.**, Wageningen UR Plant Breeding, Wageningen University & Research Centre, PO Box 16, 6700 AA Wageningen, the Netherlands.

**Perales, H.**, El Colegio de la Frontera Sur, Carretera Panamericana y Periferico Sur s/n, San Cristobal, Chiapas, Mexico 29290. E-mail: hperales@ecosur.mx

**Phillips, J.**, School of Biosciences, University of Birmingham, Birmingham, B15 2TT, UK.

**Pignone, D.**, Institute of Bioscience and Bioresources, National Research Council (CNR), Via Amendola 122/O, Bari, Italy, and Department of Biology Agriculture and Food Science, National Research Council (CNR) Monterotondo St, Rome, Italy. E-mail: domenico.pignone@cnr.it

**Pinheiro de Carvalho, M.Â.A.**, ISOPlexis Genebank, Biotechnology and Genetic Resources Group, University of Madeira, 9000-390 Funchal, Portugal, and ICAAM, University of Évora, Apartado 94, 7006-554, Évora, Portugal.

**Pink, D.A.**, Harper Adams University, Newport, Shropshire, TF10 8NB, UK.

**Pinzón, S.**, Bioversity International, Regional Office for the Americas, C/O CIAT, Apartado Aereo 6713, Cali, Colombia.

**Plasencia, F.**, International Potato Center (CIP), Avenida La Molina 1895, Apartado 1895, Lima 12, Peru.

**Polreich, S.**, International Potato Center (CIP), Avenida La Molina 1895, Apartado 1895, Lima 12, Peru.

**Poulsen, G.**, Nordic Genetic Resource Center, Box 41, SE-230 53 Alnarp, Sweden.

**Pritchard, J.**, School of Biosciences, University of Birmingham, Birmingham, B15 2TT, UK.

**Raggi L.**, Department of Agricultural, Nutritional and Environmental Sciences, University of Perugia, Borgo XX Giugno 74, 06121 Perugia, Italy.

**Rahmanian, M.**, Centre for Sustainable Development (CENESTA), No 108, Azarbaijan Street, Tehran, 13169, Iran. E-mail: maryam.rahm@gmail.com

**Ramirez, M.**, Bioversity International, Regional Office for the Americas, C/O CIAT, Apartado Aereo 6713, Cali, Colombia.

**Rapanà, N.**, Institute of Bioscience and Bioresources, National Research Council (CNR), Via Amendola 122/O, Bari, Italy.

**Rašomavičius, V.**, Institute of Botany, Nature Research Centre, Žaliųjų Ežerų g. 49, LT-08406 Vilnius, Lithuania.

**Razavi, K.**, Centre for Sustainable Development (CENESTA), Tehran, Islamic Republic of Iran.

**Rocha, F.**, Banco Português de Germoplasma Vegetal, Instituto Nacional de Investigação Agrária e Veterinária, I.P. (INIAV, I.P.), Quinta S. José, S. Pedro de Merelim, 4700-859 Braga, Portugal.

**Rodriguez, F.**, International Potato Center (CIP), Avenida La Molina 1895, Apartado 1895, Lima 12, Peru.

**Rubio Teso, M.L.**, Universidad Rey Juan Carlos, Departamento de Biología, Geología, Física y Química Inorgánica, Área de Biodiversidad y Conservación, C/Tulipán, s/n, 28933 Móstoles, Madrid, Spain. E-mail: marialuisa.rubio@urjc.es

**Saifan, S.**, National Center for Agricultural Research and Extension, PO Box 639, Baq'a 19381, Jordan.

**Sainz, H.**, Universidad Pública de El Alto (UPEA), Av. Surce esq. Pascoe, El Alto, La Paz, Bolivia.

**Salimi, M.**, Centre for Sustainable Development (CENESTA), No 108, Azarbaijan Street, Tehran, 13169, Iran.

**Santoni, S.**, UMR AGAP 1334, INRA, 2, Place P. Viala – 34060, France.

**Sensen, S.**, Federal Office for Agriculture and Food (BLE), 513- Information and Coordination Centre for Biological Diversity, Bonn, Germany.

**Sharma, G.**, School of Biosciences, University of Birmingham, Birmingham, B15 2TT, UK.

**Shibli, R.**, University of Mutah, Mu'tah, Karak Governorate, Jordan.

**Simon, A.**, Research Centre for Agrobiodiversity, Külsömezö 15, H-2766 Tápiószele, Hungary.
**Smith, C.**, Quaker United Nations Office, 13 Avenue du Mervelet, 1209 Geneva, Switzerland. E-mail: chelseasmth204@gmail.com
**Solano, W.**, Tropical Agricultural Research and Higher Education Centre (CATIE), 7170 Cartago, Turrialba, 30501 Costa Rica.
**Sosa, C.C.**, International Center for Tropical Agriculture (CIAT), Km 17, Recta Cali-Palmira, Cali, Colombia.
**Strãjeru, S.**, Vegetal Genetic Resources Bank, B-dul 1 Mai 17, PC 720224, Suceava, Romania.
**Street, K.**, International Center for Agricultural Research in the Dry Areas (ICARDA), PO Box 6299, Rabat, Morocco. E-mail: k.street@cgiar.org
**Struik, P.C.**, Centre for Crop Systems Analysis, Wageningen University, PO Box 430, 6700 AK Wageningen, the Netherlands.
**Subedi, A.**, Centre for Development Innovation, Wageningen University and Research Centre, PO Box 88, 6700 AB Wageningen, the Netherlands. E-mail: abishkar.subedi@wur.nl
**Taifour, H.**, Royal Botanic Garden, Tell Ar-Rumman, Al Balqa Governorate, Jordan.
**Tan, A.**, Soyak Mavisehir A. 14/3, Karsıyaka, Izmir, Turkey.
**Taylor, N.**, Faculty of Biological Sciences, University of Leeds, Leeds, LS2 9JT, UK.
**Tenaillon, M.**, CNRS, UMR0320/UMR8120 Génétique Végétale, Ferme du Moulon, F-91109 Gif-sur-Yvette, France.
**Terrazas, F.**, Fundación PROINPA (Promoción e Investigación de Productos Andinos), Cochabamba, Bolivia.
**Thomas, E.**, Bioversity International, Regional Office for the Americas, C/O CIAT, Apartado Aereo 6713, Cali, Colombia. E-mail: evert.thomas@gmail.com
**Thormann, I.**, Bioversity International, Via dei Tre Denari 472/a, 00057 Maccarese, Rome, Italy. E-mail: i.thormann@cgiar.org
**Thörn, E.**, Swedish University of Agricultural Sciences (SLU), Box 57, 23053 Alnarp, Sweden. E-mail: eva.thorn@slu.se
**Thornström, C.-G.**, Swedish University of Agricultural Sciences (SLU), Box 7080, 75007 Uppsala, Sweden.
**Torricelli, R.**, Department of Agricultural, Nutritional and Environmental Sciences, University of Perugia, Borgo XX Giugno 74, 06121 Perugia, Italy. E-mail: renzo.torricelli@unipg.it
**Tyagi, R.K.**, Division of Germplasm Conservation, National Bureau of Plant Genetic Resources, Pusa Campus, New Delhi 110012, India.
**van Etten, J.**, Bioversity International, C/O Tropical Agricultural Research and Higher Education Centre (CATIE), 7170 Cartago, Turrialba, 30501 Costa Rica.
**van Kaauwen, M.**, Wageningen UR Plant Breeding, Wageningen University and Research Centre, PO Box 16, 6700 AA Wageningen, the Netherlands.
**van Workum, W.T.**, ServiceXS, Plesmanlaan 1d, 2333 BZ Leiden, the Netherlands.
**van Zonneveld, M.**, Bioversity International, C/O Tropical Agricultural Research and Higher Education Centre (CATIE), 7170 Cartago, Turrialba, 30501 Costa Rica.
**Venegas, C.**, Centro de Educacion y Tecnología (CET), Chonchi Rural, Sector Notuco, Casilla 40 Chiloé, Chile.
**Veteläinen, M.**, Boreal Plant Breeding Ltd, Myllytie 10, FI-31600 Jokioinen, Finland.
**Vigouroux, Y.**, Institut de Recherche pour le Développement (IRD), UMR Diversité, Adaptation et Développement des Plantes (DIADE), 911 Avenue Agropolis, 34394 Montpellier, France Cedex 5, France.
**Vincent, H.**, School of Biosciences, University of Birmingham, Birmingham, B15 2TT, UK. E-mail: holly.vincent@gmail.com
**Virgin, I.**, Stockholm Environment Institute (SEI), Linnégatan 87D, 115 23 Stockholm, Sweden.
**Vögel, R.**, Landesamt für Umwelt, Gesundheit und Verbraucherschutz Brandenburg (LUGV) Abt. Technischer Umweltschutz (TUS) Ref. Klimaschutz, Umweltbeobachtung und -toxikologie (T2) Tramper Chaussee 2, D-16225 Eberswalde, Germany.
**Voorrips, R.**, Wageningen UR Plant Breeding, Wageningen University and Research Centre, PO Box 16, 6700 AA Wageningen, the Netherlands.

**Vosman, B.**, Wageningen UR Plant Breeding, Wageningen University and Research Centre, PO Box 16, 6700 AA Wageningen, the Netherlands. E-mail: ben.vosman@wur.nl
**Weibull, J.**, Swedish Board of Agriculture, Plant and Environment Department, Plant Regulations Division, SE-55182 Jönköping, Sweden.
**Wiersema, J.H.**, United States Department of Agriculture, Agricultural Research Service, National Germplasm Resources Laboratory, Beltsville Agricultural Research Center, Beltsville, Maryland, USA. E-mail: john.wiersema@ars.usda.gov

# 'ENHANCED GENEPOOL UTILIZATION – Capturing wild relative and landrace diversity for crop improvement' 16–20 June 2014 – Cambridge, UK

## Conference Participants

# Preface

The legacy of the Convention on Biological Diversity (CBD) and the FAO International Treaty on Plant Genetic Resources for Food and Agriculture (ITPGRFA) has been a refocusing of national and international biodiversity conservation on to *in situ* activities, and for socio-economic species this has placed a much greater emphasis on on-farm and genetic reserve conservation. However, this has thrown up a series of challenges, among them being the need for cross-community integration to underpin future food security. Far too often, the diverse stakeholders involved in: conservation planning and its implementation; *in situ* and *ex situ* conservation; conservation, characterization and sustainable use; biodiversity and agrobiodiversity conservation; local, national, regional and global conservation and use; and conservation practitioners and policy makers, have worked in virtual isolation. This volume illustrates that real progress is being made in the science of agrobiodiversity conservation and use, and although complete integration of all stakeholders remains a challenge, the challenges are being addressed and solutions to blockage are beginning to emerge, as reported in the following chapters.

In Europe, the quest to meet these challenges, especially in relation to crop wild relative (CWR) and landrace (LR) *in situ* conservation and use, has been led by the European Cooperative Programme for Plant Genetic Resources – *In Situ* and On-Farm Working Groups. The Working Groups first spawned the EC FP5-funded PGR Forum project (pgrforum.org), which made significant progress in developing *in situ* conservation methodologies, and then the EC GenRes-funded AEGRO project (aegro.jki.bund.de) developed the methodologies further for both CWR and LR and began to test their implementation. This has been followed most recently by the EC FP7-funded PGR Secure (pgrsecure.org), whose focus is more on bridging the gaps between conservation and use stakeholders, and so enhancing crop improvement, as a means of underpinning European food security in the face of climate change. The final dissemination conference, entitled 'ENHANCED GENEPOOL UTILIZATION – Capturing wild relative and landrace diversity for crop improvement', was held at the National Institute of Agricultural Botany (NIAB) Innovation Farm, Cambridge, UK, between 16 and 20 June 2014 – with the aim of showcasing novel characterization techniques and conservation strategies to identify and preserve CWR and LR traits to increase options for crop improvement. This book represents the Conference Proceedings. The aim of the conference was to provide exemplars for CWR and LR conservation, promote their germplasm use and improve the conservation and use policy context, and these aspirations are followed through in this volume. Specifically, the objectives of the book, which are synonymous with those of the conference, are:

- Characterization techniques: 'omics' techniques and predictive tools to identify traits and expedite plant breeding.
- Conservation strategies: national, regional and global CWR and LR conservation strategy development; targeted conservation to meet the needs of the plant breeding community; integration of CWR and LR diversity into existing biodiversity conservation programmes.
- Facilitating CWR and LR use: pre-breeding; meeting breeders' needs; integrating conservation and user communities; policy enhancement.
- Informatics development: characterization, trait and conservation data management and accessibility; inter-information system operability.

The research requires the integration and collaboration of policy, conservation and breeding sectors and the products will be disseminated to all appropriate stakeholders. Although PGR Secure is a European-based project, the research, the conference and this volume, although developed in Europe, are generic and therefore transferable across the globe. The 42 chapters are divided into seven sections, grouped on their content as follows: Part I Breeders' Use of Exotic Germplasm, Chapters 1–6; Part II Improving Access to PGRFA, Chapters 7–9; Part III CWR Conservation, Chapters 10–22; Part IV LR Conservation, Chapters 23–31; Part V Community-based Conservation and Use, Chapters 32–34; Part VI PGR Conservation and Use Policy, Chapters 35–39; and Part VII Conservation Informatics, Chapters 40–42. The text is designed to be of use to conservation professionals, geneticists, ecologists, agriculturalists, agro and biodiversity policy makers, academics and postgraduate students in the hope that they will use the methodologies and experiences expounded to resolve the challenges identified and enhancing crop improvement, as a means of underpinning European food security in the face of climate change.

# Acknowledgements

We acknowledge the financial support of the EU Seventh Framework Programme, THEME KBBE.2010.1.1-03 for the PGR Secure project ('Characterization of biodiversity resources for wild crop relatives to improve crops by breeding', Project Grant Agreement No 266394). Further, we acknowledge the support of the NIAB Innovation Farm, Graminor, Limagrain, the European Association for Research on Plant Breeding (EUCARPIA) and the European Cooperative Programme for Plant Genetic Resources (ECPGR) for helping facilitate the conference 'ENHANCED GENEPOOL UTILIZATION – Capturing wild relative and landrace diversity for crop improvement', from which these proceedings are drawn. Specifically, the authors of Chapters 1, 8, 10–18, 26, 29, 31, 34–37, 39 and 42 are grateful for funding provided by the EU-funded PGR Secure project.

# 1 Using Phenomics and Genomics to Unlock Landrace and Wild Relative Diversity for Crop Improvement

B. Vosman,[1*] K. Pelgrom,[1] G. Sharma,[2] R. Voorrips,[1] C. Broekgaarden,[1] J. Pritchard,[2] S. May,[3] S. Adobor,[3] M. Castellanos-Uribe,[3] M. van Kaauwen,[1] R. Finkers,[1] B. Janssen,[4] W.T. van Workum[4] and B.V. Ford-Lloyd[2]

[1]*Wageningen UR Plant Breeding, Wageningen, the Netherlands;* [2]*School of Biosciences, University of Birmingham, Birmingham, UK;* [3]*University of Nottingham, Loughborough, UK;* [4]*ServiceXS, Leiden, the Netherlands*

## 1.1 Introduction

At present, it is not possible to feed the world population without the application of insecticides, and with the predicted growth of the human population, problems are expected to increase. Worldwide yield losses caused by insects would be at least 30–50% were no insecticides used. However, the use of pesticides is hazardous to the environment and usually not very durable, as insects develop resistance against pesticides very rapidly. Another factor that may have profound effects on the interaction between insects and plants is climate change. Climate change may potentially disturb plant–insect interactions directly and indirectly. It can change population development and distribution patterns of the pest insect, as well as of its natural enemies (Pritchard *et al.*, 2014).

The genus *Brassica* contains many important vegetables such as cabbage, broccoli, cauliflower, Brussels sprout and kale, which belong to the species *Brassica oleracea*, and Chinese cabbage, pak choi and turnip, which belong to *Brassica rapa*. In Europe, *Brassica* vegetables are cultivated across approximately 430,000 ha, of which more than half are grown in Eastern Europe. *Brassica* vegetables suffer from pest insects, including cabbage thrips, root flies, aphids, caterpillars and whiteflies (Fig. 1.1a,b). In particular, Brussels sprouts, kale and Savoy cabbage suffer from the cabbage aphid (*Brevicoryne brassicae*) and whitefly (*Aleyrodes proletella*). These insects feed on the underside of leaves, where they consume sap from the phloem, leading to disrupted plant growth and chlorosis (leaf bleaching). Additionally, they secrete a sugary substance (honeydew) that serves as a good substrate for moulds. Because the cabbage aphid and whitefly feed on the underside of leaves, they are difficult to control, and growers would therefore benefit strongly from resistant varieties.

Plants can defend themselves against herbivores through physical and chemical barriers that can be present constitutively (i.e. present regardless of attack and forming a first line of defence) or induced upon herbivore attack (Schoonhoven *et al.*, 2005). Morphological and anatomical features, such as a wax layer or leaf toughness, prevent herbivores from settling or

*Corresponding author; e-mail: ben.vosman@wur.nl

 *Enhancing Crop Genepool Use: Capturing Wild Relative and Landrace Diversity for Crop Improvement* (eds N. Maxted, M. Ehsan Dulloo and B.V. Ford-Lloyd)

Fig. 1.1. (a) Kale leaf infested with cabbage whitefly (*Aleyrodes proletella*) and (b) detail of the cabbage whitefly life cycle showing adults, eggs and different larval stages on a heavily infested *Brassica* leaf.

feeding on a plant. In addition, plants can deter herbivores through the production of repellent volatile secondary metabolites and defensive compounds or proteins that directly affect herbivore performance (Broekgaarden *et al.*, 2011). Glucosinolates, a group of secondary metabolites that are found almost exclusively in *Brassica* species, are well-studied defence compounds. When plant cells are disrupted, glucosinolates are hydrolysed by the enzyme, myrosinase, resulting in the formation of a variety of toxic compounds such as isothiocyanates (Halkier and Gershenzon, 2006). However, most specialist insects have evolved enzymes to detoxify glucosinolates and/or their breakdown products (Ratzka *et al.*, 2002). Therefore, plant resistance towards specialist herbivores is probably based on defensive compounds or proteins other than glucosinolates. Several proteins/compounds and the genes encoding them have been shown to play an important role in plant resistance towards herbivores (Van Doorn and de Vos, 2013), and some genes have already been identified (Smith and Clement, 2012).

To be able to develop insect-resistant varieties, one first needs to identify resistance sources. Often, these are found in crop wild relatives (CWR) or landraces (LR). Once the accessions containing the resistance traits are known, the genes involved need to be located to facilitate efficient transfer to the crop species. Novel phenomics, genomics and transcriptomics technologies can speed up the location of such genes and the identification of molecular markers linked to them (Broekgaarden *et al.*, 2011).

In the PGR Secure project, we aimed to: (i) identify host plant resistance to the cabbage whitefly and cabbage aphid via a germplasm screen where both are specialist phloem-feeding insects that feed only on members of the *Brassicaceae* family; (ii) elucidate the resistance mechanism; and (iii) provide tools to breeders that will facilitate resistance breeding.

## 1.2 Identification of Resistant Accessions

As a first step to identifying resistant material, we screened a collection of 432 accessions, including wild material and landraces of *B. oleracea*, as well as other CWR, for cabbage whitefly and aphid resistance in the field (Pelgrom *et al.*, 2012). A series of no-choice tests in which field-grown plants were infested with whiteflies was used to eliminate obviously susceptible accessions. Using this approach, it was possible to screen a very large number of accessions in a cost-efficient way and in a relatively short time (Pelgrom *et al.*, 2015). A selected group of putatively resistant accessions was retested for whitefly resistance in the greenhouse. Resistant accessions were identified among several *B. oleracea* var. *capitata* (heading cabbage) landraces. Resistance in heading cabbage had been observed previously in cultivar (cv.) Rivera (Broekgaarden *et al.*, 2010, 2012) and the present results suggest that the resistance observed in cv. Rivera is not unique but spread widely among heading cabbage landraces. Resistance was also observed among the wild relatives *Brassica villosa*, *Brassica incana* and *Brassica montana* (Pelgrom *et al.*, 2015). Where heading cabbage resistance is only expressed in plants of at least 12 weeks old, some wild relatives were already starting to express resistance at the age of 6 weeks. Since farmers plant these crops at an age of 5–6 weeks, this earlier expression of resistance is of great practical importance. Some level of resistance to the cabbage aphid was observed in *Brassica fruticulosa* and in *B. villosa*. These results show that although resistance against whiteflies has been found in the crop species itself, wild relatives are very important for crop improvement. The resistance to whiteflies in the wild relatives is likely based on a different mechanism than the resistance found in the crop species. Some accessions of *B. villosa* and *B. incana* are densely covered with trichomes, which may play a role in the resistance observed. For the cabbage aphid, resistance was not present in the crop species itself, once again underpinning the importance of including wild relatives in germplasm collections (Maxted *et al.*, 2012).

## 1.3 Location of Resistance Factors

Plants can defend themselves against phloem-feeding insects by means of physical and chemical barriers. Resistance components can be present in the form of morphological adaptations, such as trichomes (leaf hairs) or wax layers on the surface of the leaf, but may also be present in deeper cell layers or in the phloem. The electrical penetration graph (EPG) technique, in which a phloem-feeding insect is made part of a low-voltage electrical circuit, can shed light on the location of the resistance factors in the plant (Tjallingii, 1978). From the EPG readings, information can be obtained about the insect's feeding activities on the plants, which can be translated into parameters that provide information on the presence and location of resistance factors (Tjallingii, 1978; Alvarez *et al.*, 2006). Parameters that can be used are, for instance, the time an insect spends on the plant without probing, the time it needs to reach a sustained phloem sap ingestion, or the length of the period during which it takes up the phloem sap. Using the EPG technique, we could show large differences in the feeding behaviour of cabbage aphids on different *Brassica* accessions (Fig. 1.2). During the 6 h recording period, aphids did not reach a period of sustained phloem sap ingestion on some accessions of *B. villosa*, *B. incana* and *B. montana*, whereas they had no problems reaching this stage on some *B. oleracea* accessions. When comparing the activities of the cabbage whitefly on a resistant (cv. Rivera) and susceptible (cv. Christmas Drumhead) heading cabbage accession, it was found that the resistance factors were present at the phloem level and that morphological differences between the two cultivars most likely did not play a role in the resistance (Broekgaarden *et al.*, 2012). The EPG results suggest that the resistance is probably based on reduced ingestion of phloem sap on cv. Rivera. Also, it was observed that a number of whiteflies on cv. Rivera did show xylem sap uptake, possibly triggered by dehydration as a result of insufficient phloem uptake. Such information on the location of resistance factors as uncovered by the EPG technique is important, as it may point to different resistance mechanisms present, and combining different resistance mechanisms towards insects may increase resistance as well as durability of resistance.

## 1.4 Metabolomics Analysis

Secondary metabolites can play an important role in the defence against herbivores. Well-described defence compounds in *Brassicaceae* are the glucosinolates (Hopkins *et al.*, 2009). To identify metabolites possibly involved in the resistance against whitefly, we performed metabolomics analysis on two sets of plant material. The first set was selected from 125 landrace accessions of heading cabbages differing in levels of whitefly resistance. The second set was selected from an $F_2$ population derived from a cross between two *B. oleracea* cultivars (heading cabbages) (i.e. susceptible cv. Christmas Drumhead and resistant cv. Rivera). Differences in metabolite profiles between resistant and susceptible plants may be used to identify the compounds that are linked to the resistance against cabbage whitefly. From both sets of plant material (landrace accessions and the $F_2$ population), we used the most resistant and most susceptible

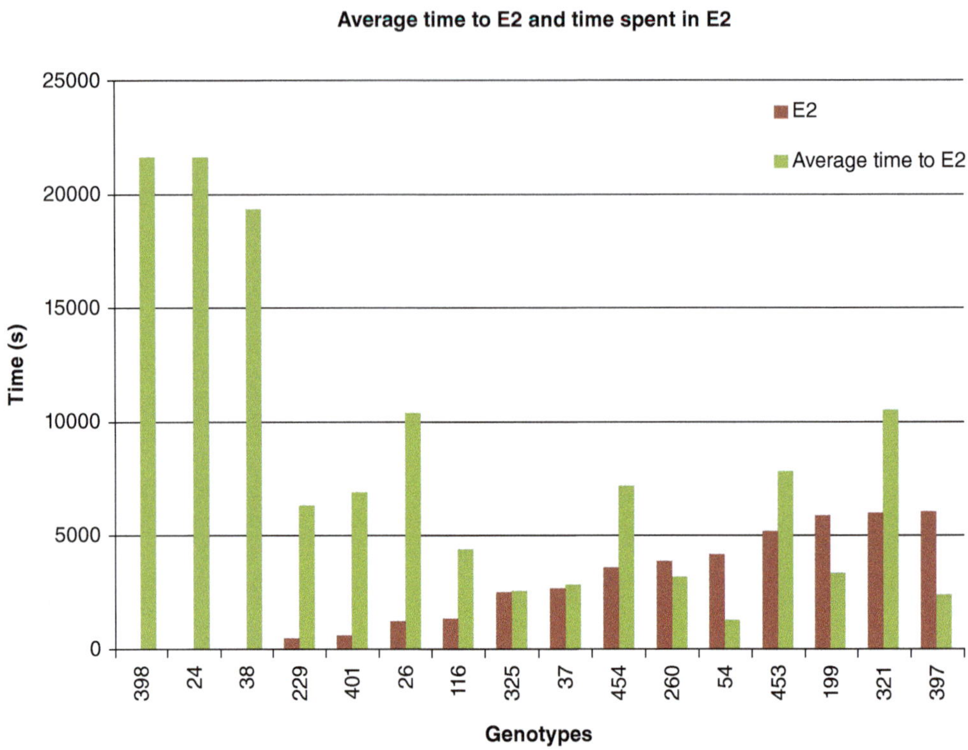


Fig. 1.2. Electrical penetration graph analysis of aphids (*Brevicoryne brassicae*) feeding on different accessions of *Brassica* spp. The average time to reach the phloem phase is shown in green and the time that the aphid is feeding from the phloem (E2) is shown in red. Numbers indicate different accessions used (24, 26: *B. incana*; 37, 38: *B. montana*; 397, 398, 401: *B. villosa*; 453, 454: *B. fruticulosa*). All others are *B. oleracea*.

plants/accessions for analysis. Two complementary metabolomics platforms were used to identify compounds related to susceptibility and resistance – gas chromatography-mass spectrometry (GC-MS) and liquid chromatography-mass spectrometry (LC-MS) – in both negative and positive ionization modes. Both the GC-MS and LC-MS data showed no significant differences in metabolites between the resistant and susceptible groups. We believe that, based on this result, it is unlikely that the resistance is based on a metabolite. Other mechanisms that may be based on a protein are more likely, although it cannot be excluded that a metabolite not detected by any of the platforms used is the causal agent.

## 1.5 Transcriptomics Analysis

We studied differential gene expression in landraces and species of *Brassica* to obtain information about candidate genes underpinning resistance factors. Genome-wide gene expression in response to aphid and whitefly infestation in *Brassica* CWR and LR was studied using the Affymetrix Arabidopsis Gene 1.0 ST array (Affymetrix, Santa Clara, California), which consisted of 600,941 probes representing 28,500 genes. This array is derived from the inbred *Arabidopsis thaliana* accession Columbia. The TAIR 10 genome annotation, containing 33,602 genes and 41,671 gene models, was used (GEO Platform reference: GPL17416).

RNA samples were prepared from the leaf tissue of 12-week-old plants that were either induced with aphids for 24 h or non-induced. Differences in gene expression were seen in the sets of material previously classified as resistant or susceptible to aphids based on EPG or field evaluation of resistance.

Different sets of genes were found to be expressed differentially when resistant/susceptible accessions were determined using field experiment results in comparison with resistance/susceptibility assessment from the feeding behaviour (EPG). This is likely to be for a number of reasons, including plants being exposed to different environments and, in the case of induced, because of differences in plant stage, time of exposure to the insect, but also most importantly because the EPG assessment is able to dissect the physiological nature of any resistance more critically.

For resistance/susceptibility assessment based on the non-penetration phase in the EPG analysis, gene ontology analysis showed that the genes differentially expressed were mostly related to membrane or cell wall activity, whereas assessment based on the pathway phase showed differentially expressed genes involved in the respiration electron transport chain, oxidation reduction processes and in intracellular signal transduction. When resistance/susceptibility assessment was based on active feeding phase E2, it was represented by genes such as PP2-A10, which were known to be associated with phloem-based defence against insect pests in plants.

A number of candidate genes for resistance have been revealed in both induced and non-induced situations, and some of these may clearly contribute in different physiological ways related to modifying the feeding behaviour of the aphids. This suggests that for future crop improvement, stacking of different resistance genes as part of the plant breeding process may increase effective resistance. The strongest candidates for further development of aphid resistance in *Brassica* crops are: (i) ESM1 – At3g142101; (ii) pectin-lyase superfamily – At1g04680; (iii) PDCB3 – At1g18650; and (iv) NPC6 – At3g48610. For resistance assessed by way of the EPG pathway, most of the highly significant differentially expressed genes code for hypothetical proteins, the value of which, in terms of resistance, is therefore unknown. However, NUDX1 – AT1G68760, which was highly significant and showed a fold change of 2, encodes the first defined nudix hydrolase in *Arabidopsis*, which may or may not be associated with disease susceptibility.

## 1.6 Next-generation Sequencing and SNP-array Development

Molecular markers are an indispensable tool in modern plant breeding. They can be used to make early selection of plant material possible and to facilitate recurrent parent selection, thus speeding up the breeding process. The marker type most widely used is the single nucleotide polymorphism (SNP). To obtain SNP markers that are informative in *B. oleracea* and its relatives, we carried out an RNAseq analysis of one representative plant of each of the taxa used and of the parents of our mapping populations. In total, the leaf RNA of 15 selected plants was sequenced, resulting in the identification of *c.*2 million SNPs. From these SNPs, a selection was made based on the position of the SNP on the *B. oleracea* reference genome (Yu *et al.*, 2013). The 90 k Affymetrix Axiom array produced contains *c.*40,000 SNPs selected from a set of broccoli varieties, 21,000 polymorphic SNPs from a set of heading cabbages, 4200 already validated *B. oleracea* SNPs and *c.*5000 SNPs that are polymorphic between *B. oleracea* and the wild relative *B. incana*, as well as 5000 that are polymorphic between *B. oleracea* and *B. montana*. The array also contains *c.*5000 SNPs that are polymorphic within *B. fruticulosa*. The distribution of the SNPs over the different pseudochromosomes and scaffolds of the *B. oleracea* genome sequence is shown in Table 1.1. The array will be very useful in a number of applications including quantitative trait loci (QTL) mapping in *B. oleracea* and CWR, association mapping in *B. oleracea*, as well as relationship analysis among species, subspecies, varieties and landraces. As an example, we used the array to study the relationships among a set of 114 *B. oleracea* accessions (landraces and varieties). These accessions were labelled according to their origin (collection site as indicated in the passport data). From the principal component analysis (PCA) (Fig. 1.3), it can be seen that most of the accessions clustered according to their origin, suggesting local adaptation. This is especially true for the accessions originating from Africa (Egypt) and the accessions from the south-east (Turkey). However, there was also substantial overlap among the groups, which may point towards exchange of genetic material. Also, some accessions were identified that showed an unexpected clustering; for example, one sample from Egypt that clustered with the samples from north-western Europe. This may point towards mislabelling, but more likely it points to exchange of accessions among regions.

**Table 1.1.** Distribution of the SNPs present on the Axiom array over the different pseudochromosomes and scaffolds of the *Brassica oleracea* genome sequence based on their origin.

| | Number of SNPs | | | | | | | |
|---|---|---|---|---|---|---|---|---|
| Chromosome | *B. oleracea* var. *capitata* | *B. oleracea* var. *italica* | *B. incana* | *B. montana* | *B. fruticulosa* | *B. napus* | CWR | Total number of chromosomes |
| C01 | 1,445 | 4,637 | 396 | 406 | 376 | 349 | 435 | 8,044 |
| C02 | 1,560 | 4,545 | 349 | 321 | 296 | 313 | 579 | 7,963 |
| C03 | 2,542 | 4,955 | 555 | 606 | 574 | 521 | 581 | 10,334 |
| C04 | 1,605 | 3,842 | 434 | 381 | 417 | 296 | 465 | 7,440 |
| C05 | 1,216 | 3,907 | 347 | 396 | 365 | 137 | 349 | 6,717 |
| C06 | 1,614 | 4,211 | 512 | 522 | 493 | 264 | 443 | 8,059 |
| C07 | 1,994 | 4,813 | 372 | 347 | 363 | 371 | 539 | 8,799 |
| C08 | 1,587 | 5,593 | 458 | 488 | 459 | 312 | 408 | 9,305 |
| C09 | 1,285 | 3,631 | 454 | 449 | 442 | 140 | 447 | 6,848 |
| Total C01–C09 | 14,848 | 40,134 | 3,877 | 3,916 | 3,785 | 2,703 | 4,246 | 73,509 |
| Scaffolds | 6,147 | 0 | 1,123 | 1,084 | 1,017 | 0 | 0 | 9,371 |
| Unmapped | 0 | 0 | 0 | 0 | 0 | 2,212 | | 2,212 |
| Total SNPs | 20,995 | 40,134 | 5,000 | 5,000 | 4,802 | 4,915 | 4,246 | 85,092 |

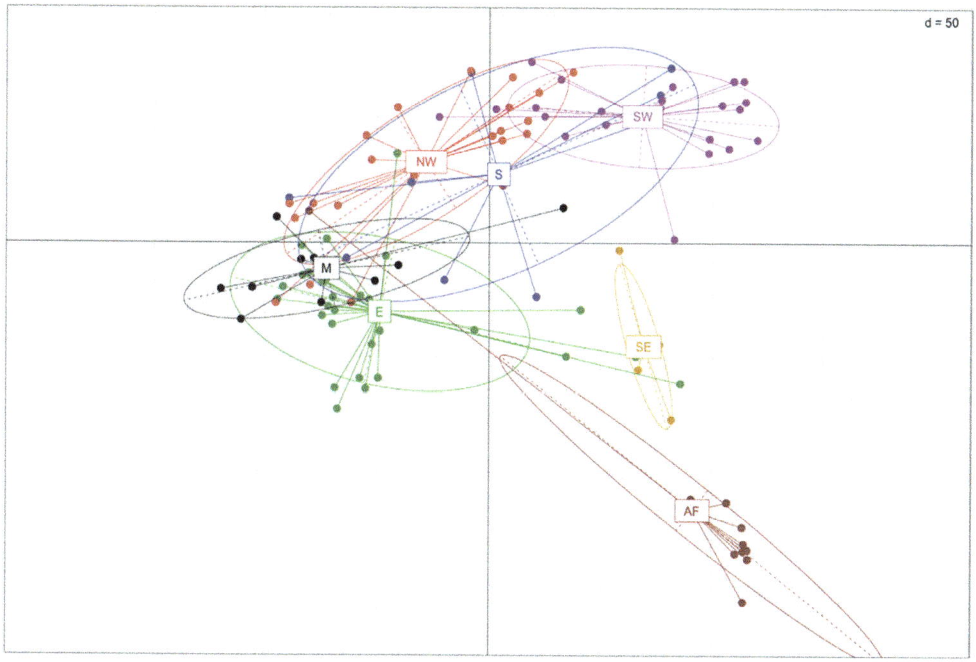


Fig. 1.3. PCA analysis of 114 landraces and varieties. The accessions were collected in northern Africa (AF, brown), Eastern Europe (E, green), south-eastern Europe (SE, yellow), Central Europe (M, black), Southern Europe (S, blue), north-western Europe (NW, red) and south-western Europe (SW, purple). The data from 62,434 polymorphic SNP markers were used for the analysis. The first axis explains 6.8% of the variation, the second axis 4.2%.

## 1.7 Mapping of Resistance Genes

Host plant resistance to phloem-feeding insects can be mediated by several mechanisms (Broekgaarden *et al.*, 2011). QTL mapping was used to identify chromosomal regions involved in whitefly resistance. Based on the germplasm screen, resistant and susceptible plants have been selected as parents for crosses. Within the heading cabbages, an $F_2$ population was obtained from a cross between the whitefly-susceptible cv. Christmas Drumhead and the resistant cv. Rivera. In this population of 180 plants, we measured whitefly adult survival and oviposition rate, as well as some morphological characteristics possibly involved in the resistance (time of head formation, leaf wax layer and leaf toughness). The population was genotyped using 150 SNP markers equally spaced over the chromosomes. QTLs were found for the whitefly resistance parameters 'adult survival' and 'oviposition rate', explaining 14% and 13% of the variance, respectively. A strong QTL was found for 'wax layer', explaining 64% of the variance. None of the measured morphological traits co-localized with adult survival and oviposition rate. Although a strong resistance towards the cabbage whitefly was observed in the heading cabbage cv. Rivera, no major QTLs were found for survival and oviposition rate. The resistance in this variety is probably based on the interaction of several genes or different resistance mechanisms.

In a second cross, we used a fully whitefly-resistant plant of the CWR *B. incana* and a susceptible *B. oleracea* cultivar. The resulting $F_1$ was backcrossed with the *B. incana* parent, as selfing of the $F_1$ plant yielded only a limited number of seeds. In the backcross population, we mapped whitefly resistance to a single locus, explaining 57% of the variance for whitefly adult survival and 82% for oviposition rate. At the same locus, we also mapped the presence/absence of trichomes. There was a strong negative correlation between the presence of trichomes and whitefly adult survival (–0.71) and oviposition rate (–0.89). This suggests that the presence of the trichomes is likely responsible for the resistance observed, although it cannot be excluded that the resistance is caused by a gene linked closely to the gene governing the presence/absence of the trichomes. Further research will be needed to elucidate this.

## 1.8 Conclusions

- The PGR Secure project has delivered valuable information on the extent of whitefly and aphid resistance in landrace accessions of *B. oleracea* var. *capitata*, as well as in wild relatives of *B. oleracea*.
- The whitefly resistance present in *B. villosa*, *B. incana* and *B. montana* is expressed both in 6- and 12-week-old plants, which indicates that this form of resistance is probably different from that already present in *B. oleracea*. These sources can be used in breeding resistant varieties.
- By combining novel phenomics, genomics and transcriptomics technologies, resistance breeding can be speeded up significantly. The developed and publicly available 90 k Affymetrix Axiom Brassica array can play an important role in this.
- The SNP markers linked to the resistance QTLs will facilitate an efficient introgression of the QTLs into high-yielding varieties.

## Acknowledgements

This research was funded by a grant from the EU Seventh Framework Programme for the PGR Secure project (Grant Agreement No 266394) and a grant from the Ministry of Economic Affairs of the Netherlands (Project Nr BO-26.03-009-001) for the contribution of the Wageningen UR Plant Breeding partners.

## References

Alvarez, A.E., Tjallingii, W.F., Garzo, E., Vleeshouwers, V., Dicke, M., *et al.* (2006) Location of resistance factors in the leaves of potato and wild tuber-bearing Solanum species to the aphid *Myzus persicae*. *Entomologia Experimentalis et Applicata* 121, 145–157.

Broekgaarden, C., Poelman, E.H., Voorrips, R.E., Dicke, M. and Vosman, B. (2010) Intraspecific variation in herbivore community composition and transcriptional profiles in field-grown *Brassica oleracea* cultivars. *Journal of Experimental Botany* 61, 807–819.

Broekgaarden, C., Snoeren, T.A.L., Dicke, M. and Vosman, B. (2011) Exploiting natural variation to identify insect-resistance genes. *Plant Biotechnology Journal* 9, 819–825.

Broekgaarden, C., Riviere, P., Steenhuis, G., del Sol Cuenca, M., Kos, M., *et al.* (2012) Phloem-specific resistance in *Brassica oleracea* against the whitefly *Aleyrodes proletella*. *Entomologia Experimentalis et Applicata* 142, 153–164.

Halkier, B. and Gershenzon, J. (2006) Biology and biochemistry of glucosinolates. *Annual Review of Plant Biology* 57, 303–333.

Hopkins, R.J., van Dam, N.M. and Van Loon, J.J.A. (2009) Role of glucosinolates in insect–plant relationships and multitrophic interactions. *Annual Review of Entomology* 54, 57–83.

Maxted, N., Kell, S., Ford-Lloyd, B., Dulloo, E. and Toledo, Á. (2012) Toward the systematic conservation of global crop wild relative diversity. *Crop Science* 52, 774–785.

Pelgrom, K., Sharma, G., Broekgaarden, C., Voorrips, R., Bas, N., *et al.* (2012) Looking for resistance to phloem feeders in *Brassica oleracea*. *Crop Wild Relatives* 8, 12–14.

Pelgrom, K.T.B., Broekgaarden, C., Voorrips, R.E., Bas, N., Visser, R.G.F., *et al.* (2015) Host plant resistance towards the cabbage whitefly in *Brassica oleracea* and its wild relatives. *Euphytica* 202, 297–306.

Pritchard, J., Broekgaarden, C. and Vosman, B. (2014) Effects of climate change on plant–insect interactions and prospects for resistance breeding using genetic resources. In: Jackson, M., Ford-Lloyd, B. and Parry, M. (eds) *Plant Genetic Resources and Climate Change*. CAB International, Wallingford, UK, pp. 270–284.

Ratzka, A., Vogel, H., Kliebenstein, D.J., Mitchell-Olds, T. and Kroymann, J. (2002) Disarming the mustard oil bomb. *Proceedings of the National Academy of Sciences of the United States of America* 99, 11223–11228.

Schoonhoven, L.M., van Loon, J.J.A. and Dicke, M. (2005) *Insect–Plant Biology*, 2nd edn. Oxford University Press, New York.
Smith, C.M. and Clement, S.L. (2012) Molecular bases of plant resistance to arthropods. *Annual Review of Entomology* 57, 309–328.
Tjallingii, W.F. (1978) Electronic recording of penetration behaviour by aphids. *Entomologia Experimentalis et Applicata* 24, 721–730.
Van Doorn, A. and de Vos, M. (2013) Resistance to sap-sucking insects in modern-day agriculture. *Frontiers in Plant Science* 4, 222.
Yu, J., Zhao, M., Wang, X., Tong, C., Huang, S., *et al.* (2013) Bolbase: a comprehensive genomics database for *Brassica oleracea*. *BMC Genomics* 14, 664.

# 2 Pre-domesticating Wild Relatives as New Sources of Novel Genetic Diversity

D. Falk*
*Department of Plant Agriculture, University of Guelph, Guelph, Ontario, Canada*

## 2.1 Introduction

Wild relatives of many domesticated crop species have been used mainly as a source of specific genes for increased disease resistance. It is often difficult to exploit these genepools efficiently because they lack a number of essential traits necessary for cultivation and management using modern agricultural technology. Many of these traits were developed in our modern crops as part of the process of domestication. These traits are usually associated with seed dispersal, seed size, seed coats and seed dormancy, which are essential to survival in the wild but undesirable in domesticated crops. By incorporating these 'domestic' traits into the wild species, they may be managed using standard farming methods. It would then be easier to select desirable gene combinations within these diverse and variable populations, and introgress these selected traits more effectively into modern, elite germplasm and cultivars. Cultivated barley (*Hordeum vulgare*) and its wild progenitor, *Hordeum spontaneum*, are used as a model to employ the proposed methodology and illustrate the increased efficiency and effectiveness of this approach to taming wild relative germplasm. By 'pre-domesticating' a series of diverse populations of *H. spontaneum* and conducting breeding and selection within these backgrounds, it is possible to exploit a much larger portion of the vast reservoir of quantitative genetic variation that is not being utilized at this time. Applying recurrent selection, via an efficient genetic male sterile system within *H. spontaneum* itself, may lead to greater genetic diversity for nutritional quality, agronomic traits, processing quality, and even yield, than currently exists within cultivated barley. This approach is essentially re-domesticating barley on a much broader scale than the original domestication event(s); and it could help address the current issue of lack of useful diversity. This concept is applicable to most crop species.

## 2.2 Domestication of Barley

Barley, *H. vulgare* L., is postulated to have been derived from the wild barley, *H. spontaneum*, about 10,000 years ago in the Fertile Crescent region of the Middle East (Harlan, 1992; Badr *et al.*, 2000). This domesticated barley spread into Europe with ancient farmers, and eventually throughout the rest of the world in relatively modern times. Barley is currently the fourth most widely grown grain crop and is grown in nearly every country in the world. It is commonly

*E-mail: dfalk@uoguelph.ca

used as animal feed, is the basis for beer production and is used as food in some cultures.

The initial 'domestication' that led to the cultivation of cereals, and eventually to modern agriculture, was based on selection of a tough rachis form of *H. spontaneum* that was much better adapted to human harvesting of the crop (Harlan, 1992). This loss of the natural seed dispersal mechanism has been shown to be a result of a mutation at the separate, but tightly linked, Bt and Bt2 loci (Nevo, 1992), which are responsible for the disarticulation of the rachis (shattering) when the individual seeds mature on the spike. There would be a high natural selection for this characteristic in a human-managed system where the seed harvested is used for sowing the subsequent crop. The plants with the wild-type fragile rachis would have few seeds remaining in the spikes for harvesting (Fig. 2.1a), whereas those with a tough rachis would retain many more seeds (Fig. 2.1b) and contribute a larger portion of the seeds that would be sown for the next crop (Harlan, 1992). This trait basically distinguishes wild barley from primitive domesticated barley. The two different loci (Bt and Bt2) involved in creating the tough rachis, which is diagnostic of domestication (Nevo, 1992), can create problems for modern breeders in that crossing cultivars of the different genotypes can result in a fragile rachis phenotype in the $F_1$ and a portion of the later-generation progeny. The resulting high rate of shattering is incompatible with modern harvesting and planting methods, in addition to reducing harvestable yield and leading to high levels of volunteer grain in the following crop. The tough rachis would have been highly selectable in any barley population that was harvested and then replanted in a cultivated scenario (Harlan, 1992; Nevo, 1992). The loss of the natural seed dispersal mechanism of the wild relatives of domesticated cultivated crop plants is a common occurrence across a wide range of crop kinds, such as barley, wheat (*Triticum* spp.), soybeans (*Glycine max*), dry beans (*Phaseolus vulgaris*), rye (*Secale cereale*), oats (*Avena sativa*), rice (*Oryza sativa*), brassicas and many other crops harvested for seed/grain.

## 2.3 Traits Developed Under Cultivation

Subsequent to the loss of the seed dispersal mechanism, which was an essential element in true domestication, leading to dependence on humans for propagation of the population, a number of other traits would also have been desirable in making a crop better suited/adapted to a cultivated management system. Traits like the tough rachis would have been selected naturally as being favourable for the genotypes that possessed them without conscious selection being practised by early farmers. The fragile rachis and the ease of threshing trait (*thresh-1*) are the two essential traits distinguishing cultivated *H. vulgare* from wild *H. spontaneum* (Schmalenbach, 2011). The *thresh-1* gene would have been favoured heavily by selection within the tough rachis background of the early cultivated crop.

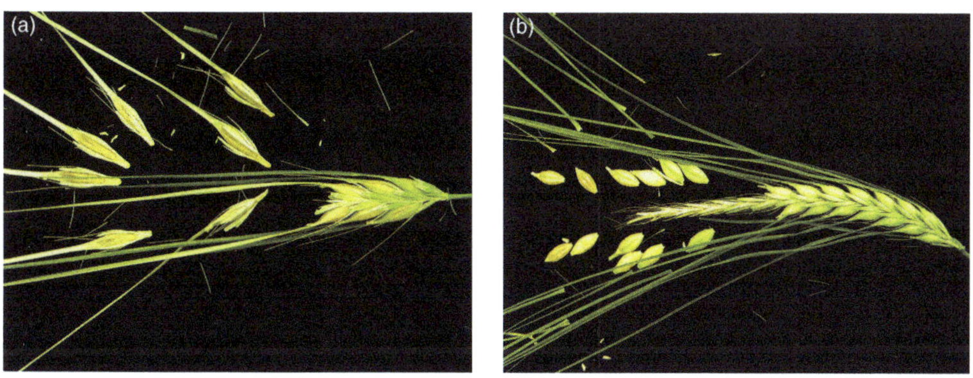

Fig. 2.1. (a) Wild barley, *H. spontaneum*, with a disarticulating rachis that remains attached to the grains and strongly attached, very rough awns for enhanced seed dispersal. (b) Domesticated barley, *H. vulgare*, with a tough rachis and awns that separate readily from the grains on threshing.

The wild-type forms of both traits are essential for survival in the wild, and the domestic version of either trait would not have been maintained in wild populations, while they were fixed quickly and multiplied under cultivation.

One of the important traits of long-term survival of a seed-propagated species in the wild is some form of seed dormancy. Under cultivation in which seed is harvested from those plants that have produced mature seed from an annual sowing cycle, there would be a very high selection against any genotypes that did not germinate rapidly and in the same season in which they were sown. Thus, loss of dormancy would be a natural response to selection under human management and would have occurred early in the domestication process. Plants with mutations toward the more desirable state of low or no dormancy would have been highly favoured in the annual cycle of sowing and harvesting, and then re-sowing the seed from the harvested plants the following season. This is essentially an unconscious phenotypic recurrent selection system with a 1-year cycle and natural selection for the desired plant characteristics under cultivation.

Another trait that would improve adaption to a cultivated environment rapidly is increase in seed size (Harlan, 1992). Since competition with other barley seedlings and with weeds would both be favoured by larger, more vigorous seedlings, there is a natural selection toward larger seed size. This would also be augmented to some extent by human harvesting and cleaning of seed by winnowing (gleaning), where the lighter seeds might be lost at harvesting or blown away in the winnowing process. Being grown in a less stressful environment under cultivation would have provided greater opportunity to fill the grains better and further promote larger seed size.

Synchronous development/ripening was favoured strongly when the crop was being cut with a sickle rather than seeds being pulled from individual plants by hand as they ripened, so as harvesting technology developed, perhaps even before full domestication, there was selection for those plants where all the seed ripened at the same time, and so was all harvested and subsequently would be viable for sowing in the next growing season. As crop management systems became more sophisticated and efficient and larger numbers of plants were harvested simultaneously, there was even greater selection in favour of synchrony of tillering and uniformity of ripening in the entire field/population. A major step forward in uniformity of ripening was the development of pure line selections from the landraces that were developed almost universally through phenotypic mass selection practised by farmers from ancient times up until the 19th century. The introduction of highly mechanized agriculture at the beginning of the 20th century necessitated greater uniformity of plant type, growth and development. All plants need to emerge uniformly in order to utilize herbicides that have narrow windows of crop safety, as well as ripening uniformly and being uniform in height and harvestability for large, mechanical harvesting equipment. Uniformity of the harvested grain was also highly desirable for large-scale processing, particularly malting and brewing of barley.

The sterile lateral spikelets giving a two-rowed spike and the attached segment of rachis were essential elements of the triplet seed dispersal mechanism(s) of wild barley (Nevo, 1992), but such seed dispersal was actually undesirable under cultivation where harvested seed was used for the next planting cycle and was replaced rapidly by the tough rachis, but still two-rowed spike. Therefore, when the spontaneous mutation to the six-rowed trait would have occurred in a domesticated population, the greater seed number per spike would have increased the proportion of this genotype in the population, which would comprise the following sowing generation (Fig. 2.2). Natural selection, based

Fig. 2.2. Wild barley (left) showing differences in maturity and seed size within the spike; cultivated two-rowed barley (centre) and cultivated six-rowed barley (right) showing uniformity of maturity and seed size within the spike.

on the increased number of progeny produced by six-rowed plants relative to two-rowed plants, would have resulted in a relatively rapid shift toward six-rowed types in the population. The larger spikes with more seed would likely have been noted by farmers and might also have been subjected to positive human selection, in addition to the natural selection occurring in the background. There is still some controversy about the occurrence and persistence of the six-rowed types of wild barley (*Hordeum agriocrithon*), but most such collections have proven to be hybrids of wild two-rowed and domesticated six-rowed types (Nevo, 2012).

Cultivation inadvertently would have created an environment where stronger straw would have been favoured. Plants that were standing upright at ripening/harvesting would have contributed more of their seed to the following generation for propagation than those whose spikes were lying on the ground or those straws that had broken off prior to harvesting with a sickle or, later, more mechanized means (Fig. 2.3). Stems breaking, or falling over prior to seed maturity, may have been a significant component of short-range seed dispersal in wild barleys (Fig. 2.4). Stronger straw would also have been selected naturally as seed size increased and as spikes became heavier with the occurrence of the six-rowed types in cultivation. The spikes had to be harvested easily in order to contribute seeds to the next planting.

Fig. 2.3 Wild barley (*H. spontaneum*) with plants tied up and mesh bags on spikes to retain seed at maturity.

Fig. 2.4 Shattered seed and lodged stems of wild barley (*H. spontaneum*) on the ground around maturing plants.

## 2.4 Human Selection Under Domestication

In addition to selection through natural means in response to the cultivated environment, other traits would have been selected by humans specifically for their desirable attributes. A principle trait that humans would have preferred was smooth awns (*rr*) over the very rough awns (*RR*) of wild barley. In the wild situation, the barbs on the awns help the seeds to move along the ground in response to wind and water action, or to catch in the fur of passing animals and disperse more widely. Under cultivation, rough awns would have been particularly annoying to farmers handling bundles of unthreshed grain by hand and carrying them from the fields to threshing areas. The awns tend to cause severe itching, and also cause the grains attached to them to move inside clothing as a person moves. The awns would have made the grains difficult to eat for both humans and domesticated animals. When spontaneous mutations to smooth awns occurred, the seed of these very desirable plants would have been saved and sown preferentially by observant farmers, and so smooth awned, much easier handled barleys would have developed. This trait, which would have been highly undesirable in a wild plant, where seed dispersal was critical to the long-term survival of the population, would have been favoured to almost the same degree in the opposite direction under cultivation, where it would have made the crop much easier to handle. This preference for smooth awns may have reversed with the advent of mechanical reaping and modern combine harvesters, as spikes with rough awns tend to dry down faster than smooth awns, due to the greater surface areas. Swaths of rough-awned barley tend to remain more open and less compacted after cutting, which allows greater air movement and faster drying down to threshing moisture.

The occurrence of a mutation to naked (*nn*), or hulless, seeds would have been noticed quickly by farmers because the hull on covered (wild type) barley is not removed easily and so generally is tolerated by necessity when using barley as food (or feed). The hull has few desirable features in the use of barley as food, even though it is quite valuable in protecting the embryo, and the seed in general, from damage during the mechanical threshing operation. Without the fibrous hull, naked barley would have been much easier to eat and would have had a higher concentration of nutrition in the form of carbohydrates (starches), for energy, and also of proteins, and even oils, from the embryo. Hulless barleys have been found in quite early cultivation of the crop and commonly have been used as human food in ancient, as well as modern, societies that use barley as a major source of food (Harlan, 1992). The genetic loss of the hull makes the embryo more vulnerable to damage during harvesting, handling and even sowing, so 'gentler' methods of seed handling and processing would be desirable with hulless barleys compared to the covered types. Covered barley is preferred for malting and brewing, where embryo damage reduces germination uniformity and reduces the quality of the final product.

With the hull removed during the threshing operation, the appearance of the grain would then have been subjected to human influence. Most wild barleys have a blue aleurone, which is associated with flavonoids that may be responsible for undesirable flavours and tastes that reduce consumption by animals in the wild populations (Nevo, 1992). Cultivated barleys would have shown some variation for seed colour when spontaneous mutations to the non-blue aleurone occurred within the hulless types being used for food. Humans who were consuming barley and barley products regularly would have noted the more desirable flavour of the non-blue types readily, so selection against the dominant blue aleurone trait/genes (*Bl1* and *Bl2*) would have been very effective in converting the population to a less colourful but better tasting food.

Selection, whether consciously by humans or by natural conditions, for the numerous mutations toward a more desirable crop would have imposed a series of bottlenecks on the domesticated cultivated crop. This would have occurred early in the cultivated history of the crop and, over time, caused a significant loss of diversity, which is recognized today as a serious limitation in further crop improvement. Because the mutations for most of these traits would occur at relatively low frequencies, and subsequently would be subjected to very high selection intensity, the few individuals descended directly from the original individuals in which the mutations occurred would dominate the population. The low frequency of outcrossing in barley would also have contributed to this problem of loss of diversity within populations/landraces.

When humans migrated in response to population increases, land degradation, climate changes and wars, they took their staple food crops with them, including barley. As barleys moved into higher latitudes and environments with different abiotic and biotic stresses, the natural variation present in many of the landraces was the source of adaptability that allowed cultivation of the crop to be conducted in many diverse environments around the world. Many barleys are functionally day neutral in the low latitudes of Mediterranean and tropical climates, whereas they are much more photoresponsive in the more temperate regions of Europe and the Americas and the short-season, very long-day regions near the polar circles in northern and southern regions of the world. Specific adaptation to each of the widely different climatic regions would have contributed further to the loss of diversity within populations.

In the early, self-sufficient farming scenarios where animals provided the power and locomotion for the farming operation, the straw of most cereals, including barley, was often used as a form of forage for animals. As modern, mechanized, fossil fuel-based crop production systems became more common, animals were no longer a part of the basic system. Cash cropping became a specialized type of agriculture where the grain was sold to generate income, which was then used to purchase plant and animal food products for consumption and manufactured products that were not produced on the individual farm. With animals, and the associated need for forages, removed from the system, the straw became a less desirable product and so there was a shift toward shorter straw varieties with a higher harvest index (ratio of desired/economic part of the crop to the total harvestable biomass). Semi-dwarf barleys have several advantages over their taller counterparts, such as better standability simply because they do not have as much leverage factor leading to lodging. Shorter straw is often a result of less cell elongation in the stems, which results in thicker, stronger cell walls, and therefore stronger straw. Less elongation also frees up some energy to be used for grain filling, and increases yield. Such shorter barleys are more tolerant of higher rates of fertilizer application, which leads to higher grain yields and avoids the severe lodging of normal tall types. The greater use of irrigation, and associated increases in yield and total biomass production, is also an environment in which shorter straw is a distinct advantage.

The morphological evolution of barley from the wild to the most modern cultivars has occurred in several reasonably distinct steps. The first step, and arguably the most significant, was the transition from the wild, self-sowing, self-reliant *H. spontaneum* to the cultivated version with the tough rachis that was dependent on humans for propagation. In the cultivated environment, these primitive domesticated barleys would have undergone natural selection for larger seeds, more vigorous and competitive seedlings and more synchronous tillering and seed ripening, to give a higher portion of fully mature seeds at the time of human harvesting. There would also have been a natural tendency for those seeds that were threshed most easily to be used most frequently for the next sowing, even without any conscious human selection. Natural selection would have also favoured those types that were the most productive under the cultivated environment, which may have included soil enrichment from animal manures, and so greater productive ability would have been selected naturally, including the development of six-rowed spikes and disease resistance. Human-mediated mass selection may have also favoured the more productive plants and the larger-seeded types.

## 2.5 Modern Loss of Diversity

Human preference would have played a dominant, and maybe the exclusive, role in selection for hulless grains and non-blue aleurone types (and possibly smooth awns) in ancient times. All of these traits are undesirable in a truly wild species. Ancient migrations would have resulted in selection for more specifically adapted types. Modern, mechanized, high-productivity-driven agriculture has favoured the development of widely adapted, photoperiod-insensitive, semi-dwarf types with high harvest indices and the ability to respond to increased inputs of irrigation, fertilizer and pesticides.

The most modern, highly domesticated barley cultivar theoretically would be/have: six-rowed, hulless, semi-dwarf, day-neutral, smooth awned, white aleurone, large seed, high, synchronous tillering ability, along with the basic tough rachis, threshability gene and lack of long-term dormancy. Different combinations of these traits are used for certain specific end uses where there are advantages to certain aspects of the crop morphology, such as covered grain and two-rowed spikes in cultivars specifically used in malting and brewing.

## 2.6 Modern Agriculture

Originally, farms produced raw commodities, both crops and animals, which were then processed minimally to become food, most of which was consumed on the same farm. The food that was not consumed fresh was preserved on the farm, or centrally, then distributed via local farmers' markets to small retail outlets, where it was purchased by an urban consumer. In urban environments, consumers produced very little of their own food. The modern trend toward specialized farming practices has resulted in farmers who also produce very little of their own food. The commodities produced on the modern farm are essentially not edible in their initial harvested state without further processing. So, the crops (and animals) are generally 'exported' from the farm to a processing facility, where they are converted into edible products and then purchased by consumers, including farmers, for preparation of meals in the home. This is in sharp contrast to early farmers, who produced nearly all of their own food, and if there were an excess of a crop or animal, they would process it into a preserved or transportable form that was then sold off the farm. Most of their time and energy went into producing what they themselves needed, and there would have been a small amount of excess that would be their net profit. Modern farmers produce almost all of their crops for export off the farm, and then purchase the vast majority of the food the family eats from local retail outlets, owned by national chains or multinational corporations, after their commodities have been processed into those food products. Shifts toward convenience foods and processed food mean that farmers, who produce the vast majority of the raw products that are processed into food, actually eat very little of their own production directly on the farm. They are now as much 'consumers' as are their urban counterparts.

## 2.7 Modern Breeding Methods

Modern breeding programmes have evolved from more traditional pedigree or bulk-type methods developed on discovering Mendel's laws of genetic patterns of segregation at the beginning of the 20th century. These early methods emphasized bi-parental crosses among known homozygous and homogeneous pure lines. The females' spikes were emasculated and pollen of a specific male genotype was applied to the stigmas of individual flowers to produce small numbers of nearly identical crossed ($F_1$) seeds. These hybrid seeds are then grown and selfed to produce an $F_2$ population, which may be subjected to selection for some basic traits. Plants are usually self-pollinated for several generations and then subjected to selection for numerous quantitatively inherited/expressed traits when a high level of homozygosity, and the associated high heritability, has been reached. Populations of doubled haploid plants may also be produced from the $F_1$ plants to capture the $F_2$ gametes in the homozygous form and then be subjected to selection. These methods rely on segregation among the alleles contributed by the original pair of parents, and are thus limited to two alleles at a locus and their recombination products.

The main limitation to diversity is the selection of parents to use in the original cross combinations. Since crossing is time-consuming and costly, a limited number of crosses are made in most programmes, and a few elite parents are used in most of the crosses. The same few elite parents are likely used in other programmes breeding for adaptation to similar regions or with similar quality or disease resistance goals. McProud (1979) postulated that, in addition to the limited founder diversity, long breeding cycle times were another major contributor to lack/loss of diversity in several barley breeding programmes that he studied.

Modern approaches using marker-assisted selection may be contributing even more to the loss of diversity by using few sources of single major genes that have identifiable markers associated with specific alleles to backcross into populations for molecular selection. Usually, a few individuals containing the desirable alleles, or allelic combinations, at a few specific loci are selected from these populations for further evaluation for additional agronomic and quality traits. Thus, breeding population size is often limited by the efficiency and effectiveness of the marker selection technique itself.

## 2.8 Proposed Methods and Materials

Loss of diversity is almost universal in modern crops developed for the commodity market by large, focused, 'industrial' breeding programmes. The ideotype is fairly narrow and generally well defined by 'the market' and regional adaptation. A different approach to breeding needs to be developed to change the trend toward ever greater loss of diversity in the future.

The basic methods being outlined in this paper are a combination of the domestication process proposed by Cowling *et al.* (2009) and the RIPE (Recurrent Introgressive Population Enrichment) system (Falk, 2002) employed as the core barley breeding method at the University of Guelph, Ontario, Canada (Falk, 2011). The first step is to specify the critical traits necessary to produce a primitive cultivated barley phenotype and identify the specific alleles necessary to obtain this phenotype. When the number of loci is determined and the segregation at each locus is known, then the process of assembling the desired genotype in a wild (*H. spontaneum*) background can be undertaken using the appropriate population sizes to ensure that the critical genotype would be produced. The pre-sowing selection of male sterile plants developed by Falk *et al.* (1982) would be used for initial intercrossing and for subsequent backcrosses.

There are two basic routes that can be taken in this project, one resulting in diverse populations of *H. spontaneum* with the basic domestication traits and the second leading to the introgression of multiple chromosome segments of *H. spontaneum* into an elite population of modern, cultivated barley that has an increased potential to produce new cultivars with novel *H. spontaneum* genes.

The first route would be simply to incorporate the minimal phenotype of the domesticated barley (tough rachis, threshability and possibly low dormancy) into a diverse genepool of *H. spontaneum* strains. This could be accomplished by crossing the *H. spontaneum* selections onto male sterile *H. vulgare* plants (based on the *sex1msg6* system of Falk *et al.*, 1982, and Falk, 2011) and then selfing the $F_1$ crosses to obtain an $F_2$ population that would be selected for the 'domestication complex'. Selected $F_3$ lines with the full domestication phenotype, which are also segregating for male sterility, would be backcrossed with a diverse set of *H. spontaneum*, followed by selfing and repeated selection for the domestication phenotype. Thus, a very diverse population of *H. spontaneum* with basic domestication traits and segregating for male sterility would be assembled. This pre-domesticated population could then be subjected to selection for numerous desirable agronomic, disease resistance and quality traits, and should have the genetic potential to respond to selection in many different directions, including basic adaptation. The ability to respond to selection should be much greater than current populations of *H. vulgare* or any single population of *H. spontaneum*. When a phenotype with the desired degree of expression of a trait is obtained, it could then be crossed with modern cultivars without the necessity to select for domestication traits, in addition to all the other segregating factors inherent in such wide crosses. By fixing the domestication traits in the *H. spontaneum* background before beginning the breeding/selection process in the diverse population(s), they could be managed like 'normal' cultivated barley (i.e. conventional planting, harvesting and processing equipment). Having these critical genes fixed prior to crossing with modern cultivated barley would also simplify the segregation in the hybrid population and give an improved probability of selecting a desired genotype in the population.

This approach would be a novel use of genetic diversity from *H. spontaneum* and could give rise to phenotypes well beyond the range of current *H. vulgare* lines. After exploitation of the tremendous genetic potential of *H. spontaneum* (Ellis *et al.*, 2000), it could be incorporated efficiently into *H. vulgare* using the male sterility already present in the derived *H. spontaneum*. The *H. spontaneum* would be much easier to manage, as it would have a tough rachis, be more threshable and would not have the seed dormancy of the true wild type. Seed size and plant type may also respond quickly to natural selection in the cultivated environment.

It should be possible to select for a number of desirable traits within this 'pre-domesticated' wild barley (Ellis *et al.*, 2000; Gong *et al.*, 2012). There should be significant variation for adaptation, seed size, straw strength and increased nutritional quality, and possibly even improved malting quality, in the diversity that could be assembled by intercrossing among different *H. spontaneum* accessions, and would contain just as much useful genetic variation. Ultimately, populations of such 'pre-domesticated' barley from the Middle East could be combined with similarly 'pre-domesticated' barleys from the Himalayas (Nevo, 2012; Sun *et al.*, 2012) and other diverse adaptation zones (Nevo, 1992) using the incorporated male sterility. This could result in far more genetic diversity than currently exists in cultivated barley, and it would be far more accessible to breeders than wild *H. spontaneum* itself. It is debatable whether the *rob1* allele (Falk, 2011) would be of value in these populations, as recombination would be the principal objective rather than the development of pure lines as cultivars. The persistence of male sterility, and the associated outcrossing in the population, would be a desirable feature.

A second approach to creating more diverse populations of *H. vulgare* involves isolating the domestication traits from a series of initial biparental crosses of male sterile cultivated barleys with a range of *H. spontaneum* lines. The first cross series would then be backcrossed to male sterile plants from the elite barley population in a fashion similar to Cowling *et al.* (2009) and the basic RIPE system of Falk (2011). The population produced from this backcross would be managed in the manner of the segregating population illustrated by Cowling *et al.* (2009) and could then be taken to 'domesticated' pure lines as $BC_1$ (75% *H. vulgare*), or backcrossed again to the elite population and the $BC_2$ (87.5% *H. vulgare*) 'domesticated' pure lines developed. By backcrossing to domesticated barley, the proportion of lines homozygous for the desired domesticated genes would increase (Cowling *et al.*, 2009) and it would be easier to isolate fully domesticated lines with some desired traits from the *H. spontaneum*. This is the general process used at present in many breeding programmes; however, it is made much more efficient through the use of the *sex1*-marked male sterile system (Falk, 2011) and the greater diversity of the *H. spontaneum* foundation lines. The *rob* allele could be introduced in the second BC, as it is in the most recent version of the RIPE system (Falk, 2011).

## 2.9 Discussion

Wild barley, *H. spontaneum*, even though it is fully compatible with domesticated barley, is used in a limited way in breeding, primarily as a source of specific genes for disease resistance. However, *H. spontaneum* populations contain a tremendous range of traits that would be useful in the domesticated crop (Nevo, 1992, 2012; Badr *et al.*, 2000). Many of these traits are determined by quantitative genes or complex inheritance and/or expression patterns and so cannot be transferred readily by simple backcrossing. If wild barleys were pre-domesticated by incorporation of the essential domestic traits of tough rachis and ease of threshability into a range of *H. spontaneum* populations, the ability to use the diversity represented by the wild barleys would be greatly simplified.

The pre-domesticated wild barleys could be crossed with domesticated barleys and new combinations of traits could be extracted readily from the recombinant populations without the issues around shattering and threshability normally associated with such wide crosses. The pre-domesticated *H. spontaneum* could be managed under 'normal' management practices and selection for numerous desirable traits could be applied easily to the populations. Thus, *H. spontaneum* potentially could be exploited directly as a source of new cultivars with unique and valuable traits, as well as used as parents in crosses with domesticated barley more efficiently and effectively than is currently possible. New combinations of genes for agronomic and quality traits, as well as disease resistance and adaptation, would lead to greater diversity of the cultivated crop and the potential for developing new cultivars with a greater expression of existing traits. This approach can lead to the re-domesticating of barley on a much greater scale than the original domestication event(s); and pre-domesticating *H. spontaneum* would address the issue of lack of useful diversity in the domesticated crop (Nevo, 2012). This concept is applicable to most crop species, particularly where the wild progenitor still exists and is fully compatible with the domesticated crop species. The fewer essential traits that separate the wild species from the cultivated crop, the easier the task of pre-domestication would be.

## References

Badr, A., Muller, K., Schafer-Pregl, R., El Rabey, H., Effgen, S., *et al.* (2000) On the origin and domestication history of barley (*Hordeum vulgare*). *Molecular Biology and Evolution* 17(4), 499–510.

Cowling, W.A., Buirchell, B.J. and Falk, D.E. (2009) A model for incorporating novel alleles from the primary gene pool into elite crop breeding programs while reselecting major genes for domestication or adaptation. *Crop and Pasture Science* 60(40), 1009–1015.

Ellis, R.P., Forster, B.P., Robinson, D., Handley, L.L., Gordon, D.C., *et al.* (2000) Wild barley: a source for crop improvement in the 21st century? *Journal of Experimental Botany* 51(342), 9–17.

Falk, D.E. (2002) Theory, methodology, and results of using recurrent selection in cereals. *EUCARPIA*, Salsomaggiorre, Italy.

Falk, D.E. (2011) Generating and maintaining diversity at the elite level in crop breeding. *Genome* 53, 982–991.
Falk, D.E., Kasha, K.J. and Reinbergs, E. (1982) Presowing selection of genetic male sterile plants to facilitate hybridization in barley. *Barley Genetics IV. Proceedings of the 4th International Barley Genetics Symposium* Edinburgh, Scotland, pp. 778–785.
Gong, X., Li, C., Zhang, G., Yan, G., Lance, R., *et al.* (2012) Novel genes from wild barley *Hordeum spontaneum* for barley improvement. In: Zhang, G., Li, C. and Liu, X. (eds) *Advances In Barley Sciences, Proceedings of the 11th International Barley Genetics Symposium*. Zhejiang University Press, Hangzhou, China, pp. 43–56.
Harlan, J.R. (1992) *Crops and Man*, 2nd edn. American Society of Agronomy, Crop Science Society, Madison, Wisconsin, pp. 284.
McProud, W. (1979) Repetitive cycling and simple recurrent selection in traditional barley breeding programs. *Euphytica* 28(2), 473–480.
Nevo, E. (1992) Origin, evolution, population genetics and resources for breeding of wild barley, *Hordeum spontaneum* in the Fertile Crescent. In: Shewry, P.R. (ed.) *Barley: Genetics, Biochemistry, Molecular Biology and Biotechnology*. CAB International, Wallingford, UK, pp. 19–41.
Nevo, E. (2012) Evolution of wild barley and barley improvement. In: Zhang, G., Li, C. and Liu, X. (eds) *Advances In Barley Sciences, Proceedings of the 11th International Barley Genetics Symposium*. Zhejiang University Press, Hangzhou, China, pp. 1–16.
Schmalenbach, I., March, T.J., Bringezu, T., Waugh, R. and Pillen, K. (2011) High-resolution genotyping of wild barley introgression lines and fine-mapping of the threshability locus *thresh-1* using the Illumina Goldengate assay. *Genes, Genomes, Genetics* 1(3), 187–196.
Sun, D., Xu, T., Zhang, G., Ling, Z., Sun, D., *et al.* (2012) Exploiting and utilizing the novel annual wild barley germplasms on the Quing-Tibet plateau. In: Zhang, G., Li, C. and Liu, X. (eds) *Advances In Barley Sciences, Proceedings of the 11th International Barley Genetics Symposium*. Zhejiang University Press, Hangzhou, China, pp. 64–71.

# 3 Unravelling Quinoa Domestication with Wild Ancestors

D. Bertero[1]* and A. Alercia[2]
[1]Faculty of Agronomy, University of Buenos Aires, Buenos Aires, Argentina;
[2]Bioversity International, Rome, Italy

## 3.1 Introduction

Little known a few years ago to the general public and to most scientists in the field of agriculture, the Andean crop quinoa (*Chenopodium quinoa* Willd.) made a sudden appearance in the international community. A combination of traits made this species known not only to consumers (because of the high nutritional quality of the seeds) but also to crop physiologists and experts in agriculture for marginal lands. The only edible seed-producing halophyte, the high tolerance of quinoa to salinity made it a model plant for research, with a number of articles published in just a few years (see Jellen *et al.*, 2013, for a more detailed revision). Quinoa also shows a very high tolerance to water deficits and low temperatures (Jacobsen *et al.*, 2003), which explains why this is the only species growing under rain-fed conditions in the southern highlands of Bolivia, where a cold and dry environment limits agriculture, lacking a frost-free period, with poor soils and frequent salinity problems and with an average rainfall of *c.*150 mm/year. These extreme environmental conditions suggest that quinoa could be a kind of 'climate change prone crop', providing food where many of the other crops fail. Evidence of this growing interest was the declaration of 2013 as the International Year of Quinoa by the United Nations and the recent publication of a book on the state of the art of quinoa by FAO (Bazile *et al.*, 2014).

Details on the process of domestication and migration of this crop, however, remain obscure, and up to now have attracted the attention of only a few researchers, mostly archaeologists. Possibly, this is a combination of different interests in the species, which first focused on botany (taxonomy, morphology), then nutrition and finally crop/plant and seed physiology, all related to practical concerns and also because some hypotheses about quinoa's domestication were considered 'facts' (for example, it is given for certain that quinoa was domesticated in the Titicaca Basin).

In this paper, we approach quinoa domestication asking three questions:

**1.** What do we know about quinoa genetic structure, relations among related species and phylogenetics?
**2.** What is known about *Chenopodium hircinum*, its putative wild ancestor?
**3.** How can research on *C. hircinum* contribute to the understanding of the quinoa domestication process and breeding for adaptation to conditions where quinoa performs poorly but *C. hircinum* has adapted?

*Corresponding author; e-mail: bertero@agro.uba.ar

## 3.2 Quinoa Genetic Structure and Relations

Based on a combination of morphological (seed and leaves) traits and isozyme variation, a wide geographical range was explored for cultivated, weedy and wild quinoa and related species (Wilson, 1990). This analysis gave origin to various discoveries and hypotheses about quinoa genetic structure and dissemination that are still valid today. Quinoa belongs to subsection *Cellulata* of the genus *Chenopodium*, and is distinguished from other members of the same genus by particular seed traits. The more closely related species to quinoa are *Chenopodium berlandieri* and *C. hircinum*. Wilson (1990) proposed the distinction of two subspecies within *C. berlandieri*: subsp. *berlandieri*, from North America (wild), and subsp. *nuttalliae* from Mexico (huazontle, cultivated). A third subspecies, not included in the analysis (subsp. *jonesianum*) was domesticated in the USA, but became extinct around 1000 years ago (Wilson, 1981; Gremillion, 1993; Smith, 2006). Wilson distinguished between *C. quinoa* subsp. *quinoa* (cultivated) and subsp. *milleanum* or var. *melanospermum* (weed escapes and wild forms). A further var. *melanospermum* has been proposed by Hunziker (1943) as a variety of quinoa. So far, Wilson has been the only author conducting a study on *C. hircinum* using the same approach (Wilson, 1988a,b). Cultivated quinoa was clearly organized into two genetic groups: Andean and coastal, this latter from central and southern Chile. The Andean forms were further subdivided into northern and southern Andes groups, but with less clear distinction than that found in the comparison with the coastal group. This Andean–coastal division had already been defined by Tapia *et al.* (1979), who proposed naming these variants as 'ecotypes', to distinguish morphological types linked to different areas of cultivation.

Based on this theory, the geographical distribution of variability (lower variability the further away from the Titicaca Basin) and the capacity for interbreeding among all these species (Wilson *et al.*, 1983; Wilson and Manhart, 1993), Wilson proposed a series of related hypotheses: (i) the ancestors of quinoa originated in North America and migrated to South America, giving origin to *C. hircinum*; (ii) quinoa was domesticated independently in the Andes (those starting in *C. berlandieri* (subsp. *nuttalliae* and *jonesianum*) being independent domestications); (iii) the coastal group is an archaic form of quinoa, as some panicle shapes ('amarantiform', as distinct from the ancestral 'glomerulate' shape) and seed traits (floury against translucent perisperm) are only found in Andean quinoa and are considered derived traits (not present in wild populations). Wilson also used the reduced variability detected by isozyme analysis in Chilean quinoa as verification of a founder effect associated with this early, non-representative, migration of some populations to southern America. As quinoa and the related species are allopolyploid, and because of the close relations between the aforementioned species, the alternative hypothesis suggesting quinoa arose in the Andes after a South American polyploidization (Gandarillas, 1979) was disregarded as highly unrealistic. It has been also suggested that an extinct tetraploid could have been involved in quinoa domestication (Wilson, 1990), but no scientific evidence to support this has been found so far.

There was a knowledge gap until the development of DNA-based molecular markers. The use of molecular techniques such as random amplified polymorphic DNA (RAPD) and amplified fragment length polymorphism (AFLP), and the subsequent development of microsatellite markers (simple sequence repeat, SSR), allowed further studies of variability and genetic structure of the species. Then, it was confirmed that the notion of reduced genetic variability the further the distance from Lake Titicaca, instrumental in the hypothesis of that being the centre of origin, was not supported by the use of these markers (del Castillo *et al.*, 2007; Fuentes *et al.*, 2009; Anabalón Rodriguez and Thomet Isla, 2009). Nevertheless, none of this research allowed for more than a very fragmented reconstruction of the genetic structure and the migration process.

Regarding other aspects involved in Wilson's hypothesis, the close relation between *C. quinoa* and *C. berlandieri* was proved by artificial (Wilson, 1980) and natural (Wilson and Manhart, 1993) hybridization between them, supporting further the identity of the subsection *Cellulata*. Wilson also tested the hypothesis that quinoa and huazontle (*nuttalliae*) were the result of a single domestication (like North and South American maize). Hybridization between both species proved that the white seed characteristics of domesticated forms were controlled by different genes in both species, thereby discarding that hypothesis (Heiser and Nelson, 1974). A note of caution has been posed in what is understood by quinoa relatives, as some germplasm collections include more distant species like *Chenopodium pallidicaule* (cañihua) and *Chenopodium ambrosioides* (paico). This loose definition of 'quinoa relatives' has appeared several times in the literature (Jacobsen and Mujica, 2002; Bazile *et al.*, 2013) but, to our knowledge, no Andean quinoa collection includes *C. berlandieri* (wild and cultivated forms). This led Jellen and co-workers to propose that germplasm collections should focus only on species with a demonstrated close relationship with quinoa (Jellen *et al.*, 2011, 2014). Dr Jellen's team recently proposed considering the *Chenopodium quinoa-hircinum-berlandieri* as a single biological species, and went one step further in identifying the diploid ancestors of this complex. One of these genomes is present in *Chenopodium ficifolium* from Eurasia, and is also present in *Chenopodium album* (Kolano *et al.*, 2011), the other genome in *Chenopodium standleyanum* from North America (Jellen *et al.*, 2014), although this is still presented as a provisional hypothesis after having explored many *Chenopodium* species from all over the world. This recent evidence strongly supports the idea of North America as the place of origin of quinoa ancestors (Wilson, 1990).

## 3.3 Hypotheses about Quinoa Domestication

Quinoa has been part of the Andean agriculture for thousands of years and supposedly as a domesticate from as early as 7000 years BP (Dillehay *et al.*, 2007). This date was argued based on information about the context and not on the seeds themselves, but cultivated forms were present ~5000 years BP in Ayacucho, Peru (Risi and Galwey, 1984), ~3000 BP in Chinchorro, Chile (Risi and Galwey, 1984), ~3000 BP in Antofagasta de la Sierra, Argentina (Andrade *et al.*, 2014) and ~3500 BP in Chiripa, Bolivia (Bruno and Whitehead, 2003). Gandarillas (1979) proposed that quinoa was domesticated in the Titicaca Basin, somewhere between Cuzco, Peru, and Lake Poopo, Bolivia, using the large phenotypic variation found around that area as the main support for his hypothesis. Wilson (1988b) commented on a proposition by Aellen (1929) stating that Ecuadorian 'ajaras' (*C. quinoa* subsp. *milleanum*) could be a link between quinoa and *C. berlandieri* subsp. *szchackeii* from North America and suggested that quinoa could have been domesticated in the northern Andes. Gandarillas also objected to the notion of North America being the place of origin of quinoa wild ancestors, basing this assertion on hybridizations he carried out using *C. quinoa*, *C. hircinum*, *Chenopodium petiolare* and *C. berlandieri* ssp. *nuttalliae* (Gandarillas, 1986). He stated that the Andean forms of *C. hircinum* were diploid plants, and that by crossing with another diploid, *C. quinoa* originated in the Andes (Gandarillas, 1984). According to Wilson (1988b), no evidence of *C. hircinum* diploid forms has been found, and recent genetic analyses show that the chances of quinoa being the result of an independent polyploidization event are minimal (Jellen *et al.*, 2014). The 'Ecuadorian domestication hypothesis' was not supported by data on genetic variability and was also discarded (Wilson, 1988a).

The domestication syndrome proposed for North American *Chenopodium* (Gremillion, 1993) was used by Bruno (Bruno and Whitehead, 2003; Bruno, 2006) to study the domestication process in Chiripa, Bolivia, close to the Titicaca lake. Her study encompassed a period from 1500 BC to 100 AD. She found a trend in seed size (bigger seeds), relative testa width (episperm to seed) and colour (black to white), indicating an increasing selection pressure against wild forms and supporting the notion of quinoa being domesticated somewhere around Lake Titicaca at that time. Nonetheless, a few years later, evidence of an independent domestication process was found near Oruro, around 300 km south of Chiripa (Langlie *et al.*, 2011). This theory was made more complex with the findings of what was regarded by the authors as another domestication event thousands of kilometres away, in central Chile (Planella *et al.*, 2005). The study corroborates that quinoa is poly- rather than monophyletic, and this matches with proof of the other two domestication events, at least in North America (Heiser and Nelson, 1974; Gremillion, 1993).

To further complicate this, no 'domesticated trait' seems to be fixed in quinoa (when variation for some traits of agronomic interest is eliminated). The crop is sympatric to wild forms over almost all its distribution and easily cross-pollinates with them. Wilson (1990) found that by using isozyme markers, quinoa populations in the Andes were more similar to their local wild counterparts than to quinoa populations from other sites, and this was also verified by using microsatellite markers (Gonzalez Marín, 2009). A comparison of local accessions in the north-west Argentina germplasm collection confirmed a wide range of variation for most domestication traits (seed size, colour, degree of branching, panicle compactness and whether or not a simple terminal – or several axillary – panicles were found within modern crops (Curti *et al.*, 2012)), contrasting in some way with Bruno's theory of simple linear temporal tendencies in quinoa. The truncate margin of the seed, a 'distinctive domestication trait' for several authors, is also found in some 'ajara' accessions (Wilson, 1988c), perhaps as a consequence of regular gene flow between cultivated and wild forms.

## 3.4 *Chenopodium hircinum sensu stricto*

Very little is known about this species. Distributed in Chaco and Pampa biomes in Argentina, Uruguay and Brazil (Wilson, 1988b,c, 1990), and also in Chile (i.e. specimens stored in the Herbarium of the University of Concepcion (www2.udec.cl/~herbconc/)), it grows in environments where there has never been proof of quinoa cultivation (Uruguay and Brazil) or where its cultivation disappeared after the XVIII century, as in central Argentina (Gilii and Xuarez, 1790: Paucke, 2010). Herbarium specimens of this species are found in several universities and research institutions, but it has rarely been included in quinoa germplasm collections (del Castillo *et al.*, 2007; Rojas *et al.*, 2014) and, in the best case, do not include representative collections from the lowlands. The most thorough analysis of variation in *C. hircinum* was made by Wilson (1988b,c), who began his research by considering the possibility that *C. hircinum* included two forms, those sympatric with quinoa ('ajaras') and those from the lowlands. After making comparisons at fruit, leaf and molecular (isozyme) levels, he proposed that Andean forms be considered either as subspecies of quinoa (*C. quinoa* subsp. *milleanum*) or to keep it as varietal adscription (*C. quinoa* var. *melanospermum*), as proposed by Hunziker (1943). This analysis also suggested lowland *C. hircinum* as a separate entity. It has well-developed lateral and basal branching, a diffuse and lax inflorescence and extremely small fruits, this latter perhaps as a consequence of selective constraints not present in domesticates (Wilson, 1988b; Sadras, 2007). Allopatry with the Andean crop–weed complex indicates reproductive isolation and specific differentiation (Wilson, 1990), although still cross fertile (Wilson *et al.*, 1983) with quinoa, as previously stated. Wilson has also emphasized that, on the basis of isozyme analysis, *C. hircinum* exhibits more polymorphic loci, more alleles and a higher level of heterozygosity than the Andean crop–weed complex (Wilson, 1988c), with Andean quinoas having a subset of alleles found in Argentinean populations of *C. hircinum*. In a recent analysis (Maughan *et al.*, 2014), SNPs from quinoa were amplified successfully (81%) in two *C. hircinum* accessions from the Argentinean humid Pampas as part of a transferability test of these markers to quinoa wild relatives (*C. berlandieri* was also included). This research was unable to detect higher polymorphism in wild versus domesticated samples. Nevertheless, these SNPs were developed to differentiate quinoa accessions and not wild species, pointing to a need to develop *C. hircinum* and *berlandieri* specific markers (R. Jellen, Washington, 2013, personal communication).

A particular remark made by Wilson is that some populations from mid-elevation altitudes in north-west Argentina have intermediate traits between *C. hircinum* and 'ajara', perhaps as a result of a contact zone or past sympatry (Wilson, 1990). No systematic seed collection of *C. hircinum* has been done there and it is part of ongoing research (R. Curti, Argentina, 2014, personal communication), and the original collection by Wilson is lost or unavailable (H. Wilson, Texas, 2011, personal communication). In addition, other areas where *C. hircinum* is found have not been explored systematically, and little research has been done on its physiology and variation of traits other than fruit and seeds

(González *et al.*, 1988; Moris *et al.*, 1996). A collection of *C. hircinum* samples (comprising around 15 accessions) was initiated by the Faculty of Agronomy of the University of Buenos Aires and the University of Salta (Argentina), including accessions from the humid Pampas (Buenos Aires and Santa Fé provinces) and Chaco environments (Córdoba, Santa Fé, Santiago del Estero and Tucumán provinces). Preliminary observations of this germplasm showed high variation in seed traits (important ranges of size and colour), leaf morphology, degree of branching and salinity tolerance.

## 3.5 Concluding Remarks

Well-founded data on quinoa variability and genetic structure appeared only a decade ago with the development of highly informative DNA-based markers. Although the former geographical distribution of variability (Gandarillas, 1986; Wilson, 1988a,b, 1990) has been challenged by this research, the main hypothetical framework, posed by Wilson (1990), remains as the reference idea guiding research about where, how and when quinoa was domesticated. Archaeological research is also fairly recent and supportive of Wilson's theory, at least for the general picture, and molecular research using wild species also provides support along the same lines. Much work remains to be done on the relationship between quinoa and its wild relatives in terms of genealogy, distribution and timing of domestication; however, the open gene flow between species and the concrete possibility that all of them are just a single species (Jellen *et al.*, 2014) also puts forward a challenge to the simple notion of just one domestication (and a bottleneck reduction in genetic diversity), followed by migration and adaptation to specific environments. It is highly likely that quinoa could have 'harvested genes' from local wild plants by recurrent introgression processes. Besides its relevance to domestication research, *C. hircinum* can be a source of adaptive traits not present in quinoa (as an example, *C. hircinum* grows in extremely hot environments in central Argentina and adapts to different soil types). A new scenario for quinoa domestication can be drawn from the relationship between quinoa and its wild relatives. Further research on *C. hircinum* could contribute to the understanding of the quinoa domestication process and to quinoa breeding and adaptation to new environmental challenges. There is a need to create a representative collection and to develop species-specific markers that will contribute to the identification of the 'domestication trait alleles'.

## References

Aellen, P. (1929) Beitrag zur Systematik der *Chenopodium* Arten Americas, vorwiegend aud Grund der Sammlung des United States National Museum in Washington, D.C. *Repertorium speciarum novarum regni vegetabilis* 26, 31–64.

Anabalón Rodriguez, L. and Thomet Isla, M. (2009) Comparative analysis of genetic and morphologic diversity among quinoa (*Chenopodium quinoa* Willd.) accessions of the south of Chile and highland accessions. *Journal of Plant Breeding and Crop Science* 1, 210–216.

Andrade, A., Babot, P., Bertero, H.D., Costa Tártara, S.M., Curti, R.N., *et al.* (2014) Argentina. In: Bazile, D., Bertero, H.D. and Nieto, C. (eds) *Estado del arte de la quinua en el mundo en 2013*. FAO and CIRAD, Santiago, Chile, and Montpellier, France, pp. 504–518.

Bazile, D., Fuentes, F. and Mujica, A. (2013) Historical perspectives and domestication. In: Barghava, A. and Srivastava, S. (eds) *Quinoa. Botany, Production and Uses*. CAB International, Wallingford, UK, pp. 16–35.

Bazile, D., Bertero, H.D. and Nieto, C. (2014) *Estado del arte de la quinua en el mundo en 2013*. FAO and CIRAD, Santiago, Chile, and Montpellier, France.

Bruno, M.C. (2006) A morphological approach to documenting the domestication of *Chenopodium* in the Andes. In: Zeder, M.A., Bradley, E.S. and Smith, B.D. (eds) *Documenting Domestication: New Genetic and Archaeological Paradigms*. University of California Press, Berkeley, California, pp. 32–45.

Bruno, M.C. and Whitehead, W.T. (2003) *Chenopodium* cultivation and formative period agriculture at Chiripa, Bolivia. *Latin American Antiquity* 14, 339–355.

Curti, R.N., Andrade, A.J., Bramardi, S., Velásquez, B. and Bertero, H.D. (2012) Eco-geographic structure of phenotypic diversity in cultivated populations of quinoa (*Chenopodium quinoa* Willd.) from North-Western Argentina. *Annals of Applied Biology* 160, 114–125.
del Castillo, C., Winkel, T., Mahy, G. and Bizouz, J.P. (2007) Genetic structure of quinoa (*Chenopodium quinoa* Willd.) from the Bolivian Altiplano as revealed by RAPDs markers. *Genetic Resources and Crop Evolution* 54, 897–905.
Dillehay, T.D., Rossen, J., Andres, T.C. and Williams, D.E. (2007) Preceramic adoption of peanut, squash and cotton in Northern Peru. *Science* 316, 1890–1893.
Fuentes, F.F., Martínez, E.A., Hinrichsen, P.V., Jellen, E.N. and Maughan, P.J. (2009) Assessment of genetic diversity patterns in Chilean quinoa (*Chenopodium quinoa* Willd.) germplasm using multiplex fluorescent microsatellite markers. *Conservation Genetics* 10, 369–377.
Gandarillas, H. (1979) Botánica. In: Tapia, M., Gandarillas, H., Alandia, S., Cardozo, A., Mujica, A., Ortiz, R., Otazu, V., Rea, J., Salas, B. and Zanabria, E. (eds) *Quinua y Cañihua, Cultivos Andinos*. CIID-IICA, Serie de Libros y Materiales educativos No. 40, Bogotá, Colombia, pp. 20–33.
Gandarillas, H. (1984) Obtención experimental de *Chenopodium quinoa* Willd. Ministerio de Asuntos Campesinos y Agropecuarios IBTA, La Paz, Bolivia.
Gandarillas, H. (1986) *Estudio anatómico de los órganos de la quinua. Estudio de caracteres correlacionados y sus efectos sobre el rendimiento. Hibridaciones entre especies de la Sub-sección Cellulata del género Chenopodium*. Ministerio de Asuntos Campesinos y Agropecuarios, Instituto Boliviano de Tecnología Agropecuaria and Centro Internacional de Investigaciones para el desarrollo, La Paz, Bolivia.
Gilii, F.L. and Xuarez, G. (1790) *Osservazioni fitologiche sopra alcune piante esotiche introdotte in Roma*. Stamperia di Arcangelo Casalatti, Rome, Italy.
González, J.A., Gallardo, M. and Israilev, L.A. (1988) Leaf flavonoids in *Chenopodium hircinum* Schrad. and *Chenopodium album* L. (*Chenopodiaceae*). *Phyton* 63, 279–281.
Gonzalez Marín, S.P. (2009) Estudio del flujo de genes en quinoa (*Chenopodium quinoa* Willd.) en campo de agricultores mediante el uso de marcadores microsatélites. MSc Thesis, Escuela Politécnica del Ejército, Salgolquí, Ecuador.
Gremillion, K.J. (1993) The evolution of seed morphology in domesticated *Chenopodium*: an archaeological case study. *Journal of Ethnobiology* 13, 149–169.
Heiser, C.B. and Nelson, D.C. (1974) On the origin of the cultivated chenopods (*Chenopodium*). *Genetics* 78, 503–505.
Hunziker, A.T. (1943) Las especies alimenticias de *Amaranthus* y *Chenopodium* cultivadas por los indios de America. *Revista Argentina de Agronomía* 30, 297–353.
Jacobsen, S.E. and Mujica, A. (2002) Genetic resources and breeding of the Andean grain crop quinoa (*Chenopodium quinoa* Willd.). *Plant Genetic Resources Newsletter* 130, 54–61.
Jacobsen, S.E., Mujica, A. and Jensen, C.R. (2003) The resistance of quinoa (*Chenopodium quinoa* Willd.) to adverse abiotic factors. *Food Reviews International* 19, 99–110.
Jellen, E.N., Kolano, B.A., Sederberg, M.C., Bonifacio, A. and Maughan, P.J. (2011) *Chenopodium*. In: Kole, C. (ed.) *Wild Crop Relatives: Genomics and Breeding Resources*. Springer, Berlin, Germany, pp. 35–61.
Jellen, E.N., Maughan, P.J., Bertero, H.D. and Munir, H. (2013) Prospects for quinoa (*Chenopodium quinoa*) improvement through biotechnology. In: Jain, S.M. and Gupta, S.D. (eds) *Biotechnology of Neglected and Underutilized Crops*. Springer, Berlin, Germany, pp. 173–201.
Jellen, E.N., Maughan, P.J., Fuentes, F. and Kolano, B.A. (2014) Botánica, filogenia y evolución. In: Bazile, D., Bertero, H.D. and Nieto, C. (eds) *Estado del arte de la quinua en el mundo en 2013*. FAO and CIRAD, Santiago, Chile, and Montpellier, France, pp. 12–25.
Kolano, B., Gardunia, B.W., Michalska, M., Bonifacio, A., Fairbanks, D., *et al.* (2011) Chromosomal localization of two novel repetitive sequences isolated from the *Chenopodium quinoa* Willd. genome. *Genome* 54, 710–717.
Langlie, B.A.S., Hastorf, C.A., Bruno, M.C., Bermann, M., Bonzani, R.M., *et al.* (2011) Diversity in Andean *Chenopodium* domestication: describing a new morphological type from La Barca, Bolivia 1300–1250 B.C. *Journal of Ethnobiology* 31, 72–88.
Maughan, P.J., Jellen, E.N. and Raney, J.A. (2014) Herramientas moleculares y genomicas para la quinua. In: Bazile, D., Bertero, H.D. and Nieto, C. (eds) *Estado del arte de la quinua en el mundo en 2013*. FAO and CIRAD, Santiago, Chile, and Montpellier, France, pp. 26–32.
Moris, M., González, J.A., Gallardo, M. and Prado, F.E. (1996) Anatomical and functional differences and nyctinastic leaf movements in *Chenopodium album* L. and *Chenopodium hircinum* Schrad. (*Chenopodiaceae*). *Botanical Journal of the Linnean Society* 12, 133–141.

Paucke, F. (2010) *Hacia allá y para acá.* Ministerio de Innovación y Cultura de la Provincia de Santa Fé, Santa Fé, Argentina.

Planella, M.T., Cornejo, L. and Tagle, B. (2005) Alero Las Morrenas 1: Evidencia de cultigenos entre cazadores-recolectores de finales del periodo Arcaico en Chile Central. *Chungara* 37, 59–74.

Risi, J.C. and Galwey, N.W. (1984) The *Chenopodium* grains of the Andes: Inca crops for modern agriculture. *Advances in Applied Biology* 10, 145–216.

Rojas, W., Pinto, M., Alanoca, C., Gomez Pando, L., Leon Lobos, P., *et al.* (2014) Estado de la conservación ex situ de los recursos genéticos de quinua. In: Bazile, D., Bertero, H.D. and Nieto, C. (eds) *Estado del arte de la quinua en el mundo en 2013.* FAO and CIRAD, Santiago, Chile, and Montpellier, France, pp. 65–94.

Sadras, V.O. (2007) Evolutionary aspects of the trade-off between seed size and numbers in crops. *Field Crops Research* 100, 125–138.

Smith, B.D. (2006) Eastern North America as an independent center of plant domestication. *Proceedings of the National Academy of Sciences* 103, 12223–12228.

Tapia, M., Gandarillas, H., Alandia, S., Cardozo, A., Mujica, A., *et al.* (1979) *Quinua y Cañihua, Cultivos Andinos.* CIID-IICA, Serie de Libros y Materiales educativos No. 40, Bogotá, Colombia.

Wilson, H.D. (1980) Artificial hybridization among *Chenopodium* species sect. *Chenopodium*. *Systematic Botany* 5, 235–263.

Wilson, H.D. (1981) Domesticated *Chenopodium* of the Ozark Bluff dwellers. *Economic Botany* 35, 233–239.

Wilson, H.D. (1988a) Quinua biosystematics I: domesticated populations. *Economic Botany* 42, 461–477.

Wilson, H.D. (1988b) Quinua biosystematics II: free-living populations. *Economic Botany* 42, 478–494.

Wilson, H.D. (1988c) Allozyme variation and morphological relationships of *Chenopodium hircinum* (s.l.). *Systematic Botany* 13, 215–228.

Wilson, H.D. (1990) Quinoa and relatives (*Chenopodium* sect. *Chenopodium* subsect. *Cellulata*). *Economic Botany* 44, 92–110.

Wilson, H.D. and Manhart, J. (1993) Crop/weed gene flow: *Chenopodium quinoa* Willd. and *C. berlandieri* Moq. *Theoretical and Applied Genetics* 86, 642–648.

Wilson, H.D., Barber, S.C. and Walters, T.W. (1983) Loss of duplicate gene expression in tetraploid *Chenopodium*. *Biochemical and Systematic Ecology* 11, 7–13.

# 4 Screening Wild *Vigna* Species and Cowpea (*Vigna unguiculata*) Landraces for Sources of Resistance to *Striga gesnerioides*

O. Oyatomi,[1]* C. Fatokun,[1] O. Boukar,[1] M. Abberton[1] and C. Ilori[2]
[1]*International Institute of Tropical Agriculture (IITA), Ibadan, Nigeria;* [2]*Department of Crop Protection and Environmental Biology, University of Ibadan, Ibadan, Nigeria*

## 4.1 Introduction

Cowpea (*Vigna unguiculata* (L.) Walp.), also known as black-eyed or southern pea, belongs to the genus *Vigna*, section *Catiang*, species *unguiculata*. It comprises four subspecies, namely: *unguiculata*, *stenophylla*, *dekindtiana* and *tenuis* (Ng and Marechal, 1985). The subspecies *unguiculata* is the only one cultivated, while the other three are wild relatives. Subspecies *unguiculata* is itself subdivided into four cultivar groups (cv-gr) namely *Unguiculata*, *Biflora*, *Sesquipedalis* and *Textilis* (Westphal, 1974). The cv-gr *unguiculata* is the most diverse of the four and is widely grown in Africa, Asia and Latin America. Cowpea is a grain legume grown in savannah regions of the tropics and subtropics.

Cowpea is a warm-season, annual, herbaceous legume. Plant types are often categorized as erect, semi-erect, prostrate (trailing) or climbing. There is much variability within the species. Growth habit ranges from indeterminate to fairly determinate, with the non-twining types tending to be more determinate. Cowpea generally is strongly taprooted.

It is grown mostly in west and central African countries. Its value lies in its high protein content, ability to tolerate drought and the fact that it fixes atmospheric nitrogen, which allows it to grow on and improve poor soils. It is an important protein supplement in the diet of many African people (Bressani, 1985). The dry seed contains approximately 51% carbohydrates, 23–25% protein, 13% fat and 3.5% minerals. It may contribute 60% of the total protein intake of the people of the western part of Nigeria (Rachie and Rawal, 1975). However, cowpea yield is affected negatively by different biotic and abiotic factors. Yield losses in cowpea associated with *Striga gesnerioides* have been reported to be as great as 83% and 100% (Cardwell and Lane, 1995). Emechebe *et al.* (1991) reported 100% yield losses on farmers' fields in the northern Guinea savannah of Nigeria in susceptible local varieties. Yield losses associated with *Alectra vogelii* (Benth.), an obligate, root-parasitic flowering plant of the family Orobanchaceae, range between 70% and 100% (Parker, 1991; Kureh *et al.*, 1999). Various control measures, including cultural practices, chemical and biological means and host plant resistance, have been suggested (Dube and Oliver, 2001; Boukar *et al.*, 2004), but no single field method seems to be fully adequate. Efforts have been made to identify

*Corresponding author; e-mail: o.oyatomi@cgiar.org

 

natural sources of genetic resistance within cowpea germplasm and in the selection and breeding of improved lines with resistance to *Striga* and *Alectra* (Singh and Emechebe, 1997).

Fatokun and Singh (1987) crossed cultivated cowpea successfully with a hairy wild relative (*Vigna pubescens*). They used tissue culture techniques to rescue the hybrid embryos, which would otherwise have shrivelled and degenerated. The $F_1$ plants were vigorous in growth but partially sterile, with only about 32% pollen viability. Cytological investigations of $F_1$ plants showed meiotic abnormalities in the pollen mother cells. These abnormalities included a few univalents and quadrivalents, suggesting some structural differentiation in the chromosomes. Barone and Ng (1990) could not obtain an interspecific cross between *V. unguiculata* and *Vigna vexillata*. They concluded that the following were barriers to crossability: lack of fertilization and collapse of fertilized ovules 5–8 days after pollination. All attempts by Fatokun (2002) to cross *V. vexillata* with various cultivated and non-cultivated cowpeas were also unsuccessful.

## 4.2 Materials and Methods

The studies reported here were conducted on the experimental field of the International Institute of Tropical Agriculture (IITA) at Minjibir, Kano State, Nigeria (Sudan savannah, 12° 08.997′ N, 8° 39.733′ E) during the planting seasons of 2012 and 2013. In 2012, 350 accessions of wild *Vigna* belonging to 45 different species and from 32 countries of origin collected from the Genetic Resources Center (GRC) of IITA were used for the study. Accessions of wild *Vigna* species were planted in August 2012 on a single row of 2 m length, with 75 cm and 20 cm between and within rows, respectively, per accession per replicate in a randomized complete block design in three replications with two susceptible and one resistant check. Seed of *S. gesnerioides* was pre-mixed with oven-dried sandy soil at the ratio of 1.0 g *S. gesnerioides* seed to 1.0 kg oven-dried sandy soil, and each planting hole was inoculated artificially with one teaspoonful of the mixture. Three seeds of each accession were sown per hill and thinned to two plants per stand 2 weeks later. All management practices such as weed control, insecticide application to prevent pre- and post-flowering insects were done at regular intervals.

In 2013, 280 accessions of cowpea landraces that were geographically co-located with the resistant wild *Vigna* species from the *Striga* screening experiment in 2012 were selected from the cowpea core collection available in the GRC. Seeds of the 280 accessions of cowpea landraces were planted following the same planting, field maintenance and data collection procedures as in the previous experiment in the same location in August 2013.

## 4.3 Data Collection

Data were collected on the number of emerged *Striga*/plot (2 m²) at 9 weeks after planting (WAP) and at harvest to assess the host support for *Striga*, and the accessions were thereafter classified using the Singh *et al.* (1997) method as follows:

Resistant (R) = no *Striga* emergence on plot.
Moderately resistant (MR) = 1–2 *Striga* emergence per plot.
Susceptible (S) = 3–4 *Striga* per plot
Highly susceptible (HS) = 5 or more *Striga* emergence per plot.

## 4.4 Results

Twenty-one genotypes from 11 wild *Vigna* species (*Vigna ambacensis*, *Vigna davyi*, *Vigna glabrescens*, *Vigna marina*, *Vigna mungo*, *Vigna oblongifolia*, *Vigna parkeri*, *Vigna racemosa*, *Vigna reticulata*, *V. vexillata* and *V. unguiculata* subsp. *dekindtiana*) originating from Democratic Republic of Congo, Ghana, Kenya, Japan, South Africa, Equatorial Guinea, Republic of Benin, Zimbabwe, Namibia, Costa Rica, Zambia, Republic of Niger, Gambia and Philippines, showed resistance (R) to *S. gesnerioides* with no *Striga* emergence observed in their plots. One hundred and twenty-eight accessions showed moderate resistance (MR), 155 accessions were susceptible (S), with *Striga* counts per plot ranging from 3 to 4, while 35 accessions with high levels of susceptibility (HS) had *Striga* counts per plot ranging from 5 to 7. The 45 wild

*Vigna* species were classified on the basis of their varying degrees of reactions to *S. gesnerioides* (Table 4.1).

Sixteen landrace accessions of cowpea *V. unguiculata*, comprising of seven from Zambia, six from Niger, two from Kenya and one from Ghana, had no *S. gesnerioides* emergence and were classified as resistant (R). Two accessions showed moderate resistance (MR), 25 accessions were susceptible (S), while 237 showed high levels of susceptibility (HS), having *Striga* counts per plot ranging from 5 to 38.

**Table 4.1.** Reaction of wild *Vigna* species to *Striga gesnerioides*.

| Species | Number resistant (R) | Number moderately resistant (MR) | Number susceptible (S) | Number highly susceptible (HS) |
|---|---|---|---|---|
| *V. adenantha* | 0 | 1 | 1 | 0 |
| *V. ambascensis* | 4 | 2 | 3 | 0 |
| *V. baoulensis* | 0 | 4 | 6 | 0 |
| *V. benuensis* | 0 | 1 | 0 | 0 |
| *V. comosa* | 0 | 0 | 2 | 0 |
| *V. davyi* | 1 | 0 | 0 | 0 |
| *V. dekindtiana* | 0 | 1 | 0 | 0 |
| *V. filicaulis* | 0 | 1 | 1 | 0 |
| *V. fischeri* | 0 | 1 | 0 | 0 |
| *V. friesiorum* | 0 | 1 | 0 | 0 |
| *V. frutescens* | 0 | 2 | 0 | 0 |
| *V. gentryi* | 0 | 1 | 0 | 0 |
| *V. glabrescens* | 1 | 1 | 0 | 0 |
| *V. gracilis* | 0 | 0 | 0 | 2 |
| *V. heterophylla* | 0 | 1 | 0 | 0 |
| *V. hosei* | 0 | 0 | 1 | 1 |
| *V. juruana* | 0 | 0 | 0 | 1 |
| *V. kirkii* | 0 | 2 | 0 | 0 |
| *V. lasiooarpa* | 0 | 1 | 0 | 0 |
| *V. laurentii* | 0 | 0 | 0 | 1 |
| *V. linearis* | 0 | 0 | 0 | 1 |
| *V. lobatifolia* | 0 | 0 | 0 | 1 |
| *V.longifolia* | 0 | 0 | 0 | 1 |
| *V. longiloba* | 0 | 0 | 0 | 1 |
| *V. luteola* | 0 | 0 | 0 | 1 |
| *V. macrosperma* | 0 | 0 | 1 | 0 |
| *V. marina* | 2 | 0 | 0 | 0 |
| *V. membranacea* | 0 | 1 | 1 | 1 |
| *V. minima* | 0 | 1 | 0 | 0 |
| *V. multinervis* | 0 | 1 | 0 | 0 |
| *V. mungo* | 1 | 0 | 0 | 0 |
| *V. nigritia* | 0 | 0 | 0 | 2 |
| *V. oblongifolia* | 4 | 2 | 0 | 0 |
| *V. parkeri* | 1 | 0 | 0 | 0 |
| *V. peduncularis* | 0 | 0 | 0 | 1 |
| *V. racemosa* | 1 | 1 | 1 | 0 |
| *V. radiata* | 0 | 0 | 1 | 0 |
| *V. reticulata* | 2 | 3 | 0 | 1 |
| *V. trilobata* | 0 | 1 | 0 | 0 |
| *V. triphylla* | 0 | 0 | 0 | 1 |
| *V. unguiculata* | 1 | 92 | 128 | 28 |
| *V. venulosa* | 0 | 0 | 1 | 0 |
| *V. vexillata* | 2 | 5 | 3 | 2 |
| *V. wittei* | 0 | 1 | 0 | 0 |

## 4.5 Discussion

The results of the experiments indicate that 21 accessions from the following wild *Vigna* species – *V. ambacensis*, *V. davyi*, *V. glabrescens*, *V. marina*, *V. mungo*, *V. oblongifolia*, *V. parkeri*, *V. racemosa*, *V. reticulata*, *V. vexillata* and *V. unguiculata* subspecies *dekindtiana* – are potential sources of genes for resistance to *S. gesnerioides*. However, due to the strong cross-incompatibility between cowpea and *V. vexillata* reported by Barone and Ng (1990) and Fatokun (2002), it may be difficult to access these resistance genes for incorporation into cowpea using conventional crossing methods. Interspecific crosses between the following resistant species *ambacensis*, *parkeri*, *mungo*, *racemosa*, *oblongifolia*, *reticulata*, *glabrescens* and *marina*, with cowpea (*V. unguiculata*) have not yet been reported. Further studies on cross-compatibility between cowpea and these wild relatives would be needed in case any of them is crossable. *Striga*-susceptible cowpea landrace accessions showed more host support for *Striga* emergence by having higher *Striga* counts per plot ranging between 1 and 38 compared to the wild *Vigna* species, with lesser *Striga* counts per plot across both the resistant and the susceptible accessions.

These results indicate that cowpea landrace genotypes that are geographically co-located with the resistant wild *Vigna* species could be sources of resistance to *S. gesnerioides*. TVu-997, TVu-8788, TVu-8453, TVu-5478, TVu-4806, TVu-15474, TVu-15016, TVu-15011, TVu-14980, TVu-13485, TVu-13297, TVu-13035, TVu-12470, TVu-12449, TVu-12431 and TVu-12430 were identified as potential sources of resistance to *S. gesnerioides*. They have the same geographical origins of Ghana, Kenya, Niger, Zambia and Zimbabwe, with the wild *Vigna* accessions TVNu-1070, TVNu-1083, TVNu-585, TVNu-1535, TVNu-1537, TVNu-1589, TVNu-1647 and TVNu-491 belonging to the following species: *ambacensis*, *parkeri*, *oblongifolia*, *unguiculata* and *reticulata*, respectively. Of all these resistant wild *Vigna* species, only TVNu-1589 (*V. unguiculata* subsp. *dekindtiana*) is potentially cross-compatible with cowpea (*V. unguiculata*).

## 4.6 Conclusions

- We identified some wild *Vigna* species and landrace germplasm lines as highly resistant to *S. gesnerioides*.
- Most of the resistant wild *Vigna* species are not members of section *Catiang*, where cultivated cowpea (*V. unguiculata*) belongs.
- Accessing the *Striga* resistance genes in these wild *Vigna* species may pose a great challenge as cross-incompatibility may be a problem.
- Studies have been initiated to establish if any of the *Striga* resistance genes found in the tested wild cowpea (*V. unguiculata* ssp. *dekindtiana*) and cowpea landrace germplasm lines are allelic to those previously used in the development of resistant cowpea varieties.

## Acknowledgements

The authors thank the International Institute of Tropical Agriculture (IITA) and the Genetic Resources Center of IITA for financing this project.

## References

Barone, A. and Ng, N.Q. (1990) Embryological study of crosses between *Vigna unguiculata* and *V. vexillata*. In: Ng, N.Q. and Monti, L.M. (eds) *Cowpea Genetic Resources*. IITA, Ibadan, Nigeria, pp.151–160.

Bressani, R. (1985) Nutritive value of cowpea. In: Singh, S.R. and Rachel, K.O. (eds) *Cowpea Research, Production and Utilization*. John Wiley and Sons, Chester, UK, pp. 353–360.

Boukar, O., Kong, L., Singh, B.B., Murdock, L. and Ohm, H.W. (2004) AFLP and AFLP-derived SCAR markers associated with *Striga gesnerioides* resistance in cowpea [*Vigna unguiculata* (L.) Walp.]. *Crop Science* 44, 1259–1264.

Cardwell, K.F. and Lane, J.A. (1995) Effect of soils, cropping system and host phenotype on incidence and severity of *Striga gesnerioides* on cowpea in West Africa. *Agriculture Ecosystems and Environment* 53, 253–262.

Dube, M.P. and Oliver, A. (2001) Le *Striga gesnerioides* et son hôte, le niébé: interaction et méthodes de lutte. *Canadian Journal of Botany* 79, 1225–1228.
Emechebe, A.M., Singh, B.B., Leleji, O.I., Atokple, I.D.K. and Adu, J.K. (1991) Cowpea *Striga* problems and research in Nigeria. In: Kim, S.K. (ed.) *Combating* Striga *in Africa*. Proceedings of an International Workshop, 1988, Ibadan, Nigeria.
Fatokun, C.A. (2002) Breeding cowpea for resistance to insect pests: attempted crosses between cowpea and *Vigna vexillata*. In: Fatokun, C.A., Tarawali, S.A., Singh, B.B., Kormawa, P.M. and Tamo, M. (eds) *Challenges and Opportunities for Enhancing Cowpea Production*. Proceedings of the World Cowpea Conference III, International Institute of Tropical Agriculture (IITA), Ibadan, Nigeria, 4–8 September 2000, pp. 52–61.
Fatokun, C.A. and Singh, B.B. (1987) Interspecific hybridization between *Vigna pubescens* and *V. unguiculata* [L.] Walp. through embryo rescue. *Plant Cell, Tissue and Organ Culture* 9, 229–233.
Kureh, I., Katung, P.D. and Orakwue, F.C. (1999) Reaction of soybean varieties to preconditioning and concentration of seed inoculum of *Alectravogelii* (Benth). Science Forum. *Journal of Pure and Applied Science* 2, 116–124.
Ng, N.Q. and Marechal, R. (1985) Cowpea taxonomy, origin and germplasm. In: Singh, S.R. and Rachie, K.O. (eds) *Cowpea Research, Production and Utilization*. John Wiley and Sons, Chichester, UK, pp.11–21.
Parker, C. (1991) Production of crops against parasitic weeds. *Crop Production* 10, 6–13.
Rachie, K.O. and Rawal, K.M. (1975) Integrated Approaches to Improving Cowpeas, *Vigna unguiculata* (L.) Walp. Technical Bulletin No. 5, IITA, Ibadan, Nigeria.
Singh, B.B. and Emechebe, A.M. (1997) Advances in research on cowpea *Striga* and *Alectra*. In: Singh, B.B., Mohan, R., Dashiel, K.E. and Jackai, L.E.N. (eds) *Advances in Cowpea Research*. IITA-JIRCAS, Ibadan, Nigeria, pp. 215–224.
Westphal, E. (1974) Pulses in Ethiopia: their taxonomy and agricultural significance. In: *Agricultural Research Report*. Center for Agricultural Publishing and Documentation, Wageningen, the Netherlands, pp. 213–232.

# 5 Wild *Lactuca saligna*: A Rich Source of Variation for Lettuce Breeding

A. Lebeda,[1]* E. Křístková,[1] M.Kitner,[1] B. Mieslerová[1] and D.A. Pink[2]
[1]*Palacký University in Olomouc, Faculty of Science, Department of Botany, Olomouc-Holice, Czech Republic;* [2]*Harper Adams University, Newport, UK*

## 5.1 Introduction

An amazing diversity of leaf forms and colours, shape of heads and suitability for various cultivation procedures can be observed in cultivated lettuce (*Lactuca sativa* L., fam. Asteraceae). This is the only cultivated species within the group of 99 species of the genus *Lactuca* (Lebeda *et al.*, 2004b), and the high degree of variation results from intense breeding. The main aim of exploitation of wild relatives of lettuce in lettuce breeding is to introduce resistance to diseases and pests; however, they can also be sources of new plant morphotypes. Willow-leaf lettuce (*Lactuca saligna* L.) is a modest looking plant, with beauty hidden in large variation of fine details on leaves, stem and flowers, with richness of internal features (e.g. secondary metabolites) and with great potential to be exploited in lettuce breeding programmes for economically important features (Lebeda *et al.*, 2007).

## 5.2 Plant Morphology

*L. saligna* L. is an annual, winter annual, to biennial, rarely perennial, herb. Its stem is 30–100 cm long, simple or branched, glabrous, rarely with trichomes; leaves are thick, with trichomes on the adaxial side, stem leaves entire or divided. The inflorescence (head) is composed of from 6 to 15 florets. Their ligules are yellow, with anthocyanin colour on the adaxial side. Anthocyanin coloration is present in the anther tube. Heads form a composed inflorescence – a narrow, spike-like panicle. The fruit (achene) is composed of the achene body, beak and white deciduous, non-setose pappus (Feráková, 1977).

Willow-leaf lettuce is morphologically a very polymorphic species, but only two varieties are recognized. Infraspecific classification follows the shape of the stem leaves. Divided leaves are attributed to *L. saligna* var. *runcinata*. Within the other variety, *L. saligna* var. *saligna* with entire leaves, there are two forms distinguished: f. *saligna* (with trichomes) and f. *wallrothii* (without trichomes) (Feráková, 1977).

## 5.3 Chromosome Number, DNA Content, Secondary Metabolites

*L. saligna* is a diploid species with chromosome number $n = 9$ (Feráková, 1977), similar to other *Lactuca* species from Europe, the Mediterranean, the Middle East, Africa and India (Lebeda *et al.*, 2007). The relative DNA content established by flow cytometry for 15 *Lactuca* species ranges

*Corresponding author; e-mail: ales.lebeda@upol.cz

from 17.96 pg to 2.02 pg, so the relative DNA content of 4.76 pg in *L. saligna* is very low (Doležalová *et al.*, 2003b). Karyotypes of *L. sativa* and *Lactuca serriola* are very similar, and they are intermediate between *L. saligna* and *Lactuca virosa*, and this leads to the postulation of an evolutionary scenario: all species having a common ancestor, but with *L. virosa* split off much earlier than *L. saligna* (Koopman *et al.*, 1993; Koopman, 1999).

Plants of the genus *Lactuca* have been shown to produce sesquiterpene lactones as their characteristic secondary metabolites. Michalska *et al.* (2009) confirmed that *L. saligna* accumulated in its roots all eight compounds analysed, i.e. lactucin-type guaianolides, 8-deoxylactucin, jacquinelin, crepidiaside B, lactucin, 11b,13-dihydrolactucin and lactucopicrin, and also vernoflexuoside 1,4glucozaluzanin C and lactuside A. In leaves, 8-deoxylactucin, jacquinelin, crepidiaside B and lactucopicrin were detected. The composition of sesquiterpene lactones in roots of *L. saligna* is similar to *L. serriola* and *Lactuca indica*: the spectra of these compounds in its leaves associate *L. saligna* more closely to *Lactuca perennis* and *Lactuca tenerrima* (Michalska *et al.*, 2009).

## 5.4 Taxonomy and Relation to Other *Lactuca* Species

In a recent taxonomic classification of the genus *Lactuca* L., species were organized into seven sections and two geographic groups, summarized by Lebeda *et al.* (2007). Section *Lactuca* L., subsection *Lactuca* L. comprises cultivated *L. sativa* and *L. saligna*, together with wild species – *Lactuca aculeata* Boiss. & Kotschy, *Lactuca azerbaijanica* Rech.f., *Lactuca altaica* Fisch. & Mey., *Lactuca dregeana* D.C., *Lactuca georgica* Grossh., *Lactuca livida* Boiss. & Reut., *Lactuca scarioloides* Boiss., *L. serriola* L. and *L. virosa* L.

Koopman *et al.* (1998) critically analysed the available data on plant morphology, SDS-electrophoresis pattern of seed proteins, isozyme analyses of foliar esterases of *L. saligna*, *L. sativa/L. serriola* and *L. virosa* and demonstrated a close relationship between *L. sativa/ L. serriola* and *L. virosa*, while *L. saligna* was more distinct. Data on crossability, karyotype, chromosome banding pattern and DNA content gave *L. sativa* and *L. serriola* an intermediate position between *L. saligna* and *L. virosa* (Koopman *et al.*, 1998). Results from amplified fragment length polymorphisms (AFLPs), restriction fragment length polymorphism (RFLP) and internal transcribed spacer (ITS-1) molecular analyses provided an unequivocal view of the relationship between the species (Koopman *et al.*, 1998). Study of sufficient newly collected plant samples is needed to elucidate relationships between *L. altaica* and its relatives, *L. serriola* and *L. saligna*.

*L. saligna* is partly interfertile with *L. sativa*, and according to the definition of genepools of Harlan and de Wet (1971), it belongs to the secondary genepool of cultivated lettuce (Lebeda *et al.*, 2007). This position is supported by the *L. saligna* profile of ITS-1 DNA sequences (Koopman *et al.*, 1998, 2001).

## 5.5 Geographic Distribution and Ecological Requirements

Willow-leaf lettuce is Eurasiatic to sub-Mediterranean, distributed around the Mediterranean, Near East, around the Caspian Sea and further eastwards, and to the south (Iraq, Iran, Saudi Arabia), and occupying scattered locations throughout North Africa. However, the species is rare in the greater part of its natural distribution area (Lebeda *et al.*, 2004b). In Europe it extends as far as 52° latitude north (Feráková, 1977). *L. saligna* has been retreating in Europe, in spite of its inclination to spread in ruderal communities. It occurs as a facultative halophyte in western France, England and Belgium, mostly at the seaside, and it is very rare in the Netherlands (Meusel and Jäger, 1992). It is considered to be absent in the Czech Republic (Grulich, 2004) and as endangered in Austria (Adler *et al.*, 1994). It is an adventive plant in North and South America, namely in the USA, Canada and Argentina, and has been introduced to Australia (Feráková, 1977; Lebeda *et al.*, 2004b).

In Europe, it grows from lowland to low montane elevations (10–300 m), exceptionally, higher; in open sunny exposures in Italy, it has been found at 1000 m, in Cyprus at 1680 m and in Turkey at 2400 m (Feráková, 1977). It prefers warm, fertile, semi-arid, slightly salty soils. This synanthropic species occurs commonly on ruderal sites, influenced by human activities: along roads, railways, borders of woods, arable fields and riverbanks.

## 5.6 Representation in Germplasm Collections, Redundancies, Gaps and Errors

Detailed analyses of wild *Lactuca* species representation in the world's genebanks show that *L. saligna* is represented by 10% of all accessions. However, the world area of *L. saligna* distribution is not covered, and the primary source of origin (country) is known for 33% of accessions only. Accessions are exchanged among institutions, and only a very limited number of newly acquired samples are from original collecting activities (Lebeda *et al.*, 2004a). Duplicates within and among collections, as well as mistakes in taxonomic determination, have been recognized also among and within *L. saligna* accessions (Doležalová *et al.*, 2004). The quality and quantity of plant material held by official genebanks is not sufficient, and any study of *L. saligna* biodiversity should focus on additional and original plant material newly collected either in areas of high diversity of this species or in marginal areas of its distribution.

## 5.7 Field Studies and Collecting Missions

The main aims of field studies and collecting missions performed by staff members of the Department of Botany, Faculty of Science, Palacký University in Olomouc, Czech Republic, are to complete our knowledge of the natural distribution of *L. saligna*, to study the features of species populations in their natural habitats and to acquire original seed material for further studies (Lebeda *et al.*, 2001, 2012a; Beharav *et al.*, 2008). In the period 1998–2013, a total of 144 seed samples of *L. saligna* were collected in Europe: Greece (2), Croatia (19), France (9), Hungary (1), Italy (57), Portugal (6) and Slovakia (31); the Near East: Israel (8) and Jordan (8); and America: the USA (3). Each seed sample is represented by achenes from one maternal plant. The number of seed samples collected from each population depends on the size of the population.

Our field observations confirmed data on *L. saligna* distribution in Europe given in floras and summarized by Feráková (1977) and Lebeda *et al.* (2004b). Lebeda *et al.* (2001) reported a high density of species populations in the Po River lowlands (Italy), and single plants mostly growing in sunny exposures along roads extending up to 300 m (Appennino Ligure) were observed as well. Willow-leaf lettuce is frequent along the seashore of the Riviera di Ponente (Italy) and the Côte d'Azur (France), in some locations occurring in mixed populations with *L. serriola* (Lebeda *et al.*, 2001). We observed similar population characteristics in coastal areas of Croatia and its islands (namely Brač, Mljet) in 2010–2014.

We did not observe this species in Belgium, northern coastal and central France, the Netherlands, Czech Republic, Germany, Switzerland, Austria or Slovenia. There are large areas in Europe where the occurrence of *L. saligna* has been reported in local floras, namely in Spain, Portugal, central and southern Italy and central France, as well as in substantial parts of Hungary, but the Balkans and Ukraine have not been surveyed yet. Lebeda *et al.* (2001) confirmed the occurrence of this species in Europe in typical habitats along roads, highways, parking sites, petrol stations, in walls and pavements in towns, in ruderal places and along fields. The population size and density of plants in locations are usually very low, especially when comparing these parameters to *L. serriola*. Huge plant density has been recorded in south-west Slovakia (Trnovec nad Váhom, district Nitra) on the meadow surrounding a source of thermal water (A. Lebeda and I. Doležalová, Czech Republic, 2006, personal communication).

*L. saligna* is a commonly occurring species in the Near East. It has been recorded throughout Israel, except for desert areas (e.g. Negev and Judean deserts) and extreme environmental/soil conditions (Dead Sea area) at various altitudes (10–1277 m) and different habitats and soils (Beharav *et al.*, 2008), very often sharing identical locations with *L. serriola* and *L. aculeata* (A. Lebeda, Czech Republic, 2007–2010, personal communication). Field studies in Georgia (South-west Asia) confirmed the presence of willow-leaf lettuce at an elevation of 1444 m in the Gombori Pass (A. Lebeda, Czech Republic, 2013, personal communication). *L. saligna*, as an adventive plant in North America, has been recorded in only one place in Salinas (California) during recent field studies in the USA, and it has not been observed in any other place across the USA, so its occurrence remains very rare in the USA (Lebeda *et al.*, 2012a).

## 5.8 Regeneration of Seed Samples

Collected seed samples are regenerated in the greenhouse (Palacký University in Olomouc) according to the international standards summarized by van Hintum and Boukema (1999). Each sample is represented by 16 plants. During regeneration, the morphological traits and phenological characteristics are described according to the descriptor lists of Doležalová *et al.* (2002, 2003a), and the taxonomic status of samples is verified or specified on the level of a lower taxonomic unit. This assessment is followed by scanning, photo-documentation and additional measurements. Regenerated seed is stored long-term at below −18°C and used in subsequent studies of morphology, molecular polymorphism and disease resistance.

## 5.9 Morphological and Phenological Variation

Variation in morphological traits and phenological characteristics either within or among populations of *L. saligna* were observed during cultivation of collected plant samples in the greenhouse (Beharav *et al.*, 2008; Křístková *et al.*, 2010, 2011; Tvardková, 2010; Lebeda *et al.*, 2012b; Matoušková, 2013; M. Kitner, Czech Republic, 2014, personal communication).

The formation of a leaf rosette was observed on samples from Europe and the USA, in contrast to the majority of samples originating from Israel and Jordan that did not develop a leaf rosette. This trait, connected to the very early beginning of bolting of these plants, and considered as an adaptation to local ecological conditions, i.e. hot and dry weather in late spring and early summer, is fixed genetically (Matoušková, 2013).

The shape of stem leaves varied from entire, narrow lanceolate, corresponding to the definition of *L. saligna* var. *saligna* by Feráková (1977), to divided, pinnately lobed to pinnately sect, which fits the definition of *L. saligna* var. *runcinata*. Geographic distribution of both varieties has not yet been resolved, and to our recent knowledge we cannot attribute specific leaf shape to specific ecogeographic conditions. Plants of *L. saligna* var. *saligna* with entire leaves have been observed only sporadically (several samples from Slovakia, Greece and Israel) (Tvardková, 2010; Matoušková, 2013).

Large variation in the shape of stem leaves was observed. The grouping of divided leaves proposed by Křístková *et al.* (2009) took into consideration the length of the apical lobe. The shape of the apex on the stem leaves varied from acute and subacute to obtuse, and was not related clearly to the depth of incisions (Matoušková, 2013).

Variation of trichome distribution on leaves and stems and the anthocyanin distribution pattern were not related clearly to specific original ecological conditions of the samples or to other morphological traits.

The spike-like panicle as a type of composed inflorescence attributed to the species *L. saligna* by Feráková (1977) has been recorded very rarely, on one sample from Israel only; the majority of plants developed a pyramidal panicle, exceptionally, a corymbose panicle.

The number of florets in the heads varied from 5 to 19. Different patterns of anthocyanin distribution (i.e. diffused, in points or stripes) and variation in intensity of anthocyanin coloration on ligules and bracts in heads were observed. The presence of anthocyanin in anther tubes is a species-specific trait (Feráková, 1977). Its absence in anther tubes in several samples from Portugal, Israel and Jordan is considered to be a spontaneous mutation (Matoušková, 2013).

The greatest variability in morphological traits within and between plant samples was observed for samples from Israel (Matoušková, 2013). Examples of morphological variation are documented in Fig. 5.1.

Developmental stages of plants (beginning of bolting and flowering) are influenced (selected for) by the original ecogeographic conditions of samples and persist when plants are cultivated in unified conditions and are fixed genetically. Plants originating from coastal areas of northern Italy (Riviera di Ponente) bolted relatively early, but the beginning of flowering was delayed.

**Fig. 5.1.** *Lactuca saligna*: (a) young plant in natural habitat – stony road in coastal area of Croatia; (b) regeneration of collected plant samples in the greenhouse (Palacký University in Olomouc, Czech Republic); (c) achenes with beaks and pappus; (d) shape of rosette leaves of *L. saligna* (♀), *L. sativa* (♂) and their hybrid ($F_1$); (e) inflorescence (head) with species-typical anthocyanin coloration of anther tubes; (f) architecture of composed inflorescence of *L. saligna* (♀); (g) their hybrid ($F_1$); (h) *L. sativa* (♂); (j) position of heads on stem; (k–n) infraspecific variation in shape of rosette leaves; (o–t) infraspecific variation in shape of stem leaves.

Plants from Central Europe (Slovakia and higher elevations of southern France (Rhône Alpes)) entered the stage of bolting later, but the period between bolting and flowering was shorter (Křístková *et al.*, 2011). A similar phenomenon was observed for *L. serriola* originating from different ecogeographical conditions (Lebeda *et al.*, 2005).

## 5.10 Molecular Polymorphism

Molecular genetic markers represent one of the most powerful tools for the analysis of genetic variation and structure of plant populations, for the selection of desirable genotypes for lettuce breeding, and the association of heritable traits with underlying genetic variation and mapping of genes and the identification and variation of resistance genes. Markers may be used in applications such as genetic linkage analysis and trait mapping, diversity analysis, association studies and marker-assisted selection, either in theoretical plant science or in modern breeding programmes (Appleby *et al.*, 2009).

Reports on various studies using molecular biology, where *L. saligna* was used, were summarized recently by Lebeda *et al.* (2014). These studies were focused mainly on inferring the relationships among *Lactuca* species (Kesseli *et al.*, 1991; Koopman *et al.*, 2001; Yang *et al.*, 2007), application and/or design of various molecular markers in a larger context of *Lactuca* spp. (Simko, 2009; van Treuren and van Hintum, 2009; Riar *et al.*, 2011; Simko *et al.*, 2011), genetic mapping and interspecific hybridization (Jeuken *et al.*, 2001, 2009; Jeuken and Lindhout, 2004; Simko *et al.*, 2010; Stoffel *et al.*, 2012; den Boer *et al.*, 2013; Rauscher and Simko, 2013).

The number of studies where *L. saligna* is the primary aim (i.e. genetic variation in natural populations of *L. saligna* or for *L. saligna* germplasm maintenance and characterization) is limited. The only study published so far is a paper by Kitner *et al.* (2008), describing differences among 67 *L. saligna* accessions collected in natural and semi-natural habitats in four European (Czech Republic, France, Italy, Portugal) and three Middle Eastern countries (Israel, Jordan, Turkey).

Recently, we performed a population genetic study comparing two natural *L. saligna* populations with two *L. serriola* and two *L. aculeata* populations from Israel by microsatellite and AFLP markers (Kitner *et al.*, unpublished). Contrary to *L. serriola* and *L. aculeata*, both *L. saligna* populations exhibited higher diversity values and a greater number of simple sequence repeat (SSR) genotypes, as well as the number of plants with a heterozygous constitution. Morphological traits on individual plants and their allelic profiles indicate that flowers on a single maternal plant can be pollinated by pollen coming from several different plants. This might point to the possibility of exchange of individuals between populations (confirmed by the higher outcrossing rate) and/or that selfing is not the strict mode of reproduction in the natural population of *L. saligna* in the Near East, e.g. due to the environmental factors. This is even more important in areas where various cross-compatible *Lactuca* species occur in the same location. However, this needs to be verified and proven in more focused research (M. Kitner *et al.*, 2014, unpublished results).

## 5.11 Resistance to Diseases and Pests

Reports on resistance to the most important pathogens and pests of cultivated lettuce observed within wild *Lactuca* species have recently been reviewed critically by Lebeda *et al.* (2014). The current information about sources of resistance detected in *L. saligna* is summarized in Table 5.1. In this contribution, we will focus in detail on different aspects of the interactions between *L. saligna* and two economically very important pathogens, lettuce downy mildew and lettuce powdery mildew.

## 5.12 Lettuce Downy Mildew

The oomycete pathogen, lettuce downy mildew (*Bremia lactucae* Regel), has a very significant economic impact on lettuce production in many lettuce producing areas, and the study of its biology and epidemiology, and sources of resistance, including wild *Lactuca* germplasm, have been a high priority of researchers and breeders (Lebeda and Blok, 1991; Lebeda *et al.*, 2014).

Currently, one of the most challenging resistances for lettuce breeders to exploit is the possible non-host resistance located in *L. saligna* (Lebeda *et al.*, 2002; Beharav *et al.*, 2006; Petrželová *et al.*, 2011). The phenomenon of non-host resistance in some *L. saligna* accessions was hypothesized for the first time by Lebeda (1986).

The phenotype of non-host resistance (e.g. in some *L. saligna* accessions and most Asteraceae species) is characterized by a lack of symptoms (Crute and Dickinson, 1976; Lebeda and Syrovátko, 1988; Lebeda and Reinink, 1994; Sedlářová *et al.*, 2001). Nevertheless, expression of macroscopic

Table 5.1. Sources of resistance to diseases and pests in *Lactuca saligna*. (From Lebeda *et al.*, 2014.)

| Name of pathogen/pest | No. of accession/sample | Type of resistance | References |
|---|---|---|---|
| Viral pathogens | | | |
| CMV | PI 261653 | R to CMV-B CMV-L2 | Edwards *et al.* (1983) |
| LNSV | PI 271940, PI 490999 | | Grube *et al.* (2005) |
| LNYV | PI 261653 | | Haley and McCreight (1990) |
| Bacterial pathogens | | | |
| Corky root (*Sphingomonas suberifaciens*) | PI 261653, PI 490999, PI 491204, PI 491206, UC83UK2, UC83US1 | n.k. | Brown and Michelmore (1988) |
| Fungal pathogens | | | |
| *Bremia lactucae* | LSA/6 | Non-host | Lebeda (1986) |
| | e.g. CGN 05310, CGN 05315, CGN 05319, CGN 05329, CGN 13326, CGN 13330, CGN 13375, LSA/7, 09-H58-0888 | Non-host | Lebeda and Zinkernagel (2003) |
| | e.g. 8-6, 11-3, 14-8, 18-4, 20-4, 20-22 | Non-host | Petrželová *et al.* (2011) |
| | 151-23, 198-12, 199-34 | R to isolates Bl-21, Isr-74 | Beharav *et al.* (2006) |
| *Golovinomyces cichoracearum* | LSA/92/1, LSA/92/2 | n.k. | Lebeda (1985) |
| | 05282, 05304, 05306, 05308, 05309, 05311, 05313, 05314, 05315, 05318,05319, 05320, 05322, 05323, 05326, 05330, 05895, 09311, 09313 | n.k. | Lebeda (1994) |
| | 09-H58-1013 | Race-specific | Lebeda *et al.* (2012c) |
| Anthracnose | LAG1 | Immune | Galea and Price (1988) |
| *Microdochium panattoniana* | UC83US1 | R to races A, B, C, D, E | Ochoa *et al.* (1987) |
| | UC83UK2 | | Ochoa *et al.* (1987) |
| | PI 273482, PI 490999 | | |
| Stemphylium leaf spot (*Stemphylium* spp.) | n.s. | R | Netzer *et al.* (1985) |
| Nematodes | | | |
| *Meloidogyne hapla* | PI 281876, PI 491000, PI 261653, PI 273582 (M-HR) | n.k. | Abawi and Robinson (1991) Kaur and Mitkowski (2010) |
| *Meloidogyne incognita* | n.s. | n.k. | Rich *et al.* (2010) |
| Aphids | | | |
| *Liriomyza langei* | PI 490999, PI 261653-1, PI 509525 | R | Mou and Liu (2003, 2004) |
| *Pemphigus bursarius* | 006186, 001627, HRIGRU1630 | R | Cole *et al.* (1991) |

*Notes*: CMV = *Cucumber mosaic virus*; LNSV = *Lettuce necrotic stunt virus*; LNYV = *Lettuce necrotic yellows virus*; M-HR = moderate-highly resistant genotype; n.k. = not known; n.s. = not specified; R = resistant.

chlorosis (Crute and Dickinson, 1976), necrosis (Norwood *et al.*, 1981; Lebeda and Reinink, 1994) or sub-epidermal necrosis (Lebeda and Reinink, 1994; Lebeda *et al.*, 2006) has also been recorded. Detailed histological studies of the pathogenesis of *B. lactucae* show that in some *L. saligna* interactions with *B. lactucae*, the formation of secondary vesicles represents the final stage of oomycete development, which is considered to be a characteristic of non-host resistance (Lebeda and Reinink, 1994; Sedlářová *et al.*, 2001; Lebeda *et al.*, 2002, 2006).

Biochemical studies reveal that the supposed non-host resistance in the *L. saligna* accession CGN 05271 is accompanied by only minor changes in the level of $H_2O_2$, the content of which is generally lower compared to the other *Lactuca* species (Sedlářová *et al.*, 2007). Non-host resistance of *L. saligna* (CGN 05271) correlated with low amounts of NO production and a relatively small-scale increase of NOS activity (M. Petřivalský, Czech Republic, 2014, personal communication). The non-host resistance in *L. saligna* (CGN 05271) is expressed before haustoria formation (Sedlářová *et al.*, 2001) and is characterized by a lack of the hypersensitive response, which might relate to the previously mentioned adjustment of oxidative processes (see reactive oxygen species (ROS), antioxidants and ROS-scavenging enzymes and nitric oxide, NO synthase and NO modulators in Sedlářová *et al.*, 2007).

Theoretically, non-host resistance should be very effective, durable and not influenced by changes in environmental conditions (Lebeda *et al.*, 2002). However, recent findings indicate that at least some *L. saligna* accessions possess race-specific resistance factors (Jeuken and Lindhout, 2002; Lebeda and Zinkernagel, 2003), in addition to possible non-host resistance to *B. lactucae* (Beharav *et al.*, 2006; Petrželová *et al.*, 2011).

Recent studies with individual plants (Lebeda and Zinkernagel, 2003; Beharav *et al.*, 2006) and populations (Petrželová *et al.*, 2007) of *L. saligna* showed a high degree of resistance to all *B. lactucae* races originating from lettuce, and also those from *L. serriola* (Lebeda, 1986; Lebeda and Boukema, 1991). Moreover, studies at the tissue, cellular and physiological levels (Lebeda and Reinink, 1994; Lebeda and Pink, 1998; Sedlářová and Lebeda, 2001; Sedlářová *et al.*, 2001, 2007; Lebeda *et al.*, 2002, 2006) confirmed that the mechanism of resistance in *L. saligna* differed significantly from the mechanisms known in *L. sativa*, *L. serriola* and *L. virosa* (Lebeda *et al.*, 2002). This is also in agreement with the conclusions of Koopman *et al.* (2001) based on the expectation that *L. virosa* is more close to the *sativa–serriola* complex than to *L. saligna*.

Analyses of the interactions between 60 samples of *L. saligna* representing 17 Israeli populations with three highly virulent Californian isolates of *B. lactucae* showed that this species was not an absolute non-host for *B. lactucae*, at least at the seedling stage, but none of the 60 samples of *L. saligna* were susceptible to lettuce downy mildew at the adult stage (Beharav *et al.*, 2014).

The quantitative genetic basis of the resistance of *L. saligna* to lettuce downy mildew has been studied using substitution lines derived from the interspecific cross of *L. saligna* and *L. sativa* and is reported by Jeuken and Lindhout (2002, 2004). Other studies related to this topic are mentioned in the section 'Interspecific Hybridization' below.

Data obtained in histological studies of resistance in wild *Lactuca* spp. suggest there is a wide range of resistance mechanisms in *Lactuca* spp. against *B. lactucae* (Lebeda *et al.*, 2008), and those identified in *L. saligna* differ to mechanisms in *L. serriola* and *L. virosa* (Lebeda *et al.*, 2002).

## 5.13 Lettuce Powdery Mildew

Lettuce powdery mildew, *Golovinomyces cichoracearum* (DC.) Heluta, infects mainly representatives of the family Asteraceae and can also be dangerous on cultivated chicory and lettuce crops (Lebeda and Mieslerová, 2011). The first authenticated report of powdery mildew infection of cultivated lettuce was in southern California (USA) (Pryor, 1941). Since then, the occurrence of powdery mildew on cultivated lettuce has been reported from various countries in Europe (Lebeda and Mieslerová, 2011). However, as recently reported, the biggest impact of the pathogen is on lettuce and chicory crops in the USA (Blancard *et al.*, 2006).

Only a limited number of studies have evaluated wild *Lactuca* species for resistance to lettuce powdery mildew (Lebeda, 1985, 1994). The results of screening more than 100 accessions of various *Lactuca* species (including *L. saligna* accessions), under natural infection, revealed high variability in resistance to *G. cichoracearum* (Lebeda, 1985, 1994). From these experiments, it is obvious that some *L. saligna* accessions are potentially useful sources of resistance, especially when they carry resistance to *B. lactucae* as well (Lebeda, 1985, 1994; Lebeda *et al.*, 2012c).

Two accessions of *L. saligna* (09-H58-1013 and 09-H58-1010) were studied in detail. These accessions are part of the differential set of 13 *Lactuca* spp. genotypes that have been used for study of intraspecific variability of *G. cichoracearum* genotypes. In general, accession *L. saligna* 09-H58-1013 showed a great variability in response to inoculation of various *G. cichoracearum* isolates. Out of the whole differential set, this genotype was found to be 'resistant' to the tested isolates most frequently. However, it did not show complete resistance to all tested isolates (Lebeda *et al.*, 2013).

From these results, it is clear that resistance in *L. saligna* (as well as some other wild *Lactuca* spp.) is characterized by race specificity, but the genetic background of resistance is poorly understood (Lebeda, 1994; Lebeda and Mieslerová, 2011).

Knowledge on the interaction between *L. saligna* accessions and *G. cichoracearum* has now improved. A study of the development of isolate GC 1/11 on both *L. saligna* genotypes (09-H58-1013 and 09-H58-1010) found that germ tube development and sporulation was reduced on both *L. saligna* genotypes; however, the intensity of hypersensitive reaction was not high (Filová, 2012, 2014).

## 5.14 Interspecific Hybridization

*L. saligna* is predominantly self-pollinating, but its flower morphology does allow transmission of alien pollen. A high rate of outcrossing in *L. saligna* natural populations has been reported (Kitner *et al.*, 2014, unpublished results). There is a possibility of natural interspecific hybridization of *L. saligna* with some other closely related wild *Lactuca* species, and this is supported by the results of hybridization experiments, where fully fertile hybrids raised from the cross of *L. saligna* with *L. serriola* and partly fertile hybrids from the crosses with *L. sativa* (Rulkens, 1987; de Vries, 1990) have been produced. Zohary (1991) reported successful hybridization of *L. altaica* × *L. saligna*.

The proposition of the phenomenon of non-host resistance to *B. lactucae* in *L. saligna* (Bonnier *et al.*, 1992; Lebeda and Reinink, 1994) initiated intense attempts to cross it with cultivated lettuce. To map the downy mildew resistance in *L. saligna*, two different independent $F_2$ populations of *L. saligna* × *L. sativa* were generated by Jeuken *et al.* (2001), one from the cross of *L. saligna* CGN 5271 and *L. sativa* 'Olof', the second from *L. saligna* CGN 11341 and *L. sativa* 'Norden'. Jeuken and Lindhout (2004) developed backcross inbred lines in which chromosome segments from *L. saligna* CGN 5271 were introgressed into cultivated lettuce cv. 'Olof', and subsequent genetic mapping of lines has led to efficient quantitative trait loci (QTL) detection for non-host resistance to lettuce downy mildew in wild *L. saligna* (Jeuken *et al.*, 2008; Zhang *et al.*, 2009a,b) and better understanding of the genetics of resistance to *Bremia lactucae* (den Boer *et al.*, 2014).

Controlled interspecific hybridization experiments of *L. saligna* × *L. sativa* realized by the authors of this chapter at the Department of Botany (Palacký University in Olomouc) in 2009 and 2010 confirmed the results of Pink and Keane (1993), that *L. saligna* was known to produce hybrids with *L. sativa* and *L. serriola* when used as the female parent. The most significant morphological traits of $F_1$ hybrids were localized on leaves and flowers. Their hybrid character was confirmed by isozymes.

## 5.15 Known Examples of Exploitation of *L. saligna* in Lettuce Breeding

The exploitation of wild *Lactuca* species in lettuce breeding is strategically important for breeding companies. Data on the use of wild *Lactuca* spp. in lettuce breeding are incomplete in Europe because such pedigrees are not released into the public domain. Nevertheless, from sporadic and unofficial sources, we may surmise that breeding companies expend significant effort connected to the exploration and exploitation of wild *Lactuca* germplasm. The pedigree analysis of 328 proprietary and publicly developed lettuce cultivars registered in the USA from 1970 to 2004 showed that three wild *Lactuca* species, *L. serriola*, *L. saligna* and *L. virosa*, were involved in this process (Mikel, 2007).

*L. saligna* PI 261653, originally collected in Portugal and provided to the USA national germplasm collection by the Experiment Station for Growing Vegetables under Glass, Naaldwijk (the Netherlands) in 1959, is resistant to several viruses of lettuce, corky root, nematode *Meloidogyne hapla* and aphids *Liriomyza langei* (Table 5.1). It was crossed with the Iceberg type of cultivated lettuce by R.W. Robinson, who developed a cultivar 'Salad Crisp', the first lettuce cultivar with genetic material from *L. saligna* (Provvidenti *et al.*, 1980). Currently, the cultivar 'Salad Crisp' is maintained within the USDA National Genetic Resources Program in their plant germplasm collection under the accession number PI 635076 (Anon., 2014). Tamaki *et al.* (1995) reviewed a series of lettuce breeding lines raised from various combinations of crosses of *L. saligna* PI 261653 with cultivar 'Montello', which does not have any wild lettuce in its pedigree, and with 'Vanguard', which was bred from commercial lettuce crossed with *L. virosa*, and pointed out that the high sesquiterpene lactone content in *L. saligna* and *L. virosa*, whether or not associated with virus resistance, might be of concern to lettuce breeders and consumers.

Cultivar 'Salad Bibb' (HXP 3550) is an open pollinated butterhead-type lettuce with multiple disease resistance, developed by Hasib S. Humaydan (Harris Moran Seed Co.) through a series of crosses of unspecified *L. saligna* with several *L. sativa* cvs. (PVPA 8500060) (Mikel, 2007; Ryder and McCreight, 2014).

As a result of studies in the 1990s, a new lettuce cultivar 'Titan' (Sluis & Groot), with the race-specific gene *Dm6* plus resistance derived from *L. saligna*, was released in the Netherlands (Lebeda *et al.*, 2002). However, this resistance is no longer effective (Lebeda and Zinkernagel, 2003).

Unfortunately, knowledge of the use of *L. saligna* as a donor of resistance genes, not only against *B. lactucae* but also against other diseases and pests in specific lettuce cultivars, is limited due to a lack of information on cultivar pedigrees.

## 5.16 Conclusion

Our knowledge about the origin, process of domestication, diversification and spread of lettuce around the globe is still fragmentary (Lebeda *et al.*, 2009). The taxonomy of wild *Lactuca* species and related genera is currently unclear (Lebeda *et al.*, 2007; Funk *et al.*, 2009) and needs more detailed elaboration at the level of the genus, involving all known described species. Collection, conservation and evaluation of wild *Lactuca* germplasm is incomplete. During the past 70 years, the genus *Lactuca* has been studied intensively with respect to exploitation of wild relatives in commercial lettuce breeding.

*L. saligna* is a good example of a widely distributed wild relative of cultivated lettuce, with great potential to serve a model plant species for ecogeographic and evolutionary studies and for the process of lettuce domestication, but also for exploitation in terms of economically important traits useful in lettuce breeding.

## Acknowledgements

The research was supported by grant MSM 6198959215 and by internal grant PrF_2016_001 from Palacký University in Olomouc. The authors thank Dr Alex Beharav (Institute of Evolution, University of Haifa, Israel) for providing some seed samples of wild *Lactuca* spp.

## References

Abawi, G.S. and Robinson, R.W. (1991) Reaction of selected lettuce germplasm to artificial inoculation by *Meloidogyne hapla* in the greenhouse. *Journal of Nematology* 23, 519.

Adler, W., Oswald, K. and Fischer, R. (1994) *Exkursionsflora von Őstereich*. Verlag Eugen Ulmer, Stuttgart, Germany.

Anon. (2014) United States Department of Agriculture, Agriculture Research Service. Available at: www.ars-grin.gov (accessed 18 August 2014).

Appleby, N., Edwards, D. and Batley, J. (2009) New technologies for ultra-high throughput genotyping in plants. *Methods in Molecular Biology* 513, 19–39.
Beharav, A., Lewinsohn, D., Lebeda, A. and Nevo, E. (2006) New wild *Lactuca* genetic resources with resistance against *Bremia lactucae*. *Genetic Resources and Crop Evolution* 53, 467–474.
Beharav, A., Ben-David, R., Doležalová, I. and Lebeda, A. (2008) Eco-geographical distribution of *Lactuca saligna* natural populations in Israel. *Israel Journal of Plant Sciences* 56, 195–206.
Beharav, A., Ochoa, O. and Michelmore, M. (2014) Resistance in natural populations of three wild *Lactuca* species from Israel to highly virulent Californian isolates of *Bremia lactucae*. *Genetic Resources and Crop Evolution* 61, 603–609.
Blancard, D., Lot, H. and Maisonneuve, B. (2006) *A Color Atlas of Diseases of Lettuce and Related Salad Crops. Observation, Biology and Control*. Academic Press, Burlington, Massachusetts.
Bonnier, F.J.M., Reinink, K. and Groenwold, R. (1992) Genetic analysis of *Lactuca* accessions with new major gene resistance to lettuce downy mildew. *Phytopathology* 84, 462–468.
Brown, P.R. and Michelmore, R.W. (1988) The genetics of corky root resistance in lettuce. *Phytopathology* 78, 1145–1150.
Cole, R.A., Sutherland, R.A. and Riggall, W.E. (1991) The use of polyacrylamide gradient gel electrophoresis to identify variation in isozymes as markers for *Lactuca* species and resistance to the lettuce root aphid *Pemphigus bursarius*. *Euphytica* 56, 237–242.
Crute, I.R. and Dickinson, C.H. (1976) The behaviour of *Bremia lactucae* on cultivars of *Lactuca sativa* and other composites. *Annals of Applied Biology* 82, 433–450.
de Vries, I.M. (1990) Crossing experiments of lettuce cultivars and species (*Lactuca* sect. *Lactuca*, Compositae). *Plant Systematics and Evolution* 171, 233–248.
den Boer, E., Zhang, W.N., Pelgrom, K., Visser, R.G.F., Niks, R.E., *et al.* (2013) Fine mapping quantitative resistances to downy mildew in lettuce revealed multiple sub-QTLs with plant stage dependent effects reducing or even promoting the infection. *Theoretical and Applied Genetics* 126, 2995–3007.
den Boer, E., Pelgrom, E.T.B., Zhang, N.W., Visser, R.G.F., Niks, R.E., *et al.* (2014) Effects of stacked quantitative resistances to downy mildew in lettuce do not simply add up. *Theoretical and Applied Genetics* 127, 1805–1816.
Doležalová, I., Křístková, E., Lebeda, A. and Vinter, V. (2002) Description of morphological characters of wild *Lactuca* L. spp. genetic resources (English-Czech version). *Horticultural Science (Prague)* 29, 56–83.
Doležalová, I., Křístková, E., Lebeda, A., Vinter, V., Astley, D., *et al.* (2003a) Basic morphological descriptors for genetic resources of wild *Lactuca* spp. *Plant Genetic Resources Newsletter* 134, 1–9.
Doležalová, I., Lebeda, A., Dziechciarková, M., Křístková, E., Astley, D., *et al.* (2003b) Relationships among morphological characters, isozymes polymorphism and DNA variability – the impact on *Lactuca* germplasm taxonomy. *Czech Journal of Genetics and Plant Breeding* 39, 59–67.
Doležalová, I., Lebeda, A., Tiefenbachová, I. and Křístková, E. (2004) Taxonomic reconsideration of some *Lactuca* spp. germplasm maintained in world genebank collections. *Acta Horticulturae* 634, 193–201.
Edwards, M.C., Gonsalves, D. and Provvidenti, R. (1983) Genetic analysis of cucumber mosaic virus in relation to host resistance: location of determinants for pathogenicity to certain legumes and *Lactuca saligna*. *Phytopathology* 73, 269–273.
Feráková, V. (1977) *The genus* Lactuca L. *in Europe*. Univerzita Komenského, Bratislava, Czechoslovakia.
Filová, D. (2012) Study of the development of powdery mildew (*Golovinomyces cichoracearum*) on selected representatives of genus *Lactuca*. Bc thesis. Palacký University in Olomouc, Olomouc, Czech Republic (in Czech).
Filová, D. (2014) Study of the development of powdery mildew (*Golovinomyces cichoracearum*) on selected representatives of genus *Lactuca* and family Asteraceae. MSc thesis. Palacký University in Olomouc, Olomouc, Czech Republic (in Czech).
Funk, V.A., Susanna, A., Stuessy, T.F. and Bayer, R.J. (eds) (2009) *Systematics, Evolution and Biogeography of Compositae*. International Association for Plant Taxonomy, Vienna, Austria.
Galea, V.J. and Price, T.V. (1988) Resistance of lettuce and related species to anthracnose (*Microdochium panattonianum*) in Australia. *Plant Pathology* 37, 363–372.
Grube, R.C., Hayes, R., Mou, B. and McCreight, J.D. (2005) *Lettuce breeding, USDA-ARS. California Lettuce Research Board Annual Report, 2004–2005*. California Lettuce Research Board, California.
Grulich, V. (2004) *Lactuca* L. – locika. In: Slavík, B. and Štěpánková J. (eds) *Květena České republiky, díl 7*. Academia, Praha, Czech Republic, pp. 487–497.
Haley, V. and McCreight, J.D. (1990) Resistance of wild lettuce (*Lactuca saligna* L.) to lettuce infectious yellows virus. *HortScience* 25, 1163.

Harlan, J.R. and de Wet, J.M.J. (1971) Towards a rational classification of cultivated plants. *Taxon* 20, 509–517.

Jeuken, M. and Lindhout, P. (2002) *Lactuca saligna*, a non-host for lettuce downy mildew (*Bremia lactucae*), harbors a new race-specific Dm gene and three QTLs for resistance. *Theoretical and Applied Genetics* 105, 384–391.

Jeuken, M. and Lindhout, P. (2004) The development of lettuce backcross inbred lines (BILs) for exploitation of the *Lactuca saligna* (wild lettuce) germplasm. *Theoretical and Applied Genetics* 109, 394–401.

Jeuken, M., van Wijk, R., Peleman, J. and Lindhout, P. (2001) An integrated interspecific AFLP map of lettuce (*Lactuca*) based on two *L. sativa* × *L. saligna* $F_2$ populations. *Theoretical and Applied Genetics* 103, 638–647.

Jeuken, M., Pelgrom, K., Stam, P. and Lindhout, P. (2008) Efficient QTL detection for nonhost resistance in wild lettuce: backcross inbred lines versus $F_2$ population. *Theoretical and Applied Genetics* 116, 845–857.

Jeuken, M.J.W., Zhang, N.W., McHale, L.K., Pelgrom, K., den Boer, E., *et al.* (2009) Rin4 causes hybrid necrosis and race-specific resistance in an interspecific lettuce hybrid. *Plant Cell* 21, 3368–3378.

Kaur, P. and Mitkowski, N.A. (2010) Evaluation of *Lactuca* germplasm for resistance to the northern root-knot nematode (*Meloidogyne hapla* Chitwood). *International Journal of Vegetable Science* 17, 26–36.

Kesseli, R., Ochoa, O. and Michelmore, R. (1991) Variation at RFLP loci in *Lactuca* spp. and origin of cultivated lettuce (*L. sativa*). *Genome* 34, 430–436.

Kitner, M., Lebeda, A., Doležalová, I., Maras, M., Křístková, E., *et al.* (2008) AFLP analysis of *Lactuca saligna* germplasm collections from four European and three Middle Eastern countries. *Israel Journal of Plant Sciences* 56, 185–193.

Koopman, W.J.M. (1999) Plant systematics as a useful tool for plant breeders: examples from lettuce. In: Lebeda, A. and Křístková, E. (eds) *Eucarpia Leafy Vegetables '99. Proceedings of the Eucarpia Meeting on Leafy Vegetables Genetics and Breeding, Olomouc, Czech Republic, 8–11 June, 1999.* Palacký University in Olomouc, Olomouc, Czech Republic, pp. 95–105.

Koopman, W.J.M., De Jong, J.H. and De Vries, I.M. (1993) Chromosome banding pattern in Lettuce species (*Lactuca* sect. *Lactuca*, *Compositae*). *Plant Systematics and Evolution* 185, 249–257.

Koopman, W.J.M., Guetta, E., van de Wiel, C.C.M., Vosman, B. and van den Berg, R.G. (1998) Phylogenetic relationships among *Lactuca* (Asteraceae) species and related genera based on ITS-1 DNA sequences. *American Journal of Botany* 85, 1517–1530.

Koopman, W.J.M., Zevenbergen, M.J. and van den Berg, R.G. (2001) Species relationships in *Lactuca* s.l. (Lactuceae, Asteraceae) inferred from AFLP fingerprints. *American Journal of Botany* 88, 1881–1887.

Křístková, E., Doležalová, I. and Lebeda, A. (2009) Morphological grouping of *Lactuca saligna* germplasm originating from Italy and France. In: Meglič, V. and Bastar, M.T. (eds) *Book of Abstracts of 19th EUCARPIA Conference, Genetic Resources Section, Ljubljana, Slovenia, May 26–29, 2009.* Kmetijski Institut Slovenije, Ljubljana, Slovenia, p. 46.

Křístková, E., Lebeda, A., Doležalová, I., Kitner, M., Petrželová, I., *et al.* (2010) Willow-leaf lettuce – distribution and variation. In: Dolenc Koce, J., Vodnik, D. and Pongrac, P. (eds) *Book of Abstracts, 5th Slovenian Symposium on Plant Biology with International Participation, September 6–8, 2010, Ljubljana, Slovenia.* The Slovenian Society of Plant Biology, Ljubljana, Slovenia, p. 47.

Křístková, E., Tvardková, M. and Lebeda, A. (2011) Characterization of developmental stages in *Lactuca saligna* germplasm from Europe and USA. In: *Eucarpia Leafy Vegetables 2011, August 24–26, 2011.* Université Lille Nord de France, Villneuve d´Ascq, France, p. 78.

Lebeda, A. (1985) Differences in resistance of wild *Lactuca* species to natural infection of lettuce powdery mildew (*Erysiphe cichoracearum*). *Euphytica* 34, 521–523.

Lebeda, A. (1986) Specificity of interactions between wild *Lactuca* spp. and *Bremia lactucae* isolates from *Lactuca serriola*. *Journal of Phytopathology* 117, 54–64.

Lebeda, A. (1994) Evaluation of wild *Lactuca* species for resistance of natural infection of powdery mildew (*Erysiphe cichoracearum*). *Genetic Resources and Crop Evolution* 41, 55–57.

Lebeda, A. and Blok, I. (1991) Race-specific resistance genes to *Bremia lactucae* in new Czechoslovak lettuce cultivars and location of resistance in a *Lactuca serriola* × *Lactuca sativa* hybrid. *Archiv für Phytopathologie und Pflanzenschutz* 27, 65–72.

Lebeda, A. and Boukema, I.W. (1991) Further investigation of the specificity of interactions between wild *Lactuca* spp. and *Bremia lactucae* isolates from *Lactuca serriola*. *Journal of Phytopathology* 133, 57–64.

Lebeda, A. and Mieslerová, B. (2011) Taxonomy, distribution and biology of lettuce powdery mildew (*Golovinomyces cichoracearum* sensu stricto). *Plant Pathology* 60, 400–415.

Lebeda, A. and Pink, D.A.C. (1998) Histological aspects of the response of wild *Lactuca* spp. and their hybrids, with *L. sativa* to lettuce downy mildew (*Bremia lactucae*). *Plant Pathology* 47, 723–736.

Lebeda, A. and Reinink, K. (1994) Histological characterization of resistance in *Lactuca saligna* to lettuce downy mildew (*Bremia lactucae*). *Physiological and Molecular Plant Pathology* 44, 125–139.
Lebeda, A. and Syrovátko, P. (1988) Specificity of *Bremia lactucae* isolates from *Lactuca sativa* and some Asteraceae plants. *Acta Phytopathologica et Entomologica Hungarica* 23, 39–48.
Lebeda, A. and Zinkernagel, V. (2003) Characterization of new highly virulent German isolates of *Bremia lactucae* and efficiency of resistance in wild *Lactuca* spp. germplasm. *Journal of Phytopathology* 151, 274–282.
Lebeda, A., Doležalová, I., Křístková, E. and Mieslerová, B. (2001) Biodiversity and ecogeography of wild *Lactuca* spp. in some European countries. *Genetic Resources and Crop Evolution* 48, 153–164.
Lebeda, A., Pink, D.A.C. and Astley, D. (2002) Aspects of the interactions between wild *Lactuca* spp. and related genera and lettuce downy mildew (*Bremia lactucae*). In: Spencer-Phillips, P.T.N., Gisi, U. and Lebeda, A. (eds) *Advances in Downy Mildew Research*. Kluwer, Dordrecht, the Netherlands, pp. 85–117.
Lebeda, A., Doležalová, I. and Astley, D. (2004a) Representation of wild *Lactuca* spp. (Asteraceae, Lactuceae) in world genebank collections. *Genetic Resources and Crop Protection* 51, 167–174.
Lebeda, A., Doležalová, I., Feráková, V. and Astley, D. (2004b) Geographic distribution of wild *Lactuca* species (Asteraceae, Lactucae). *The Botanical Review* 70, 328–356.
Lebeda, A., Doležalová, I., Křístková, E. and Novotná, A. (2005) Comparative study of variation of some morphological and developmental characteristics of *Lactuca serriola* germplasm collected in Central Europe (Czech Republic) and the British Isles (England, UK) In: Bullitta, S. (ed.) *Plant Genetic Resources of Geographical and Other Islands. (Conservation, Evaluation and Use for Plant Breeding). Proceedings of the XVII Eucarpia Genetic Resources Section Meeting, Castelsardo, Italy, 30 March–2 April 2005*. CNR-ISPAAM, Sassari, Italy, p. 46.
Lebeda, A., Sedlářová, M., Lynn, J. and Pink, D.A.C. (2006) Phenotypic and histological expression of different genetic backgrounds in interactions between lettuce, wild *Lactuca* spp., *L. sativa* × *L. serriola* hybrids and *Bremia lactucae*. *European Journal of Plant Pathology* 115, 431–441.
Lebeda, A., Ryder, E.J., Grube, R., Doležalová, I. and Křístková, E. (2007) Lettuce (Asteraceae; *Lactuca* spp.). In: Singh, R.J. (ed.) *Genetic Resources, Chromosome Engineering, and Crop Improvement, Vol 3, Vegetable Crops*. CRC Press, Taylor and Francis Group, Boca Raton, Florida, pp. 377–472.
Lebeda, A., Sedlářová, M., Petřivalský, M. and Prokopová, J. (2008) Diversity of defence mechanisms in plant–oomycete interactions: a case study of *Lactuca* spp. and *Bremia lactucae*. *European Journal of Plant Pathology* 122, 71–89.
Lebeda, A., Doležalová, I., Křístková, E., Kitner, M., Petrželová, I., *et al.* (2009) Wild *Lactuca* germplasm for lettuce breeding: recent status, gaps and challenges. *Euphytica* 170, 15–34.
Lebeda, A., Doležalová, I. and Novotná, A. (2012a) Wild and weedy *Lactuca* species, their distribution, ecogeography and ecobiology in USA and Canada. *Genetic Resources and Crop Evolution* 59, 1805–1822.
Lebeda, A., Křístková, E., Doležalová, I., Kitner, M., Petrželová, I., *et al.* (2012b) Biodiversity of *Lactuca saligna* and its exploitation in lettuce breeding. In: Bedö, Z. and Láng, L. (eds) *Plant Breeding for Future Generations, Proceedings of the 19th Eucarpia General Congress, 21–24 May 2012. Budapest, Hungary*. Agricultural Institute, Centre for Agricultural Research, Academy of Sciences, Martonvásár, Hungary, p. 257.
Lebeda, A., Mieslerová, B., Petrželová, I., Korbelová, P. and Česneková, E. (2012c) Patterns of virulence variation in the interaction between *Lactuca* spp. and lettuce powdery mildew (*Golovinomyces cichoracearum*). *Fungal Ecology* 5, 670–682.
Lebeda, A., Mieslerová, B., Petrželová, I. and Korbelová, P. (2013) Host specificity and virulence variation in populations of lettuce powdery mildew pathogen (*Golovinomyces cichoracearum* s. str.) from prickly lettuce (*Lactuca serriola*). *Mycological Progress* 12, 533–545.
Lebeda, A., Křístková, E., Kitner, M., Mieslerová, B., Jemelková, M., *et al.* (2014) Wild *Lactuca* species, their genetic diversity, resistance to diseases and pests, and exploitation in lettuce breeding. *European Journal of Plant Pathology* 138, 597–640.
Matoušková, Z. (2013) Morphological traits of *Lactuca saligna* L. from selected countries in Europe, Asia and America. MSc thesis. Palacký University in Olomouc, Olomouc, Czech Republic (in Czech).
Meusel, H. and Jäger, E.J. (1992) *Vergleichende Chorologie der Zentraleuropäischen Flora*. Gustav Fischer Verlag, Jena, Germany.
Michalska, K., Stojakowska, A., Malarz, J., Doležalová, I., Lebeda, A., *et al.* (2009) Systematic implications of sesquiterpene lactones in *Lactuca* species. *Biochemical Systematics and Ecology* 37, 174–179.
Mikel, M.A. (2007) Genealogy of contemporary North American lettuce. *HortScience* 42, 489–493.
Mou, B. and Liu, Y. (2003) Leafminer resistance in lettuce. *HortScience* 38, 570–572.

Mou, B. and Liu, Y. (2004) Host plant resistance to leafminers in lettuce. *Journal of the American Society for Horticultural Science* 129, 383–388.
Netzer, D., Globerson, D., Weintal, C. and Elyassi, R. (1985) Sources and inheritance of resistance to *Stemphylium* leaf spot of lettuce. *Euphytica* 34, 393–396.
Norwood, J.M., Crute, I.R. and Lebeda, A. (1981) The location and characteristics of novel sources of resistance to *Bremia lactucae* Regel (downy mildew) in wild *Lactuca* L. species. *Euphytica* 30, 659–668.
Ochoa, O., Delp, B. and Michelmore, R.W. (1987) Resistance in *Lactuca* spp. to *Microdochium panattoniana* (lettuce anthracnose). *Euphytica* 36, 609–614.
Petrželová, I., Lebeda, A., Nevo, E. and Beharav, A. (2007) Variation of response against *Bremia lactucae* in natural populations of *Lactuca saligna*. In: Lebeda, A. and Spencer-Phillips, P.T.N. (eds) *Advances in Downy Mildew Research, Vol* 3. Palacký University in Olomouc and JOLA, v.o.s., Kostelec na Hané, Czech Republic, pp. 169–173.
Petrželová, I., Lebeda, A. and Beharav, A. (2011) Resistance to *Bremia lactucae* in natural populations of *Lactuca saligna* from some Middle Eastern countries and France. *Annals of Applied Biology* 159, 442–455.
Pink, D.A.C. and Keane, E.M. (1993) Lettuce: *Lactuca sativa* L. In: Kalloo, G. and Bergh, B.O. (eds) *Genetic Improvement of Vegetable Crops*. Pergamon Press, Oxford, UK, pp. 543–571.
Provvidenti, R., Robinson, R.W. and Shail, J.W. (1980) A source of resistance to a strain of cucumber mosaic virus in *Lactuca saligna* L. *HortScience* 15, 528–529.
Pryor, D.E. (1941) A unique case of powdery mildew on lettuce in the field. *Plant Disease Report* 25, 74.
Rauscher, G. and Simko, I. (2013) Development of genomic SSR markers for fingerprinting lettuce (*Lactuca sativa* L.) cultivars and mapping genes. *BMC Plant Biology* 13, 11.
Riar, D.S., Rustgi, S., Burke, I.C., Gill, K.S. and Yenish, J.P. (2011) EST-SSR development from 5 *Lactuca* species and their use in studying genetic diversity among *L. serriola* biotypes. *Journal of Heredity* 102, 17–28.
Rich, J., Brito, J., Ferrell, J. and Kaur, R. (2010) Weed hosts of root-knot nematodes common to Florida. University of Florida, IFAS, ENY-060. Available at: ufdc.ufl.edu/IR00004190/00001 (accessed 19 October 2015).
Rulkens, A.J.H. (1987) *De CGN sla-collectie: inventarisatie, paspoortgegevens en enkele richtlijnen voor de toekomst. CGN Report: CGN-T 1987-1*. Centre for Genetic Resources (CGN), Wageningen, the Netherlands.
Ryder, E.J. and McCreight, J.D. (2014) Vegetable Cultivar Descriptions for North America, Lettuce (M–Z). Available at: cuke.hort.ncsu.edu/cucurbit/wehner/vegcult/lettucemz.html (accessed 11 June 2014).
Sedlářová, M. and Lebeda, A. (2001) Histochemical detection and role of phenolic compounds in the defense response of *Lactuca* spp. to lettuce downy mildew (*Bremia lactucae*). *Journal of Phytopathology* 149, 693–697.
Sedlářová, M., Lebeda, A. and Pink, D.A.C. (2001) The early stages of interaction between effective and non-effective race-specific genes in *Lactuca sativa*, wild *Lactuca* spp. and *Bremia lactucae* (NL16). *Journal of Plant Disease and Protection* 108, 477–489.
Sedlářová, M., Luhová, L., Petřivalský, M. and Lebeda, A. (2007) Localisation and metabolism of reactive oxygen species during *Bremia lactucae* pathogenesis in *Lactuca sativa* and wild *Lactuca* spp. *Plant Physiology and Biochemistry* 45, 607–616.
Simko, I. (2009) Development of EST-SSR markers for the study of population structure in lettuce (*Lactuca sativa* L.). *Journal of Heredity* 100, 256–262.
Simko, I., Pechenick, D.A., McHale, L.K., Truco, M.J., Ochoa, O.E., *et al.* (2010) Development of molecular markers for marker-assisted selection of dieback resistance in lettuce (*L. sativa*). *Acta Horticulturae (ISHS)* 859, 401–408.
Simko, I., Hayes, R.J., Truco, M.J. and Michelmore, R.W. (2011) Mapping a dominant negative mutation for triforine sensitivity in lettuce and its use as a selectable marker for detecting hybrids. *Euphytica* 182, 157–166.
Stoffel, K., van Leeuwen, H., Kozik, A., Caldwell, D., Ashrafi, H., *et al.* (2012) Development and application of a 6.5 million feature Affymetrix Genechip® for massively parallel discovery of single position polymorphisms in lettuce (*Lactuca* spp.). *BMC Genomics* 13, 185.
Tamaki, H., Robinson, R.W., Anderson, J.L. and Stoewsand, G.S. (1995) Sesquiterpene lactones in virus-resistant lettuce. *Journal of Agricultural and Food Chemistry* 43, 6–8.
Tvardková, M. (2010) Morphologic traits of willow-leaf lettuce (*Lactuca saligna* L.) from France, Italy, Slovakia and the USA. BSc thesis. Palacký University in Olomouc, Olomouc, Czech Republic (in Czech).

van Hintum, T.J.L. and Boukema, I. (1999) Genetic resources of leafy vegetables. In: Lebeda, A. and Křístková, E. (eds) *Eucarpia Leafy Vegetables '99. Proceedings of the Eucarpia Meeting on Leafy Vegetables Genetics and Breeding, Olomouc, Czech Republic, 8–11 June, 1999*. Palacký University in Olomouc, Olomouc, Czech Republic, pp. 59–71.

van Treuren, R. and van Hintum, T.J.L. (2009) Comparison of anonymous and targeted molecular markers for the estimation of genetic diversity in ex situ conserved *Lactuca*. *Theoretical and Applied Genetics* 119, 1265–1279.

Yang, T.J., Jang, S.W. and Kim, W.B. (2007) Genetic relationships of *Lactuca* spp. revealed by RAPD, Inter-SSR, AFLP, and PCR-RFLP analyses. *Journal of Crop Science and Biotechnology* 10, 29–34.

Zhang, N.W., Lindhout, P., Niks, R.E. and Jeuken, M.J.W. (2009a) Genetic dissection of *Lactuca saligna* nonhost resistance to downy mildew at various lettuce developmental stages. *Plant Pathology* 58, 923–932.

Zhang, N.W., Pelgrom, K., Niks, R.E., Visser, R.G.F. and Jeuken, M.J.W. (2009b) Three combined quantitative trait loci from nonhost *Lactuca saligna* are sufficient to provide complete resistance of lettuce against *Bremia lactucae*. *Molecular Plant Microbe Interactions* 22, 1160–1168.

Zohary, D. (1991) The wild genetic resources of cultivated lettuce (*Lactuca sativa* L.). *Euphytica* 53, 31–35.

# 6 Capturing Wild Relative and Landrace Diversity for Crop Improvement Using a New Selection Tool to Exploit Genetic Resources in Durum Wheat (*Triticum durum* Desf.)

**D. Pignone,[1,2]* D. De Paola,[1] N. Rapanà[1] and M. Janni[1]**
[1]*Institute of Bioscience and Bioresources, National Research Council (CNR), Bari, Italy;* [2]*Department of Biology Agriculture and Food Science, National Research Council (CNR), Rome, Italy*

## 6.1 Introduction

Wheat is one of the oldest crops, domesticated at the very early beginning of agriculture. Originating in the Middle East within the Fertile Crescent, it moved along with human migrations to colonize almost the entire world. For centuries, the combined effect of adaptation to different environments and selection operated empirically on the basis of cultural factors brought the fixation of new alleles and traits in the genetic pool of this crop. Moreover, the wheat genepool is constituted by different 'species' with three levels of ploidy, i.e. 2×, 4× and 6×, which each correspond to a different breakthrough of the domestication process (Heun *et al.*, 1997; Lev-Yadun *et al.*, 2006).

In the Mediterranean, tetraploid wheat belonging mostly to the species *Triticum durum* Desf. (syn. *Triticum turgidum* L. subsp. *durum*) is traditionally grown under rain-fed conditions; the majority of durum is produced in Italy, Greece and France. Italy accounted for nearly 58% of EU durum production in 2010, followed by Greece (21%) and France (12%) (Taylor and Koo, 2011).

Drought and high temperature stress at the end of the growing season usually constrain crop yield potential as these stresses coincide with the grain-filling period in the Mediterranean region (Al-Karaki, 2012). Although predictive models differ in their details, they generally agree that the climate change effect in the Mediterranean will be more severe than in many other regions, and also that the Mediterranean region will become hotter and drier over the next century. In fact, a rise in the mean temperature and a decrease in annual precipitation are forecast by the Intergovernmental Panel on Climate Change (IPCC, 2012). The challenge for durum wheat will be to overcome this climatic variation while maintaining the present-day high standards in production and quality.

New variability can be found in all the durum wheat accessions available all over the world in genebanks. The Eurisco database (eurisco.ecpgr.org) reports a total of over 17,000 durum wheat accessions held in different

*Corresponding author; e-mail: domenico.pignone@cnr.it

 

genebanks, but only some 6800 of them are traditional varieties and/or landraces. This material is of incontestable usefulness to mine for new alleles and adaptation traits to changing climatic conditions to meet novel requirements by the agrofood industry and to provide the grounds for the developing genomic technologies for wheat breeding (Habash *et al.*, 2009). Unfortunately, the great number of available accessions is at the same time the major resource and the strongest barrier to the possibility of utilizing the traits present in the genepool.

New approaches are therefore needed to overcome the constraints inherent in analysing almost 7000 accessions, which implies an even larger number of genotypes, since most landraces are actually a mixture of different genotypes. A possible way to explore durum wheat germplasm better could be the development of a subset of genotypes, representative of the overall variation of the species, taking advantage of the cleistogamic reproduction habit of this plant.

In the past, a few attempts have been made to reduce the number of accessions while maintaining at the same time most of the genetic variation, such as the core collection approach. Nevertheless, the variable nature of landraces did not allow a complete exploitation of these efforts (Spagnoletti *et al.*, 1993). A different approach to reduce the number of genotypes while maintaining a good representation of the genetic background is the constitution of a working durum wheat collection obtained by means of a single seed descent (SSD) selection.

## 6.2 Materials and Methods

### 6.2.1 Plant material

The seedbank of the Institute of Bioscience and Bioresources (IBBR) of the National Research Council of Italy in Bari hosts more than 5600 accessions of durum wheat, mostly constituted by landraces, collected from the whole range of the species. They represent approximately 40 countries, although samples from the Mediterranean region and other centres of diversity (*sensu* Vavilov, see Dvorak *et al.*, 2011) are more abundantly represented (Fig. 6.1). A total of 452 accessions, including three reference varieties reputed to be characteristic of this durum wheat collection, were selected according to their origin, trying to retain a good representation of all agroclimatic regions where the plants were grown or collected. One seed per accession was collected at random and used to start the SSD lines. The process of using a single seed per plant in each sowing season was repeated for three generations to obtain the S3 generation.

### 6.2.2 Production of the Single Seed Descent genotypes collection

The Single Seed Descent (SSD; Brim, 1966) technique was originally developed with the aim to increase breeding efficiency, while at the same time maximizing the variation in the segregating generations of a cross between elite varieties (Snape, 1976). The goal of the SSD process is, in theory, to produce a set of inbred lines that represent the entire range of variation in the original recombinant population. In our approach, instead of starting from an $F_2$ population, we started from a collection of landraces, aimed at establishing a set of genotypes that represented the highest range of variation present in the germplasm collection used. In order to obtain a working collection represented by highly homozygous genotypes, 500 genotypes were randomly selected, on the basis of country of origin, from the IBBR seedbank. This genebank stores 5659 samples identified as *T. durum* and the samples to be used identified and treated as described in this material and methods section. In the $S_3$ generation, 452 genotypes were obtained due to some losses caused by absence of germination, destruction by birds or by severe parasitic attack, and moreover, some contaminants of *Triticum aestivum*, often present in durum wheat landraces, were discovered.

### 6.2.3 Genotyping

DNA was extracted from young leaves of the S3 generation (Zadoks *et al.*, 1974). The DNA samples were analysed using the Illumina Infinium iSelect 90 k wheat chip at TraitGenetics (Gatersleben, Germany), containing 81,587 molecular markers,

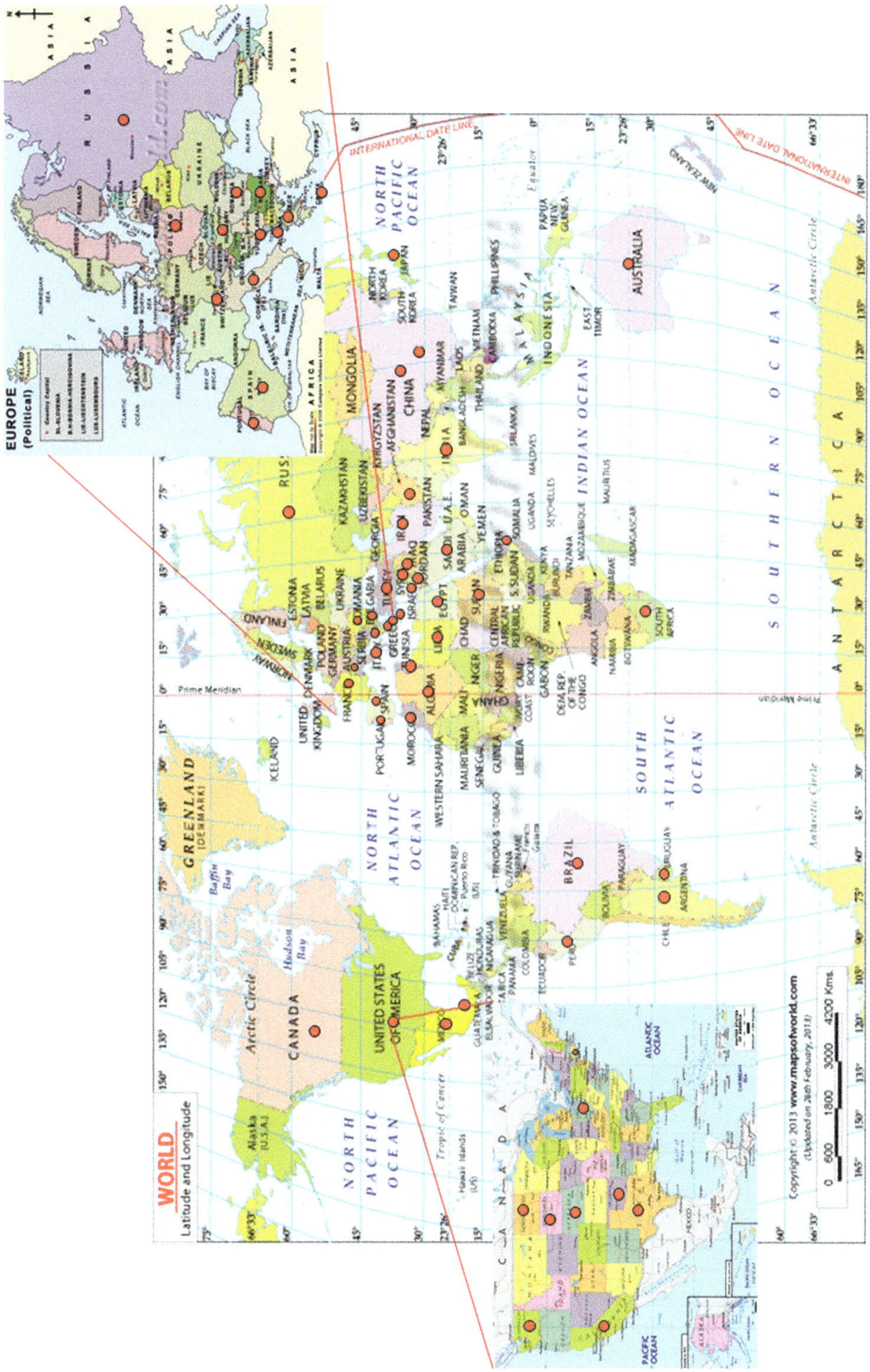


Fig. 6.1. Map of representativeness of the SSD collection.

used to survey single nucleotide polymorphism (SNP) variation in the sample of 452 SSD genotypes, including three durum wheat varieties ('Langdon', 'Svevo' and 'Cappelli') as reference. SNP allele clustering and genotype calling for all wheat samples was performed with GenomeStudio Software using a specific durum wheat cluster file developed at TraitGenetics. The genotyping data were analysed to assess mean heterozygosity and genetic diversity based on pairwise comparison of variation over all SNP loci. As a result of this analysis, a distance matrix of differences was generated. Principal coordinate analysis (PCoA) was used to visualize the patterns of genetic relationships contained in the matrix of differences.

### 6.2.4 Field and postharvest phenotyping

The SSD genotype collection in S3 was evaluated phenotypically for both qualitative and quantitative traits at two phenological stages according to the Zadoks scale (Zadoks *et al.*, 1974): at hard-dough development (Z87) on the plant and at complete ripeness (Z94) in postharvest. The collection was evaluated for both qualitative and quantitative traits. Three qualitative characters were chosen: awn colour (Awn C), spike density (SpD) and seed colour (SC) (IBPGR, 1985). Six quantitative traits were also recorded: plant height (PH), number of spikes per plant (NSPP), number of spikelets per spike (NSPS), number of kernels per spike (NKPS), kernel yield per spike (KYPS) and thousand kernel weight (TKW).

## 6.3 Results and Discussion

### 6.3.1 Genotyping of the SSD collection

Identifying DNA variants that contribute to determine traits of interest is a central goal of modern breeding to improve crop yield and quality and increase crop resistance to diseases and tolerance to biotic and abiotic stresses. Considerable progress has been made technologically to assay for genetic variation in the eukaryotic genomes of both plants and animals. The most impacts have been those based on the development of the next-generation sequencing (NGS) technologies (Koboldt *et al.*, 2013; Pabinger *et al.*, 2014). This advance in sequencing technologies proceeds with other high-throughput technologies that allow genotyping of thousands of loci per run using array-based approaches. A genome-wide genotyping chip for wheat was available from Illumina Infinium iSelect 90 k and was used to genotype our SSD collection.

The analysis of the 90 k SNP chip on the 452 genotypes revealed an average heterozygosity around 0.4%, within the expected figure since the outcrossing rate in durum wheat is less than 1%. Based on pairwise comparison of all samples, a distance matrix of the number of differences was generated. Total number of differences ranged from 0 (reference varieties) to 10,429 (12.8%), with a mean value of 6222 (7.6%) different SNPs. The codes used to describe the SNPs are reported in Table 6.1.

Despite the low level of heterozygosity (Table 6.2), these results confirm that the SSD population retains a high level of diversity to exploit in breeding programmes. As a measure of relatedness between different SSD genotypes,

**Table 6.1.** Codes used to identify the nucleotides in the output of the 90 k array.

| | |
|---|---|
| A | Adenine |
| C | Cytosine |
| G | Guanine |
| T | Thymine |
| R | Guanine/adenine (purine) |
| Y | Cytosine/thymine (pyrimidine) |
| K | Guanine/thymine |
| M | Adenine/cytosine |
| S | Guanine/cytosine |
| W | Adenine/thymine |

**Table 6.2.** Observed heterozygote loci. The percentage has been calculated over all loci and all genotypes.

| Observed heterozygote | IUB code | Number | Per cent over all observations |
|---|---|---|---|
| A/G | R | 106,726 | 0.31817369 |
| C/T | Y | 112,227 | 0.334573382 |
| G/T | K | 22,509 | 0.067104282 |
| A/C | M | 21,973 | 0.065506348 |
| G/C | S | 829 | 0.002471431 |
| A/T | W | 230 | 0.000685681 |

the genetic distance over all 81,587 SNP marker loci was quantified based on the total number of differences. The results are summarized via principal coordinates analysis in Fig. 6.2. The first three axes explain almost 70% of the observed variation (first axis 31.38%, second axis 21.50% and third axis 16.40% of the total variation). This analysis demonstrated a mainstream uniform distribution of the variation and some 'islands' with a peculiar distribution of the genetic diversity, as shown by the scatterplot of the first two principal coordinates (Fig. 6.2).

Genotypic data demonstrate that the collection of SSD genotypes retains a high level of diversity, and at the same time, each SSD genotype is genetically uniform with a very low level of heterozygosity. There is a wide range of distances between pairs of SSD genotypes. Based on these data, a dendrogram can be constructed giving a representation of the distances among genotypes. Because of the complexity of such a dendrogram, a clear-cut delimitation of the branches into discrete subgroups is not feasible, supporting the view that this collection has an even distribution of the genetic variation.

Nevertheless, some grouping, although not clear, emerges. This can be the result of human influence on the diffusion of selected favoured genotypes. A question emerges from this observation: does the SSD collection also include indicators of the co-evolution of the wheat–human complex? A better understanding of this issue might arise from analyses based on a Bayesian approach (Stephens and Balding, 2009).

One important outcome of the present study is that, applying a correct phenotyping procedure, possibly based on automated systems, the high number of markers can help anchor a wide range of phenotypic data on previously produced genetic maps. Existing genetic maps are based mostly on crosses between elite varieties and may therefore be based on a limited genetic base. The SSD collection, being based on a representation of the whole genepool, will allow enriching the existing maps with new, potentially useful traits.

### 6.3.2 Pre- and postharvest phenotyping

The SSD genotypes in S3 were evaluated phenotypically for both qualitative and quantitative traits at two phenological stages according to the scale of Zadoks *et al.* (1974): at hard-dough development (Z87) on the plant and at complete ripeness (Z94) in postharvest. The collection was evaluated through three qualitative traits – awn colour (Awn C), spike density (SpD) and seed colour (SC) (IBPGR, 1985) – and six quantitative traits – plant height (PH), number of spikes per plant (NSPP), number of spikelets per spike (NSPS), number of kernels per spike (NKPS), kernel yield per spike (KYPS) and thousand kernel weight (TKW). In 2013, the SSD genotypes

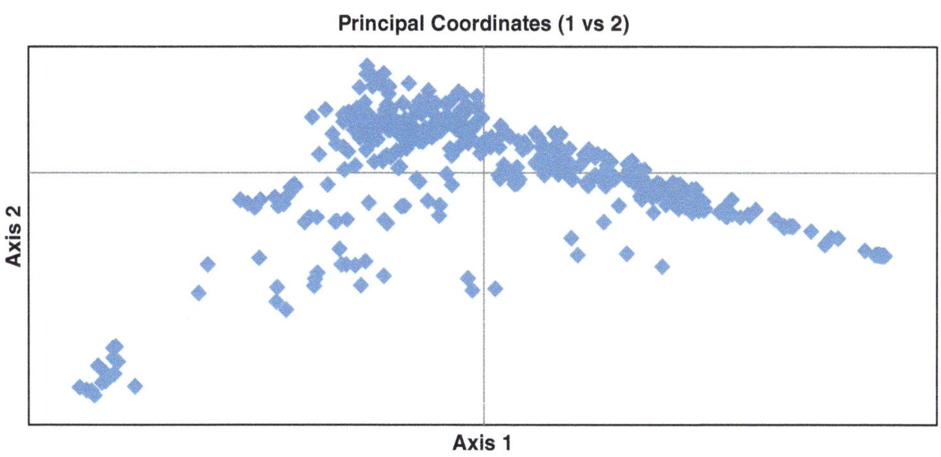


Fig. 6.2. Principal coordinates analysis (PCoA). Plot of the first two coordinates using genotypic data for 81,587 SNPs. In order to view the figure more easily, individual names are not included.

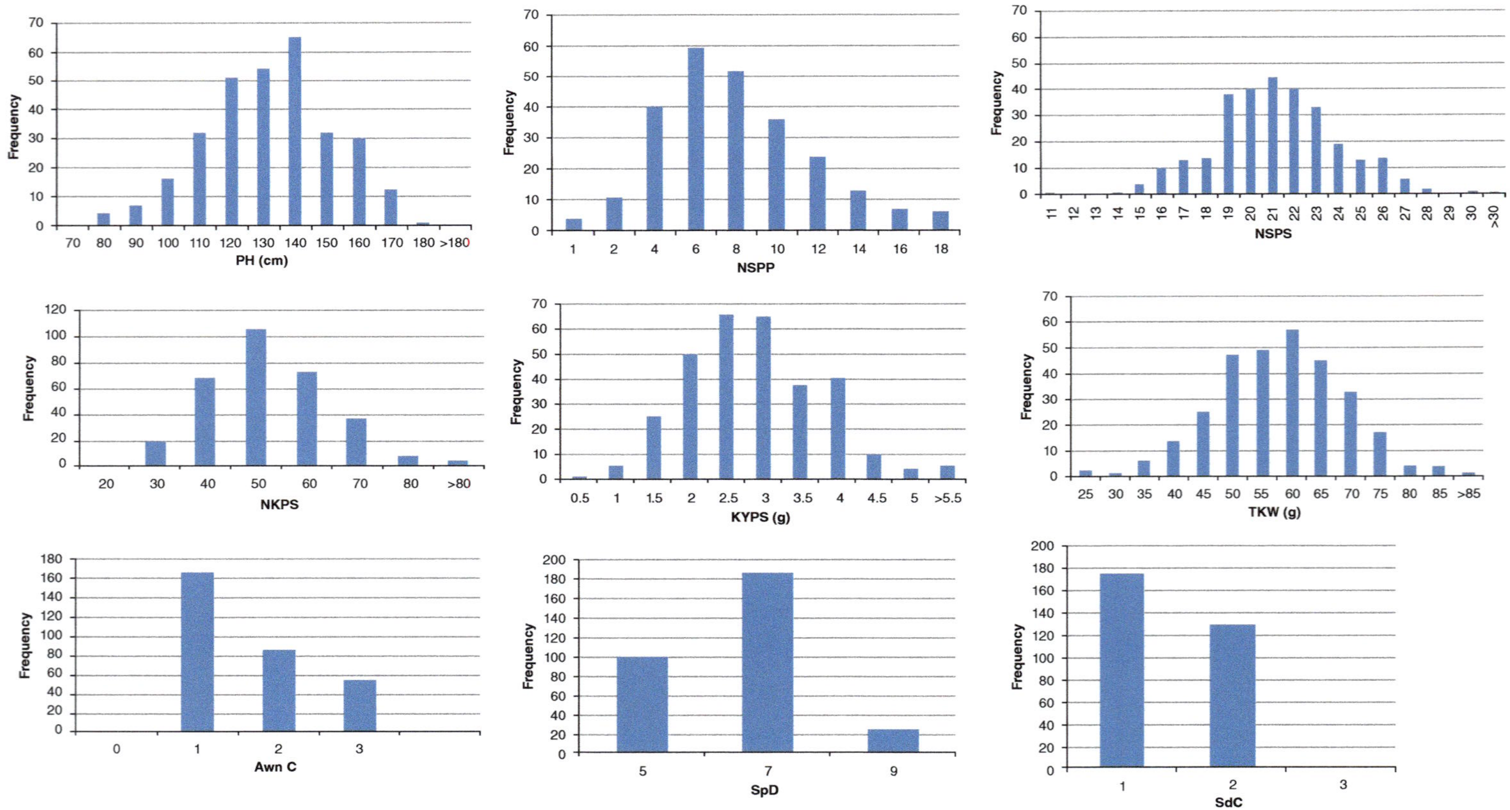


Fig. 6.3. Frequency distribution of phenotypic traits analysed (PH = plant height (cm); NSPP = number of spikes per plant; NSPS = number of spikelets per spike; NKPS = number of kernels per spike; KYPS = kernel yield per spike; TKW = thousand kernel weight; awn C = awn colour; SpD = spike density; SC = seed colour).

were scored for qualitative and quantitative phenotypic traits. Mean and standard deviation were calculated for each quantitative trait, as well as for the proportion of the different classes for each character; frequency distribution was also calculated. These results highlight that the data points for all traits that scored significantly fit a normal distribution curve (Fig. 6.3).

## 6.4 Conclusions

One of the major problems in exploiting the potential of germplasm stored in *ex situ* genebanks is often the overwhelming number of samples and genotypes they contain. This is particularly true for major crops. Several methods have been proposed in the past for reducing this amount to a comfortable number of samples yet still retaining a large amount of the total genetic variation of that species. The SSD method originally developed to facilitate the selection of recombinant lines has been applied in durum wheat to obtain a set of genotypes potentially retaining most of the variation of this species. The morphological and genetic analyses performed on these genotypes strongly support the successful outcome of a limited scale working collection of genotypes, useful for both theoretical studies and practical applications such as: gene mining, genotype × environment (G × E) interaction, genotype–phenotype association, breeding for specific traits, etc. These results demand further studies and this SSD genotype collection is proposed to the scientific community as a genetic tool for all scientists and students interested in durum wheat.

## References


Al-Karaki, G.N. (2012) Phenological development–yield relationships in durum wheat cultivars under late-season high-temperature stress in a semiarid environment. *ISRN Agronomy* 2012, Article ID 456856.

Brim, C.A. (1966) A modified pedigree method of selection in soybeans. *Crop Science* 16, 220.

Dvorak, J., Luo, M.-Ch. and Akhunov, E.D. (2011) N.I. Vavilov's theory of centres of diversity in the light of current understanding of wheat diversity, domestication and evolution. *Czech Journal of Genetics and Plant Breeding* 47, S20–S27.

Habash, D.Z., Kehel, Z. and Nachit, M. (2009) Genomic approaches for designing durum wheat ready for climate change with a focus on drought. *Journal of Experimental Botany* 60, 2805–2815.

Heun, M., Schäfer-Pregl, M., Klawan, D., Castagna, R., Accerbi, M., *et al.* (1997) Site of Einkorn wheat domestication identified by DNA fingerprinting. *Science* 278, 1312–1314.

IBPGR (International Board for Plant Genetic Resources) (1985) *Descriptors for Wheat* (revised). IBPGR, Rome, pp. 15.

IPCC (Intergovernmental Panel on Climate Change) (2012) *Managing the Risks of Extreme Events and Disasters to Advance Climate Change Adaptation. A Special Report of Working Groups I and II of the Intergovernmental Panel on Climate Change*. Field, C.B., Barros, V., Stocker, T.F., Qin, D., Dokken, D.J., *et al.* (eds) Cambridge University Press, UK.

Koboldt, D.C., Steinberg, K.M., Larson, D.E., Wilson, R.K. and Mardis, E.R. (2013) The next-generation sequencing revolution and its impact on genomics. *Cell* 155(1), 27–38.

Lev-Yadun, S., Gopher, A. and Abbo, S. (2006) How and when was wild wheat domesticated? *Science* 313, 296.

Pabinger, S., Dander, A., Fischer, M., Snajder, R., Sperk, M., *et al.* (2014) A survey of tools for variant analysis of next-generation genome sequencing data. *Briefings in Bioinformatics* 15(2), 256–278.

Snape, J.W. (1976) A theoretical comparison of diploidized haploid and single seed descent populations. *Heredity* 36, 275–277.

Spagnoletti Zeuli, P.L. and Qualset, C.O. (1993) Evaluation of five strategies for obtaining a core subset from a large genetic resource collection of durum wheat. *Theoretical and Applied Genetics* 87, 295–304.

Stephens, M. and Balding, D.J. (2009) Bayesian statistical methods for genetic association studies. *Nature Reviews – Genetics* 10, 681–690.

Taylor, R.D. and Koo, W.W. (2011) Outlook of the US and World Wheat Industries, 2010–2020. *Agribusiness and Applied Economics Reports* 715. Available at: purl.umn.edu/157807 (accessed 5 February 2014).

Zadoks, J.C., Chang, T.T. and Konzak, C.F. (1974) A decimal code for the growth stages of cereals. *Weed Research* 14, 415–421.

# 7 How the Focused Identification of Germplasm Strategy (FIGS) is Used to Mine Plant Genetic Resources Collections for Adaptive Traits

**K. Street,[1]* A. Bari,[1] M.Mackay[2] and A. Amri[1]**

*[1]International Center for Agricultural Research in the Dry Areas (ICARDA), Rabat, Morocco; [2]Queensland Alliance for Agriculture and Food Innovation (QAAFI), The University of Queensland, Brisbane St Lucia, Australia*

## 7.1 Introduction

With the world population estimated to reach 9.3 billion people by 2050 (United Nations, 2011), a 70% increase in global food production will be required (FAO, 2009). To put this into context, the demand for cereals alone will increase by nearly 1 billion more tonnes than the 2.1 billion tonnes required in 2009 (FAO, 2009). What is more, our agroecosystems will face significant challenges from factors such as climate change, land degradation and water availability, as well as evolving pest and disease virulence. There are two viable options to address these challenges: better agronomic practices and plant improvement (Fischer, 2009).

The International Maize and Wheat Improvement Center (CIMMYT) wheat breeding programme, for example, has achieved yield improvement rates of up to 2–3% per year (Lantican *et al.*, 2003), due to the selection of several genetic traits such as phenological adaptation, efficient use of water, resistance to biotic stresses, tolerance to abiotic stresses and superior grain quality (Khush, 2001; Fischer, 2009). These improvements are due to the utilization of naturally occurring genetic diversity (Linden, 1991). In this context, genebanks, which are a rich source of diversity, are of key importance to help meet the increasing demand for food (Hoisington *et al.*, 1999; FAO, 2003; Acquaah, 2007).

However, searching genebanks for useful traits can be problematic. A lack of adequate evaluation data along with the large numbers of accessions conserved have been highlighted as significant issues hindering their effective use in plant improvement (Gollin *et al.*, 2000; Koo and Wright, 2000; FAO, 2010). In fact, identifying novel and often rare alleles from large genebanks has been likened to searching for a needle in a haystack. Given that evaluation for useful traits is often expensive and research budgets are usually restrictive, an efficient technique is required that can identify relatively small subsets of germplasm that will have a high probability of containing the sought-after traits.

The core collection approach was suggested by Frankel (1984) as a means to improve the utilization of genebanks by attempting to capture the majority of genetic variation within a genebank into smaller subsets of accessions (5–10% of the original collection) that can be evaluated

*Corresponding author; e-mail: k.street@cgiar.org

routinely for desirable traits. How core sets are developed varies considerably and has received much attention in the literature (see Brown, 1989a,b; Hodgkin *et al.*, 1995). By and large, though, they are of static size and membership, constructed to maximize measures of diversity based on plant attributes such as morphology, physiology, phenology, ecogeographic origin and, more recently, molecular data. Variations and refinements of the core collection approach include mini-cores that further aggregate genetic variation into even smaller subsets (Upadhyaya *et al.*, 2009). Core collections have been adopted enthusiastically by the genetic resource community and have become industry standard, perhaps even dogma. However, the value of core collections as a tool to uncover specific traits is based on the assumption that some of the variation measured is actually useful. In this context, the point has been made that rare and adaptive alleles, most of which are thought to be functional, may not be captured by core collections (Brown and Spillane, 1999; Polignano *et al.*, 2001; Gepts, 2006; Dwivedi *et al.*, 2008; Pessoa-Filho *et al.*, 2010; Xu, 2010). In fact, in terms of uncovering useful genes, the return on investment in core collections does not seem to have been demonstrated convincingly. Rather, the literature suggests much of the focus has been on methods (or sampling strategies) to establish core collections (Holbrook *et al.*, 1993; Ortiz *et al.*, 1998; Hu *et al.*, 2000; Malosetti and Abadie, 2001).

FIGS (Focused Identification of Germplasm Strategy) is an alternative approach to searching germplasm accessions that aims to maximize the likelihood of capturing specific adaptive traits in subsets, flexible in size and make-up, sourced from one or more genebanks (Mackay, 1990; Street *et al.*, 2008; Glaszmann *et al.*, 2010; Kilian and Graner, 2012). FIGS is based on the premise that many accessions in *ex situ* collections, or genebanks, have been collected from the environment in which they evolved, often over millennia. Therefore, their adaptive traits will have been determined largely by the selection pressures that were present in the environment of origin, which has been shown to influence gene flow, and thus spatial/geographic differentiation (Vavilov, 1957; Lin and Bradshaw, 1975; Spieth, 1979). Nikolai Ivanovich Vavilov was one of the first pioneers to recognize the importance of ecoclimatic conditions when searching for the source material to include in plant improvement (Vavilov, 1932, 1957; Dorofeyev, 1992; Kurlovich *et al.*, 2000). Vavilov used the term 'climatic analogy' for the selection of suitable strains guided by climate and soil data. His 'differential phytogeographical method' also has elements that link the morphophysiological trait characters of species and strains to specific environments (Vavilov, 1920, 1922, 1935). FIGS extends this work by, for example, utilizing recent advances in geographic information systems (GIS) technology to develop environmental profiles for germplasm collection sites with pinpoint accuracy. This paper explores how FIGS works and ways in which it has been deployed to identify useful trait variation within genetic resource collections.

## 7.2 Genebank Material Suitable for FIGS Analysis

Since FIGS relies on a linkage between adaptive traits within *in situ* populations and long-term environmental conditions, only georeferenced landraces, crop wild relatives, pastures and rangeland species are suitable for the FIGS approach. While there are approximately 4.5 million accessions classified as landraces and wild relatives conserved in genebanks worldwide (FAO, 2010), a large proportion of them, especially older ones, lack collection site latitude and longitude data. This does not mean the data necessary to capture georeferences is not available. Many collection missions, even before global positioning system (GPS) devices were available, were documented rigorously with good descriptions of collection sites; the NI Vavilov Institute (VIR) collection is a good example of this, with records dating back to the 1920s. Using gazetteers (e.g. the Getty Thesaurus of Geographic Names and the GEOnet Names Server) and various GIS software packages, reasonably accurate coordinates can be determined from collection mission notes. Efforts to develop software tools to automate georeferencing have also been made. The BioGeomancer Project is one such initiative (Guralnick *et al.*, 2006). Various methods have also been developed to check the accuracy of captured coordinates. For example, FloraMap can be used to identify accessions with climatically discordant localities (Jones and Gladkov, 1999).

Whatever the tools used to capture geo-coordinates, it is a painstaking and tedious process that requires a significant commitment of resources. However, with recent advances in GIS software and the availability of agroclimatic and soil GIS surfaces, the utility of capturing collection site geo-coordinates is becoming obvious and more efforts are being made to georeference collections.

Of the global collections documented in the Genesys database (available at: www.genesys-pgr.org), 75% do not have collection site geo-coordinates; it must be noted, though, that a significant percentage of these are not landraces or wild relatives. Despite this, that still leaves 692,489 accessions that are georeferenced (M.C. Mackay, Queensland, 2015, personal communication), which is an adequate number to make FIGS viable. Further, efforts by genebanks to georeference their collections are not always uniform, whereby site coordinates might have been captured for significant collections but not for others. For example, in a recent exercise aimed at searching for reproductive frost tolerance in bread wheat, 41,538 georeferenced wheat accessions conserved at the International Center for Agricultural Research in the Dry Areas (ICARDA), VIR, the US Department of Agriculture (USDA) and the Australian Winter Cereals Collection (AWCC) were made available to the FIGS process. Likewise, of the 206,800 accessions recorded in the worldwide barley register, approximately 75,690 are georeferenced (J. Konopka, Aleppo, 2009, personal communication), which provides more than enough data to work with.

## 7.3 Environmental Data

By and large, the more classes of environmental data at a fine-grained resolution, the better. That is, FIGS can use a whole range of data classes, including anecdotal data such as farmer information about crop attributes, binary data such as irrigated or non-irrigated, scalar data such as aspect or slope, percentage data such as cloud cover and continuous data such as precipitation or temperature. But as indicated above, the finer the data resolution, the more efficient the process is likely to be. For example, climatic data can range from hourly to yearly and can be at a scale from just a few square metres to 30 km$^2$. For the FIGS approach, daily data at a few metres resolution would be considerably more useful than yearly averages at a 30 km$^2$ scale.

FIGS has been deployed successfully to find novel alleles, including new sources of resistance to powdery mildew (Bhullar *et al.*, 2009, 2010), Sunn pest (El-Bouhssini *et al.*, 2009, 2012), Russian wheat aphid (El-Bouhssini *et al.*, 2011, 2012), drought resistance traits in faba bean (Khazaei *et al.*, 2013), stem rust resistance in wheat (Bari *et al.*, 2012) and net blotch in barley (Endresen *et al.*, 2011) using long-term average monthly data at a 1 km$^2$ grid size. The climatic variables included monthly average minimum temperature, monthly average maximum temperature, monthly average precipitation, monthly average evapotranspiration and monthly average aridity index as point data extracted from continuous GIS surfaces modelled from data collection sites, as described by De Pauw (2008). Further, boron tolerance in bread wheat was found using a FIGS approach that relied on coarse-grained world soil map point data to guide the filtering process (Mackay, 1995).

Despite these successes, it is envisaged that significant improvements can be made in the FIGS process by using climatic data at a daily resolution presented to the algorithms deployed in a way that emphasizes the conditions during the crop cycle. This necessitates that daily climatic data for each collection site are arranged in databases so that the onset of the growing season is expressed as Day 1 instead of 1 January. This allows within-season climatic comparisons between collection sites, no matter what the geographic location and the actual site-specific date. This approach also facilitates screening for specific selection pressures at critical crop development stages; for example, heat or cold during the reproductive phase.

However, this refinement has posed two significant challenges. First, an accurate model is required to calculate the onset of growing period at a minimum 1 km$^2$ resolution for any given geo-coordinate. ICARDA created a model that used a combination of monthly temperature and moisture data to determine an onset date for a given geo-coordinate (De Pauw, 2008). However, there are considerable inconsistencies between actual observed dates and modelled values, particularly in humid environments.

Work is ongoing to improve the accuracy of the onset date model.

The second challenge is to generate accurate continuous surfaces for daily climatic data for the whole of the northern hemisphere. An exercise of this scale, which requires considerable computing power and storage capacity, had not been attempted before ICARDA recently generated over 285 terabytes of daily data for a range of climatic parameters by downscaling the 0.25 degree (25 $km^2$) European Centre for Medium-Range Weather Forecasts reanalysis data to 1 $km^2$ grid sizes. This daily data, coupled with upgraded onset data, will allow us to calculate more accurately the timing of crop development phases in relation to climatic conditions. This, in turn, will allow the detection of environment–trait relationships with a greater resolution.

Soil conditions also exert selection pressures on *in situ* populations for important traits such as salinity, acid, aluminium, boron and waterlogging tolerances, as well as increasingly important factors such as nutrient use efficiencies. While large-scale soil maps are typically coarse grained, the FIGS approach has been successful in using such maps to identify both salinity and boron tolerance (Mackay, 1995).

The soil data source currently used by ICARDA is the Harmonized World Soil Database (FAO, 2015), the results of an international collaboration, which is a 30 arc-second raster database with over 15,000 different soil mapping units that combines existing regional and national updates of soil information with the information contained in the 1:5,000,000 scale FAO-UNESCO Soil Map of the World (FAO-UNESCO, 1971–1981). The database includes 32 fields, describing the physical and chemical attributes of the top- and subsoils.

## 7.4 FIGS Pathways

It must be emphasized here that FIGS is an approach that uses information about the collection sites of genetic resources to help identify accessions containing specific adaptive traits. Beyond this there are no hard or fast rules as to how FIGS is deployed. What follows is a description of two pathways used to identify useful traits by the ICARDA FIGS scientists and their collaborators.

### 7.4.1 Filtering

The most straightforward approach is to apply filters to collection site data that are designed to 'mimic' selection pressures for specific traits. For example, using digital maps that detailed the probability of encountering saline soils, a set of 200 bread wheat landraces with possible tolerance to high soil salinity were identified in accessions from VIR, ICARDA and the AWCC. A second core set of 200 accessions was constructed using published methods. Both sets were tested for salinity tolerance in a phytotron constructed at VIR using the method of Munns and James (2003) to measure the ability of plants to exclude sodium from their leaves. It was found that 21% of the FIGS salinity set were tolerant compared to only 2% of the core set (I. Koserava, St Petersburg, 2010, personal communication).

A simple filtering approach was also used to identify accessions of bread wheat containing new genes for Sunn pest (El-Bouhssini *et al.*, 2009, 2012) and Russian wheat aphid (El-Bouhssini *et al.*, 2009, 2012) tolerance in wheat. Using the Sunn pest study as an example, a database containing passport and average monthly climatic data for the ICARDA, VIR and AWCC bread wheat collections was assembled (over 16,000 accessions). The following filters were applied:

**1.** From the 16,000 accessions available, only those from between 30–45 latitude and 35–80 longitude were selected because this was historically the broad geographic 'square' within which Sunn pest had been reported.
**2.** Accessions from China, Pakistan and India were excluded because of only recent reports of Sunn pest in these countries. This left 6328 accessions.
**3.** Randomly selected one accession per collection site, which left 2830 accessions.
**4.** Excluded particularly dry environments with annual precipitation below 280 mm.
**5.** Excluded material where the winter minimum temperatures fell below −10°C, which left 1502 accessions residing in three genebanks.

The 534 accessions conserved at ICARDA were evaluated for Sunn pest resistance.

Previously, thousands of ICARDA bread wheat accessions had been evaluated for resistance to

Sunn pest, with no tolerance or resistance identified. Of the 534 accessions tested in this instance, 9 accessions exhibiting juvenile resistance were identified. It can be seen from this example that even a very simple filter, using no more than geographic location, annual rainfall and monthly minimum temperature, was sufficient to uncover rare genetic variation.

A slightly more involved filtering process was used by Khazaei *et al.* (2013) to uncover drought resistance traits in faba bean landraces from the ICARDA collection. Two sets of germplasm were compared for leaf physiology and morphology, one originating from water-limiting environments, the other from wetter locations. The dry set of 201 entries was constructed as follows:

**1.** Accessions from collection sites where the annual rainfall was below 300 mm or greater than 550 mm were not considered.
**2.** Of the remaining accessions, one accession per collection site was chosen at random.
**3.** A hierarchical cluster analysis was performed on the remaining collection sites using the following parameters: long-term averages for yearly precipitation, yearly aridity, minimum yearly temperature and maximum yearly temperature, extracted from the ICARDA in-house database and the BIO4, BIO15, BIO16 and BIO19 variables extracted from the WorldClim databases (Hijmans *et al.*, 2005). The climate variables were chosen to combine temperature and precipitation factors that would influence the length of growing season and seasonal moisture availability. Twenty clusters were identified.
**4.** Accessions contained in six clusters were dropped because the average aridity index for the cluster was above 0.6 or below 0.1 (indicating irrigated sites).
**5.** For each of the remaining clusters, the accessions were sorted according to the BIO15 climate variable (a measure of the variation in seasonal moisture variation) for their respective collection sites (Hijmans *et al.*, 2005). Highly variable sites were ranked higher.
**6.** Any accession with a score of 50 or lower was discarded.
**7.** The remaining accessions within each cluster were ranked based on collection site long-term yearly precipitation.
**8.** A set of 201 accessions were chosen by selecting the lowest ranked accession in each cluster and the process was repeated until the desired set size was achieved.

When tested, even under well-watered conditions, the dry set exhibited traits that would allow plants to maintain favourable water balances under dry conditions. Further research also revealed that the dry set had more prolific root systems.

Just recently, a more sophisticated filtering process was used successfully to find tolerance to *barley yellow dwarf virus* (BYDV) (A. Amri, Rabat, 2014, personal communication) and indications of tolerances for leaf miner (*Liriomyza cicerina* R.) and pod borer (*Helicoverpa armigera* Hubn.) in chickpea (M. El Bouhssini, Rabat, 2014, personal communication), all of which were considered rare traits. In the case of the latter two traits, there are no reports of any tolerances having been found previously. The advance here was in the use of daily climatic data for temperature, precipitation and aridity, derived by extrapolating values from monthly data, and estimates of the onset of growing period date for each collection site as described above. For each collection site in the ICARDA database, a day-degree accumulation model for crop development was used to determine when in the growing season specific development phases were most likely to occur. This allowed filters to be applied within specific phases of the crop's development for any given set of locations.

The BYDV example is used here to demonstrate how a filtering strategy was imposed on collection sites once the timing of crop development was estimated. Here, candidate collection sites were chosen based on environmental conditions most likely to favour high populations of the vector aphid within the growing season, and also conditions conducive to population maintenance between cropping seasons, within stands of volunteer cereals and both annual and perennial pasture grasses and wild grasses. Thus, during the cropping season, temperature optima for aphid population development and virus infection are likely to be the most important filters, while rainfall during the off-season would also be important.

With the above in mind, an index was developed by combining variable rank values for in-season and off-season long-term values for collection site precipitation, average temperature, a count of the days where maximum temperature fell between 18 and 28°C and where minimum temperature fell between 8 and 22°C. Different weights were given to the parameter rankings depending on their perceived influence on aphid population densities and virus infection rates. For example, in-season temperature values were given a higher weighting than off-season temperatures, while off-season average rainfall was given a higher ranking than in-season averages. These weighted variable rankings were combined to create the final index used to select 200 barley accessions for evaluation. Three accessions were subsequently identified as resistant and 13 as moderately resistant. This is considered to be an excellent result given the rarity of resistance found so far in barley populations.

The above filtering strategy could have been improved significantly with comprehensive data to support decisions on appropriate weighting levels for the climatic parameters. In this case, the decisions were largely subjective, based on what limited literature was available on environmental conditions conducive to BYDV epidemics. The point here is that FIGS does not necessarily require precise data to obtain a desirable result. Clearly, though, higher-quality data will lead to superior results.

### 7.4.2 Modelling

Where evaluation data exist for a trait whose emergence in an environment is likely to be influenced by multiple climatic factors, a modelling approach can be used to identify 'best-bet' FIGS sets. In an EcoTILLING study to search for allelic variation for the powdery mildew *Pm3* gene in bread wheat, linear models were used to identify candidate environments likely to yield resistant accessions. The following approach was used:

**1.** Site geo-coordinates, elevation and the long-term monthly climatic averages (for temperature, precipitation and humidity) of the collection sites of 400 bread wheat landraces with known resistance to powdery mildew were combined with those of 16,000 unevaluated landrace accessions.
**2.** A hierarchical cluster analysis was performed on all the collection site data.
**3.** Unevaluated accessions were retained from sites that clustered the same as sites from where resistant accessions were collected.
**4.** A principal component analysis was performed on the agroclimatic data for the collection sites of the resistant accessions and the non-evaluated accessions retained in step 3.
**5.** Accessions were retained if their collection site x, y, z component scores were all sufficiently close to those of a site that yielded resistant accessions.

Accessions (1320) chosen in this way were infected with four powdery mildew isolates that were avirulent or virulent to known *Pm3* alleles. From 45% of the sites identified in the FIGS analysis, 212 accessions were resistant to one or other of the isolates. The study went on to reveal seven new resistance alleles for the *Pm3* gene, doubling the known functional allelic diversity at this locus (Bhullar *et al.*, 2009, 2010).

Linear models such as those used above are limited because the data must conform to certain assumptions. It has been suggested that non-linear recursive partitioning algorithms may be more robust and easier to use since they have fewer assumptions and thus require less data preparation when compared to linear approaches (Drake *et al.*, 2006; Tirelli *et al.*, 2009). Bari *et al.* (2012, 2014) have experimented with the use of machine learning approaches using Random Forest (RF), Neural Networks (NN) and Support Vector Machine (SVM) models to identify relationships between long-term monthly average climatic data and resistance to yellow and stem rust in bread wheat. Relationships determined in this way can then be used to select FIGS sets for resistance to these diseases in the search for allelic variation.

The modelling framework used by Bari *et al.* (2012, 2014) defines the potential trait–environment relationship within a non-linear framework where recursive learning techniques are used to overcome the problem of restrictive parametric paradigms on the one hand and the prerequisite distribution assumptions on the other (Drake *et al.*, 2006). The studies proceeded in two steps: first, a learning phase trains a predictive model using collection site environmental data and disease scores for a subset of accessions that had been evaluated previously. This process quantifies the relationship, if any exists, between the disease score and a suite of environmental variables in an interactive process where the model parameters are tuned until the rate of false classifications are minimized. In the second phase, the 'trained' model is supplied with only the collection site climatic data of a further set of accessions that had also been evaluated for, in this case, *stem rust* or *yellow rust* resistance. The model then uses the relationships determined in the training phase to classify the second set of accessions as potentially resistant or susceptible. The results of these predictions were then validated by the disease scores and the models evaluated using a variety of accuracy metrics.

The three models used were able to identify sites yielding rust resistant accessions with a high degree of accuracy. In both studies, high values for the model performance metrics were obtained, which quantify the agreement between the models' predictions and the actual disease scores. This demonstrates a strong environmental component in the geographic distribution of resistance genes for the two diseases. This study also concurs with the work done by Endresen *et al.* (2011, 2012), who used different non-linear models to demonstrate a relationship between disease resistance and the long-term climatic conditions at sites of origin for genebank accessions. These studies clearly validate the utility of a modelling approach where evaluation data exist for adaptive traits.

The studies described above used long-term average monthly data. It is expected that using daily data extracted from relevant intervals within the crop development cycle will improve the predictive power of the models. This was found to be the case when monthly, a calendar year of daily climatic data, and daily climatic data from within the growing season were presented to the NN, RP and SVM models to classify accessions for resistance to yellow rust. The accuracy metrics were significantly higher for the daily data and higher still for daily data from within the growing season (A. Bari, Rabat, 2014, personal communication).

## 7.5 Implications for Genebank Management

Since 2012, the ICARDA genebank has used FIGS subsetting to select germplasm when responding to seed requests. More than 58 FIGS subsets were constructed for various traits in bread wheat, durum wheat, barley, lentil, chickpea, faba bean and grass pea. In this context, FIGS has helped the ICARDA genebank to rationalize the distribution of the genetic resources by sending smaller sets of germplasm whilst maximizing the user's likelihood of receiving the trait variation required. In the long-term, this will reduce the resources required to maintain adequate supplies of seed for requested accessions.

## 7.6 Response by the User Community

It is curious to note that the response of genetic resource users to FIGS has been considerably more enthusiastic than that of the genetic resources community. Since FIGS has been presented at international conferences, the small FIGS team at ICARDA has been overwhelmed with requests for FIGS sets, and have been invited to collaborate on a range of ambitious trait-mining projects such as the hunt for reproductive frost tolerance in bread wheat. In this respect, FIGS has been responsible for a more vibrant relationship between the ICARDA genebank and its user community. The perceived value of the FIGS approach to the end-user community has been demonstrated best by the Grains Research and Development Corporation (GRDC) in Australia, who have funded FIGS development consistently from its inception. GRDC is a statutory authority, funded both by the Australian Government and by Australian grain growers, the latter being directly involved in formulating research needs and evaluating research project outcomes.

## 7.7 Conclusion

While FIGS to date has shown considerable potential as an effective means of mining genebanks for adaptive traits, it is still very much a technology at its early stages of development, with scope for significant improvement. To date, only a small team of workers has contributed to FIGS development and so input by other scientists is likely to enhance the approach. The areas that could be investigated or improved on include:

- The use of a wider range of environmental and ecological variables.
- Incorporation of long-term remote sensing data into the suite of environmental data.
- Refinement of growing season onset date models.
- Assembly of a greater range of evaluation data for georeferenced material that can be used in modelling for key traits.
- Development and use of biological diversity hotspot maps for key traits.
- Capture of collection site geo-coordinates for collections where there are adequate data to do so.
- Development of decision support software to walk researchers through the FIGS approach and to automate as much of the selection process as possible – such software could be linked to global genetic resource databases.

## Acknowledgements

We gratefully acknowledge the financial support of the Grains Research and Development Corporation, who have supported the development of FIGS from the beginning.

## References

Acquaah, G. (2007) *Principles of Plant Genetics and Breeding*. Blackwell Publishing, Malden, Massachusetts.

Bari, A., Street, K.A., Mackay, M.C., Endresen, D.T.F., De Pauw, E., *et al.* (2012) Focused identification of germplasm strategy (FIGS) detects wheat stem rust resistance linked to environmental variables. *Genetic Resources and Crop Evolution* doi:10.1007/s10722-011-9775-5 (published online 3 December 2011).

Bari, A., Amri, A., Street, K.A., Mackay, M.C., De Pauw, E., *et al.* (2014) Predicting resistance to stripe (yellow) rust (*Puccinia striiformis*) in wheat genetic resources using focused identification of germplasm strategy. *The Journal of Agricultural Science* 152(06), 906–916.

Bhullar, N.K., Street, K., Mackay, M., Yahiaoui, N. and Keller, B. (2009) Unlocking wheat genetic resources for the molecular identification of previously undescribed functional alleles at the Pm3 resistance locus. *Proceedings of the National Academy of Sciences of the United States of America* 106, 9519–9524.

Bhullar, N.K., Zhang, Z., Wicker, T. and Keller, B. (2010) Wheat gene bank accessions as a source of new alleles of the powdery mildew resistance gene Pm3: a large scale allele mining project. *BMC Plant Biology* doi:10.1186/1471-2229-10-88.

Brown, A.H.D. (1989a) Core collections – a practical approach to genetic-resources management. *Genome* 31, 818–824.

Brown, A.H.D. (1989b) The case for core collections. In: Brown, A.H.D., Frankel, O.H., Marshall, D.R. and Williams, J.T. (eds) *The Use of Plant Genetic Resources*. Cambridge University Press, Cambridge, UK.

Brown, A.H.D. and Spillane, C. (1999) Implementing core collections – principles, procedures, progress, problems and promise. In: Johnson, R.C. and Hodgkin, T. (eds) *Core Collections for Today and Tomorrow*. Crop Science Society of America, Madison, Wisconsin, pp. 1–10.

De Pauw, E. (2008) *Climatic and Soil Datasets for the ICARDA Wheat Genetic Resource Collections of the Eurasia Region: Explanatory Notes*. ICARDA GIS Unit, Aleppo, Syria.

Dorofeyev, V.F. (ed.) (1992) N.I. Vavilov, *Origin and Geography of Cultivated Plants*. Cambridge University Press, Cambridge, UK.

Drake, J.M., Randin, C. and Guisan, A. (2006) Modelling ecological niches with support vector machines. *Journal of Applied Ecology* 43, 424–432.

Dwivedi, S.L., Upadhyaya, H.D., Thomas Stalker, H., Blair, M.W., Bertioli, D.J., *et al.* (2008) Enhancing crop gene pools with beneficial traits using wild relatives. *Plant Breeding Reviews* 30, 180–230.

El-Bouhssini, M., Street, K., Joubi, A., Ibrahim, Z. and Rihawi, F. (2009) Sources of wheat resistance to Sunn pest, *Eurygaster integriceps* Puton, in Syria. *Genetic Resources and Crop Evolution* 56(8), 1065–1069.

El Bouhssini, M.E., Street, K., Amri, A., Mackay, M., Ogbonnaya, F.C., *et al.* (2011) Sources of resistance in bread wheat to Russian wheat aphid (*Diuraphis noxia*) in Syria identified using the Focused Identification of Germplasm Strategy (FIGS). *Plant Breeding* 130, 96–97.
El Bouhssini, M., Ogbonnaya, F.C., Chen, M.S., Lhaloui, S., Rihavvi, F., *et al.* (2012) Sources of resistance in primary synthetic hexaploid wheat (*Triticum aestivum* L.) to insect pests: Hessian fly, Russian wheat aphid and Sunn pest in the Fertile Crescent. *Genetic Resources and Crop Evolution* 60, 621–627, doi:10.1007/s10722-012-9861-3.
Endresen, D.T.F., Street, K., Mackay, M., Bari, A. and De Pauw, E. (2011) Predictive association between biotic stress traits and ecogeographic data for wheat and barley landraces. *Crop Science* 51, 2036–2055.
Endresen, D.T.F., Street, K., Mackay, M., Bari, A., Amri, A., *et al.* (2012) Sources of resistance to stem rust (Ug99) in bread wheat and durum wheat identified using Focused Identification of Germplasm Strategy (FIGS). *Crop Science* 52, 764–773.
FAO (Food and Agriculture Organization) (2003) Unlocking the Water Potential of Agriculture. FAO, Rome. Available at: www.fao.org/docrep/006/Y4525E/y4525e00.htm (accessed 5 August 2015).
FAO (2009) *High Level Expert Forum – How to Feed the World in 2050*. FAO, Rome.
FAO (2010) *The Second Report on the State of the World's Plant Genetic Resources for Food and Agriculture*. FAO, Rome.
FAO (2015) Harmonized World Soil Database. Food and Agriculture Organisation of the UN, Rome, Italy. Available at: www.fao.org/publications/card/en/c/2fa14e5e-ae97-516e-9dd2-24bc7abbc823/ (accessed 5 August 2015).
FAO-UNESCO (1971–1981) *Soil Map of the World 1:5000000*. UNESCO, Paris.
Fischer, R.A. (2009) Farming systems of Australia: exploiting the synergy between genetic improvement and agronomy. In: Victor, O., Calderini, D. and Calderini, S. (eds) *Crop Physiology – Applications for Genetic Improvement and Agronomy*. Academic Press, Sydney, Australia, pp. 23–54.
Frankel, O.H. (1984) Genetic perspective of germplasm conservation. In: Arber, W., Limensee, K., Peacock, W.J. and Stralinger, P. (eds) *Genetic Manipulations: Impact on Man and Society*. Cambridge University Press, Cambridge, UK, pp. 161–170.
Gepts, P. (2006) Plant genetic resources conservation and utilization: the accomplishments and future of a societal insurance policy. *Crop Science* 46, 2278–2292.
Glaszmann, J.C., Kilian, B., Upadhyaya, H.D. and Varshney, R.K. (2010) Accessing genetic diversity for crop improvement: current opinion. *Plant Biology* 13, 1–7.
Gollin, D., Smale, M. and Skovmand, B. (2000) Searching an *ex situ* collection of wheat genetic resources. *American Journal of Agricultural Economics* 82(4), 812–27.
Guralnick, R.P., Wieczorek, J., Beaman, R. and Hijmans, R.J. (2006) The BioGeomancer Working Group: automated georeferencing to map the world's biodiversity data. *Public Library of Sciences Biology* 4(11), e381, doi:10.1371/journal.pbio.0040381.
Hijmans, R.J., Cameron, S.E., Parra, J.L., Jones, P.G. and Jarvis, A. (2005) Very high resolution interpolated climate surfaces for global land areas. *International Journal of Climatology* 25(15), 1965–1978.
Hodgkin, T., Brown, A.H.D., van Hintum, T.J.L. and Morales, E.A.V. (eds) (1995) *Core Collections of Plant Genetic Resources*. John Wiley & Sons, Chichester, UK.
Hoisington, D., Khairallah, M., Reeves, T., Ribaut, J.-M., Skovmand, B., *et al.* (1999) Plant genetic resources: What can they contribute toward increased crop productivity? *Proceedings of the National Academy of Sciences of the United States of America* 96, 5937–5943.
Holbrook, C.C., Anderson, W.F. and Pittman, R.N. (1993) Selection of a core collection from the United States germplasm collection of peanut. *Crop Science* 33, 859–861.
Hu, J., Zhu, J. and Xu, H.M. (2000) Methods of constructing core collections by stepwise clustering with three sampling strategies based on the genotypic values of crops. *Theoretical and Applied Genetics* 101, 264–268.
Jones, P.G. and Gladkov, A. (1999) *FloraMap Version 1. A Computer Tool for Predicting the Distribution of Plants and Other Organisms in the Wild*. CD-ROM and Manual. Centro Internacional de Agricultura Tropical, Cali, Colombia.
Khazaei, H., Street, K., Bari, A., Mackay, M. and Stoddard, F.L. (2013) The FIGS (Focused Identification of Germplasm Strategy) approach identifies traits related to drought adaptation in *Vicia faba* genetic resources. *Public Library of Sciences* ONE 8(5), e63107, doi:10.1371/journal.pone.0063107.
Khush, G.S. (2001) Green revolution: the way forward. *Nature Reviews, Genetics* 2, 815–822.
Kilian, B. and Graner, A. (2012) NGS technologies for analyzing germplasm diversity in genebanks. *Briefings in Functional Genomics* 11, 38–50.

Koo, B. and Wright, B.D. (2000) The optimal timing of evaluation of genebank accessions and the effects of biotechnology. *American Journal of Agricultural Economics* 82(4), 797–811.
Kurlovich, B.S., Rep'ev, S.I., Petrova, M.-V., Buravtseva, T.V., Kartuzova, L.T., *et al.* (2000) The significance of Vavilov's scientific expeditions and ideas for development and use of legume genetic resources. *Plant Genetic Resources Newsletter* 124, 23–32.
Lantican, M.A., Pingali, P.L. and Rajaram, S. (2003) Is research on marginal lands catching up? The case for unfavourable wheat growing environments. *Agricultural Economics* 29, 353–361.
Lin, W. and Bradshaw, A.D. (1975) The potential for evolution of heavy metal tolerance in plants: the rapid evolution of copper tolerance in *Agrostis stolonifera. Heredity* 34(2), 165–187.
Linden, E. (1991) Will we run low on food? *Time Magazine*, 19 August.
Mackay, M.C. (1990) Strategic planning for effective evaluation of plant germplasm. In: Srivastava, J.P. and Damania, A.B. (eds) *Wheat Genetic Resources: Meeting Diverse Needs*. John Wiley & Sons, Chichester, UK, pp. 21–25.
Mackay, M.C. (1995) One core collection or many? In: Hodgkin, T., Brown, A.H.D., van Hintum, T.J.L. and Morales, E.A.V. (eds) *Core Collections of Plant Genetic Resources*. John Wiley & Sons, New York, pp. 199–210.
Malosetti, M. and Abadie, T. (2001) Sampling strategy to develop a core collection of Uruguayan maize landraces based on morphological traits. *Genetic Resources and Crop Evolution* 48, 381–390.
Munns, R. and James, R.A. (2003) Screening methods for salinity tolerance: a case study with tetraploid wheat. *Plant and Soil* 253, 201–218.
Ortiz, R., Ruiz-Tapia, E.N. and Mujica-Sanchez, A. (1998) Sampling strategy for a core collection of Peruvian quinoa germplasm. *Theoretical and Applied Genetics* 96, 475–483.
Pessoa-Filho, M., Rangel, P.H.N. and Ferreira, M.E. (2010) Extracting samples of high diversity from thematic collections of large gene banks using a genetic-distance based approach. *BMC Plant Biology* 10, 127.
Polignano, G.B., Uggenti, P. and Scippa, G. (2001) Diversity analysis and core collection formation in Bari faba bean germplasm. *FAO/Bioversity Plant Genetic Resources Newsletter* 125, 33–38.
Spieth, P.T. (1979) Environmental heterogeneity: a problem of contradictory selection pressures, gene flow, and local polymorphism. *American Naturalist* 113(2), 247–260.
Street, K., Mackay, M., Zuev, E., Kaul, N., El Bouhssini, M., *et al.* (2008) Diving into the genepool: a rational system to access specific traits from large germplasm collections. In: Appels, R., Eastwood, R., Lagudah, R., Langridge, E. and Mackay, M. (eds) *Proceedings of the 11th International Wheat Genetics Symposium, Brisbane, Australia*. Sydney University Press, Australia, pp. 28–31.
Tirelli, T., Pozzi, L. and Pessani, D. (2009) Use of different approaches to model presence/absence of *Salmo marmoratus* in Piedmont (Northwestern Italy). *Ecological Informatics* 4, 234–242.
United Nations (2015) *2015 Revision of the World Population Prospects*. Population Division, United Nations Department of Economic and Social Affairs, United Nations Secretariat, New York. Available at: esa.un.org/unpd/wpp/index.htm (accessed 8 December 2015).
Upadhyaya, H.D., Pundir, R.P.S., Dwivedi, S.L. and Gorda, C.L.L. (2009) Mini core collections for efficient utilization of plant genetic resources. In: *Crop Improvement Programs. Information Bulletin 78*. International Crops Research Institute for Semi-Arid Tropics, Patancheru, Andhra Pradesh, India, p. 52.
Vavilov, N.I. (1920) Zakon gomologicheskikh ryadov v nasledstvennoj izmenchivosti. (The law of homologous series in variation). *Proceedings of the III All-Russian plant breeding conference*. Saratov, p. 16 (in Russian). (Cited in Loskotov, 1999, 82.)
Vavilov, N.I. (1922) The law of homologous series in variation. *Journal of Genetics* 12(1), 47–89.
Vavilov, N.I. (1932) Problema Novykh Kul'tur (Problems concerning new crops). Sel'chozgiz, Moscow-Leningrad, (USSR). In: Dorofeyev, V.F. (ed.) (1992) *Origin and Geography of Cultivated Plants*. Cambridge University Press, Cambridge, UK, pp. 256–285.
Vavilov, N.I. (1935) The phyto-geographical basis for plant breeding. Studies of the original material used for plant breeding. (First published in 1935 in Teoreticheskie osnovy selektsii (Theoretical basis of plant breeding). Moscow-Leningrad, USSR). In: Dorofeyev, V.F. (ed.) (1992) *Origin and Geography of Cultivated Plants*. Cambridge University Press, Cambridge, UK, pp. 316–366.
Vavilov, N.I. (1957) *Mirovye resurcy sortov chlebnych zlakov, zernovych bobovych, l'na i ich ispol'zovanie v selekcii. Opyt agroklimatičecskogo obozrenija važnejšich polevych kultur. (World Resources of Cereals, Grain Leguminous Crops and Flax and their Utilization in Plant Breeding. Agroecological Survey of the Principal Field Crops)*. Izdatel'stvo Akademii Nauk SSR, Moskva, Leningrad, pp. 463 (in Russian).
Xu, Y. (2010) Plant genetic resources: management, evaluation and enhancement. In: Xu, Y. (ed.) *Molecular Plant Breeding*. CAB International, Wallingford, UK, pp. 151–194.

# 8 Predictive Characterization Methods for Accessing and Using CWR Diversity

I. Thormann,[1*] M. Parra-Quijano,[2] M.L. Rubio Teso,[3] D.T.F. Endresen,[4] S. Dias,[1] J.M. Iriondo[3] and N. Maxted[5]

[1]*Bioversity International, Rome, Italy;* [2]*International Treaty on Plant Genetic Resources for Food and Agriculture, FAO, Rome, Italy;* [3]*Universidad Rey Juan Carlos, Madrid, Spain;* [4]*UiO Natural History Museum, University of Oslo, Oslo, Norway;* [5]*School of Biosciences, University of Birmingham, Birmingham, UK*

## 8.1 Introduction

The growing interest in crop wild relatives (CWR) – wild progenitors or wild plant species closely related to crops – as a source of adaptive genetic diversity in crop improvement and research has led to: (i) gap analyses for targeted *ex situ* and *in situ* conservation (Maxted *et al.*, 2004; Ramirez-Villegas *et al.*, 2010; Parra-Quijano *et al.*, 2012a; Vincent *et al.*, 2012; Shehadeh *et al.*, 2013); (ii) a growing number of national CWR inventories (FAO, 2010; Phillips *et al.*, 2014; Dulloo *et al.*, 2015); (iii) a global inventory of priority CWR of 173 major food crops (Vincent *et al.*, 2013); and (iv) the development of a global network for *in situ* conservation of CWR, currently being discussed by the Commission on Genetic Resources for Food and Agriculture (FAO, 2013). To enhance access to and use of CWR diversity, CWR accessions in genebanks and CWR populations growing in and outside protected areas need to be characterized and evaluated to allow researchers, breeders and farmers to focus their resources on the most promising accessions and populations. Characterization and evaluation (C&E) data held by genebanks is often incomplete or lacking, in particular for CWR. Indeed, the lack of C&E data continues to be reported as one major limitation to the use of *ex situ* conserved plant genetic resources (PGR) (FAO, 2010), and traditional characterization and evaluation methods cannot catch up with the growing number of accessions. Even fewer data are available for CWR populations conserved *in situ*. *In situ* C&E is not a routine activity of protected area managers, as their implementation is complex and resource intensive (Guarino *et al.*, 2002). Biodiversity conservationists and managers of protected areas generally see conservation as the end goal and will not focus on collecting the C&E data that are needed for systematic utilization of this material in breeding programmes. However, if CWR conservation is itself to be sustainable, then the conservation–use link needs to be explicit and the greater the demonstrated value of the conserved CWR populations, the more likely that resources will be allocated to their conservation. Novel approaches to enhance the characterization of plant genetic resources are being developed, such as those that use genomics, transcriptomics, metabolomics and high-throughput phenotyping, as well as less resource-intensive predictive characterization

*Corresponding author; e-mail: i.thormann@cgiar.org

techniques, which build on geographic location and agroecological data to optimize the search for populations and accessions with adaptive traits and characteristics.

Collectively, predictive characterization comprises a set of approaches that use geographic and environmental data along with geographic information systems (GIS) analysis to search for particular traits in a normally large set of populations under consideration. They represent a more cost-effective approach than phenotypic C&E of the complete germplasm collection. They build on the hypothesis that the different environments exert divergent selective pressures on plant populations, increasing the probability of finding specific traits under certain circumstances.

Predictive characterization methods are predictive in the sense that they assign a probability of trait presence to uncharacterized germplasm (either *ex situ* or *in situ*) using (i) biotic and abiotic characteristics and/or ecogeographical information associated with the location where the germplasm has been collected or occurs, and (ii) previously recorded C&E data of trait occurrence associated with a set of locations different from those where the germplasm has been collected or occurs (compare schema 1 in Thormann *et al.*, 2014). In each case, a predictor is used to build the hypothesis that germplasm from a particular location will be differentiated genetically depending on the environmental conditions that are normally experienced in that site.

One of the first systematic applications of finding a predictive link between a resistance/tolerance trait and a set of environmental parameters, named the Focused Identification of Germplasm Strategy (FIGS) (Mackay and Street, 2004; Street *et al.*, 2008), used the biotic and abiotic matching techniques. FIGS was developed further at the International Centre for Agricultural Research in the Dry Areas (ICARDA), based on early work by Michael Mackay in the 1980s and 1990s (Mackay 1986, 1990, 1995). The first FIGS studies used a series of filters based on scientific expert knowledge for matching environmental profiles that were known to be suitable for adaptations leading to the target trait properties in landraces growing in such locations. The method can be illustrated using the work by El Bouhssini *et al.* (2009) on Sunn pest resistance in wheat. Starting with 16,000 wheat landraces from different genebank collections, germplasm material collected in China, Pakistan and India was excluded based on expert knowledge, arguing that the Sunn pest had only recently been reported there and thus providing too little time for adaptive resistance in the germplasm to evolve. This reduced the candidate set down to 6328 landraces. The next filter excluded environments that were too dry for the Sunn pest insects to thrive (less than 280 mm precipitation/year) and environments with too low winter temperatures (below −10°C), reducing the candidate set further down to 1502 landraces. Finally, 534 landraces were tested and 9 new and formerly uncharacterized resistant landraces were identified by ICARDA in field experiments during 2007 and 2008. A previous extensive series of field experiments conducted at ICARDA from 2000 to 2006 including more than 2000 wheat landraces had not been able to identify any wheat landraces with significant resistance.

Previous FIGS studies were mainly applied to major crops, in particular to wheat and barley (Kaur *et al.*, 2008; Bhullar *et al.*, 2009; El Bouhssini *et al.*, 2009, 2011; Endresen, 2010; Endresen *et al.*, 2011, 2012; Bari *et al.*, 2012, 2014) and recently also to faba bean (Khazaei *et al.*, 2013). Building on the foundation of the FIGS approach, further studies that used ecogeographical information or previously recorded C&E data were developed and tested for their applicability to CWR within the context of the PGR Secure project 'Novel characterization of crop wild relative and landrace resources as a basis for improved crop breeding' (www.pgrsecure.org) (Thormann, 2012; Thormann *et al.*, 2015). These studies for CWR material have explored the so-called ecogeographical filtering and calibration methods.

The results of the application of the ecogeographical filtering and calibration methods to CWR of the genera *Avena*, *Beta*, *Brassica* and *Medicago*, which were the target genera of the PGR Secure project, are reported and discussed here.

## 8.2 Ecogeographical Filtering Method

The ecogeographical filtering method filters occurrence records, combining the spatial distribution of the target taxon on an ecogeographical land characterization map (ELC) (Parra-Quijano *et al.*, 2012b,c) with the ecogeographical characterization of those environments that are likely to impose selection pressure for the adaptive trait investigated. In the predictive characterization context, a taxon-specific ELC map is used, aiming at representing the adaptive scenarios that are present over the territory studied.

### 8.2.1 Implementation of the ecogeographical filtering method

The process followed for the implementation of the ecogeographical filtering method, using a procedure that can be achieved with low and cost-effective resources, is described in the following steps (Thormann *et al.*, 2014):

**1.** The occurrence data of the taxa of interest were compiled. The European occurrences of *Avena, Beta, Brassica* and *Medicago* were compiled from the GBIF (Global Biodiversity Information Facility, www.gbif.org), the crop-specific databases of the European Cooperative Programme for Plant Genetic Resources (ECPGR), GRIN (Germplasm Resources Information Network of the USA, www.ars-grin.gov) and SINGER (System-wide Information Network for Genetic Resources; data now available from Genesys – Gateway to Genetic Resources, www.genesys-pgr.org/welcome). Occurrences where latitude and longitude were missing but location description provided were then georeferenced. The quality of all georeferences was checked and ranked according to the georeferencing quality evaluation procedure developed by Parra-Quijano *et al.* (2014) and implemented in the CAPFITOGEN tool GEOQUAL (Parra-Quijano *et al.*, 2014). This allowed excluding records with low precision of site location. As predictive characterization methods build on the link between plant adaptation and environmental conditions of the collecting/occurrence site, the quality of coordinates becomes a key factor since the higher the quality of coordinates, the higher the precision of the extracted ecogeographical information.

**2.** Taxon-specific ELC maps were developed, according to the procedures described and implemented in the CAPFITOGEN tool, ELCmapas (Parra-Quijano *et al.*, 2014). They were developed based on the variables most relevant for adaptation and distribution of the taxon, which were identified through expert consultations and literature reviews.

**3.** The environmental profile of the habitat that was likely to promote the selection for the adaptive trait that was studied was described and the most appropriate variables (geophysical, edaphic and/or bioclimatic) that describe the environment profile and their respective threshold values were defined based on literature review and consultation with experts.

**4.** The R-script for the ecogeographical filtering method developed in the PGR Secure project (Thormann *et al.*, 2014) was run with the data and ELC maps obtained in steps 1–3 to determine the subset of occurrences most likely to harbour resistance/tolerance to the trait of interest. Through the R-script, the ecogeographical category from the ELC map was assigned to each occurrence according to its coordinates and the records were grouped according to their ELC map category. The script then selected occurrences from each group that complied with the specific ecogeographical variables determined in step 3 and generated an ecogeographical core set. To obtain the final subset, the ecogeographical variables were used to rank the ecogeographical core set and to select, based on the previously identified threshold value, the records with the highest or lowest values of the variable.

### 8.2.2 Application to the four project genera

The occurrence data set compiled in the PGR Secure project contained 31,745 records of passport data of European landrace and CWR occurrences and their geo-locations for all four genera. This included 3905 records of *Avena*, 1461 of *Beta*, 1384 of *Brassica* and 2153 of *Medicago* CWR. Literature reviews and consultations with breeders and experts about important traits of interest in the four project genera and variables describing the respective environmental profiles resulted in the traits, variables and thresholds summarized in Table 8.1.

#### *Results from the* Avena *case study*

Aluminium (Al) toxicity exists in many soils throughout the world and is a limiting factor for

Table 8.1. Traits of interest to breeders and the variables and thresholds used to describe environments in which these traits might have greater probability of development.

| Genus | Identified trait | Identified variable(s) | Threshold value |
|---|---|---|---|
| *Avena* | Aluminum toxicity | Soil pH; | pH <5.5 |
| | | Soil organic carbon content (T_OC) | T_OC <1.2% |
| *Beta* | Drought | De Martonne aridity index (De Martonne, 1926), calculated based on temperature and precipitation of the three driest months (July, August, September) | IaR-DM <10 |
| *Brassica* | Drought | De Martonne aridity index | IaR-DM <10 |
| | Salinity | Topsoil salinity (TSS) measured as electrical conductivity in dS/m (deciSiemens/metre) | TSS >4 dS/m |
| | | Mean temperature values for the driest months | Highest values in records with TSS >4 |
| *Medicago* | Frost | BIOCLIM 11 | <–2°C |

full expression of the yield potential of crops. Acid subsoils reduce root penetration into this region, thereby diminishing the water supply, in particular in soils with low water retention and soils cultivated without irrigation. The development of Al toxicity tolerant genotypes in combination with liming is probably the most feasible alternative to overcome Al toxicity for cultivation on soils with acid subsoils (Fageria *et al.*, 1988; de Oliveira *et al.*, 2005). It is impossible to stipulate a general threshold value above which Al phytotoxicity appears. There is considerable variation in Al tolerance between species and between varieties of the same species (Andersson, 1988). It has been shown that the Al toxicity tolerance trait in *Avena* is inheritable (Sanchez-Chacon *et al.*, 2000; Wagner *et al.*, 2001; de Oliveira *et al.*, 2005; Nava *et al.*, 2006).

Availability and phytotoxicity of Al depend on a number of abiotic and biotic factors. Biotic factors that influence Al toxicity are dependent on the plant species itself and the individual genetic properties of the plant. The main abiotic soil-dependent characteristic influencing Al toxicity is the pH value (Andersson, 1988; Rout *et al.*, 2001). Toxicity is more severe below pH 5.0 but can occur even at values above pH 5.5 (de Oliveira *et al.*, 2005). The solubility of Al increases exponentially as the pH decreases below 4.5 (Andersson, 1988). Additional major factors determining Al toxicity are the organic matter content and the dominant clay mineral as they influence the pH. High organic matter content reduces Al availability to plants as organic matter and organic acids are capable of complexing Al (Haynes and Mokolobate, 2001). Clay content, depending on the dominant clay mineral, can increase toxicity, as soils rich in clay have a large Al fraction, which can dissolve on acidification. For example, clay content together with base saturation and pH were the soil factors that offered the best estimate for the uptake of Al by *Anemone memorosa* L. in acid forest soils (Tyler *et al.*, 1987).

To identify and describe environment profiles that may generate Al toxicity tolerance traits, the pH value was considered as the primary criterion and organic matter content as the second most important factor. The organic matter content is provided as topsoil organic carbon (T_OC; percentage of organic carbon contained in the soil) in soil databases. These two variables together represent the best simple indicator of the health status of soil, according to the Harmonized World Soil Database (HWSD; www.iiasa.ac.at/Research/LUC/External-World-soil-database). With reference to the HWSD classes of soils and organic carbon content, environments with a soil pH <5.5 and soil organic matter content below 1.2% were selected as potentially interesting environments for the development of Al toxicity tolerance. The ELC map developed for the genus *Avena* was based on the variables listed in Table 8.2.

The final set of occurrences selected from the PGR Secure data set according to the variables and thresholds described above, and with a georeference quality rank ≥50, included 104 *Avena* CWR records. Geographical distribution is shown in Fig. 8.1.

Table 8.2. Environmental variables used to generate the ELC map for the *Avena* species.

| Bioclimatic variables | Edaphic variables | Geophysical variables |
|---|---|---|
| BIOCLIM 1 | Drainage class | Elevation |
| BIOCLIM 4 | Total exchangeable bases in topsoil (T_TEB) | Latitude (POINT_Y) |
| BIOCLIM 6 | Topsoil organic carbon (T_OC) | Longitude (POINT_X) |
| BIOCLIM 11 | Topsoil sand fraction (T_SAND) | Eastness |
| BIOCLIM 19 | Topsoil salinity (T_ECE) | Northness |

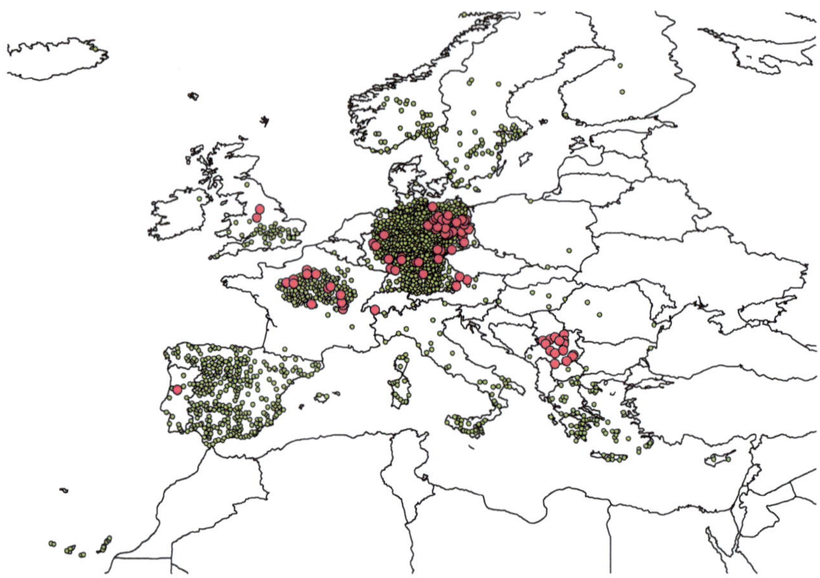

Fig. 8.1. Location of wild *Avena* occurrences included in the PGR Secure data set. Occurrences selected through ecogeographical filtering are highlighted.

### *Results from the* Beta *case study*

Precipitation as a single variable does not explain plant water availability. For instance, low annual precipitation in a particular location does not necessarily mean that plants may be subject to selective pressures for drought tolerance if precipitation is distributed appropriately throughout the plant life cycle, the ability of the soil to retain the water is good and the distribution of temperatures throughout the period does not increase the need for water. Aridity indexes include precipitation as well as temperature as variables and can be used to evaluate plant water availability, and consequently selective pressures in favour of drought tolerance. A great number of aridity indices have been formulated throughout time. Some of them are quite simple, involving essentially the temperature and precipitation variables for a given month (e.g. Köppen index, Gaussen index), whereas others are more complex and involve variables such as number of months, mean maximum temperatures (e.g. Emberger index, Walter index or De Martonne index) (Wallén, 1967) or solar radiation, wind speed and air humidity, along with temperature and precipitation (Penman-Monteith evapotranspiration (PM-ETo); Monteith, 1965). The lack of availability of these variables throughout the territory under study and the search for simplicity favoured agreement on the De Martonne index (De Martonne, 1926), which had been shown as being suitable for the measurement of aridity applied to biological studies (Wallén, 1967; Paltineanu *et al.*, 2007a,b). The

De Martonne index (Iar-DM) provides aridity values for a given year according to:

$$\text{Iar-DM} = P/(tm + 10)$$

where $P$ is the annual mean precipitation (mm) and $tm$ is the annual mean temperature (degrees Celsius); 10 is a constant to avoid negative values (Almorox, 2003). Aridity classes based on the De Martonne index are provided in Table 8.3.

To search for drought tolerance in *Beta* CWR accessions and populations, the De Martonne aridity index was calculated using the temperature and precipitation of the three driest months (July, August, September), when precipitation was the lowest and temperature the highest, so plant water availability was at its lowest. Environment profiles for habitats of CWR likely to contain tolerance to drought would correspond to those with De Martonne aridity values below 10, including both the arid and extremely arid categories.

The genus-specific *Beta* ELC map used in PGR Secure had been developed previously in the AEGRO project (aegro.jki.bund.de/aegro/) based on the variables listed in Table 8.4.

The final set of occurrences selected from the PGR Secure data set according to the variables and thresholds described above, and with a georeference quality rank ≥50, included 31 *Beta* CWR records, most of them originating from Greece, as shown in Fig. 8.2.

Table 8.3. Classification of zones according to the De Martonne index. (From Almorox, 2003.)

| Iar-DM value | Zone classification |
|---|---|
| 0–5 | Extremely arid (desert) |
| 5–10 | Arid (steppic) |
| 10–20 | Semi-arid (Mediterranean) |
| 20–30 | Subhumid |
| 30–60 | Humid |
| >60 | Perhumid |

### Results from the *Brassica* case study

SALT TOLERANCE. Salinity is one of the factors that limit plant growth and affect yields in crops worldwide (Zhu, 2001; Parida and Das, 2005; Yamaguchi and Blumwald, 2005). High salt concentration generates an osmotic gradient that makes it difficult for the plants to take water from the soil (Jordán López, 2005), and therefore can cause similar effects to those produced by drought. Salinity is a common problem for crops, related not only to high salt concentration in soils but also to factors such as type of water used for irrigation, temperature, drainage, etc. On the other hand, salinity tolerance is a trait controlled by many genes (Shannon, 1997), and it is measured with different agronomic parameters (i.e. yield, survival, plant height, leaf area, leaf injury, relative growth rate and relative growth reduction) or biochemical indicators (antioxidants and organic solutes such as glycine betaine and proline) (Ashraf and Harris, 2004). These characteristics make it difficult to select the appropriate landrace accessions or CWR populations for breeding.

For the selection of variables to filter the environmental profiles, measuring for topsoil salinity (TSS) is considered as it provides electrical conductivity (EC) values (deciSiemens/metre) and is used as the most common indicator of salt concentration (Seelig, 2000). The TSS variable already takes into account the soil water content, its availability and possible soil washing due to precipitation, so utilization of the 'excess salt' variable was considered redundant in relation to

Table 8.4. Environmental variables used to generate the ELC map for *Beta* species in the AEGRO project.

| Bioclimatic variables | Edaphic variables | Geophysical variables |
|---|---|---|
| BIOCLIM 3, 6 and 12 | Mineralogical profile of soil (MINERALOGY) | Northness |
| PRECIP 2, 6, 7 and 8 | World reference base for soil resources (WRB) coder for soil typological unit (STU) (WRBCODESTU) | Elevation |
| TMED 1, 3 and 11 | Depth to rocks (DEPTHTOROC) | Global irradiation on an optimal inclination (SOLRADOP) |
| TMIN 1 and 12 | Dominant parent material (obstacle to roots) (DOMPARMAT) | |

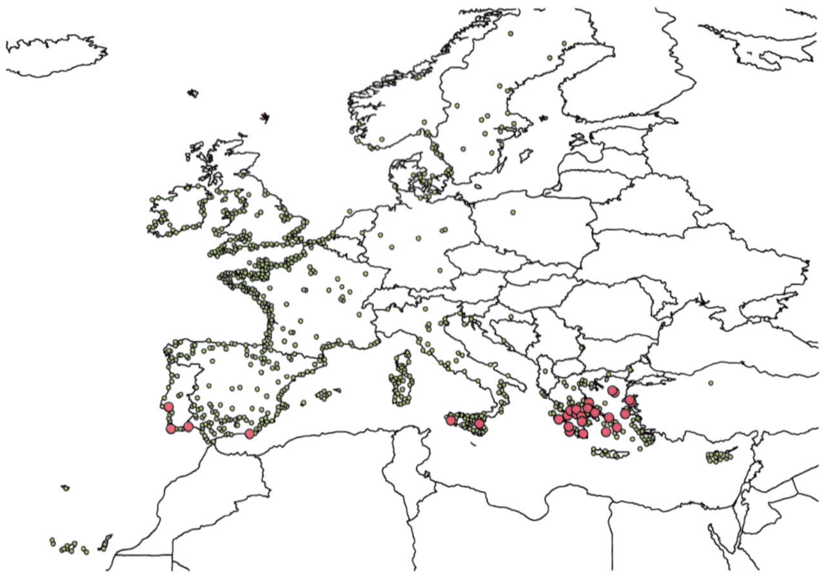

Fig. 8.2. Location of wild *Beta* occurrences included in the PGR Secure data set. Occurrences selected through ecogeographical filtering are highlighted.

the TSS. On the other hand, temperature was contemplated as a variable that could modulate the intensity of selection for salinity tolerance (for a particular TSS value, the higher the temperature, the higher the stress), so temperatures of the driest months were included as limiting factors in the analysis.

Soils can be classified into four classes, from 'non-saline' to 'strongly saline' according to its EC, as is shown in Table 8.5.

The environmental profiles most likely to generate salt tolerance would correspond to those in which EC values are greater than 4 dS/m. Once the accessions and populations were ranked according to TSS, mean temperature values for the driest months were included in the study. The accessions and populations holding the mean temperature values belonging to the highest quartile would be selected. No record within the *Brassica* data set met these conditions.

Table 8.5. Classification of soil salinity according to electrical conductivity (EC). (From Jordán López, 2005.)

| EC (dS/m) | Type of soil | Crop behaviour |
|---|---|---|
| <2 | Non-saline | Crop plants are not affected |
| 2–4 | Slightly saline | Some crop species may show growth problems |
| 4–8 | Saline | Most crop species are affected in their development |
| 8–16 | Strongly saline | Only tolerant species can be cultivated |

DROUGHT TOLERANCE. The same variables and thresholds as those described in the previous section for drought tolerance in *Beta* CWR were used to filter for *Brassica* populations. The De Martonne index was calculated according to the same parameters and variables (temperature and precipitation in the driest months). The ELC map developed for the genus *Brassica* was based on the variables listed in Table 8.6.

The final set of *Brassica* occurrences selected from the PGR Secure data set for drought tolerance, and with a georeference quality rank ≥50, included 128 CWR records. Geographical distribution is shown in Fig. 8.3.

### *Results from the* Medicago *case study*

Climate changes are expected to give more unpredictable and unstable winter climates, with

Table 8.6. Environmental variables used to generate the ELC map for *Brassica* species.

| Bioclimatic variables | Edaphic variables | Geophysical variables |
|---|---|---|
| BIOCLIM 1 | Total exchangeable bases in topsoil (T_TEB) | Global irradiation on an optimal inclination (SOLRADOP) |
| BIOCLIM 2 | Topsoil pH (T_PH_H2O) | Latitude (POINT_Y) |
| BIOCLIM 6 | Topsoil texture (T_TEXTURE) | Longitude (POINT_X) |
| BIOCLIM 11 | Topsoil organic carbon (T_OC) | Eastness |
| BIOCLIM 12 | Topsoil salinity (T_ECE) | Northness |
| PRECIP 5 | | Slope degree (SLOPEDG) |

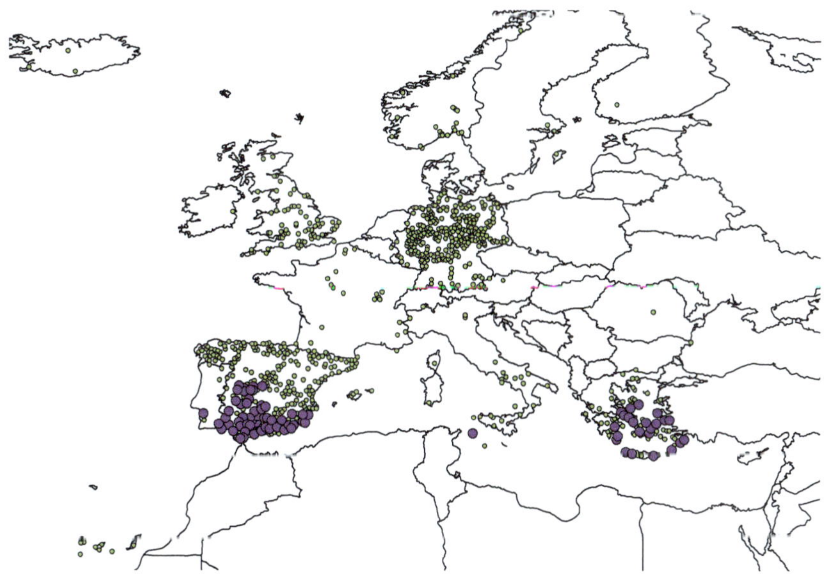

Fig. 8.3. Location of wild *Brassica* occurrences included in the PGR Secure data set. Occurrences selected through ecogeographical filtering are highlighted.

more cycles of freezing and thawing and shorter-lasting snow cover. These climates pose new combinations of winter stresses and increased stresses for perennial plants in many regions (Sandve *et al.*, 2011).

Of nearly one-third of a million plant species that exist on the planet, only 10–15% can tolerate any significant frost, and only a very small fraction can survive extreme cold (Dansereau, 1957). Boyer (1982) highlights that low temperature is one of the most important factors that limits the growth, development and distribution of plants in the world. In natural conditions, a phenomenon known as cold acclimatization allows plants to increase their tolerance to low, non-freezing temperatures (Levitt, 1980; Sakay and Larcher, 1987; Thomashow, 1999).

It is impossible to design a single approach that will satisfy the requirements of all low-temperature tolerance breeding programmes. For example, the criteria used to select for low-temperature tolerance are different for actively growing plants that must survive short periods of late spring frost compared to plants that must withstand exposure to long periods of below-freezing winter temperatures.

Overall, the capacity for acclimatization or prevention of low-temperature stress evidently varies between species and crop growth stages.

Table 8.7. Classification of freeze temperatures. (From Koss *et al.*, 1988.)

| Freeze/frost classification | Temperature thresholds | Effect |
|---|---|---|
| Light freeze | −2°C to 0°C | Tender plants are killed, with little destructive effect on other vegetation. |
| Moderate freeze | −4°C to −2°C | Widely destructive effect on most vegetation, with heavy damage to fruit blossoms and tender and semi-hardy plants. |
| Severe freeze | −4°C and less | Heavy damage to most plants. At these temperatures, the ground freezes solid, with the depth of the frozen ground dependent on the duration and severity of the freeze, soil moisture and soil type. |

Table 8.8. Environmental variables used to generate the ELC map for *Medicago* species.

| Bioclimatic variables | Edaphic variables | Geophysical variables |
|---|---|---|
| BIOCLIM 5 | Drainage class | Latitude (POINT_Y) |
| BIOCLIM 6 | Subsoil pH (S_PH_H2O) | Longitude (POINT_X) |
| BIOCLIM 10 | Topsoil pH (T_PH_H2O) | Northness |
| BIOCLIM 12 | Topsoil salinity (T_ECE) | Slope degree (SLOPEDG) |
| TMED 11 | Topsoil sand fraction (T_SAND) | – |
| TMAX 5 | – | – |

From the available studies and existing models, it was observed that an effective value to take into consideration would be the mean temperature for the coldest quarter, provided as bioclimatic variable BIO 11 in the WorldClim data set (Hijmans *et al.*, 2005). We considered −2°C as the threshold value, in order to focus on areas with moderate to severe freeze (see Table 8.7), providing for an increased likelihood of harbouring plants with tolerant traits.

The ELC map developed for the genus *Medicago* was based on the variables listed in Table 8.8.

The final set of occurrences selected from the PGR Secure data set based on the above criteria, and with a georeference quality rank ≥45, included 45 *Medicago* CWR accessions. Geographical distribution of these accessions is shown in Fig. 8.4.

## 8.3 Calibration Method

The calibration method uses a training set with existing C&E data for the trait of interest, together with ecogeographical data specific to the environment at the sites from where these accessions were collected, to identify existing relationships between the trait and the environment. This training set is required to calibrate a predictive model based on these relationships. This predictive model is then applied to other non-evaluated accessions to identify those that, according to this model, are likely to have a higher probability of genetic adaptation for a target trait property. The model therefore aims to identify a subset that is more likely to show the target trait property than a subset merely selected randomly. The use of the calibration method has been described in recent studies on morphological and agricultural traits in barley (Endresen, 2010; Endresen *et al.*, 2011) and wheat stem rust (Endresen *et al.*, 2011, 2012; Bari *et al.*, 2012).

Calibration of such predictive models can lead easily to very specific models that are overfitted to the training set. Such models are too complex and too closely fitted to the training set. To find an appropriate model complexity, the iterative model calibration should be stopped when a generic pattern is identified. Through an independent test set of similar accessions with known trait data, it is evaluated if the model is sufficiently generic to perform well on other similar accessions. If a separate and independent test set is not available, a (dependent) test set can be created by using a set of records from the training set.

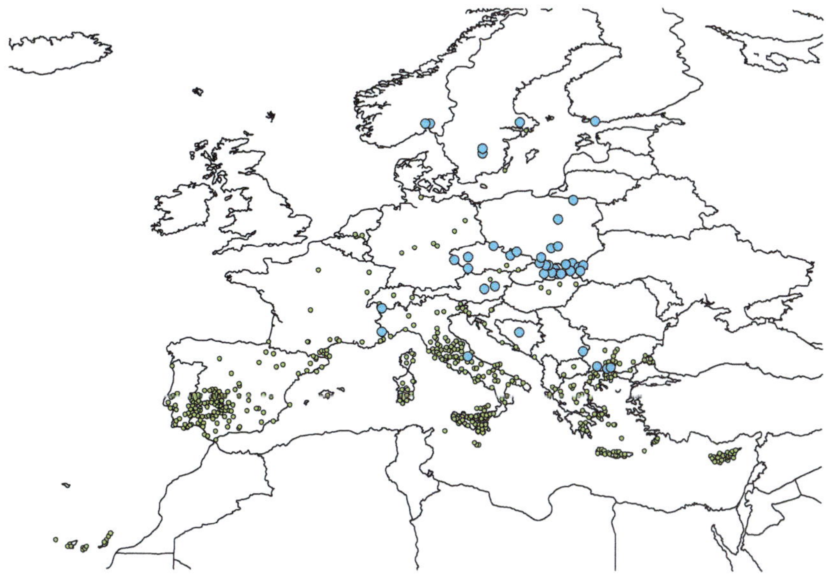

Fig. 8.4. Location of wild *Medicago* occurrences included in the PGR Secure data set. Occurrences selected through ecogeographical filtering are highlighted.

R-scripts for performing a predictive characterization study using the calibration method for classification have been developed using algorithms such as the Random Forest (Breiman, 2001) for calibration of the model (Thormann *et al.*, 2014). Random Forest was one of the methods that performed well in previous FIGS studies (Bari *et al.*, 2012). The R-script was developed using an existing data set with occurrence and trait data for stem rust in wheat made available by USDA GRIN (public domain) and environment data from the WorldClim data set (www.worldclim.org) (Hijmans *et al.*, 2005), which are the same stem rust data and modelling approaches that were used by Endresen *et al.* (2011) and Bari *et al.* (2012).

The amount of evaluation data available for CWR of *Avena*, *Beta*, *Brassica* and *Medicago* from relevant databases has been insufficient for the application of the calibration method to the data sets of the PGR Secure project.

## 8.4 General Considerations

Predictive characterization does not replace actual field trials, but considerably reduces the size of field trials by reducing the set of candidate accessions that the breeder needs to screen before finding novel alleles for a target trait. It is also true that it is a significantly cheaper option than applying molecular techniques to identify accessions with desired traits. This is achieved through less expensive pre-screening against an environment profile. The predictive methods therefore help to utilize more efficiently the limited and costly land area and human working time for field screening.

In theory, better responses to abiotic stresses could be expected in the populations with the lowest values below the identified thresholds, e.g. the lowest De Martonne indices when looking for drought tolerance, but these populations might not be the best to use for breeding. On the one hand, marginal populations being at the edge of the species distribution conditions are likely to have a small population size (Brown, 1984) and may have low genetic diversity (Aronson *et al.*, 1990; Lynch and Lande, 1993; Sagarin and Gaines, 2002). On the other hand, selection of the most marginal populations might be justified, regardless of their population size or genetic diversity, as we are interested in only one trait (drought tolerance in this case). Low diversity levels would not be a major problem in breeding programmes as long as the desired trait is, indeed, found in the selected populations. However, difficulties may arise if the desired traits are associated with unwanted traits (such as fixed deleterious alleles as a result of inbreeding or small plant size) or have multigenic controlled inheritance. Populations able to withstand the most extreme abiotic conditions for one particular trait could, in fact, be maladapted to the rest of the desired characteristics of the crop. From this perspective, populations subject to slightly less abiotic stress (probably with larger population sizes and greater genetic diversity) might provide a better source of adaptive traits. It might therefore be useful to establish if populations included in the subsets selected through predictive characterization are located at the geographic margins of distribution, and if their population size is large enough to support mean population genetic diversity levels, avoiding severe inbreeding and genetic drift effects.

The calibration method may be said to have an advantage over the ecogeographical filtering method in that the former is an evidence-based approach, the evidence being the explicit link to past C&E data, so it does take into account the genetic fixation of adaptive traits. So if there is only a partial relationship between site locality and the presence of adaptive traits, it should be more successful at predicting sites with the desired adaptive traits than the ecogeographical method. Obviously, where the basic assumption of predictive characterization is false and there is no correlation between environmental conditions and patterns of genetic diversity for adaptive traits, both methods will be equally likely to fail in predicting localities where populations with desired adaptive traits might be found, but this would equally hold true for any logical method the breeder might apply to select accessions for C&E trials.

The ecogeographical filtering method is the method better suited for CWR, as it only requires coordinates from collecting sites. For the calibration method to be applied, there does need to be previous C&E data available for the target taxon, and this may rarely be the case for CWR taxa. None of the methods are appropriate for modern cultivars, as they are not expected to show this association between traits and the environment. This is because their traits have not arisen as a result of natural selection but have been selected artificially to provide a high yield under a wide range of environmental conditions.

The two predictive characterization methods presented here have not been tested with genetic data as a variable. Further research is required to know how the inclusion of genetic diversity data might support predictive characterization. The relationship between neutral genetic marker diversity and population fitness, as well as heritability of quantitative traits, has been found to be very weak (Reed and Frankham, 2001; McKay and Latta, 2002), and low differentiation at neutral markers does not necessarily mean a corresponding lack of adaptive differentiation (Conner and Hartl, 2004). It therefore should not be used as a predictor variable for adaptive genetic differentiation, but its use in the development of ELC maps or the generation of the ecogeographical core set might be investigated. Data about adaptive genetic diversity or diversity of loci linked to quantitative traits could probably serve as variables in the calibration method in a similar way in which C&E data are used, or support the description of environment profiles. Considering the decreasing costs for genotyping and the continuous improvement of technologies, more data that can be used in this research are expected to be generated.

## Acknowledgements

This research was funded by a grant from the EU Seventh Framework Programme for the PGR Secure project (Grant Agreement No 266394).

## References

Almorox, J. (2003) *Climatología aplicada al Medio Ambiente y Agricultura*. UPM. ETSI Agrónomos, The Universidad Politecnica de Madrid, Madrid.

Andersson, M. (1988) Toxicity and tolerance of aluminium in vascular plants. *Water, Air, and Soil Pollution* 39, 439–462.

Aronson, J.A., Kigel, J. and Shmida, A. (1990) Comparative plants sizes and reproductive strategies in desert and Mediterranean populations of ephemeral plants. *Israel Journal of Botany* 39, 413–430.
Ashraf, M. and Harris, P.J.C. (2004) Potential biochemical indicators of salinity tolerance in plants. *Plant Science* 166(1), 3–16.
Bari, A., Street, K.A., Mackay, M.C., Endresen, D.T.F., De Pauw, E., *et al.* (2012) Focused Identification of Germplasm Strategy (FIGS) detects wheat stem rust resistance linked to environmental variables. *Genetic Resources and Crop Evolution* doi:10.1007/s10722-011-9775-5 (published online 3 December 2011).
Bari, A., Amri, A., Street, K.A., Mackay, M.C., De Pauw, E., *et al.* (2014) Predicting resistance to stripe (yellow) rust (*Puccinia striiformis*) in wheat genetic resources using focused identification of germplasm strategy. *The Journal of Agricultural Science* 152(06), 906–916.
Bhullar, N.K., Street, K., Mackay, M., Yahiaoui, N. and Keller, B. (2009) Unlocking wheat genetic resources for the molecular identification of previously undescribed functional alleles at the Pm3 resistance locus. *Proceedings of the National Academy of Sciences of the United States of America* 106, 9519–9524.
Boyer, J.S. (1982) Plant productivity and environment. *Science* 218, 443–448.
Breiman, L. (2001) Random forests. *Machine Learning* 45(1), 5–32.
Brown, J.H. (1984) On the relationship between abundance and distribution of species. *The American Naturalist* 124(2), 255–279.
Conner, J.K. and Hartl, D.L. (2004) *A Primer of Ecological Genetics.* Sinauer Associates, Inc, Sunderland, Massachusetts.
Dansereau, P. (1957) *Biogeography, An Ecological Perspective*. The Ronald Press Co, New York.
De Martonne, E. (1926) Une nouvelle fonction climatologique: l'indice d'aridité. *La Meteorologie* 2, 449–458.
De Oliveira, P.H., Fererizzi, L.C., Kothe Milach, S.C., Gotuzzo, C. and Taizo Sawasato, J. (2005) Inheritance in oat (*Avena sativa* L.) of tolerance to aluminium toxicity. *Crop Breeding and Applied Biotechnology* 5, 302–309.
Dulloo, M.E., Fiorino, E. and Thormann, I. (2015) Research on crop wild relatives. In: Redden, R., Yadav, S.S., Maxted, N., Dulloo, M.E., Guarino, L. and Smith, P. (eds) *Crop Wild Relatives and Climate Change*. Wiley-Blackwell, USA, pp. 108–129.
El Bouhssini, M.E., Street, K., Joubi, A., Ibrahim, Z. and Rihawi, F. (2009) Sources of wheat resistance to Sunn pest, *Eurygaster integriceps* Puton, in Syria. *Genetic Resources and Crop Evolution* 56(8), 1065–1069.
El Bouhssini, M.E., Street, K., Amri, A., Mackay, M., Ogbonnaya, F.C., *et al.* (2011) Sources of resistance in bread wheat to Russian wheat aphid (*Diuraphis noxia*) in Syria identified using the Focused Identification of Germplasm Strategy (FIGS). *Plant Breeding* 130, 96–97.
Endresen, D.T.F. (2010) Predictive association between trait data and ecogeographical data for Nordic barley landraces. *Crop Science* 50(6), 2418–2430.
Endresen, D.T.F., Street, K., Mackay, M., Bari, A. and de Pauw, E. (2011) Predictive association between biotic stress traits and ecogeographical data for wheat and barley landraces. *Crop Science* 51(5), 2036–2055.
Endresen, D.T.F., Street, K., Mackay, M., Bari, A., Amri, A., De Pauw, E., *et al.* (2012) Sources of resistance to stem rust (Ug99) in bread wheat and durum wheat identified using Focused Identification of Germplasm Strategy (FIGS). *Crop Science* 52, 764–773.
Fageria, N.K., Ballgar, V.C. and Wright, R.J. (1988) Aluminium toxicity in crop plants. *Journal of Plant Nutrition* 11(3), 303–319.
FAO (Food and Agriculture Organization) (2010) *The Second Report on the State of the World's Plant Genetic Resources for Food and Agriculture*. FAO of the United Nations, Rome.
FAO (2013) *Report of the Fourteenth Regular Session of the Commission on Genetic Resources for Food and Agriculture*. FAO of the United Nations, Rome. CGRFA 14/13/report. Available at: www.fao.org/nr/cgrfa/cgrfa-meetings/cgrfa-comm/fourteenth-reg/en/ (accessed 19 October 2015).
Guarino, L., Jarvis, A., Hijmans, R.J. and Maxted, N. (2002) Geographical information systems (GIS) and the conservation and use of plant genetic resources. In: Engels, J.M.M., Ramanatha Rao, V., Brown, A.H.D. and Jackson, M.T. (eds) *Managing Plant Genetic Diversity*. CAB International, Wallingford, UK, pp. 387–404.
Haynes, R.K. and Mokolobate, M.S. (2001) Amelioration of Al toxicity and P deficiency in acid soils by additions of organic residues: a critical review of the phenomenon and the mechanisms involved. *Nutrient Cycling in Agroecosystems* 59, 47–63.

Hijmans, R.J., Cameron, S.E., Parra, J.L., Jones, P.G. and Jarvis, A. (2005) Very high resolution interpolated climate surfaces for global land areas. *International Journal of Climatology* 25(15), 1965–1978.

Jordán López, A. (2005) Suelos salinos y sódicos. In: Jordán López, A. *Manual de edafología. Dpto. De Cristalografía, Mineralogía y Química Agrícola de la Universidad de Sevilla*. Escuela Universitaria Técnica Agrícola, Universidad de Sevilla, Seville, Spain pp. 135–144.

Kaur, N., Street, K., Mackay, M., Yahiaoui, N. and Keller, B. (2008) Allele mining and sequence diversity at the wheat powdery mildew resistance locus Pm3. In: Appels, R., Eastwood, R., Lagudah, E., Langridge, P., Mackay, M., Mcintyre, L. and Sharp, P. (eds) *11th International Wheat Genetics Symposium*. Sydney University Press, Brisbane, Australia, pp. 32–34.

Khazaei, H., Street, K., Bari, A., Mackay, M. and Stoddard, F.L. (2013) The FIGS (focused identification of germplasm strategy) approach identifies traits related to drought adaptation in *Vicia faba* genetic resources. *Public Library of Science ONE* 8(5), e63107.

Koss, J.W., Owenby, J.R., Steurer, P.M. and Ezell, D.S. (1988) Freeze/Frost Data. *Climatography of the U.S.* No 20, supplement No 1, I-3–I-5.

Levitt, J. (1980) *Response to Plant Environmental Stress*, 2nd edn. Academic Press, New York.

Lynch, M. and Lande, R. (1993) Evolution and extinction in response to environmental change. In: Kareiva, P., Kingsolver, J. and Huey, R. (eds) *Biotic Interactions and Global Change*. Sinauer Associates, Inc, Sunderland, Massachusetts, pp. 234–250.

McKay, J.K. and Latta, R.G. (2002) Adaptive population divergence: markers, QTL and traits. *Trends in Ecology and Evolution* 17(6), 285–291.

Mackay, M.C. (1986) Utilizing wheat genetic resources in Australia. In: McLean, R. (ed.) *Proceedings of the 5th Assembly of the Wheat Breeders' Society in Australia*. 18–22 August 1986, Merredin. Western Australian Department of Agriculture, Perth, Australia, pp. 56–61.

Mackay, M.C. (1990) Strategic planning for effective evaluation of plant germplasm. In: Srivastava, J.P. and Damania, A.B. (eds) *Wheat Genetic Resources: Meeting Diverse Needs.* John Wiley & Sons, Chichester, UK, pp. 21–25.

Mackay, M.C. (1995) One core collection or many? In: Hodgkin, T., Brown, A.H.D., van Hintum, T.J.L. and Morales, A.A.V. (eds) *Core Collections of Plant Genetic Resources. Proceedings from the IBPGR/CGN/CENARGEN Workshop on 'Core Collections: Improving the Management and Use of Plant Germplasm Collections'*, Brasilia, August 1992. John Wiley & Sons, Chichester, UK, pp. 199–210.

Mackay, M.C. and Street, K. (2004) Focused identification of germplasm strategy – FIGS. In: Black, C.K., Panozzo, J.F. and Rebetzke, G.J. (eds) *Cereals 2004. Proceedings of the 54th Australian Cereal Chemistry Conference and the 11th Wheat Breeders' Assembly, 21–24 September 2004, Canberra, Australian Capital Territory (ACT).* Cereal Chemistry Division, Royal Australian Chemical Institute, Melbourne, Australia, pp. 138–141.

Maxted, N., Mabuza-Dlamini, P., Moss, H., Padulosi, S., Jarvis, A., *et al.* (2004) An ecogeographic survey: African *Vigna*. *Systematic and Ecogeographic Studies of Crop Genepools 10*. IPGRI, Rome, pp. 1–468.

Monteith, J.L. (1965) Evaporation and environment. In: Fogg, G.E. (ed.) *The State and Movement of Water in Living Organisms*. 19th Symposium, Society of Experimental Biology. Cambridge University Press, Cambridge, UK, pp. 205–234.

Nava, I.C., Delatorre, C.A., de Lima Duarte, I.T., Teixeira Pacheco, M. and Federizzi, L.C. (2006) Inheritance of aluminium tolerance and its effects on grain yield and grain quality in oats (*Avena sativa* L.). *Euphytica* 148, 353–358.

Paltineanu, C., Tanasescu, N., Chitu, E. and Mihailescu, I.F. (2007a) Relationships between the De Martonne aridity index and water requirements of some representative crops: a case study from Romania. *International Agrophysics* 21, 81–93.

Paltineanu, C., Mihailescu, I.F., Seceleanu, I., Dragota, C. and Vasenciuc, F. (2007b) Using aridity indices to describe some climate and soil features in Eastern Europe: a Romanian case study. *Theoretical and Applied Climatology* 90, 263–274.

Parida, A.K. and Das, A.B. (2005) Salt tolerance and salinity effects on plants: a review. *Ecotoxicology and Environmental Safety* 60, 324–349.

Parra-Quijano, M., Iriondo, J.M. and Torres, E. (2012a) Improving representativeness of genebank collections through species distribution models, gap analysis and ecogeographical maps. *Biodiversity and Conservation* 21, 79–96.

Parra-Quijano, M., Iriondo, J.M. and Torres, E. (2012b) Ecogeographical land characterization maps as a tool for assessing plant adaptation and their implications in agrobiodiversity studies. *Genetic Resources and Crop Evolution* 59(2), 205–217.

Parra-Quijano, M., Iriondo, J.M., Frese, L. and Torres, E. (2012c) Spatial and ecogeographical approaches for selecting genetic reserves in Europe. In: Maxted, N., Dulloo, M.E., Ford-Lloyd, B.V., Frese, L.,

Iriondo, J. and Pinheiro de Carvalho, M.A.A. (eds) *Agrobiodiversity Conservation: Securing the Diversity of Crop Wild Relatives and Landraces*. CAB International, Wallingford, UK, pp. 20–28.
Parra-Quijano, M., Torres, E., Iriondo, J.M. and López, F. (2014) CAPFITOGEN tools user manual Version 1.2. International Treaty on Plant Genetic Resources for Food and Agriculture, FAO, Rome, 138 p. Available from: www.agrobiodiversidad.org/blog/?p=1189 (accessed 12 September 2014) and ftp.fao.org/ag/agp/planttreaty/publi/2014/capfitogen_ENG92014.pdf (accessed 12 September 2014).
Phillips, J., Kyratzis, A., Christoudoulou, C., Kell, S. and Maxted, N. (2014) Development of a national crop wild relative conservation strategy for Cyprus. *Genetic Resources and Crop Evolution* 61(4), 817–827.
Ramirez-Villegas, J., Khoury, C., Jarvis, A., Debouck, D.G. and Guarino, L. (2010) A gap analysis methodology for collecting crop genepools: a case study with *Phaseolus* beans. *PLoS ONE* 5(10), e13497, doi:10.1371/journal.pone.0013497.
Reed, D.H. and Frankham, R. (2001) How closely correlated are molecular and quantitative measures of genetic variation? A meta-analysis. *Evolution* 55(6), 1095–1103.
Rout, G.R., Samantaray, S. and Das, P. (2001) Aluminium toxicity in plants: a review. *Agronomie* 2, 3–21.
Sagarin, R.D. and Gaines, S.D. (2002) The 'abundant centre' distribution: to what extent is it a biogeographical rule? *Ecology Letters* 5, 137–147.
Sakay, A. and Larcher, W. (1987) *Frost Survival of Plants: Responses and Adaptation to Freezing Stress*. Springer-Verlag, Berlin.
Sanchez-Chacon, C.D., Federizzi, L.C., Milach, S.C.K. and Pacheco, M.T. (2000) Variabilidad e genetica e heranca da tolerancia a toxicidade do aluminio em aveia. *Pesquisa Agropecuaria Brasileira* 35, 1797–1808.
Sandve, S.R., Kosmala, A., Rudi, H., Fjellheim, S., Rapacz, M., *et al*. (2011) Molecular mechanisms underlying frost tolerance in perennial grasses adapted to cold climates. *Plant Science* 180(1), 69–77.
Seelig, B.D. (2000) *Salinity and Sodicity in North Dakota Soils*. EB 57. North Dakota State University Extension Service, Fargo, North Dakota.
Shannon, M.C. (1997) Adaptation of plants to salinity. *Advances in Agronomy* 60, 75–120.
Shehadeh, A., Amri, A. and Maxted, N. (2013) Ecogeographic survey and gap analysis of *Lathyrus* L. species. *Genetic Resources and Crop Evolution* 60, 2101–2113, doi:10.1007/s10722-013-9977-0.
Street, K., Mackay, M., Zuev, E., Kaul, N., El Bouhssini, M., *et al*. (2008) Swimming in the genepool – a rational approach to exploiting large genetic resource collections. Available at hdl.handle.net/2123/3390 (accessed 12 September 2014).
Thomashow, M.F. (1999) Plant cold acclimatization: freezing tolerance genes and regulatory mechanisms. *Annual Review of Plant Physiology and Molecular Biology* 50, 571–599.
Thormann, I. (2012) Applying FIGS to crop wild relatives and landraces in Europe. *Crop Wild Relative*, Issue 8, 14–16. Available at: www.pgrsecure.bham.ac.uk/sites/default/files/documents/newsletters/CWR_Issue_8.pdf (accessed 12 September 2014).
Thormann, I., Parra-Quijano, M., Endresen, D.T.F., Rubio Teso, M.L., Iriondo, M.J., *et al*. (2014) *Predictive Characterization of Crop Wild Relatives and Landraces. Technical Guidelines Version 1*. Bioversity International, Rome.
Thormann, I., Parra-Quijano, M., Iriondo, J.M., Rubio-Teso, M.L., Endresen, D.T., Maxted, M., Dias, S. and Dulloo M.E. (2015) Two predictive characterization approaches to search for target traits in crop wild relatives and landraces. *Crop Wild Relative* 10, 16–18. ISSN 1742-3694 (online), www.pgrsecure.bham.ac.uk/sites/default/files/documents/newsletters/CWR_Issue_10.pdf
Tyler, G., Berggren, D., Bergkvist, B., Falkengreen-Grerup, U., Folkeson, L., *et al*. (1987) Soil acidification and metal solubility in forests of southern Sweden. In: Hutchinson, T.C. and Meema, K.M. (eds) NATO ASI Series G16. *Effects of Atmospheric Pollutants on Forests, Wetlands and Agricultural Ecosystems*. Springer-Verlag, Berlin, pp. 347–359.
Vincent, H., von Bothmer, R., Knüpffer, H., Amri, A., Konopka, J., *et al*. (2012) Genetic gap analysis of wild *Hordeum* taxa. *Plant Genetic Resources, Characterization and Utilization* 10, 242–253.
Vincent, H., Wiersema, J., Kell, S.P., Dobbie, S., Fielder, H., *et al*. (2013) A prioritized crop wild relative inventory to help underpin global food security. *Biological Conservation* 167, 265–275.
Wagner, C.W., Milach, S.C.K. and Federizzi, L.C. (2001) Genetic inheritance of aluminum tolerance in oat. *Crop Breeding and Applied Biotechnology* 1, 22–26.
Wallén, C.C. (1967) Aridity definitions and their applicability. Geografiska Annaler. Series A, Physical Geography, Vol 49, No 2/4, *Landscape and Processes: Essays in Geomorphology*. Swedish Society for Anthropology and Geography. Wiley, pp. 367–384.
Yamaguchi, T. and Blumwald, E. (2005) Developing salt-tolerance crop plants: challenges and opportunities. *Trends in Plant Science* 10(12), 615–620.
Zhu, J.K. (2001) Plant salt tolerance. *Trends in Plant Science* 6(2), 66–71.

# 9 Keeping a Finger on the Pulse: Monitoring the Use of CWR in Crop Improvement

C. Smith*
*Quaker United Nations Office, Geneva, Switzerland*

## 9.1 Introduction

Fellow contributors to this issue present recent examples of crop wild relatives (CWR) contributing to crop improvement efforts and highlight how advancements in the field are facilitating increased use today. These cases are part of a larger trend in the use of CWR – one that is exceedingly difficult to quantify but that is otherwise fairly intuitive. Technological advancements in breeding, coupled with increasing environmental pressures, have incited crop scientists and breeders to look to more distantly related species for sources of required genetic variation. The CWR conservationist community has described this dynamic for nearly two decades (see, for example, Maxted *et al.*, 1997a,b). This trend is likely to continue in light of the burgeoning field of applied genomics, the availability of tools for predictive trait mining and improved collaboration in the context of ever more acute climate pressures.

Efforts have been made to quantify use and place a value on the genetic contributions from wild to cultivated crop genepools (for examples see Prescott-Allen and Prescott-Allen, 1986; Pimentel *et al.*, 1997; Hein and Gatzweiler, 2006; Hunter and Heywood, 2011; PricewaterhouseCoopers LLD, 2013, commissioned by Kew Millennium Seed Bank). These efforts are useful communication tools but are limited by the extent to which they capture the breadth of ongoing work. The value of CWR is commonly estimated by focusing on the final stage in the long process of breeding: variety release. Such *downstream* benefits take a very long time to accrue and do not reflect the host of other benefits derived from ongoing work with CWR *upstream* in the forms of basic research and applied pre-breeding. Less tangible benefits include enhanced understanding of gene–trait relationships and compilation of the data required to make CWR usable in breeding programmes.

This paper takes a macro perspective of the state of the use of CWR in crop improvement programmes under the auspices of the Consultative Group on International Agricultural Research (CGIAR), with some supplementary perspectives from outside of the CG family where required. It asks, *to what extent are CWR being used by crop scientists and pre-breeders today?* A series of interviews was conducted with experts across 19 crop communities (Table 9.1) between February and April 2014. The results are compared with those reported in 1986 and 2007 by Prescott-Allen and Prescott-Allen and Hajjar and Hodgkin, respectively.

*E-mail: chelseasmth204@gmail.com

Table 9.1. Status of CWR use.

| Crop | Mid-1980s | 1986–2007 | 2007–2014 | Post-2014 |
|---|---|---|---|---|
| Wheat | Extensive | Extensive | **Extensive use:** | A1 |
| Rice | – | Extensive | Many varieties and introgression | B1 |
| Tomato[a] | Extensive | Extensive | lines released | B1 |
| Oat[a] | Extensive | – | Many different types of traits | B1 |
| Barley[b] | – | Considerable | transferred from wild species | A2 |
| Soybean[b] | Modest | Modest | | A2 |
| Banana | – | Considerable | **Considerable use:** | A3 |
| Potato[c] | | Considerable | Few varieties released | A3 |
| Maize | | Moderate | **Moderate use:** | C1 |
| Chickpea | | Considerable | Varieties not yet released | C3 |
| Cassava | | Considerable | New and valuable traits identified | C2 |
| Lentil | | Modest | Ongoing efforts to characterize | A2 |
| Sorghum | | Moderate | and evaluate collections | B2 |
| Sweet potato | | – | Techniques to overcome crossing | C5 |
| Yam | | – | barriers being employed | A4 |
| Pearl millet | | Moderate | **Modest use:** | A5 |
| Finger millet | | – | Very little exploration | C5 |
| Bean | | Moderate | of wild genepools | A2 |
| Cowpea | | Modest | Crosses achieved but no | B2 |
| | | | varieties released | |

**Need to look to CWR for required genetic variation:**
A = insufficient variation in LR and crop genepools
B = sufficient variation for some traits
C = sufficient variation for most traits

**Genomics resources:**
Reference genome and CWR genomes available = 1
Reference genome available; re-sequencing under way = 2
Reference genome available = 3
Reference genome sequencing under way = 4
No resources available; no sequencing efforts reported = 5

*Notes:* [a]High prevalence of wild germplasm in all new varieties released; [b]extensive use outside of CGIAR, several varieties released; [c]difficult to track CWR contribution.

The results highlight: (i) a positive trend in the overall use of CWR; (ii) a growing emphasis on their role in adapting cultivated crops to climate change; and (iii) that the majority of ongoing work with CWR has yet to result in quantifiable breeding outputs. Measuring use simply by counting the number of cultivated varieties released with traits from wild species underestimates the current use of CWR. By extension, valuation exercises that focus exclusively on the economic value of varieties underestimates the use and potential use values of CWR.

Looking at (i), the genomics resources currently available to breeders, and (ii) the relative need for breeders to look outside of cultivated crops and landrace genepools for required greater genetic diversity, this paper suggests that today's technological advancements coupled with more extreme environmental pressures will incite greater use of CWR in the coming years. Findings are relevant to the CWR conservation community advocating for increased investment into enhanced wild genepool conservation and use.

## 9.2 Quantifying the Use of CWR

Monitoring trends is no easy feat. An immense body of literature documents the identification and successful transfer of beneficial traits from CWR to crop genepools. Reviews of this literature and variety release reports are useful for gaining an overall picture of the use of CWR (Harlan, 1976; Prescott-Allen and Prescott-Allen, 1981; Maxted and Kell, 2009; Hunter and

Heywood, 2011). Together, they communicate a positive trend in use, measured by the volume of studies published.

Published literature is an imperfect proxy for use. Studies are difficult to compare with one another, and advancements reported cannot be aggregated easily. Terms such as *tolerance*, *resistance* and *immunity* are used to communicate varying degrees of success and improvements, and are not always quantified. Different modes of resistance have varying effects. Durable versus transient resistance and polygenic versus monogenic resistance indicate varying degrees of effectiveness against variants of pathogens and overall robustness of the resistance (Prescott-Allen and Prescott-Allen, 1986).

Moreover, the majority of publications come from a basic research perspective rather than an applied breeding perspective. Research into the basic biology of species, gene–trait relationships and the use of particular wild species as model organisms is published to a greater extent than ongoing work to develop breeding lines with wild germplasm. The breeding lines developed within university institutions are often underused in applied breeding programmes because the parents selected do not reflect the breeding goals of the given programme and target environment. Furthermore, successful transfers of traits do not always result in improved commercial cultivars. This leaves the reviewer with an impression of what *can* be done, rather than what *is* being done (H. Dempewolf, 2014, personal Skype communication). As Prescott-Allen and Prescott-Allen (1986) emphasized, while 'the number of crops that currently benefit from wild genetic resources is not small ... it is substantially smaller than might be inferred from the literature'.

A case can also be made that use is underestimated in the literature. Pre-breeders and breeders within the CGIAR generally underpublish on their ongoing work and prioritize reporting once materials are ready to be released, in the form of variety release reports (C. Khoury; S. Hearne, 2014, personal Skype communication). Contributions from CWR are also difficult to trace as materials get passed between users without associated pedigree information (E. Weltzien-Rattunde; M. Bonierbale, 2014, personal Skype communication). The wealth of ongoing research is insufficiently captured. Particularly in the case of the private sector, where breeding efforts are commercially sensitive, information is not made readily available (R. Eastwood, 2014, personal Skype communication).

Variety release is only part of the story. The majority of efforts to use CWR take place *upstream*, i.e. trait identification and the transfer of novel genes into cultivated crop backgrounds. This work may never translate directly into cultivars released *downstream*, but is indispensable for facilitating future use of CWR. Basic research into species' taxonomies, ploidy levels and flowering times, and the advancing study of gene–trait relationships and genotype–environment interactions is a prerequisite for making use of wild species. Applied pre-breeding efforts to identify, select and transfer traits of interest from wild species in cultivated crop backgrounds and develop pre-breeding lines from this material is likewise essential – wild genetic variation would otherwise remain inaccessible to breeders developing new crop varieties. Pre-breeding also entails screening CWR accessions and conducting field trials for pre-breeding lines, resulting in large amounts of characterization and evaluation information. These data will facilitate the increased use of CWR in years to come, but do not translate directly into quantifiable outputs.

## 9.3 Benchmarks for Comparison

In 1986, Prescott-Allen and Prescott-Allen published a major review of the use of CWR in breeding in Canada and the USA, including the first effort to quantify the economic value of all wild species. This continues to be a widely referenced benchmark in the field. The list of crops grown or imported by Canada and the US that had been improved significantly by wild species includes tomato, bread wheat, potato, oat, sunflower, sugarbeet, blueberry, lucerne, sweet clover, cotton, tulip, hops, lettuce, sugarcane, palm oil, strawberry, bell pepper, tobacco and cacao (Prescott-Allen and Prescott-Allen, 1986).

Hajjar and Hodgkin (2007) undertook the first comprehensive review of the use of wild species in breeding since that published by the Prescott-Allens, aimed at portraying how far breeders had come in the 20 years prior. They reported a steady increase in the release of cultivars containing wild genes, although their list of crops differed substantially, selected according to the mandate crops of the CGIAR. They measured CWR use in terms of the number of varieties released and the number of different traits transferred successfully from wild to those released species, with some references made to varieties under development. Approximately double the number of wild species were used for nine of the crops common to the two reviews, and over 100 beneficial traits had been transferred successfully to released varieties. The crops that had benefitted the most from CWR until the mid-1980s continued to dominate, namely wheat and tomato (Hajjar and Hodgkin, 2007).

Both the Prescott-Allens and Hajjar and Hodgkin reported that although significant, use was more modest than had been anticipated. Contributions of CWR 'remain less than might have been expected given improved procedures for intercrossing species from different genepools, advances in molecular methods for managing backcrossing programmes, increased numbers of wild species accessions in genebanks, and the substantial literature on beneficial traits associated with wild relatives' (Hajjar and Hodgkin, 2007).

## 9.4 Limitations to Use

There are a host of factors limiting the use of CWR in crop improvement programmes today despite significant advancements in the fields of genetics and genomics. These include, among others, persistent biological barriers to crossing, limited passport, characterization and evaluation information on accessions held in genebanks and for pre-breeding materials (materials under development), and human capacity and resource constraints. An in-depth look at the bottlenecks in the process of using CWR is beyond the scope of this paper. Instead, the emphasis here is that there is a significant amount of ongoing work with CWR despite these limitations, the bulk of which cannot be readily translated into quantifiable breeding outputs.

Breeding with CWR is also a long-term endeavour. The process of transferring genes from wild species to domesticated crop backgrounds and repeatedly backcrossing with the cultivated parent to reduce linkage drag commonly takes one decade or longer, depending on the complexity of the desired trait and the size of linkage blocks inherited. So while Koebner and Ortiz (2013), summarizing recent developments discussed during the European Association for Research on Plant Breeding held in June 2013, assert that the perceived constraints of using CWR in plant breeding programmes are outdated, these advancements take time to percolate down into applied breeding practice. This is particularly the case for crops with smaller pre-breeding programmes.

## 9.5 Findings

### 9.5.1 More extensive use of CWR

The steady increase in the use of CWR reported by Hajjar and Hodgkin (2007) has continued until today. Pre-breeders and genebank managers report a range of advancements, including new varieties released and new sources of traits identified. Experts in barley, chickpea, lentil, rice, soybean, tomato, wheat and yam breeding explicitly describe a strong positive trend in the use of CWR. Experts in all other crops, with the exceptions of cowpea and common bean, highlight ongoing explorations, advancements and great potential for future use.

Table 9.1 presents the state of use across four time intervals: until the mid-1980s, from 1986 to 2007, from 2007 until the present, and projected use post-2014. *Extensive use* describes the successful development and release of many new cultivars with genetic contributions from wild species; *considerable use* denotes that few varieties have been released; *moderate use* describes crop communities which have yet to release varieties with wild germplasm but which have made significant gains in terms of identifying sources of new traits, successfully transferring traits to cultivated crop backgrounds and overcoming crossing barriers; and *modest use* denotes that very little exploration into wild genepools has taken place.

The use of CWR remains highly crop dependent. Crops with long histories of breeding with CWR continue to benefit the most from wild progenitors, having well-established breeding programmes with relatively secure sources of funding. Rice, tomato and wheat in particular have well-established pre-breeding programmes that focus specifically on CWR and more advanced tools at the disposal of pre-breeders for identifying traits of interest and making crosses. Looking at rice as one example, since 2007, new breeding lines have been developed and a number of wild species' genomes have been re-sequenced to understand the function of genes and how genes control various types of resistances and tolerances to different stress conditions (K.K. Jena, 2014, personal Skype communication).

Experts report more ongoing work to borrow abiotic stress tolerances from wild species than in the past. Extensive efforts are under way to identify and transfer drought, heat and soil salinity tolerances within barley, cassava, chickpea, lentil, potato, rice, soybean, sweet potato, tomato, wheat and yam pre-breeding programmes. Progress has been achieved in understanding the genetic base of polygenic traits and the mechanisms that control complex environmental stress tolerances. This is consistent with the message from the conservation community: CWR not only have the *potential* to contribute to food security in the context of climate change, but it is *imperative* that they play a feature role in adaptation and mitigation efforts (for examples see Heywood *et al.*, 2007; Ford-Lloyd *et al.*, 2011; McCouch, 2013; Dempewolf *et al.*, 2014). According to Breithaupt (2008), CWR are better placed than even landrace varieties to lend 'climate ready' traits to domesticated crops, being adapted to more marginal environments themselves.

Looking only at progress achieved between 2007 and 2014, this trend does not stand out. Only barley and wheat varieties have been released with improved abiotic stress tolerances since 2007. Column 'Post-2014' in Table 9.2 depicts ongoing work with CWR and varieties at various stages of development. Each entry represents an interesting story told by pre-breeders but not yet captured by those reporting on the use and potential use values of CWR. Using rice again as an example, genes associated with saline soil tolerance have been transferred successfully from a wild halophyte species to cultivated rice. Varieties will be developed and released in the coming years (K.K. Jena, 2014, personal Skype communication).

It is also striking that biotic traits are being borrowed to the benefit of every crop surveyed, with the sole exception of bean. Biotic and abiotic stresses are intimately linked. Crops face changing and intensifying pressures from insects and pathogens as environmental conditions change

**Table 9.2.** Types of traits taken from CWR and used in varieties released.

| | Mid-1980s | 1986–2007 | 2007–2014 | Post-2014 |
|---|---|---|---|---|
| Banana | Unreported | β | β | β |
| Barley | Unreported | α | α | β α ✓↑ |
| Bean | Unreported | β α ↑ | None | |
| Cassava | Unreported | β ✓ | None | β α ✓ |
| Chickpea | Unreported | α | None | β α |
| Cowpea | Unreported | None | None | β |
| Finger millet | Unreported | Unreported | None | β |
| Lentil | Unreported | None | None | β α ✓ |
| Maize | Unreported | None | None | β |
| Oat | β | Unreported | β | β ✓ |
| Pearl millet | Unreported | β ψ | None | β ψ |
| Potato | β | β | β ↑ | β α ✓↑ |
| Rice | β | β α ψ ↑ | β ↑ | β α ✓↑ |
| Sorghum | Unreported | None | None | β |
| Soybean | None | None | β | β α |
| Sweet potato | Unreported | Unreported | None | β α |
| Tomato | β ✓ | β α ✓ | β | β α ✓ |
| Wheat | β ↑ | β ✓↑ | β α ✓↑ | β α ✓ ψ ↑ |
| Yam | Unreported | Unreported | None | β α ✓ |

*Notes*: β = biotic; α = abiotic; ✓ = quality; ψ = CMS (cytoplasmic male sterility); ↑ = yield.

(N. Maxted, 2014, personal Skype communication), thus biotic resistances will continue to be a top priority for breeding programmes under climate change.

## 9.6 Predicting Future Use

There are many factors that together determine the extent of CWR use. These include, among others, the availability of funding, characterization and evaluation data, the expertise and capacity of breeders, the persistence of crossing barriers and the availability of sufficient genetic variation within crop genepools. Two indicators are used here to predict future use: (i) the relative size of cultivated crops' genepools, and (ii) the genomics resources currently available. The genetic base of crops represents the *need*, as identified by experts, to look outside of landrace and cultivated genepools for traits required to adapt to changing environmental conditions. Availability of genomics resources is a proxy for *investment and technological advances* facilitating the use of CWR.

Based on these two indicators, it is predicted that today's technological advancements, coupled with more extreme environmental pressures, will incite greater use of CWR in the coming years.

### 9.6.1 Indicator 1: The relative size of cultivated crop genepools

Tanksley and McCouch (1997) famously describe the domestication bottleneck process whereby cultivated species' genepools lose genetic diversity through recurrent selection. The result for many crops has been a narrowing of the genetic base. An important determining factor in which crop communities make use of CWR is the *need* to look to outside of landrace and crop genepools for the required genetic variability, denoted by the letters 'A', 'B' and 'C' in the column 'Post-2014' (Table 9.1).

The letter 'A' identifies crops with narrow genetic bases where breeders have been incited increasingly to look to wild species as sources for novel traits and more extreme tolerances; 'B' identifies crops with sufficient variation for *some* traits of interest; and 'C' identifies crops with relatively wide genetic bases where breeders have found sufficient genetic variation for *most* traits of interest. Sufficient variation for most traits has been found in crop genepools for maize, chickpea, cassava, sweet potato and finger millet without going to distantly related species, where crossing barriers and linkage drag are concerns. There is a much greater imperative for breeders to look to wild genepools of wheat, barley, soybean, banana, potato, lentil, yam, pearl millet and the common bean.

It is important to note that the value of CWR for crops that have retained wide genetic bases is still widely recognized. First of all, exploring wild genepools contributes to a better understanding of the history of cultivated crops – valuable for understanding the co-development of hosts and pathogens and finding sources of resistance. Second, the potential use value of genetic diversity is immense, given the difficulty in predicting future climate change scenarios and which traits will be needed. Third, the greater adaptive capacity of wild species will likely be required as climate pressures become more extreme. Using cassava as an example, many cultivated varieties currently exist in dry climates and breeders have not yet been compelled to seek out sources of more extreme drought tolerance from the wild. However, even greater variability for drought tolerance will likely be found within wild genepools, and breeders may need to explore this diversity in the coming years (C. Hersey, 2014, personal Skype communication). Experts across the crop communities emphasize the potential use value of CWR, regardless of the genetic base of cultivated crops.

### 9.6.2 Indicator 2: Availability of genomics resources

The rapidly developing field of genomics promises to help researchers identify useful regions of re-sequenced wild genomes with much more precision, reduce linkage drag and the time required for transferring genes into cultivated crop backgrounds. Whole genome sequences have been released for all crops except sweet potato, yam, finger and pearl millets. Efforts to re-sequence CWR for the remainder of crops have been at least initiated, with the exceptions of potato, banana and chickpea.

The availability of genomic resources is summarized in column 'Post-2014', Table 9.1. A score of 1 denotes a high level of availability of genomics resources (both reference genome and re-sequencing data for the CWR available). A score of 5 denotes no available resources or ongoing sequencing efforts reported. There is a sliding scale in-between.

As with previous advancements, it will take time for these resources to yield tangible breeding outputs for most crops. While this technology hosts immense potential for the future, it does not account for the extent of the increased use of CWR over the past decade. Genomic resources on their own are insufficient for predicting the breeding value of species, and utility is predicated on the availability of extensive phenotypic data for genebank collections. High-throughput genotyping approaches, namely genotyping-by-sequencing, will then be used to link re-sequenced genomes with phenotypes of interest to breeders. The insufficiency of characterization and evaluation data for CWR accessions has prompted experts from some crop communities to question the immediate benefit of genomic resources (Global Crop Diversity Trust (GCDT), unpublished consultation reports).

Strategies for mapping allelic variation using genomic data will facilitate predictive trait mining of core collections based on where accessions are collected from (see Mackay and Street, 2004; Khazaei *et al.*, 2013, for discussion on the application of the Focused Identification of Germplasm Strategy (FIGS)). There is an increasing expectation that genebank managers keep more detailed passport data on all new accessions coming in, and develop core collections that adequately represent the genetic variation available in crop genepools (R. Sackville Hamilton, 2014, personal Skype communication). Current progress being made in predictive trait mining will likely result in increased breeding outputs in the future, although numbers of varieties released do not yet reflect this work.

As strategies for increasing coordination and information sharing among genebanks, pre-breeders and breeders improve, CWR will become exponentially more usable. Online databases and portals such as the Plant Genetic Resources Diversity Gateway (pgrdiversity.bioversityinternational.org) and the DivSeek initiative (www.divseek.org) are being developed. These resources will increase access to CWR accessions, pre-breeding materials in development, associated characterization and evaluation data and genomics resources among crop scientists along the pre-breeding/breeding chain.

## 9.7 Interpreting Results

If these two indicators have predictive value, there will continue to be extensive use of CWR in wheat improvement programmes (A1) and modest use in finger millet (C5). These two extreme examples fit with expert projections of future use (H.J. Braun; H. Ojulong, 2014, personal Skype communications). Interpretation becomes less straightforward with other crops. In the case of maize (C1), for example, the availability of genomics resources will not *necessarily* incite greater use of CWR, as sufficient genetic variability for most traits has been found within the crop's wide genetic base. This fits with the story told by maize experts (S. Hearne; M. Banziger, 2014, personal Skype communication).

The predictive value of these indicators is perhaps best exemplified by the cases of lentil (A2), sorghum (B2) and yam (A4). Varieties have not yet been released with contributions from CWR for these three crops. However, experts have highlighted tremendous potential for future contributions in each case, and report a significant amount of ongoing exploration that is not otherwise reflected in the quantitative progress made since 2007.

On the other hand, very little advancement is predicted in sweet potato (C5), while C. Yencho, applied sweet potato breeder at North Carolina State University, emphasizes their potential. Wild species represent a 'potentially very fertile territory that, as we are beginning to develop new tools and methodologies, we need to look to as a source of new traits and a source of understanding of cultivated germplasm' (2014, personal Skype communication). In the case of bean (A2), the high potential highlighted in Table 9.1 is not currently reflected in crop improvement efforts (B. Raatz, 2014, personal Skype communication). The underlying message throughout this contribution remains: current use and use values are difficult to quantify, and future use and potential use values are difficult to predict.

## 9.8 Conclusion

This study supplies evidence that efforts to quantify the use of CWR based on cultivars released underestimates the breadth of ongoing use. By extension, valuation efforts undertaken to date underestimate the total use and potential use values of CWR by focusing exclusively on the economic contributions of certain key species. The story is more complex. The vast majority of ongoing work *upstream* with wild species has yet to be realized *downstream* in the form of new varieties. It will take time for pre-breeding work – characterizing and evaluating wild accessions and developing lines useful to breeders – to translate into quantifiable breeding outputs. And ongoing research into the basic biology of wild species and relationships between genes and traits of interest and genetic–environmental interactions will surely facilitate the future use of CWR, while outputs from this work can be quantified to an even lesser extent.

An analysis of expert opinions on the current use of CWR reveals a positive trend, as well as a greater emphasis on using CWR to develop varieties adapted to predicted climate change scenarios. These trends will likely continue, if not intensify, in light of new tools and methods emerging from the burgeoning field of genomics and more extreme environmental pressures threatening the viability of crop production systems. All evidence suggests that crop scientists and pre-breeders will be exploring wild genepools increasingly for traits of interest in the coming decades.

The policy implications of these findings include a growing imperative to invest in CWR conservation both *in situ* and *ex situ*, and to facilitate greater access and sharing of wild germplasm, materials under development and associated characterization and evaluation information among crop improvement programmes. The resilience of agricultural systems around the world will depend upon the availability and accessibility of genetic variation, in all of its breadth.

## References

Breithaupt, H. (2008) Up to the challenge? Rising prices for food and oil could herald a renaissance of plant science. *EMBO Reports* 9, 832–834.

Dempewolf, H., Eastwood, R.J., Guarino, L., Khoury, C.K. and Müller, J.V. (2014) Adapting agriculture to climate change: a global initiative to collect, conserve, and use crop wild relatives. *Agroecology and Sustainable Food Systems* 38, 369–377.

Ford-Lloyd, B.V., Schmidt, M., Armstrong, S.J., Barazani, O., Engels, J., *et al.* (2011) Crop wild relatives – undervalued, underutilized, and under threat? *Bioscience* 61(7), 559–565.

Hajjar, R. and Hodgkin, T. (2007) The use of wild relatives in crop improvement: a survey of developments over the last 20 years. *Euphytica* 156(1–2), 1–13.

Harlan, J.R. (1976) Genetic resources in wild relatives of crops. *Crop Science* 16, 329–333.

Hein, L. and Gatzweiler, F. (2006) The economic value of coffee (*Coffea arabica*) genetic resources. *Ecological Economics* 60(1), 176–185.

Heywood, V., Casas, A., Ford-Lloyd, B., Kell, S. and Maxted, N. (2007) Conservation and sustainable use of crop wild relatives. *Agriculture, Ecosystems & Environment* 121(3), 245–255.

Hunter, D. and Heywood, V. (2011) *Crop Wild Relatives: A Manual for In Situ Conservation.* Earthscan, London.

Khazaei, H., Street, K., Bari, A., Mackay, M. and Stoddard, F.L. (2013) The FIGS (focused identification of germplasm strategy) approach identifies traits related to drought adaptation in *Vicia faba* genetic resources. *Public Library of Science One* 8(5), e63107.

Koebner, R. and Ortiz, R. (2013) Fishing in the gene pool – how useful was the catch? *Plant Genetic Resources* 11(3), 283–287.

McCouch, S., Baute, G.J., Bradeen, J., Bramel, P., Bretting, P.K., *et al.* (2013) Agriculture: feeding the future. *Nature* 499(7456), 23–24.

Mackay M.C. and Street, K. (2004) Focused identification of germplasm strategy – FIGS. In: Black, C.K., Panozzo, J.F. and Rebetzke, G.J. (eds) *Cereals 2004. Proceedings of the 54th Australian Cereal Chemistry Conference and the 11th Wheat Breeders' Assembly,* 21–24 September 2004, Canberra, Australian Capital Territory. Royal Australian Chemical Institute, Melbourne, Australia, pp. 138–141.

Maxted, N. and Kell, S. (2009) *Establishment of a Global Network for the In Situ Conservation of Crop Wild Relatives: Status and Needs*. FAO Commission on Genetic Resources for Food and Agriculture, Rome.
Maxted, N., Ford-Lloyd, B.V. and Hawkes, J.G. (eds) (1997a) *Plant Genetic Conservation: The In Situ Approach*. Chapman and Hall, London.
Maxted, N., Ford-Lloyd, B.V. and Hawkes, J.G. (1997b) Complementary conservation strategies. In: Maxted, N., Ford-Lloyd, B.V. and Hawkes, J.G. (eds) *Plant Genetic Conservation: The In Situ Approach.* Chapman and Hall, London, pp. 20–55.
Pimentel, D., Wilson, C., Mccullum, C., Huang, R. and Dwen, P. (1997) Economic and environment benefits of biodiversity. *BioScience* 47(11), 747–757.
Prescott-Allen, R. and Prescott-Allen, C. (1981) In situ *Conservation of Crop Genetic Resources: A Report to the International Board for Plant Genetic Resources*. IBPGR, Rome.
Prescott-Allen, R. and Prescott-Allen, C. (1986) *The First Resource: Wild Species in the North American Economy*. Yale University Press, New Haven, Connecticut.
PricewaterhouseCoopers LLD (2013) Crop wild relatives: A valuable resource for crop development. Available at: pwc.blogs.com/files/pwc-seed-bank-analysis-for-msb-0713.pdf (accessed 5 July 2015).
Tanksley, S.D. and McCouch, S.R. (1997) Seed banks and molecular maps: unlocking genetic potential from the wild. *Science* 277(5329), 1063–1066.

# 10 Joining Up the Dots: A Systematic Perspective of Crop Wild Relative Conservation and Use

N. Maxted,[1*] A. Amri,[2] N.P. Castañeda-Álvarez,[1,3] S. Dias,[4] M.E. Dulloo,[4] H. Fielder,[1] B.V. Ford-Lloyd,[1] J.M. Iriondo,[5] J. Magos Brehm,[1,6] L.-B. Nilsen,[7] I. Thormann,[4] H. Vincent[1] and S.P. Kell[1]
[1]*School of Biosciences, University of Birmingham, Birmingham, UK;* [2]*International Center for Agricultural Research in the Dry Areas (ICARDA), Rabat, Morocco;* [3]*International Center for Tropical Agriculture (CIAT), Cali, Colombia;* [4]*Bioversity International, Rome, Italy;* [5]*Universidad Rey Juan Carlos, Madrid, Spain;* [6]*Centro de Biologia Ambiental, Faculdade de Ciências da Universidade de Lisboa, Lisbon, Portugal;* [7]*Swiss Federal Institute of Technology Zurich (ETHZ), Center for Development and Cooperation (NADEL), Zurich, Switzerland*

> … it appears strange to me that so many of our cultivated plants should still be unknown or only doubtfully known in the wild state.
>
> (Darwin, 1868)

## 10.1 Introduction

The growing human population, the increasing demand for food, scarcer non-renewable resources for agricultural production and the likely adverse impacts of climate change on food production present a potential 'perfect storm' for the global community in the 21st century. The human population size is currently 7.3 billion and is predicted to rise to 9.5 billion in 2050 (86% living in developing countries, primarily Africa, where malnourishment is already most severe) (United Nations, 2015). Although some models do show population increase beginning to slow, it can be argued that the planet is already beyond its human carrying capacity, as evidenced by the widespread over-exploitation of our natural resources and unsustainable environmental management practices. To adequately meet the food demand required for the global population in 2050, it is likely that food supplies will need to increase globally by 60% and in developing countries by 100% above the levels of 2005 (FAO, 2011a). Furthermore, climate change is predicted to reduce agricultural production by 2% per decade, resulting in up to 40% yield reduction by 2050 (Porter *et al.*, 2014). Not taking preventive action in the face of such challenges inevitably will result in increased food insecurity and human malnourishment, which in turn is likely to give rise to social instability and unrest globally, as is already evidenced in some regions of the world (Kelley *et al.*, 2015).

*Corrresponding author; e-mail: nigel.maxted@dial.pipex.com

 

Ensuring the availability of food for the coming decades will require a suite of carefully targeted actions that sustain and enhance food production and delivery. One of these is the more effective utilization of the largely untapped wealth of genetic diversity found in crop wild relative (CWR) species. Plant breeders increasingly are looking for traits beyond the traditional pool of diversity towards the use of more 'exotic' germplasm (Tanksley and McCouch, 1997; Vollbrecht and Sigmon, 2005; McCouch *et al.*, 2013). The importance of CWR for crop improvement was further emphasized by Feuillet *et al.* (2008), who questioned the ability of breeders to increase or simply sustain crop yield and quality in the face of dynamic biotic and abiotic threats without more systematic exploitation of the diversity contained in CWR populations. CWR contain greater breadth of diversity than crops, because they have not been subject to the genetic bottleneck of domestication or human directed selection. They are a vital source of novel genetic diversity, particularly that required for climate change mitigation (for examples see Table 10.1), and new technology for incorporating traits from 'exotic' germplasm into commercial cultivars has made their use for crop improvement easier (Maxted and Kell, 2009).

Historically, many plant breeders have been cautious of exploiting CWR diversity because of difficulties associated with cross incompatibility, hybrid sterility and linkage drag (Stebbins, 1958; Zeven *et al.*, 1983). Recent technological advances have improved the ease of transfer of traits between distantly related species and increased the value of CWR by expanding their usefulness into the secondary and tertiary crop genepools (Meilleur and Hodgkin, 2004; Jones *et al.*, 2014). Breeders have often relied on searching for specific beneficial traits associated with particular CWR taxa rather than searching more generally for beneficial genes, and have avoided transfer into polyploid crops where transfer was more difficult (e.g. rice, sorghum and sweet potato). They had also seldom in the past required trait diversity from outside of the crop itself, so their crop-based working collections had the diversity they needed in-house. However, this is no longer the case given current and future predicted environmental instability and the need for a substantive increase in crop production. Therefore, breeders are now turning increasingly to CWR as the technical issues over their use are becoming resolved and the need for 'exotic' germplasm grows (Tanksley and McCouch, 1997; Dwivedi *et al.*, 2008). The broad conclusion of these authors is that the vast majority of CWR diversity remains as yet untapped for utilization for both major and minor crops. Some may argue that the use of closely related species in breeding has been outmoded by the rapid development of biotechnological techniques, and that in the future any traits from any donor organism will be available to the genetic engineer. It is difficult to argue against this theoretical proposition. However, in practice, this may not be entirely the case in the foreseeable future because:

- Although GM technology is now well advanced for gene transfers across wide genepools, it is unlikely to be adopted universally because the perception of the public in many parts of the world does not support its application.
- The majority of the world's breeders, particularly those working in developing countries and/or on minor crops, are unlikely to have access to GM technology, at least in the short to medium term.
- GM engineering tends to focus on qualitative traits regulated by single loci, and many traits associated with biotic and abiotic factors are regulated by the interaction of multiple genes located in different regions of the genome.
- Adaptive traits are more likely to be located in taxa related to crops than in unrelated taxa – this is why the original crops were domesticated from these genepools, as they were 'predisposed' for domestication.
- Breeders cannot predict which traits they will need in the long-term or in which taxa these traits will be found.

Therefore, pragmatically, it seems likely that for the foreseeable future gene introductions will continue to be easiest, most effective and more accepted from the wild species that are close relatives of crops, or are even direct ancestors of them, and which are known to provide a significant contribution to our provisioning ecosystem services. The fact that CWR (like other wild species) are threatened by rapid changes in land use (Brummitt *et al.*, 2015a) and climate change

Table 10.1. Examples of the transfer of climate change mitigation traits from CWR to crops. (From Maxted and Kell, 2009.)

| Crop | CWR | Application(s) | Reference(s) |
|---|---|---|---|
| Chickpea (*Cicer arientinum*) | *C. reticulatum* | Drought and temperature tolerance | Hajjar and Hodgkin (2007) |
| Sunflower (*Helianthus annuus*) | *H. paradoxus* | Salt tolerance | Lexer *et al.* (2004) |
| Barley (*Hordeum vulgare*) | *H. spontaneum* | Drought and temperature tolerance | Hadjichristodoulou (1993), Eglinton *et al.* (2001), Baum *et al.* (2003), Forster *et al.* (2004), Talame *et al.* (2004), Hajjar and Hodgkin (2007), Chen *et al.* (2008) |
| Tomato (*Lycopersicon esculentum*) | *L. cheesmanii*, *L. pennellii*, *L. peruvianum* | Various forms of resistance, drought and salinity tolerance, soluble solids, insect resistance | Stevens and Rick (1986), Rick and Chetelat (1995) |
| Cassava (*Manihot esculenta*) | *M. angustiloba*, *M. davisiae*, *M. rubricaulis* | Drought tolerance, adaptation to high altitudes and cool temperatures | Jennings (1995) |
| Banana and plantain (*Musa acuminata*, *Musa balbisiana*) | *M. balbisiana*, *M. itinerans*, *M. nagensium* | Drought resistance, waterlogging resistance | INIBAP/IPGRI (2006) |
| Rice (*Oryza sativa*) | *O. longistaminata* | Drought resistance and yield increase | Brar (2005) |
| | *O. rufipogon* | Drought resistance | Zhang *et al.* (2006) |
| Wheat (*Triticum aestivum*) | *Aegilops cylindrica*, *Thinopyrum junceum* | Salt tolerance | Farooq *et al.* (1992, 1995), Wang *et al.* (2003) |
| | *Aegilops* spp., *Thinopyrum* spp., *Elytrigia elongata*, *Hordeum* spp. | Salt tolerance | Colmer *et al.* (2006) |
| | *Ae. speltoides*, *T. turgidum* | Salt tolerance | Noori (2005) |
| | *Ae. tauschii*, *T. turgidum* | Waterlogging tolerance | Villareal *et al.* (2001) |
| | *Agropyron elongatum* | Drought tolerance | Goodman *et al.* (1987) |
| | *Thinopyrum bessarabicum* | Salt resistance | King *et al.* (1997a,b) |

(Jarvis *et al.*, 2008; Warren *et al.*, 2013) means we need to ensure that they are conserved effectively and are more widely available for use. In fact, it can be argued that only through recognition of their value as gene donors for crops and their wider use by breeders can long-term genetic conservation of CWR be ensured.

It has been estimated that the contribution of CWR to improving food production has an annual value of US$115 billion worldwide (Pimentel *et al.*, 1997), and in the individual case of producing sweeter tomatoes for the US market, a single gene from *Solanum chmielewskii* increased sales by US$5–8 million/year (Iltis, 1988). The estimated current annual value of the use of CWR of major crops is US$120 billion (PwC, 2013). Given the significant current and potential value of CWR, it is surprising that many are threatened with extinction or significant genetic erosion – a situation that is being exacerbated by the negative impacts of human-mediated change on wild populations. The Sampled Red List Index for Plants project (Brummitt *et al.*, 2015b) has estimated that 20% of all plants are currently threatened with extinction. However, this statistic does not take into account the potential impact of climate change, which is also predicted to have a further and substantial impact on wild plant populations. For example, Jarvis *et al.* (2008) used distribution modelling and three migrational scenarios to evaluate changes in the potential range size and fragmentation of climatically suitable areas for groundnut (*Arachis* spp.), potato (*Solanum* spp.) and cowpea (*Vigna* spp.) genepools. For these genepools, they found that: (i) 16–22% of these CWR species would be extinct by 2055; (ii) the majority of species were expected to lose greater than 50% distributional range; and (iii) the populations that will remain will be highly fragmented, placing the species under greater threat of genetic erosion or extinction. The authors found that the impact of climate change varied depending on the crop genepool. They showed that while *Vigna* spp. were likely to be least impacted, with 0–2 (out of 48) species predicted to be extinct and 8–41 predicted to lose more than 50% of their current geographic range, 24–31 (out of 50) species of wild *Arachis* were predicted to be extinct, with 85–94% range decreases for the remaining species. Similar modelling of climate change scenarios in Mexico by Lira *et al.* (2009) highlighted that most of the eight wild cucurbit taxa studied were predicted not to survive under accepted climate change models. In past periods of climatic change, plant species had the ability to migrate to areas where the climate was more moderate, but today the global anthropogenic influence over ecosystem change is occurring faster than potential migration rates and there are fewer free niches for migrating species to occupy, and so the choice is either to adapt or to become extinct.

The most commonly applied means of assessing the threatened status of wild taxa is the application of the *IUCN Red List Categories and Criteria* (IUCN, 2001), a standardized system for the classification of taxa according to their extinction risk (or threatened status). In the most comprehensive assessment of the threatened status of CWR to date, 572 European species in 25 economically important crop genepools/groups were assessed (Bilz *et al.*, 2011; Kell *et al.*, 2012). The results of this study show that at least 11.5% (66) of the species are threatened, with 3.3% (19) of them being critically endangered (CR), 4.4% (22) endangered (EN) and 3.8% (25) vulnerable (VU). A further 4.5% (26) of the species are classified as near threatened, and one species (*Allium jubatum* J.F. Macbr.) is regionally extinct. The remaining species are regionally assessed as data deficient (DD) (29%) or least concern (LC) (54.7%). However, of the species assessed as LC, around one-third are threatened at national level (Kell *et al.*, 2012). The same authors report that of the 25 crop genepools/groups for which the European CWR have been assessed, at least 14 contain regionally and/or globally threatened (CR, EN or VU) or near threatened (NT) species (92 species in total, of which 65 are endemic to Europe), the highest number occurring in the *Brassica* complex, which in total contains 137 species native to, and with a significant proportion of the global population in, Europe. At least 8–50% of the species assessed in each of these crop genepools/groups are threatened or NT, and these percentages are likely to increase when the DD species are re-evaluated.

Kell *et al.* (2012) also analysed the factors threatening CWR diversity and reported 31 distinct threats, the most frequent being 'livestock farming and ranching', 'tourism and recreation areas' and 'housing and urban areas'. However, the authors note that we should not conclude that farming *per se* is threatening CWR diversity – in fact, farmed areas (including arable land and pasture) are one of the primary habitats of CWR species. Rather, it is unsustainable farming practices, such as severe overgrazing, conversion of land to monocultures and the heavy application of fertilizers and herbicides, that are the major threats to CWR that grow in agricultural areas (Kell *et al.*, 2012). Continued overgrazing, expansion of industrialized agriculture and the adverse effects of climate change might not allow the survival of populations found in natural and semi-natural habitats, even for species adapted to the harshest environments.

Further, as pointed out by Maxted *et al.* (1997a), the loss of genetic diversity must always be greater than the loss of species diversity because there will be genetic erosion from the species that remain extant, so it should be recognized that species distribution modelling significantly underestimates the loss of genetic diversity from CWR taxa. A potential further specific threat to CWR diversity is that many CWR of major crops are found in disturbed, pre-climax communities (Jain, 1975), which themselves are the habitats that are most likely to be subject to increasing levels of anthropogenic change and destruction as a result of global change. This was corroborated by a recent study of CWR distribution in the UK (Jarvis *et al.*, 2015), who found that CWR were most abundant in grassland habitats, although CWR related to food crops were most common in cropped and weedy areas, fertile grassland and lowland woodland. They found that CWR were associated significantly with linear features (e.g. hedgerows, roadsides, field boundaries and field margins) and disturbed habitats and landscape features that are not normally a conservation priority under site-based conservation measures, emphasizing the need for a reappraisal of *in situ* wild plant species conservation in the CWR context.

Increasing recognition of the value of and threats to CWR diversity, together with the fact that the natural distributions of CWR seldom follow political or administrative borders, have resulted in their recognition as a special case for conservation in a number of global policy instruments. However, currently, there are no mechanisms in place to organize technically coordinated, effective and efficient *in situ* conservation actions globally for CWR, and interest in their conservation has only recently been addressed by policy makers. There is some embryonic but growing interest in the conservation of CWR by the nature conservation community, largely as an exemplar of ecosystem services, demonstrating the potential economic value of protected area conservation to funders (Stolton *et al.*, 2008; Hopkins and Maxted, 2010; Hunter *et al.*, 2012; IUCN, 2014).

At the global level, the growing perceived value of CWR and the requirement for more effective CWR conservation is recognized in a number of policy instruments. The Second Global Plan of Action for Plant Genetic Resources for Food and Agriculture (Second GPA) (FAO, 2011b) puts great emphasis on CWR in several of its priority activities, and Article 5 of the International Treaty on Plant Genetic Resources for Food and Agriculture (FAO, 2001) also promotes *in situ* conservation of CWR, including in protected areas. The Conference of the Parties (COP) to the Convention on Biological Diversity (CBD) underlined the importance of CWR in Aichi Target 13 of the CBD Strategic Plan (CBD, 2010a) which states that 'by 2020 the status of crop and livestock genetic diversity in agricultural ecosystems and of wild relatives has been improved' and that '*in situ* conservation of wild relatives of crop plants could be improved inside and outside protected areas', as well as in the CBD Global Strategy for Plant Conservation 2011–2020 (CBD, 2010b), in which Target 9 is '70 per cent of the genetic diversity of crops including their wild relatives and other socio-economically valuable plant species conserved'. These explicit goals can be achieved through the identification of gaps in current *in situ* and *ex situ* CWR conservation, combined with improved recognition of CWR in national and regional policy, ultimately leading to systematic conservation and use of CWR diversity.

## 10.2 The State of the Art

The aim of genetic conservation of CWR is to maximize the maintenance of genetic diversity, but it is explicitly utilitarian in that it makes the link between CWR genetic diversity, its conservation and its utilization (Fig. 10.1). The model for plant genetic resources conservation proposed by Maxted *et al.* (1997b) divides the process into three elements: conservation planning, conservation and utilization.

### 10.2.1 CWR conservation planning

In recent years, significant progress has been made in CWR conservation planning. Progress has been made not only in the development of techniques to assist the prioritization of CWR taxa and in diversity and gap analysis but also in the development of systematic conservation planning itself, through the development of CWR conservation strategies commonly focused on the conservation of CWR diversity at the national level.

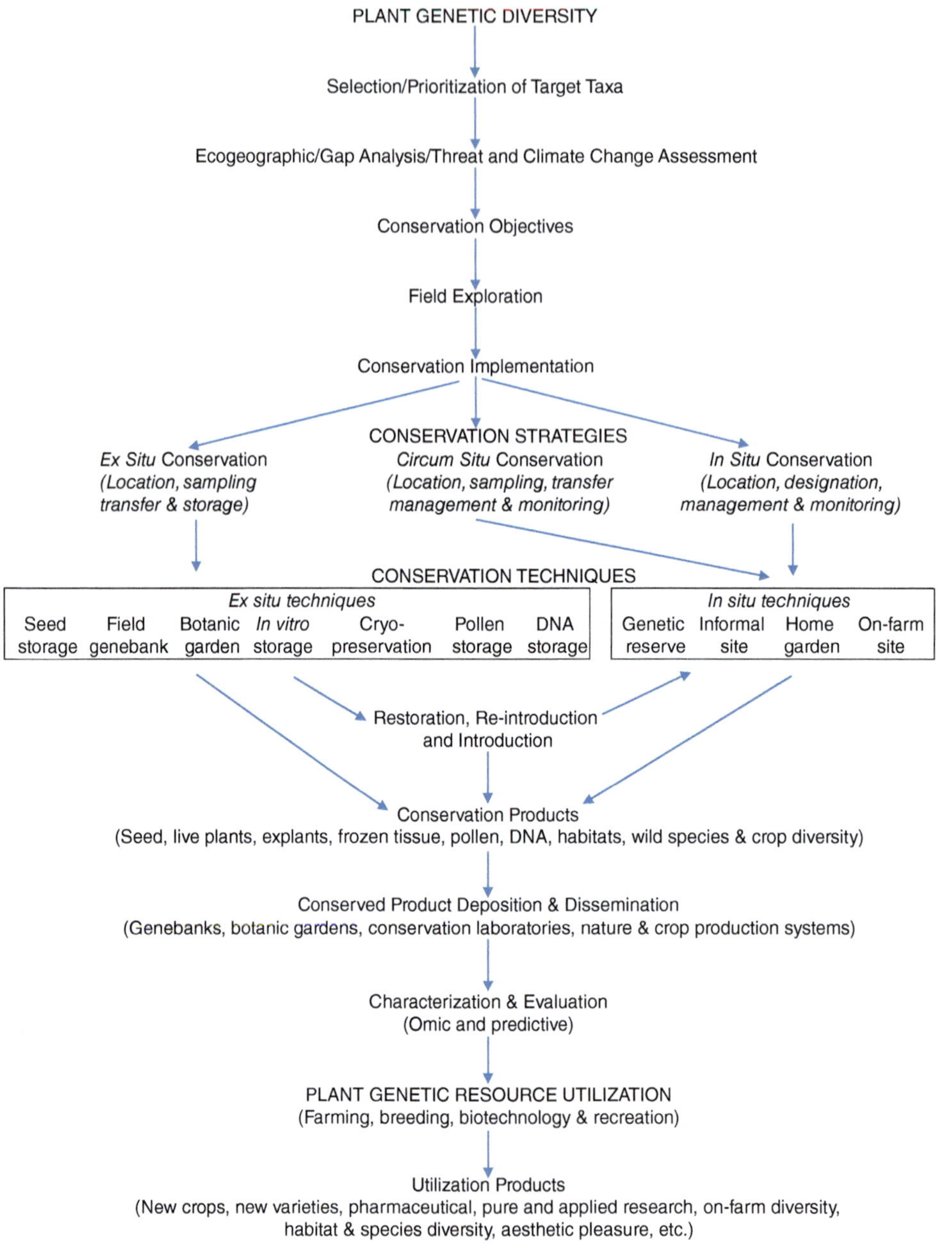


**Fig. 10.1.** Model for plant genetic conservation. (From Maxted *et al.*, 1997b.)

### *Prioritization of target taxa*

Conservation of CWR is always likely to be limited by the financial, temporal and technical resources available, and with an estimate of 50–60,000 CWR existing globally (Maxted and Kell, 2009), we are forced to set priorities and to select those taxa on which to focus scarce conservation resources. Each CWR species has a perceived value, and the establishment of its value is aided by considering some or all of the following criteria (Maxted *et al.*, 1997a):

- current conservation status;
- socio-economic use;
- perceived threat;
- taxonomic and genetic distinction;
- ecogeographic distinction;
- biological importance;
- cultural importance;
- relative cost of conservation;
- conservation sustainability;
- legislation;
- ethical and aesthetic considerations; and
- conservation agency priorities.

All CWR have specific socio-economic use value associated with their genetic potential for contributing traits to crops, and therefore contribute to provisioning ecosystem services. This is alongside their ecosystem service value as wild plant species *per se*, for example, providing a feeding source for insects that pollinate crops or, less closely associated with agriculture, their value in chemically neutralizing air pollutants (McNeely *et al.*, 2009).

Within the context of prioritizing CWR species, a combination of the conservation criteria listed above have been used (Maxted *et al.*, 2007; Magos Brehm *et al.*, 2010; Idohou *et al.*, 2012; Kell *et al.*, 2012; Khoury *et al.*, 2013; Phillips *et al.*, 2014) – however, the three important criteria for CWR are (Kell *et al.*, 2015):

- The socio-economic value of the related crop (Ford-Lloyd *et al.*, 2008), which can be based on economic and food supply statistics (e.g. using FAOSTAT or regional or national databases).
- Potential for utilization, which can be assessed in two ways (Maxted and Kell, 2009; Maxted *et al.*, 2012a): (i) by defining the relative genetic relationship of the wild relative to the crop using the Gene Pool and Taxon Group concepts (Maxted *et al.*, 2006) to provide an indication of the relative ease with which crosses may be made with the crop; and (ii) by documenting the known actual or potential use of a CWR for crop improvement (i.e. highlighting taxa that have already been used for crop improvement or that contain traits of potential use for crop improvement). Details on the potential utilization value of CWR that are related to 173 food crops have been reported by Vincent *et al.* (2013) and can be found at www.cwrdiversity.org/checklist/ and via USDA GRIN at www.ars-grin.gov/~sbmljw/cgi-bin/taxcwr.pl?language=en.
- Relative threat status (Ford-Lloyd *et al.*, 2008; Maxted and Kell, 2009), which is commonly assessed using the *IUCN Red List Categories and Criteria* (IUCN, 2001) or based on the known geographic distribution range of the taxon and/or threats to the habitats in which it occurs if a Red List assessment of a species has not been undertaken or published.

In general, native CWR populations are given priority over those that are outside their native range, because it is the responsibility of nations to conserve the biodiversity within their jurisdiction and there is an inherent drive for countries to preserve their native resources. The occurrence status of populations is often unclear, but those that are considered native are usually archaeophytes (populations of taxa introduced before 1500 AD), while neophytes (introduced after 1500 AD) or diaphytes (populations of taxa repeatedly introduced and only established as a non-permanent casual (Kornas, 1990)) are considered as non-native. However, in terms of CWR conservation, it can be argued that the occurrence status of populations is irrelevant because they may contain genetic diversity that is valuable for crop improvement, especially because many taxa are able to adapt rapidly to new environments (Ford-Lloyd *et al.*, 2014).

### Ecogeographic/gap analysis

As noted by Maxted *et al.* (1995), conservationists do not throw dice when selecting where to collect or conserve germplasm – conservation is based on existing evidence and CWR are conserved *in situ* or *ex situ* from locations where they are found. The means of locating where to conserve CWR diversity involves a combination of ecogeographic and gap analyses. Ecogeographic analysis is defined by Castañeda Álvarez *et al.* (2011) as the process of gathering and synthesizing information on ecological, geographical, taxonomic and genetic diversity. The results are predictive and can be used to assist in the formulation of complementary *in situ* and *ex situ* conservation priorities. Routinely, this generalized collation of background information on a species is accompanied by the application of an explicit gap analysis methodology (Maxted *et al.*, 2012b). This involves a comparison of the CWR diversity that exists in nature with the sample that is actively conserved, the identification of key elements of diversity that are under-conserved, and therefore the location of 'gaps' in the current conservation strategy. This process may be broken down into the following steps: (i) circumscription of target taxon and target area; (ii) assessment of natural diversity through a review of intrinsic taxonomic, genetic and ecogeographic diversity combined with threat assessment; (iii) assessment of current complementarity between *in situ* and *ex situ* conservation strategies; and (iv) reformulation of the conservation strategy through analysis of the differences between the pattern of natural, intrinsic diversity and the elements of that diversity already effectively represented by existing *in situ* and *ex situ* conservation actions.

There are now numerous examples of how ecogeographic/gap analysis techniques have been applied for diverse groups of CWR globally (e.g. African *Vigna* species – Maxted *et al.*, 2004; *Aegilops* species – Maxted *et al.*, 2008a; 12 globally important food crop groups – Maxted and Kell, 2009; *Phaseolus* species – Ramírez-Villegas *et al.*, 2010; and temperate forage and pulse legume species – Maxted *et al.*, 2012c). The most extensive use of ecogeographic/gap analysis techniques is currently being applied by the Global Crop Diversity Trust (GCDT) initiative 'Adapting Agriculture to Climate Change: Collecting, Protecting, and Preparing Crop Wild Relatives' (Dempewolf *et al.*, 2014). Using published Gene Pool and Taxon Group concepts, the project has developed a CWR inventory for 173 crops, containing 1392 priority species (Vincent *et al.*, 2013). Ecogeographic data have been collated for these species and *ex situ* and *in situ* conservation priorities identified using the Maxted *et al.* (2008b) and Ramírez-Villegas *et al.* (2010) methodology. The key results of the two analyses of *ex situ* (Castañeda-Álvarez *et al.*, 2016) and *in situ* (Vincent *et al.*, 2016) conservation priorities are shown in Figs 10.2 and 10.3, respectively, and discussed in detail in Vincent *et al.* (2016) and Chapter 12, this volume, and in Castañeda-Álvarez *et al.* (2016) and Chapter 13, this volume.

### *National CWR strategies*

Meilleur and Hodgkin (2004) observed, in the context of *in situ* CWR conservation, that there were 'weak links between the "site-selection and/or management-recommendations" process and the "official-protected-site and/or management-change-designation" process'. This remains the case today, but in an attempt to bridge the gap between conservation planning and conservation implementation, the focus has moved from solely identifying individual *in situ* and *ex situ* conservation actions to the development of national CWR conservation strategies for the comprehensive complementary conservation of all priority CWR within a country, the country commonly being the natural administrative unit for conservation planning and implementation.

Maxted *et al.* (2007) proposed a model for how such a strategy might be developed (Fig. 10.4), and this model has now been applied in the UK (Maxted *et al.*, 2007), Venezuela (Berlingeri and Crespo, 2012), Benin (Idohou *et al.*, 2012), the USA (Khoury *et al.*, 2013), Finland (Fitzgerald, 2013), Spain (Rubio Teso *et al.*, 2013), Italy (Panella *et al.*, 2014) and Cyprus (Phillips *et al.*, 2014). Progress in national CWR conservation strategy planning has also been made in Albania, Armenia, Azerbaijan, Belarus, Bolivia, Bulgaria, Czech Republic, Greece, India, Ireland, Madagascar, Mauritius, Mexico, Norway, Peru, South Africa, Sri Lanka, Sweden, Uzbekistan and Zambia.

Applying the national CWR conservation planning methodology has two clear advantages over *ad hoc* CWR conservation planning. First, it ensures that a holistic view of national CWR conservation is taken, so that alternative actions are compared and the most effective ones employed to conserve CWR diversity systematically with the resources available. Second, it promotes participatory decision making – for example, breeders can be involved in selecting the CWR to be actively conserved; botanists can confirm the correct nomenclature for candidate CWR taxa and help in identifying sources of ecogeographic data, as well as providing information on the threats facing CWR taxa and populations; protected area managers and environmental agencies will be involved in nominating and establishing *in situ* genetic reserves; and germplasm end-users

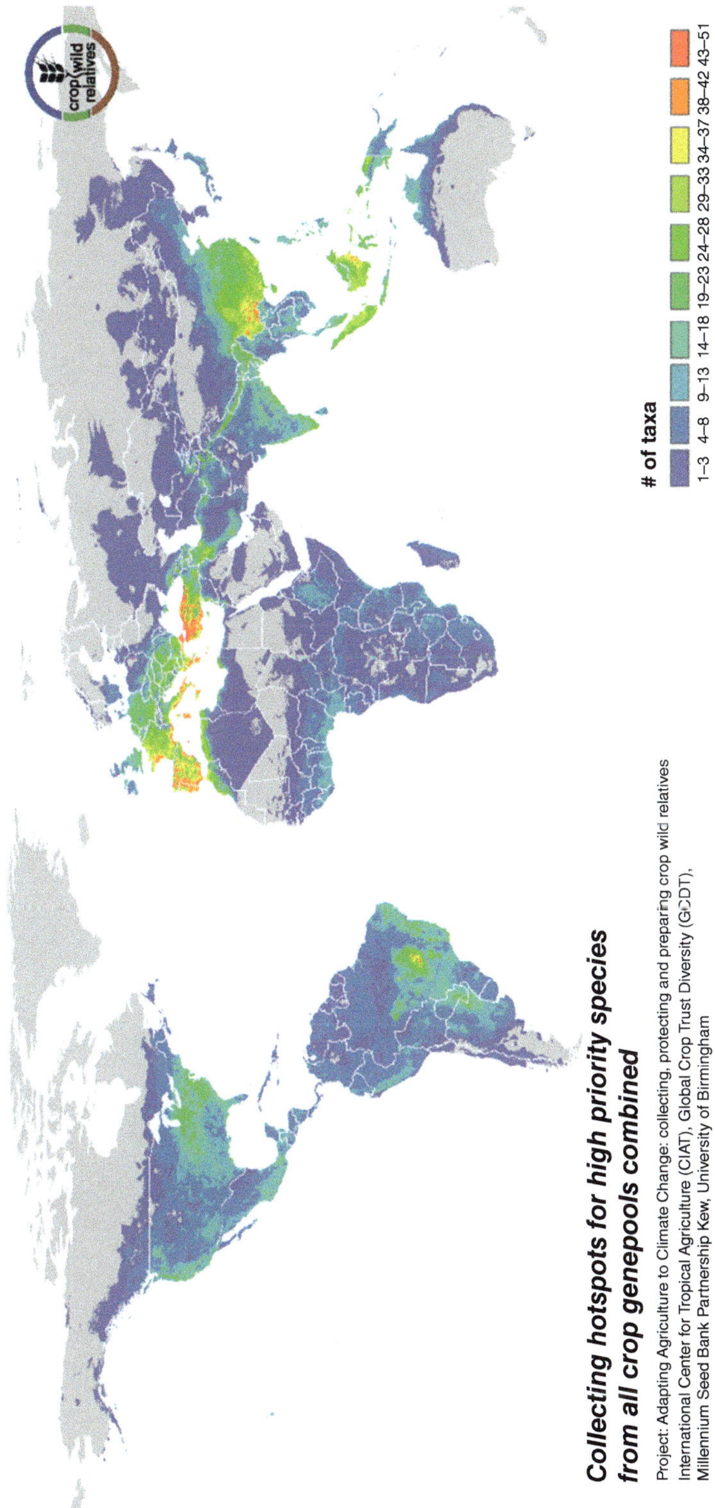


**Fig. 10.2.** Priorities for global *ex situ* CWR conservation. (From Castañeda Álvarez *et al.*, 2016a.)

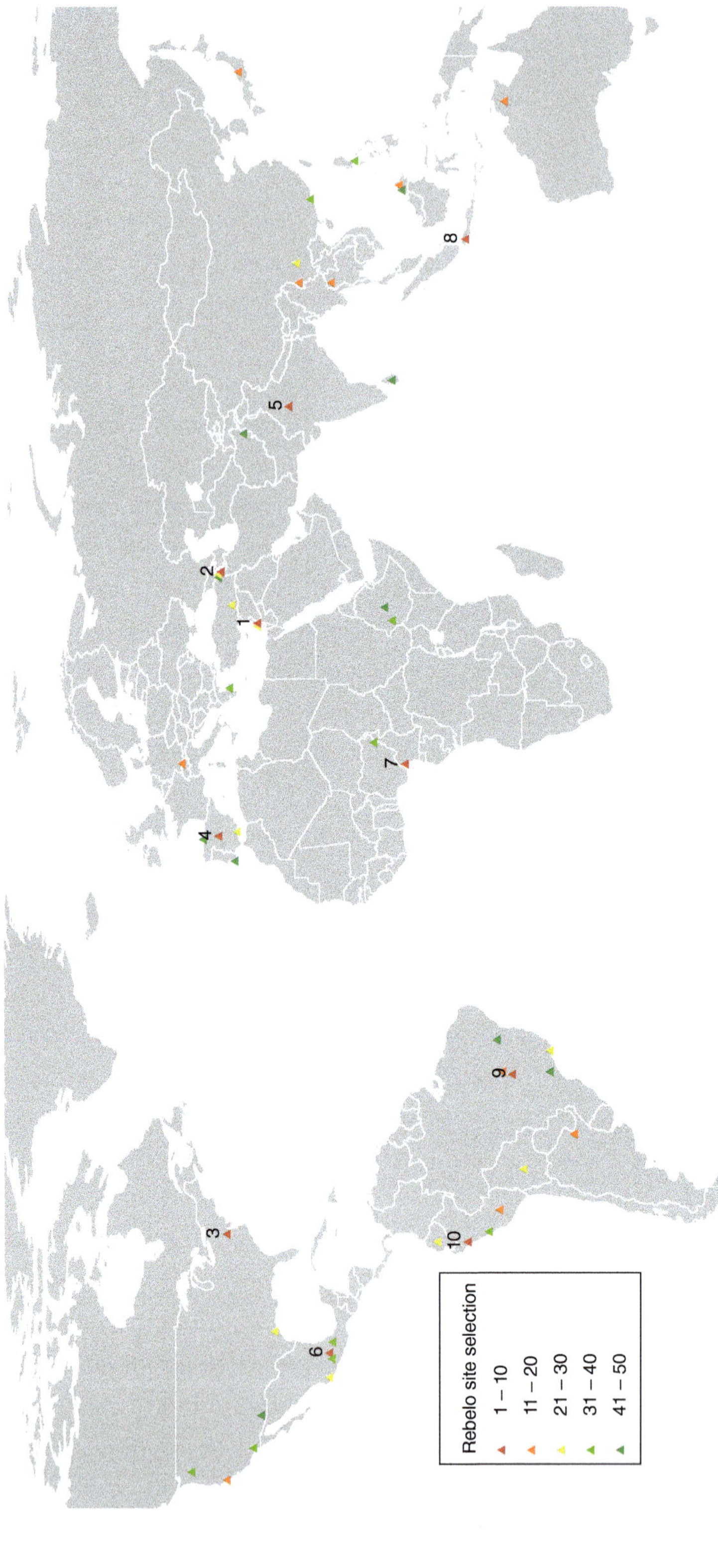


**Fig. 10.3.** Priorities for global *in situ* CWR conservation. (From Vincent *et al.*, 2016a.)

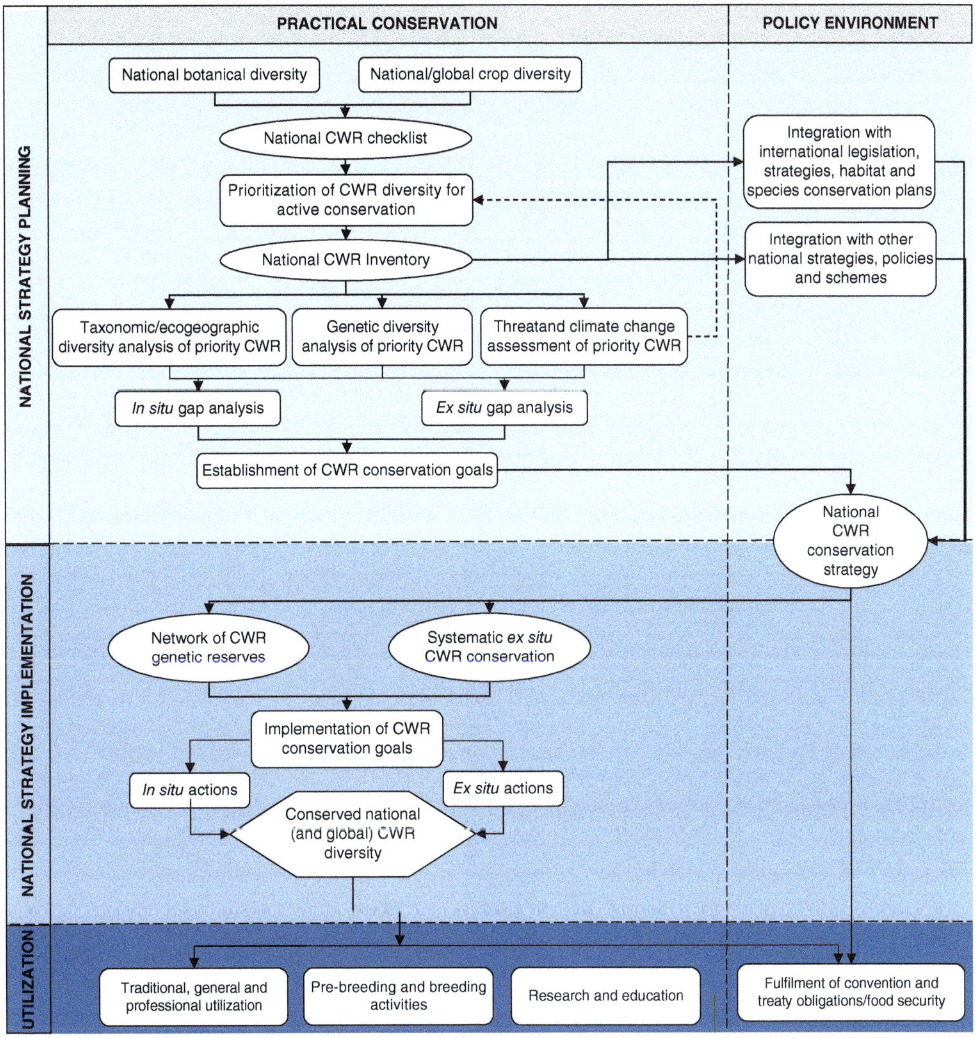


Fig. 10.4. Model for implementation of national CWR conservation strategies. (From Maxted *et al.*, 2007, 2013.)

(primarily breeders) will be involved in accessing the conserved diversity. Experience has shown that for full implementation, it is critical to obtain buy-in from the broadest possible stakeholder community – particularly protected area managers and agencies for the establishment of *in situ* genetic reserves. In a broader context, FAO (2015) has also prepared Guidelines for Developing National Strategies for Plant Genetic Resources for Food and Agriculture (PGRFA) as an aid to enhance conservation and use of plant genetic resources, including CWR. It should be stressed that national CWR conservation planning should address the conservation of the full range of genetic diversity present in the CWR taxa prioritized, that both native and historically introduced populations are included and both *in situ* and *ex situ* conservation techniques are employed to maximize the chances of conserving the various traits required by germplasm users.

### 10.2.2 Conservation implementation

#### *Ex situ conservation*

The quotation from Darwin (1868) at the start of this chapter could be brought up to date and paraphrased to indicate that it appears strange that so many of our cultivated plants' wild relatives should still be under- or only partially conserved. Given their value and threatened status, it might be expected that CWR would have been the focus of specific, systematic *in situ* and *ex situ* conservation action, but this is far from the case. For several decades, the focus has been on the *ex situ* conservation of PGRFA in genebanks. However, few *ex situ* conservation programmes have focused intentionally on the active conservation of CWR as a resource for crop improvement and future food security. An analysis of European genebank collections data stored in EURISCO (eurisco.ecpgr.org) revealed that only 4% of accessions were of wild species (Dias and Gaiji, 2005; Dias *et al.*, 2012), and while this has been estimated more recently at 7% (S. Dias, Rome, 2013, personal communication), the breadth of coverage of crop genepools is limited (Kell *et al.*, 2008, 2012). Further, Kell *et al.* (2012) found that most European priority CWR species were represented by very few *ex situ* accessions, were reported by only one genebank, and had been collected from only a small part of the species' range. At the global level, FAO (2010), based on data provided by WIEWS (World Information and Early Warning System on PGRFA; apps3.fao.org/wiews/wiews.jsp), reported that in 2009, 18% of accessions in germplasm collections were of wild species and that this was an increase of 3% since 1996. However, this increase was comparable with the increase in numbers of research/breeding materials and landrace accessions, so may simply represent an increase in the number of accessions reported overall or a general increase in the size of collections of all types of germplasm. Further, the percentage of accessions alone is not an adequate indication of the *ex situ* conservation status of CWR. Information on the number of species, crop genepool coverage and the genetic representativeness of the collections is needed in order to undertake a comprehensive assessment. A recent initiative (Dempewolf *et al.*, 2014) aims to redress the requirement for systematic *ex situ* CWR conservation by identifying global priority CWR, developing and implementing an *ex situ* conservation action plan for priority species and promoting the use of the conserved diversity in crop improvement programmes. However, the data held in EURISCO and the assessment made as part of the report on the State of the World's PGRFA refers to agriculturally based genebanks only – botanic gardens also often have genebanks of wild species, and these are likely to hold some CWR material. Maxted *et al.* (2010) reviewed European *ex situ* seed collections held in botanic gardens via the ENSCONET portal and found that CWR taxa accounted for 61.8% of total germplasm holdings and that the 5756 CWR species included represented 33% of the 17,495 CWR species found in Europe. Although initially it would appear that botanic garden genebanks are more effective at conserving CWR diversity, CWR taxa are conserved here not because of their CWR status but simply because they are wild plant taxa. The result is also skewed because the list of CWR used is broad, based on any species found in the same genus as a crop rather than those priority CWR most closely related to crops and most likely to be used (see further discussion in Kell *et al.*, 2008). It is likely that the CWR taxa held in crop genebanks are more likely to include high-priority CWR species from the primary and secondary genepools because these are more likely to have been requested by users. Further, as Heywood (1999) notes, taxa conserved in botanic garden genebanks are likely to have few accessions per taxon, to be small in numbers of seed conserved and genetically poorly sampled. Therefore, it remains likely that, globally, the Consultative Group for International Agricultural Research centres (CGIAR – www.cgiar.org) holds the most significant collections of CWR diversity of their mandate crops, but these still do not contain genetically representative samples of all their mandated CWR taxa (Maxted and Kell, 2009). This may be explained in part by breeders' historic focus on the exploitation of genetic diversity existing within crops, but it is still surprising given that most diversity in crop genepools is located in the related wild species (Tanksley and McCouch, 1997; Maxted *et al.*, 2008c).

The global initiative 'Adapting Agriculture to Climate Change: Collecting, Protecting, and Preparing Crop Wild Relatives' (Dempewolf *et al.*, 2014) has generated a global inventory of priority CWR taxa and identified regions where collecting for *ex situ* conservation should be focused. The Harlan and de Wet CWR Inventory contains baseline information on taxonomy, nomenclature, distribution, genebank storage behaviour, herbaria likely to contain specimens and relationship to crop for 173 food crop genepools (www.cwrdiversity.org/checklist/; Vincent *et al.*, 2013). This resource was created to provide a foundation for global *ex situ* CWR conservation planning, but will prove equally useful for planning *in situ* conservation, as will the database of 5 million locations of the 1667 global priority CWR taxa representing 173 food crops. These resources have been used as the basis of *ex situ* gap analyses of 81 crop genepools, and national PGRFA programmes are now working with the Millennium Seed Bank, Kew (UK), to fill the identified gaps. The collected germplasm will be stored in relevant national and international genebanks, with further safety duplication for long-term security at the Svalbard Global Seed Vault in Norway. In 2014, collecting focused on Cyprus, Georgia, Italy, Portugal and Vietnam, and will be extended to other countries in subsequent years of the project, as well as using CWR diversity for the pre-breeding of rice, sunflower and other crops.

Parallel to global *ex situ* initiatives, there is a need for complementary regional and national CWR *ex situ* conservation. At the regional level, no such systematic sampling and *ex situ* conservation of germplasm is in place. Further, although national CWR conservation strategies have been developed for several countries, only limited progress has been made in filling the *ex situ* gaps identified.

### In situ *conservation*

Given the poor state of *ex situ* CWR conservation, which has historically been the main focus of PGRFA conservation, it is hardly surprising that the situation regarding *in situ* CWR conservation is less advanced. While most of the world's national parks and other protected areas contain CWR populations, the protected areas were established to conserve particular habitats and charismatic or threatened species – thus, CWR are only conserved passively or because they happen to be found in the same location as charismatic taxa or are characteristic of threatened habitats, and therefore individual CWR populations could decline or become extinct without site managers being aware (Maxted *et al.*, 1997b). Although sites that conserve CWR *in situ* are sometimes recognized (Table 10.2), in many cases even these sites are not managed in the most appropriate manner to conserve the genetic diversity of CWR populations. These sites would not meet the set of quality standards for CWR genetic reserves proposed by Iriondo *et al.* (2012), and their designation has been *ad hoc* and opportunistic rather than as a result of any deliberate scientific decision. Also, they are independent of each other and so do not together constitute a desired global network of genetic reserves. The FAO Commission on Genetic Resources for Food and Agriculture (CGRFA) recently held a Technical Workshop to debate the means and to explore the options for the establishment of a global network for *in situ* PGRFA conservation (FAO, 2013). Such a network would provide the necessary platform to raise awareness of the social and economic value of *in situ* conservation (and on-farm management) in partnerships with national and regional level activities, but the development of such a network requires further debate, a governance structure, an evidence research base and financial security.

It is normally assumed that *in situ* CWR conservation will take place in existing protected areas because: (i) these sites already have an associated long-term conservation ethos and are less prone to hasty management changes associated with private land or roadsides where conservation value and sustainability is not a consideration; (ii) it is relatively easy to amend the existing site management to facilitate the genetic conservation of CWR species; and (iii) it means that creating novel conservation sites can be avoided, negating the possibly prohibitive cost of acquiring previously non-conservation managed land (Maxted *et al.*, 2008d). Yet there is limited evidence of active *in situ* CWR conservation globally, and the fact that ecosystems are becoming increasingly unstable has increased the attraction of moving samples of populations to genebanks or other *ex situ* repositories where the germplasm is under firmer scientific control and where breeders can access it more easily. However, such a strategy halts evolutionary adaptation, known to occur naturally and often rapidly *in situ* (Ford-Lloyd *et al.*, 2014), and therefore *ex situ* conservation should not be the only or preferred approach used, particularly given the sheer numbers of species involved and the need to conserve multiple populations of each to maximize the genetic diversity available.

Table 10.2. Examples of CWR conserved *in situ* in protected areas.

| CWR | Protected area | Country | References |
|---|---|---|---|
| Teosinte (*Zea diploperennis*) | MAB Sierra de Manantlán Biosphere Reserve | Mexico | Sanchez-Velasquez (1991) |
| Wild emmer wheat (*Triticum turgidum* subsp. *dicoccoides*) | Ammiad, Galilee | Israel | Anikster *et al.* (1997); Safriel *et al.* (1997) |
| Wild coffee (*Coffea mauritiana*, *C. macrocarpa*, *C. myrtifolia*) | Black River Gorges National Park | Mauritius | Dulloo *et al.* (1999) |
| Wild wheats (*Triticum turgidum* subsp. *dicoccoides*, *T. monococcum*, *Ae. tauschii*, *Ae. speltoides*) | Ceylanpinar | Turkey | Karagöz (1998) |
| Chestnut (*Castanea sativa*), wild plum (*Prunus cerasifera* var. *divaricata*) | Kazdağ | Turkey | Kűçűk *et al.* (1998) |
| Wild onions (*Allium columbianum*, *A. geyeri* and *A. fibrillum*) | Great Basin, Washington State | USA | Hannan and Hellier (1999, unpublished results); Hellier (2000) |
| Wild grapevine (*Vitis rupestris*, *V. shuttleworthii*, *V. monticola*) | Witchita Mountains and Ouachita National Forest, Oklahoma, Clifty Creek, Missouri | USA | Pavek *et al.* (2003) |
| Wild bean populations (*Phaseolus* spp.) | Central Valley | Costa Rica | Baudoin *et al.* (2008) |
| *Medicago* spp., *Vicia* spp., *Trifolium* spp., *Lathyrus* spp., *Lens* spp., *Triticum* spp., *Avena* spp., *Hordeum* spp., *Aegilops* spp., *Allium* spp., *Amygdalus* spp., *Prunus* spp., *Pyrus* spp., *Pistacia* spp. and *Olea* spp. | Abu Taha, Sale-Rsheida, Ajloun, Wadi Sair | Lebanon, Syria, Jordan, Palestinian Territories | Al-Atawneh *et al.* (2008) |
| Wild wheats (*Triticum boeoticum*, *T. urartu*, *T. araraticum*) | Erebuni | Armenia | Avagyan (2008) |
| Wild beet (*Beta patula*) | Desertas Islands | Portugal | Pinheiro de Carvalho *et al.* (2012) |
| Wild *Solanum* spp. | Laguna de los Pozuelos Natural Monument and Los Cardones National Park | Argentina | Marfil *et al.* (2015) |

Therefore, a complementary approach involving active *in situ* and *ex situ* conservation is required. This is emphasized in the Second Report on the State of the World's Plant Genetic Resources for Food and Agriculture (FAO, 2010), which recognizes the global importance of CWR diversity and acknowledges the fact that it remains seriously under-conserved, both *in situ* and *ex situ*.

### 10.2.3 CWR use

The Russian botanist, Nikolai Ivanovich Vavilov, recognized and championed the potential of CWR for crop improvement in the 1920s and 1930s; for example, referring to the use of wild *Aegilops*, *Secale*, *Haynaldia* and *Agropyron* species in wheat breeding (Vavilov, 1949). Subsequently, CWR were routinely used by agricultural scientists to improve major crops in the 1940s and 1950s, and by the 1960s and 1970s, this practice had resulted in some significant breeding improvements (Borlaug, 1983; Meilleur and Hodgkin, 2004). There has been increasing use of CWR in breeding programmes in recent years, and it seems likely that this has arisen from increased awareness and recognition that (Hajjar and Hodgkin, 2007; Maxted and Kell, 2009):

**1.** If crops are to maintain or increase production levels, there is a need for new trait diversity outside that which historically has been used by farmers and plant breeders – CWR offer the necessary novel genetic diversity that can enhance crop productivity or commodity improvement, promote disease and pest resistance and increase tolerance of adverse or marginal environments.
**2.** Globally, agriculture is being practised in more adverse or marginal environments, whether due to human degradation of habitats, climate change, or the demand for food forcing the expansion of agricultural lands – the desired traits necessary to grow crops in these environments may be found in CWR species.
**3.** There is a continuous and growing demand for novel diversity by breeders to be used in the development of new varieties, due to the relatively short-term commercial lifespan of modern cultivars (usually 5–10 years).
**4.** Conventional and biotechnological breeding techniques have advanced dramatically in recent years, enabling more precise identification of desirable traits, relatively easy transfer to the crop and less problems with the transfer of unwanted characteristics. Additionally, these techniques are becoming cheaper, and thus increasingly accessible to researchers globally.

The widest survey of CWR use was undertaken by Maxted and Kell (2009), who looked at the recorded use of CWR in crop improvement for 29 major crops and found 234 references reporting the identification of useful traits in 183 CWR taxa. It should be noted that the primary data sources were published journal papers, and we do not know how closely this correlates with the actual use of CWR in commercial crop breeding (a useful trait might be present in a CWR that could be transferred to a crop, but this does not mean that a novel variety was the result). It may be difficult to give a precise estimate of CWR use by breeders, because the information is likely to be commercially sensitive and therefore not readily available. Bearing this in mind, the following points were made (Maxted and Kell, 2009):

- The degree to which breeders use CWR species varies between crops. CWR use is particularly prominent in barley, cassava, potato, rice, tomato and wheat. Rice and wheat are the crops in which CWR have been most widely used, both in terms of number of CWR taxa used and in terms of successful attempts to introgress traits from the CWR to the crop that resulted in significant gains in production and productivity of these crops.
- The exploitation of the potential diversity contained in CWR species remains hit and miss, as the approach by breeders to CWR use has not been systematic or comprehensive; therefore, the vast majority of CWR diversity remains untapped for utilization. In fact, there are very few pre-breeding programmes around the world focusing on transferring genes from CWR to crops.
- The number of publications of papers detailing the use of CWR in breeding has increased gradually over time, presumably as a result of technological developments for trait transfer,

with 2% of citations recorded prior to 1970, 13% in the 1970s, 15% in the 1980s, 32% in the 1990s, and 38% after 1999.

- The most widespread CWR use has been and remains in the development of disease and pest resistance, with the references citing disease resistance objectives accounting for 39%, pest and disease resistance 17%, abiotic stress 13%, yield increase 10%, cytoplasmic male sterility and fertility restorers 4%, quality improvers 11% and husbandry improvement 6% of the reported interspecific trait transfers.
- Since 2000, the number of attempts to improve quality, husbandry and end-product commodities using CWR diversity has increased substantially.

It should also be remembered that in addition to their potential use in breeding, CWR could also be crucial in the rehabilitation/restoration of degraded ecosystems, either by using accessions stored in genebanks or by introducing populations based on homoclinal environment similarities.

### *Omics approaches*

Advances in molecular biology have opened up new possibilities for identifying genes and genotypes of interest for crop improvement, but progress has been somewhat limited for poorly understood complex traits such as drought tolerance. The application of new molecular technologies has also been very limited in the context of identifying patterns of genetic diversity *in situ* and planning for *in situ* conservation, where both traditional population genetic and landscape genetic approaches could benefit from high-throughput molecular genetic data if such approaches were not held back by the need for massive computational power (Ford-Lloyd *et al.*, 2014). Current initiatives such as DivSeek (www.divseek.org) should alleviate this problem by increasing accession characterization information and by making the genotypic and phenotypic data more easily accessible to users in the coming years.

The time and cost of whole-genome sequencing is being reduced rapidly, and together with a massive capability for identifying and employing single nucleotide polymorphisms (SNPs), this is enabling technology for a range of more or less traditional and novel applications. These can include identifying meaningful SNP variation in genes that are expressed (gene mining) (Edwards and Batley, 2010). SNPs can be used for producing high-density genetic maps, for fine mapping of traits and for identifying marker–trait associations by way of 'association genetics' (Waugh *et al.*, 2009). The Beijing Genomics Institute is currently sequencing 10,000 rice genotypes, aiming for high-throughput screening of phenotypic traits, especially those that are relevant to improving rice production or quality (Ford-Lloyd *et al.*, 2014).

Advances encompassed in second- and third-generation sequencing technologies may soon lead to enhanced gene discovery and the location of genes underpinning traits of interest, and greater capacity to follow these genes through generations of breeding programmes will speed up the plant breeding process (Stamp and Visser, 2012). Inevitably, new problems will arise in terms of the masses of data generated. The issue will not just be the ease with which data analysis can be undertaken, but also how to store such quantities of data. But conquering these problems will mean that as well as data mining, genetic information could be used to plan both *in situ* and *ex situ* conservation and in mining the genetic diversity conserved under both conditions. Simple population genetic analysis carried out on a scale that is an order of magnitude greater than that which we are used to could be a very powerful tool for use in *in situ* CWR conservation (Ford-Lloyd *et al.*, 2014).

There are increasingly good examples of the use of these advancing technologies. Although focused on *Arabidopsis*, genome-wide association studies (GWAS) have been used to identify candidate loci for lifetime fitness and local adaptation (Fournier-Level *et al.*, 2011). Climatic signals of local adaptation were particularly evident, suggesting a way forward for a greater understanding of how plants will or will not be able to respond to changes in climate. CWR, even those that are currently poorly described, may be put to greater use with a better understanding of their involvement in domestication processes. This has been achieved by the high-throughput sequencing of whole genomes (Malory *et al.*, 2011). 'Massively parallel sequencing' (Tucker *et al.*, 2009) has provided the opportunity to characterize homologues of known genes of rice in one of its distant relatives, *Microlaena stipoides*. Of the 18 genes where homologues were identified, several were related to heading date, which could have significance in terms of drought avoidance (Malory *et al.*, 2011).

Studies of whole-genome expression analysis, by whatever means, have yielded valuable results already. The protein kinase gene, *Pstol1* (phosphorus-starvation tolerance 1), discovered in the rice landrace 'Kasalath', conferring tolerance to phosphorus deficiency, has in part been identified using studies of whole-genome expression and is a clear example of the value of conserving and exploring traditional germplasm (Gamuyao *et al.*, 2012). RNA-Seq can also be used to identify the molecular mechanisms for adaptation to both biotic and abiotic stresses, and its advantage is that it can be applied to any species, not just those whose genomes have currently been sequenced. Studies of the adaptation of eucalyptus to water deficit conditions has painted a complex picture, but one which will be valuable in combatting climate change (Villar *et al.*, 2011) in the future.

## 10.3 Beyond the State of the Art

### 10.3.1 Improving CWR conservation

A review of the current state of the art shows a mixed picture of improving CWR conservation planning on the one hand, but poor implementation of planned *in situ* and *ex situ* CWR conservation on the other. It is the link between the planning and implementation that is largely missing. Poor implementation may be due to lack of financial resources and technical skills (particularly for *in situ* or on-farm conservation), lack of specific policy and legal instruments, and because implementation usually involves diverse actors that historically have not collaborated – for example, agencies responsible for nature conservation *in situ* and those responsible for agricultural diversity conservation, primarily *ex situ*. Also, even though there is increasing interest in the use of CWR by end-users, actual use is still limited by the CWR diversity currently available. Fundamental to improving the state of the art is the development of an agreed definition of a CWR (Maxted *et al.*, 2006), which has enabled the identification of the breadth of taxonomic CWR diversity, as well as providing the foundation for prioritizing CWR taxa for conservation action. Despite this, both CWR conservation and use is still almost entirely *ad hoc*, often being based on the personal interests of individuals. However, the focus of the 'Adapting Agriculture to Climate Change: Collecting, Protecting, and Preparing Crop Wild Relatives' project is global priority CWR diversity, and there is a need for complementary action to fill regional and national gaps, as well as to meet the complementary requirement for *in situ* conservation action for the bulk of CWR diversity, which is not practical or possible to conserve *ex situ*.

#### *Bridging gaps between communities*

Current CWR conservation is too often hindered by a serious lack of collaboration and integration, which ultimately will continue to limit actual use if left unresolved. Integration is required between the:

- planning and actual implementation of conservation priorities;
- conservation (*in situ* and *ex situ*), characterization and sustainable use of CWR diversity;
- existing sectorial networking between biodiversity and agrobiodiversity stakeholders;
- implementation of local, national, regional and global CWR conservation;
- conservation of CWR diversity and its supply and end-user application; and
- awareness, valuation, governance and policy related to CWR conservation and use.

Integrating the disparate elements involved in agrobiodiversity conservation and use will require 'joining up the dots', and only once this is achieved will we have a better chance of food and nutritional security.

#### *A renaissance for genebank conservation*

In recent years, the global economic turndown has led many governments to reconsider their investment in agrobiodiversity conservation, and genebank funding has come under threat or has actually declined. However, environmental instability, food insecurity and the growing human population has forced genebank funding back up the list of governmental priorities. Recently, with initiatives such as 'Adapting Agriculture to Climate Change: Collecting, Protecting, and Preparing Crop Wild Relatives' (www.cwrdiversity.org), DivSeek (www.divseek.org) and Public–Private Partnerships (PPP) for plant (pre-)breeding – (e.g. see www.nordgen.org/index.php/en/content/view/ full/1907) – there is evidence of a renaissance in *ex situ* conservation. Genebanks are moving away from being primarily repositories of 'frozen' genetic diversity to becoming more dynamic digital genetic resource information centres; for example, bringing together genotypic and phenotypic information, together with cultural and environmental information, in a form that meets users' requirements. But even this new role perhaps undersells their potential value. Why not improve the systematic representation of genetic diversity held *ex situ*, or let the genebanks act more directly as the conduit between diversity found in nature or on-farm and the end-users? Genebanks could, for example, investigate the genetic distance between existing, ecogeographically diverse accessions identified using ecogeographic land characterization (ELC) maps, and if significant genetic distance is found between the sampled accessions, use gap analysis techniques to identify additional priority sites to be sampled and collect additional accessions to fill the diversity gaps. If insignificant genetic distance is found between the sampled accessions, then it can be assumed that the existing collections already represent maximum diversity and no further collection is required. It also seems unlikely that breeders or farmers in the future will access material with desired traits directly from *in situ* or on-farm sites, therefore providing a role for genebank staff to be more proactive in collecting to meet specific users' demands. Good genebanks already have regular meetings with their users and suggest material from their collections most likely to meet the users' needs. This could be extended by using predictive characterization methods to identify candidate populations in nature or on-farm, which genebanks could then collect specifically to serve the user community better.

### *Un-complementary conservation?*

There are few who would argue against the need for complementary conservation, but this does require the implementation of both *in situ* and *ex situ* actions. However, this is not the *status quo*. Currently, if we compare implementation of the active *in situ* and *ex situ* conservation of agrobiodiversity globally, over 99% of funding or technical resources are spent on *ex situ* alone; therefore, although we pay lip service to the concept of implementing complementary conservation for agrobiodiversity, there is no balanced implementation. Historically, the reasons for this imbalance include: (i) the science of *ex situ* conservation is well advanced, straightforward and relatively inexpensive to implement compared to *in situ* conservation; (ii) Hawkes (1991) concluded that *in situ* conservation was in its infancy, and although this is no longer the case, practical *in situ* conservation actions remain limited and experimental; (iii) *in situ* conservation often requires the collaboration of diverse actors who historically have not collaborated; therefore, establishing CWR *in situ* reserves is complex and time-consuming; and (iv) users have been able to gain relatively easy access to the germplasm and traits they require from genebanks. Although each of these points remains at least in part valid, it can be argued that climate change has altered the game and that relying on the relatively small sample of agrobiodiversity held in genebanks is unnecessarily limiting utilization, with users requiring greater adaptive trait diversity, which they should be able to access whether or not it exists in a genebank. Therefore, there is a requirement for a more balanced range of *in situ* actions to complement existing *ex situ* actions.

### *The birth of systematic* in situ *CWR conservation*

Such a level of integration will need a novel, holistic approach to CWR conservation and use, which is applicable at global, regional, national and local geographic scales. This requirement was acknowledged by the 13th Regular Session of the CGRFA (FAO, 2011c), which recognized the importance of establishing a global network for *in situ* conservation and on-farm management of PGRFA and requested FAO to elaborate on the means and opportunities for this. A similar requirement was specifically recognized in Europe in the 13th meeting of the ECPGR Steering Committee (2012), which acknowledged the importance of *in situ* conservation and recommended the development of a concept for *in situ* conservation of CWR in Europe, and which was subsequently proposed by Maxted *et al.* (2015). Both recommend the establishment of a broad, decentralized and participatory approach to the conservation of CWR diversity, and possibly the establishment of a global network that could support the coordination of efforts, help raise resources and create more awareness of the value and necessity of the *in situ* conservation of agrobiodiversity. FAO initiated a consultation process and held a technical workshop, 'Towards the establishment of a global network for *in situ* conservation and on-farm management of PGRFA', in November 2012 to identify options, ways and means for establishing a global network (FAO, 2014), and its implementation is still being actively discussed.

EMPLOYING EXISTING NETWORKS OR ESTABLISHMENT OF A NOVEL CWR *IN SITU* NETWORK? There are currently no existing initiatives and networks that actively implement global *in situ* CWR conservation. However, global, regional and national networks that have related functions do exist:

- Related existing global networks:
  - FAO Globally Important Agricultural Heritage Systems
  - UNESCO Man and Biosphere Programme
  - UNESCO World Heritage Sites
  - CBD Programme of Work on Protected Areas
  - IUCN Key Biodiversity Areas
  - IUCN Crop Wild Relative Specialist Group
  - Plantlife Important Plant Areas
- Related existing regional or national networks
  - ECPGR Wild Species Conservation in Genetic Reserves Working Group (see www.ecpgr.cgiar.org/working-groups/wild-species-conservation/)
  - Natura 2000 Network (see www.natura.org/)
  - EUROPARC Federation (see www.europarc.org/)
  - National plant genetic resource programmes
  - National protected area networks.

Several of these networks at present tangentially support CWR conservation, but their roles could be clarified and efforts enhanced. Of the existing regional or national networks, the ECPGR Wild Species Conservation in Genetic Reserves Working Group has the closest remit, but it does not actually manage a network for *in situ* CWR conservation, rather a network of experts in conservation. While at the national level virtually every country has a national PGR programme and a national protected area network, the national PGR programmes tend to focus primarily on *ex situ* PGRFA conservation, while National Protected Area Networks almost exclusively focus on habitat, mega-fauna or *in situ* conservation of rare and/or threatened species. So although existing initiatives and networks do relate to the objectives of the proposed global CWR network, none of them duplicate its scope or functionality. Therefore, part of establishing a global network will be to strengthen regional and national *in situ* conservation networks and the linkages/collaboration between them. There are other global conservation organizations such as the International Treaty on Plant Genetic Resources for Food and Agriculture (ITPGRFA), the CGIAR Centres and the global Crop Trust (www.croptrust.org) that have a significant interest in PGRFA conservation, and which may also be interested in collaboration in *in situ* CWR conservation. The CGIAR, within its Strategic Research Framework and research programmes, has recognized the importance of collective efforts for *ex situ* and *in situ* conservation of CWR.

GLOBAL CWR *IN SITU* NETWORK FUNCTIONALITY. The function and purpose of establishing a global *in situ* conservation network needs to be debated by the stakeholder communities. In the FAO technical workshop 'Towards the establishment of a global network for *in situ* conservation and on-farm management of PGRFA', a national/local approach to implementing the global network was proposed. However, this is realistically the only possible approach, given that each individual CWR population will be found in a particular locality and its *in situ* management will come under the jurisdiction of a national agency, with local conservation staff and communities managing the site. Therefore, whether the *in situ* CWR conservation site has country, national, continental or global importance, it is inevitable, given the principles of the ITPGRFA and CBD, that each country has national sovereignty over its own genetic resources and that the *in situ* populations will need to be managed at the national geopolitical level. However, the establishment of a global networking mechanism on *in situ* conservation could help to enhance coordination and collaboration among stakeholders at various levels, and thereby contribute to the implementation of the Second GPA, the ITPGRFA, and to meeting the Aichi Biodiversity Targets of the CBD (CBD, 2010b).

In terms of national *in situ* CWR conservation, priorities are likely to vary among countries, and may also differ from continental or global priorities. For example, some countries may only assign priority to CWR of nationally important crops and not to those that may have regional or global importance. Thus, some populations of regionally or globally important CWR will not be included in national strategies. Conversely, some nationally important taxa may not be considered a priority at the regional or global level. Therefore, national/local collaboration and coordination within a global network would ensure systematic CWR diversity coverage and maximize the benefits for long-term food security.

Establishing a global network will necessarily involve strengthening regional and national *in situ* management networks as collaborators in the establishment of the global network and in providing a firm foundation for the global network. The support that a global network could provide regional and national networks would help to underpin both higher and lower priority CWR conservation, ensuring greater breadth of CWR diversity is conserved. The global network for CWR conservation will need to take this into account, and one important function will be to initiate a dialogue between the farmer and the plant breeder, to ensure breeders do meet farmer's needs, as well as awareness raising of the value of CWR population diversity and its continued evolution. The core functions of a global network for *in situ* CWR conservation are listed in Box 10.1.

GLOBAL CWR *IN SITU* NETWORK STRUCTURE. The establishment of a global network would necessarily need to be complemented and built on *in situ* activities at the national and regional geographic scales, so *de facto* these would form a 'network of networks' – networks of national sites contributing to regional networks, and in turn regional networks contributing to the global network. Furthermore, the development of a 'network of networks' approach enables a more efficient way of conserving CWR resources at the national, regional, continental or global geographic scales rather than the global network being simply the sum of all nationally established *in situ* conservation sites.

There is a choice to be made over individual site management for any sites included in the global network: should they be (i) managed directly through a global network secretariat, or (ii) a virtual network based on nationally implemented and managed sites and networks? The costs of establishing a novel, stand-alone network would be significant, and so it would appear that

**Box 10.1.** Core functions of a global network for *in situ* CWR conservation.

- Awareness raising at global, regional and national levels of the critical importance of PGRFA existing outside the *ex situ* conservation realm. This function would apply to several socio-economic domains, including food and nutritional security, safeguarding of the environment, income generation and improved livelihoods.
- Sharing of information and experience is an essential activity that will aid the establishment of an evidence base of best practice as a resource for those undertaking *in situ* CWR conservation.
- Strengthening of partnerships and linkages that could foster viable 'communities of practice' for the conservation, management and sustainable use of PGRFA outside genebanks.
- Generation, coordination and implementation of tools, initiatives and interventions that could be relevant to *in situ* conservation.
- Implementation of evidence-based interventions in the areas of *in situ* conservation, with a view to demonstrating positive impacts at the country level.
- Mainstreaming the conservation of PGRFA in nature reserves and other protected areas.

(Text adapted from FAO, 2014.)

option (ii) is the only viable option. Were this to be agreed, then the national, regional and continental networks might be viewed as a virtual network of networks, with individual, nationally implemented and managed sites providing their foundation.

The proposed concept for agreeing 'most suitable' locations for the *in situ* conservation of CWR diversity is summarized in Fig. 10.5 (after Maxted *et al.*, 2015). In essence, effective and systematic global *in situ* conservation of CWR diversity will be achieved via three interrelated geographic, or, more precisely, geopolitical levels of conservation strategy planning: (i) national (Fig. 10.5 in light green); (ii) regional (Fig. 10.5 in blue); and (iii) global (Fig. 10.5 in orange). National governments will be at the core of the establishment of the network, and their support will be essential to its success. Therefore, whether a particular site is provided with national only, national and regional, or national, regional and global designation as part of a CWR network, its inclusion is justified by it containing significant CWR populations that have respective national, regional or global value, and the national authority has precedence over this decision.

The use of combined bottom-up and top-down approaches to the identification of priority sites containing CWR diversity of global importance for inclusion in the global network will require research by national, regional and global authorities to identify localities where national, regional and globally important CWR populations are found. These could be recommended to individual countries as sites within their borders where genetic reserves might be established (Fig. 10.5 in grey). The actual establishment of the regionally or globally identified genetic reserves would be at the discretion of individual nations, although policy and resource support to encourage their establishment could be forthcoming from the international community. Therefore, the final integrated global CWR *in situ* network would contain both genetic reserves identified by individual countries (bottom-up) and those initially identified by regional or global research (top-down), but the latter's nomination for inclusion in the integrated global network would be by individual countries and the genetic reserve conservation planning, practical management and monitoring would necessarily be implemented at national level, though potentially with international support. The global CWR *in situ* network is therefore integrated because it is composed of both the bottom-up and top-down elements.

The integrated global CWR *in situ* network would be driven by international, regional and national policy on conservation and sustainable utilization of PGRFA (Fig. 10.5 in red) and implemented at national level (Fig. 10.5 in green). The purpose of the integrated strategy is to preserve CWR genetic resources for use in crop improvement and to maintain cultivar development options – in particular, to provide a wide pool of diversity as insurance against the negative impacts of climate change on crop production. Therefore, a fundamental element is making conserved CWR germplasm available to the user community (Fig. 10.5 in purple), and to achieve this, the interface between *in situ*, *ex situ* and use of CWR conservation needs to be strengthened (Maxted *et al.*, 2015). As indicated by the cyclical flow of the related strategies in Fig. 10.5, planning and implementing global *in situ* conservation of CWR will be an iterative process requiring periodic review and updating as CWR conservation and utilization policy, science and practice develops. Promoting awareness of the value of CWR to food and economic security, as well as raising additional funding, will be critical to support this process and to ensure long-term *in situ* CWR conservation (Maxted *et al.*, 2015).

CWR POPULATION INCLUSION IN A GLOBAL NETWORK. To ensure the scientific validity of the global CWR *in situ* conservation network and to maximize the CWR diversity conserved *in situ*, it is advisable that populations nominated for inclusion meet a number of criteria. In the context of establishing a European CWR *in situ* conservation network, Maxted *et al.* (2015) proposed such a set of criteria that could be adapted for global use, as follows. The CWR population:

- is native at that location, or if introduced, has existed at that location for at least 15 generations;
- contains distinct or complementary genetic or ecogeographic diversity or specific traits of interest (e.g. an example of high importance to the CWR user community is *Beet necrotic yellow vein virus* (BNYVV) resistance in *Beta vulgaris* subsp. *maritima*

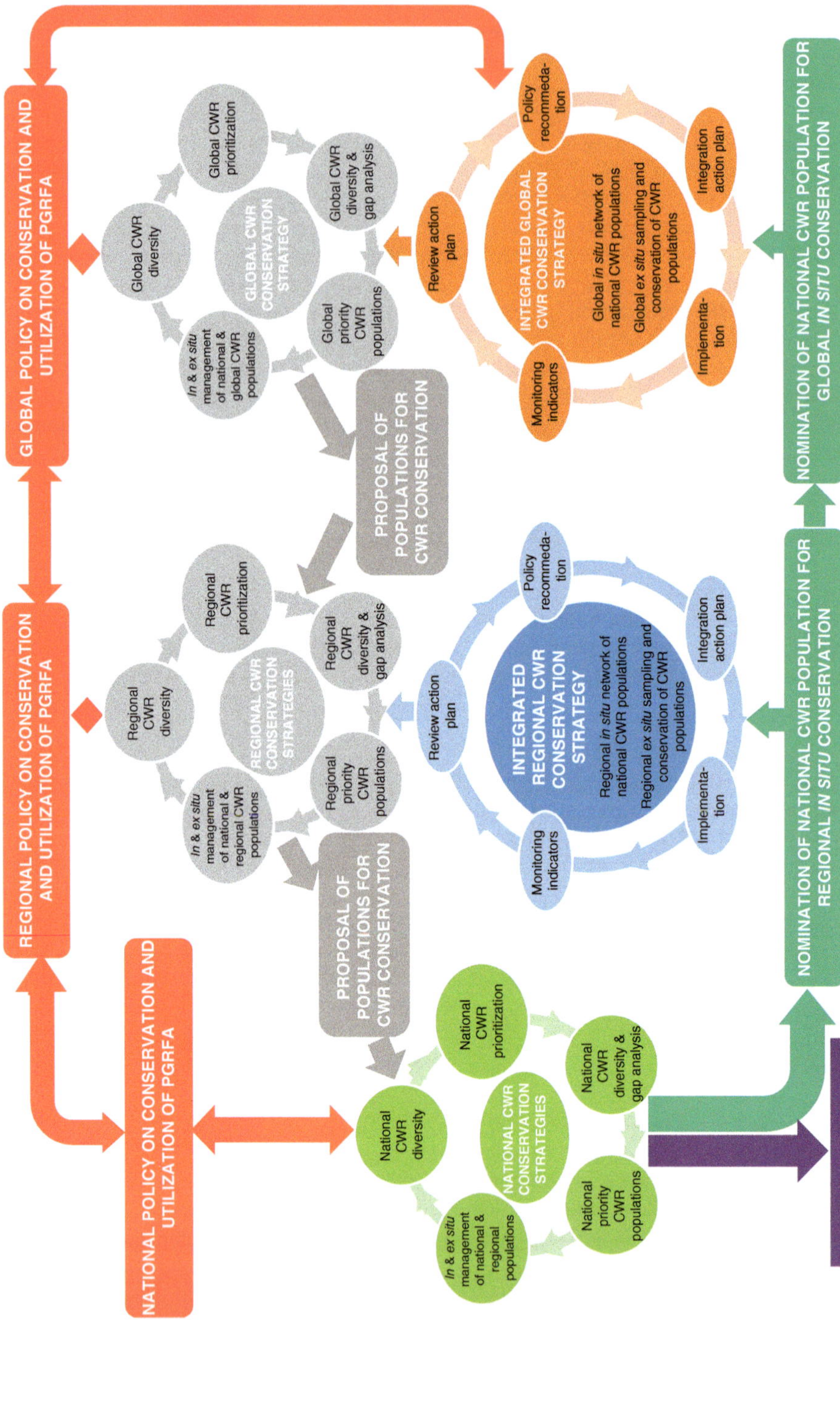


**Fig. 10.5.** Schematic representation of the structure of a potential integrated global CWR *in situ* network. (From Maxted *et al.*, 2015.)

populations from the Kalundborg Fjord area, Denmark) that enhance the overall value of the network;
- should not be threatened, so there is a good chance of long-term survival (conventionally thought to mean a 99% survival probability over 1000 years; Shaffer, 1981), and threats such as from development or climate change will need to have been assessed/modelled and found to be negligible;
- is actively and sustainably managed as a long-term *in situ* conservation resource according to the minimum quality standards for genetic reserve conservation (see Iriondo *et al.*, 2012, for full discussion and summary in Box 10.2);
- is sampled and held in a backup *ex situ* facility every 15 generations;
- is accessible for research or utilization in accordance with the international treaty from a specified *ex situ* facility as part of the multilateral system; and
- is nominated through an appropriate national agency.

Once national agencies in the country nominate a site for inclusion in the global network, it would be the responsibility of the authority overseeing the global network to review whether the selection criteria for inclusion are met and a decision made as to whether the nominated site is added. Sites included within the network would need to be reviewed periodically to ensure they continue to meet the inclusion criteria, and where this is not the case, recommendations for changes would be made and the sanction of de-selection from the global network would be available.

Individual CWR reserves will pragmatically often contain multiple priority species selected using gap analysis techniques, and designated sites may often be found in CWR hotspots. However, they will also need to be complementary, and may in certain cases contain single CWR populations to ensure the breadth of CWR diversity coverage. Ideally, designated global CWR *in situ* conservation sites would occur within formally designated protected areas, but many CWR populations of value occur outside protected areas, so a site may occur within or outside a protected area. In both cases, active and sustained *in situ* CWR conservation management commitment is pivotal to the long-term success of the global CWR *in situ* conservation network.

**Box 10.2.** The criterion related to the minimum quality standards for genetic reserve conservation of designated CWR populations proposed by Iriondo *et al.* (2012):

- Location
  - located following a rigorous scientific process
  - located in a protected area network
- Spatial structure
  - clear boundaries of the genetic reserve should be defined
  - sufficient extent to conserve CWR populations and natural processes
- Target taxa
  - genetic reserves are designed to capture maximum genetic diversity
  - demographic survey of target CWR taxa
- Populations
  - population sizes are large enough to sustain long-term populations
- Management
  - site recognized by the appropriate national agencies
  - management plan formulated
  - monitoring plans are designed and implemented
  - local community involved in site management
  - clearly defined procedure to regulate the use of genetic material
- Quality standards for the protected areas selected for the establishment of genetic reserves
  - site has legal foundation
  - site management plan recognizes the existence of the genetic reserve and addresses its objectives.

A KEY ROLE FOR NATIONAL CWR *IN SITU* CONSERVATION PROGRAMMES. For a global CWR *in situ* conservation network to be effective in conserving CWR diversity, it needs to build on and be integrated with national and regional CWR *in situ* activities. However, many of the sites currently recognized as conserving CWR *in situ* may not be managed in the most appropriate manner to conserve the genetic diversity of the CWR populations and would not meet the set of quality standards for CWR genetic reserves proposed by Iriondo *et al.* (2012).

The EC-funded FP7 project, PGR Secure (www.pgrsecure.org) developed previously conceived approaches to the creation and implementation of national CWR conservation strategies (Maxted *et al.*, 2007), which involved the establishment of a putative network of national CWR genetic reserves across Europe. At least three sites meet the set of quality standards for CWR genetic reserves proposed by Iriondo *et al.* (2012), and their designation has been based on scientific selection principles. Although they are independent and do not together constitute a desired European network of CWR genetic reserves, they can be seen as the core of such a network that could be expanded with time and resources. They are:

- Portugal: Ilhéu do Desembarcadouro, a small island in the east of Madeira, and Ilhéu Chão, one of the Desertas Islands located 11 miles south-east of Madeira. Both contain small populations of the wild beet relative, *Beta patula* Aiton. Pinheiro de Carvalho *et al.* (2012) proposed that these sites should be recognized as protected areas within the Natural Park of Madeira specifically to conserve *B. patula*.
- Norway: The Færder National Park in south-east Norway will be the site of the first proposed *in situ* CWR genetic reserve in the country. The area covers 340 km$^2$, of which 15 km$^2$ is land and home to at least 45 CWR taxa.
- England: The Lizard Peninsula is the most southerly point of the British mainland and home to around 100 CWR taxa. It is envisaged that this CWR-rich site will become the first *in situ* CWR genetic reserve in the UK.

These sites have yet to be recognized formally, but it seems likely that they will be designated formally by the appropriate national authorities in the near future, providing a legacy for PGR Secure but also exemplars that other countries and protected areas may follow. Further, the experience of planning for the establishment of these genetic reserves has shown it is possible to meet the quality standards proposed by Iriondo *et al.* (2012), and in doing so bridge the division between the biodiversity and agrobiodiversity conservation communities, bringing the two communities together to conserve the genetic diversity of CWR *in situ*. There are, as acknowledged above, many other protected areas that might in the future meet these quality standards.

GLOBAL CWR *IN SITU* NETWORK GOVERNANCE. The establishment of a global CWR *in situ* network will require that an authority takes geopolitical and administrative responsibility for overseeing its establishment and long-term effective operation. There are no existing international agencies that have a remit to facilitate the coordination of a network responsible for *in situ* conservation of global CWR diversity. However, several organizations have related mandates that might be extended to address such a requirement, including FAO, through its Globally Important Agricultural Heritage Systems and International Treaty, CGIAR, UNESCO, through its Man and Biosphere Programme and World Heritage Sites, the Secretariat of the CBD, through its Programme of Work on Protected Areas, and IUCN, through its Key Biodiversity Areas. It may be one organization takes the lead but that management decisions are made by a consortium of organizations that are involved to ensure all stakeholders are represented in policy and governance decisions.

The coordinating agency would have management and coordination responsibility for the virtual network and would need to: (i) ensure that the minimum criteria for inclusion in the global network are met; (ii) provide necessary training and expertise in *in situ* conservation; (iii) promote access to *in situ* conserved populations linked to benefit sharing; (iv) increase awareness of the value of CWR for agriculture and the environment; and (v) work with national PGRFA coordinators to nominate and maintain healthy CWR populations.

Some may argue that it is premature to discuss the financing of a global CWR *in situ* conservation network during the initial discussion of its inception, but one argument that may inhibit that discussion is lack of clarity over the magnitude of costs associated with its establishment. The costs may be subdivided between: (i) establishment of the network by the international coordinating agency; (ii) running costs of the global network; (iii) production of national CWR conservation strategies; (iv) establishment of national genetic reserves to join the global network; and (v) running costs of national genetic reserves within the global network. It is estimated that a ballpark figure of US$19.3 million would be required to research, establish and run the global CWR *in situ* network for 10 years. Following establishment, the recurrent costs for a 10-year period would be in the region of US$2.5 million global costs and US$2.0 million national costs to maintain the network. A significant figure, but surely not too high a price to pay compared to the US$5.3 billion that it took to collect and conserve the world's germplasm held *ex situ* in genebanks and the annual running cost of US$30.5 million of maintaining that material (Hawkes *et al.*, 2000), as well as taking into account the annual value of the use of CWR of as much as US$115 billion (Pimentel *et al.*, 1997).

A FURTHER LOOK AT *IN SITU* CONSERVATION. It has been argued that in terms of *in situ* conservation of CWR species, genetic reserve conservation is the most appropriate conservation technique (Maxted *et al.*, 2008c), and this remains the logical focus of CWR *in situ* activities. However, this assumes that CWR diversity is adequately represented in the existing protected area networks, and to an extent, this assumption can be questioned. CWR, particularly those of the major food crops, are often ruderal species found in disturbed habitats (Jain, 1975; Jarvis *et al.*, 2015), and therefore conventional protected areas may not be the only appropriate locations for their *in situ* conservation. Therefore, as well as establishing genetic reserves within existing protected areas, and possibly, in rare cases of high priority, establishing new formal protected areas, other, less formal initiatives should be explored (Maxted *et al.*, 2008e; Heywood, 2011). For example, the signing of long-term agreements with landowners outside of protected areas, to manage their cultivated field margins, fallow land, pastures, orchards or woodland in a way that promotes CWR population maintenance over an extended period. Such maintenance agreements for less formal CWR *in situ* conservation have been experimented with in, for example, the Global Environment Facility funded Dryland Agrobiodiversity Project (Conservation and Sustainable Use of Dryland Agrobiodiversity in Jordan, Lebanon, Syria and the Palestinian Authority), but it is uncertain how sustainable they have been in the long-term (Al-Atawneh *et al.*, 2008). As with much of the development of PGR conservation science, we can learn from other, broader biodiversity applications. In this context, Heywood (2011) reviews the application of conservation easements, which are voluntary agreements between conservation agencies and landowners that restrict or limit the development and management of a site. Such agreements may allow the conservation agency to manage or influence the management of the site, the agreement is legally binding, and in return, the landowner may get tax benefits (Merenlender *et al.*, 2004). The US Nature Conservancy (TNC) plays a leading role in conservation easements implementation. For example, in the mid-1970s, TNC established a conservation easement on 1800 acres in the Blackfoot River Valley in Montana, and this area has now been expanded to cover 30,254 acres. The TNC currently manages more than 2 million acres via conservation easements in the USA, and their model is being employed increasingly in Canada, Latin America and Australia (TNC, 2015).

There is a clear need for research into similar approaches to the conservation of CWR diversity. Therefore, although the simplest way to conserve CWR *in situ* is likely to be to establish the interrelated network of networks of national, regional and global genetic reserves within existing protected areas, it should not be assumed that this is the only possible option. CWR *in situ* conservation on privately owned land using schemes such as in conservation easements or, even more informally, through the raising of awareness of the value of CWR among the general public and their inclusion in more community-based conservation schemes should be considered.

### 10.3.2 Improving CWR use

CWR use by those developing new crop varieties is expanding but often inhibited by lack of characterization and evaluation (C&E) data. The First Report on the State of the World's PGRFA (FAO, 1998) found that 35% of countries reported lack of C&E data as a major constraint on germplasm use, and the Second Report on the State of the World's PGRFA (FAO, 2010) stated that 'country reports were virtually unanimous in suggesting the most significant obstacle for greater use of PGRFA is the lack of adequate C&E data'. It is obvious that conventional field C&E of crop material has failed to meet demand, and for CWR it seems likely that even less C&E information is available. Novel approaches will be required to improve CWR use, including omics, predictive characterization and improved informatics to make CWR genetic diversity available to the user community.

#### *Novel omics approaches to characterization and evaluation*

We know from Vosman *et al.* (Chapter 1, this volume) that the diversity of crops where omics approaches are being investigated (tomato, brassicas, barley, quinoa, rice, oats, cowpea, faba bean, wheat, pearl millet, potato), in itself represents widely differing challenges to the scientific community. Hence, a range of novel characterization techniques can, and may, be employed, depending on the crop.

If we just consider quantitative trait loci (QTL) mapping or marker-assisted selection (MAS) as well established techniques, more novel techniques are needed to overcome effectively such problems as are encountered because of having to move from one ploidy level to another (oats), or looking for relatively uncommon, complex insect resistance traits, as in brassicas (Borém and Fritsche-Neto, 2014). Crops with large genomes, as is the case with barley, present their own problems, which are gradually being overcome, while improving our understanding of the domestication route for some crops and their 'redomestication' are beginning to reveal a range of useful traits previously unknown (Ortiz *et al.*, 2007). Overcoming problems that may or may not be crop specific by way of various omics approaches can lead to success in the identification and transfer of a range of traits in breeding programmes, including insect and pathogen resistance, environmental adaptability and abiotic stress tolerances. Many of these techniques increasingly require knowledge of whole-genome DNA sequences. While this is 'relatively' straightforward by means of next-generation whole-genome sequencing of crops for which reference genomes are available, it is much more problematic in those crops with large genomes and/or where there is as yet no reference genome. But even here, techniques focused on exome sequencing or RNA-Seq rather than whole-genome sequencing provide a means of moving forward (Hübner *et al.*, 2015).

So in the context of rapidly developing technology, what issues now need to be addressed? While DNA sequence data are now made publicly available – something that is more or less mandatory for the scientific publication of research – this is not the case for phenotypic data in the same way. Documentation has been discussed and debated for the past 40 years, so it could be argued that it is disappointing to have reported a clear lack of publicly available information for phenotypic characterization, and something needs to be done to rectify this situation. This has been taken up by current initiatives such as DivSeek (www.divseek.org) and PPP for plant (pre-)breeding – (e.g. see www.nordgen.org/index.php/en/content/view/full/1907) – which plan to gather and provide access to data via the web phenotypic and genetic characterization data.

The other, more positive, issue is the increasing emphasis on the use of DNA sequence data focusing on enhancing the use of PGRs, and CWR in particular, and this is clearly exemplified in the chapters of this book. Is it now appropriate to propose a challenge to the CWR community of sequencing/resequencing as many as 200 genomes of the 1392 priority CWR identified by Vincent *et al.* (2013)? The '3000 Rice Genomes Project' (Chinese Academy of Agricultural Sciences (CAAS), International Rice Research Institute (IRRI) and Beijing Genomics Institute (BGI)) clearly demonstrates what can be done with the will and the resources. The challenge to the CWR community is by no means as great, but would be of immense value for informing both conservation actions and sustainable utilization of germplasm. Who will take up this challenge, and where will the funding come from? The benefits not only for enhancing the use of genetic resources but also for informing conservation strategies are clear, and this has now become an achievable objective.

### *Predictive characterization for mining genetic resources*

Predictive characterization is a technique that uses Geographical Information System (GIS) analysis to predict the C&E traits of accessions based on the geographic provenance of collection localities (Thormann *et al.*, 2014, and Chapter 8, this volume). The first application of the concept of predictive characterization that identified a link between a resistance trait and a set of environmental parameters was developed by the International Center for Agricultural Research in the Dry Areas (ICARDA) as the Focused Identification of Germplasm Strategy (FIGS) (Mackay and Street, 2004; Street *et al.*, 2008), which used biotic and abiotic matching techniques. There are three current approaches identified: (i) matching of biotic and abiotic characteristics associated with a collecting site (Kaur *et al.*, 2008); (ii) ecogeographical information associated with a collecting site (Parra-Quijano *et al.*, 2012; Khoury *et al.*, 2015a,b); and (iii) previously recorded C&E data of trait occurrences associated with a set of locations different from those where the germplasm being examined had been collected (Endresen, 2010). In each case, an attempt is made to predict which traits might be present in germplasm from a particular location (see Fig. 10.6). Examples of the predictive characterization of CWR of *Avena*, *Beta*, *Brassica* and *Medicago* are discussed in Chapter 8, this volume.

There is now a growing literature (Bhullar *et al.*, 2009; El Bouhssini *et al.*, 2009, 2011; Endresen, 2010; Endresen *et al.*, 2011, 2012; Bari *et al.*, 2012, 2014; Khazaei *et al.*, 2013) that is showing the value and relevance of predictive characterization approaches to mine genetic resources conserved *ex situ*, *in situ*, or simply found in nature. This approach seems to offer significant enhancement for the use of CWR diversity because (i) it applies a relatively simple computer-based approach, and (ii) the costs of application are significantly less than conventional and genomic approaches to C&E, easing its adoption and application worldwide. If the promise of predictive characterization is fulfilled, it might be imagined that a sorghum breeder in Uganda, or even a rice farmer in Thailand, might be able to use a simple package to predict which accessions he or she wishes to experiment with. Predictive characterization does not replace actual field trials, but it does reduce the number of candidate accessions that the breeder needs to screen before finding novel alleles for a target trait. Further, predictive characterization could be routinely used as a precursor to genomic-based characterization to enable a more cost-effective selection of accessions to be characterized. Even more strikingly, the breeder may be able to use a predictive characterization package to decide which *in situ* CWR population he or she wishes to obtain material from for inclusion in his or her crossing programme, the local genebank staff liaising with protected area managers to provide the desired CWR germplasm to the breeder.

### *End-user-orientated informatics*

It would appear that the lack of available information on conserved agrobiodiversity is as much an impediment to utilization as the *ad hoc* nature of the agrobiodiversity conserved. For example, if I am a plant breeder looking for accessions with known adaptive traits, it is difficult, if not impossible, to find which accessions have the desired adaptive trait, and further, which genebanks have material of those accessions. Why is this the case when it is such an obvious end-user requirement? The problem is not simply that the data may be unknown (it is often known), but it is that the knowledge is currently inaccessible to the user community, and this is undoubtedly limiting the utilization of conserved diversity.

However, the problem is recognized, and one example where it is beginning to be addressed is in the development of the prototype Plant Genetic Resources Diversity Gateway (the PGR Diversity Gateway – pgrdiversity.bioversityinternational.org; Dias, 2015), which was developed during the PGR Secure project. The aims of the PGR Diversity Gateway are to promote and facilitate the use of agrobiodiversity by end-users in breeding and crop improvement by providing trait and QTL information, and in conservation by providing information on agrobiodiversity

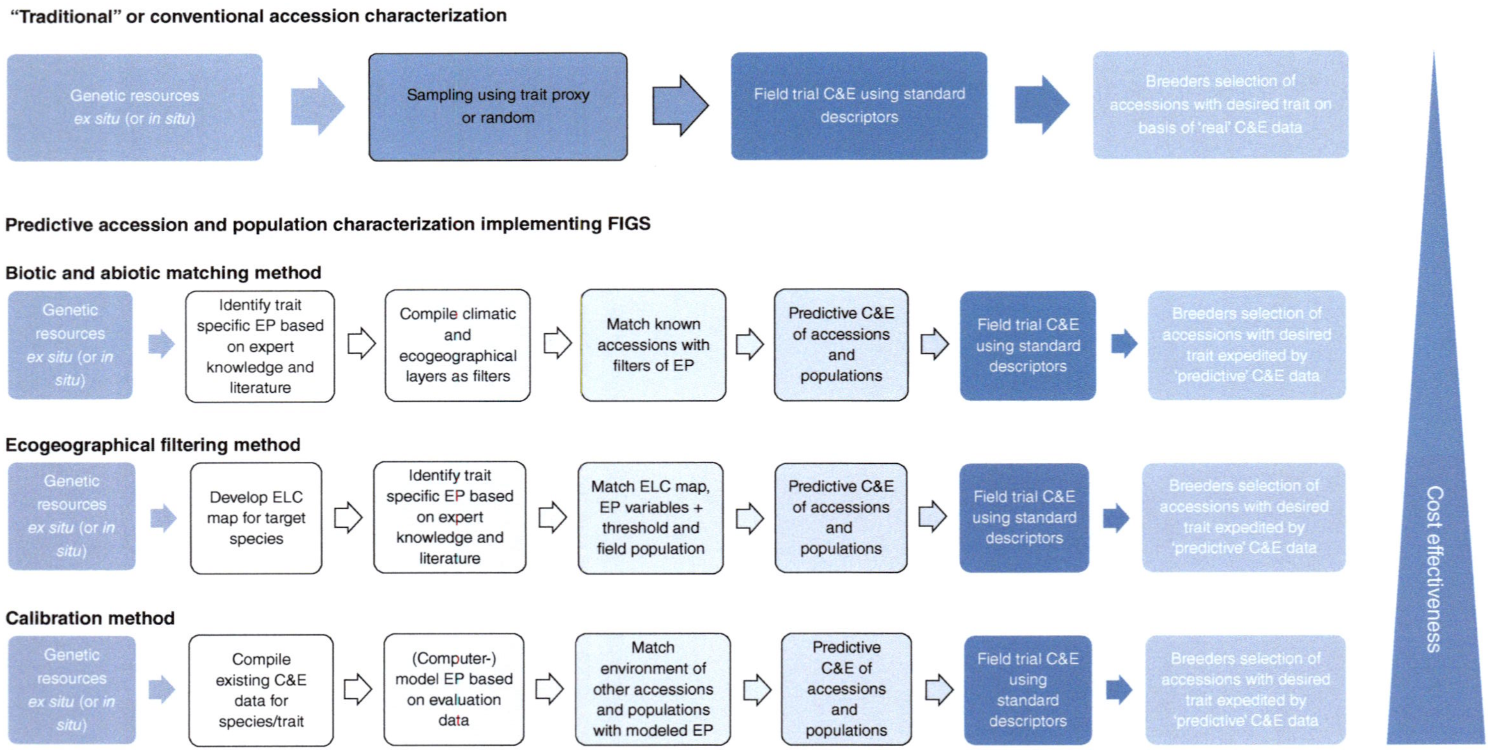


Key: C&E = Characterization and Evaluation; EP = Environmental Profile; ELC = Ecogeographical Land Characterization; FIGS = Focused Identification of Germplasm Strategy

**Fig. 10.6.** Schematic representation of conventional versus predictive characterization and evaluation methods. (From Thormann *et al.*, 2014.)

inventories and conservation strategies. Further, the PGR Diversity Gateway, or a future similar initiative, will link existing tools such as EURISCO (www.ecpgr.cgiar.org/resources/germplasm-databases/eurisco-catalogue/) more effectively to accessions known to contain specific adaptive traits and enable users to request the accession from a genebank in a single process.

It can be conceptualized that the process of end-user agrobiodiversity information acquisition is analogous to a sunflower structure, with the capitulum providing the foundation for two types of floret: ray florets on the outer rim and disc florets in the centre. In the analogy with end-user agrobiodiversity information acquisition, the central disc florets represent the information associated with agrobiodiversity accessions or populations and the outer ray florets are the various end-user communities, with the capitulum being the informatics base that links agrobiodiversity knowledge and provides tools that link users to the knowledge they require (Fig. 10.7). We currently have a reasonable understanding of the central disc florets containing agrobiodiversity information and the outer ray florets made up of the various end-user communities, but we have yet to develop the comprehensive capitulum informatics base that links different kinds of agrobiodiversity information and makes the link between this information and the specific requirements of diverse end-user communities. Elements of the capitulum exist, for example EURISCO, pcGRIN, GENESYS, etc., but the remaining challenge is to join up these existing elements in a unified system and to develop new elements where required – only when this is done will conserved agrobiodiversity be utilized efficiently.

## 10.4 Conclusion

CWR have an increasing and significant value for food security and human well-being, but today they are threatened, under-conserved and underutilized. Together with increased recognition of CWR value has come increased global,

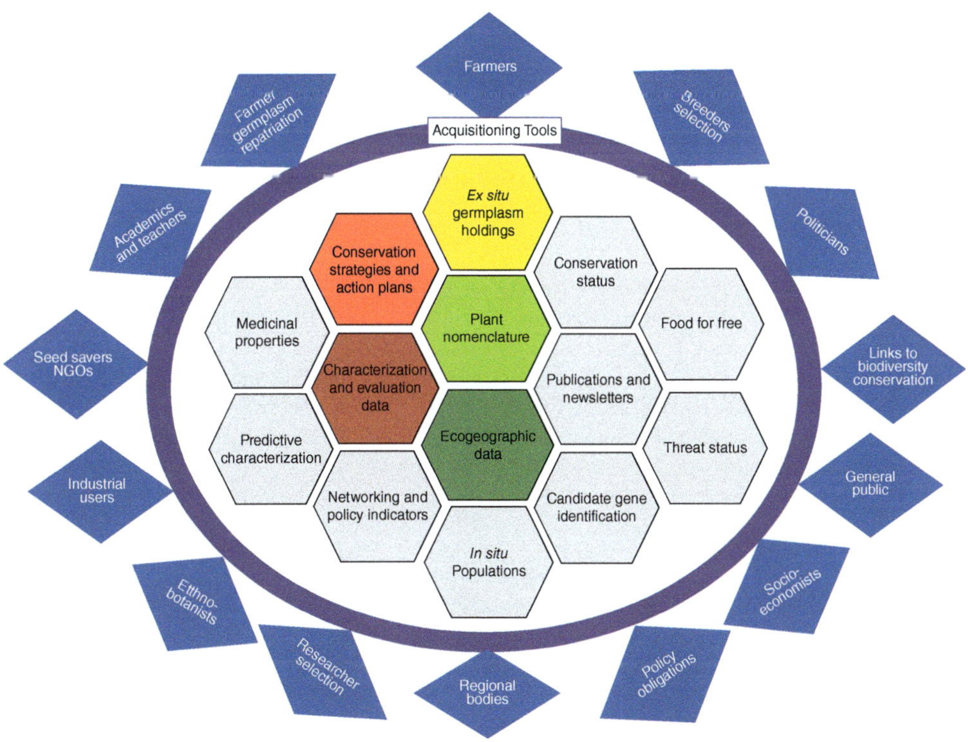


Fig. 10.7. Sunflower schematic diagram of end-user agrobiodiversity information acquisition.

regional and national attention to their complementary *in situ* and *ex situ* conservation, along with the need to link conserved diversity to use, and it is widely recognized that use provides the only long-term guarantee of sustained conservation. However, to achieve this ideal holistic conservation and use of CWR diversity, there is a requirement for the integration of: (i) conservation (*in situ*/*ex situ*), characterization and sustainable use; (ii) biodiversity and agrobiodiversity interventions; (iii) local, national, regional and global implementation; and (iv) improved awareness, valuation, governance and policy to promote CWR conservation and use. This is a complex task, given that many of the stakeholders have previously worked in isolation.

Along with a review of the state of the art of CWR conservation and use, arguments have been put forward to move beyond the current state of the art by:

**1.** Improving integration between CWR conservation and the broader conservation and CWR user communities.
**2.** Looking again at the potential role of genebanks beyond the conventional storage of germplasm and providing them with a more dynamic, facilitating, entrepreneurial role.
**3.** Ensuring a more appropriate balance between *ex situ* and *in situ* CWR conservation actions.
**4.** The commencement of global, regional and nationally integrated systematic *in situ* CWR conservation, both via a network of genetic reserves and less formally designated *in situ* conservation sites.
**5.** The continuing development and implementation of the growing range of omics techniques to locate adaptive genes that meet the end-users' requirements.
**6.** The application of predictive characterization approaches to mine genetic resources conserved *ex situ*, *in situ*, or simply found in nature.
**7.** The provision of an end CWR user-orientated informatics base that truly facilitates end-users' access to the CWR germplasm they need, whether actively conserved or not and whether they themselves have a clear image of what they require.

Central to the way forward is the proposal to establish a global, decentralized *in situ* network of networks for CWR conservation that strengthens existing partnerships and makes novel linkages between communities, but particularly links CWR conservation to germplasm users. Given that this is the desired long-term outcome, we do not have to wait for the full *in situ* network of networks to be implemented, we can take action now. In the short term, we need to raise awareness of the value of CWR diversity for future food security, build capacity and work at individual national levels to promote active CWR conservation and use, strengthening the already existing networks and programmes and their integration, in our home countries, where we often know personally those that might or should be involved.

Finally, it is clear from this review that significant progress has and is being made to refute Darwin's assertion that CWR are either 'unknown or only doubtfully known', but if real progress is to be made in joining up the dots of CWR conservation and use, we need to redouble our efforts to ensure the full picture is in focus.

## Acknowledgements

This research was partially funded by a grant from the EU Seventh Framework Programme for the PGR Secure project (Grant Agreement No 266394).

## References

Al-Atawneh, N., Amri, A., Assi, R. and Maxted, N. (2008) Management plans for promoting *in situ* conservation of local agrobiodiversity in the West Asia centre of plant diversity. In: Maxted, N. Ford-Lloyd, B.V., Kell, S.K., Iriondo, J.M., Dulloo, M.E. and Turok, J. (eds) *Crop Wild Relative Conservation and Use*. CAB International, Wallingford, UK, pp. 338–361.

Anikster, Y., Feldman, M. and Horovitz, A. (1997) The Ammiad experiment. In: Maxted, N., Ford-Lloyd, B.V. and Hawkes, J.G. (eds) *Plant Genetic Conservation: The In Situ Approach*. Chapman & Hall, London, pp. 239–253.

Avagyan, A. (2008) Crop wild relatives in Armenia: diversity, legislation and conservation issues. In: Maxted, N., Ford-Lloyd, B.V., Kell, S.K., Iriondo, J.M., Dulloo, M.E. and Turok, J. (eds) *Crop Wild Relative Conservation and Use*. CAB International, Wallingford, UK, pp. 58–68.

Bari, A., Street, K., Mackay, M., Endresen, D.T.F., de Pauw, E., *et al.* (2012) Focused identification of germplasm strategy (FIGS) detects wheat stem rust resistance linked to environmental variables. *Genetic Resources and Crop Evolution* 59(7), 1465–1481.

Bari, A., Amri, A., Street, K., Mackay, M., de Pauw, E., *et al.* (2014) Predicting resistance to stripe (yellow) rust (*Puccinia striiformis*) in wheat genetic resources using focused identification of germplasm strategy. *Journal of Agricultural Science* 152, 906–916.

Baudoin, J.P., Rocha, O.J., Degreef, J., Zoro Bi, I., Ouédraogo, M., Guarino, L. and Toussaint, A. (2008) *In situ* conservation strategy for wild Lima bean (*Phaseolus lunatus* L.) populations in the Central Valley of Costa Rica: a case study of short-lived perennial plants with a mixed mating system. In: Maxted, N., Ford-Lloyd, B.V., Kell, S.K., Iriondo, J.M., Dulloo, M.E. and Turok, J. (eds) *Crop Wild Relative Conservation and Use*. CAB International, Wallingford, UK, pp. 364–379.

Baum, M., Grando, S., Backes, G., Jahoor, A., Sabbagh, A., *et al.* (2003) QTLs for agronomic traits in the Mediterranean environment identified in recombinant inbred lines of the cross Arta x H. *spontaneum* 41–1. *Theoretical and Applied Genetics* 107, 1215–1225.

Berlingeri, C. and Crespo, M.B. (2012) Inventory of related wild species of priority crops in Venezuela. *Genetic Resources and Crop Evolution* 59, 655–681.

Bhullar, N.K., Street, K., Mackay, M., Yahiaoui, N. and Keller, B. (2009) Unlocking wheat genetic resources for the molecular identification of previously undescribed functional alleles at the Pm3 resistance locus. *Proceedings of the National Academy of Sciences of the United States of America* 106, 9519–9524.

Bilz, M., Kell, S.P., Maxted, N. and Lansdown, R.V. (2011) *European Red List of Vascular Plants*. Publications Office of the European Union, Brussels, Luxembourg. Available at: ec.europa.eu/environment/nature/conservation/species/redlist/downloads/European_vascular_plants.pdf (accessed 6 March 2015).

Borém, A. and Fritsche-Neto, R. (eds) (2014) *Omics in Plant Breeding*. John Wiley and Sons, Inc, Chichester, UK.

Borlaug, N.E. (1983) Contributions of conventional plant breeding to food production. *Science* 219(4585), 689–693.

Brar, D.S. (2005) Broadening the gene pool of rice through introgression from wild species. In: Toriyama, K., Heong, K.L. and Hardy, B. (eds) *Rice is Life: Scientific Perspectives for the 21st Century*. Proceedings of the World Rice Research Conference held in Tsukuba, Japan, 4–7 November 2004. International Rice Research Institute, Philippines, pp. 157–160.

Brummitt, N., Bachman, S.P., Aletrari, E., Chadburn, H., Griffiths-Lee, J., *et al.* (2015a) The Sampled Red List Index for Plants, phase II : ground-truthing specimen-based conservation assessments. *Philosophical Transactions of the Royal Society of London: Biological Sciences* 370(1662), 20140015, doi:10.1098/rstb.2014.0015.

Brummitt, N.A., Bachman, S.P., Griffiths-Lee, J., Lutz, M., Moat, J.F., *et al.* (2015b) Green plants in the red: a baseline global assessment for the IUCN Sampled Red List Index for Plants. *PLoS ONE* 10(8), e0135152, doi:10.1371/journal.pone.0135152.

Castañeda Álvarez, N.P., Vincent, H.A., Kell, S.P., Eastwood, R.J. and Maxted, N. (2011) Ecogeographic surveys. In: Guarino, L., Ramanatha Rao, V. and Goldberg, E. (eds) *Collecting Plant Genetic Diversity: Technical Guidelines. 2011 update*. Bioversity International, Rome. Available at: cropgenebank.sgrp.cgiar.org/index.php?option=com_content&view=article&id=679 (accessed 6 March 2015).

Castañeda-Álvarez, N.P., Khoury, C.K., Achicanoy, H.A., Bernau, V., Dempewolf, H., *et al.* (2016) Global priorities for crop wild relative conservation for food security (in preparation).

CBD (Convention on Biological Diversity) (2010a) *Global Strategy for Plant Conservation*. Secretariat of the Convention on Biological Diversity, Montreal, Canada.

CBD (2010b) *Strategic Plan for Biodiversity 2011–2020*. Secretariat of the Convention on Biological Diversity, Montreal, Canada.

Chen, G., Li, C., Shi, Y. and Nevo, E. (2008) Wild barley, *Hordeum spontaneum*, a genetic resource for crop improvement in cold and arid regions. *Sciences in Cold and Arid Regions* 1, 0115–0124.

Colmer, T.D., Flowers, T.J. and Munns, R. (2006) Use of wild relatives to improve salt tolerance in wheat. *Journal of Experimental Botany* 57, 1059–1078.

Darwin, C. (1868) *The Variation of Animals and Plants under Domestication*. John Murray Publishers, London.

Dempewolf, H., Eastwood, R.J., Guarino, L., Khoury, C.K., Müller, J.V., *et al.* (2014) Adapting agriculture to climate change: a global initiative to collect, conserve, and use crop wild relatives. *Agroecology and Sustainable Food Systems* 38, 369–377.

Dias, S. (2015) Plant Genetic Resources Diversity Gateway for the conservation and use of crop wild relative and landrace traits. *Crop Wild Relative* 10, 29–30.
Dias, S. and Gaiji, S. (2005) EURISCO: a window on Europe's plant genetic diversity: overview of European CWR collections. Unpublished presentation given at the First International Conference on Crop Wild Relative Conservation and Use, Agrigento, Italy, September 2005.
Dias, S., Dulloo, M.E. and Arnaud, E. (2012) The role of EURISCO in promoting use of agricultural biodiversity. In: Maxted, N., Dulloo, M.E., Ford-Lloyd, B.V., Frese, L., Iriondo, J.M. and Pinheiro de Carvalho, M.A.A. (eds) *Agrobiodiversity Conservation: Securing the Diversity of Crop Wild Relatives and Landraces*. CAB International, Wallingford, UK, pp. 270–277.
Dulloo, M.E., Guarino, L., Engelmann, F., Maxted, N., Newbury, H.J.N., *et al.* (1999) Complementary conservation strategies for the genus *Coffea* with special reference to the Mascarene Islands. *Genetic Resources and Crop Evolution* 45, 565–579.
Dwivedi, S.L., Upadhyaya, H.D., Thomas Stalker, H., Blair, M.W., Bertioli, D.J., *et al.* (2008) Enhancing crop gene pools with beneficial traits using wild relatives. *Plant Breeding Reviews* 30, 180–230.
ECPGR (European Cooperative Programme for Plant Genetic Resources) (2012) *Report of the Thirteenth Meeting of the ECPGR Steering Committee Meeting.* Federal Ministry of Agriculture, Forestry, Environment and Water Management Austria, 4–7 December 2012. Available at: www.ecpgr.cgiar.org/about-ecpgr/steering-committee/13th-sc-meeting/ (accessed 6 March 2015).
Edwards, D. and Batley, J. (2010) Plant genome sequencing: applications for crop improvement. *Plant Biotechnology Journal* 8, 2–9.
Eglinton, J.K., Evans, D.E., Brown, A.H.D., Langridge, P., McDonald, G., *et al.* (2001) The use of wild barley (*Hordeum vulgare* spp. *spontaneum*) in breeding for quality and adaptation. Proceedings of the 10th Barley Technical Symposium, Canberra, 16–20 September 2001. Available at: www.regional.org.au/au/abts/1999/eglinton.htm (accessed 4 August 2008).
El Bouhssini, M.E., Street, K., Joubi, A., Ibrahim, Z. and Rihawi, F. (2009) Sources of wheat resistance to Sunn pest, *Eurygaster integriceps* Puton, in Syria. *Genetic Resources and Crop Evolution* 56(8), 1065–1069, doi:10.1007/s10722-009-9427-1.
El Bouhssini, M.E., Street, K., Amri, A., Mackay, M., Ogbonnaya, F.C., *et al.* (2011) Sources of resistance in bread wheat to Russian wheat aphid (*Diuraphis noxia*) in Syria identified using the focused identification of germplasm strategy (FIGS). *Plant Breeding* 130(1), 96–97.
Endresen, D.T.F. (2010) Predictive association between trait data and ecogeographic data for Nordic barley landraces. *Crop Science* 50(6), 2418–2430.
Endresen, D.T.F., Street, K., Mackay, M., Bari, A. and de Pauw, E. (2011) Predictive association between biotic stress traits and eco-geographical data for wheat and barley landraces. *Crop Science* 51(5), 2036–2055.
Endresen, D.T.F., Street, K., Mackay, M., Bari, A., Amri, A., *et al.* (2012) Sources of resistance to stem rust (Ug99) in bread wheat and durum wheat identified using focused identification of germplasm strategy. *Crop Science* 52(2), 764–773.
FAO (Food and Agriculture Organization of the United Nations) (1998) The State of the World's Plant Genetic Resources for Food and Agriculture. FAO, Rome. Available at www.fao.org/agriculture/crops/thematic-sitemap/theme/seeds-pgr/sow/en/ (accessed 6 March 2015).
FAO (2001) International Treaty on Plant Genetic Resources for Food and Agriculture. FAO, Rome. Available at: www.planttreaty.org/ (accessed 6 March 2015).
FAO (2010) The Second Report on the State of the World's Plant Genetic Resources. FAO, Rome. Available at: www.fao.org/agriculture/seed/sow2/en/ (accessed 6 March 2015).
FAO (2011a) World Agriculture Towards 2030/2050: The 2012 Revision. ESA E Working Paper No 12-03. FAO, Rome. Available at: www.fao.org/economic/esa/esag/en (accessed 6 March 2015).
FAO (2011b) Second Global Plan of Action for Plant Genetic Resources for Food and Agriculture. FAO, Rome. Available at: www.fao.org/docrep/015/i2624e/i2624e00.htm (accessed 6 March 2015).
FAO (2011c) Thirteenth Regular Session of the Commission on Genetic Resources for Food and Agriculture, CGRFA-13/11/Report. FAO, Rome. Available at: www.fao.org/docrep/meeting/024/mc192e.pdf (accessed 6 March 2015).
FAO (2013) Towards the establishment of a global network for *in situ* conservation and on-farm management of PGRFA. Report of Technical Workshop held in Rome, 13 November, 2012. FAO, Rome. Available at: www.fao.org/agriculture/crops/core-themes/theme/seeds-pgr/itwg/6th/technical-workshop/en/ (accessed 5 April 2013).

FAO (2014) Concept note on global networking on *in situ* conservation and on-farm management of plant genetic resources for food and agriculture. Information document to the 7th Session of the Intergovernmental Working Group on Plant Genetic Resources for Food and Agriculture (CGRFA/WG-PGR-7/14/Inf.3), Commission on Genetic Resources for Food and Agriculture, Rome. Available at: www.fao.org/3/a-ml477e.pdf (accessed 6 March 2015).
FAO (2015) Draft Guidelines for Developing a National Strategy for Plant Genetic Resources for Food and Agriculture: Translating the Second Global Plan of Action for Plant Genetic Resources for Food and Agriculture into National Action. FAO, Rome. Available at: www.fao.org/3/a-mm566e.pdf (accessed 6 March 2015).
Farooq, S., Iqbal, N., Asghar, M. and Shah, T.M. (1992) Intergeneric hybridization for wheat improvement. VI. Production of salt tolerant germplasm through crossing wheat (*Triticum aestivum* L.) with *Aegilops cylindrica* and its significance in practical agriculture. *Journal of Genetics and Breeding* 46, 125–132.
Farooq, S., Asghar, M., Iqbal, N., Asian, E., Arif, M., *et al.* (1995) Production of salt tolerant wheat germplasm through crossing cultivated wheat with *Aegilops cylindrica*. II. Field evaluation of salt tolerant germplasm. *Cereal Research Community* 23, 275–282.
Feuillet, C., Langridge, P. and Waugh, R. (2008) Cereal breeding takes a walk on the wild side. *Trends in Genetics* 24, 24–32.
Fitzgerald, H. (2013) The National Crop Wild Relative Strategy Report for Finland. MTT, Report 121. 95 pp. Available at: jukuri.mtt.fi/bitstream/handle/10024/481549/mttraportti121.pdf (accessed 6 March 2015).
Ford-Lloyd, B.V., Kell, S.P. and Maxted, N. (2008) Establishing conservation priorities for crop wild relatives. In: Maxted, N., Ford-Lloyd, B.V., Kell, S.P., Iriondo, J., Dulloo, E. and Turok, J. (eds) *Crop Wild Relative Conservation and Use*. CAB International, Wallingford, UK, pp. 110–119.
Ford-Lloyd, B., Engels, J.M.M. and Jackson, M. (2014) Genetic resources and conservation challenges under the threat of climate change. In: Jackson, M., Ford-Lloyd, B. and Parry, M. (eds) *Plant Genetic Resources and Climate Change*. CAB International, Wallingford, UK, pp. 16–37.
Forster, B.P., Ellis, R.P., Moir, J., Talam, V., Sanguineti, M.C., *et al.* (2004) Genotype and phenotype associations with drought tolerance in barley tested in North Africa. *Annals of Applied Biology* 144, 157–168.
Fournier-Level, A., Korte, A., Cooper, M.D., Nordborg, M., Schmitt, J., *et al.* (2011) A map of local adaptation in *Arabidopsis thaliana*. *Science* 334, 86–89.
Gamuyao, R., Chin, J.H., Pariasca-Tanaka, J., Pesaresi, P., Catausan, S., *et al.* (2012) The protein kinase Pstol1 from traditional rice confers tolerance of phosphorus deficiency. *Nature* 488, 535–539.
Goodman, R.M., Hauptli, H., Crossway, A. and Knauf, V.C. (1987) Gene transfer in crop improvement. *Science* 236, 48–54.
Hadjichristodoulou, A. (1993) The use of wild barley in crosses for grain production under dryland conditions. *Euphytica* 69, 211–218.
Hajjar, R. and Hodgkin, T. (2007) The use of wild relatives in crop improvement: a survey of developments over the last 20 years. *Euphytica* 156, 1–13.
Hawkes, J.G. (1991) International workshop on dynamic *in situ* conservation of wild relatives of major cultivated plants: summary of final discussion and recommendations. *Israel Journal of Botany* 40, 529–536.
Hawkes, J.G., Maxted, N. and Ford-Lloyd, B.V. (2000) *The* ex situ *Conservation of Plant Genetic Resources*. Kluwer, Dordrecht, the Netherlands.
Hellier, B.C. (2000) Genetic, morphologic, and habitat diversity of two species of *Allium* native to the Pacific Northwest, USA and their implications for *in situ* seed collection for the National Plant Germplasm System. MSc thesis. Washington State University, Pullman, Washington.
Heywood, V.H. (1999) The role of botanic gardens in *ex situ* conservation of agrobiodiversity. In: Gass, T., Frese, L., Begemann, F. and Lipman, E. (eds) *Implementation of the Global Plan of Action in Europe – Conservation and Sustainable Utilization of Plant Genetic Resources for Food and Agriculture*. Proceedings of the European Symposium, 30 June–3 July 1988. Braunschweig, Germany. International Plant Genetic Resources Institute, Rome, pp. 102–107.
Heywood, V.H. (2011) Conservation strategies for species/populations occurring outside of protected areas. In: Hunter, D. and Heywood, V.H. (eds) *Crop Wild Relatives: A Manual of in situ Conservation*. Earthscan, UK, pp. 253–274.
Hopkins, J. and Maxted, N. (2010) *Crop Wild Relatives: Plant Genetic Conservation for Food Security*. Natural England, Peterborough, UK.

Hübner, S., Korol, A.B. and Schmid, K.J. (2015) RNA-Seq analysis identifies genes associated with differential reproductive success under drought-stress in accessions of wild barley *Hordeum spontaneum*. *BMC Plant Biology* 15, 134.

Hunter, D., Maxted, N., Heywood, V.H., Kell, S. and Borelli, T. (2012) Protected areas and the challenge of conserving crop wild relatives. *Parks* 18(1), 87–98.

Idohou, R., Assogbadjo, A.E., Fandohan, B., Gouwakinnou, G.N., Kakai, R.L.G., *et al.* (2012) National inventory and prioritization of crop wild relatives: case study for Benin. *Genetic Resources and Crop Evolution* 60(4), 1337–1352.

Iltis, H.H. (1988) Serendipity in the exploration of biodiversity. What good are weedy tomatoes? In: Wilson, E.O (ed.) *Biodiversity*. National Academy Press, Washington, DC, pp. 98–105.

INIBAP/IPGRI (2006) *Global Conservation Strategy for Musa (Banana and Plantain).* INIBAP/IPGRI, Montpellier/Rome.

Iriondo, J.M., Maxted, N., Kell, S.P., Ford-Lloyd, B.V., Lara-Romero, C., *et al.* (2012) Quality standards for genetic reserve conservation of crop wild relatives. In: Maxted, N., Dulloo, M.E., Ford-Lloyd, B.V., Frese, L., Iriondo, J.M. and Pinheiro de Carvalho, M.A.A. (eds) *Agrobiodiversity Conservation: Securing the Diversity of Crop Wild Relatives and Landraces*. CAB International, Wallingford, UK, pp. 72–77.

IUCN (International Union for Conservation of Nature) (2001) *The IUCN Red List of Threatened Species. 2001 Categories and Criteria (version 3.1).* IUCN Species Survival Commission. IUCN, Gland, Switzerland, and Cambridge, UK. Available at: www.iucnredlist.org/static/categories_criteria_3_1 (accessed 6 March 2015).

IUCN (2014) Plants for People: Conserving the world's most important trees, crop wild relatives, and medicinal plants. IUCN, Gland, Switzerland, and Cambridge, UK. Available at: www.iucn.org/about/work/programmes/species/our_work/plants/plants_projects_initiatives/plants_for_people_/ (accessed 6 March 2015).

Jain, S.K. (1975) Genetic reserves. In: Frankel, O.H. and Hawkes, J.G. (eds) *Crop Genetic Resources for Today and Tomorrow*. Cambridge University Press, Cambridge, UK, pp. 379–396.

Jarvis, A., Lane, A. and Hijmans, R.J. (2008) The effect of climate change on crop wild relatives. *Agriculture, Ecosystems and Environment* 126(1), 13–23.

Jarvis, S.G., Fielder, H., Hopkins, J., Maxted, N. and Smart, S. (2015) Distribution of crop wild relatives of conservation priority in the UK landscape. *Biological Conservation* 191, 444–451.

Jennings, D.L. (1995) Cassava, *Manihot esculenta* (Euphorbiaceae). In: Smartt, J. and Simmonds, N.W. (eds) *Evolution of Crop Plants*. Longman Group, Harlow, Essex, UK, pp. 128–132.

Jones, J.D.G., Witek, K., Verweij, W., Jupe, F., Cooke, D., *et al.* (2014) Elevating crop disease resistance with cloned genes. *Philosophical Transactions of the Royal Society B* 369, 20130087.

Karagöz, A. (1998) *In situ* conservation of plant genetic resources in the Ceylan State Farm. In: Zencirci, N., Kaya, Z., Anikster, Y. and Adams, W.T. (eds) *The Proceedings of International Symposium on in situ Conservation of Plant Diversity*. Central Research Institute for Field Crops, Ankara, pp. 87–92.

Kaur, N., Street, K., Mackay, M., Yahiaoui, N. and Keller, B. (2008) Allele mining and sequence diversity at the wheat powdery mildew resistance locus Pm3. In: Appels, R., Eastwood, R., Lagudah, E., Langridge, P., Mackay, M., Mcintyre, L. and Sharp, P. (eds) *11th International Wheat Genetics Symposium*. Sydney University Press, Brisbane, Australia, pp. 32–34.

Kell, S.P., Knüpffer, H., Jury, S.L., Ford-Lloyd, B.V. and Maxted, N. (2008) Crops and wild relatives of the Euro-Mediterranean region: making and using a conservation catalogue. In: Maxted, N., Ford-Lloyd, B.V., Kell, S.P., Iriondo, J., Dulloo, E. and Turok, J. (eds) *Crop Wild Relative Conservation and Use*. CAB International, Wallingford, UK, pp. 69–109.

Kell, S.P., Maxted, N. and Bilz, M. (2012) European crop wild relative threat assessment: knowledge gained and lessons learnt. In: Maxted, N., Dulloo, M.E., Ford-Lloyd, B.V., Frese, L., Iriondo, J.M. and Pinheiro de Carvalho, M.A.A. (eds) *Agrobiodiversity Conservation: Securing the Diversity of Crop Wild Relatives and Landraces*. CAB International, Wallingford, UK, pp. 218–242.

Kell, S., Qin, H., Chen, B., Ford-Lloyd, B.V., Wei, W., *et al.* (2015) China's crop wild relatives: diversity for agriculture and food security. *Agriculture, Ecosystems and Environment* 209, 138–154, doi.org/10.1016/j.agee.2015.02.012.

Kelley, C.P., Mohtadib, S., Canec, M.A., Seagerc, R. and Kushnirc, Y. (2015) Climate change in the Fertile Crescent and implications of the recent Syrian drought. *Proceedings of the National Academy of Sciences of the United States of America* 112(11), 3241–3246. Available at: www.pnas.org/content/112/11/3241 (accessed 4 March 2015).

Khazaei, H., Street, K., Bari, A., Mackay, M. and Stoddard, F.L. (2013) The FIGS (focused identification of germplasm strategy) approach identifies traits related to drought adaptation in *Vicia faba* genetic resources. *PLoS ONE* 8(5), e63107, doi:10.1371/journal.pone.0063107.

Khoury, C.K., Greene, S., Wiersema, J., Maxted, N., Jarvis, A., *et al.* (2013) An inventory of crop wild relatives of the United States. *Crop Science* 53, 1496–1508.

Khoury, C.K., Castañeda-Álvarez, N.P., Achicanoy, H.A., Sosa, C.C., Bernau, V., *et al.* (2015a) Crop wild relatives of pigeonpea [*Cajanus cajan* ( L .) Millsp.]: Distributions, *ex situ* conservation status, and potential genetic resources for abiotic stress tolerance. *Biological Conservation* 184, 259–270, doi:10.1016/j.biocon.2015.01.032

Khoury, C.K., Heider, B., Castañeda-Álvarez, N.P., Achicanoy, H., Sosa, C.C., *et al.* (2015b) Distributions, *ex situ* conservation priorities, and genetic resource potential of crop wild relatives of sweetpotato [*Ipomoea batatas* (L.) Lam., I. series Batatas]. *Frontiers in Plant Science* 6, 251, doi:10.3389/fpls.2015.00251.

King, I.P., Forster, B.P., Law, C.C., Cant, K.A., Orford, S., *et al.* (1997a) Introgression of salt tolerance genes from *Thinopyrum bessarabicum* into wheat. *New Phytologist* 137, 75–81.

King, I.P., Law, C.C., Cant, K.A., Orford, S., Gorham, J., *et al.* (1997b) *Tritipyrum*, a potential new salt-tolerant cereal. *Plant Breeding* 116, 127–132.

Kornas, J. (1990) Plant invasions in Central Europe: historical and ecological aspects. In: Di Castri, F., Hansen, A.J. and Debussche, M. (eds) Biological Invasions in Europe and the Mediterranean Basin. *Monographic Biology* 65, 19–36.

Kűçűk, S.A., Tan, A.Ş., Sabanci, C.O., Cinsoy, A.S., Őnul, K., *et al.* (1998) Ecogeographic and floristic differentiation of chestnut gene management zone in Kazdağ. In: Zencirci, N., Kaya, Z., Anikster, Y. and Adams, W.T. (eds) *The Proceedings of International Symposium on in situ Conservation of Plant Diversity*. Central Research Institute for Field Crops, Ankara, pp. 135–148.

Lexer, C., Lai, Z. and Rieseberg, L.H. (2004) Candidate gene polymorphisms associated with salt tolerance in wild sunflower hybrids: implications for the origin of *Helianthus paradoxus*, a diploid hybrid species. *New Phytologist* 161, 225–233.

Lira, R., Téllez, O. and Dávila, P. (2009) The effects of climate change on the geographic distribution of Mexican wild relatives of domesticated Cucurbitaceae. *Genetic Resources and Crop Evolution* 56, 691–703.

McCouch, S., Baute, G.J., Bradeen, J., Bramel, P., Bretting, P.K., *et al.* (2013) Agriculture: feeding the future. *Nature* 499 (7456), 23–24.

Mackay, M.C. and Street, K. (2004) Focused identification of germplasm strategy – FIGS. In: Black, C.K., Panozzo, J.F. and Rebetzke, G.J. (eds) *Cereals 2004*. Proceedings of the 54th Australian Cereal Chemistry Conference and the 11th Wheat Breeders' Assembly, 21–24 September 2004, Canberra, Australian Capital Territory (ACT). Cereal Chemistry Division, Royal Australian Chemical Institute, Melbourne, Australia, pp. 138–141.

McNeely, J.A., Mittermeier, R.A., Brooks, T.M., Boltz, F. and Ash, N. (2009) *The Wealth of Nature: Ecosystem Services, Biodiversity and Human Well-being*. Cemex, Arlington, Washington.

Magos Brehm, J., Maxted, N., Martins-Loução, M.A. and Ford-Lloyd, B.V. (2010) New approaches for establishing conservation priorities for socio-economic important plant species. *Biodiversity and Conservation* 19(9), 2715–2740.

Malory, S., Shapter, F.M., Elphinstone, M.S., Chivers, I.H. and Henry, R.J. (2011) Characterizing homologues of crop domestication genes in poorly described wild relatives by high-throughput sequencing of whole genomes. *Plant Biotechnology Journal* 9, 1131–1140.

Marfil, C.F., Hidalgo, V. and. Masuelli, R.W. (2015) *In situ* conservation of wild potato germplasm in Argentina: example and possibilities. *Global Ecology and Conservation* 3, 461–476.

Maxted, N. and Kell, S.P. (2009) Establishment of a Global Network for the In Situ Conservation of Crop Wild Relatives: Status and Needs. Commission on Genetic Resources for Food and Agriculture. FAO, Rome. Available at: www.fao.org/docrep/013/i1500e/i1500e18a.pdf (accessed 6 March 2015).

Maxted, N., Van Slageren, M.W. and Rihan, J. (1995) Ecogeographic surveys. In: Guarino, L., Ramanatha Rao, V. and Reid, R. (eds) *Collecting Plant Genetic Diversity: Technical Guidelines*. CAB International, Wallingford, UK, pp. 255–286.

Maxted, N., Hawkes, J.G., Guarino, L. and Sawkins, M. (1997a) Towards the selection of taxa for plant genetic conservation. *Genetic Resources and Crop Evolution* 44, 337–348.

Maxted, N., Ford-Lloyd, B.V. and Hawkes, J.G. (1997b) Complementary conservation strategies. In: Maxted, N., Ford-Lloyd, B.V. and Hawkes, J.G. (eds) *Plant Genetic Conservation: The in situ Approach*. Chapman & Hall, London, pp. 20–55.

Maxted, N., Mabuza-Diamini, P., Moss, H., Padulosi, S., Jarvis, A., *et al.* (2004) *Systematic and Ecogeographic Studies on Crop Genepools 11: An Ecogeographic Study African* Vigna. International Plant Genetic Resources Institute, Rome.

Maxted, N., Ford-Lloyd, B.V., Jury, S., Kell, S.P. and Scholten, M.A. (2006) Towards a definition of a crop wild relative. *Biodiversity and Conservation* 15(8), 2673–2685.

Maxted, N., Scholten, M.A., Codd, R. and Ford-Lloyd, B.V. (2007) Creation and use of a national inventory of crop wild relatives. *Biological Conservation* 140, 142–159.

Maxted, N., White, K., Valkoun, J., Konopka, J. and Hargreaves, S. (2008a) Towards a conservation strategy for *Aegilops* species. *Plant Genetic Resource: Characterization and Utilization* 6(2), 126–141.

Maxted, N., Dulloo, M.E., Ford-Lloyd, B.V., Iriondo, J. and Jarvis, A. (2008b) Genetic gap analysis: a tool for more effective genetic conservation assessment. *Diversity and Distributions* 14, 1018–1030.

Maxted, N., Ford-Lloyd, B.V. and Kell, S.P. (2008c) Crop wild relatives: establishing the context. In: Maxted, N., Ford-Lloyd, B.V., Kell, S.P., Iriondo, J., Dulloo, E. and Turok, J. (eds) *Crop Wild Relative Conservation and Use*. CAB International, Wallingford, UK, pp. 3–30.

Maxted, N., Iriondo, J., Dulloo, E. and Lane, A. (2008d) Introduction: The integration of PGR conservation with protected area management. In: Iriondo, J.M., Maxted, N. and Dulloo, E. (eds) *Plant Genetic Population Management*. CAB International, Wallingford, UK, pp. 1–22.

Maxted, N., Iriondo, J., De Hond, L., Dulloo, E., Lefèvre, F., *et al.* (2008e) Genetic reserve management. In: Iriondo, J.M., Maxted, N. and Dulloo, E. (eds) *Plant Genetic Population Management*. CAB International, Wallingford, UK, pp. 65–87.

Maxted, N., Kell, S.P., Toledo, A., Dulloo, M.E., Heywood, V., *et al.* (2010) A global approach to crop wild relative conservation: securing the gene pool for food and agriculture. *Kew Bulletin* 65, 561–576.

Maxted, N., Castañeda Álvarez, N.P., Vincent, H.A. and Magos Brehm, J. (2012a) Gap analysis: A tool for genetic conservation. In: Guarino, L., Ramanatha Rao, V. and Goldberg, E. (eds) *Collecting Plant Genetic Diversity: Technical Guidelines. 2011 update*. Bioversity International, Rome. Available at: cropgenebank.sgrp.cgiar.org/index.php?option=com_content&view=article&id=678 (accessed 6 March 2015).

Maxted, N., Kell, S.P., Ford-Lloyd, B.V., Dulloo, M.E. and Toledo, A. (2012b) Toward the systematic conservation of global crop wild relative diversity. *Crop Sciences* 52(2), 774–785.

Maxted, N., Hargreaves, S.P., Kell, S.P., Amri, A., Street, K., *et al.* (2012c) Temperate forage and pulse legume genetic gap analysis. *Bocconea* 24, 5–36.

Maxted, N., Magos Brehm, J. and Kell, S.P. (2013) *Resource Book for the Preparation of National Plans for Conservation of Crop Wild Relatives and Landraces.* Commission on Genetic Resources for Food and Agriculture. FAO, Rome, 457 pp. Available at: www.fao.org/fileadmin/templates/agphome/documents/PGR/PubPGR/ResourceBook/TEXT_ALL_2511.pdf (accessed 3 November 2015).

Maxted, N., Avagyan, A., Frese, L., Iriondo, J.M., Magos Brehm, J., *et al.* (2015) Preserving diversity: a concept for *in situ* conservation of crop wild relatives in Europe: version 2. Wild Species Conservation in Genetic Reserves Working Group, European Cooperative Programme for Plant Genetic Resources, Rome. Available at: www.pgrsecure.org/documents/Concept_v2.pdf (accessed 6 March 2015).

Meilleur, B.A. and Hodgkin, T. (2004) *In situ* conservation of crop wild relatives. *Biodiversity and Conservation* 13, 663–684.

Merenlender, A.M., Hutsinger, L., Guthey, G. and Fairfax, S.K. (2004 Land trusts and conservation easements: who is conserving what for whom? *Conservation Biology* 18, 67–75.

Noori, S.A.S. (2005) Assessment for salinity tolerance through intergeneric hybridization: *Triticum durum* x *Ae. speltoides. Euphytica* 146, 149–155.

Ortiz, R., Trethowan, R., Ortiz Ferrara, G., Iwanaga, M., Dodds, J.H., *et al.* (2007) High yield potential, shuttle breeding, genetic diversity, and a new international wheat improvement strategy. *Euphytica* 157(3), 365–384.

Panella, L., Landucci, S., Torricelli, R., Gigante, D., Donnini, D., *et al.* (2014) The national crop wild relative strategy for Italy: first steps to be taken. PGR Secure [online] Available at: www.pgrsecure.bham.ac.uk/sites/default/files/documents/public/National_CWR_Conservation_Strategy_Italy.pdf (accessed 6 March 2015).

Parra-Quijano, M., Iriondo, J.M. and Torres, E. (2012) Ecogeographical land characterization maps as a tool for assessing plant adaptation and their implications in agrobiodiversity studies. *Genetic Resources and Crop Evolution* 59, 205–217.

Pavek, D.S., Lamboy, W.F. and Garvey, E.J. (2003) Selecting *in situ* conservation sites for grape genetic resources in the USA. *Genetic Resources and Crop Evolution* 50, 165–173.

Phillips, J., Kyratzis, A., Christoudoulou, C., Kell, S. and Maxted, N. (2014) Development of a national crop wild relative conservation strategy for Cyprus. *Genetic Resources and Crop Evolution* 61(4), 817–827.

Pimentel, D., Wilson, C., McCullum, C., Huang R., Dwen, P., *et al.* (1997) Economic and environmental benefits of biodiversity. *BioScience* 47, 747–757.
Pinheiro de Carvalho, M.Â.A., Nóbrega, H., Freitas, G., Fontinha, S. and Frese, L. (2012) Towards the establishment of a genetic reserve for *Beta patula* Aiton. In: Maxted, N., Dulloo, M.E., Ford-Lloyd, B.V., Frese, L., Iriondo, J.M. and Pinheiro de Carvalho, M.A.A. (eds) *Agrobiodiversity Conservation: Securing the Diversity of Crop Wild Relatives and Landraces.* CAB International, Wallingford, UK, pp. 36–44.
Porter, J.R., Xie, L., Challinor, A., Cochrane, K., Howden, M., *et al.* (2014) Food Security and Food Production Systems. In: Field, C.B., Barros, V.R., Dokken, D.J., Mach, K.J., Mastrandrea, M.D., Bilir, T.E., Chatterjee, M., Ebi, K.L., Estrada, Y.O., Genova, R.C., Girma, B., Kissel, E.S., Levy, A.N., MacCracken, S., Mastrandrea, P.R. and White, L.L. (eds) *Climate Change 2014: Impacts, Adaptation, and Vulnerability. Part A: Global and Sectoral Aspects. Contribution of Working Group II to the Fifth Assessment Report of the Intergovernmental Panel on Climate Change.* Cambridge University Press, Cambridge, UK, and New York, pp. 485–533.
PwC (2013) *Crop Wild Relatives: A Valuable Resource for Crop Development.* PwC Valuations, PwC, London.
Ramírez-Villegas, J., Khoury, C., Jarvis, A., Debouck, D.G. and Guarino, L. (2010) A gap analysis methodology for collecting crop gene pools: a case study with *Phaseolus* beans. *PLoS ONE* 5(10), 1–18.
Rick, C.M. and Chetelat, R. (1995) Utilization of related wild species for tomato improvement. First International Symposium on Solanaceae for Fresh Market. *Acta Horticulturae* 412, 21–38.
Rubio Teso, M.L., Iriondo, J.M., Parra, M. and Torres, E. (2013) National strategy for the conservation of crop wild relatives of Spain. Available at: www.pgrsecure.bham.ac.uk/sites/default/files/documents/public/National_CWR_Conservation_Strategy_Spain.pdf (accessed 6 March 2015).
Safriel, U.N., Anikster, Y. and Waldman, M. (1997) Management of nature reserves for conservation of wild relatives and the significant of marginal populations. *Bocconea* 7, 233–239.
Sanchez-Velasquez, L.R. (1991) *Zea diploperennis*: Mejoramiento genetico del maiz, ecologia y la conservación de recursos naturales. *Tiempos de ciencia 24 Qulio-septiembre:* 1–8. University of Guadalajara, Jalisco, Mexico.
Shaffer, M.L. (1981) Minimum population sizes for species conservation. *Bioscience* 31(2), 131–134.
Stamp, P. and Visser, R. (2012) The twenty-first century, the century of plant breeding. *Euphytica* 186, 585–591.
Stebbins, G.L. (1958) The inviability, weakness, and sterility of interspecific hybrids. *Advances in Genetics* 9, 147–215.
Stevens, M.A. and Rick, C.M. (1986) *The Tomato Crop. A Scientific Basis for Improvement.* Chapman and Hall, New York.
Stolton, S., Maxted, N., Kell, S.P., Ford-Lloyd, B. and Dudley, N. (2008) Protected areas and plant agrobiodiversity. In: Secretariat of the Convention on Biological Diversity, *Protected Areas in Today's World: Their Values and Benefits for the Welfare of the Planet.* CBD Technical Series No 36, Montreal, pp. 42–49.
Street, K., Mackay, M., Zuev, E., Kaul, N., El Bouhssini, M., *et al.* (2008) Diving into the genepool – a rational system to access specific traits from large germplasm collections. Available at hdl.handle.net/2123/3390 (accessed 12 September 2014).
Talame, V., Sanguineti, M.C., Chiapparino, E., Bahri, H., Salem, B.M., *et al.* (2004) Identification of *Hordeum spontaneum* QTL alleles improving field performance of barley grown under rainfed conditions. *Annals of Applied Biology* 144, 309–319.
Tanksley, S.D. and McCouch, S.R. (1997) Seed banks and molecular maps: unlocking genetic potential from the wild. *Science* 277, 1063–1066.
Thormann, I., Parra-Quijano, M., Endresen, D.T.F., Rubio Teso, M.L., Iriondo, M.J., *et al.* (2014) Predictive characterization of crop wild relatives and landraces. Technical guidelines version 1. Bioversity International, Rome. Available at: www.bioversityinternational.org/index.php?id=244&tx_news_pi1%5Bnews%5D=4967&cHash=7cd3c6c2b8360927b83fa6ef7cc28d99 (accessed 6 March 2015).
TNC (The Nature Conservancy) (2015) Private Lands Conservation. The Nature Conservancy, Arlington, Virginia. Available at: www.nature.org/about-us/private-lands-conservation/conservation-easements/index.htm (accessed 6 March 2015).
Tucker, T., Marra, M. and Friedman, J.M. (2009) Massively parallel sequencing the next big thing in genetic medicine. *American Journal of Human Genetics* 85(2), 142–54.
United Nations (2015) *World Population Prospects: The 2015 Revision.* Population Division, Population Estimates and Projections Section, United Nations Department of Economic and Social Affairs, New York. Available at: esa.un.org/unpd/wpp/index.htm (accessed 3 November 2015).
Vavilov, N.I. (1949) The origin, variation, immunity and breeding of cultivated plants. *Chronika Botanica* 13, 1–54.

Villar, E., Klopp, C., Noirot, C., Novaes, E., Kirst, M., *et al.* (2011) RNA-Seq reveals genotype-specific molecular responses to water deficit in eucalyptus. *BMC Genomics* 12, 538, doi:10.1186/1471-2164-12-538.

Villareal, R.L., Sayre, K., Banuelos, O. and Mujeeb-Kazi, A. (2001) Registration of four synthetic hexaploid wheat (*Triticum turgidum*/*Aegilops tauschii*) germplasm lines tolerant to waterlogging. *Crop Science* 41, 274.

Vincent, H., Wiersema, J., Kell, S., Fielder, H., Dobbie, S., *et al.* (2013) A prioritized crop wild relative inventory to help underpin global food security. *Biological Conservation* 167, 265–275.

Vincent, H., Castañeda-Álvarez, N.P. and Maxted, N. (2016) Global priorities for *in situ* conservation of wild plant genetic resources: towards the establishment of a global network of crop wild relative reserves (in preparation).

Vollbrecht, E. and Sigmon, B. (2005) Amazing grass: developmental genetics of maize domestication. *Biochemical Society Transactions* 33, 1502–1506.

Wang, W., Vinocur, B. and Altaian, A. (2003) Plant responses to drought, salinity and extreme temperatures: towards genetic engineering for stress tolerance. *Planta* 218, 1–14.

Warren, R., VanDerWal, J., Price, J., Welbergen, J.A., Atkinson, I., *et al.* (2013) Quantifying the benefit of early climate change mitigation in avoiding biodiversity loss. *Nature Climate Change* 3(5), 1–5.

Waugh, R., Jannink, J., Muehlbauer, G.J. and Ramsay, L. (2009) The emergence of whole genome association scans in barley. *Current Opinion in Plant Biology* 12, 218–222.

Zeven, A.C., Knott, D.R. and Johnson, R. (1983) Investigation of linkage drag in near isogenic lines of wheat by testing for seedling reaction to races of stem rust, leaf rust, and yellow rust. *Euphytica* 32, 319–327.

Zhang, X., Zhou, S., Fu, Y., Su, Z., Wang, X.K., *et al.* (2006) Identification of a drought tolerant introgression line derived from Dongxiang common wild rice (*Oryza rufipogon* Griff.). *Plant Molecular Biology* 62, 247–259.

# 11 Europe's Crop Wild Relative Diversity: From Conservation Planning to Conservation Action

**S.P. Kell,* B.V. Ford-Lloyd and N. Maxted**
*School of Biosciences, University of Birmingham, Birmingham, UK*

## 11.1 Introduction

### 11.1.1 The value of Europe's crop wild relative diversity

Europe has a wealth of native and endemic diversity of wild species related to crops of regional and global socio-economic importance. Heywood and Zohary (1995) drew attention to the significance of crop wild relative (CWR) diversity in the region, highlighting the 'rich wild gene pools' (p. 375) of several cereals, food legumes, fruit crops and vegetables, as well as aromatic plants, ornamentals and forestry crops. Examples include the native wild relative diversity of oats (*Avena sativa* L.), sugarbeet (*Beta vulgaris* L.), carrot (*Daucus carota* L.), apple (*Malus domestica* Borkh.), annual meadow grass (*Festuca pratensis* Huds.), perennial rye grass (*Lolium perenne* L.) and white clover (*Trifolium repens* L.). Many minor crop species have significant wild relative diversity in the region, including asparagus (*Asparagus officinalis* L.), lettuce (*Lactuca sativa* L.), sage (*Salvia officinalis* L.), raspberries and blackberries (*Rubus* spp.), as well as herbs and aromatic plants such as mints (*Mentha* spp.) and chives (*Allium* spp.) (Maxted *et al.*, 2008). Europe is also an important region for forest genetic resources, such as pine, poplar and sweet cherry (*Pinus*, *Populus* and *Prunus* spp.), and ornamental plants, such as sweet pea (*Lathyrus odoratus*), sweet pinks (*Dianthus* spp.) and violets (*Viola* spp.) (Kell *et al.*, 2008). The recent creation of the Harlan and de Wet inventory of globally important CWR taxa (www.cwrdiversity.org/checklist/) and the associated geographic analysis of their distribution (Vincent *et al.*, 2013) indicated that southern Europe was globally significant in terms of its richness of species related to economically important crops.

Today, agricultural production is challenged by climate change. Although food production in Europe is likely to be less affected by climate change in the first half of the 21st century than some other regions of the world (IPCC, 2007), an increase in extreme weather events due to climate change can have far-reaching impacts. For example, in 2003, temperatures reached 6°C above long-term averages and there were rainfall shortages of up to 300 mm (Trenberth *et al.*, 2007). These extreme climatic changes resulted in estimated economic losses in the European Union (EU) agriculture sector of €13 billion (Létard *et al.*, 2004). In a review of the implications of climate change for food plant production in Europe, Kovats *et al.* (2014) highlight the significant negative effects of summer heatwaves and drought on grain yields, particularly in southern Europe. In an examination of climate change impacts on crop production under different

*Corresponding author; e-mail: s.kell@bham.ac.uk

 

climate change scenarios, the authors underline the fact that while negative impacts on crop yields in central and northern Europe will likely be less significant than in the south of the region, increased climatic variability could have marked effects in the north and greater crop damage could result from insect pests and pathogens in response to rising temperatures.

One option for mitigating the impacts of climate change on food production is to develop crop varieties with increased resistance to heat shock, drought, pests and diseases (Easterling *et al.*, 2007). The authors observe that the many climate change adaptation studies on wheat, rice and maize crops indicate that this option alone, or combined with other adaptations such as changes in planting times and locations and improved water management, has the potential to provide an average of 10% increase in yield across all regions, all crops and different temperature regimes. Kovats *et al.* (2014) also draw attention to the fact that at 'the high range of the projected temperature changes, only plant breeding aimed at increasing yield potential jointly with drought resistance and adjusted agronomic practices may reduce risks of yield shortfall'. Given the problem of increased climate variability, the authors also emphasize the need for greater use of between- (as well as within-) species genetic diversity in farming systems. This places even greater emphasis on the need to conserve a broad range of plant genetic diversity both within and between species in order to provide options for the adaptation of a wide range of crop species as an insurance against climate variability.

Plant breeders and farmers are therefore in need of a continuous supply of diverse and novel sources of genetic diversity to produce new crop varieties able to cope with the impacts of changing growing conditions (Hawkes *et al.*, 2000; FAO, 2008; Feuillet *et al.*, 2008; Maxted *et al.*, 2012, 2014, 2015). Due to the breadth of genetic diversity inherent in CWR populations, which are adapted to a wide range of environmental conditions and constantly changing in response to biotic and abiotic pressures (Maxted and Kell, 2009; Kell *et al.*, 2012a), they are likely to become increasingly important as sources of genetic diversity to produce crop varieties able to cope in the altered environmental conditions induced by climate change (FAO, 2008; Feuillet *et al.*, 2008; Maxted and Kell, 2009; Kell *et al.*, 2012a; Maxted *et al.*, 2012, 2013, 2014, 2015). CWR are therefore a fundamental component of plant genetic resources for food and agriculture (PGRFA) and may contribute significantly to future food security (FAO, 2008; Maxted and Kell, 2009; Maxted *et al.*, 2011, 2012, 2014, 2015; Kell *et al.*, 2012a).

### 11.1.2 What do we know about European CWR diversity?

Kell *et al.* (2005) created the first comprehensive catalogue of CWR for Europe and the Mediterranean using a broad definition of a CWR (i.e. any species in the same genus or closely related genera to any type of cultivated plant species), and found that approximately 80% of the flora of the region consisted of crops and their wild relatives. Further analysis revealed that more than 15,000 species are native to Europe, of which at least half are endemic (Kell *et al.*, 2008). The authors found that four countries contain more than 20% of the species in the Euro-Mediterranean region: Turkey, Spain, Italy and France, which is consistent with the overall proportions of the flora of the region that occur in these countries. However, per unit area, Greece has the highest concentration of CWR diversity in the region. A high percentage of CWR species occur on the EU's oceanic islands, such as the Canary Islands (Spain) and the Azores (Portugal), but also other islands such as Sicily (Italy) and Malta, and Corsica (France). For example, around 10% of the crop and CWR taxa of the Spanish territories occur in the Canary Islands – taxa that are not found in mainland Spain (Kell *et al.*, 2008). This is not surprising, since islands exhibit high levels of endemism due to their isolation from continental areas, so they are natural reservoirs of unique genetic diversity (Dulloo *et al.*, 2002).

In the most comprehensive assessment of the Red List status of CWR to date, 572 European species in 25 economically important crop genepools/groups were assessed (Bilz *et al.*, 2011; Kell *et al.*, 2012b). The results of this study show that at least 11.5% (66) of the species are threatened, with 3.3% (19) of them being Critically Endangered (CR), 4.4% (22) Endangered (EN) and 3.8% (25) Vulnerable (VU). A further 4.5% (26) of the species are classified as Near

Threatened and one species (*Allium jubatum* J.F. Macbr.) is Regionally Extinct. The remaining species are regionally assessed as Data Deficient (DD) (29%) or Least Concern (LC) (54.7%). However, of the species assessed as LC, around one-third are threatened at national level (Kell *et al.*, 2012b). The same authors reported that of the 25 crop genepools/groups for which the European CWR were assessed, at least 14 contain regionally and/or globally threatened (CR, EN or VU) or Near Threatened (NT) species (92 species in total, of which 65 are endemic to Europe), the highest number occurring in the brassica complex, which in total contains 137 species native to and with a significant proportion of the global population in Europe. At least 8–50% of the species assessed in each of these crop genepools/groups are threatened or NT, and these percentages are likely to increase when the DD species are re-evaluated. Crop genepools/groups of particular concern in terms of the percentage of regionally threatened wild species are brassica, beet, lettuce, wheat and allium (Kell *et al.*, 2012b).

Kell *et al.* (2012b) also analysed the factors threatening CWR diversity and reported 31 distinct threats – the most frequent being 'livestock farming and ranching', 'tourism and recreation areas' and 'housing and urban areas'. However, the authors note that we should not conclude that farming *per se* is threatening CWR diversity – in fact, farmed areas (including arable land and pasture) are one of the primary habitats of CWR species. It is unsustainable farming practices, such as severe overgrazing, conversion of land to monocultures and the heavy application of fertilizers and herbicides that are the major threats to CWR that grow in agricultural areas (Kell *et al.*, 2012b). Although there is sufficient evidence that CWR diversity is threatened by climate change (Maxted *et al.*, 2014), IUCN Red List assessments do not reflect this directly, as the impacts are often less direct and so cannot be attributed unequivocally to climate change. What is actually noted is overgrazing, increased threat from fires or competition from alien species, each of which may have at its foundation changes in the biotic or abiotic environment themselves attributable to climate change (Kell *et al.*, 2012b).

Given their value and threatened status, it might be expected that CWR would have been the focus of specific, systematic conservation efforts, but this is far from the case, either in Europe or elsewhere (Hoyt, 1988; Maxted *et al.*, 1997a, 2011, 2013, 2015; Maxted, 2003; FAO, 2010; Kell *et al.*, 2012a). For example, based on data available in EURISCO, only around 9% of total germplasm accessions in European genebanks are of wild origin (Dias *et al.*, 2012). Further, the ratio of the number of accessions of cultivated species to wild species is striking at 12:1, with an average of 167 for each cultivated species and 14 for each wild species. This may be explained in part by breeders' historic focus on the exploitation of genetic diversity existing within crops, but it is still surprising given that most diversity in crop genepools is located in the related wild species (Maxted *et al.*, 2008). The situation is even less satisfactory for *in situ* CWR conservation. Recently, a set of standards for genetic reserve *in situ* CWR conservation was established, but it is thought that no European protected areas currently meet these standards and only a few claim to be actively conserving CWR diversity *in situ* (Anikster *et al.*, 1997; Tan and Tan, 2002; Avagyan, 2008; Pinheiro de Carvalho *et al.*, 2012).

Although CWR have been used by plant breeders to broaden crop breeding pools since the early 20th century, the conservation of CWR has only been addressed by policy makers relatively recently. However, no mechanisms currently exist to organize technically coordinated, effective and efficient *in situ* conservation actions for CWR across political borders in Europe, therefore, a systematic regional approach to *in situ* CWR genetic diversity conservation is required (Maxted, 2003, Maxted *et al.*, 2013, 2015). Critically, the same authors note that while there has been some embryonic but growing interest in the conservation of CWR by the nature conservation community, collaboration between the environmental and agricultural sectors at all geographic levels still needs to be improved – a situation that seems to have changed little since the need for an integrated, multidisciplinary approach to CWR conservation was highlighted by Heywood (1997).

### 11.1.3 Progress in implementing national CWR conservation strategies in Europe

The concept of developing national CWR conservation strategies proposed by Maxted *et al.* (2007) to help ensure more systematic CWR conservation has been widely adopted by the international PGRFA community as the standard framework for national CWR conservation planning. However, although some CWR taxa may have been included incidentally in national biodiversity strategies and action plans, national conservation strategies specifically for CWR have yet to be developed in most European countries. The EU-funded Seventh Framework Programme PGR Secure project (2011–2014; www.pgrsecure.org) provided the opportunity to extend the national strategy approach to a significant number of European countries, and has resulted in a sea change in CWR conservation in the region. National strategy development was supported with project funding in Finland, Italy, Spain and the UK, with additional funding allocated for the provision of technical support for the development of strategies in Albania, Bulgaria, Cyprus, Czech Republic and Norway. Substantial project resources were also committed to provide training and ongoing web-based and individual expert support to national PGR programmes throughout Europe (see Kell *et al.*, 2012c, and www.pgrsecure.org/helpdesk), which have resulted in the initiation of national strategies in a number of other countries, including Greece, Lithuania, Sweden and Turkey.

The development of national CWR conservation strategies in several European countries throughout the region has provided an opportunity to test the national strategy planning model and analyse variation in its application between countries. Initial comparisons indicate close correlation between countries with regard to the choice of prioritization criteria, as well as approaches to diversity and gap analyses, although there is some variation in the application of the criteria, as well as in the level and scope of the diversity analyses undertaken – the latter which may be guided largely by the availability of technical expertise and resources (see Iriondo *et al.*, Chapter 14, this volume). What is clear at this stage is that while much progress has been made in planning CWR conservation, few countries have actually implemented the strategy – *in situ* and *ex situ* conservation actions specifically for national CWR resources are in place in only a few rare cases and on an *ad hoc* basis.

A critical issue that has been highlighted by a number of European national PGR programmes is the lack of clarity with regard to the coordinating/implementing body responsible for CWR conservation. CWR often fall in the 'gaps' between the remits of the environmental protection (or nature conservation) agencies and the agricultural administration agencies (Maxted *et al.*, 2008). *In situ* populations of CWR require protection and active conservation in the same way as other wild species, but the focus of nature conservation agencies tends to be on habitat and rare or threatened species conservation. Incidentally, some of these species may be CWR and some target habitats may contain CWR, but the *in situ* conservation of priority national CWR diversity is rarely (in all but a few cases) a specific objective (Maxted, 2003). In terms of *ex situ* PGRFA (including CWR) conservation, this is mainly the responsibility of the agricultural sector. However, *ex situ* conservation of European CWR diversity is currently inadequate because the focus of PGRFA conservation has historically been mainly on the collection and conservation of crop germplasm (Maxted *et al.*, 2008). Furthermore, some national programmes have reported a lack of coordination, or even competition, between the environment protection agencies and agricultural administration when it comes to developing a policy on CWR conservation and management.

It is clear that the lack of national policy dedicated to CWR conservation and use is undermining progress in practical implementation of the CWR conservation needs identified in national CWR conservation strategy documents. The development of regional policy on CWR conservation is needed to obligate member states to fill the CWR conservation 'gaps' and take action to implement national CWR conservation priorities, thus driving forward CWR conservation across the region (Maxted *et al.*, 2013, 2015).

## 11.2 A Regional Approach to CWR Conservation in Europe

### 11.2.1 Why a regional CWR conservation strategy?

A national approach to CWR conservation (see Maxted *et al.*, 2007, 2013) is important because it is the responsibility of individual nations to conserve genetic resources within their jurisdiction, and they are required by the Second Global

Plan of Action (GPA) for Plant Genetic Resources for Food and Agriculture, the International Treaty on Plant Genetic Resources for Food and Agriculture (ITPGRFA) and the Convention on Biological Diversity (CBD) to report on their activities. However, national priorities vary between nations and are likely to be distinct from regional priorities. For example, some countries only assign priority to CWR of nationally important crops and not to those of regionally important crops that may occur in their country, or some prioritize only endemic CWR taxa. Thus, some populations of regionally important CWR are not included in national strategies. Conversely, some nationally important taxa are not considered a priority at the regional level, therefore, by producing a regional list of priority taxa based on national priorities, taxa that are not in need of immediate conservation action throughout their entire range would be included and resources misspent. This point was illustrated by Maxted *et al.* (1997b) with the example of *Vicia bithynica* L. It is native throughout central and southern Europe and the Mediterranean Basin, but the extreme north-western edge of its distribution is the south coast of Britain, where the species is found at a few localities, all of which are threatened by increased levels of tourism and natural coastal erosion. Here the populations are being genetically eroded and the species may in the near future even become extinct in the UK. However, it is thriving in its centre of diversity and is clearly not threatened at the international level. Therefore, while *in situ* conservation of *V. bithynica* is a priority in the UK, it is not a priority at the regional level.

In addition to national strategies, a Europe-wide CWR conservation strategy is thus needed to: (i) ensure that regionally important CWR resources are targeted for conservation action across their full range; (ii) ensure that resources are not allocated incorrectly to conservation of taxa that are nationally but not regionally important; and (iii) provide a framework for directing European policy on the conservation of PGRFA. Furthermore, in a context of limited resources assigned to CWR conservation by nations and European institutions, the development of a regional CWR conservation strategy may enable a more efficient way of conserving CWR resources than simply having the sum of the national strategies of the European countries.

The national and regional approaches to CWR conservation in Europe may be considered as top-down and bottom-up, respectively, but what is critical is that the two approaches are not viewed as independent of one another – rather, that they are harmonized and implemented in a coordinated way towards an integrated European CWR conservation strategy (Maxted *et al.*, 2013, 2015).

## 11.2.2 Developing the regional strategy

### *Baseline taxon data*

The baseline taxon data for the Europe-wide CWR conservation strategy is provided by the *Crop Wild Relative Catalogue for Europe and the Mediterranean* (the CWR Catalogue), originally published by Kell *et al.* (2005) as an output of the EU-funded Framework 5 project, PGR Forum (www.pgrforum.org). An outline of the methodology for creating the CWR Catalogue and the results of data analysis can be found in Kell *et al.* (2008).

The CWR Catalogue provides an overview of the breadth of crop and CWR diversity in the European region and the baseline data for conservation planning at the regional scale. Further, national CWR checklists have been extracted and provided to each European country for use in the national PGR programmes to form the basis of national checklists, inventories and, subsequently, national CWR conservation strategies and action plans. However, for the development of a Europe-wide CWR conservation strategy, it is necessary to select regional priority species – those with the greatest potential to contribute to food and economic security in the region.

### *Selection of priority CWR species at the regional level*

The role of prioritization in the conservation planning process is widely recognized. Three main criteria are of greatest relevance when assigning priorities to CWR species in the context of conservation planning:

**1.** The socio-economic value of the crop to which they are related (Ford-Lloyd *et al.*, 2008).
**2.** Their potential ease of use or known value in crop improvement programmes (Maxted and Kell, 2009; Maxted *et al.*, 2012).

**3.** Their relative threat status (Ford-Lloyd *et al.*, 2008; Maxted and Kell, 2009).

In general, priority is likely to be given to native species. However, depending on how long they have been present, some introduced CWR populations may harbour important genetic diversity, especially because many taxa are able to adapt rapidly to new environments (Ford-Lloyd *et al.*, 2014). Therefore, introduced populations may be considered for inclusion in CWR conservation plans once the policy to support the initial strategy is in place and actions implemented.

In 2009, European CWR species were prioritized on the basis of their socio-economic importance and native status in Europe in the context of the production of a European Red List (Bilz *et al.*, 2011; Kell *et al.*, 2012b). The initial selection was limited to the native wild relatives of food and forage/fodder crops due to their importance for food security. Within these major groups, crop genepools were selected on the basis of (i) production quantity and economic value in the region averaged over a five year period, and/or (ii) their inclusion in Annex I of the ITPGRFA – a list of PGRFA established according to criteria of food security and interdependence which includes 78 genera containing human or animal food crops, 59 of which contain food CWR or wild populations of forage/fodder species native to Europe. Species in the same genera as the priority crops or in closely related genera were selected, and this resulted in a list of 591 CWR species in 58 genera and 25 crop genepools/crop groups.[1] The Red List assessment of the initial list of 591 priority CWR species provided a snapshot of the threat status of CWR in Europe. However, the list was produced several years ago and required updating in line with the latest production quantity and economic value statistics, revised taxonomic classifications and some new prioritization concepts developed recently in the context of prioritizing China's CWR (Kell *et al.*, 2015). To this end, the following steps were taken to redefine Europe's priority CWR:

**1.** FAO crop production statistics (FAO, 2014) were consulted to obtain the annual production values of human food crops cultivated in Europe over the ten year period 2002–2011. Human food crops with an average annual value of more than US$500 million over this period that have CWR native to Europe were identified and the native wild relatives of these crops selected from the CWR Catalogue for Europe and the Mediterranean to create a base list of regionally important CWR based on the economic importance of the associated crops. The regional value of human food crops in terms of average annual energy supply per capita over the ten year period 2000–2009 was calculated from FAO food supply statistics (FAO, 2014) for Europe to highlight crops of particular regional value for food security.
**2.** By consulting the data available on the degree of relationship between the crop species and the wild relatives in the crop genepool and/or the known value of CWR in crop improvement programmes (Vincent *et al.*, 2013), taxa in the base priority list that are likely to have greater use value for crop improvement were identified.
**3.** The European Red List of Vascular Plants (Bilz *et al.*, 2011) was used to identify threatened and Near Threatened taxa in the base priority list, and those endemic to Europe were identified based on data in the CWR Catalogue for Europe and the Mediterranean. Using occurrence data in the CWR Catalogue for Europe and the Mediterranean, countries containing regionally and globally threatened or Near Threatened CWR were identified.
**4.** Lists of priority CWR at each of these three levels were compiled to create the European CWR Inventory with the purpose of providing the foundations for regional CWR conservation planning in Europe.

Using this prioritization approach, the species selected on the basis of criteria 1 and 2 combined are considered priorities for regional conservation planning, regardless of their regional (for native species) or global (for endemic species) Red List status. While some of these species are relatively widespread and not threatened at the taxon level, individual populations not only may be threatened but also may contain unique genetic diversity that could be valuable for crop improvement (Kell *et al.*, 2012b). An example is *Beta vulgaris* subsp. *maritima*, a primary wild relative of cultivated beets which grows in coastal areas and is widely distributed across much of Europe. If the Red List status of this taxon (which is Least Concern) was applied as a prioritization criterion, it might be overlooked in conservation planning at the regional scale. However, one of the main objectives of *in situ* CWR conservation is the maintenance of the optimum amount of genetic diversity of a genepool in nature, which is prerequisite for the evolutionary processes that generate novel genetic variation (Kell *et al.*, 2012b). On this basis, ten candidate populations of *B. vulgaris* subsp. *maritima* have been proposed for *in situ* conservation management because they are known to harbour genetic variation for traits useful for plant breeding (Kell *et al.*, 2012a).

### *The European CWR Inventory as a basis for regional conservation planning*

The purpose of identifying the highest priority crop genepools in Europe and those that contain taxa of greatest potential use for crop improvement and/or threatened or Near Threatened taxa is to produce a list of high priority taxa to inform regional CWR conservation planning. On the basis of the process described above, 192 CWR species native to Europe were identified as the highest priority taxa requiring immediate conservation status assessment and complementary conservation planning. The initial results of diversity and gap analyses have revealed that the responsibility for conserving these priority species is Europe-wide, with some 30 countries containing native, wild populations of 20 or more species. Alarmingly, less than half of the species are represented in genebank collections, and of these, around 50% are represented by only eight accessions or less (Kell *et al.*, 2014).

## 11.3 An Integrated European CWR Conservation Strategy

The integrated European CWR conservation strategy brings together the national and regional approaches to maximize the active conservation of priority populations of CWR taxa throughout the region (Fig. 11.1). In summary, the concept (which has been endorsed by the Steering Committee of the European Cooperative for Plant Genetic resources (ECPGR) – see Maxted *et al.*, 2015) is as follows (Maxted *et al.*, 2013, 2015):

**1. National CWR conservation strategy:** each country has its own national CWR conservation strategy implemented through *in situ* and *ex situ* activities undertaken by national agencies.

**2. Regional CWR conservation strategy:** this comprises a regional network of *in situ* CWR conservation populations backed up by germplasm collection and *ex situ* management, without consideration of national borders. A regional expert authority (e.g. the ECPGR Wild Species Conservation in Genetic Reserves Working Group) provides leadership on the identification of conservation targets, oversees the technical aspects of the implementation of the strategy, and provides support to the relevant national agencies responsible for the realization of the identified targets.

**3. Integrated European CWR conservation strategy:** the two distinct strategic levels are married into one coherent integrated whole:

- *National–regional integration:* priority national populations are nominated by the national PGR coordinator for inclusion in the integrated European CWR conservation strategy for formal recognition as part of the European network of priority *in situ* CWR populations.
- *Regional–national integration:* individual CWR conservation populations identified in the regional CWR conservation strategy are implemented at the national level as detailed in 2 above.

Critically, while the focus of *in situ* CWR conservation has historically been on the designation of genetic reserves, this concept proposes a paradigm shift away from the conservation of sites to the designation of priority CWR conservation populations (most appropriate crop wild relative populations – MAWPs). This new concept is appropriate given that: (i) many CWR populations occur outside of formally designated protected areas; (ii) environmental change in response to future climate scenarios is expected to cause range shifts in some species and a great deal of uncertainty with regard to the stability and suitability of the sites in which they occur; and (iii) if in extreme circumstances *in situ* conservation is unfeasible, greater emphasis can be placed on *ex situ* conservation and the possible introduction of populations to more stable localities. This change also reflects the reality of *in situ* conservation – in designating a genetic reserve for CWR conservation, it was not the site itself that

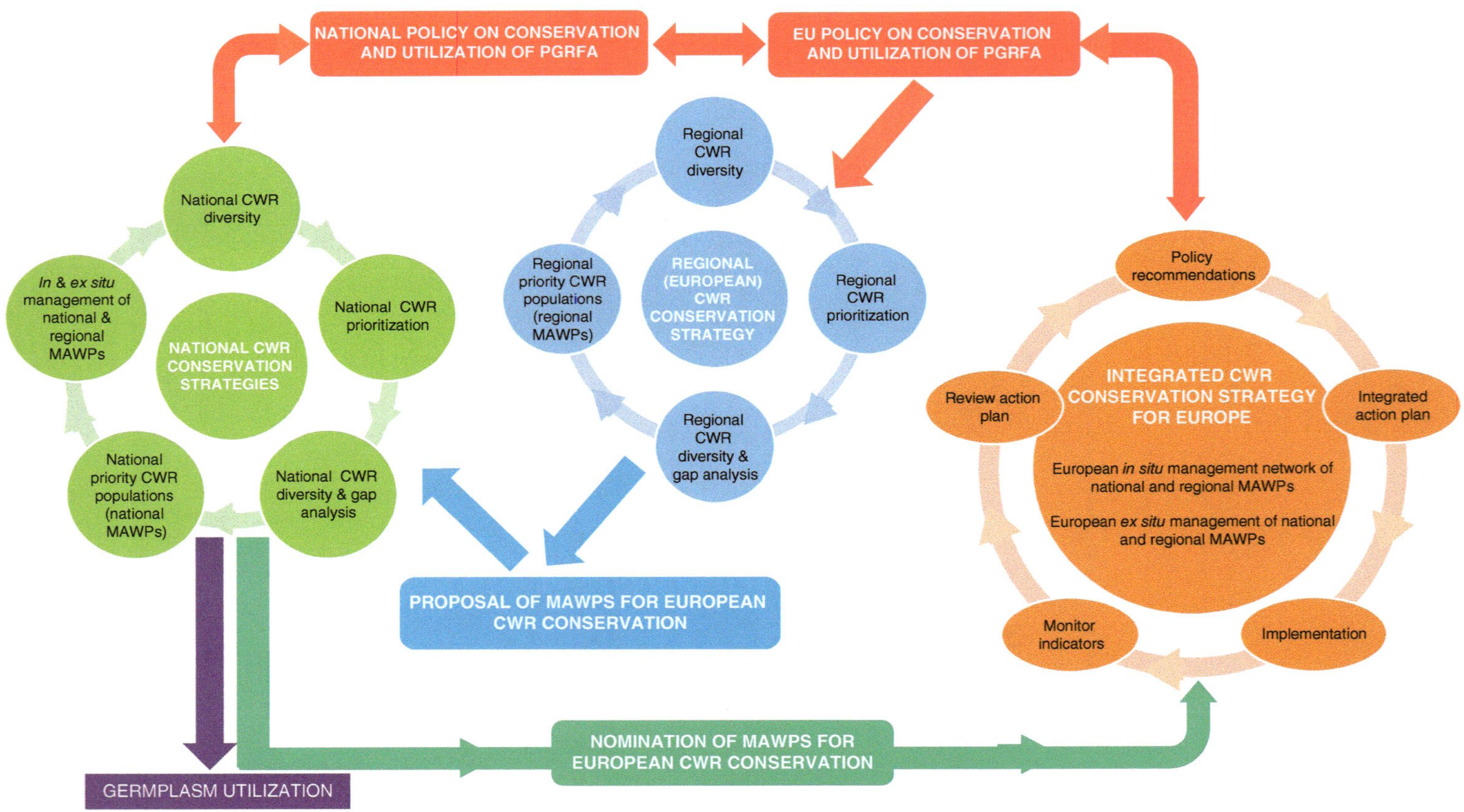


**Fig. 11.1.** Schematic representation of the concept for *in situ* conservation of CWR in Europe. (From Maxted *et al.*, 2013, 2015). *Note*: MAWPs = most appropriate crop wild relative populations.

was being conserved but the populations of the target taxa within that locality, so it was more accurate to actually refer to the populations themselves rather than the site.

As well as ensuring the conservation of national CWR diversity, the national network of MAWPs may also contribute to the European (regional) network of MAWPs if they contain CWR diversity of regional importance. In turn, the European network of MAWPs may also contribute to the global network if they contain CWR diversity of global importance. Thus, some national MAWPs, particularly those in Vavilov centres of diversity (Vavilov, 1926), may also be designated as MAWPs of international importance and be part of the regional and global networks. Conversely, it is logical that each MAWP included in the regional or global network is also nominated as part of a country's national CWR diversity network. As a result of the integration of national and regional strategies, there are likely to be cases where a number of site designations or target populations for *ex situ* conservation might duplicate conservation efforts. In this case, it may be necessary to eliminate the redundancy of multiple countries proposing similar conservation targets, with a corresponding increase in efficiency in the process (Maxted *et al.*, 2013).

The integrated European CWR conservation strategy will require periodic review and updating according to future developments in CWR conservation and utilization science and practice, as well as regional agrobiodiversity conservation policy. For example, while the initial focus is on the highest priority crop genepools to support the European agricultural industry and food security, in the future the strategy may be developed to include a wider range of socio-economically important crop genepools, particularly when a number of national CWR conservation strategies are available for review and comparison and in which particular crop genepools may be highlighted as priorities across the region. In this regard, the planning and implementation of the initial strategy can act as a blueprint for the inclusion of further crop genepools in the future. Continual long-term monitoring of the implementation of the integrated strategy will also be required to highlight aspects requiring adaptation in the future. Triennial national reports and a ten year review cycle have been proposed (Maxted *et al.*, 2013).

## 11.4 A New Policy Paradigm for CWR Conservation in Europe

A critical aspect of the strategy is the integration of national and regional CWR conservation actions. This requires the inclusion of regional priority species in national CWR conservation planning. European nations should have an obligation to monitor/conserve populations of these species, whether nationally threatened or not. This approach will require a regional authoritative body to oversee its implementation, therefore, the practicalities of implementing this integration need to be addressed and incorporated into European policy on agrobiodiversity conservation. As no European legislation with a focus on CWR conservation currently exists, there is at present no means of enforcing this obligation on EU member states or those European countries not currently within the EU. Emphasis therefore needs to be placed on the development of a clear regional policy on CWR conservation with buy-in from national PGR programmes throughout the region. For the regionally important CWR species that are included in Annex I of the ITPGRFA, the Treaty may be used as leverage for obligating European nation states to actively conserve CWR genetic resources within their jurisdiction. However, as the Treaty does not cover all European priority crop genepools, it is vital that EU legislation with a specific focus on CWR is developed.

This legislation could be achieved using a combination of approaches. First, a specific EU Directive on PGR could be enacted that would contemplate the protection of priority CWR populations within existing European biodiversity protection infrastructures. Second, the inclusion of priority CWR (if not already included) in the annexes of the EU Habitats Directive would place an obligation on EU member states to conserve populations of the species within their jurisdiction. However, since the species for inclusion in the Habitats Directive have to be proposed by nation states, buy-in on regional policy with respect to CWR conservation is needed from the onset of discussions in this area. A possible complication in terms of changes to the Habitats Directive is that it falls under the remit of the European Commission (EC) Directorate General (DG) for Environment, while PGRFA issues fall under the remit of the EC DG for Agriculture and Rural Development. As already highlighted, CWR frequently fall in a 'responsibility gap' between environment and agriculture agencies, both at national and regional levels – it is critical that this issue is resolved and the gap closed.

In terms of policy in support of CWR conservation in Europe, there are a number of other issues to consider, including how to ensure the success of conservation actions that depend on cross-border cooperation and the need for a central coordinating body to collect reports on the implementation of the integrated strategy. These policy-related issues will require discussion between the relevant actors (e.g. the EC, European Environment Agency and ECPGR) once the technical aspects of the integrated European CWR conservation strategy have been developed, agreed and finalized.

## Acknowledgements

This research was funded by a grant from the EU Seventh Framework Programme for the PGR Secure project (Grant Agreement No 266394).

## Note

[1] In this context, the term 'crop group' is used to refer to genera containing multiple crops (e.g. onion, leek, garlic, etc., in the genus *Allium*), crop complexes such as the brassica complex which contains multiple crops within multiple genera, or crops grouped according to their category of use (e.g. legume forages), as listed in Annex I of the ITPGRFA.

## References

Anikster, Y., Feldman, M. and Horovitz, A. (1997) The Ammiad experiment. In: Maxted, N., Ford-Lloyd, B.V. and Hawkes, J.G. (eds) *Plant Genetic Conservation: The In Situ Approach*. Chapman and Hall, London, pp. 239–253.

Avagyan, A. (2008) Crop wild relatives in Armenia: diversity, legislation and conservation issues. In: Maxted, N., Ford-Lloyd, B.V., Kell, S.P., Iriondo, J., Dulloo, M.E. and Turok, J. (eds) *Crop Wild Relative Conservation and Use*. CAB International, Wallingford, UK, pp. 58–68.

Bilz, M., Kell, S.P., Maxted, N. and Lansdown, R.V. (2011) *European Red List of Vascular Plants*. Publications Office of the European Union, Luxembourg. Available at: ec.europa.eu/environment/nature/conservation/species/redlist/downloads/European_vascular_plants.pdf (accessed 7 July 2015).

Dias, S., Dulloo, M.E. and Arnaud, E. (2012) The role of EURISCO in promoting use of agricultural biodiversity. In: Maxted, N., Dulloo, M.E., Ford-Lloyd, B.V., Frese, L., Iriondo, J.M. and Pinheiro de Carvalho, M.A.A. (eds) *Agrobiodiversity Conservation: Securing the Diversity of Crop Wild Relatives and Landraces*. CAB International, Wallingford, UK, pp. 270–277.

Dulloo, M.E., Kell, S.P. and Jones, C.G. (2002) Impact and control of invasive alien species on small islands. *International Forestry Review* 4(4), 277–285, doi.org/10.1505/ifor.4.4.277.40525.

Easterling, W.E., Aggarwal, P.K., Batima, P., Brander, K.M., Erda, L., *et al.* (2007) Food, fibre and forest products. In: Parry, M.L., Canziani, O.F., Palutikof, J.P., van der Linden, P.J. and Hanson, C.E. (eds) *Climate Change 2007: Impacts, Adaptation and Vulnerability. Contribution of Working Group II to the Fourth Assessment Report of the Intergovernmental Panel on Climate Change*. Cambridge University Press, Cambridge, UK, pp. 273–313.

FAO (Food and Agriculture Organization) (2008) Climate Change and Biodiversity for Food and Agriculture. FAO, Rome. Available at: www.fao.org/uploads/media/FAO_2008a_climate_change_and_biodiversity_02.pdf (accessed 7 July 2015).

FAO (2010) The Second Report on the State of the World's Plant Genetic Resources. FAO, Rome. Available at: www.fao.org/agriculture/seed/sow2/en/ (accessed 8 July 2015).

FAO (2014) FAOSTAT. Available at: faostat3.fao.org/home/E (accessed 11 May 2014).

Feuillet, C., Langridge, P. and Waugh, R. (2008) Cereal breeding takes a walk on the wild side. *Trends in Genetics* 24(1), 24–32.

Ford-Lloyd, B.V., Kell, S.P. and Maxted, N. (2008) Establishing conservation priorities for crop wild relatives. In: Maxted, N., Ford-Lloyd, B.V., Kell, S.P., Iriondo, J., Dulloo, E. and Turok, J. (eds) *Crop Wild Relative Conservation and Use*. CAB International, Wallingford, UK, pp. 110–119.

Ford-Lloyd, B.V., Engels, J.M.M. and Jackson, M. (2014) Genetic resources conservation challenges under the threat of climate change. In: Jackson, M., Ford-Lloyd, B.V. and Parry, M.L. (eds) *Plant Genetic Resources and Climate Change*. CAB International, Wallingford, UK, pp. 16–37.

Hawkes, J.G., Maxted, N. and Ford-Lloyd, B.V. (2000) *The Ex Situ Conservation of Plant Genetic Resources*. Kluwer Academic Publishers, Dordrecht, the Netherlands, 250 pp.

Heywood, V.H. (1997) Conservation of the wild relatives of European cultivated plants: conclusions. In: Valdés, B., Heywood, V.H., Raimondo, F.M. and Zohary, D. (eds) *Proceedings of the Workshop on 'Conservation of the Wild Relatives of European Cultivated Plants'*. Azienda Foreste Demaniali della Regione Siciliana, Sicily, pp. 445–452.

Heywood, V.H. and Zohary, D. (1995) A catalogue of wild relatives of cultivated plants native to Europe. *Flora Mediterranea* 5, 375–415.

Hoyt, E. (1988) *Conserving the Wild Relatives of Crops*. IBPGR, IUCN, WWF, Rome and Gland, Switzerland.

IPCC (Intergovernmental Panel on Climate Change) (2007) *Fourth Assessment Report Climate Change 2007: Synthesis Report*. IPCC, Geneva, Switzerland.

Kell, S.P., Knüpffer, H., Jury, S.L., Maxted, N. and Ford-Lloyd, B.V. (2005) *Catalogue of Crop Wild Relatives for Europe and the Mediterranean*. Available online via the PGR Forum Crop Wild Relative Information System (CWRIS – www.pgrforum.org/cwris/cwris.asp) and on CD-ROM. © University of Birmingham, UK.

Kell, S.P., Knüpffer, H., Jury, S.L., Ford-Lloyd, B.V. and Maxted, N. (2008) Crops and wild relatives of the Euro Mediterranean region: making and using a conservation catalogue. In: Maxted, N., Ford-Lloyd, B.V., Kell, S.P., Iriondo, J., Dulloo, M.E. and Turok, J. (eds) *Crop Wild Relative Conservation and Use*. CAB International, Wallingford, UK, pp. 69–109.

Kell, S.P., Maxted, N., Frese, L. and Iriondo, J.M. (2012a) *In situ* conservation of crop wild relatives: a strategy for identifying priority genetic reserve sites. In: Maxted, N., Dulloo, M.E., Ford-Lloyd, B.V., Frese, L., Iriondo, J.M. and Pinheiro de Carvalho, M.A.A. (eds) *Agrobiodiversity Conservation: Securing the Diversity of Crop Wild Relatives and Landraces*. CAB International, Wallingford, UK, pp. 7–19.

Kell, S.P., Maxted, N. and Bilz, M. (2012b) European crop wild relative threat assessment: knowledge gained and lessons learnt. In: Maxted, N., Dulloo, M.E., Ford-Lloyd, B.V., Frese, L., Iriondo, J.M. and Pinheiro de Carvalho, M.A.A. (eds) *Agrobiodiversity Conservation: Securing the Diversity of Crop Wild Relatives and Landraces*. CAB International, Wallingford, UK, pp. 218–242.

Kell, S.P., Negri, V., Torricelli, R., Maxted, N., Maggioni, L. and Fielder, H. (compilers) (2012c) Conservation Strategies for European Crop Wild Relative and Landrace Diversity. Report of the Joint PGR Secure/ECPGR Workshop, 7–9 September 2011, Palanga, Lithuania. Available at: www.pgrsecure.bham.ac.uk/sites/default/files/meetings/palanga/CWR_and_LR_Workshop_Report_FINAL.pdf (accessed 8 July 2015).

Kell, S., Iriondo, J., Magos Brehm, J., Ford-Lloyd, B.V., Harris, C., *et al.* (2014) Europe's Crop Wild Relatives: From Conservation Planning to Conservation Action. Oral communication, international conference, 'Enhanced genepool utilization – Capturing wild relative and landrace diversity for crop improvement', NIAB Innovation Farm, Cambridge, UK, 16–20 June 2014. Available at: www.pgrsecure.bham.ac.uk/sites/default/files/documents/public/Conference_presentations/Kell_etal.pdf (accessed 8 July 2015).

Kell, S., Qin, H., Chen, B., Ford-Lloyd, B.V., Wei, W., *et al.* (2015) China's crop wild relatives: diversity for agriculture and food security. *Agriculture, Ecosystems and Environment* 209, 138–154, doi.org/10.1016/j.agee.2015.02.012.

Kovats, R.S., Valentini, R., Bouwer, L.M., Georgopoulou, E., Jacob, D., *et al.* (2014) Europe. In: Barros, V.R., Field, C.B., Dokken, D.J., Mastrandrea, M.D., Mach, K.J., Bilir, T.E., Chatterjee, M., Ebi, K.L., Estrada, Y.O., Genova, R.C., Girma, B., Kissel, E.S., Levy, A.N., MacCracken, S., Mastrandrea, P.R. and White, L.L. (eds) *Climate Change 2014: Impacts, Adaptation, and Vulnerability. Part B: Regional Aspects. Contribution of Working Group II to the Fifth Assessment Report of the Intergovernmental Panel on Climate Change*. Cambridge University Press, Cambridge, UK, and New York, pp. 1267–1326.

Létard, V., Flandre, H. and Lepeltier, S. (2004) Information Report No 195 (2003–2004) done on behalf of the joint mission of information (1) 'France et les Français Face à la Canicule: Les Leçons d'une Crise' (France and the French Facing a Heat Wave: The Lessons of a Crisis), 391 pp. Available at: www.senat.fr/rap/r03-195/r03-195.html (accessed 7 July 2015).

Maxted, N. (2003) Conserving the genetic resources of crop wild relatives in European protected areas. *Biological Conservation* 113, 411–417.

Maxted, N. and Kell, S. (2009) Establishment of a Network for the *In Situ* Conservation of Crop Wild Relatives: Status and Needs. Commission on Genetic Resources for Food and Agriculture. Food and Agriculture Organization of the United Nations, Rome, 211 pp. Available at: www.fao.org/3/a-i1500e/i1500e18d.pdf (accessed 8 July 2015).

Maxted, N., Ford-Lloyd, B.V. and Hawkes, J.G. (1997a) Complementary conservation strategies. In: Maxted, N., Ford-Lloyd, B.V. and Hawkes, J.G. (eds) *Plant Genetic Conservation: the In Situ Approach*. Chapman and Hall, London, pp. 15–40.

Maxted, N., Hawkes, J.G., Guarino, L. and Sawkins, M. (1997b) The selection of taxa for plant genetic conservation. *Genetic Resources and Crop Evolution* 44, 337–348.

Maxted, N., Scholten, M.A., Codd, R. and Ford-Lloyd, B.V. (2007) Creation and use of a national inventory of crop wild relatives. *Biological Conservation* 140, 142–159.

Maxted, N., Ford-Lloyd, B.V. and Kell, S.P. (2008) Crop wild relatives: establishing the context. In: Maxted, N., Ford-Lloyd, B.V., Kell, S.P., Iriondo, J., Dulloo, E. and Turok, J. (eds) *Crop Wild Relative Conservation and Use*. CAB International, Wallingford, UK, pp. 3–30.

Maxted, N., Kell, S. and Magos Brehm, J. (2011) Options to Promote Food Security: On-Farm Management and In Situ Conservation of Plant Genetic Resources for Food and Agriculture. Commission on Genetic Resources for Food and Agriculture, Food and Agriculture Organization of the United Nations (FAO), Rome, 27 pp. Available at: www.fao.org/3/a-am489e.pdf (accessed 8 July 2015).

Maxted, N., Kell, S.P., Ford-Lloyd, B.V., Dulloo, M.E. and Toledo, A. (2012) Toward the systematic conservation of global crop wild relative diversity. *Crop Sciences* 52(2), 774–785.

Maxted, N., Avagyan, A., Frese, L., Iriondo, J.M., Kell, S.P., *et al.* (2013) Preserving Diversity: *In Situ* Conservation of Crop Wild Relatives in Europe – the Background Document. Rome, Italy: *In Situ* and On-farm Conservation Network, European Cooperative Programme for Plant Genetic Resources. Available at: www.pgrsecure.org/documents/Background_Document.pdf (accessed 7 July 2015).

Maxted, N., Kell, S. and Magos Brehm, J. (2014) Crop wild relatives and climate change. In: Jackson, M., Ford-Lloyd, B.V. and Parry, M.L. (eds) *Plant Genetic Resources and Climate Change*. CAB International, Wallingford, UK, pp. 114–136.

Maxted, N., Avagyan, A., Frese, L., Iriondo, J.M., Magos Brehm, J., *et al.* (2015) ECPGR Concept for *In Situ* Conservation of Crop Wild Relatives in Europe. Wild Species Conservation in Genetic Reserves Working Group, European Cooperative Programme for Plant Genetic Resources, Rome, Italy. Available at: www.ecpgr.cgiar.org/fileadmin/templates/ecpgr.org/upload/WG_UPLOADS_PHASE_IX/WILD_SPECIES/Concept_for_in__situ_conservation_of_CWR_in_Europe.pdf (accessed 8 July 2015).

Pinheiro de Carvalho, M.A.A., Nóbrega, H., Freitas, G., Fontinha, S. and Frese, L. (2012) Towards the establishment of a genetic reserve for *Beta patula* Aiton. In: Maxted, N., Dulloo, M.E., Ford-Lloyd, B.V., Frese, L., Iriondo, J.M. and Pinheiro de Carvalho, M.A.A. (eds) *Agrobiodiversity Conservation: Securing the Diversity of Crop Wild Relatives and Landraces*. CAB International, Wallingford, UK, pp. 20–28.

Tan, A. and Tan, A.S. (2002) *In situ* conservation of wild species related to crop plants: the case of Turkey. In: Engels, J.M.M., Ramantha Rao, V., Brown, A.H.D. and Jackson, M.T. (eds) *Managing Plant Genetic Diversity*. CAB International, Wallingford, UK; International Plant Genetic Resources Institute (IPGRI), Rome, pp. 195–204.

Trenberth, K.E., Jones, P.D., Ambenje, P.G., Bojariu, R., Easterling, D.R., *et al.* (2007) Observations: surface and atmospheric climate change. In: Solomon, S., Qin, D., Manning, M., Chen, Z., Marquis, M., Averyt, K.B., Tignor, M. and Miller, H.L. (eds) *Climate Change 2007: The Physical Science Basis. Contribution of Working Group I to the Fourth Assessment Report of the Intergovernmental Panel on Climate Change*. Cambridge University Press, Cambridge, UK, pp. 235–336.

Vavilov, N.I. (1926) Studies on the origins of cultivated plants. *Bulletin of Applied Botany and Plant Breeding* 16, 1–245.

Vincent, H., Wiersema, J., Kell, S., Fielder, H., Dobbie, S., *et al.* (2013) A prioritized crop wild relative inventory to help underpin global food security. *Biological Conservation* 167, 265–275.

# 12 An Approach for *In Situ* Gap Analysis and Conservation Planning on a Global Scale

H. Vincent,[1]* N.P. Castañeda-Álvarez[1,2] and N. Maxted[1]
[1]*School of Biosciences, University of Birmingham, Birmingham, UK;*
[2]*International Center for Tropical Agriculture (CIAT), Cali, Colombia*

## 12.1 Introduction

With the human population estimated to reach 9.6 billion by 2050 (United Nations, 2013), the increase in demand for food, combined with climate change, predicted increases in extreme weather events, reduced availability of natural resources and an increasingly animal-based diet globally is likely to overwhelm the current agricultural system (Godfray *et al.*, 2010; Kastner *et al.*, 2012). Furthermore, as climate change is predicted to reduce food crop production by up to 2% per decade until the end of the 21st century (Porter *et al.*, 2014), existing agricultural practices will need to adapt to ensure future food security (FAO *et al.*, 2013). These adaptations include reducing food wastage, effective sustainable management of natural resources, sustainable intensification of current food systems and improving current crop varieties to withstand changing climates and prevent reduction in yield (Godfray *et al.*, 2010; Tilman *et al.*, 2011).

Crop wild relatives (CWR) are a vast repository of genetic traits that have been used successfully to improve traditional crops and could, if used more systematically, help to diversify and adapt current crop varieties to ensure future food security (McCouch *et al.*, 2013). CWR are being utilized increasingly as gene donors in plant breeding programmes (Maxted and Kell, 2009), having been used to improve a number of crop attributes, including: nutritional quality (Sebolt *et al.*, 2000; Nassar, 2003); resistance to biotic and abiotic stresses (Panella and Lewellen, 2007); and yield (Jordan *et al.*, 2004; Brar, 2005). However, despite their importance and increasing popularity within crop improvement, CWR are poorly conserved globally (Heywood and Dulloo, 2005; Maxted and Kell, 2009; Dias *et al.*, 2012) and, like the majority of wild plants, are threatened with increased risk of extinction due to damaging anthropogenic activities, including human-induced climate change (Jarvis *et al.*, 2008; Kell *et al.*, 2011).

With both the Convention on Biological Diversity (CBD) and the Food and Agriculture Organization (FAO) explicitly stating the importance of global conservation and utilization of CWR for agriculture in Aichi Target 13 (CBD, 2011), the International Treaty on Plant Genetic Resources for Food and Agriculture (ITPGRFA) (FAO, 2001) and the CBD Global Strategy for Plant Conservation 2011–2020 (CBD, 2010), the need for a systematic conservation strategy for the wild relatives of the world's most important crops is paramount. To address this conservation need, a 10-year initiative – 'Adapting agriculture to climate change: collecting, protecting

*Corresponding author; e-mail: holly.vincent@gmail.com

 

and preparing the crop wild relatives' – was launched, led by the Global Crop Diversity Trust (GCDT) and the Millennium Seed Bank, Kew (GCDT, 2011). The initiative involves:

- identifying priority CWR related to the world's most important crops (Vincent *et al.*, 2013);
- *ex situ* gap analysis to determine novel potential areas worldwide for priority CWR sampling (see Castañeda-Álvarez *et al.*, 2016, and Chapter 13, this volume);
- *in situ* gap analysis to clarify key regions and existing protected areas for the development of multi-crop CWR genetic reserves;
- focused CWR germplasm collection and storage in seed genebanks;
- characterization of novel germplasm and development of pre-breeding lines for testing and evaluation; and
- dissemination of knowledge and genetic material to plant breeders, researchers and other users of plant genetic resources (PGR) (Dempewolf *et al.*, 2014).

*In situ* CWR conservation plays a vital role in complementary conservation, alongside *ex situ* activities, as it allows wild populations to continue adapting to a changing environment and to evolve new traits that might prove useful in plant breeding. Historically, *in situ* conservation has been overlooked due to the larger upfront costs involved, longer-term project commitments, lack of funding and difficulties securing and monitoring genetic reserves, but hopefully, with increased advocacy by major biodiversity organizations and growing awareness of the importance of complementary conservation of these species and their habitats, CWR *in situ* conservation will receive the attention it deserves.

Gap analysis is an established and widely used technique in CWR conservation that enables thorough assessment of the efficacy of the current conservation framework and identification of taxa, intra-taxon genetic diversity and geographical sites that are under-represented and require additional conservation action (Maxted *et al.*, 2008a, 2011a; Maxted and Kell, 2009; Ramírez-Villegas *et al.*, 2010; Parra-Quijano *et al.*, 2012a). However, the majority of gap analysis techniques developed for PGR and CWR conservation are targeted at *ex situ* conservation, and comparatively *in situ* gap analysis has been left somewhat underdeveloped. This chapter addresses this gap by providing an overview of the methodologies and research associated with *in situ* gap analysis of global priority CWR taxa; readers are referred to Vincent *et al.* (2016) for full details.

## 12.2 Methodology

Successful *in situ* conservation planning consists of the following basic steps: (i) circumscription of target taxon and target area; (ii) evaluating current conservation actions; (iii) setting conservation targets; (iv) ensuring persistence in targeted taxa through conservation efforts; (v) selecting and designing a network of reserves; and (vi) setting priorities for action among conservation areas (Groves, 2003). This general framework has been adapted and developed over recent years to apply to the specific needs of CWR conservation (Iriondo *et al.*, 2008; Maxted *et al.*, 2008a, 2011b; Hunter and Heywood, 2011; Vincent *et al.*, 2016).

The methodology utilized for the global CWR *in situ* gap analysis was further expanded to eight steps: (i) determining target taxa and scope; (ii) ecogeographic conspectus; (iii) analysis of genetic diversity data; (iv) evaluating current *in situ* conservation actions; (v) analysis of projected climate change impacts; (vi) reserve selection; (vii) setting priorities among conservation areas; and (viii) expert feedback. The detailed methodology description presented below is extracted from Vincent *et al.* (2016).

### 12.2.1 Determining target taxa and scope

Setting a clear goal for conservation efforts is an essential first step in gap analysis and conservation planning in general. This often requires meeting with project partners and stakeholders to discuss fully the project requirements and goals. The taxa and geographic area intended for inclusion in gap analysis can be clarified in many ways, from being directly stated as part of a specific programme to conserve a particular species, to the need for further elucidation as part of a broader project. Depending on the scope (i.e. local, national, regional or global), various sources of information should be consulted in order to determine the CWR taxa that will be considered in the gap analysis; for example, analyses at the local and national scale can benefit from published CWR inventories, as is the case for Benin (Idohou *et al.*, 2013), Portugal (Magos Brehm *et al.*, 2008), USA (Khoury *et al.*, 2013) and Venezuela (Berlingeri and Crespo, 2012). Online sites, like the Crop Wild Relatives Global Portal (2015), which contains information, reports and books from previous projects that deal with the identification of CWR taxa at the national level in multiple countries, can also help to elucidate target taxa for a particular country. In addition, this information can be complemented with The Harlan and de Wet Inventory (www.cwrdiversity.org/checklist/), which lists the CWR associated with the world's most important food crops (Vincent *et al.*, 2013).

Prioritization is a key technique in defining a set of target CWR taxa for conservation, especially when working from a national inventory or a large geographical area of interest. Maxted *et al.* (1997) and Magos Brehm *et al.* (2010) describe several criteria, along with the benefits and pitfalls of each, for CWR prioritization; these include economic value, legislation, ethnobotanical value, threat assessment and current conservation status. Furthermore, local and national analyses may only consider native CWR; thus, it is necessary to check the native distribution for each taxon. This information can be found in The Harlan and de Wet Inventory for CWR associated with globally important crops, along with other sources, such as the Germplasm Resources Information Network (GRIN, 2015) and the World Checklist of Selected Plant Families (WCSP, 2015).

### 12.2.2 Ecogeographic conspectus

Once the final list of target taxa is ready for inclusion in the gap analysis, it is necessary to begin accumulating data for each CWR. Passport data for target taxa occurrences can be obtained from a wide range of sources, including major herbariums worldwide, seed genebank databases, experts' personally collected data and online databases. The data should then be collated and organized in a single database, preferably following standards that will ensure the interoperability of the data set with external requesters (Castañeda Álvarez *et al.*, 2011). The quality of the data gathered needs to be assessed and checked thoroughly to remove obvious misplaced records, calculate coordinates where locality descriptions are available and of good quality, and standardize CWR nomenclature following a particular taxonomic system.

Examples of online repositories where occurrence records can be obtained are: the Global Biodiversity Information Facility (www.gbif.org); Genesys – the gateway to genetic resources (www.genesys-pgr.org); the European Cooperative Programme for Plant Genetic Resources database (eurisco.ipk-gatersleben.de); and the global CWR occurrence data set, which will be available for download soon (Global Consortium of Crop Wild Relative Occurrence Data Providers, 2015, unpublished results).

Information on breeding systems of target taxa should also be collected either from experts or the literature, as it can help to plan how many populations will be targeted in the reserve selection software to ensure persistence. Finally, it is important to gather any genetic diversity information for each target taxon, so that the reserve selection software can aim to conserve as much of the infraspecific diversity of each priority CWR as possible.

### 12.2.3 Analysis of genetic diversity data

Before the current conservation status of the target CWR is evaluated, it is important to examine any genetic diversity data that may be available. Individual germplasm and herbarium records may have this data, as is the case for tomato (www.kazusa.or.jp/tomato), rice (www.oryzasnp.org), maize (www.panzea.org) and wheat (www.cerealsdb.uk.net). These resources enable *in silico* analysis that potentially can increase understanding of the genetic structure of CWR populations, thus identifying geographical areas with high genetic diversity and/or unique genes that can be targeted for conservation. Ideally, monitoring the genetic composition of CWR populations should be undertaken to help to detect whether genetic erosion processes are taking place, and information obtained fed back into conservation efforts (Marfil *et al.*, 2015).

Despite the usefulness of genetic information to inform conservation needs, many CWR taxa are still lacking this kind of information. In such cases, environmental adaptation can be used as a proxy for genetic diversity. Ecogeographic land characterization (ELC) maps cluster edaphic, geophysical and climatic variables to qualify land into categories that represent different environmental niches. When overlaid with taxa occurrences, it is possible to determine how many environmental niches each taxon exists in, which can act as a substitute for infraspecific diversity. Parra-Quijano *et al.* (2012a) developed this technique for CWR and applied it successfully to collect germplasm of six *Lupinus* species located in Spain from various environmental niches identified by an ELC map (Parra-Quijano *et al.*, 2012b). This helped to increase both the genetic and geographic representativeness of the genebank collections of those species. This method can also be applied to *in situ* conservation, as reserve selection algorithms can be set to conserve instances of each CWR–environmental niche combination. To aid conservation planners, Parra-Quijano *et al.* (2014) have developed 'ELC Mapas'; a program that creates an ELC map for a defined geographical area via user-defined environmental inputs.

### 12.2.4 Evaluating current *in situ* conservation actions

To assess current *in situ* conservation actions, it is recommended that the distribution and passport data of each target taxon be compared to the spatial extent of existing protected areas (PAs). There are many available sources of PA geospatial data at varying levels of detail, examples of which include:

- Global: Protected Planet, previously the World Database on Protected Areas (WDPA) (www.protectedplanet.net).
- Regional: European Environmental Agency (www.eea.europa.eu/data-and-maps/data/natura-5).
- National: Protected Areas Database of the United States (PAD-US) (gapanalysis.usgs.gov/padus/data/download).
- National: Protected Areas of South Africa (bgis.sanbi.org/protectedareas/protectedAreas.asp).

Non-terrestrial PAs should be removed and the extent of each remaining PA overlaid with georeferenced accession and herbarium passport data. Then, using a geographic information system (GIS), overlaps between taxon occurrence points and PAs can be recorded to determine the current conservation status of each taxon and priority CWR as a whole. Also, following on from step 12.2.3, it will be possible to determine how much genetic diversity of each individual taxon is currently represented in the existing PA network, particularly when using ELC maps.

As this process only identifies taxa that may be being conserved passively, it is advisable to check with the management plans of each PA containing target taxa records to ascertain whether they are mentioned explicitly. However, management plans for sites worldwide are rarely available to researchers, so it is usual practice to assume only passive conservation is occurring.

### 12.2.5 Analysis of projected climate change impacts

As *in situ* conservation, in the majority of cases, is a long-term commitment, it is important to ensure that the populations of target taxa chosen for conservation are likely to persist. Therefore, before establishing a reserve, it is advisable to model the threats that compromise the long-term persistence of taxa populations, such as mining, urbanization, animal grazing and climate change. The impact of most threats that may affect CWR taxa habitats negatively can be reduced by implementing management plans collectively with stakeholder participation, and even introducing protective legislation on a national or global level. This is not the case for climate change, which cannot be mitigated locally; therefore, understanding the potential impacts of climate change on CWR distributions is a critical step during reserve planning to ensure the CWR populations conserved are those most likely to persist over time. Jarvis *et al.* (2008) modelled the current and future potential distributions of CWR associated with groundnut (*Arachis hypogaea* L.), potato (*Solanum tuberosum* L.) and *Vigna* species using the Bioclim climate envelope (Busby, 1991). These models, when compared, predicted high fragmentation in the current habitats of CWR and major reductions in habitat range, thus highlighting the need for climate change mitigation within conservation strategies.

Climate change projections are freely available for different future dates (i.e. 2020, 2050, 2100), for the four Representativeness Concentration Pathways (RCPs), as per the IPCC Fifth Assessment Report (cmip-pcmdi.llnl.gov/cmip5), and for multiple General Circulation Models (GCMs). For instance, the WorldClim website holds freely available future climate data, downscaled and calibrated using the current scenario of Worldclim (Hijmans *et al.*, 2005), including raster files of Bioclim variables (www.worldclim.org/CMIP5). The future climate models can be used to assess whether particular grid cells/sites will still be stable long term for target taxa or to assess taxa data points individually for significant change that would mean the population would likely perish.

### 12.2.6 Reserve selection

Before selecting sites for genetic reserves, it is important to set targets for conservation. This can be a percentage of a certain ecosystem, a fixed number of target taxa occurrences, or a proportion of a taxon's distributional range. One method of setting targets is to analyse the breeding system of each taxon and determine the minimum number of populations needed to ensure persistence *in situ*; expert advice would be beneficial in this case. A more simple method suggested by Brown and Briggs (1991), and supported by Dulloo *et al.* (2008), is to conserve a minimum of five populations per taxon throughout its range.

When designing a network of reserves, it is typical to run the georeferenced occurrence records through a reserve selection algorithm to produce an ideal group of sites that fulfil all conservation targets, which can serve as a starting point for discussions with experts and land managers. There are many reserve selection algorithms available online that can assist in choosing sites for conservation, while meeting complex conservation targets, several of which are free to use. One widely used algorithm in CWR conservation planning is the Rebelo complementarity analysis (Rebelo and Siegfried 1990, 1992) feature in DIVA-GIS (Hijmans *et al.*, 2001), which selects taxon-rich sites based on the taxa that have not yet been selected by previous iterations. However, this algorithm lacks flexibility in terms of setting targets for individual taxa, and there are other algorithms and programs that can offer more thorough data analysis. These sophisticated programs include:

- Marxan (www.uq.edu.au/marxan/get-marxan-software);
- Zonation (cbig.it.helsinki.fi/software/zonation); and
- ConsNet (uts.cc.utexas.edu/~consbio/ Cons/consnet_home.html).

Marxan is a flexible tool that attempts to solve the minimum set reserve problem using simulated annealing; it can be run with both taxon coordinate occurrence data and polygons of species distribution. Marxan can also analyse different scenarios while offering options to weight species due to rarity or relative importance, set targets for the minimum number of occurrences for each taxon to be achieved in the whole network and set costs for each potential planning unit. Zonation is similar to Marxan, but operates by iteratively removing the least valuable cell/site while matching this with the overall complementarity and connectivity of the total solution. Zonation also allows the use of point data and raster data as inputs. ConsNet is software that uses smart heuristic searches to find reserve solutions that best meet the criteria of the conservation planners. It has many possible user inputs, such as costings of cells/sites, and different search options, such as minimum area search and maximum representation search. All of these programs require some experimentation with different user settings to produce significant results.

When using reserve selection algorithms, many projects run only the complete data set; however, it is necessary to run several scenarios on subsets of the data. It is particularly important to analyse the taxon occurrence points that fall within PAs separately from those not protected, or at least factor the presence of PAs into site costs in full data runs. This is due to it being easier to set up genetic reserves in existing PAs, rather than acquire land for new sites.

### 12.2.7 Setting priorities among conservation areas

Usually, it is not possible to establish a complete network of reserves due to economic, political or other restraints. Therefore, it is necessary to prioritize sites from the ideal network for *in situ* conservation action. One method for prioritizing sites is to compare complementarity runs against the complete reserve network, to determine which sites will provide the most CWR richness. Further sites can be prioritized based on the cost of site, threats and priority species (such as Annex 1 food crops of the ITPGRFA); further prioritization techniques are described by Magos Brehm *et al.* (2010). The presence of other important biodiversity or overlap with Myers or Vavilov hotspots may also be good rationale for prioritizing sites, as this may help persuade conservation agencies and relevant authorities to set up sites if there is overall high biodiversity, not just CWR diversity.

### 12.2.8 Expert feedback

If the goal is to establish a complete network of reserves, experience has shown this will only be possible with full stakeholder buy-in (Maxted *et al.*, 2008b); therefore, it is important to involve stakeholders throughout each stage of conservation planning. However, the first complete 'ideal' run of a reserve selection programme is a key time to get stakeholder feedback and expert input. This will provide a basis for discussion on potential sites, priorities and whether they are practically feasible sites for genetic reserves. These discussions can lead back to further reserve planning software runs, as there will often be a need to compromise between the best possible solution as determined by the reserve selection algorithm and the requirements of the stakeholders.

## 12.3 Summary of Global *In Situ* Gap Analysis[1]

A global *in situ* gap analysis for the CWR of 173 crop genepools, as listed in The Harlan and de Wet Inventory (Vincent *et al.* 2013), was undertaken to identify an ideal worldwide network of genetic reserves that maximizes the number of priority CWR conserved *in situ*. The global CWR occurrence database (Global Consortium of Crop Wild Relative Occurrence Data Providers, 2015) was edited to include only records related to target CWR; non-native and poorly georeferenced records were also removed, leaving occurrence data for 1158 priority species. A generalized ELC map was produced for the whole terrestrial world at the 20 km × 20 km resolution, using the ELC Mapas tool developed by Parra-Quijano *et al.* (2014). The georeferenced data were overlaid on the ELC map to extract the environmental niche for each occurrence point, resulting in 6236 unique CWR–environmental niche combinations. PA data were obtained from the World Database on Protected Areas (WDPA) as a geospatial database (IUCN and UNEP-WCMC, 2012) and a shapefile for the Natura 2000 sites was also obtained (European Environment Agency, 2013), as this important data set was not included in the WDPA. The PA files were edited to remove non-terrestrial PAs, then overlaps between taxon points and PAs were recorded to determine the current, albeit passive, conservation status of each population, taxon and priority CWR as a whole. During the time frame of this study, it was not possible to contact all the PA managers where CWR appeared to exist. Therefore, it was assumed there was no active management of CWR in these PAs. Nineteen GCMs for 2070 with RCP 45 were downloaded from the Worldclim website as Bioclim rasters and combined to determine which taxa would be most likely to survive predicted climate change. Marxan was used to model different scenarios for conserving subsets of the priority CWR taxa, with the aim of discovering the best sites for multiple crop CWR conservation, with priority for sites that were potentially tolerant to climate change and overlapped with PAs. Target numbers of populations to conserve were set as five per individual taxon, as recommended by Dulloo *et al.* (2008). However, during analysis of the ELC categories per taxon, a target of only one per ELC–individual species combination was studied, due to their large number. Marxan scenarios were run for:

- all taxa;
- all Genepool 1B (GP1B) taxa;
- all taxa related to ITPGRFA Annex1 food crops;
- all GP1B taxa related to ITPGRFA Annex1 food crops;
- all ELC–taxa combinations;
- all GP1B ELC–taxa combinations;
- all ELC–taxa combinations related to ITPGRFA Annex1 food crops; and
- all GP1B ELC–taxa combinations related to ITPGRFA Annex1 food crops.

Finally, Rebelo complementarity analysis was run on the final Marxan reserve solutions to further prioritize a set of the top 50 sites worldwide that would conserve maximal amounts of taxa diversity.

The gap analysis found that 403 of the 1158 studied species, including four whole genera, had fewer than ten good-quality occurrence records, which was the minimum number of records needed to produce a gap analysis assessment reliably. Out of the remaining 755 CWR species, the gap analysis found that 129 CWR were not found in a single PA, with 20 of these being primary genepool species. Additionally, over 50% of the assessable species had less than 10% of their occurrence records within PAs, suggesting a poor level of overlap between CWR and protected area occurrence, and therefore, at least, passive *in situ* conservation. In terms of genetic diversity, 434 of the assessable species had less than 50% of their associated environmental niches found in PAs; however, 84 species had all of their niches represented in the existing PA network. Worryingly, only 104 of the 1158 priority CWR have been assessed by the IUCN Red List (IUCN, 2015), leaving the conservation status of these species unknown.

Figure 12.1 shows the best Marxan solution for all priority CWR of the 173 globally important crops studied in the gap analysis, while Fig. 12.2 highlights the top 50 complementary sites from the same scenario. The richest complementary CWR site was found on the Israel/Jordan/Palestine/Syria border and contained 61 unique CWR species from 20 genera. The remaining top ten sites of importance were as follows: the southern Armenia/Azerbaijan border (53 CWR from 23 genera); Virginia, USA (30 CWR from 11 genera); southern Madrid, Spain (52 CWR from 25 genera); Rajasthan, India (24 CWR from 17 genera); Mexico state, Mexico (24 CWR from 12 genera); Moungo, Cameroon (17 CWR from 9 genera); Java, Indonesia (20 CWR from 10 genera); Brasilia region, Brazil (18 CWR from 6 genera); and Cajamarca, Peru (16 CWR from 2 genera). Out of the top ten complementary sites, all overlap with PAs except those in Rajasthan, India, and Java, Indonesia; these two sites require urgent ground truthing to confirm the presence of CWR and to formulate *in situ* conservation strategies outside of the PA network. All of the top ten sites should be a priority for establishing new reserves, or at the very least, careful and regular monitoring of populations, their health and threats.

## Acknowledgements

Data gathering and analyses were undertaken as part of the initiative 'Adapting Agriculture to Climate Change: Collecting, Protecting and Preparing Crop Wild Relatives', which is financed by the Government of Norway. The project is managed by the Global Crop Diversity Trust with the Millennium Seed Bank of the Royal Botanic Gardens, Kew, UK, and implemented in partnership with national and international genebanks and plant breeding institutions around the world. Further information can be found at www.cwrdiversity.org.

## Note

[1] *In situ* gap analysis results are summarized here, but readers are referred to Vincent *et al.* (2016) for full details.

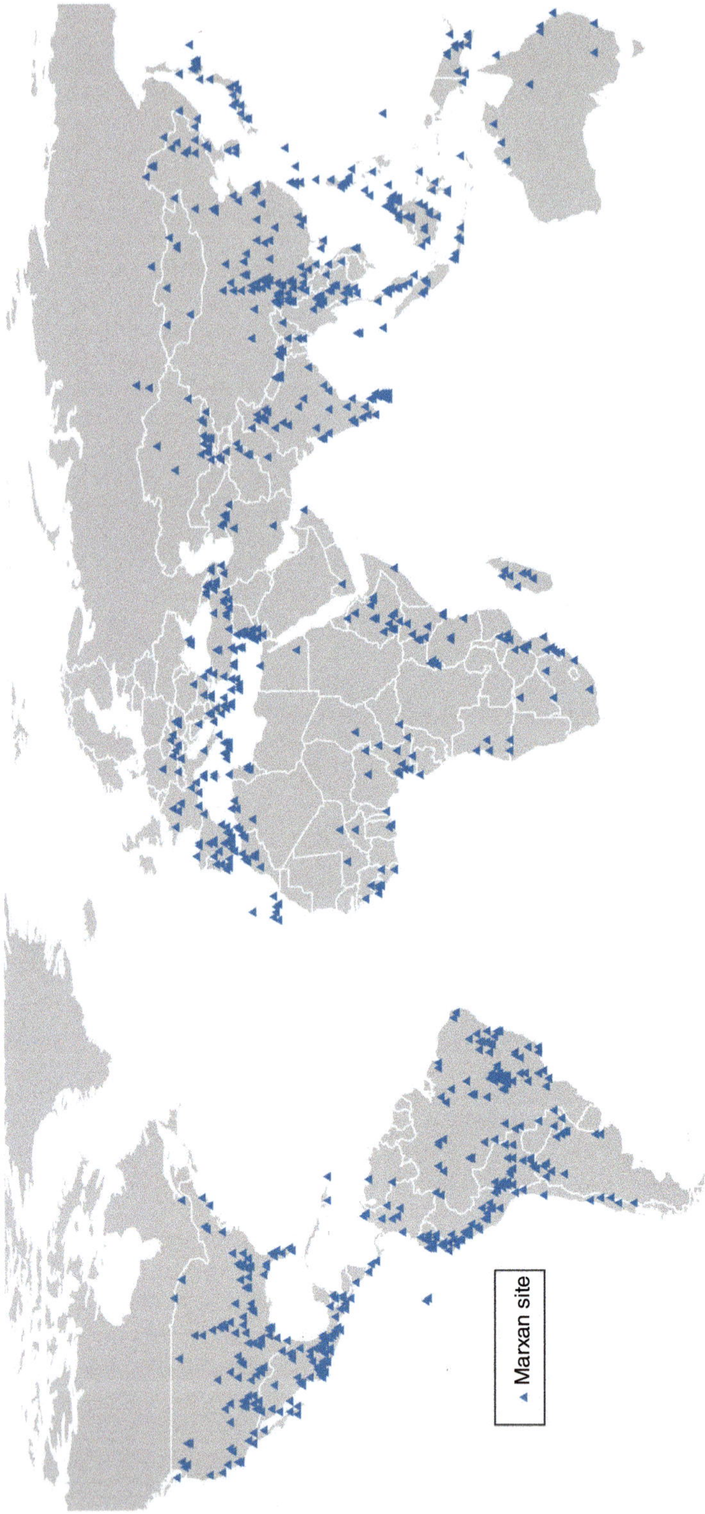


Fig. 12.1. Best Marxan solution for all priority CWR of 173 globally important crops.

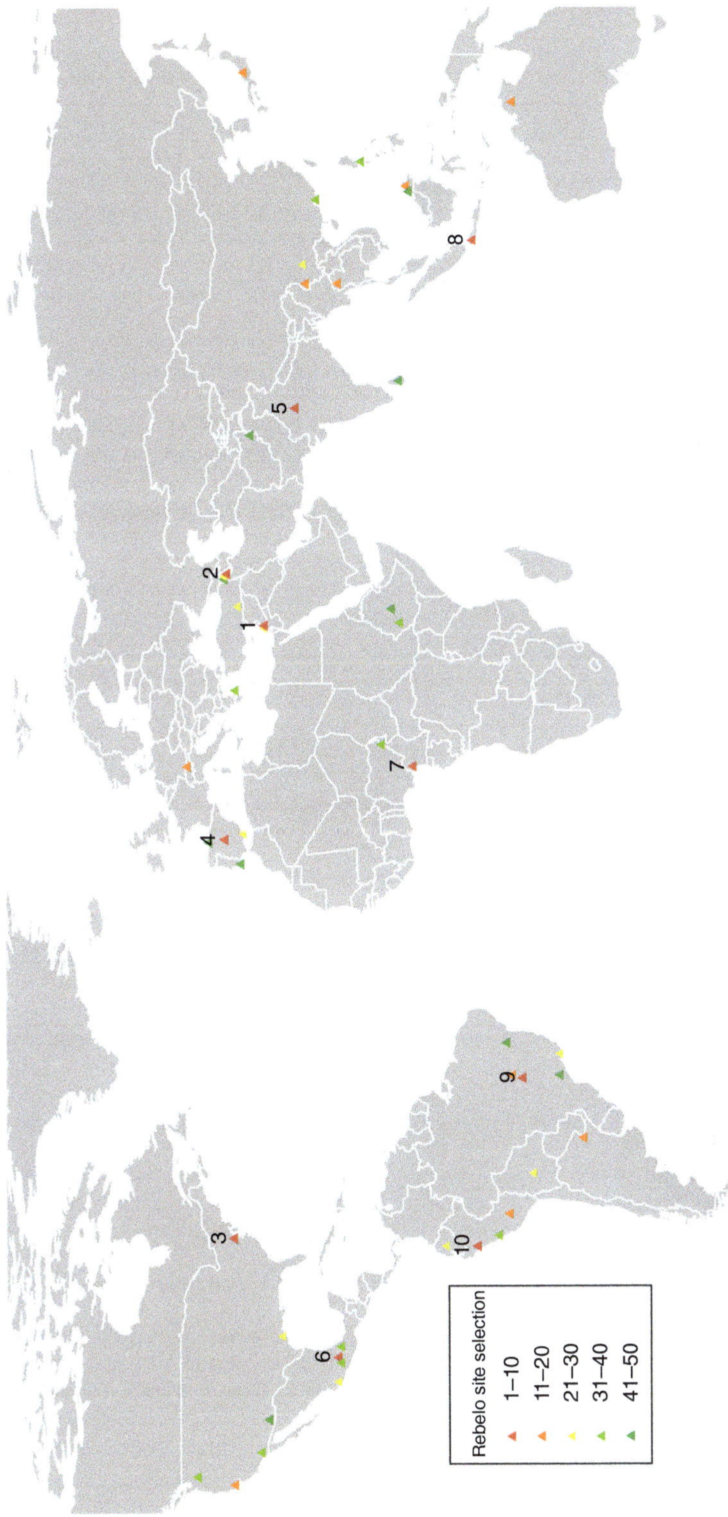


**Fig. 12.2.** Top 50 complementary sites from the 'all priority CWR' best solution.

## References

Berlingeri, C. and Crespo, M.B. (2012) Inventory of related wild species of priority crops in Venezuela. *Genetic Resources and Crop Evolution* 59(5), 655–681.

Brar, D. (2005) Broadening the genepool and exploiting heterosis in cultivated rice. In: Toriyama, K., Heong, K.L. and Hardy, B. (eds) *Rice is Life: Scientific Perspectives for the 21st Century*. Proceedings of the World Rice Research Conference, Tokyo and Tsukuba, Japan, 4–7 November 2004. International Rice Research Institute (IRRI), Los Baños, Philippines, and Japan International Research Center for Agricultural Sciences (JIRCAS), Tsukuba, Japan, pp. 157–160.

Brown, A.H.D. and Briggs, J.D. (1991) Sampling strategies for genetic variation in *ex situ* collections of endangered plant species. In: Falk, D.A. and Holsinger, K.E. (eds) *Genetics and Conservation of Rare Plants*. Oxford University Press, New York, pp. 99–119.

Busby, J.R. (1991) BIOCLIM – a bioclimatic analysis and prediction system. In: Margules, C.R. and Austin, M.P. (eds) *Nature Conservation: Cost-effective Biological Surveys and Data Analysis*. CSIRO, Canberra, pp. 64–68.

Castañeda Álvarez, N.P., Vincent, H., Kell, S., Eastwood, R. and Maxted, N. (2011) Ecogeographic surveys. In: Guarino, L., Ramanatha Rao, V. and Goldberg, E. (eds) *Collecting plant genetic diversity: Technical guidelines. 2011 update*. Bioversity International, Rome. Available at: cropgenebank.sgrp.cgiar.org/index.php?option=com_content&view=article&id=390&Itemid=557 (accessed 29 January 2015).

Castañeda-Álvarez, N.P., Khoury, C.K., Achicanoy, H.A., Bernau, V., Dempewolf, H., *et al.* (2016) Global priorities for crop wild relative conservation for food security. (In preparation.)

CBD (Convention on Biological Diversity) (2010) *Global Strategy for Plant Conservation*. Secretariat of the Convention on Biological Diversity, Montreal, Canada.

CBD (2011) *Conference of the Parties Decision X/2: Strategic plan for biodiversity 2011–2020*. Secretariat of the Convention on Biological Diversity, Montreal. Available at: www.cbd.int/decision/cop/?id=12268 (accessed 1 October 2014).

Crop Wild Relatives Global Portal (2015) Crop wild relatives in national inventories. Available at: www.cropwildrelatives.org/inventories-and-strategies (accessed 15 January 2015).

Dempewolf, H., Eastwood, R.J., Guarino, L., Khoury, C.K., Müller, J.V., *et al.* (2014) Adapting agriculture to climate change: a global initiative to collect, conserve, and use crop wild relatives. *Agroecology and Sustainable Food Systems* 38(4), 369–377, doi:10.1080/21683565.2013.870629.

Dias, S., Dulloo, M.E. and Arnaud, E. (2012) The role of EURISCO in promoting use of agricultural biodiversity. In: Maxted, N., Dulloo, M.E., Ford-Lloyd, B.V., Frese, L., Iriondo, J.M. and Pinheiro de Carvalho, M.A.A. (eds) *Agrobiodiversity Conservation: Securing the Diversity of Crop Wild Relatives and Landraces.* CAB International, Wallingford, UK, pp. 270–277.

Dulloo, M.E., Labokas, J., Iriondo, J.M., Maxted, N., Lane, A., *et al.* (2008) Genetic reserve location and design. In: Iriondo, J.M., Maxted, N. and Dulloo, M.E. (eds) *Conserving Plant Genetic Diversity in Protected Areas.* CAB International, Wallingford, UK, pp. 23–64.

European Environment Agency (2013) Natura 2000 data – the European network of protected sites. Available at: www.eea.europa.eu/data-and-maps/data/natura-5 (accessed 10 September 2014).

FAO (Food and Agriculture Organization) (2001) International Treaty on Plant Genetic Resources for Food and Agriculture. FAO, Rome. Available at: www.planttreaty.org/content/texts-treaty-official-versions (accessed 20 January 2015).

FAO, IFAD and WFP (2013) *The State of Food Insecurity in the World 2013. The Multiple Dimensions of Food Security*. FAO, Rome.

GCDT (Global Crop Diversity Trust) (2011) *$50 Million Pledge for Climate Change Adaptation*. Global Crop Diversity Trust, Rome.

Godfray, C.J., Beddington, J.R., Crute, I.R., Haddad, L., Lawrence, D., *et al.* (2010) Food security: the challenge of feeding 9 billion people. *Science* 327, 812–818.

GRIN (Germplasm Resources Information Network) (2015) Germplasm Resources Information Network – (GRIN) (Online Database). USDA, ARS, National Genetic Resources Program. National Germplasm Resources Laboratory, Beltsville, Maryland. Available at: www.ars-grin.gov/cgi-bin/npgs/html/tax_search.pl?language=en (accessed 29 January 2015).

Groves, C.R. (2003) *Drafting a Conservation Blueprint: A Practitioner's Guide to Planning for Biodiversity*. Island Press, Washington, DC.

Heywood, V.H. and Dulloo, E. (2005) In situ *Conservation of Wild Plant Species – A Critical Global Review of Good Practices*. IPGRI Technical Bulletin No 11, FAO and IPGRI, Rome.

Hijmans, R.J., Cruz, M., Rojas, E. and Guarino, L. (2001) *DIVA-GIS Manual. Version 1.4. A Geographic Information System for the Management and Analysis of Genetic Resources Data*. International Potato Center, (CIP), Lima.

Hijmans, R.J., Cameron, S.E., Para, J.L., Jones, P.G. and Jarvis, A. (2005) Very high resolution interpolated climate surfaces for global land areas. *International Journal of Climatology* 25, 1965–1978.

Hunter, D. and Heywood, V. (eds) (2011) *Crop Wild Relatives: A Manual of In Situ Conservation*. Earthscan, London.

Idohou, R., Assogbadjo, A.E., Fandohan, B., Gouwakinnou, G.N., Glèlè Kakaï, R.L., *et al.* (2013) National inventory and prioritization of the crop wild relatives: case study for Benin. *Genetic Resources and Crop Evolution* 60, 1337–1352.

Iriondo, J.M., Maxted, N. and Dulloo, M.E. (eds) (2008) *Conserving Plant Genetic Diversity in Protected Areas*. CAB International, Wallingford, UK.

IUCN (International Union for Conservation of Nature) (2015) 2015 IUCN Red List of Threatened Species. Available at: www.iucnredlist.org (accessed 21 July 2015).

IUCN and UNEP-WCMC (2012) The World Database on Protected Areas (WDPA) (Online). UNEP-WCMC, Cambridge, UK. Available at: www.protectedplanet.net (accessed 12 December 2012).

Jarvis, A., Lane, A. and Hijmans, R.J. (2008) The effect of climate change on crop wild relatives. *Agriculture, Ecosystems & Environment* 126, 13–23.

Jordan, J., Butler, D., Henzell, B., Drenth, J. and McIntyre, L. (2004) Diversification of Australian sorghum using wild relatives, new directions for a diverse planet: *Proceedings of the 4th International Crop Science Congress*, Brisbane, Australia, 26 September–1 October 2004.

Kastner, T., Rivas, M.J.I., Koch, W. and Nonhebel, S. (2012) Global changes in diets and the consequences for land requirements for food. *Proceedings of the National Academy of Sciences* 109, 6868–6872.

Kell, S.P., Maxted, N. and Bilz, M. (2011) European crop wild relative threat assessment: knowledge gained and lessons learnt. In: Maxted, N., Dulloo, M.E., Ford-Lloyd, B.V., Frese, L., Iriondo, J.M. and Pinheiro de Carvalho, M.A.A. (eds) *Agrobiodiversity Conservation: Securing the Diversity of Crop Wild Relatives and Landraces*. CAB International, Wallingford, UK, pp. 218–242.

Khoury, C.K., Greene, S., Wiersema, J., Maxted, N., Jarvis, A., *et al.* (2013) An inventory of crop wild relatives of the United States. *Crop Science* 53(4), 1496.

McCouch, S., Baute, G.J., Bradeen, J., Bramel, P., Bretting, P.K., *et al.* (2013) Feeding the future. *Nature* 499, 23–24.

Magos Brehm, J., Maxted, N., Ford-Lloyd, B.V. and Martins-Loução, M.A. (2008) National inventories of crop wild relatives and wild harvested plants: case-study for Portugal. *Genetic Resources and Crop Evolution* 55, 779–796.

Magos Brehm, J., Maxted, N., Martins-Loução, M.A. and Ford-Lloyd, B.V. (2010) New approaches for establishing conservation priorities for socio-economically important plant species. *Biodiversity Conservation* 19, 2715–2740.

Marfil, C.F., Hidalgo, V. and Masuelli, R.W. (2015) *In situ* conservation of wild potato germplasm in Argentina: example and possibilities. *Global Ecology and Conservation* 3, 461–476.

Maxted, N. and Kell, S. (2009) *Establishment of a Global Network for the In Situ Conservation of Crop Wild Relatives: Status and Needs*. FAO, Rome.

Maxted, N., Hawkes, J.G., Guarino, L. and Sawkins, M. (1997) The selection of taxa for plant genetic conservation. *Genetic Resources and Crop Evolution* 44, 337–348.

Maxted, N., Dulloo, E., Ford-Lloyd, B., Iriondo, J. and Jarvis, A. (2008a) Gap analysis: a tool for complementary genetic conservation assessment. *Diversity and Distributions* 14, 1018–1030.

Maxted, N., Iriondo, J., De Hond, L., Dulloo, E., Lefèvre, F., *et al.* (2008b) Genetic reserve management. In: Iriondo, J.M., Maxted, N. and Dulloo, E. (eds) *Plant Genetic Population Management*. CAB International, Wallingford, UK, pp. 65–87.

Maxted, N., Castañeda Álvarez, N.P., Vincent, H. and Magos Brehm, J. (2011a) Chapter 41: Gap analysis: a tool for genetic conservation. In: Guarino, L., Ramanatha Rao, V. and Goldberg, E. (eds) *Collecting Plant Genetic Diversity: Technical Guidelines – 2011 Update*. Bioversity International, Rome. Available at: cropgenebank.sgrp.cgiar.org/index.php?option=com_content&view=article&id=390 &Itemid=557 (accessed 29 January 2015).

Maxted, N., Dulloo, M.E., Ford-Lloyd, B.V., Frese, L., Iriondo, J.M., *et al.* (eds) (2011b) *Agrobiodiversity Conservation: Securing the Diversity of Crop Wild Relatives and Landraces*. CAB International, Wallingford, UK.

Nassar, N.M.A. (2003) Gene flow between cassava, *Manihot esculenta* Crantz, and wild relatives. *Genetics and Molecular Research* 2(4), 334–347.

Panella, L. and Lewellen, R.T. (2007) Broadening the genetic base of sugar beet: introgression from wild relatives. *Euphytica* 154, 383–400.

Parra-Quijano, M., Iriondo, J.M. and Torres, E. (2012a) Ecogeographical land characterization maps as a tool for assessing plant adaptation and their implications in agrobiodiversity studies. *Genetic Resources and Crop Evolution* 59, 205–217.

Parra-Quijano, M., Iriondo, J.M. and Torres, E. (2012b) Improving representativeness of genebank collections through species distribution models, gap analysis and ecogeographical maps. *Biodiversity and Conservation* 21, 79–96.

Parra-Quijano, M., Torres, E., Iriondo, J.M. and López, F. (2014) *CAPFITOGEN Tools User Manual Version 1.2.* International Treaty on Plant Genetic Resources for Food and Agriculture, FAO, Rome.

Porter, J.R., Xie, L., Challinor, A.J., Cochrane, K., Howden, S.M., *et al.* (2014) Food security and food production systems. In: Field, C.B., Barros, V.R., Dokken, D.J., Mach, K.J., Mastrandrea, M.D., Bilir, T.E., Chatterjee, M., Ebi, K.L., Estrada, Y.O., Genova, R.C., Girma, B., Kissel, E.S., Levy, A.N., MacCracken, S., Mastrandrea, P.R. and White, L.L. (eds) *Climate Change 2014: Impacts, Adaptation and Vulnerability. Part A: Global and Sectoral Aspects. Contribution of Working Group II to the Fifth Assessment Report of the Intergovernmental Panel on Climate Change.* Cambridge University Press, Cambridge, UK, and New York, pp. 485–533.

Ramírez-Villegas, J., Khoury, C., Jarvis, A., Debouck, D.G. and Guarino, L. (2010) A gap analysis methodology for collecting crop genepools: a case study with *Phaseolus* beans. *PLoS ONE* 5(10), e13497, doi:10.1371/journal.pone.0013497.

Rebelo, A.G. and Siegfried, W.R. (1990) Protection of Fynbos vegetation: ideal and real-world options. *Biological Conservation* 54(1), 15–31.

Rebelo, A.G. and Siegfried, W.R. (1992) Where should nature reserves be located in the Cape Floristic Region, South Africa? Models for the spatial configuration of a reserve network aimed at maximizing the protection of floral diversity. *Conservation Biology* 6(2), 243–252.

Sebolt, A.M., Shoemaker, R.C. and Diers, B.W. (2000) Analysis of a quantitative trait locus allele from wild soybean that increases seed protein concentration in soybean. *Crop Science* 40, 1438–1444.

Tilman, D., Balzer, C., Hill, J. and Befort, B.L. (2011) Global food demand and the sustainable intensification of agriculture. *Proceedings of the National Academy of Sciences of the United States of America* 108, 20260–20264.

United Nations (2013) *World Population Prospects: The 2012 Revision.* Population Division, Dept of Economic and Social Affairs, United Nations, New York.

Vincent, H., Wiersema, J., Kell, S., Fielder, H., Dobbie, S., *et al.* (2013) A prioritized crop wild relative inventory to help underpin global food security. *Biological Conservation* 167, 265–275.

Vincent, H., Castañeda-Álvarez, N.P. and Maxted, N. (2016) Global priorities for *in situ* conservation of wild plant genetic resources: towards the establishment of a global network of crop wild relative reserves. (In preparation.)

WCSP (2015) World Checklist of Selected Plant Families. Facilitated by the Royal Botanic Gardens, Kew. Available at: apps.kew.org/wcsp/ (accessed 29 January 2015).

# 13 The Distributions and *Ex Situ* Conservation of Crop Wild Relatives: A Global Approach

**N.P. Castañeda-Álvarez,[1,2]* C.K. Khoury,[1,3] C.C. Sosa,[1] H.A. Achicanoy,[1] V. Bernau,[1] H. Vincent,[2] A. Jarvis,[1,4] P.C. Struik[3] and N. Maxted[2]**

*[1]International Center for Tropical Agriculture (CIAT), Cali, Colombia; [2]School of Biosciences, University of Birmingham, Birmingham, UK; [3]Centre for Crop Systems Analysis, Wageningen University, Wageningen, the Netherlands; [4]CGIAR Research Program on Climate Change, Agriculture and Food Security (CCAFS), Cali, Colombia*

## 13.1 Introduction

In the coming decades, we will need to produce more food for a growing population, with less arable land, irrigation water and phosphorus available, while at the same time reducing the negative impacts of agriculture to the environment (Rosegrant and Cai, 2001; Godfray *et al.*, 2010). This situation is worsened by the reduction of crop yields as a consequence of increasing temperatures and frequency of droughts, and of extreme weather events due to climate change (Porter *et al.*, 2014). Multiple strategies have been proposed to overcome such challenges, including the synthesis of more productive, resilient and nutritious crop varieties (FAO, 2008; Foley *et al.*, 2011). For this, the broadest range of plant genetic resources (PGRs), including crop wild relatives (CWR), are required as sources of diverse traits for crop breeding programmes (Hawkes *et al.*, 2000; McCouch *et al.*, 2013).

Increasingly, CWR are recognized as sources of novel traits in crop breeding, many of which are not found in cultivated species as CWR have not experienced the genetic bottlenecks of domestication, dispersal and modern plant breeding (Maxted *et al.*, 2008a; van de Wouw *et al.*, 2010). CWR have been used successfully in crop improvement; for example, to increase resistance against pests and diseases and tolerance to abiotic constraints, improve yields, enhance nutritional content and quality and, more generally, to broaden the genetic base of important crops in order to enhance food security (Hajjar and Hodgkin, 2007; Maxted and Kell, 2009). Other important uses of CWR include direct sources of food (wild gathered food plants), especially in times of shortage, and providers of ecosystem services (Brehm *et al.*, 2010). Given the advances and growing accessibility of biotechnological tools that enable the understanding of the mechanisms behind hybridization (Tanksley and McCouch, 1997) and predictive tools for identifying CWR with specific desired traits (Thormann *et al.*, 2014), it is expected that the use of CWR for producing new crop varieties will increase through conventional and molecular crop breeding. Therefore, it is paramount to make the diversity of CWR

*Corresponding author; e-mail: n.p.castaneda@cgiar.org

available and accessible to crop breeders and this is achieved conventionally via genebanks or other *ex situ* collections.

Taking action to conserve CWR is urgently needed (CBD, 2010a,b). Currently, less than 10% of the germplasm accessions held in publicly accessible genebanks correspond to CWR taxa (FAO, 2010; Dias *et al.*, 2012). In addition, CWR are exposed to threats like extreme changes in the current land cover and increasing pressures on natural habitats due to anthropogenic activities (Brummitt and Bachman, 2010; Kell *et al.*, 2012). Similarly, the rapid increase of global temperature and abrupt changes in rainfall patterns are expected to reduce the environmental niches of some of these taxa (Jarvis *et al.*, 2008; Ureta *et al.*, 2012), unless CWR are able to adapt rapidly to the new climatic conditions, migrate towards areas with suitable environmental conditions, or assisted migration plans are implemented.

Prioritizing taxa for urgent conservation and the geographical regions where such taxa are likely to be found is key for directing resources devoted to collecting CWR. Under the umbrella of the 'Adapting agriculture to climate change: collecting, protecting and preparing the crop wild relatives' project (Dempewolf *et al.*, 2013), more than 1000 CWR taxa have been assessed to determine how well represented they currently are in genebanks that are accessible to the global plant breeding community. Likewise, the regions where such taxa are likely to occur have been mapped, identifying geographic hotspots of CWR species richness, as well as locations where further collecting of plant material is highly recommended (Castañeda-Álvarez *et al.*, in press). The main outputs of this assessment helped to guide decisions regarding which regions and countries should be selected to collect samples of CWR evaluated as high priorities for *ex situ* conservation (Dempewolf *et al.*, 2013). The methodology, results and discussion of this study are summarized in the following sections; readers are referred to Castañeda-Álvarez *et al.* (in press) for full details.

## 13.2 Methodology

Gap analyses are frequently used in conservation science as an approach to identify the taxa and/or regions that are outside the boundaries of established conservation efforts like reserves and natural parks and therefore help in setting conservation priorities (Rodrigues *et al.*, 2004; Langhammer *et al.*, 2007). The general gap analysis methodology was adapted for PGR use by Maxted *et al.* (2008b) and further refined for *ex situ* application by Ramírez-Villegas *et al.* (2010), and this was modified and adapted by Castañeda-Álvarez *et al.* (in press) to assess the representativeness of more than 1000 CWR in *ex situ* genebanks. The methodology includes five steps:

**1. Definition of crops and their associated wild relatives:** the closest CWR taxa of 81 crops relevant for global food security and income generation were chosen for the analysis (Table 13.1). This included all the crops in Annex I of the International Treaty on Plant Genetic Resources (FAO, 2009), and the list of CWR taxa was retrieved from The Harlan and de Wet Crop Wild Relative Inventory (www.cwrdiversity.org/checklist/ – Vincent *et al.*, 2013). The CWR defined as closest to their associated crop belong to the primary genepool (GP-1b) and secondary genepool (GP-2) according to the classification proposed by Harlan and de Wet (1971), the primary and secondary genetic-relative classes (Wiersema *et al.*, 2012), taxon groups TG1b, TG2 and TG3 (Maxted *et al.*, 2006), or the provisional primary and secondary groups (Vincent *et al.*, 2013). Additional taxa from more distantly related genepools (or equivalent grouping) were included when confirmed and potential uses had been reported in the literature.

**2. Data inputs required by the gap analysis methodology:** gap analysis relies on geographic inputs to model the distributions of CWR and to determine the extent of their representativeness in genebanks. The main inputs required for the analyses are:

- *Germplasm occurrence data*: having a comprehensive data set representing most of the accessions available for plant breeders was critical for the gap analysis. These data were obtained from genebanks databases. These databases contain relevant information of each of the accessions deposited, such as the unique identifier of the accession, identification, collector, date of collecting and genebank acquisition, details of the site where the accession

**Table 13.1.** List of crop genepools included in the gap analysis.

| | | |
|---|---|---|
| Adzuki beans [*Vigna angularis* (Willd.) Ohwi & H. Ohashi][a] | Garlic (*Allium sativum* L.) | Plums (*Prunus domestica* L.) |
| Almonds [*Prunus dulcis* (Mill.) D. A. Webb] | Grapefruit (*Citrus paradisi* Macfad.)[a] | Potatoes (*Solanum tuberosum* L.)[a] |
| Apples (*Malus domestica Borkh.*)[a] | Grapes (*Vitis vinifera* L.) | Pumpkins (*Cucurbita maxima* Duchesne) |
| Apricots (*Prunus armeniaca* L.) | Grasspea (*Lathyrus sativus* L.)[a] | Quinoa (*Chenopodium quinoa* Willd.) |
| Asparagus (*Asparagus officinalis* L.)[a] | Groundnut (*Arachis hypogaea* L.) | Rapeseed (*Brassica napus* L.)[a] |
| Aubergine (*Solanum melongena* L.)[a] | Lagos yam (*Dioscorea cayennensis* Lam.)[a] | African rice (*Oryza glaberrima* Steud.)[a] |
| Bambara [*Vigna subterranea* (L.) Verdc.][a] | Leeks (*Allium ampeloprasum* L.) | Rice (*Oryza sativa* L.)[a] |
| Bananas (*Musa acuminata* Colla)[a] | Lemons [*Citrus limon* (L.) Osbeck] | Rye (*Secale cereale* L.)[a] |
| Plantains (*Musa balbisiana* Colla)[a] | Lentils (*Lens culinaris* Medik.)[a] | Safflower (*Carthamus tinctorius* L.) |
| Barley (*Hordeum vulgare* L.)[a] | Lettuce (*Lactuca sativa* L.) | Sorghum [*Sorghum bicolor* (L.) Moench][a] |
| Beans (*Phaseolus vulgaris* L.)[a] | Lima beans (*Phaseolus lunatus* L.)[a] | Soybean [*Glycine max* (L.) Merr.] |
| Black mustard [*Brassica nigra* (L.) K.Koch][a] | Lucerne (*Medicago sativa* L.)[a] | Spinach (*Spinacia oleracea* L.) |
| Breadfruit [*Artocarpus altilis* (Parkinson ex F.A. Zorn) Fosberg][a] | Maize (*Zea mays* L.)[b] | Strawberries [*Fragaria* × *ananassa* (Duchesne ex Weston) Duchesne ex Rozier][a] |
| Broom millet [*Setaria italica* (L.) P. Beauv.] | Mango (*Mangifera indica* L.) | *Sugarbeet* (*Beta vulgaris* L.)[a] |
| Cabbages (*Brassica oleracea* L.)[a] | Melons (*Cucumis melo* L.) | Sugarcane (*Saccharum officinarum* L.) |
| Carrots (*Daucus carota* L.)[a] | Mung peas [*Vigna radiata* (L.) R. Wilczek][a] | Sunflower (*Helianthus annuus* L.)[a] |
| Cassava (*Manihot esculenta Crantz*)[c] | Mustard [*Brassica juncea* (L.) Czern.][a] | Sweetpotato [*Ipomoea batatas* (L.) Lam.][a] |
| Cherries [*Prunus avium* (L.) L.] | Oats (*Avena sativa* L.)[a] | Tomatoes (*Solanum lycopersicum* L.) |
| Chickpeas (*Cicer arietinum* L.)[a] | Onions (*Allium cepa* L.) | Turnip (*Brassica rapa* L.)[a] |
| Chillies (*Capsicum annuum* L.) | Oranges [*Citrus sinensis* (L.) Osbeck][a] | Urd beans [*Vigna mungo* (L.) Hepper][a] |
| Cocoa (*Theobroma cacao* L.) | Papayas (*Carica papaya* L.) | Vetches (*Vicia sativa* L.)[a] |
| Cottonseed (*Gossypium hirsutum* L.) | Peaches [*Prunus persica* (L.) Batsch] | Water yam (*Dioscorea alata* L.)[a] |
| Cowpeas [*Vigna unguiculata* (L.) Walp.][a] | Pearl millet [*Pennisetum glaucum* (L.) R. Br.][a] | Watermelons [*Citrullus lanatus* (Thunb.) Matsum. & Nakai] |
| Cucumbers (*Cucumis sativus* L.) | Pears (*Pyrus communis* L.) | Wheat (*Triticum aestivum* L.)[a] |
| Faba beans (*Vicia faba* L.)[a] | Peas (*Pisum sativum* L.)[a] | White guinea yam (*Dioscorea* rotundata Poir.)[a] |
| Finger millet [*Eleusine coracana* (L.) Gaertn.][a] | Pigeonpeas [*Cajanus cajan* (L.) Millsp.][a] | Yautia [*Xanthosoma sagittifolium* (L.) Schott][a] |
| Foxtail millet (*Panicum miliaceum* L.) | Pineapples [*Ananas comosus* (L.) Merr.] | Zucchini (*Cucurbita pepo* L.) |

*Notes*: [a]Crop genepool and wild relatives are part of Annex I (FAO, 2009); [b]only the crop is part of Annex I plus genus *Tripsacum*; [c]only crop genepool is covered by Annex I.

was collected (latitude, longitude, locality description, elevation), whether or not the accession is available, and biological status of the accession (whether it is a wild sample, weed, landrace, breeding material, or an advanced cultivar). Many individual genebanks have their databases available online (e.g. The World Vegetable Center (avrdc.org), the International Center for Tropical Agriculture (isa.ciat.cgiar.org/urg), the International Potato Center (germplasmdb.cip.cgiar.org), Centre for Genetic Resources, the Netherlands (documents.plant.wur.nl/cgn/Website/downloads/download.htm)), and existing online catalogues bring together multiple genebank databases in a single outlet, as is the case for EURISCO (eurisco.ipk-gatersleben.de), which gathers information from the European genebanks, and GENESYS (www.genesys-pgr.org), which centralizes the passport data from the collections of CGIAR centres and more than 400 institutions globally. For some isolated cases, databases are reached by direct requests to the corresponding genebank or data manager.

- *Reference occurrence data*: this category includes herbarium specimens, observations reported in the literature and germplasm accessions classified as not available for distribution. The underlying assumption for using this information is that it can provide details of localities where the taxa have been previously recorded but have not yet been collected, or need to be re-collected. Germplasm occurrence data and reference occurrence data are useful to determine the potential distributions of each taxon, by estimating its environmental niches. As for the case of germplasm occurrence data, some individual herbaria have their specimens digitized and available online. Likewise, global initiatives like the Global Biodiversity Information Facility (www.gbif.org) and Global Plants on JSTOR (plants.jstor.org) are virtual resources where occurrence records and digitized plant specimens are made available to the global community. Various large collections have yet to undergo the digitizing process; therefore, personal visits to key herbaria were required to obtain details on the biology and locality descriptions where the taxa had been recorded.
- *Environmental layers*: spatial-explicit environmental layers serve as drivers of distribution in environmental niche modelling. For the CWR global analysis, the 19 bioclimatic variables described by Nix (1986) and calculated using the Worldclim database (Hijmans *et al.*, 2005) were used at a resolution of 2.5 min (~4.6 km at the equator). Maxent was the modelling algorithm selected for producing the environmental niche models of each CWR, due to its wide use and recognized performance (Elith *et al.*, 2006).

**3. Preparing and standardizing all data inputs:** data organization and standardization were performed on all 158 data sets that were gathered in step 2, as they followed different standards and were not interoperable. For this, an extended list of ecogeographic data descriptors was designed (Castañeda-Álvarez *et al.*, 2011), and all data sets were prepared to follow these descriptors. Subsequently, all data records were curated, especially for the fields containing information on the identification, locality and coordinates where the taxon had been observed/collected. The nomenclatural identification of each record was revised and validated using three tools: 'Taxonomic Name Resolution Service – TNRS' (Boyle *et al.*, 2013), the R package 'taxonstand' (Cayuela *et al.*, 2012) and GRIN taxonomy (USDA, 2015). Manual revisions took place when discrepancies were found among the Latin names suggested by these online services. The coordinates of the records were also examined, by identifying mismatching coordinates (e.g. coordinates falling in countries different from those described in the passport data) and data points falling in water bodies, and recalculating coordinates when these issues were detected and when the record contained enough locality and geographical descriptions of the site where the taxon was recorded but no coordinates were available. This data set is accessible to the public at: www.cwrdiversity.org/checklist/cwr-occurrences.php.

**4. Applying the methodology:** the gap analysis was applied in R, with the associated packages maptools (Bivand and Lewin-Koh, 2014), rgdal (Bivand *et al.*, 2014), SDMTools (VanDerWal *et al.*, 2014), raster (Hijmans, 2014), sp (Pebesma and Bivand, 2005; Bivand *et al.*, 2013) and dismo (Hijmans *et al.*, 2014). This methodology quantifies three metrics: the Sampling Representativeness Score (SRS), the Geographic Representativeness Score (GRS) and the Ecological Representativeness Score (ERS). Environmental niche models are required to calculate the GRS and ERS, and the ERS utilizes an extra layer to determine the existing ecological units where the taxon is likely to be found. The 'Terrestrial Ecoregions of the World' map was used for this case (Olson *et al.*, 2001). For future implementations, users can include equivalent layers, such as the 'Ecogeographical Land Characterization' maps (Parra-Quijano *et al.*, 2011). Two main outputs are obtained from the gap analysis: (i) an equally weighted score is calculated with the SRS, GRS and ERS, and is used to identify taxa requiring high, medium, low and very low priority for *ex situ* conservation; and (ii) a map indicating the collecting hotspots or places where further efforts for collecting high priority taxa should be addressed.

**5. Capturing expert opinions:** with the purpose of assessing how well the gap analysis captured known conservation priorities of CWR, a survey was completed by 44 experts in the fields of botany, plant breeding and PGR conservation. The survey asked the experts to score the prioritization results for the taxa of their expertise, to assess the quality of the occurrence data used as inputs of the analysis and the distribution models produced, and to provide a prioritization score based on their experience on probable threats that may affect CWR in their habitats and/or uses of CWR for breeding. The answers obtained were used to fine-tune analyses (when required) and to assess the level of correspondence between the gap analysis results and experts' knowledge. Individual crop genepool results of this evaluation are available at: www.cwrdiversity.org/expert-evaluation/.

## 13.3 Results

### 13.3.1 A global database of CWR occurrence records

The initial effort of gathering, standardizing and georeferencing occurrence records of CWR for producing environmental niche models and applying the gap analysis methodology resulted in a large database, perhaps the largest of its kind, containing over 5.6 million occurrence records, and these were used to generate a CWR distribution map (Fig. 13.1). The geographic coverage of the database comprises 96% of the countries of the world. Two-thirds of the records in the database correspond to CWR species, and the remaining one-third corresponds to records of cultivated species.

### 13.3.2 CWR Global Atlas

An interactive global atlas of CWR was also prepared for presenting current recorded distributions, potential distributions, potential areas for further collecting (or collecting priorities map) and collecting prioritization category (Fig. 13.2). In the same way, maps presenting species richness and hotspots for further collecting were prepared per crop genepool. This atlas is available at: www.cwrdiversity.org/distribution-map.

### 13.3.3 Conservation priorities

CWR are alarmingly under-represented in *ex situ* genebanks, as 838 (out of 1182) CWR taxa have been categorized as high priorities for further collecting, and all the crop groups, including 81 crop genepools (Fig. 13.3). Only a few crop genepools are well represented in genebanks in terms of associated CWR: grass pea, wheat, chickpea, lentil, barley and tomato (Castañeda-Álvarez *et al.*, in press). In terms of the degree of relatedness of CWR and their associated crops, most of the taxa ranked as high priorities for further conservation belong to the secondary genepool (395 taxa), followed by the primary genepool (155 taxa) and more distantly related taxa from the tertiary genepool (138 taxa). In general, collecting efforts aiming to increase the representativeness of high-priority taxa should focus on the Mediterranean basin, the Near East, South-east and East Asia and South America (Fig. 13.4).

## 13.4 Discussion

Urgent actions to improve the current representativeness of CWR in *ex situ* genebanks are

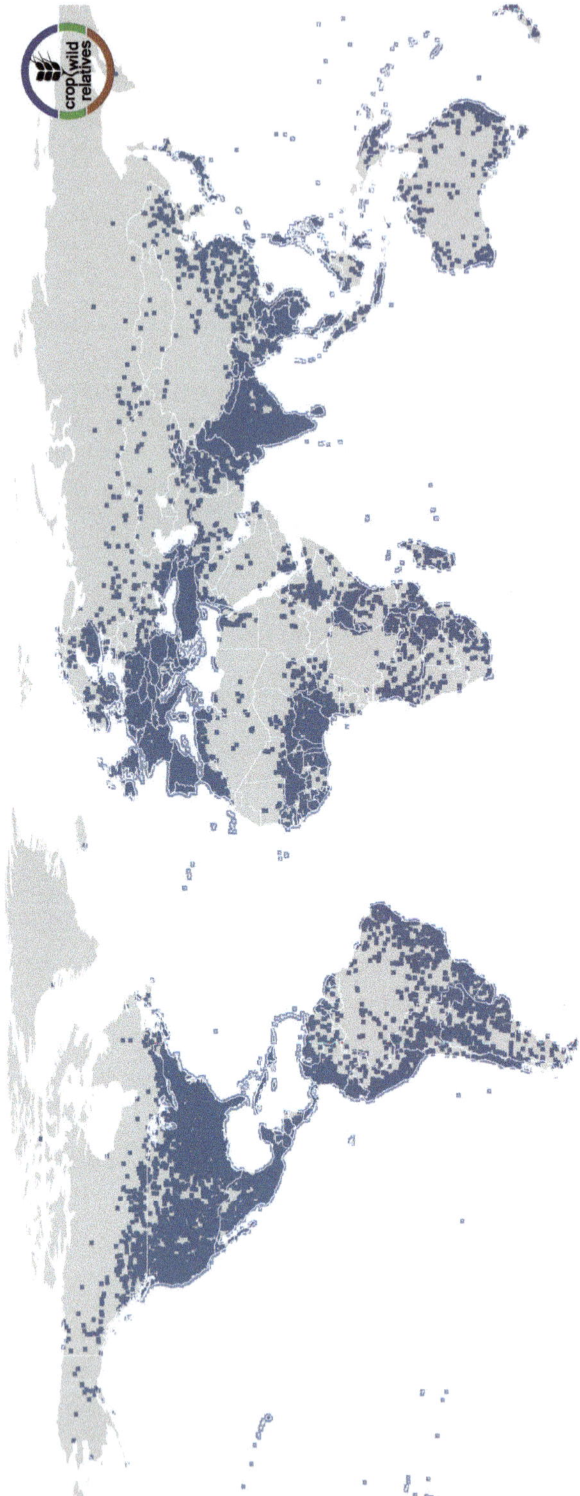


**Fig. 13.1.** Records with geographic coordinates held in the Crop Wild Relative Global Occurrence Database. (From Castañeda-Álvarez *et al.*, 2014.)

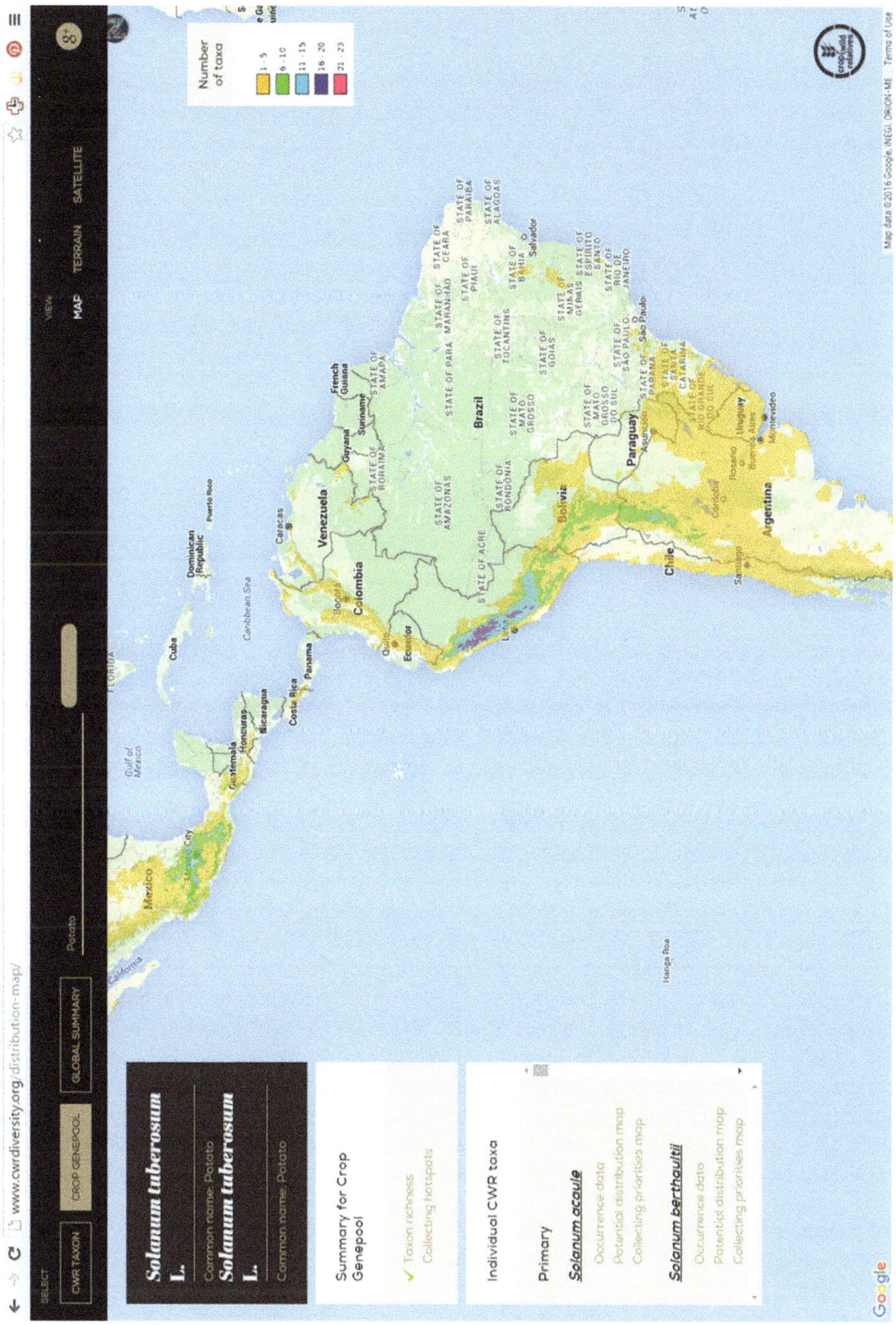


**Fig. 13.2.** Screen capture of the CWR Global Atlas, showcasing the potential species richness of the CWR of potato.

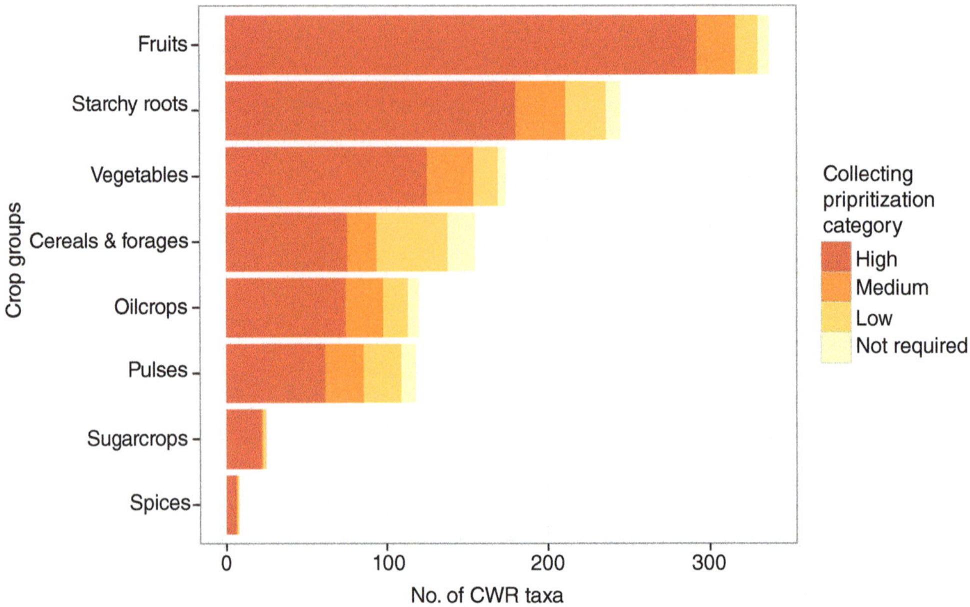


**Fig. 13.3.** Count of CWR taxa per collecting prioritization category and clustered by crop group.

needed, given the potential of CWR for agricultural improvement and adaptation, the threats that are likely to affect CWR populations in their natural habitats and the overall low extent of representativeness in genebanks. Collections of CWR taxa have already started in partnership with local institutions under the global initiative 'Adapting agriculture to climate change: collecting, protecting and preparing the crop wild relatives'. Despite this initiative being the largest of its kind, additional resources are required to enhance the conservation of CWR *ex situ*, especially for the wild relatives of crops not yet listed in Annex I, like fruit and vegetables. Due to the spatial resolution used for this study, the results obtained can be used to back further global, national and regional strategies to foster the effective conservation and use of CWR.

Taxon richness maps were also created as an intermediate product of the gap analysis, for each crop genepool and for all genepools considered in the analysis together, leading to a global CWR richness map. These outputs in particular can help to identify CWR-rich regions, and therefore guide *in situ* conservation strategies, which is discussed further by Vincent *et al.* (see Chapter 12, this volume) and Vincent *et al.* (2016). Likewise, identifying CWR hotspots can also serve for a diversity of analyses, including understanding if the current protected area networks are encompassing such hotspots within their areas or identifying ongoing conservation initiatives that may be willing to include management plans to conserve CWR within their areas of impact.

Basic research on taxonomy and understanding the crossability between CWR and associated crops is still needed for many groups. Further efforts to continue making CWR-related data retrievable, accessible, reusable, open and interoperable will improve our current understanding of the nature of CWR, their distributions and conservation status, and therefore lead to better decisions for conserving and using CWR. Increased availability and use of data on morphological traits and molecular diversity will also be key to produce refined prioritization studies of areas for *in situ* conservation and seed collecting planning, as shown by van Zonneveld *et al.* (2012).

## Acknowledgements

This work was undertaken as part of the initiative 'Adapting agriculture to climate change: collecting,

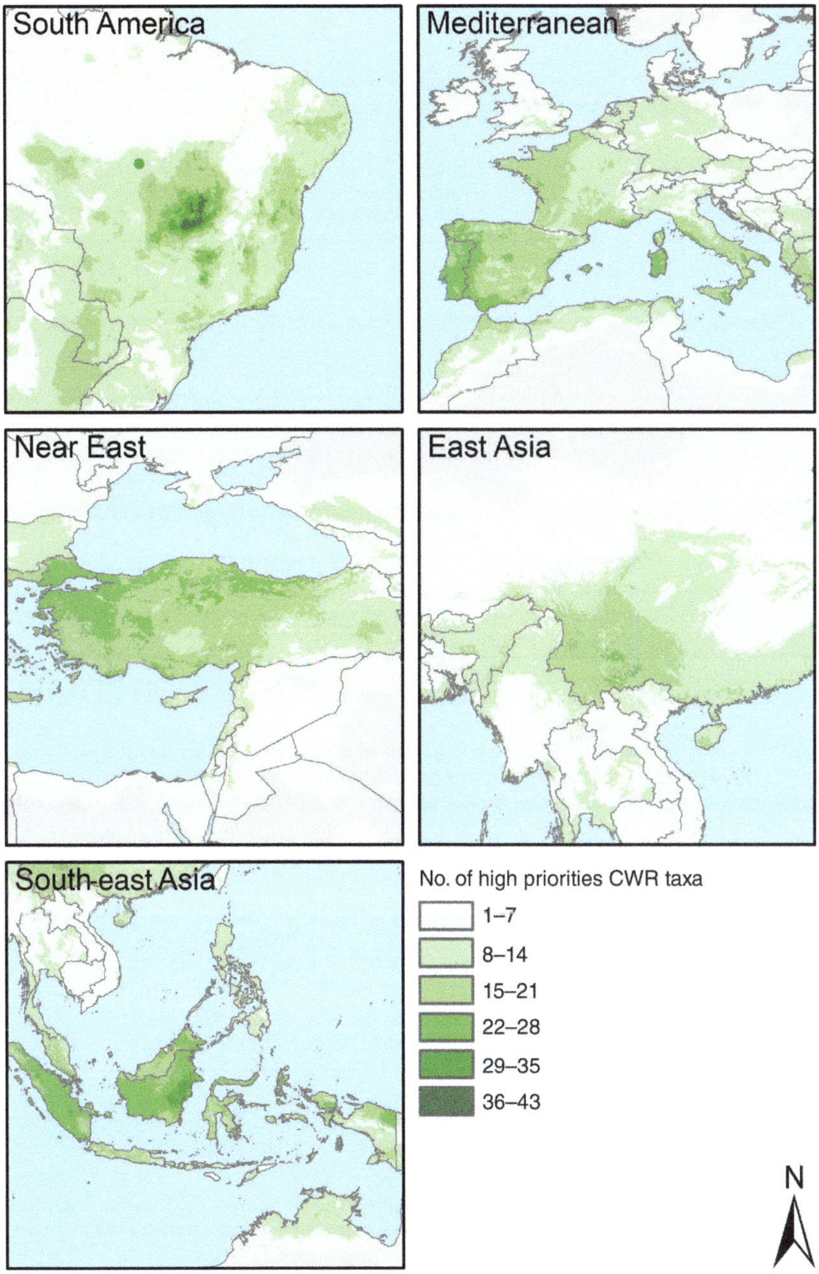


**Fig. 13.4.** Collecting hotspots of CWR categorized as high priorities for collecting.

protecting and preparing crop wild relatives', which is supported by the Government of Norway. The project is managed by the Global Crop Diversity Trust and the Millennium Seed Bank of the Royal Botanic Gardens, Kew, and implemented in partnership with national and international genebanks and plant breeding institutes around the world. For further information, go to the project website: www.cwrdiversity.org. Funding was also provided by the CGIAR Research Program on Climate Change, Agriculture, and Food Security, Cali, Colombia.

## References


Bivand, R. and Lewin-Koh, N. (2014) maptools: Tools for reading and handling spatial objects. Available at: cran.r-project.org/package=maptools (accessed 5 August 2015).

Bivand, R.S., Pebesma, E. and Gomez-Rubio, V. (2013) *Applied Spatial Data Analysis with R.* 2nd edn, Springer, New York, doi:10.1007/978-1-4614-7618-4.

Bivand, R., Keitt, T. and Rowlingson, B. (2014) rgdal: Bindings for the Geospatial Data Abstraction Library. Available at: cran.r-project.org/package=rgdal (accessed 5 August 2015).

Boyle, B., Hopkins, N., Lu, Z., Raygoza Garay, J.A., Mozzherin, D., *et al.* (2013) The taxonomic name resolution service: an online tool for automated standardization of plant names. *BMC Bioinformatics* 14(1), 1–16.

Brehm, J.M., Maxted, N., Martins-Loução, M.A. and Ford-Lloyd, B.V. (2010) New approaches for establishing conservation priorities for socio-economically important plant species. *Biodiversity and Conservation* 19(9), 2715–2740.

Brummitt, N. and Bachman, S. (2010) *Plants Under Pressure – A Global Assessment. The First Report of the IUCN Sampled Red List Index for Plants*. London, UK.

Castañeda-Álvarez, N.P., Vincent, H., Kell, S.P., Eastwood, R.J. and Maxted, N. (2011) Ecogeographic surveys. In: Guarino, L., Ramanatha, R.V. and Goldberg, E. (eds) *Collecting Plant Genetic Diversity: Technical Guidelines – 2011 Update*. Rome. Available at: cropgenebank.sgrp.cgiar.org/index.php?option= com_content&view=article&id=390&Itemid=557 (accessed 5 August 2015).

Castañeda-Álvarez, N.P., Khoury, C.K., Sosa, C.C., Eastwood, R., Harker, R., *et al.* (2014) A global database for the distributions of crop wild relatives. doi:10.6084/m9.figshare.1187065

Castañeda-Álvarez, N.P., Khoury, C.K., Achicanoy, H.A., Bernau, V., Dempewolf, H., *et al.* (in press) Global conservation priorities for crop wild relatives. *Nature Plants*.

Cayuela, L., Granzow-de la Cerda, Í., Albuquerque, F.S. and Golicher, D.J. (2012) taxonstand: An R package for species names standardisation in vegetation databases. *Methods in Ecology and Evolution* 3(6), 1078–1083.

CBD (Convention on Biological Diversity) (2010a) *Global Strategy for Plant Conservation 2011–2020*. Secretariat of the Convention on Biological Diversity, Montreal, Quebec, Canada.

CBD (2010b) *The Strategic Plan for Biodiversity 2011–2020 and the Aichi Targets*. Secretariat of the Convention on Biological Diversity, Montreal, Quebec, Canada.

Dempewolf, H., Eastwood, R.J., Guarino, L., Khoury, C.K. and Müller, J.V. (2013) Adapting agriculture to climate change: a global initiative to collect, conserve, and use crop wild relatives. *Agroecology and Sustainable Food Systems* 38, 369–377.

Dias, S., Dulloo, M.E. and Arnaud, E. (2012) The role of EURISCO in promoting use of agricultural biodiversity. In: Maxted, N., Dulloo, M.E., Ford-Lloyd, B.V., Frese, L., Iriondo, J.M., and Pinheiro de Carvalho, M. (eds) *Agrobiodiversity Conservation: Securing the Diversity of Crop Wild Relatives and Landraces*. CAB International, Wallingford, UK, pp. 270–277.

Elith, J., Graham, C.H., Anderson, R.P., Dudík, M., Ferrier, S., *et al.* (2006) Novel methods improve prediction of species' distributions from occurrence data. *Ecography* 29(2), 129–151.

FAO (Food and Agriculture Organization) (2008) Climate change and biodiversity for food and agriculture. *Technical background document from the expert consultation held on 13 to 14 February 2008*. Commission on Genetic Resources for Food and Agriculture, Food and Agriculture Organization of the United Nations, Rome, p.10.

FAO (2009) *International Treaty on Plant Genetic Resources for Food and Agriculture*. FAO, Rome.

FAO (2010) *The second report on the state of the world's plant genetic resources for food and agriculture*. Commission on Genetic Resources for Food and Agriculture, FAO, Rome.

Foley, J.A., Ramankutty, N., Brauman, K.A., Cassidy, E.S., Gerber, J.S., *et al.* (2011) Solutions for a cultivated planet. *Nature* 478(7369), 337–342.

Godfray, H.C.J., Beddington, J.R., Crute, I.R., Haddad, L., Lawrence, D., *et al.* (2010). Food security: the challenge of feeding 9 billion people. *Science* 327(5967), 812–818.

Hajjar, R. and Hodgkin, T. (2007) The use of wild relatives in crop improvement: a survey of developments over the last 20 years. *Euphytica* 156(1–2), 1–13.

Harlan, J.R. and de Wet, J.M.J. (1971) Toward a rational classification of cultivated plants. *Taxon* 20(4), 509–517.

Hawkes, J.G., Maxted, N. and Ford-Lloyd, B.V. (2000) *The ex situ Conservation of Plant Genetic Resources*. Kluwer Academic Publishers, Dordrecht, the Netherlands.

Hijmans, R.J. (2014) raster: Geographic data analysis and modeling. Available at: cran.r-project.org/package=raster (accessed 5 August 2015).
Hijmans, R.J., Cameron, S.E., Parra, J.L., Jones, P.G. and Jarvis, A. (2005) Very high resolution interpolated climate surfaces for global land areas. *International Journal of Climatology* 25, 1965–1978.
Hijmans, R.J., Phillips, S., Leathwick, J. and Elith, J. (2014) dismo: Species distribution modeling. Available at: cran.r-project.org/package=dismo (accessed 5 August 2015).
Jarvis, A., Lane, A. and Hijmans, R.J. (2008) The effect of climate change on crop wild relatives. *Agriculture, Ecosystems & Environment* 126, 13–23.
Kell, S.P., Maxted, N. and Bilz, M. (2012) European crop wild relative threat assessment: knowledge gained and lessons learnt. In: Maxted, N., Dulloo, M.E., Ford-Lloyd, B.V., Frese, L., Iriondo, J.M., and Pinheiro de Carvalho, M. (eds) *Agrobiodiversity Conservation: Securing the Diversity of Crop Wild Relatives and Landraces*. CAB International, Wallingford, UK, pp. 218–242.
Langhammer, P.F., Bakarr, M.I., Bennun, L.A., Brooks, T.M. and Clay, R.P. (2007) *Identification and Gap Analysis of Key Biodiversity Areas: Targets for Comprehensive Protected Area Systems*. IUCN, Gland, Switzerland.
McCouch, S., Baute, G.J., Bradeen, J., Bramel, P., Bretting, P.K., *et al.* (2013) Feeding the future. *Nature* 499, 23–24.
Maxted, N. and Kell, S.P. (2009) *Establishment of a Global Network for the In Situ Conservation of Crop Wild Relatives: Status and Needs.* Commission on Genetic Resources for Food and Agriculture. FAO, Rome, 211 pp. Available at: www.fao.org/docrep/013/i1500e/i1500e18a.pdf (accessed 5 August 2015).
Maxted, N., Ford-Lloyd, B.V., Jury, S., Kell, S. and Scholten, M. (2006) Towards a definition of a crop wild relative. *Biodiversity and Conservation* 15(8), 2673–2685.
Maxted, N., Ford-Lloyd, B.V., Kell, S.P., Iriondo, J., Dulloo, E., *et al.* (eds) (2008a) *Crop Wild Relative Conservation and Use*. CAB International, Wallingford, UK.
Maxted, N., Dulloo, E., Ford-Lloyd, B.V., Iriondo, J.M. and Jarvis, A. (2008b) Gap analysis: a tool for complementary genetic conservation assessment. *Diversity and Distributions* 14(6), 1018–1030.
Nix, H.A. (1986) A biogeographic analysis of Australian elapid snakes. In: Longmore, R. (ed.) *Atlas of Elapid Snakes of Australia*. Australian Government Publishing Service, Canberra, pp. 4–15.
Olson, D.M., Dinerstein, E., Wikramanayake, E.D., Burgess, N.D., Powell, G.V., *et al.* (2001) Terrestrial ecoregions of the world: a new map of life on Earth. *Bioscience* 51(11), 933–938.
Parra-Quijano, M., Iriondo, J.M. and Torres, E. (2011) Ecogeographical land characterization maps as a tool for assessing plant adaptation and their implications in agrobiodiversity studies. *Genetic Resources and Crop Evolution* 59(2), 205–217.
Pebesma, E.J. and Bivand, R.S. (2005) Classes and methods for spatial data in R. R News 5 (2). Available at: cran.r-project.org/doc/Rnews/ (accessed 5 August 2015).
Porter, J.R., Xie, L., Challinor, A.J., Cochrane, K., Howden, S.M., *et al.* (2014) Food security and food production systems. In: Field, C.B., Barros, V.R., Dokken, D.J., Mach, K.J., Mastrandrea, M.D., Bilir, T.E., Chatterjee, M., Ebi, K.L., Estrada, Y.O., Genova, R.C., Girma, B., Kissel, E.S., Levy, A.N., MacCracken, S., Mastrandrea, P.R. and White, L.L. (eds) *Climate Change 2014: Impacts, Adaptation, and Vulnerability. Part A: Global and Sectoral Aspects. Contribution of Working Group II to the Fifth Assessment Report of the Intergovernmental Panel on Climate Change.* Cambridge University Press, Cambridge, UK, and New York, pp. 485–533.
Ramírez-Villegas, J., Khoury, C., Jarvis, A., Debouck, D.G. and Guarino, L. (2010) A gap analysis methodology for collecting crop genepools: a case study with *Phaseolus* beans. *PloS ONE* 5(10), p.e13497.
Rodrigues, A., Akçakaya, H.R, Andelman, S.J., Bakarr, M.I. and Boitani, L. (2004) Global gap analysis : priority regions for expanding the global protected-area network. *BioScience* 54(12), 1092–1100.
Rosegrant, M.W. and Cai, X. (2001) Water scarcity and food security: alternative futures for the 21st century. *Water Science and Technology* 43(4), 61–70.
Tanksley, S.D. and McCouch, S.R. (1997) Seed banks and molecular maps: unlocking genetic potential from the wild. *Science* 277, 1063–1066.
Thormann, I., Parra-Quijano, M., Endressen, D.T.F., Rubio Teso, M.I., Iriondo, M.J., *et al.* (2014) *Predictive Characterization of Crop Wild Relatives and Landraces. Technical Guidelines Version 1*. Bioversity International, Rome.
Ureta, C., Martínez-Meyer, E., Perales, H.R. and Álvarez-Buylla, E.R. (2012) Projecting the effects of climate change on the distribution of maize races and their wild relatives in Mexico. *Global Change Biology* 18(3), 1073–1082.

USDA (US Department of Agriculture) (2015) *USDA ARS* National Genetic Resources Program, Germplasm Resources Information Network – GRIN. National Germplasm Resources Laboratory, Beltsville, Maryland. Available at: www.ars-grin.gov/~sbmljw/cgi-bin/taxcrop.pl?language=en (accessed 22 July 2015).
van de Wouw, M., Van Hintum, T., Kik, C., Van Treuren, R. and Visser, B. (2010) Genetic diversity trends in twentieth century crop cultivars: a meta analysis. *TAG. Theoretical and Applied Genetics* 120(6), 1241–1252.
van Zonneveld, M., Scheldeman, X., Escribano, P., Viruel, M.A., *et al.* (2012) Mapping genetic diversity of cherimoya (*Annona cherimola* Mill.): application of spatial analysis for conservation and use of plant genetic resources. *PLoS ONE* 7(1), e29845.
VanDerWal, J., Falconi, L., Januchowski, S. and Luke, S. (2014) SDMTools: Species Distribution Modelling Tools: Tools for processing data associated with species distribution modelling exercises. Available at: cran.r-project.org/package=SDMTools (accessed 5 August 2015).
Vincent, H., Wiersema, J., Kell, S., Fielder, H., Dobbie, S., *et al.* (2013) A prioritized crop wild relative inventory to help underpin global food security. *Biological Conservation* 167, 265–275.
Vincent, H., Castañeda-Álvarez, N.P. and Maxted, N. (2016) Global priorities for *in situ* conservation of wild plant genetic resources: towards the establishment of a global network of crop wild relative reserves. (In preparation.)
Wiersema, J.H., León, B. and Garvey, E.J. (2012) Identifying wild relatives of subtropical and temperate fruit and nut crops. In: Aradhya, M.K. and Kluepfel, D.A. (eds) *ISHS Acta Horticulturae 948: I International Symposium on Wild Relatives of Subtropical and Temperate Fruit and Nut Crops*. International Society for Horticultural Science, The Hague, the Netherlands, pp. 285–288.

# 14 National Strategies for the Conservation of Crop Wild Relatives

J.M. Iriondo,[1*] H. Fielder,[2] H. Fitzgerald,[3] S.P. Kell,[2] J. Labokas,[4] J. Magos-Brehm,[2] V. Negri,[5] J. Phillips,[2] M.L. Rubio Teso,[1] S. Sensen,[6] N. Taylor[7] and N. Maxted[2]

[1]*Universidad Rey Juan Carlos, Madrid, Spain;* [2]*University of Birmingham, Birmingham, UK;* [3]*Finnish Museum of Natural History, University of Helsinki, Finland;* [4]*Nature Research Centre, Vilnius, Lithuania;* [5]*University of Perugia, Perugia, Italy;* [6]*Federal Office for Agriculture and Food (BLE), Bonn, Germany;* [7]*University of Leeds, Leeds, UK*

## 14.1 Introduction

Most conservation policies are designed and implemented at the national level, including those that are executed in response to obligations contracted under international agreements. Because of this, the success of the conservation of crop wild relatives (CWR) greatly depends on the development of sound national strategies across many countries. While conservation policies for threatened species have been implemented routinely in most countries for the past three decades, the development of conservation actions for CWR at the national level is still in its preliminary stages.

During the past decade, scientists have raised concerns about the impacts of climate change on agriculture and food production (Olesen and Bindi, 2002; Turral *et al.*, 2011; IPCC, 2014). These concerns have attracted the interest of social media, and policy makers and society in general are beginning to realize the magnitude of this problem. One of the main limitations of modern cultivars to adapt to climate change lies in their narrow genetic base (Cooper *et al.*, 2001; van de Wouw *et al.*, 2010). Since CWR tend to contain greater genetic diversity than crops, some of which code for traits that can be used to improve crops, they have become a matter of great interest, and the need for conserving their genetic diversity is becoming more and more urgent (Maxted and Kell, 2009; Maxted *et al.*, 2012a). This indicates that the time is ripe for the design of national strategies for the conservation and management of CWR and the systematic implementation of conservation actions for selected CWR populations (Maxted *et al.*, 2015).

A series of international projects that have been carried out in recent years have also laid the foundations for taking action at the national level. For instance, this is the case of the projects 'European Crop Wild Relative Diversity Assessment and Conservation Forum' (PGR Forum), 'An Integrated European *In Situ* Management Work Plan: Implementing Genetic Reserves and On Farm Concepts' (AEGRO), 'Novel Characterization of Crop Wild Relative and Landrace Resources as a Basis for Improved Crop Breeding' (PGR Secure), and '*In situ* Conservation and Use of Crop Wild Relatives in Three ACP Countries of the SADC Region', co-funded by the European

*Corresponding author; e-mail: jose.iriondo@urjc.es

 

Commission; and the global project, '*In-situ* Conservation of Wild Crop Relatives through Enhanced Information Management and Field Application', developed by UNEP (United Nations Environment Programme) and funded by GEF (Global Environment Facility), uniting national agencies in Armenia, Bolivia, Madagascar, Sri Lanka and Uzbekistan.

A first attempt at providing an overview of the status of CWR conservation practices at the national level was performed at the Symposium 'Towards the Establishment of Genetic Reserves for Crop Wild Relatives and Landraces in Europe' that took place in Funchal (Madeira, Portugal) in September 2010 as the final dissemination conference of the AEGRO Project. In this symposium, a questionnaire was distributed among participants and the results were published by Maxted *et al.* (2012b). The resulting database was subsequently maintained and updated during the PGR Secure project (www.pgrsecure.bham.ac.uk/sites/default/files/documents/helpdesk/Progress_national_CWR_and_LR_conservation_Europe.pdf).

An important landmark in this process was the workshop 'Conservation Strategies for European Crop Wild Relative and Landrace Diversity' that was jointly organized by the PGR Secure project and ECPGR (the European Cooperative Programme for Plant Genetic Resources) in Palanga, Lithuania, in September 2011 (the Palanga workshop). This workshop hosted participants from 38 European countries and the USA, and provided information and training concerning the development of national strategies for CWR. In parallel, the PGR Secure project established a helpdesk that was made available online (www.pgrsecure.org/helpdesk) and which offered all relevant information regarding this subject. Moreover, additional technical support and advice were provided from the PGR Secure project via e-mail and at in-country meetings organized on demand.

The PGR Secure project funded the development of national strategies for the conservation of CWR in Finland, Italy, Spain and the UK, and used them as case studies for the development of national strategies in other countries. As a result of the Palanga workshop and other interactions between PGR Secure project leaders and representatives from different European countries, some seed funding was raised, discussions were held with different stakeholder groups and advances were made in the development of national strategies for Albania, Azerbaijan, Belarus, Bulgaria, Cyprus, Czech Republic, Greece, Lithuania, Norway, Poland, Sweden and Turkey.

In the context of other parallel projects outside of Europe, the FAO Commission of PGRFA (CGRFA) recognized the importance of the development of national strategies for agrobiodiversity conservation, including the need to conserve the full range of CWR diversity found in a country, as underlined in the Convention on Biological Diversity (CBD), the International Treaty on Plant Genetic Resources for Food and Agriculture (ITPGRFA) and the Second Global Plan of Action for Plant Genetic Resources for Food and Agriculture (Second GPA). The CGRFA have recently published Draft Technical Guidelines concerning the National Level Conservation of Crop Wild Relatives, and these are available at: www.fao.org/3/a-mm542e.pdf. Further studies or initiatives in additional countries have made progress in their national CWR strategies across the world, including Benin, Madagascar, Mauritius, South Africa and Zambia (Africa), Bolivia, Guatemala, Mexico, Peru, USA and Venezuela (America), Armenia, China, India, Jordan, Uzbekistan and Sri Lanka (Asia), and Germany, Hungary, Ireland, Portugal and Switzerland (Europe).

In this context of flourishing CWR conservation studies and initiatives at the national level, the purpose of this study was to review the current status of development and implementation of national CWR conservation strategies worldwide, and to perform a comparative study of the different approaches, challenges and solutions provided in each case.

## 14.2 Methods

Multiple resources are available on methodologies that can be applied to the development of national CWR conservation strategies, such as the CWR *In Situ* Strategy Helpdesk (aegro.jki.bund.de/aegro/index.php?id=188) of the AEGRO Project, the FAO Resource Book for the Preparation of National Plans for Conservation of Crop

Wild Relatives and Landraces (www.fao.org/fileadmin/templates/agphome/documents/PGR/PubPGR/ResourceBook/TEXT_ALL_2511.pdf), the Concept for *In Situ* Conservation of Crop Wild Relatives in Europe developed by ECPGR (Maxted *et al.*, 2015), and the PGR Secure project Helpdesk (www.pgrsecure.org/helpdesk).

Maxted *et al.* (2007) identify a series of steps that must be completed in the process of designing a national strategy for CWR (Fig. 14.1). Essentially, they involve: (i) the generation of a national checklist of CWR; (ii) the prioritization of the checklist to narrow it down to a manageable number of taxa on which conservation actions can be devised and implemented; (iii) the characterization of priority taxa to collect information that can be useful to assess the utility of the taxon for crop improvement, as well as its threat status and vulnerability, and the extent and quality of its genetic diversity; (iv) the assessment of the conservation status through *in situ* and *ex situ* gap analyses; (v) the definition of conservation goals for priority taxa; and (vi) the prescription and implementation of conservation actions on priority taxa focused on the achievement of selected goals.

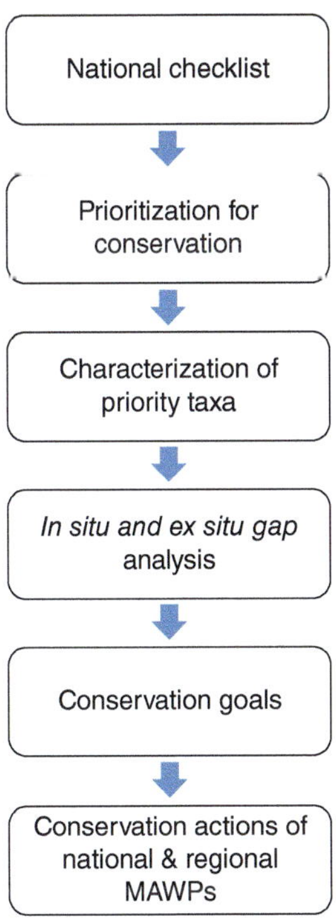


Fig. 14.1. Steps in the development of a national strategy for CWR. (From Maxted *et al.*, 2007.)
*Note*: MAWPs = most appropriate wild populations.

In this study, we have used these steps as landmarks to review the progress that has been made in different countries. Thus, using this perspective, we have summarized the contents of the database that originated from the results of the questionnaire presented at the Symposium 'Towards the Establishment of Genetic Reserves for Crop Wild Relatives and Landraces in Europe' and the PGR Secure project, further updated with communications with the National Focal Points (Maxted *et al.*, 2012b). Subsequently, we have proceeded in the same way, gathering information from various CWR conservation initiatives that are taking place in different countries, with special emphasis on Europe. The list of countries for which data were obtained from the published literature and/or direct contact with the people responsible for the studies includes China, Cyprus, Czech Republic, Finland, Germany, Italy, Jordan, Lithuania, Norway, Portugal, Spain, UK and USA (Maxted *et al.*, 2007; Magos-Brehm *et al.*, 2008; Rubio Teso *et al.*, 2012, 2015; Fitzgerald, 2013; Fitzgerald *et al.*, 2013; Khoury *et al.*, 2013; Landucci *et al.*, 2014; Phillips *et al.*, 2014; Fitzgerald and Korpelainen, 2015; Kell *et al.*, 2015; Fielder *et al.*, 2015).

## 14.3 Results and Discussion

The most up-to-date results of the database that originated from the questionnaire developed and distributed at the Symposium 'Towards the Establishment of Genetic Reserves for Crop Wild Relatives and Landraces in Europe' (Maxted *et al.*, 2012a), are available at: www.pgrsecure.bham.ac.uk/sites/default/files/documents/helpdesk/Progress_national_CWR_and_LR_conservation_Europe.pdf. Some of the most relevant information is summarized in Table 14.1. The actions that were carried out more frequently involved the generation of a national CWR checklist, a list of priority taxa and the implementation of *ex situ* conservation actions. On the other hand, the actions that were carried out less frequently concerned the development of *in situ* and *ex situ* gap analyses, *in situ* conservation actions and

Table 14.1. Progress made by European countries on different conservation actions and analysis concerning the conservation of crop wild relatives according to the results of a questionnaire developed and distributed at the Symposium 'Towards the Establishment of Genetic Reserves for Crop Wild Relatives and Landraces in Europe' that took place in Funchal (Madeira, Portugal) in 2010 and further completed subsequently by others through the PGR Secure project (raw data obtained from PGR Secure website: http://www.pgrsecure.bham.ac.uk/sites/default/files/documents/helpdesk/Progress_national_CWR_and_LR_conservation_Europe.pdf).

| Progress | Number of countries (%) |
|---|---|
| National checklist | 34 (100%) |
| CWR priority taxa | 18 (53%) |
| Threat assessment | 17 (50%) |
| *In situ*/*ex situ* gap analysis | 8 (24%) |
| National strategy | 13 (38%) |
| Information system | 10 (29%) |
| *In situ* conservation actions | 9 (26%) |
| *Ex situ* conservation actions | 18 (53%) |
| Legislation | 9 (26%) |

the enactment of legislation specific to CWR. The accuracy of the data gathered through this questionnaire may have been somewhat conditioned by lack of precise knowledge on the status of CWR conservation in some of the respondents' countries, as well as by an involuntary bias towards a too optimistic perspective of the status of CWR conservation in their countries. Nevertheless, it provided a first overall perspective of the status of CWR conservation in Europe.

### 14.3.1 National checklists, prioritization criteria and inventories

To aid the development of national CWR strategies, in the recent literature a distinction has been made between national CWR checklists and inventories (Maxted *et al.*, 2013). A checklist is a list of taxon names and authorities of all CWR taxa found in a certain geographic area, while an inventory is a prioritized list of CWR taxa present in a certain geographic area that has additional information for each taxon that will aid their conservation and use, such as nomenclature, biology, ecogeography, populations, uses, threats and conservation.

The PGR Forum project provided the first European catalogue of crop wild relatives, which made it possible to generate national checklists for all European countries (Kell *et al.*, 2005, 2008). This is why the results from the database that originated from the Funchal questionnaire indicated that 34 European countries had produced a national checklist. Nevertheless, the number of CWR national checklists that have been refined according to nationally agreed nomenclature and which are either published or easily available through the Internet is close to 10 at the European level and not greater than 20 worldwide. Different approaches have been followed in the generation of the national checklists, although in Europe many of them stem from the above-mentioned work elaborated in the context of the PGR Forum project (Kell *et al.*, 2005, 2008).

Depending on the idiosyncrasies of each country (e.g. plant species richness and diversity, organization of public administration concerning the management of agriculture, forestry and wildlife conservation and conservation priorities), different criteria have been followed concerning the initial delimitation of species in a national checklist of CWR. Thus, countries such as the Czech Republic, Finland, Norway, Portugal, UK and USA consider both native and naturalized species, whereas in Cyprus, Germany, Jordan, Lithuania and Spain naturalized taxa are not included. Furthermore, in some countries, forestry species are also excluded, because they are managed separately by the public administration. Consequently, the number of taxa present in the checklists does not depend only on species richness in each country. According to Kell *et al.* (2008), Spain has the second largest number of CWR species in the Euro-Mediterranean region after Turkey. However, the number of taxa included in the Spanish national CWR checklist is severely reduced, because it only considers the wild relatives of a subset of the most important crops. Therefore, although checklists are typically inclusive of all CWR taxa in a country, important differences among countries are found in terms of the initial delimitation of species. Thus, the number of taxa found in the national CWR checklists analysed is very variable, ranging from 941 in Spain to

7032 in Italy, with a median value of 2262 (Table 14.2).

Prioritization is a critical landmark in the process of designing a national strategy for the conservation of CWR. The reasons are clear. In a world with limited economic resources, countries cannot implement conservation actions on all taxa comprising the national CWR checklist. Thus, these actions will only be considered on those species that are prioritized following a series of criteria. There is, in general, a consensus on the main criteria to be used in the prioritization process (Kell *et al.*, 2012, 2015) – the socio-economic importance of the related crop and threat status (Ford-Lloyd *et al.*, 2008; Maxted and Kell, 2009), and potential use for crop improvement (Maxted and Kell, 2009; Maxted *et al.*, 2012b), although some differences arise in the way they are conceived and implemented.

Additional criteria or subcriteria used by some countries include distribution (Czech Republic, Lithuania and Portugal), native status (Czech Republic, Italy, Finland and Portugal), national endemic status (Czech Republic), having the centre of diversity in the country (Cyprus), stakeholder priorities, especially breeders' needs (Norway and Germany), and use categories (Czech Republic, Lithuania, Portugal, Spain and UK).

The number of taxa included in the studied national lists of priority CWR ranged from 20 (Portugal) to 821 (USA). However, in most cases, they ranged between 100 and 300, with a median of 207 (Table 14.2). In addition to the richness of CWR in a country, the number of prioritized taxa depends to a great extent on the particular way in which prioritization was carried out – just as for checklist delimitation.

Table 14.2. Number of taxa in the national CWR checklists and CWR priority lists in surveyed countries. Additional taxa considered at a second level of priority are shown in brackets.

| Country | National CWR checklist | Number of priority CWR taxa |
|---|---|---|
| Cyprus | 1722 | 178 |
| Czech Republic | 3240 | 212 |
| Finland | 1905 | 209 |
| Germany | 2874 | 545 |
| Italy | 7032 | 129 (+85 +904) |
| Jordan | 2005 | 100 |
| Lithuania | 1188 | 180 |
| Norway | 2538 | 205 |
| Portugal | 2262 | 20 |
| Spain | 941 | 580 |
| UK | 2109 | 223 |
| USA | 4600 | 821 (+1435) |

### 14.3.2 Characterization of priority taxa

To help inform conservation decisions, the priority lists should be annotated with data about the constituent taxa, such as the related crop and its use(s), distribution, threat status and inclusion in national and/or international legislation. Most of the countries included in this survey have characterized priority CWR taxa in terms of threat status and endemicity (Table 14.3). Finland, Portugal and Czech Republic were the three countries with the highest ratio of nationally threatened CWR taxa (71, 65 and 54%, respectively). Similarly, Portugal and Spain were the countries with the highest percentage of endemics (65 and 40%, respectively). These high values, always greater than the mean values obtained with the whole national floras, are typically a direct consequence of the inclusion of threat status and endemicity as selection criteria in the prioritization of CWR taxa.

### 14.3.3 Assessment of conservation status

An important part of the development of a national CWR conservation strategy is the assessment

Table 14.3. Percentage of endemic or threatened taxa at a national scale on the CWR priority lists.

| Country | Endemic (%) | Threatened (%) |
|---|---|---|
| Cyprus | 3 | 9 |
| Czech Republic | 13 | 54 |
| Finland | 0 | 71 |
| Germany | 10 | 16 |
| Jordan | 10 | 32 |
| Lithuania | 1 | 16 |
| Norway | 0 | 13 |
| Portugal | 65 | 65 |
| Spain | 40 | 24 |
| UK | 1 | 12 |

of the conservation status of priority taxa through *in situ* and *ex situ* gap analyses (Maxted *et al.*, 2012c). *In situ* gap analysis involves overlaying the distribution data of each priority taxon on a layer comprising all the protected areas of the territory. In this way, it is possible to estimate the percentage of the known populations that are subject to passive conservation because they occur in protected areas. This can be supplemented with the identification of hotspots and a complementarity analysis that provides the network that comprises the greatest number of taxa in a minimum number of sites. Complementarity analysis typically gives a more efficient conservation network, as it avoids the redundancy associated with hotspots. Therefore, it was the method most used in the proposals of *in situ* networks of genetic reserves (e.g. Maxted *et al.*, 2007; Phillips *et al.*, 2014). Most countries that have so far performed gap analysis have done so using distribution data at a resolution of 1 × 1 km or 10 × 10 km grids. There is a trade-off in the use of one or the other. The latter usually provides information on a greater number of occurrences, but they are of lower accuracy. Thus, the selection of the working resolution has been made depending on the quantity and quality of the set of species distribution data in each country.

There was not a standard parameter used when presenting the results of *in situ* gap analysis. This made it difficult to establish comparisons among countries. The UK and Lithuania found that 31 and 50% of the occurrence records of their priority CWR taxa fell within protected areas, respectively. In Cyprus, Norway and Czech Republic, 67, 88 and 99% of the taxa had at least one record in a protected area, respectively. Moreover, in Finland, 33% of the priority CWR taxa had over 60% of their populations in protected areas. In an effort to assess the genetic diversity of adaptive value that is being passively conserved in protected areas, the *in situ* gap analysis in Spain included a third layer, with the territory structured in distinct ecogeographic units. In this study, it was found that 37% of CWR priority taxa had populations covering at least 75% of their ecogeographic units in protected areas. In addition to the various ways of measuring the coverage of protected areas on the *in situ* conservation of CWR, it is worthy to remark that, up to now, no attempts have been made to establish a threshold that would correspond to an adequate passive *in situ* conservation status of priority CWR taxa. In the future, it would be desirable to discuss the best way of measuring the *in situ* conservation status of CWR and to discuss whether it would be desirable to set a threshold under this measurement type, which would set a minimum goal that all countries should aim to achieve. It should be noted that this would just be a step towards a final goal based on the implementation of active *in situ* conservation.

Complementarity analyses carried out in nine countries provide an interesting outlook on the coverage that can be obtained by selecting a limited number of sites for the *in situ* conservation of CWR (Table 14.4). In most countries, a network of between 10 and 20 genetic reserves was selected. Depending on the country, this allowed conservation of between 60 and 100% of priority CWR taxa. The coverage of the network depends to a great extent on the richness of CWR taxa in each country, but also on other factors such as the environmental heterogeneity of the country. For instance, in Finland, a network of just five sites was able to cover 60% of priority taxa, whereas in Spain, 122 sites were needed to include 100% of priority taxa. It should also be stressed that in each case, the establishment of a small number of highest-priority sites will ensure the majority of priority CWR are actively conserved, and each additional reserve will have

**Table 14.4.** Complementarity analysis carried out in different countries to determine the maximum percentage of CWR priority taxa that would be conserved *in situ* by a network of a specified number of optimally located genetic reserves.

| Country | Number of genetic reserves | Priority taxa covered by network of genetic reserves (%) |
|---|---|---|
| Finland | 5 | 60 |
| Spain | 20 | 66 |
| Cyprus | 10 | 75 |
| Czech Republic | 7 | 85 |
| Portugal | 10 | 90 |
| Lithuania | 30 | 90 |
| UK | 27 | 94 |
| Jordan | 16 | 100 |
| Norway | 20 | 100 |
| Czech Republic | 24 | 100 |
| Spain | 122 | 100 |

a diminishing contribution to priority CWR conservation. Further, it should be considered that the goal is not to conserve a single population of priority CWR species but at least five populations to represent the range of genetic diversity within the species; so this also should be built into the site selection algorithm. In many cases, the sites selected for inclusion in the national CWR network through complementarity analysis were not under any type of existing protection.

Another relevant part of any study on the conservation status of CWR is to determine how well these taxa are being conserved in genebanks. The percentage of priority CWR taxa that are currently represented in genebanks varies significantly, depending on the country under assessment (Table 14.5). In general terms, the coverage is not very good, although there are countries like Jordan where around 80% of their priority CWR taxa are represented in genebanks. It should be noted that since we are interested in preserving the genetic diversity that is present in CWR taxa, genebanks should be conserving seeds that constitute an adequate sampling of the genetic diversity of each CWR. This type of assessment was much harder to carry out. In any case, if we consider that conserving seed accessions from over five different populations can provide an adequate sampling of the genetic diversity of a particular CWR (Brown and Briggs, 1991; Dulloo *et al.*, 2008), the percentage of priority CWR taxa that was optimally conserved ranged from 3% to 35% for Finland and Jordan, respectively (Table 14.5). These figures show that the current *ex situ* conservation status of CWR is deficient, and that great effort is needed to improve this situation before it is too late. Nevertheless, haste should not preclude careful, systematic collection. In this context, the project 'Adapting Agriculture to Climate Change: Collecting, Protecting and Preparing Crop Wild Relatives', led by the Global Crop Diversity Trust and the Millennium Seed Bank of the Royal Botanic Gardens Kew, UK, and supported by the Government of Norway, is providing a definite impulse to improve this situation. However, it just focuses on CWR prioritized at the global level.

Table 14.5. Percentage of priority CWR taxa that are being conserved in genebanks and percentage of priority CWR taxa that have more than five accessions.

| Country | CWR priority taxa in genebanks (%) | CWR priority taxa >5 accessions (%) |
|---|---|---|
| Cyprus | 56 | – |
| Czech Republic | 28 | 11 |
| Finland | 27 | 3 |
| Germany | 30 | 20 |
| Jordan | 78 | 35 |
| Lithuania | 33 | 12 |
| Portugal | 15 | 0 |
| Spain | 71 | 23 |
| UK | 67 | 14 |
| USA | 55 | 18 |

### 14.3.4 Conservation goals

A review of the conservation goals that are being set in the studies carried out in different countries reveals great convergence. Most studies include: (i) the design and implementation of recovery plans of threatened CWR species; (ii) the establishment of genetic reserves in a network of areas that complement each other; (iii) the addition of priority CWR taxa to protected areas' management plans; (iv) providing special attention to priority CWR taxa with low coverage in protected areas; (v) looking for alternatives for the management of priority CWR populations outside protected areas; and (vi) collection of under-represented priority CWR taxa for *ex situ* conservation.

### 14.3.5 Conservation actions

A look at the type of conservation actions that have been taken to conserve CWR shows that most conservation initiatives traditionally have taken place outside Europe in sites related to Vavilov's Centres of Diversity. This is the case of the *in situ* conservation actions taken for: *Triticum* in Erebuni, Yerevan, Armenia (Avagyan, 2008) and in Ammiad, Eastern Galilee, Israel (Anikster and Noy-Meir, 1991); *Aegilops* in Ceylanpinar, south-east Turkey (Karagöz, 1998); *Zea perennis* in the Sierra de Manantlan, Mexico (Benz, 1988); *Citrus*, *Oryza* and *Alocasia* in Vietnam (Ngoc Hue *et al.*, 2005); *Solanum* in Pisac, Cusco,

Peru (IUCN, 2003); and *Coffea* in Yayu Forest Biosphere Reserve, Ethiopia (Hoyt, 1988). Depending on the site, conservation actions typically included inventorying, research, implementation of management recommendations, and genetic reserve creation, management and monitoring. Nevertheless, in recent years, several interesting initiatives have also been taking place in Europe, including: the establishment of a genetic reserve for *Beta patula* in Madeira, Portugal (Pinheiro de Carvalho, 2012); the establishment of a monitoring system for the *in situ* conservation of CWR in Brandenburg, Germany, completed in 2010 (pgrdeu.genres.de/insitu); securing the viability of wild grape in the old Rheinaue wetlands through targeted *in situ* management, Germany (www.genres.de/en/cultivated-and-wild-plants/conservation/in-situ-conservation/); and the preparation of a genetic reserve on the Lizard Peninsula in Cornwall, UK (Fielder *et al.*, 2015).

Concerning *ex situ* conservation, the already-mentioned CWR project led by the Global Crop Diversity Trust is currently one of the main engines for improving the *ex situ* conservation status of CWR prioritized at the global level. Nevertheless, global initiatives must be complemented by regional and national actions. In most European countries, national PGR genebanks are starting to collect and preserve the seeds of CWR related to their target crops. In parallel, botanical gardens are also promoting wild plant seed collecting and preserving initiatives that in most cases involve CWR (Maxted and Kell, 2013). An interesting step forward is the formulation of specific CWR seed collections in genebanks, as is the case in the UK (www.kew.org/discover/news/adapting-agriculture-changing-climate), or even the creation of genebanks focused specifically on CWR, as is the case in Germany (www.genbank-wel.uni-osnabrueck.de).

### 14.3.6 Limitations

The assessments made concerning the development of national strategies for the conservation of CWR have also been able to identify a set of limitations that condition the accuracy of their studies. The most cited ones include: (i) lack of systematic accurate georeferenced data on CWR populations (Cyprus, Germany, Italy, Jordan, Lithuania, Norway, Portugal, Spain); (ii) lack of digitized data (Cyprus, Germany, Jordan, Portugal); and (iii) difficulties associated with harmonizing different taxonomic treatments (Cyprus, UK). The difficulties in accessing genepool and taxon group concept information have also been mentioned in some studies, although the work being done in the context of the Global Diversity Trust project (Vincent *et al.*, 2013) and the development of the genetic relative concept by GRIN Taxonomy (USDA, 2015) is contributing substantially to the elimination of this problem. In essence, the most important limitation is not having a clear accessible picture of what we have and where it is located. CWR conservation status assessments rely on the chorological information that is available and has a minimum quality. We must be aware that there are in most cases many other CWR populations that remain unrecorded, and that the picture of the assessments might change with greater and better chorological information. Even species distribution models, which may be used to fill chorological gaps, must be based on good-quality georeferenced observation data.

It is also worth stressing the point made by Meilleur and Hodgkin (2004) that there are 'weak links between the "site-selection and/or management-recommendations" process and the "official-protected-site and/or management-change-designation" process'. Unfortunately, this still remains the case today, even for many of the studies listed above. However, in an attempt to bridge the gap between conservation planning and conservation implementation, the conservation focus has moved from solely identifying individual *in situ* and *ex situ* conservation actions to the broader development of national CWR conservation strategies.

### 14.3.7 Challenges and future directions

Concerning the concepts and methods that are currently being used in the assessments of CWR conservation status, using gap analysis techniques, perhaps the most important challenge derives from the need to stress the conservation of the genetic diversity associated with CWR. In fact, CWR are being conserved with the objective of being able to use their genetic diversity in the development of new crop varieties that are better adapted to emerging biotic and abiotic environmental conditions in the context of global change. Taking this into consideration, there may be a need to move the target conservation unit from CWR species to the genetic provenances or ecotypes of the adaptive value present in each CWR species. In this sense, the combined use of gap analysis and complementarity analysis associated with ecogeographical approaches to measuring the genetic diversity of adaptive value may provide a way to formulate genetic diversity focused conservation assessments (see Parra-Quijano *et al.*, 2012 and 2014, for further discussion).

Other future directions for the improvement of national CWR conservation strategy planning involves further refining and improving the use of ecogeographical approaches as a proxy of the genetic diversity of adaptive value, establishing a standard for measuring the *in situ* conservation status of CWR, and greater progress in the development of multi-species genetic reserves to maximize the efficiency of conservation actions. Additionally, it is important to incorporate the effect of climate change in the formulation of national CWR strategies, namely in the establishment of *ex situ* and *in situ* conservation priorities. This involves assessing both the vulnerability of priority CWR taxa and potential genetic reserves to climate change.

Regarding policy aspects, some of the most urgent challenges involve: (i) improvement in the collaboration between agricultural, PGR, biodiversity protected areas and academic sectors, as it is necessary for the establishment of CWR genetic reserves; (ii) reviewing the national policy and legal framework for CWR conservation and providing appropriate funding; (iii) improving the coordination between *in situ* and *ex situ* conservation approaches; (iv) the integration of CWR conservation data in the national PGR information infrastructure; and (v) increasing cooperative conservation efforts with neighbouring countries and working at the national level in close coordination with regional CWR strategies. To achieve these challenges, a specific EU Directive on PGRFA could be enacted that would contemplate these types of initiatives (Maxted *et al.*, 2015).

## Acknowledgements

This research was funded by a grant from the EU Seventh Framework Programme for the PGR Secure project (Grant Agreement No 266394).

## References

Anikster, Y. and Noy-Meir, I. (1991) The wild-wheat field laboratory at Ammiad. *Israel Journal of Botany* 40, 351–362.

Avagyan, A. (2008) Crop wild relatives in Armenia: diversity, legislation and conservation issues. In: Maxted, N., Ford-Lloyd, B.V., Kell, S.K., Iriondo, J.M., Dulloo, M.E. and Turok, J. (eds) *Crop Wild Relative Conservation and Use*. CAB International, Wallingford, UK, pp. 58–68.

Benz, B. (1988) *In situ* conservation of the genus *Zea* in the Sierra de Manantlan Biosphere Reserve. *Recent Advances in the Conservation and Utilization of Genetic Resources: Proceedings of the Global Maize Germplasm Workshop*. CIMMYT, Mexico DF, pp. 59–69.

Brown, A.H.D. and Briggs, J.D. (1991) Sampling strategies for genetic variation in *ex situ* collections of endangered plant species. In: Falk, D.A. and Holsinger, K.E. (eds) *Genetics and Conservation of Rare Plants*. Oxford University Press, New York, pp. 99–119.

Cooper, H.D., Spillane, C. and Hodgkin, T. (2001) Broadening the genetic base of crops: an overview. In: Cooper, H.D., Spillane, C. and Hodgkin, T. (eds) *Broadening the Genetic Base of Crops*. CAB International, FAO and IPGRI, Wallingford, UK, pp. 1–24.

Dulloo, M.E., Labokas, J., Iriondo, J.M., Maxted, N., Lane, A., *et al.* (2008) Genetic reserve location and design. In: Iriondo, J.M., Dulloo, M.E. and Maxted, N. (eds) *Conserving Plant Genetic Diversity in Protected Areas. Population Management of Crop Wild Relatives*. CAB International, Wallingford, UK, pp. 23–64.

Fielder, H., Brotherton, P., Hosking, J., Hopkins, J.J., Ford-Lloyd, B., *et al.* (2015) Enhancing the conservation of crop wild relatives in England. *PLos ONE* 10(6), e0130804.

Fitzgerald, H. (2013) *The National Crop Wild Relative Strategy Report for Finland*. MTT Report 121, Jokioinen, Finland.

Fitzgerald, H. and Korpelainen, H. (2015) Discovering Finnish crop wild relative hotspots. *Crop Wild Relative* 10, 21–23.

Fitzgerald, H., Korpelainen, H. and Veteläinen, M. (2013) Prioritization of crop wild relatives in Finland. *Crop Wild Relative* 9, 10–12.

Ford-Lloyd, B.V., Kell, S.P. and Maxted, N. (2008) Establishing conservation priorities for crop wild relatives. In: Maxted, N., Ford-Lloyd, B.V., Kell, S.P., Iriondo, J., Dulloo, E. and Turok, J. (eds) *Crop Wild Relative Conservation and Use*. CAB International, Wallingford, UK, pp. 110–119.

Hoyt E. (1988) *Conserving the Wild Relatives of Crops*. IPGRI/IUCN/WWF, Rome.

IPCC (Intergovernmental Panel on Climate Change) (2014) *Climate Change 2014: Impacts, Adaptation, and Vulnerability. Part A: Global and Sectoral Aspects. Contribution of Working Group II to the Fifth Assessment Report of the Intergovernmental Panel on Climate Change*. Cambridge University Press, Cambridge, UK, and New York.

IUCN (International Union for Conservation of Nature) (2003) 'Potato park' aims to help save genetic resources in Peru. Feature Story PR, Gland, Switzerland, 10 April 2003.

Karagöz, A. (1998) *In situ* conservation of plant genetic resources in the Ceylanpinar State Farm. In: Zencirci, N., Kaya, Z., Anikster, Y. and Adams, W.T. (eds) *Proceedings of International Symposium on In Situ Conservation of Plant Diversity.* Central Research Institute for Field Crops, Ankara, pp. 87–92.

Kell, S., Qin, H., Chen, B., Ford-Lloyd, B.V., Wei, W., *et al.* (2015) China's crop wild relatives: diversity for agriculture and food security. *Agriculture, Ecosystems and Environment* 209, 138–154, doi.org/10.1016/j.agee.2015.02.012.

Kell, S.P., Knüpffer, H., Jury, S.L., Maxted, N. and Ford-Lloyd, B.V. (2005) Catalogue of Crop Wild Relatives for Europe and the Mediterranean. Available at: www.pgrforum.org/cwris/cwris.asp (accessed 29 October 2015), and on CD-ROM, University of Birmingham, UK.

Kell, S.P., Knüpffer, H., Jury, S.L., Ford-Lloyd, B.V. and Maxted, N. (2008) Crops and wild relatives of the Euro-Mediterranean region: making and using a conservation catalogue. In: Maxted, N., Ford-Lloyd, B.V., Kell, S.P., Iriondo, J.M., Dulloo, M.E. and Turok, J. (eds) *Crop Wild Relative Conservation and Use*. CAB International, Wallingford, UK, pp. 69–109.

Kell, S.P., Maxted, N. and Bilz, M. (2012) European crop wild relative threat assessment: knowledge gained and lessons learnt. In: Maxted, N., Dulloo, M.E., Ford-Lloyd, B.V., Frese, L., Iriondo, J.M. and Pinheiro de Carvalho, M.A.A. (eds) *Agrobiodiversity Conservation: Securing the Diversity of Crop Wild Relatives and Landraces*. CAB International, Wallingford, UK, pp. 218–242.

Khoury, C.K., Greene, S., Wiersema, J., Maxted, N., Jarvis, A., *et al.* (2013) An Inventory of Crop Wild Relatives of the United States. *Crop Science* 53, 1–13.

Landucci, F., Panella, L., Lucarini, D., Gigante, D., Donnini, D., *et al.* (2014) A prioritized inventory of crop wild relatives and wild harvested plants of Italy. *Crop Science* 54, 1628–1644.

Magos-Brehm, J., Maxted, N., Ford-Lloyd, B. and Martins-Loução, M.A. (2008) National inventories of crop wild relatives and wild harvested plants: case-study for Portugal. *Genetic Resources and Crop Evolution* 55, 779–796.

Maxted, N. and Kell, S. (2009) Establishment of a Global Network for the *In Situ* Conservation of Crop Wild Relatives: Status and Needs. Commission on Genetic Resources for Food and Agriculture. FAO, Rome. Available at: www.fao.org/3/a-i1500e/i1500e18d.pdf (accessed 29 October 2015).

Maxted, N. and Kell, S.P. (2013) A role for botanic gardens in CWR conservation for food security. *BGjournal* 10(2), 32–35.

Maxted, N., Scholten, M., Codd, R. and Ford-Lloyd, B. (2007) Creation and use of a national inventory of crop wild relatives. *Biological Conservation* 140, 142–159.

Maxted, N., Kell, S., Ford-Lloyd, B., Dulloo, E. and Toledo, Á. (2012a) Toward the systematic conservation of global crop wild relative diversity. *Crop Science* 52, 774–785.

Maxted, N., Akparov, Z.I., Aronsson, M., Asdal, Å., Avagyan, A., *et al.* (2012b) Current and future threats and opportunities facing European crop wild relative and landrace diversity. In: Maxted, N., Dulloo, M.E., Ford-Lloyd, B.V., Frese, L., Iriondo, J.M. and Pinheiro de Carvalho, M.A.A. (eds) *Agrobiodiversity Conservation. Securing the Diversity of Crop Wild Relatives and Landraces*. CAB International, Wallingford, UK, pp. 333–354.

Maxted, N., Castañeda Álvarez, N.P., Vincent, H.A. and Magos Brehm, J. (2012c) Gap analysis: a tool for genetic conservation. In: Guarino, L., Ramanatha Rao, V. and Goldberg, E. (eds) *Collecting Plant*

*Genetic Diversity: Technical Guidelines. 2011 update*. Bioversity International, Rome. Available at: cropgenebank.sgrp.cgiar.org/index.php?option=com_content&view=article&id=678 (accessed 29 October 2015).

Maxted, N., Magos Brehm, J. and Kell, S.P. (2013) *Resource Book for Preparation of National Conservation Plans for Crop Wild Relatives and Landraces.* Commission on Genetic Resources for Food and Agriculture. FAO, Rome. Available at: www.fao.org/fileadmin/templates/agphome/documents/PGR/PubPGR/ResourceBook/TEXT_ALL_2511.pdf (accessed 29 October 2015).

Maxted, N., Avagyan, A., Frese, L., Iriondo, J.M., Magos Brehm, J., *et al.* (2015) Preserving diversity: a concept for *in situ* conservation of crop wild relatives in Europe: version 2. Wild Species Conservation in Genetic Reserves Working Group, European Cooperative Programme for Plant Genetic Resources, Rome. Available at: www.pgrsecure.org/documents/Concept_v2.pdf (accessed 6 March 2015).

Meilleur, B.A. and Hodgkin, T. (2004) *In situ* conservation of crop wild relatives: status and trends. *Biodiversity and Conservation* 13, 663–684.

Ngoc Hue, N.T., Tuyen, T.V., Canh, N.T., Hien, P.V., Chuong, P.V., *et al.* (eds) (2005) *In situ Conservation of Agricultural Biodiversity On-farm: Lessons Learned and Policy Implications*. Proceedings of Vietnamese National Workshop, 30 March–1 April 2004, Hanoi, Vietnam. International Plant Genetic Resources Institute, Rome.

Olesen, J.E. and Bindi, M. (2002) Consequences of climate change for European agricultural productivity, land use and policy. *European Journal of Agronomy* 16, 239–262.

Parra-Quijano, M., Iriondo, J.M., Frese, L. and Torres, E. (2012) Spatial and ecogeographical approaches for selecting genetic reserves in Europe. In: Maxted, N., Dulloo, M.E., Ford-Lloyd, B.V., Frese, L., Iriondo, J.M. and Pinheiro de Carvalho, M.A.A. (eds) *Agrobiodiversity Conservation: Securing the Diversity of Crop Wild Relatives and Landraces*. CAB International, Wallingford, UK, pp. 20–28.

Parra-Quijano, M., Torres, E., Iriondo, J.M. and López, F. (2014) *CAPFITOGEN Tools User Manual Version 1.2.* International Treaty on Plant Genetic Resources for Food and Agriculture, FAO, Rome.

Phillips, J., Kyratzis, A., Christoudoulou, C., Kell, S. and Maxted, N. (2014) Development of a national crop wild relative conservation strategy for Cyprus. *Genetic Resources and Crop Evolution* 61(4), 817–827.

Pinheiro de Carvalho, M.Â.A., Nóbrega, H., Freitas, G., Fontinha, S. and Frese, L. (2012) Towards the establishment of a genetic reserve for *Beta patula* Aiton. In: Maxted, N., Dulloo, M.E., Ford-Lloyd, B.V., Frese, L., Iriondo, J.M. and Pinheiro de Carvalho, M.A.A. (eds) *Agrobiodiversity Conservation: Securing the Diversity of Crop Wild Relatives and Landraces*. CAB International, Wallingford, UK, pp. 36–44.

Rubio Teso, M.L., Torres, M.E., Parra-Quijano, M. and Iriondo, J.M. (2012) Prioritization of crop wild relatives in Spain. *Crop Wild Relative* 8, 18–21.

Rubio Teso, M.L., Ronquillo Ferrero, C., Nebreda Trejo, A., Parra Quijano, M., Torres Lamas, E., *et al.* (2015) *In situ* conservation of CWR in Spain. Present and future. *Crop Wild Relative* 10, 24–26.

Turral, H., Burke, J. and Faurès, J.M. (2011) *Climate Change, Water and Food Security*. FAO Water Reports 36. FAO, Rome.

USDA (US Department of Agriculture), ARS (Agricultural Research Service), National Genetic Resources Program (2015) Germplasm Resources Information Network – GRIN (online database). Available at: www.ars-grin.gov/~sbmljw/cgi-bin/taxcwr.pl?language=en (accessed 7 March 2015).

van de Wouw, M., Kik, C., van Hintum, T., van Treuren, R. and Visser, B. (2010) Genetic erosion in crops: concept, research results and challenges. *Plant Genetic Resources: Characterization and Utilization* 8(1), 1–15.

Vincent, H., Wiersema, J., Kell, S., Fielder, H., Dobbie, S., *et al.* (2013) A prioritized crop wild relative inventory to help underpin global food security. *Biological Conservation* 167, 265–275.

# 15 Crop Wild Relatives: A Priority in Jordan? Developing a National Strategy for the Conservation of Plant Diversity in Jordan Using a Participatory Approach

J. Magos Brehm,[1,2]* S. Saifan,[3] H. Taifour,[4] K. Abulaila,[3] A. Al-Assaf,[5] A. El-Oqlah,[6] F. Al-Sheyab,[3] R. Bani-Hani,[7] S. Ghazanfar,[8] N. Haddad,[9] R. Shibli,[10] T. Abu Taleb,[4] B. bint Ali,[4] and N. Maxted[1]

*[1]School of Biosciences, University of Birmingham, Birmingham, UK, [2]Centro de Biologia Ambiental, Faculdade de Ciências da Universidade de Lisboa, Lisbon, Portugal; [3]National Center for Agricultural Research and Extension, Baq'a, Jordan; [4]Royal Botanic Garden, Tell Ar-Rumman, Jordan; [5]Faculty of Agriculture, University of Jordan, Amman, Jordan; [6]Yarmouk University, Irbid, Jordan; [7]Nature Protection Directorate, Amman, Jordan; [8]Royal Botanic Gardens, Kew, UK; [9]West Asia Regional Program (WARP) ICARDA, Amman, Jordan; [10]Department of Horticulture and Crop Sciences, Faculty of Agriculture, University of Jordan, Amman, Jordan*

## 15.1 Introduction

### 15.1.1 Plant diversity in Jordan

The Flora of Jordan is relatively rich and diverse compared to the total number of recorded vascular plant species on earth (250,000–300,000) (CB2, 2013). Jordan hosts over 2600 species, in 114 families and 907 genera (Ghazanfar *et al.*, 2016), which is about 1% of the total world flora. Danin (2001) has even estimated a plant species richness of 9.1 species/100 km². Jordanian flora includes 106 endemic and near-endemic species (e.g. *Crocus moabiticus* Bornm. & Dinsm., *Crucianella transjordanica* Rech. fil., *Iris petrana* Dinsm., *Micromeria sinaica* Bentham, *Scrophularia nabataerum* Eig, *Withania obtusifolia* Tackh.) (Magos Brehm *et al.*, 2015), as well as several rare species (e.g. *Ajuga chamaepytis* (Bentham) P.H. Davis, *Equisetum ramosissimum* Desf., *Lavandula pubescens* Decne., *Teucrium leucocladum* Sieber ex Heldr.) (GCEP, 1998). There are seven species endemic to Jordan and around 100 near endemics (endemic to Jordan and an adjacent country or region, such as Palestine, Israel, Jordan, Saudi Arabia and Sinai, in one geographical entity) (Ghazanfar *et al.*, 2016). There are around 375 rare taxa (GCEP, 1998; CB2, 2013), one regionally extinct and 106 threatened (Taifour and El-Oqlah, 2015). Such diversity is due mainly to the great variation in topography and climate that creates a wide range of ecological

*Corresponding author; e-mail: joanabrehm@gmail.com

 *Enhancing Crop Genepool Use: Capturing Wild Relative and Landrace Diversity for Crop Improvement* (eds N. Maxted, M. Ehsan Dulloo and B.V. Ford-Lloyd)

habitats and microhabitats. Altitude ranges from –400 m at the surface of the Dead Sea up to 1750 m at Jebel Rum. Climate varies from semi-humid Mediterranean conditions, with more than 500 mm of rainfall, to arid conditions with less than 200 mm, over only 100 km distance. Jordanian flora is thus ecologically and genetically adapted to local conditions, and most species are even adapted to the drier parts of the country. Four biogeographical regions have been recognized in Jordan – Irano-Turanian, Mediterranean, Saharo-Arabian (also referred to as Eastern Desert or 'Badia') and Sudanian (Fig. 15.1; Kasapligil, 1956) – and 13 vegetation types have been identified, namely the *Acacia* and rocky vegetation, deciduous oak forest, evergreen oak forest, Hammada vegetation, juniper forest, Mediterranean non-forest vegetation, pine forest, sandy dunes, steppe vegetation, tropical vegetation, halophytic vegetation, hydrophytic vegetation and mud flats (Al-Eisawi, 1996; Fig. 15.2).

In addition, Jordan is located at the south-western end of the Fertile Crescent of the Near East, which is the cradle of temperate agriculture origin and is also where two Vavilov centres of crop origin abut, the Asia Minor and the Mediterranean centres (Fig. 15.3). The Asia Minor

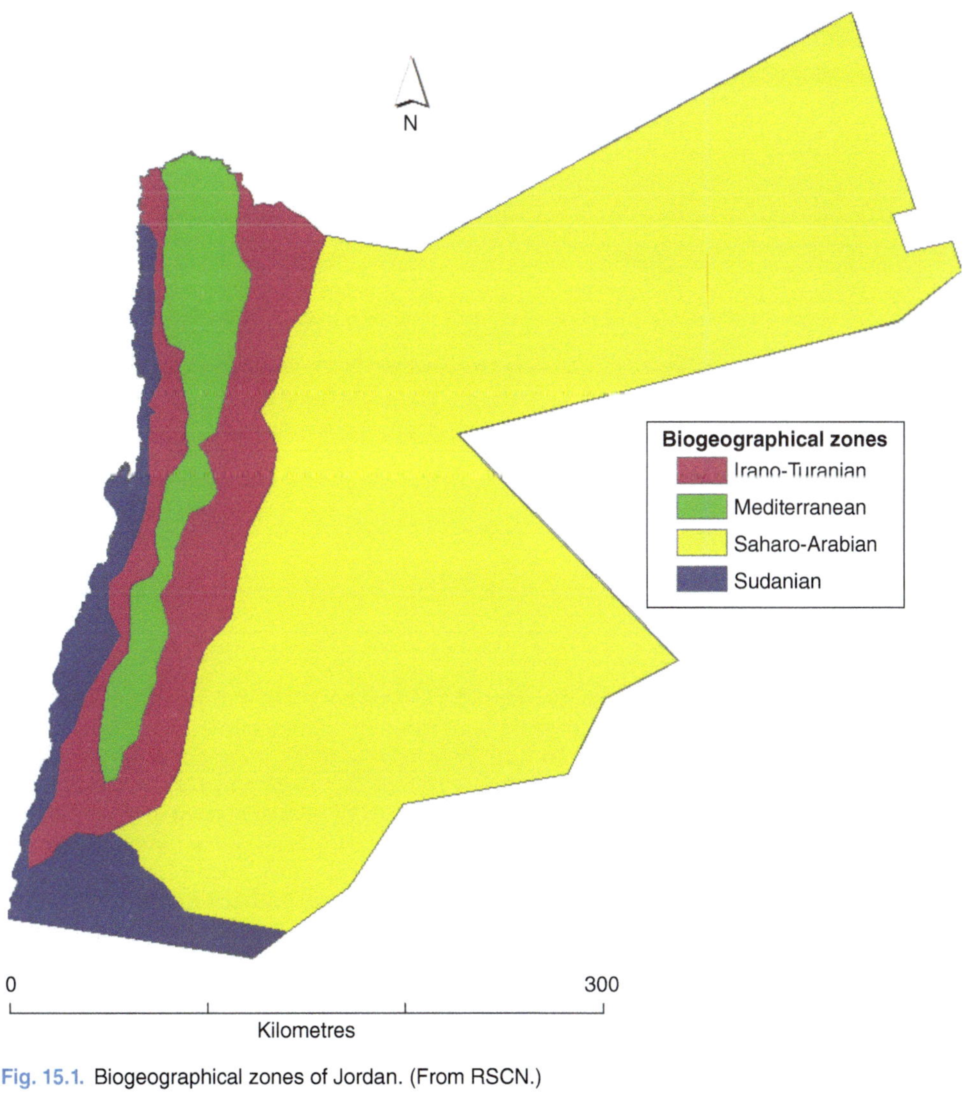


Fig. 15.1. Biogeographical zones of Jordan. (From RSCN.)

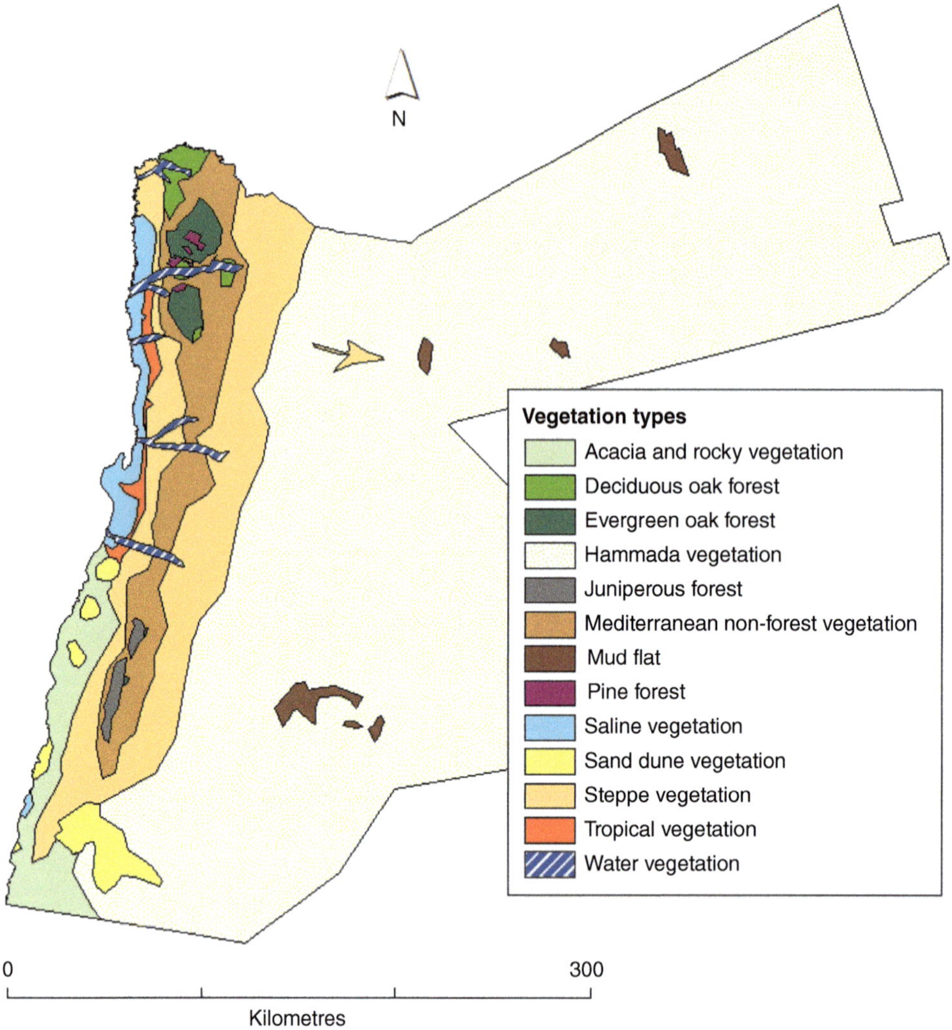


Fig. 15.2. Vegetation types of Jordan. (From RSCN, after Al-Eisawi, 1996.)

centre is the crop origin centre for wheats (*Triticum monococcum* L., *Triticum durum* Desf., *Triticum turgidum* L. and *Triticum aestivum* L.), barley (*Hordeum vulgare* L.), rye (*Secale cereale* L. and *Secale montanum* Guss.), oat (*Avena sativa* L.), chickpea (*Cicer arietinum* L.) and lentil (*Lens culinaris* Medikus and *Lens orientalis* (Boiss. Hand.-Mazz.), as well as various forages (*Medicago sativa* L., *Trifolium resupinatum* L., *Trigonella foenum-graecum* L., *Onobrychis* spp., *Lathyrus cicera* L. and several *Vicia* spp.), oil-producing plants (*Linum*, *Brassica*, *Camelina* and *Eruca* spp.), melons (*Cucumis* and *Citrullus* spp.), vegetables (*Lepidium*, *Brassica*, *Daucus*, *Eruca*, *Allium*, *Petroselinum*, *Lactuca* and *Portulaca* spp.), fruit crops (*Malus*, *Pyrus*, *Ficus*, *Cydonia*, *Cerasus*, *Amygdalus*, *Vitis* and *Pistacia*) and dye plants (*Crocus sativus* L. and *Rubia tinctorum* L.), whereas the Mediterranean centre not only also originated some of the crop genepools of Asia Minor but also some additional crops such as *Vicia faba* L., *Vicia sativa* L., *Lathyrus ochrus* (L.) DC., *Hedysarum coronarium* L., *Ornithopus sativus* Brot., *Olea europaea* L., *Ceratonia siliqua* L., *Beta vulgaris* L., *Brassica oleracea* L., *Brassica rapa* L. and *Brassica napus* L., *Portulaca oleracea* L., *Allium*, *Asparagus*, *Lactuca*, *Pastinaca* and *Tragopogon* spp. (Vavilov, 1926). A significant diversity of landraces and wild relatives of these

Fig. 15.3. Fertile Crescent and the Vavilov centres of crop origin.

crops can thus be found in Jordan. More than 14 crops of global significance have been domesticated over the past 10,000 years in the region and have spread progressively all over the world. As such, Jordan is possibly in the most important region for global food security. A recent study of global patterns of crop wild relative (CWR) diversity that included 108 highest priority CWR genera has found that the Fertile Crescent contains 254 priority taxa (Vincent *et al.*, 2013) and therefore is the region globally with the highest concentration of CWR species per unit area, and its wealth of agricultural diversity is crucial to the maintenance of global food security.

### 15.1.2 Threats to diversity

Many factors threaten the diversity of wild plant taxa in Jordan. These include: increasing development pressures combined with population growth, tourism development (infrastructure development, increasing number of visitors, effluent discharge, litter accumulation and extensive water extraction, roads development and encroachment by local population, including intensive use of resources), climate change, uncontrolled urban expansion (including transformation of forests into agricultural and urbanized areas), overgrazing and extensive woodcutting, intensive and extensive agricultural practices, unplanned mining and quarrying, uncontrolled vehicle movement, unbalanced water use and unplanned water extraction, pollution of surface and underground water resources, introduction and spread of invasive species, weak enforcement of laws and, finally, emerging pressures related to refugees, related mostly to the regional political crisis (including pressures on water resources, agricultural areas, rangelands and open spaces, as well as on ecosystem goods and services) (Ministry of the Environment, 2014). Many of these threats have led to habitat erosion and fragmentation, ecosystem degradation and pollution. Financial and management constraints, such as limited access to credit and weak agricultural extension services, have suppressed the appropriate and effective measures to counterbalance these negative factors that seriously threaten populations of wild wheat, barley, chickpeas, lentil, almond, the historical Roman olive trees, the wild relatives of many fruit trees and wild medicinal plants (NCARTT, 2007). The situation may be compounded further by contamination from genetic engineered crops, if it is allowed to be practised, and potentially with climate change. Jordan already faces irregular cycles of rainfall and drought, and Harrison (2009) has further highlighted that year-round change in temperature and consequent increased drought by the end of the 21st century will be one of the country's major problems that biodiversity is expected to face. He analysed seven projections of regional climates across Jordan in response to a mean global warming of 4°C by the end of the 21st century and found a uniform warming across the country of about 3 ± 0.5°C in winter and 4.5 ± 1°C in summer, though the level of annual precipitation is unlikely to change. Climate change is thus a major threat to the persistence of species.

Particularly for CWR, which are of such global importance in Jordan, there have been very few studies that have investigated genetic erosion in the region. Keiša *et al.* (2008) has found that local agricultural diversity is being eroded rapidly in Syria, and this is likely to be further impacted adversely by climate change. Finally, the wealth of Jordan's CWR is still largely ignored in land-use planning and implementation, economic development and biodiversity conservation endeavours. This means there is an urgent need for the development of a regional CWR conservation strategy for the Fertile Crescent, and Jordan itself, if global food security is to be safeguarded.

This chapter provides a brief overview on how the National Strategy for the Conservation of Plant Diversity for Jordan (National Strategy for short) (Magos Brehm *et al.*, 2015) has been developed using a participatory approach. Also, as CWR are identified as a priority in Jordan, the methodology can be applied specifically to CWR in other countries.

## 15.2 Developing a National Strategy

The flora of Jordan contains valuable plant species that can be used for human livelihoods, animal feed, in the rehabilitation and restoration of degraded ecosystems, and which holds diverse and globally important CWR that are most likely to contain important genes that can be used to improve crops adapted to drought and salinity (and other abiotic and biotic stresses) and to climate change. These national genetic resources need to be maintained and strategically conserved because they are not only nationally important but also crucial to underpin the world's food security. The actual implementation of the National Strategy is expected to guide national efforts in the conservation of plant genetic resources, both *in situ* and *ex situ*, as well as to encourage their fair and sustainable utilization to improve livelihoods in the rural communities that are the custodians of these resources.

A National Strategy was then developed following a participatory approach involving a wide spectrum of national stakeholders (Fig. 15.4). As a major requirement in the process of developing the National Strategy, the Royal Botanic Garden (RBG), Jordan, the International Union for the Conservation of Nature Regional Office of West Asia (IUCN ROWA), the Royal Botanic Gardens, Kew, and the University of Birmingham, UK, collated the available information on all Jordanian plant taxa from local, national and international institutions. These data served three purposes: (i) to develop an updated Jordanian flora checklist; (ii) to undertake the IUCN Red List of priority of Jordanian flora; and (iii) to develop the National Strategy.

### 15.2.1 National flora checklist and IUCN Red List

Both the national flora checklist development and the IUCN Red List received input from various national stakeholders. The 10-day workshop on the national 'Regional Red List Assessment for Plants of Jordan' (Amman, March 2013) gathered botanists from RBG, the Royal Society for the Conservation of Nature (RSCN), National Center for Agricultural Research and Extension (NCARE), Jordan University of Science and Technology

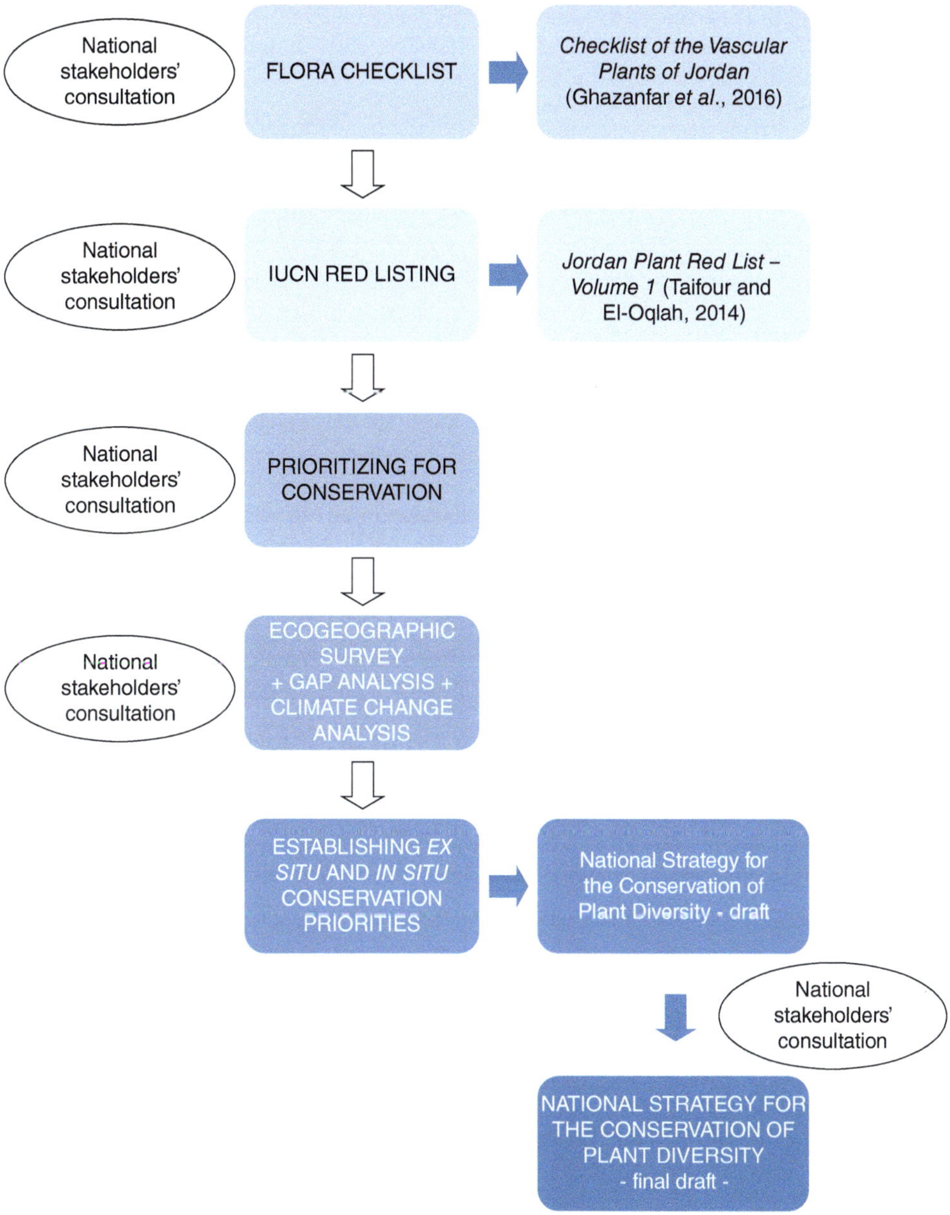


Fig. 15.4. Development of the National Strategy for Plant Conservation for Jordan.

(JUST), Yarmouk University and IUCN ROWA in order to: (i) initiate the production of an updated national flora checklist; and (ii) assess Jordan's flora species according to the *IUCN Red List Categories and Criteria* (2001).

The *Checklist of the Vascular Plants of Jordan* (Ghazanfar *et al.*, 2016) was based mainly on the list of Jordan's vascular plants prepared by Al-Eisawi (1982) and complemented with records from national databases, herbaria, genebanks, universities, surveys' checklists, scientific research, monitoring programmes, published books (such as *Flora Palestina* by Zohary, 1972) and accepted publications or records from individual botanists.

Species names were then validated using The Plant List (2010). The final version of the national flora checklist contains more than 2625 taxa belonging to 907 genera in 114 families (Ghazanfar *et al.*, 2016). This checklist includes both native and naturalized higher plants of Jordan and is organized using the Angiosperm Phylogeny Group III classification system (APG III system; Angiosperm Phylogeny Group, 2009).

About 40% of the Jordanian flora has been assessed using the 2001 *IUCN Red List Categories and Criteria* (IUCN, 2001), and the *Jordan Plant Red List – Volume 1* (Taifour and El-Oqlah, 2014) has been published. The triage methodological approach proposed and selection of the initial species to be assessed was based on a preliminary study prepared by Maxted (2011), which identified existing knowledge gaps in species information as being a major limitation for rapid threat assessment and so suggested splitting the flora into three groups based on the available information: assessment phase 1 species, where there were already existing data for Red Listing (including both known common and rare species); phase 2 species, where some data for Red Listing existed; and phase 3 species, where significant field study would be needed prior to Red Listing. Subsequent work with relevant stakeholders and survey of the available data updated the list provided by Maxted (2011), and 1072 taxa were finally selected for threat assessment. At the 'Regional Red List Assessment for Plants of Jordan' national workshop, the species-by-species regional assessments were carried out in a three-step process. First, assessors determined which taxa and which regional populations to assess. Next, the regional populations for each taxon were evaluated according to the *IUCN Red List Categories and Criteria* (IUCN, 2001) and a preliminary category was assigned. Finally, the effect of populations of the same taxon in neighbouring regions on the regional populations was then considered, and the preliminary category was up- or down-listed, if appropriate, using the *Application of IUCN Red List Criteria at Regional and National Levels* (IUCN, 2010). Out of the 1072 species assessed, 147 were found to be Not Applicable for IUCN Red List assessment and one was Data Deficient. One was considered Regionally Extinct, 106 threatened at the national level (19 Critically Endangered, 54 Endangered, 33 Vulnerable), 11 Near Threatened and the remaining 806 Least Concern.

### 15.2.2 Prioritizing for conservation

Prioritizing plant taxa for conservation in Jordan took two main phases. The first included a selection based on data availability for each taxon that resulted in 1072 taxa (see above), which were further reduced to 916 due to lack of occurrence data. The second was undertaken to prioritize further the taxa obtained in the first step. This second prioritization was based on the prioritization criteria and methodology agreed in the technical workshop on 'Establishing species conservation priorities in Jordan' (Amman, September 2012), which involved relevant Jordanian stakeholders (RBG, representatives of the Ministry of Environment of Jordan, Ministry of Agriculture of Jordan, NCARE, RSCN, University of Jordan, Hashemite University, Yarmouk University, JUST and Jordan National Center for Research and Development) and the University of Birmingham. Prior to the workshop, the participants were sent a background document on species prioritization for conservation and were asked to select the most relevant criteria for prioritizing species for conservation in Jordan, as well as the prioritization method to be used. The workshop participants voted and agreed on a point-scoring procedure based on the following criteria: whether the species was a CWR, crop use category (in case of CWR), threat status, native status, global distribution and Jordanian distribution. In addition to the criteria agreed in the workshop, a few relevant criteria were added in relation to CWR: inclusion of the genera in the International Treaty for Plant Genetic Resources for Food and Agriculture (ITPGRFA); inclusion of the related crops in national statistics; and genepool (Harlan and de Wet, 1971) or taxon group (Maxted *et al.*, 2006) concepts as the workshop participants clearly selected CWR as a major group of taxa to conserve.

Out of a total of 2625 taxa that occurred in Jordan (Ghazanfar *et al.*, 2016), 916 were selected in the first prioritization phase, from where 100 were finally prioritized for conservation. Out of the 100 priority taxa that were obtained automatically through the agreed criteria and prioritization process, invasive species were excluded and a total of 98 were left as priorities for conservation in Jordan. The 98 priority taxa are all CWR, which is not surprising given that over 75% of the Jordanian flora are CWR and that most prioritization criteria used are related to CWR. Priority taxa for conservation for Jordan include wild species of the crop genera *Astragalus, Pistacia, Pisum, Sorghum, Trifolium, Triticum* and *Vicia*, among others (Magos Brehm *et al.*, unpublished results).

### 15.2.3 Ecogeographic survey, gap and climate change analysis

Ecogeographic data were collected from more than 20 different sources that included herbaria, genebanks, bibliographic references and several experts on Jordanian flora. More than 30,000 records were compiled, which were then verified and standardized. Data verification included: (i) standardization of nomenclature to use the accepted Jordanian checklist name; (ii) georeferencing (using www.geocomm.com and geographic information systems (GIS) software); (iii) coordinate conversion to decimal degrees; (iv) a check for erroneous or outlier localities; and (v) a check for duplicates. Records with digitized localities with two or more decimals in one of the coordinates were selected for subsequent analyses to ensure geographic precision.

Ecogeographic data analyses included the creation of maps of species richness, maps of sampling bias, species distribution maps and predicted distribution maps using DIVA-GIS software (www.diva-gis.org). In addition, an individual species gap analysis was undertaken for the top 98 priorities, as well as for all 916 taxa, and further sophisticated analyses were performed for the top 20 priority taxa only. These analyses included the identification of ecogeographical gaps of priority taxa in *in situ* and *ex situ* conservation and the assessment of climate change impact on species distribution. The identification of areas for active *in situ* conservation was also undertaken via a complementarity analysis of the top 100 priority taxa (see Section 15.3.3, this chapter).

#### *Individual species gap analysis*

Through an individual species gap analysis, the species not conserved *ex situ* or *in situ* in Jordan, and species passively conserved *in situ* in Jordan, were identified. It was observed that about 75% of all 916 taxa were conserved *ex situ*, and that 22% of the 98 priority taxa had not been conserved *ex situ*, including important *Aegilops, Allium, Astragalus, Daucus, Origanum, Sorghum* and *Vicia* CWR species (Magos Brehm *et al.*, unpublished results). It was also concluded that no active *in situ* conservation of Jordanian CWR had currently been undertaken. Species monitoring has been performed for some key species in some conservation areas, but the results obtained have not been used in assisting changes in population management to ensure their persistence over time. Nevertheless, 56% of the 100 priority taxa occur within the boundaries of Jordan's existing conservation areas. Accordingly, this implies that these species are currently passively protected, and active conservation could be undertaken with the agreement of the site managers (Magos Brehm *et al.*, unpublished results).

#### *Ecogeographic gap analysis*

*Ex situ* and *in situ* ecogeographic gap analyses were undertaken for the top 20 priority species in order to detect whether the whole ecogeographic range of the priority taxa was being adequately conserved *ex situ* and *in situ*, and if not, which populations should be targeted for *ex situ* and *in situ* actions (Magos Brehm *et al.*, unpublished results), it being assumed that ecogeographic diversity was an indicator of genetic diversity and that the conservation of maximum ecogeographic diversity would result in the conservation of maximum genetic diversity. The ecogeographic gap analysis was adapted from the methodology suggested by Parra-Quijano *et al.* (2012a,b) and used an ecogeographic land characterization map (ELC) produced by Mauricio Parra-Quijano (Universidad Politécnica de Madrid, Spain, and consultant for FAO-ITPGRFA) with the CAPFITOGEN tools (Parra-Quijano *et al.*, 2014) using the Medoides method and based on the variables selected by a group of Jordanian experts. These identified the most relevant variables that portrayed different adaptive scenarios of the flora of Jordan. A total of 26 variables were used (11 bioclimatic, 11 edaphic and 4 geophysic) and produced a total of 47 ecogeographic categories (Magos Brehm *et al.*, unpublished results).

### *Climate change analysis*

Given that it is predicted that Jordan will be affected seriously by climate change (Harrison, 2009), assessing the potential effect of changes in climate on the distribution of species is of utmost importance so that adequate conservation actions can be undertaken and plant diversity can be safeguarded efficiently for future generations. The potential impact of climate change on species distribution was thus undertaken not only to help further prioritize species for conservation but also to select populations of the top 20 priority taxa for *in situ* and *ex situ* conservation. MaxEnt v. 3.3.3k (Phillips *et al.*, 2006, freely available from www.cs.princeton.edu/~schapire/maxent) was used to model species distribution under current and future climatic conditions. Most default settings were used, as they had been reported to achieve good performance with presence-only data (Phillips and Dudík, 2008). The models were validated either by data partitioning (25% data set used for validation) (when >25 records) or cross-validation (when ≤25 records). Model validation was carried out by estimating the area under curve (AUC) of the receiver operating characteristic (ROC) (Hanley and McNeil, 1982) to assess the agreement between the presence records and the predictions.

Edaphic, land cover and climate data were used to build species distribution models. Edaphic data were the SOTER-based soil parameter estimates for Jordan (ver. 1.0) (Batjes *et al.*, 2003). Land cover datum was the Global Land Cover 2000 for Jordan (forobs.jrc.ec.europa.eu/products/glc2000/glc2000.php). Current climate data were from WorldClim v. 1.3 (October 2004) at a resolution of 30 s (± 1 km) (Hijmans *et al.*, 2005; worldclim.org). Future climatic data from 2080 were downloaded from the GCM Data Portal (www.ccafs-climate.org). The A2 (more extreme) and B1 (moderate) $CO_2$ emissions scenarios, as well as the BCCR-BCM 2.0 (cooler) and MIROC 3.2 medres (hotter) global circulation models (GCMs) were selected based on the results obtained by Thorne (2012). The four possible future distribution models that resulted from the combination of the two GCMs with the two $CO_2$ emission scenarios were then combined using the ensemble approach suggested by Araújo and New (2007).

In order to identify the priority species most affected by climate change, the species range change (SRC) was calculated for each modelled species as the 'difference between the number of sites gained by the species (i.e. sites where the species could be present in the future but where it is absent at the moment) and the number of sites lost (i.e. sites where the species could be absent in the future but where it is currently present) relative to the number of sites currently occupied' (Buisson *et al.*, 2010). Out of the top 20 priority taxa, only 6 were modelled, as the other 14 did not have sufficient records that would produce robust distribution models. Out of these six, *Trifolium boissieri* Guss. ex Soy.-Will. & Godr. and *Triticum dicoccoides* Schrank ex Schübl. were affected positively by climate change and the other four were affected negatively (Magos Brehm *et al.*, unpublished results).

Finally, populations impacted negatively by climate change were identified for urgent *ex situ* conservation actions, whereas those where climate change was predicted to have a null or low impact could be conserved *in situ*.

## 15.3 Establishing *Ex Situ* and *In Situ* Conservation Priorities

The establishment of *in situ* and *ex situ* conservation priorities was determined by the combination of the individual species gap analysis, the ecogeographic gap analysis and the climate change impact on species distribution.

### 15.3.1 Priority species for *ex situ* and *in situ* conservation

Priority species were ranked and prioritized for *ex situ* conservation by combining the results from the individual species *ex situ* gap analysis, the ecogeographic gap analysis and the climate change impact on species distribution (see Section 15.2.3, this chapter) into five prioritization levels. For instance, priority species for *ex situ* conservation were those simultaneously impacted negatively by climate change and not conserved *ex situ* (priority 1) or with ecogeographic gaps in the *ex situ* conservation (priority 2); priority 3 included those species simultaneously not conserved *ex situ* and not affected negatively by climate change (or the climate change impact was unknown). Out of the taxa analysed, none were priority 1, only four ranked priority 2 and 28 were priority 3, and the remaining ranked the other lower priority levels (Magos Brehm *et al.*, unpublished results).

Priority species were ranked and prioritized for *in situ* conservation by combining the results from the individual species *in situ* gap analysis and the climate change impact on species distribution into four prioritization levels. Priority species for *in situ* conservation were those that were impacted negatively by climate change and, at the same time, did not occur in existing protected areas (PAs) (priority 1) or occurred passively within these (priority 2); these were followed by those species that simultaneously were not affected negatively by climate change (or the climate change impact was unknown) and either did not occur (priority 3) or occurred passively in existing PAs (priority 4). Out of the priority species, three were priority 1, one was priority 2, 40 were priority 3 and the remaining were priority 4 (Magos Brehm *et al.*, unpublished results).

### 15.3.2 Priority populations for *ex situ* collecting conservation

Populations for *ex situ* conservation were also ranked, by combining the frequency of the ecogeographic categories (ECs) in which they occurred and the likely negative impact of climate change, into eight prioritization levels. The first two priority levels corresponded to those populations that were impacted negatively by climate change and occurred in either low (priority 1) or medium-low (priority 2) frequency ECs (see Parra-Quijano *et al.*, 2012b for details). Priority 3 and 4 corresponded to those populations that were not affected negatively by climate change (or the climate change impact was unknown) and occurred in either low (priority 3) or medium-low (priority 4) frequency ECs. This analysis was only possible for the top 20 priority taxa, and about 80% of these taxa had populations with priorities between 1 and 4 (Magos Brehm *et al.*, unpublished results).

### 15.3.3 Priority localities for *in situ* conservation

Sites for the active *in situ* conservation were selected via two alternative approaches:

**1.** Simple complementarity analysis. DIVA-GIS version 7.5 (www.diva-gis.org) was used to perform a complementarity analysis for the top 98 priority taxa, which aimed to identify sets of grid cells that were complementary to each other; that is, which captured a maximum amount of diversity in as few cells as possible (see Rebelo, 1994a,b; Rebelo and Sigfried, 1992). The results showed that only 16 grids (10 × 10 km) were required to conserve *in situ* at least one population of each of the top 98 national priorities (Fig. 15.5).
**2.** Combination of complementarity analysis of existing PAs and outside PAs. Out of the top 98 priorities, 55 had at least one population in existing PAs and 43 did not occur in any existing PA. This analysis included:

- Selection of existing PAs, to actively conserve the 55 priority taxa that occurred in existing PAs, by complementarity analysis using an iterative procedure in Excel 2010. With this approach, the conservation of the top 98 priority taxa that occurred in existing PAs was ensured. Eight existing PAs were needed to conserve at least one population of each of the 55 priority species (only the top 98 priority taxa that occurred in existing PAs) (Fig. 15.6).
- Selection of new areas or extension of the borders of existing PAs for the conservation of the 43 top priority taxa that did not occur in any existing PA, by complementarity analysis using DIVA-GIS version 7.5. With this approach, the conservation of the top 98 priority species that did not occur in any existing PA was ensured (either the borders of existing PAs can be extended or new genetic reserves can be established). The complementarity analysis identified 12 grid cells (10 × 10 km) as the minimum number of sites needed to conserve at least one population of each 43 taxa that did not occur in existing PAs (Fig. 15.7).

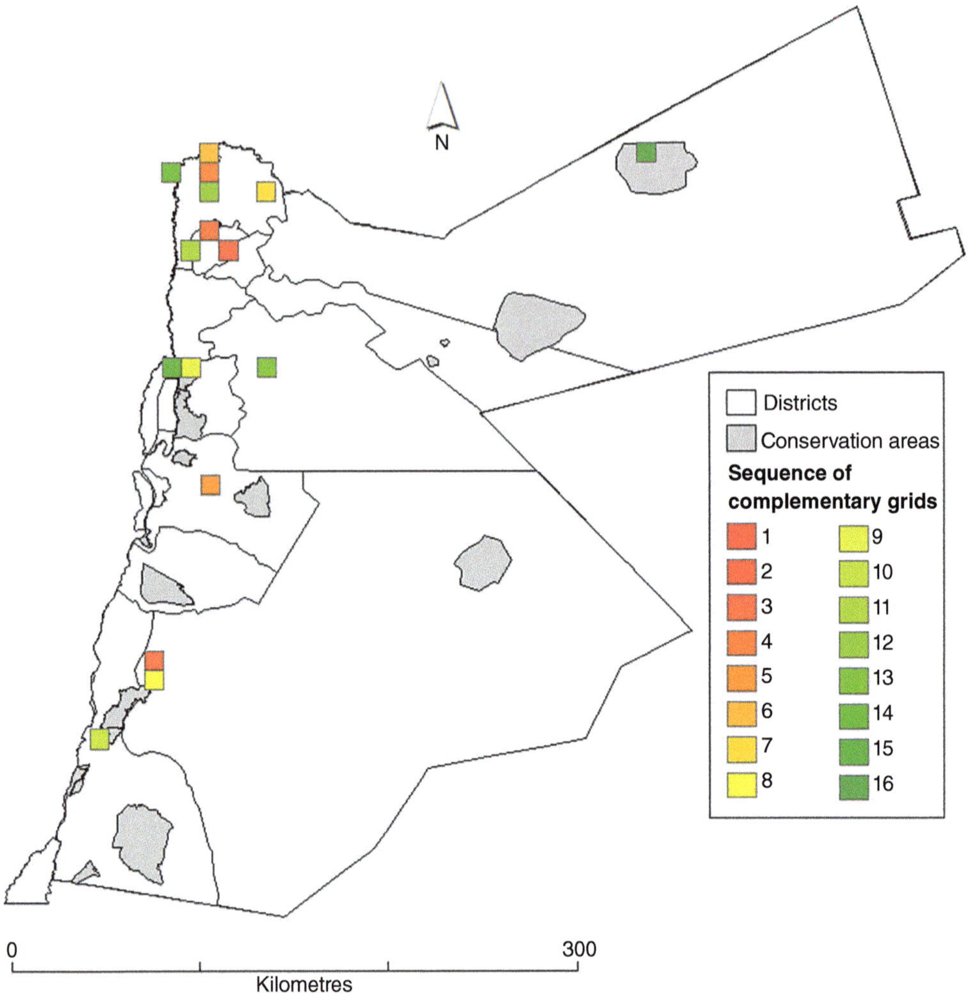


Fig. 15.5. Network of complementary grids for the conservation of the top 98 priority taxa (resolution of 10 × 10 km).

## 15.4 National Strategy for the Conservation of Plant Diversity for Jordan

A National Strategy for the Conservation of Plant Diversity for Jordan (Magos Brehm *et al.*, 2015) was then compiled, drafted and presented at the 'Wild Socio-economic Plant Conservation Strategy for Jordan' national workshop (Amman, April 2014), attended by representatives of academic societies (University of Jordan, Yarmouk University), NGOs (RBG, ASEZA–Australian Volunteers International of Jordan, RSCN, IUCN ROWA, the Oryx Conservation project in Wadi Rum), governmental bodies like the Ministry of Agriculture, Ministry of Environment and NCARE, and the University of Birmingham. The participants were sent the drafted strategy prior to the workshop and given the opportunity to provide feedback during the workshop. Based on suggestions and inputs from stakeholders, additions were made to the document. The final version of the National Strategy (Magos Brehm *et al.*, 2015) consists of over 350 pages and has been structured according to Fig. 15.8. It is expected that it will be submitted to the Biodiversity National Committee to be endorsed and subsequently

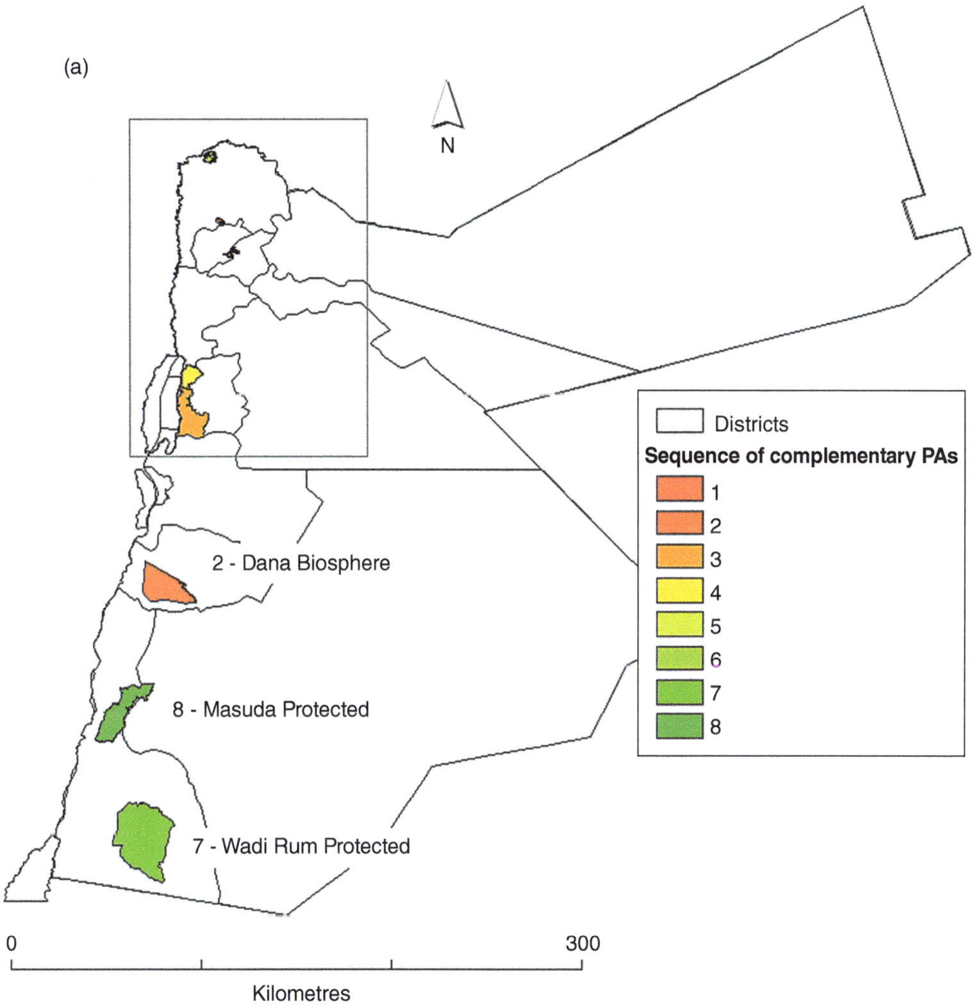


Fig. 15.6a and b. Network of complementary protected areas (PAs) for the 55 priority taxa that occur in the existing protected areas.

raised to the Ministry of Environment for adoption and implementation.

Despite its value to Jordan, as well as globally, plant diversity, and namely CWR, in Jordan is seriously threatened. However, systematic conservation of these resources is still lacking, and a concerted and strategic conservation plan should be implemented immediately. A National Strategy for the Conservation of Plant Diversity for Jordan has been developed with the participation of relevant national stakeholders, with the purpose of guiding the conservation of these resources: priorities for conservation and recommendations for active *in situ* and *ex situ* conservation have been established. It is hoped that the implementation of the National Strategy for the Conservation of Plant Diversity for Jordan leads to an efficient conservation of these resources in order to make them available not only to future generations of Jordanians but also to humankind.

## Acknowledgements

The National Strategy for Plant Conservation for Jordan has been produced as part of the

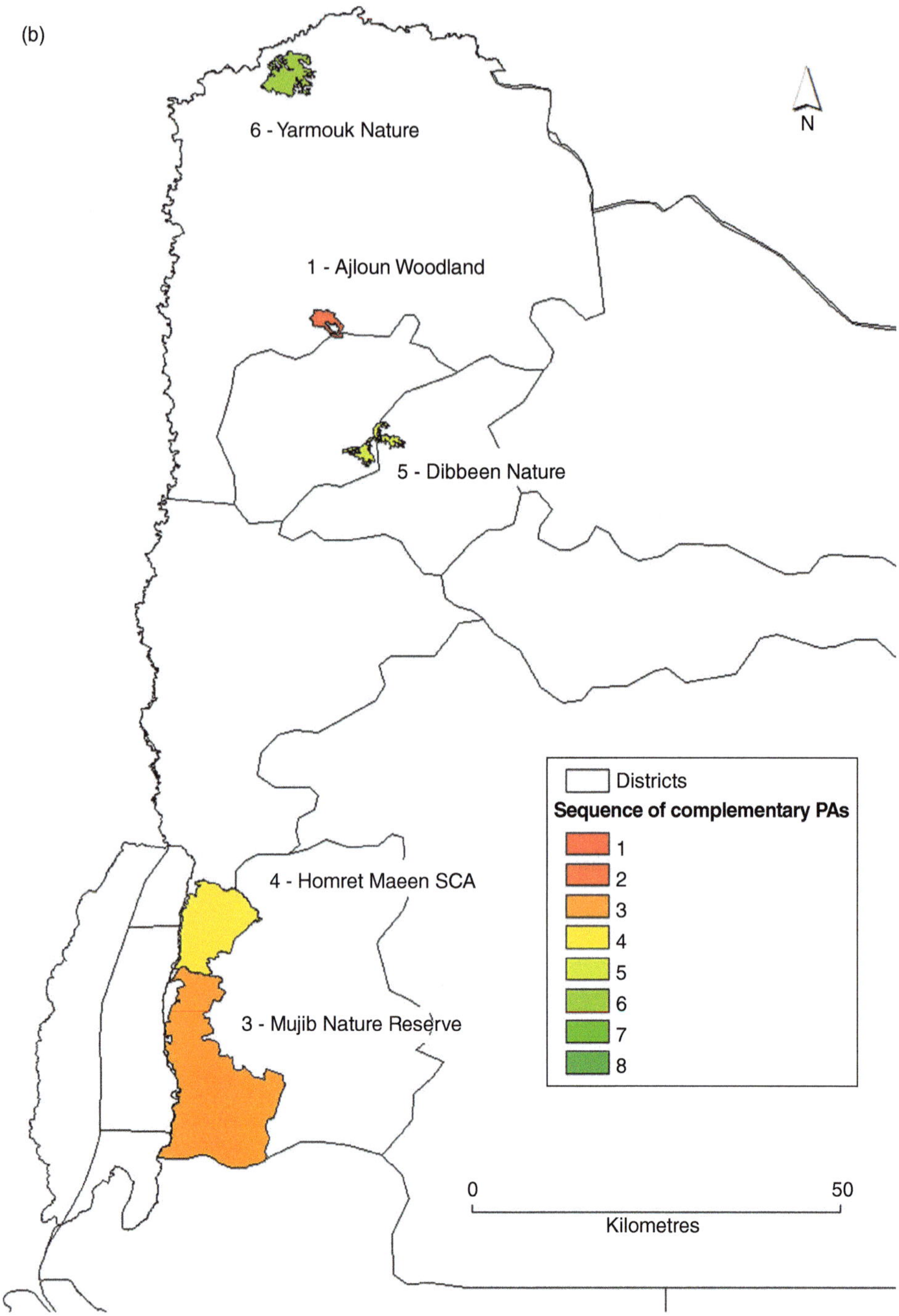


Fig. 15.6a and b. Continued.

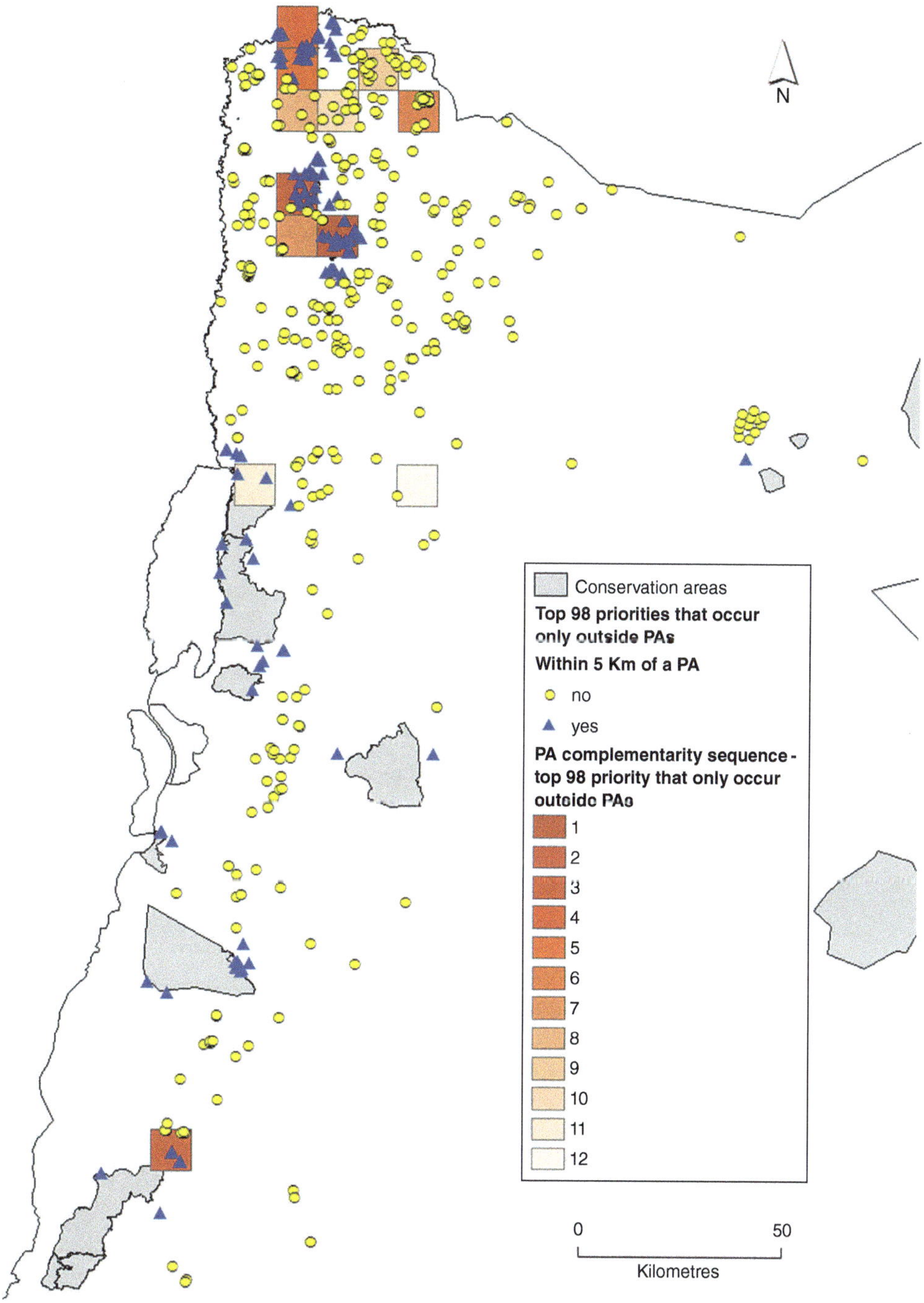


Fig. 15.7. Complementarity analysis of the 43 priority taxa that do not occur in any existing protected area (squares), together with the information on whether they occur within 5 km of an existing protected area (blue triangles) or not (yellow dots). Darker squares indicate higher priority for conservation, as they include a larger number of these taxa.

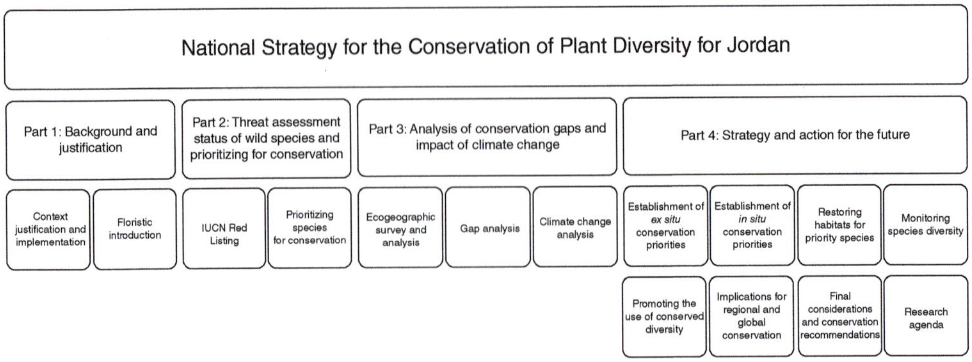


**Fig. 15.8.** Structure of the National Strategy for the Conservation of Plant Diversity for Jordan.

BOT-ERA – reinforcing cooperation between the Royal Botanic Garden of Jordan and the European Research Area project, funded by the EU Seventh Framework Programme, THEME [INCO.2011-6.2 KBBE], Grant Agreement No 294416. The authors would like to acknowledge the contribution made to the development of the National Strategy for Plant Conservation for Jordan by the IUCN ROWA, the curators of herbaria and genebanks (JUST, Al Albait University, Yarmuk University, Badia Research and Development Program, Royal Botanic Garden Edinburgh, Hebrew University of Jerusalem, NCARE), RSCN, the United Nations Compensation Committee, ICARDA, individual experts (Hala Barakat, Nora Castañeda, Avinoam Danin, Ori Fragman-Sapir, James Hearsum, Claus Holzapfel, Shelagh Kell, Lytton John Musselman, Mauricio Parra-Quijano, Liz Radford, Avi Shmida, Ziad Tehabsem, Imke Thormann), as well as all the participants of the 'Establishing Species Conservation Priorities in Jordan' workshop (held in 2012), the 'Regional Red List Assessment for Plants of Jordan' workshop (in 2013) and the 'Wild Socioeconomic Plant Conservation Strategy for Jordan' workshop (in 2014), namely ASEZA, Hashemite University, IUCN ROWA, Jordan National Center for Research and Development, JUST, Ministry of Environment of Jordan, Ministry of Agriculture of Jordan, NCARE, RSCN, University of Jordan and Yarmouk University. Finally, the authors would also like to thank the BOT-ERA project partners and the project Steering Committee members.

## References

Al-Eisawi, D.M. (1982) List of Jordan vascular plants. *Mitteilungen der Botanischen Staatssummlung München* 18, 79–182.

Al-Eisawi, D.M. (1996) *Vegetation of Jordan*. UNESCO (ROSTAS), Cairo.

Angiosperm Phylogeny Group (2009) An update of the Angiosperm Phylogeny Group classification for the orders and families of flowering plants: APG III. *Botanical Journal of the Linnean Society* 161(2), 105–121.

Araújo, M.B. and New, M. (2007) Ensemble forecasting of species distributions. *Trends in Ecology and Evolution* 22, 42–47.

Batjes, N.H., Rawajfih, Z. and Al-Adamat, R. (2003) Soil data derived from SOTER for studies of carbon stocks and change in Jordan (ver. 1.0; GEFSOC Project). Report 2003/04. ISRIC–World Soil Information, Wageningen. Available at: www.isric.org/data/soter-based-soil-parameter-estimates-jordan-ver-10 (accessed February 2013).

Buisson, L., Thuiller, W., Casajus, N., Lek, S. and Grenouillet, G. (2010) Uncertainty in ensemble forecasting of species distribution. *Global Change Biology* 16, 1147–1157.

CB2 Project (2013) Developing policy-relevant capacity for implementation of the global environmental conventions in Jordan. Assignment 'Developing policy-oriented research guidelines, procedures, and tools to support implementing Rio conventions in Jordan'. UNDP-MoE, Jordan.

Danin, A. (2001) Near east ecosystem and plant diversity. *Encyclopedia of Biodiversity* 4, 1–12.
GCEP (General Cooperation for Environment Protection) (1998) Jordan country study on biological diversity. The General Cooperation for Environment Protection, Amman.
Ghazanfar, S.A., Taifour, H., El-Oqlah, A., Borosova, R. and Osborne, J. (2016) *Checklist of the Vascular Plants of Jordan*. Kew Publishing, London (in press).
Hanley, J.A. and McNeil, B.J. (1982) The meaning and use of the area under a Receiver Operating Characteristic (ROC) curve. *Radiology* 143, 29–36.
Harlan, J.R. and de Wet, J.M.J. (1971) Toward a rational classification of cultivated plants. *Taxon* 20(4), 509–517.
Harrison, S.P. (2009) Future climate change in Jordan: an analysis of state-of-the-art climate model simulations. Report submitted in fulfilment of contract 'Technical Assistance for Climate Change, International Specialist' between the Royal Society for the Conservation of Nature, RSCN on behalf of the Ministry of Planning and International Cooperation/Integrated Ecosystem Management in the Jordan Rift Valley Project and Professor Sandra Patricia Harrison, consultant, Bristol, UK.
Hijmans, R.J., Cameron, S.E., Parra, J.L., Jones, P.G. and Jarvis, A. (2005) Very high resolution interpolated climate surfaces for global land areas. *International Journal of Climatology* 25, 1965–1978.
IUCN (International Union for Conservation of Nature) (2001) *IUCN Red List Categories and Criteria*. Version 3.1. IUCN Species Survival Commission. IUCN, Gland, Switzerland, and Cambridge, UK, ii + 30 pp. Available at: www.iucnredlist.org/documents/redlist_cats_crit_en.pdf (accessed February 2013).
IUCN (2010) *Guidelines for Application of IUCN Red List Criteria at Regional and National Levels*. Version 4.0. IUCN, Gland, Switzerland, and Cambridge, UK, iii + 41 pp. Available at: www.iucnredlist.org/documents/reg_guidelines_en.pdf (accessed February 2013).
Kasapligil, B. (1956) Report to the government the Hashemite Kingdom of Jordan. Ecological survey of the vegetation in relation to the forestry and grazing. FAO report 549, Rome.
Keiša, A., Maxted, N. and Ford-Lloyd, B.V. (2008) The assessment of biodiversity loss over time: wild legumes in Syria. *Genetic Resources and Crop Evolution* 55, 603–612.
Magos Brehm, J., Saifan, S., Taifour, H., Abulaila, K., Al-Assaf, A., *et al.* (2015) *Wild Socioeconomic Plant Conservation Strategy for Jordan*. Royal Botanic Garden, Tell Ar-Rumman, Jordan.
Maxted, N. (2011) Preparation of 'IUCN Red List of threatened plants of Jordan'. Consultant's report.
Maxted, N., Ford-Lloyd, B.V., Jury, S., Kell, S.P. and Scholten, M.A. (2006) Towards a definition of a crop wild relative. *Biodiversity and Conservation* 15(8), 2673–2685.
Ministry of the Environment (2014) Fifth national report on implementation of the Convention on Biological Diversity. Ministry of the Environment, Amman.
NCARTT (National Center for Agricultural Research and Technology Transfer) (2007) The second country report on the state of plant genetic resources for food and agriculture. Amman (online). Available at: www.fao.org/docrep/013/i1500e/Jordan.pdf (accessed 25 February 2016).
Parra-Quijano, M., Iriondo, J.M., Cruz, M. and Torres, E. (2012a) Spatial and ecogeographical approaches for selecting genetic reserves in Europe. In: Maxted, N., Dulloo, M.E., Ford-Lloyd, B.V., Frese, L., Iriondo, J.M. and Pinheiro de Carvalho, M.A.A. (eds) *Agrobiodiversity Conservation: Securing the Diversity of Crop Wild Relatives and Landraces*. CAB International, Wallingford, UK, pp. 20–28.
Parra-Quijano, M., Iriondo, J.M. and Torres, E. (2012b) Improving representativeness of genebank collections through species distribution models, gap analysis and ecogeographical maps. *Biodiversity and Conservation* 21(1), 79–96.
Parra-Quijano, M., Torres, E., Iriondo, J.M. and López, F. (2014) *CAPFITOGEN Tools User Manual, Version 1.2*. International Treaty on Plant Genetic Resources for Food and Agriculture, FAO, Rome.
Phillips, S.J. and Dudık, M. (2008) Modelling of species distributions with MaxEnt: new extensions and a comprehensive evaluation. *Ecography* 31, 161–175.
Phillips, S.J., Anderson, R.P. and Schapire, R.E. (2006) Maximum entropy modelling of species geographic distributions. *Ecological Modelling* 190, 231–259.
Rebelo, A.G. (1994a) Iterative selection procedures: centres of endemism and optimal placement of reserves. *Strelitzia* 1, 231–257.
Rebelo, A.G. (1994b) Iterative selection procedures: centres of endemism and optimal placement of reserves. In: Huntley, B.J. (ed.) *Botanical Diversity in Southern Africa*. National Botanical Institute, Pretoria, South Africa, pp. 231–257.
Rebelo, A.G. and Sigfried, W.R. (1992) Where should nature reserves be located in the Cape Floristic Region, South Africa? Models for the spatial configuration of a reserve network aimed at maximising the protection of diversity. *Conservation Biology* 6, 243–252.

Taifour, H. and El-Oqlah, A. (2014) *Jordan Plant Red List*, Vol 1. Royal Botanic Garden, Jordan.
The Plant List (2010) Version 1 (online). Available at: www.theplantlist.org (accessed February 2013).
Thorne, J.H. (2012) Climate change strategy report. Recommended approaches to RSCN. Prepared for IEM Project. Available at: www.thegef.org/gef/sites/thegef.org/files/gef_prj_does/GEFProjectDocuments/MandE/TE/FY2014/World%20Bank/G001214/1214_WB_ICR.pdf (accessed 25 February 2016).
Vavilov, N.I. (1926) *Studies in the Origin of Cultivated Plants*. Institute of Applied Botany and Plant Breeding, Leningrad, Russia.
Vincent, H., Wiersema, J., Kell, S.P., Dobbie, S., Fielder, H., *et al.* (2013) A prioritised crop wild relative inventory as a first step to help underpin global food security. *Biological Conservation* 167, 265–275.
Zohary, M. (1972) *Flora Palaestina*. The Israel Academy of Sciences and Humanities, Jerusalem.

# 16 Establishing Systematic Crop Wild Relative Conservation in the UK

H. Fielder,* B. Ford-Lloyd and N. Maxted
*School of Biosciences, University of Birmingham, Birmingham, UK*

## 16.1 Introduction

Conservation in the UK, as with other countries globally, can be divided between two communities: biodiversity and agrobiodiversity conservation, with the former characterized as rare/threatened species conservation and the latter focusing on crop varieties (Maxted, 2003). Crop wild relatives (CWR) fall between these two communities, and this has resulted in a lack of conservation management or monitoring for these highly valuable resources. Though some incidental or 'passive' conservation may occur where CWR fall within protected area boundaries or have had seeds stored in genebanks as part of collection programmes for other purposes, this is not sufficient to ensure that the range of genetic diversity within them is safeguarded in the long term. It is the conservation of this breadth of genetic diversity that is crucial, as it will enable plant breeders to exploit this to produce improved crop varieties that are able to withstand the impacts of a changing climate.

In recognition of the role of CWR in aiding crop adaptation to climate change, and therefore helping to underpin future food security, CWR conservation is being placed higher on international and regional agendas with the ratification of the Convention on Biological Diversity (CBD, 1992), the more recent Global Strategy for Plant Conservation and Strategic Plan for Biodiversity (CBD, 2010a,b), the International Treaty on Plant Genetic Resources for Food and Agriculture (ITPGRFA) (FAO, 2001), as well as the European Strategy for Plant Conservation (Planta Europa, 2008) and Europe's own Biodiversity Strategy to 2020 (European Union, 2011). All of these policy documents emphasize the importance of the conservation of CWR, the genetic diversity they contain and their role in improving future food security. The UK is a contracting party to both the CBD and ITPGRFA, and as such CWR conservation should be made a priority.

The effective conservation of UK CWR diversity can be achieved by developing and implementing a conservation strategy for CWR. This should take a multi-taxon approach, providing comprehensive and systematic conservation of CWR. The implementation of active monitoring and management of CWR populations and the collection of seed for storage would take significant steps to securing this resource in the UK. The PGR Secure project focused on improving the conservation and use of European CWR. This required the development of conservation strategies for Spain, Italy, Finland and the UK. In response to this, our aims were to:

- develop conservation strategies for each devolved administration of the UK (England,

*Corresponding author; e-mail: hannahfielder88@gmail.com

 *Enhancing Crop Genepool Use: Capturing Wild Relative and Landrace Diversity for Crop Improvement* (eds N. Maxted, M. Ehsan Dulloo and B.V. Ford-Lloyd) 

Wales and Scotland) and an additional strategy for the UK as a whole;
- investigate the habitat preferences of CWR in Great Britain; and
- carry out a genetic analysis of CWR populations on the Lizard Peninsula in Cornwall, UK, with a view to establishing the first UK genetic reserve for CWR.

## 16.2 Conservation Strategies

There is a need for countries to establish national policies that answer the call of regional and international policy documents to establish CWR conservation. In the UK, conservation is managed separately within each devolved administration by separate conservation organizations and coordinated at the national level by the Joint Nature Conservation Committee (JNCC). For this reason, CWR conservation strategies were developed to mirror this hierarchy, with a strategy for each administration and one national-level strategy. Methodologies were then developed jointly between the University of Birmingham and each administration's conservation organization (Natural England, Natural Resources Wales and Scottish Natural Heritage). In taking this inclusive approach in the UK, it is more likely that policy will result through collaboration with key national organizations.

The methodology used to develop UK conservation strategies built on the framework (Fig. 16.1) set out by Maxted *et al.* (2007). This framework begins with identifying a checklist of CWR taxa within the region of study. Due to limited resources, it is often not feasible to establish active conservation for all CWR in a checklist and so a process of prioritization is carried out to identify a subset of CWR in most urgent need of active conservation. With the priority CWR identified, it is then possible to undergo a gap analysis. Here, a complementary approach is taken where gaps in current *in situ* and *ex situ* conservation efforts for priority CWR are determined and a strategy produced which ensures these gaps are filled through targeted and systematic conservation. The end product of this framework should be the effective conservation of CWR diversity, which is available for use in crop improvement. By following this framework and involving relevant stakeholders, this methodology is widely applicable to any region or country as an easily replicable model that can encourage systematic CWR conservation and the policy to support it.

The first step in developing the four conservation strategies in the UK was to create a checklist of all the CWR known to occur within each administration. This was achieved using the Vice County Census Catalogue (VCCC) and CWR Information System (CWRIS). Great Britain is divided into 112 vice-counties, which are regions of similar sizes and whose boundaries are fixed, making them suitable for biological recording over time (Stace *et al.*, 2003). Each vice-county is assigned its own 'Recorder', whose responsibility it is to survey their vice-county and to ensure that a comprehensive catalogue of all plants is established and maintained in the long term. From this database, checklists of the flora of each administration were extracted. Alongside this, a checklist of CWR for the whole of the UK was extracted from the CWRIS, developed as part of the PGR Forum project (Kell *et al.*, 2005). These two checklists were then matched against one another to create a checklist containing just the CWR occurring within each devolved administration. The resulting CWR checklists contained 1471 CWR taxa in England, 1251 CWR taxa in Wales, 1259 CWR taxa in Scotland and 2109 CWR taxa in the UK as a whole (Fielder *et al.*, 2015a, b, in press, in preparation a). These large numbers of taxa highlight the need for prioritization to identify those CWR in most urgent need of conservation.

Criteria on which to base the prioritization were selected jointly between each administration's conservation agency and the University of Birmingham. A summary of the selected criteria is shown in Table 16.1. The primary use of a CWR is by definition genetic improvement of crop plants; therefore, the value of any particular crop will infer the current relative value of its wild relatives (Maxted *et al.*, 1997a; Ford-Lloyd *et al.*, 2008; Maxted and Kell, 2009; Magos Brehm *et al.*, 2010; Kell *et al.*, 2012). Considering this, each organization chose to include the economic value of the crops to which the CWR were related. Furthermore, it is seen as a reasonable assumption that the closer the relationship between a CWR and its crop, the higher the priority for conservation, due to the greater ease of gene transfer for crop improvement. As a result, the degree of relatedness was another criterion selected by all

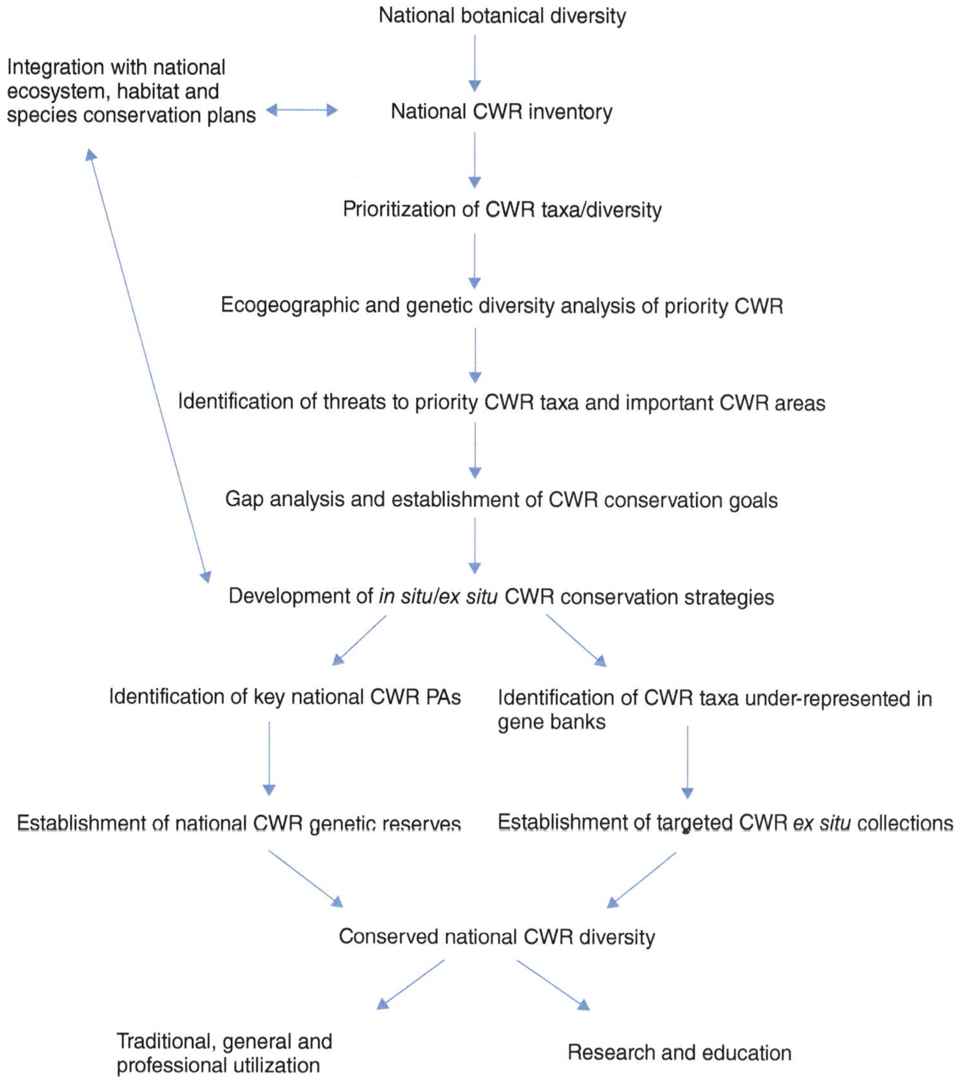


Fig. 16.1. Framework for the development of national CWR conservation strategies. (From Maxted *et al.*, 2007.)

conservation organizations, based on their genepool and taxon group classifications (Harlan and de Wet, 1971; Maxted *et al.*, 2006).

There were also a number of differences between the criteria selected by each organization and the way in which they were applied. Natural Resources Wales was the only organization to include tree species in their prioritized inventory, citing the increasing threats from new pests and diseases as the justification for this decision (Fielder *et al.*, in press). Both native and introduced (naturalized) CWR taxa were included in the UK inventory, with organizations highlighting that there was likely to have been sufficient time for unique adaptations to have occurred among introduced taxa, potentially providing a wider genepool of genetic diversity in wild relatives (Fielder *et al.*, in preparation a). Finally, various combinations of criteria relating to the threatened status of CWR were selected by each administration. Natural England chose to look at the changes in the ranges of populations of CWR in Great Britain between two survey years (1987 and 2004) (Braithwaite *et al.*,

**Table 16.1.** Criteria selected for prioritization of CWR in each devolved administration and the UK as a whole, and an explanation as to which CWR were prioritized to create four separate CWR inventories.

| Criteria | England | Wales | Scotland | UK |
|---|---|---|---|---|
| Use of related crop | CWR related to crops cultivated for food, forage or fodder | CWR related to crops cultivated for food, forage or fodder and also including trees | CWR related to crops cultivated for food, forage or fodder | CWR related to crops cultivated for food, forage or fodder |
| Native status | Native and archaeophyte CWR only | Native and archaeophyte CWR only | Native and archaeophyte CWR only | Native, archaeophyte and neophyte (excluding casual) CWR |
| Economic value of related crop | CWR related to crops listed in global, European or national agricultural statistics | CWR related to crops listed in global, European or national agricultural statistics | CWR related to crops listed in global, European or national agricultural statistics | CWR related to crops listed in global, European or national agricultural statistics |
| Degree of relatedness to crop | CWR in genepools 1b or 2 of crop or taxon groups 1b–3 | CWR in genepools 1b or 2 of crop or taxon groups 1b–3 | CWR in genepools 1b or 2 of crop or taxon groups 1b–3 | CWR in genepools 1b or 2 of crop or taxon groups 1b–3 |
| Change in population range | CWR showing a population decline between 1987 and 2004 | Not selected | Not selected | Not selected |
| Threat assessment | Not selected | CWR listed as Critically Endangered, Endangered, Vulnerable or Near Threatened according to the *IUCN Red List Categories and Criteria* (IUCN, 2001) | CWR listed as Critically Endangered, Endangered, Vulnerable or Near Threatened according to the *IUCN Red List Categories and Criteria* (IUCN, 2001) | CWR listed as Critically Endangered, Endangered, Vulnerable or Near Threat-ened according to the *IUCN Red List Categories and Criteria* (IUCN, 2001) |
| Other conservation designations | Not selected | Not selected | CWR listed as having any other rarity, threat or legal designation in the UK, independently of IUCN threat assessments | CWR listed as having any other rarity, threat or legal designation in the UK, independently of IUCN threat assessments |

2006; Fielder *et al.*, 2015a), whereas all other organizations chose to use the *IUCN Red List Categories and Criteria* (IUCN, 2001). In Scotland and the UK, any additional threat/rarity designations were also considered as justification for prioritization (Fielder *et al.*, 2016, in preparation a). The differences in taxa included in the CWR checklist and the ways in which the prioritization criteria were applied resulted in different species pools being considered for conservation in each administration.

For the priority CWR in each administration, a complementary conservation approach was then taken. *In situ* gap analyses were first carried out in order to identify hotspots of CWR taxon diversity and potential sites for the establishment of genetic reserves (Maxted *et al.*, 2012). This was achieved using geographic information systems (GIS), DIVA-GIS 7.5.0 (Hijmans *et al.*, 2012) and ArcMap 10.0 (ESRI, 2011). Figure 16.2 shows the taxon richness results for the UK, with hotspots concentrated in the south and east of the country, particularly along the south coast of the UK (e.g. Dorset) and areas of Cornwall in the south-west including the Lizard Peninsula (Fielder *et al.*, in preparation a). Similar analyses were carried out in each administration, revealing further hotspots in Wales and Scotland (e.g. the Gower Peninsula in south Wales) (Fielder *et al.*, in press). Complementarity analysis was used to identify sites suitable for CWR genetic reserves. This method, described by Rebelo (1994), makes a selection of complementary grid squares that together contain all priority CWR. Key sites again included the south coast of the UK, the Lizard Peninsula in Cornwall, the Gower Peninsula in Wales and also sites between the cities of Glasgow and Edinburgh in Scotland (Fielder *et al.*, in preparation a, 2015a, in press, 2016 respectively). Richness and complementarity

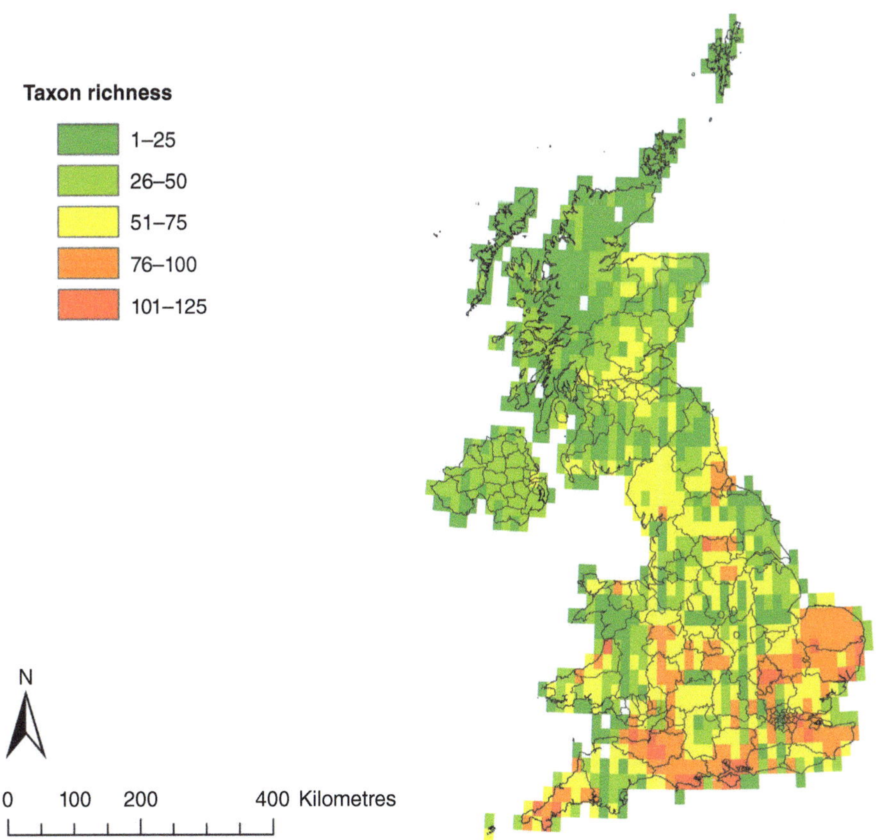


Fig. 16.2. Taxon richness of the 223 priority CWR in the UK. (From Fielder *et al.*, in preparation a.)

analyses enabled recommendations to be made as to the broad locations across the UK, and in each devolved administration, as to where the establishment of active CWR conservation would be appropriate.

It is important to note that conservation should not be focused solely on *in situ* methods. Though *in situ* conservation allows CWR to continue to adapt to their environment, it is equally important that material is collected and stored in genebanks to safeguard this genetic material (Maxted *et al.*, 1997b). Storage not only provides protection against loss of material but also it provides a point of access for plant breeders to use the material in crop improvement. As part of the UK and devolved administration conservation strategies, an *ex situ* gap analysis was also carried out. This allowed the identification of taxa for which current genebank collections insufficiently represented their full geographic range, and so highlighting where further collection was required. Significant gaps in *ex situ* collections were identified across the UK and in each administration with a particular bias in all four regions for collection of the forage CWR, *Lolium perenne* L. (Fielder *et al.*, in preparation a, 2015a, in press, 2016 respectively). It is known that the Millennium Seed Bank, Kew, stores close to 94% of all native, orthodox plant species (Stephanie Miles, 2014, personal communication); however, the gaps noted in the current study are largely for taxa that are classified as archaeophytes and neophytes (having been introduced into the UK prior to 1500 AD and post 1500 AD, respectively). In light of this, Fielder *et al.* (in preparation a) highlight that the priorities for CWR collection can differ from general plant collection priorities. Any CWR containing unique genetic diversity could provide beneficial genes for crop improvement and as such, their collection should not be limited to native species but should also include archaeophytes and neophytes, which are equally likely to contain novel diversity. A focus on collecting accessions representative of the range of geographic and, where possible, genetic diversity of all priority CWR is essential.

Overall, very similar methodologies were followed by each administration in the UK. All agreed on the overarching framework whereby priority CWR were determined from a checklist of all CWR in the country in question, and gap analyses were carried out to aid the targeting of appropriate and efficient conservation of these priority CWR. The differences in the methodologies stemmed from decisions made by each individual administration according to their own conservation priorities; for example, whether to focus solely on CWR related to human food and animal forage/fodder or whether to expand this to include tree species, etc. The opportunity to collaborate with each organization has enabled the development of tailored strategies that identify both *in situ* and *ex situ* conservation priorities for CWR in the UK.

## 16.3 CWR Habitat Preferences

The complementarity gap analysis discussed above is an effective means of identifying broad locations that contain high concentrations of CWR. Finding these broad locations is an essential step in beginning the process of introducing more fine-scale targeted conservation efforts. In order to find more specific locations suitable for CWR conservation, it is necessary to understand the basic habitat preferences associated with CWR. This will lead conservationists to be able to target their efforts not only in broad locations where there are high concentrations of CWR but also, more specifically, within those particular habitats in which CWR are most likely to be found.

Jarvis *et al.* (2015) have conducted a study of CWR habitat preferences in Great Britain using data from the Countryside Survey (CS). The CS was carried out in 2007 (Carey *et al.*, 2008) and consisted of a full vegetation survey across Great Britain. A total of 11,685 quadrats were surveyed from a stratified random sample of 589 1-km squares. From this comprehensive data set, it was possible to extract occurrence data for the priority CWR identified by Fielder *et al.* (in preparation a) and model their habitat preferences using a generalized linear mixed model with the proportion of priority CWR taxa as the response variable.

The results obtained by Jarvis *et al.* (2015) demonstrated that CWR showed a strong preference for landscape features associated with high levels of anthropogenic disturbance, including hedgerows, arable field margins and road verges. This highlights the importance of considering a wide variety of sites for *in situ* conservation. As much as it is not sufficient to establish one genetic reserve and assume that CWR genetic diversity is fully represented, the same is true of a network of genetic reserves. In addition to conservation within established protected areas, Jarvis *et al.* (2015) provide a convincing argument that the conservation of CWR is not complete without their management outside of protected areas, particularly in disturbed habitats. Establishing conservation outside of protected areas will require a different approach to that of establishing genetic reserves, whereby landowners and local stakeholders will need to be consulted. Agreements must be made between conservation organizations and these landowners to ensure there is a commitment to the long-term monitoring of areas with high concentrations of CWR. To encourage these agreements, incentives could be provided through subsidy schemes such as the Rural Development Programme (Fielder *et al.*, 2016).

Fig. 16.3. Sea beet (*Beta vulgaris* subsp. *maritima*) growing along the South West Coast Path on the western side of the Lizard Peninsula. (Photograph courtesy of Hannah Fielder.)

Fig. 16.4. Sea radish (*Raphanus raphanistrum* subsp. *maritimus*) at Spernic Cove on the eastern side of the Lizard Peninsula. (Photograph courtesy of Hannah Fielder.)

## 16.4 Genetic Diversity Analysis

The European Strategy for Plant Conservation (ESPC) is encouraging the 'establishment of 25 European crop wild relative genetic reserves covering major hotspots of species and genetic diversity' (Planta Europa, 2008). However, there are still no genetic reserves in the UK. In order to address this, a network of sites with potential for being established as genetic reserves, i.e. having high concentrations of CWR, has been identified across the UK (Fielder *et al.*, in preparation a, 2015a, in press, 2016 respectively). The proposed English and UK genetic reserve networks identified the Lizard Peninsula as a priority site, due to its diversity in terms of ecogeography and number of CWR taxa.

Knowing that the Lizard Peninsula is a hotspot of CWR taxonomic diversity, it was also necessary to investigate whether it was also a hotspot of CWR genetic diversity, in order to justify the site as suitable for a genetic reserve. To achieve this, Fielder *et al.* (in preparation b) conducted a genetic analysis of eight CWR (wild asparagus, wild garlic, wild chive, sea beet, sea carrot, sea radish, white clover and western clover, Figs 16.3 and 16.4) using amplified fragment length polymorphisms (AFLPs) in order to measure genetic diversity and distance parameters. Multiple populations were sampled across the peninsula for each CWR, to enable the investigation of patterns of genetic diversity (Fig. 16.5). Populations of the same CWR were also sampled from sites across the south-west of the UK to assess whether there was any distinction between genetic diversity on the Lizard and elsewhere.

The results presented in Fielder *et al.* (in preparation b) suggest that, in comparison with the south-west of the UK, the Lizard is a hotspot of genetic diversity in CWR. Equally, the diversity contained in the Lizard is distinct from that found in other regions of the south-west. This has demonstrated that this site is suitable for the first UK genetic reserve, as it covers a hotspot of both species and genetic diversity, a requirement

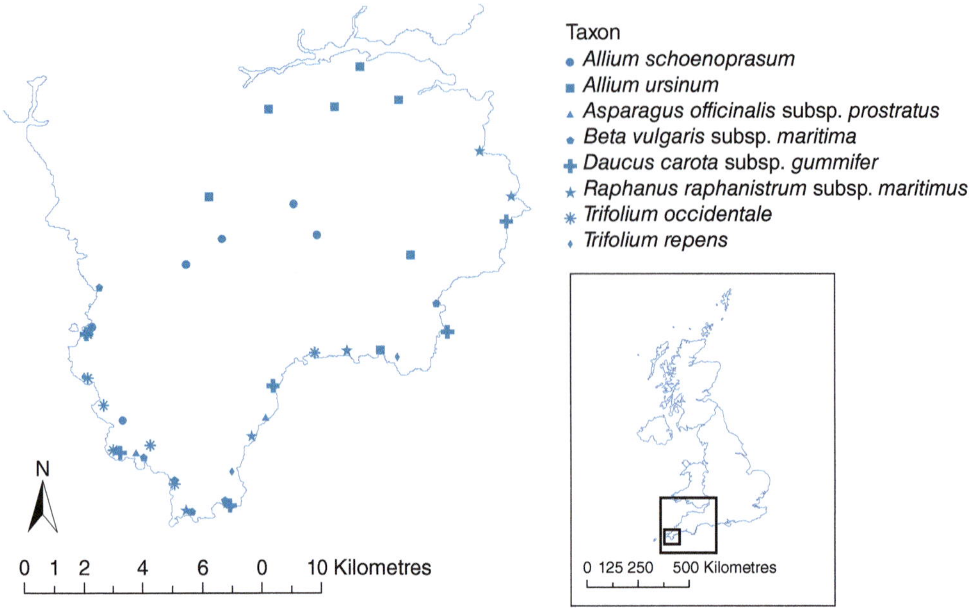


**Fig. 16.5.** The Lizard Peninsula, Cornwall, showing leaf collection sites for all eight CWR for which an AFLP analysis was carried out. (From Fielder *et al.*, in preparation b)

stated in the ESPC. However, before the site can be declared a genetic reserve, it will first have to meet a set of criteria based on Iriondo *et al.* (2012) and recently revised (Fielder *et al.*, 2015b). Natural England is working to ensure that each standard is met on the Lizard. This involves ensuring that CWR are represented in protected area management plans, that a monitoring regime for populations is established and actively maintained, and that CWR material is made available via genebanks for use by plant breeders (Fielder *et al.*, 2015b).

## 16.5 Conclusion

The PGR Secure project has provided a platform to encourage and facilitate the development of national conservation strategies in a range of countries from across Europe. In the UK, this has enabled the development of conservation strategies for each devolved administration, as well as an overarching national strategy, reflecting the way in which conservation is delivered in the UK. These four conservation strategies combine broad approaches, including gap analyses and the identification of CWR habitat preferences, with more fine-scale approaches such as the genetic diversity analysis demonstrated on the Lizard Peninsula. In combining these approaches, it will enable conservation organizations in each devolved administration to establish systematic, long-term conservation of CWR in the most appropriate locations. Equally, the national and 'subnational' approaches used in the UK and devolved administrations can be applied to other countries where conservation is managed in a similar way, both inside and outside of Europe.


## Acknowledgements

This research was funded by the EU Seventh Framework Programme funded project, PGR Secure 'Characterization of biodiversity resources for wild crop relatives to improve crops by breeding' (Grant Agreement No 266394) and Natural England (Grant Agreement No ECM 6269 ITT 748).

## References


Braithwaite, M.E., Ellis, R.W. and Preston, C.D. (2006) *Change in the British Flora 1987–2004*. Botanical Society of the British Isles, London.

Carey, P., Wallis, S., Chamberlain, P., Cooper, A., Emmett, B., *et al.* (2008) Countryside Survey: UK Results from 2007 (online). Available at: www.countrysidesurvey.org.uk (accessed 7 November 2015).

CBD (Convention on Biological Diversity) (1992) *Convention on Biological Diversity*. Secretariat of the Convention on Biological Diversity, Montreal, Canada.

CBD (2010a) Global Strategy for Plant Conservation. Secretariat of the Convention on Biological Diversity, Montreal, Canada (online). Available at: www.cbd.int/gspc/strategy.shtml (accessed 7 November 2015).

CBD (2010b) Strategic Plan for Biodiversity 2011–2020. Secretariat of the Convention on Biological Diversity, Montreal, Canada (online). Available at: www.cbd.int/decision/cop/?id=12268 (accessed 7 November 2015).

ESRI (Environmental Systems Research Institute) (2011) *ArcGIS Desktop Release 10*. Environmental Systems Research Institute, Redlands, California.

European Union (2011) The EU Biodiversity Strategy to 2020. European Commission (online). Available at: ec.europa.eu/environment/nature/info/pubs/docs/brochures/2020%20Biod%20brochure%20final%20lowres.pdf (accessed 7 November 2015).

FAO (Food and Agriculture Organization) (2001) International Treaty on Plant Genetic Resources for Food and Agriculture. FAO, Rome (online). Available at: www.planttreaty.org (accessed 7 November 2015).

Fielder, H., Brotherton, P., Hosking, J., Hopkins, J.J., Ford-Lloyd, B., *et al.* (2015a) Enhancing the conservation of crop wild relatives in England. *PLoS ONE* 10, e0130804, doi:10.1371/journal.pone.e0130804.

Fielder, H., Brotherton, P., Clitherow, J., Dulloo, E., Hosking, J., *et al.* (2015b) *Quality Standards for Genetic Reserve Conservation of Crop Wild Relatives*. University of Birmingham, Birmingham, UK.

Fielder, H., Smith, C., Ford-Lloyd, B. and Maxted, N. (2016) Enhancing the conservation of crop wild relatives in Scotland. *Journal for Nature Conservation* 29, 51–61.

Fielder, H., Jones, A., Burrows, C., Woodman, J., Ford-Lloyd, B., *et al.* (in press) Enhancing the conservation of crop wild relatives in Wales. *New Journal of Botany*. Accepted for publication.

Fielder, H., Brotherton, P., Hosking, J., Hopkins, J., Jones, A., *et al.* (in preparation a) Enhancing the conservation of crop wild relatives in the United Kingdom.

Fielder, H., Brotherton, P., Hopkins, J., Hosking, J., Ford-Lloyd, B. and Maxted, N. (in preparation b) A genetic assessment of the crop wild relatives of the Lizard Peninsula, Cornwall and its suitability to become the first UK genetic reserve.

Ford-Lloyd, B.V., Maxted, N., Kell, S.P., Mitchell, M., Scholten, M., *et al.* (2008) Establishing conservation priorities for crop wild relatives, In: Maxted, N., Dulloo, E., Ford-Lloyd, B., Kell, S.P. and Iriondo, J.M. (eds) *Crop Wild Relative Conservation and Use*. CAB International, Wallingford, UK, pp. 110–119.

Harlan, J.R. and de Wet, J.M.J. (1971) Towards a rational classification of cultivated plants. *Taxon* 20, 509–517.

Hijmans, R.J., Guarino, G. and Macathur, P. (2012) DIVA-GIS Version 7.5. Manual (online). Available at: www.diva-gis.org/documentation (accessed 7 November 2015).

Iriondo, J.M., Maxted, N., Kell, S.P., Ford-Lloyd, B.V., Lara-Romano, C., *et al.* (2012) Quality standards for genetic reserve conservation of crop wild relatives. In: Maxted, N., Dulloo, M.E., Ford-Lloyd, B.V., Frese, L., Iriondo, J.M. and Pinheiro de Carvalho, M.A.A. (eds) *Agrobiodiversity Conservation: Securing the Diversity of Crop Wild Relatives and Landraces*. CAB International, Wallingford, UK, pp. 72–77.

IUCN (International Union for Conservation of Nature) (2001) *IUCN Red List Categories and Criteria: Version 3.1*. IUCN Species Survival Commission, Gland, Switzerland, and Cambridge, UK (online). Available at: www.iucnredlist.org/static/categories_criteria_3_1 (accessed 7 November 2015).

Jarvis, S., Fielder, H., Hopkins, J., Maxted, N. and Smart, S. (2015) Distribution of crop wild relatives of conservation priority in the UK landscape. *Biological Conservation* 191, 444–451, doi:10.1016/j.biocon.2015.07.039.

Kell, S.P., Knüpffer, H., Jury, S.L., Maxted, N. and Ford-Lloyd, B.V. (2005) Catalogue of Crop Wild Relatives for Europe and the Mediterranean. Available online via the PGR Forum Crop Wild Relative Information System (CWRIS – www.pgrforum.org/cwris/cwris.asp; accessed 7 November 2015) and on CD-ROM. © University of Birmingham, UK.

Kell, S.P., Maxted, N. and Bilz, M. (2012) European crop wild relative threat assessment: knowledge gained and lessons learnt. In: Maxted, N., Dulloo, M.E., Ford-Lloyd, B.V., Frese, L., Iriondo, J.M. and Pinheiro de Carvalho, M.A.A. (eds) *Agrobiodiversity Conservation: Securing the Diversity of Crop Wild Relatives and Landraces*. CAB International, Wallingford, UK, pp. 218–242.

Magos Brehm, J., Maxted, N., Martins-Loução, M.A. and Ford-Lloyd, B.V. (2010) New approaches for establishing conservation priorities for socio-economically important plant species. *Biodiversity and Conservation* 19(9), 2715–2740.

Maxted, N. (2003) Conserving the genetic resources of crop wild relatives in European Protected Areas. *Biological Conservation* 113, 411–417.

Maxted, N. and Kell, S. (2009) *Establishment of a Network for the In Situ Conservation of Crop Wild Relatives: Status and Needs. Commission on Genetic Resources for Food and Agriculture*. FAO, Rome, 211 pp.

Maxted, N., Hawkes, J.G., Guarino, L. and Sawkins, M. (1997a) The selection of taxa for plant genetic conservation. *Genetic Resources and Crop Evolution* 44, 337–348.

Maxted, N., Ford-Lloyd, B.V. and Hawkes, J.G. (1997b) Complementary conservation strategies. In: Maxted, N., Ford-Lloyd, B.V. and Hawkes, J.G. (eds) *Plant Genetic Conservation: The In Situ Approach*. Chapman and Hall, London, pp. 545–592.

Maxted, N., Ford-Lloyd, B.V., Jury, S., Kell, S.P. and Scholten, M.A. (2006) Towards a definition of a crop wild relative. *Biodiversity and Conservation* 15(8), 2673–2685.

Maxted, N., Scholten, M.A., Codd, R. and Ford-Lloyd, B.V. (2007) Creation and use of a national inventory of crop wild relatives. *Biological Conservation* 140, 142–159.

Maxted, N., Castañeda Álvarez, N.P., Vincent, H.A. and Magos Brehm, J. (2012) Gap analysis: a tool for genetic conservation. In: Guarino, L., Ramanatha Rao, V. and Goldberg, E. (eds) *Collecting Plant Genetic Diversity: Technical Guidelines. 2011 update*. Bioversity International, Rome. Available at: cropgenebank.sgrp.cgiar.org/index.php?option=com_content&view=article&id=678 (accessed 7 November 2015).

Planta Europa (2008) *A Sustainable Future for Europe; the European Strategy for Plant Conservation 2008–2014*. Plantlife International (Salisbury, UK) and the Council of Europe (Strasbourg, France).

Rebelo, A.G., (1994) Iterative selection procedures: centres of endemism and optimal placement of reserves. In: Huntley, B.J. (ed.) *Botanical Diversity in Southern Africa*. National Botanical Institute, Pretoria, pp. 231–257.

Stace, C.A., Ellis, R.G., Kent, D.H. and McCosh, D.J. (eds) (2003) *Vice-County Census Catalogue of the Vascular Plants of Great Britain, the Isle of Man and the Channel Islands*. Botanical Society of the British Isles, London.

# 17 Optimized Site Selection for the *In Situ* Conservation of Forage CWR: A Combination of Community- and Genetic-level Perspectives

**M.L. Rubio Teso,* K. Kinoshita Kinoshita and J.M. Iriondo**
*Universidad Rey Juan Carlos, Departamento de Biología, Geología, Física y Química Inorgánica, Madrid, Spain*

## 17.1 Introduction

Climate change has been studied by now from many points of view: macro- and microecological impacts, changes to biodiversity at all levels, economical variations, sanitary threats, etc., all of them crucial to humankind. The impacts of global change on food security and crop production are topics that concern governments and scientists all over the world. The impact of climate change on crop production and yields has been studied for some time now and an increase in food insecurity, especially in countries with ongoing problems of undernourishment and hunger, has been forecast (Rosenzweig and Parry, 1994; Wheeler and Brown, 2013). Threats to food security include an increase in the frequency of extreme climate events (such as frosts, high intensity of rains and droughts) and the incidence of new weeds, pests and pathogens that can decrease crop yields dramatically. The development of new adaptations and the improvement of the adjustment capacity of crops will be the key factor in the next century for safeguarding food security (Lozte-Campen, 2011). According to the last report of the Intergovernmental Panel on Climatic Change (IPCC) in 2014, the benefits of adapting cultivated plants could make it possible to impede yield reductions by 10–15%, but this capacity would be surpassed if temperature increases at low latitudes were higher than 3°C. Several steps to overcome the challenge of safeguarding food security by adapting crops have taken place over the past years, especially in terms of breeding research, looking for crops resistant to biotic and abiotic stresses and the implementation of new breeding techniques (see Lusser *et al.*, 2011). Plant breeding can be achieved successfully when a wide range of genetic diversity related to the targeted crop is available, so currently there is global concern about the importance of preserving plant genetic resources and their diversity.

Crop wild relatives (CWR) are wild species closely related to crops to which they can contribute genetic material, but, unlike cultivated plants, they have not been domesticated (Heywood *et al.*, 2007). The genetic diversity contained in CWR is far greater than the diversity found in crops (Lidder and Sonnino, 2012), so their contribution in widening crop genetic diversity to sustain crop breeding is essential to succeed in

*Corresponding author; e-mail: marialuisa.rubio@urjc.es

 

securing food supply. Thus, CWR constitute an important source of genes for crop breeding and have been shown to be useful in broadening the genepools of many crops (Dwivedi *et al.*, 2007). There are increasing examples of this interest in selecting, prioritizing and conserving those CWR that can contribute useful traits to crops. This is, for example, evidenced by the creation of national inventories and checklists of CWR (Maxted *et al.*, 2007; Barazani *et al.*, 2008; Magos Brehm *et al.*, 2008; Scholten *et al.*, 2011; Berlingeri and Crespo, 2012; Fitzgerald, 2013; Khoury *et al.*, 2013; Rubio *et al.*, 2013; Vincent *et al.*, 2013; Panella *et al.*, 2014), as well as the development of plans to establish CWR genetic reserves (Fielder *et al.*, 2012; Pinheiro de Carvalho *et al.*, 2012). Thereby, the preservation of CWR has lately become the centre of attention for plant genetic resource (PGR) scientists. Maxted (2003) pointed out the neglected conservation of CWR and the increment of threats menacing them, causing genetic erosion of their valuable diversity. The *ex situ* and *in situ* conservation of CWR are complementary actions that need to be implemented to ensure proper conservation of these resources and halt their loss. According to this, not only the diversity of CWR species but also their genetic diversity must be the targets when planning conservation actions. Consequently, the implementation of effective measures that ensure the conservation of both aspects becomes not only a priority but also a challenge.

## 17.2 Strategies for CWR Conservation

### 17.2.1 Selection and prioritization of species: methodologies for conservation

One of the objectives of CWR conservation strategies, especially at national levels, should be the long-term preservation of the target species, protecting their taxonomic diversity as much as their genetic diversity (Maxted *et al.*, 2013). For the development of national strategies, a first step involves the creation of a checklist of resources in the country concerned. Second, a priority list of CWR is produced based on a number of criteria, such as threat status, endemicity, value of the crop, etc. An inventory list is then produced where additional associated information is compiled. An *ex situ* and *in situ* conservation assessment of the priority list is recommended to set the conservation needs clearly and to propose efficient conservation objectives.

The establishment of effective *in situ* conservation goals usually involves the identification of suitable hotspot areas with high concentration of CWR species where conservation actions can be implemented. A species richness analysis can be meaningless on its own for selecting efficient conservation areas. A complementarity analysis, which allows defining a network that would cover all the species in the minimum number of sites, is a preferred approach (Rebelo, 1994). A species-oriented method has been proposed for identifying areas for conservation and has been used in the formulation of CWR conservation strategies for the UK (Maxted *et al.*, 2007), Spain (Rubio *et al.*, 2013) or Finland (Fitzgerald, 2013). Nevertheless, this method does not take into account the genetic diversity contained in each species or the special adaptations species may have in certain areas due to particular environments. Thus, it is possible that the genetic diversity is not preserved in an efficient manner. At present, it is difficult to characterize the genetic diversity of all populations, although attempts at using neutral markers have traditionally been used to estimate genetic diversity. However, these neutral markers do not reflect the divergent selection patterns due, for example, to environmental pressures (Kirk and Freeland, 2011), and thus adaptive traits of interest for breeding may not be reflected. There is a dire need to develop novel techniques that allow us to preserve not only a broad number of species but also all their genetic diversity, especially that of adaptive value. In the near future, massive sequencing of target taxa would help in identifying the genes involved in adaptive traits of interest and to characterize this genetic diversity in the target populations directly.

## 17.3 Effective Conservation of the Genetic Diversity of CWR

### 17.3.1 Use of ELC maps

The estimation of genetic diversity contained in plant populations has been approached with the design of maps prepared to capture the specific environmental characteristics of the area under study (Parra-Quijano *et al.*, 2012a). These environmental land characterization (ELC) maps are developed for a particular area using climatic, edaphic and geophysical variables. They can be used as a proxy to gauge the genetic diversity of wild populations of CWR (Maxted *et al.*, 2012) and have been shown to be useful in improving the representativeness of seed collections in germplasm banks (Parra-Quijano *et al.*, 2012b). Thus, ELC maps can be a valuable tool to assess the potential genetic diversity of adaptive value. ELC maps have been used successfully in the studies for the development of a national strategy of CWR conservation in Spain (Rubio *et al.*, 2013). In this work, apart from the traditional gap analysis, all known occurrences of the species under study were overlaid with the ELC map developed for the Iberian Peninsula and Balearic Islands (Parra-Quijano *et al.*, 2012a). This analysis gave an idea of how many different environments hosted the prioritized Spanish CWR species. The concept of local adaptation (Williams, 1966) indicates that genotypes from a given environment will have better fitness in their own habitat than in other habitats, as well as better fitness in their habitat than alien genotypes. Thus, it is expected that various patterns of adaptation will be observed in different habitats. The addition of ecogeographical information to existing distribution data can be helpful in identifying the potential differences in some desirable traits, such as drought or frost resistance.

### 17.3.2 Target conservation units

The use of ELC maps together with the spatial distribution information for a given species allows the creation of a new target conservation unit. This conservation unit consists of one population that represents all populations of a species occurring in a particular ecogeographic area. Thus, using this conservation unit, if a particular CWR species is distributed over 14 different ecogeographic areas in a country, the target would be to conserve one population of this species from each of the 14 ecogeographical areas. Assuming that the different ecogeographic areas generate different types of genetic adaptation through divergent selection, the use of these conservation units would maximize the genetic diversity of the adaptive value of the species. In addition to this, the use of these conservation units could make it possible to select specific locations for the establishment of genetic reserves in areas with potential special adaptation patterns.

## 17.4 Effective *In Situ* Conservation of Multiple CWR Taxa

### 17.4.1 Phytosociological associations approach

The increase in the number of target conservation units can make conservation actions difficult to design and manage. A better approach would be to manage together all the CWR species that live in the same plant community. The use of phytosociological associations provides the opportunity to work with various species at a time.

The study of plant communities and the repetitive patterns of species aggregation have attracted the attention of plant ecologists and geobotanists since the beginning of the 20th century. The study of plant communities, co-occurrence of species, their relationships and environmental dependence is known as phytosociology (Braun Blanquet, 1979). Phytosociological classification – known as syntaxonomy – groups plant species according to the fidelity of the species to the community and is based on inventories or *relevés* listing all plant taxa occurring in a given area. The comparison of these inventories and repetition patterns of the presence of some species together allows phytosociologists to establish community relationships from co-occurrences and design the basic syntaxonomical unit – the association.

### 17.4.2 Information available and distribution data

The long tradition of phytosociology in many European countries has generated thousands of *relevés* that are dispersed in databases across Europe. Phytosociological databases have been developed and are also available for other non-European countries, as shown in the Global Index of Vegetation-Plot Databases (Dengler *et al.*, 2011), which registered 204 databases with 3,011,537 vegetation plots (www.givd.info/559.html, last accessed 18 September 2014). This large amount of data on European vegetation is difficult to manage (Peet and Roberts,

2013); thus, some tools, such as the TURBOVEG software (Hennekens and Schaminée, 2001) or initiatives such as the European Vegetation Archive (EVA) project or the Braun-Blanquet project (Jiménez-Alfaro *et al.*, 2013), have been developed to deal with the large amount of information.

## 17.5 Forage Use Category Group as an Example of Targeting Phytosociological Communities with Economic Importance and Use

Although the phytosociological approach has been developed for the study of wild plants and is not focused on specific groups of use, it can, nonetheless, be applied to CWR species, especially for CWR taxa classified in forage categories, as many of them are known to grow together in natural plant communities. Thus, the prioritized CWR forage species in the Spanish National Inventory provided a good case study to test the utility of using phytosociological associations in the search for CWR plant communities that could be used as conservation units.

The reference database for phytosociological associations in Spain is hosted at the Iberian and Macaronesian Plant Vegetation System (SIVIM) website (www.sivim.info/sivi) developed under the Iberiveg project funded by the National Ministry of Education and Science. In this database, more than 145,000 georeferenced inventories are available and searches by taxon or syntaxon can be performed.

Starting from the list of prioritized forage species in the Spanish National Inventory of CWR, a search for associations containing these species was carried out. Following a similar process to complementarity analysis (Rebelo, 1994), associations containing at least two prioritized forage species with high fidelity to the association were selected until the maximum number of taxa were covered using the minimum number of associations. Then, distribution data for the associations were downloaded (10 × 10 km accuracy) and overlaid with distribution data of the targeted species (1 × 1 km accuracy) obtained during the development of the national strategy of CWR conservation in Spain in the EU-funded PGR Secure project. This allowed the selection of those associations' occurrences only where the presence of the taxa could be confirmed by data from individual species occurrence of populations, provided that species with high fidelity to the association could be found in a particular plant community. In conclusion, this provided a set of sites where phytosociological associations containing the priority CWR forage species occurred.

In order to select the sites of plant communities that would better represent the genetic diversity of the adaptive value of the priority CWR species, these distribution data were overlaid with the ELC map developed for Spain (Parra-Quijano *et al.*, 2012a). With this filtered distribution data, a new conservation unit, composed of a phytosociological association in a particular ecogeographic area (PSA-EA), was created. Subsequently, a complementarity analysis for these target conservation units was performed using a 10 × 10 km grid. Finally, a gap analysis of the selected sites was carried out with the layer of Natura 2000 Sites of Community Interest, using DIVA-GIS, v 7.5.

The complementarity analysis resulted in the selection of places encompassing the maximum number of PSA-EAs in the minimum number of sites, and the gap analysis pointed out those sites already included in the Natura 2000 network and thus subject to passive protection. The results obtained with this approach (Rubio *et al.*, 2015, unpublished results) can be used to propose areas for the establishment of genetic reserves of CWR species that co-occur in certain plant communities, assuring that not only a high number of species would be protected in a reduced number of sites but also that the genetic diversity of adaptive value would potentially be well represented.

## 17.6 General Considerations and Discussion. Value of the Phytosociological Association–Ecogeographic Area Conservation Unit

The first and main advantage of the use of the phytosociological association in the creation of the target conservation unit is that it provides information on various target taxa at a time, which is considered by plant conservation managers as a desirable characteristic when planning conservation actions. Thus, it reduces the human and economic resources required for the establishment and monitoring of new genetic reserves. Furthermore, the fact that phytosociological associations have been used to delimit many of the habitats in the habitats directive (Council Directive 92/43/EEC of 21 May 1992, eur-lex.europa.eu/legal-content/EN/TXT/?uri=CELEX:01992L0043-20070101) can encourage and reinforce the discussions with conservation managers, providing an added value to the habitats containing the priority CWR taxa. The most innovative step in this approach is the combination of the ecogeographic area with the phytosociological associations to generate the PSA-EA conservation unit. The PSA-EA not only has the same advantages of the PSA itself, but also the added benefit of assigning different patterns of genetic diversity and adaptations to the different environments in which a PSA has evolved. This should be helpful in the selection process in defining the best options for the establishment of genetic reserves of CWR. Besides, the joint use of the PSA-EA target conservation units can even assist if specific genetic reserves want to be established for a particular use, enhancing the access of breeders to their traits of interest in CWR. For example, this could be the case for forage genetic reserves in ecogeographical areas that correspond to arid areas in which CWR would be expected to show higher probabilities of drought resistance, and they would thus be the best targets to start the search for drought traits for plant breeding.

On the other hand, there are also some caveats that may lessen the advantages and possibilities of this approach. The obtaining of good quality georeferencing data, both for the species and the phytosociological associations, is the first obstacle to overcome. In addition to the lack of good quality geographic information for many CWR species, the accuracy of the geographic phytosociological data is usually coarser than the data that may be obtained for particular species and adds uncertainty to the process. Furthermore, uncertainty is increased due to the contingency of the presence of the species in a given PSA, even if the species is highly faithful to the association. This makes crossing the species distribution data with the PSA distribution data unavoidable, or checking the inventories attached to a given association in a given locality. Finally, another issue to consider is the possible bias that might be found depending on the CWR group of interest. A forage species group provided a good example of the advantages of using PSA-EA as target units for conservation, as these species are known to co-occur in the same plant communities. But, are there other groups of interest for breeders that fit with the requirements to use the PSA-EA approach? Cereals or legumes (as independent groups combining the forage and food categories) or even mixed categories of interest could provide good candidates to test this approach and propose areas for the establishment of genetic reserves.

The use of the PSA-EA target conservation unit approach for the optimization of site selection for genetic reserves finds an advantageous scenario in Spain, due to the existing rich flora and the long tradition of phytosociological studies, making available vast information on syntaxonomy and its distribution throughout the country. On the other hand, the ELC map for the Iberian Peninsula and Balearic Islands has already shown its value in optimizing plant collecting missions in Spain (Parra-Quijano *et al.*, 2012b). Although this ELC map can be considered as a generalist map and may not optimally represent the different environments shaping the genetic diversity of particular species, specifications for sets of species can now be undertaken with the new tools for the creation of ecogeographical maps (*ELCmapas*) developed in 2013 (Parra-Quijano *et al.*, 2014).

Considering the urgency for the development of innovative approaches to enhance the conservation of plant genetic resources, it is time to merge the tradition of European phytosociological studies, as a vast source of information, and the innovation of informatics and GIS, to advance the search for the best methodologies aiming at the preservation of CWR. The phytosociological–ecogeographical target conservation unit approach (PSA-EA) discussed is a worthy attempt in this direction.

## Acknowledgements

This research was funded by a grant from the EU Seventh Framework Programme for the PGR Secure project (Grant Agreement No 266394).

## References

Barazani, O., Perevolotsky, A. and Hadas, R. (2008) A problem of the rich: prioritizing local plant genetic resources for *ex situ* conservation in Israel. *Biological Conservation* 141, 596–600.

Berlingeri, C. and Crespo, M.B. (2012) Inventory of related wild species of priority crops in Venezuela. *Genetic Resources and Crop Evolution* 59(5), 655–681.

Braun Blanquet, J. (1979) *Fitosociología. Bases para el estudio de las comunidades vegetales*. H. Blume ediciones, Madrid.

Dengler, J., Jansen, F., Glöckler, F., Peet, R.K., De Cáceres, M., *et al.* (2011) The Global Index of Vegetation-Plot Databases (GIVD): a new resource for vegetation science. *Journal of Vegetation Science* 22(4), 582–597.

Dwivedi, S.L., Crouch, J.H., Mackill, D.J., Xu, Y., Blair, M.W., *et al.* (2007) The molecularization of public sector crop breeding: progress, problems, and prospects. *Advances in Agronomy* 95, 163–318.

Fielder, H., Hopkins, J., Smith, C., Kell, S., Ford-Lloyd, B., *et al.* (2012) UK wild species to underpin food security: species selection, genetic reserves and targeted collection. *Crop Wild Relative* 8, 24–27. University of Birmingham, Birmingham, UK.

Fitzgerald, H. (2013) The National Crop Wild Relative Strategy Report for Finland. MTT Report 121. Available at: www.mtt.fi/mttraportti/pdf/mttraportti121.pdf (accessed 21 November 2015).

Hennekens, S.M. and Schaminée, J.H.J. (2001) TURBOVEG, a comprehensive data base management system for vegetation data. *Journal of Vegetation Science* 12, 589–591.

Heywood, V., Casas, A., Ford-Lloyd, B., Kell, S. and Maxted, N. (2007) Conservation and sustainable use of crop wild relatives. *Agriculture, Ecosystems and Environment* 21, 245–255.

Jiménez-Alfaro, B., Apostolova, I., Čarni, A., Chytrý, M., Csiky, J., *et al.* (2013) Unifying and analyzing vegetation-plot databases in Europe: the European Vegetation Archive (EVA) and the Braun-Blanquet project. In: Walker, D.A., Breen, A.L., Raynolds, M.K. and Walker, M.D. (eds) *Arctic Vegetation Archive (AVA)* Workshop, Krakow, Poland, 14–16 April 2013. CAFF Proceedings Report No 10, Akureyri, Iceland.

Khoury, C.K., Greene, S., Wiersema, J., Maxted, N., Jarvis, A., *et al.* (2013) An inventory of crop wild relatives of the United States. *Crop Science* 53, 1496–1508.

Kirk, H. and Freeland, J.R. (2011) Applications and implications of neutral *vs* non-neutral markers in molecular ecology. *International Journal of Molecular Sciences* 12, 3966–3988.

Lidder, P. and Sonnino, A. (2012) Biotechnologies for the management of genetic resources for food and agriculture. In: Goodwin, S.F., Friedmann, T. and Dunlap, J.C. (eds) *Advances in Genetics*, Vol 78. Elsiever, Philadelphia, Pennsylvania, pp. 1–112.

Lozte-Campen, H. (2011) Climate change, population growth, and crop production: an overview. In: Yadav, S.S., Redden, R.J., Hatfield, J.L., Lozte-Campen, H., Hall, A.E., *et al.* (eds) *Crop Adaptation to Climate Change*. Wiley-Blackwell, West Sussex, UK, pp. 1–11.

Lusser, M., Parisi, C., Plan, D. and Rodríguez-Cerezo, E. (2011) *New Plant Breeding Techniques. State-of-the-Art and Prospects for Commercial Development*. Publications Office of the European Union, Luxembourg.

Magos Brehm, J., Maxted, N., Ford-Lloyd, B.V. and Martins-Louçao, M.A. (2008) National inventories of crop wild relatives and wild harvested plants: case-study for Portugal. *Genetic Resources and Crop Evolution* 55, 779–796.

Maxted, N. (2003) Conserving the genetic resources of crop wild relatives in European Protected Areas. *Biological Conservation* 113, 411–417.

Maxted, N., Scholten, M., Codd, R. and Ford-Lloyd, B. (2007) Creation and use of a national inventory of crop wild relatives. *Biological Conservation* 140, 142–159.

Maxted, N., Kell, S., Ford-Lloyd, B., Dulloo, E. and Toledo, Á. (2012) Toward the systematic conservation of global crop wild relative diversity. *Crop Science* 52, 774–785.

Maxted, N., Magos Brehm, J. and Kell, S. (2013) *Resource Book for Preparation of National Conservation Plans for Crop Wild Relatives and Landraces*. Commission on Genetic Resources for Food and Agriculture, FAO, Rome.

Panella, L., Landucci, S., Torricelli, R., Gigante, D., Donnini, D., *et al.* (2014) The National Crop Wild Relative Strategy for Italy: First Steps To Be Taken. PGR Secure project. Available at: www.pgrsecure.bham.ac.uk/sites/default/files/documents/public/National_CWR_Conservation_Strategy_Italy.pdf (accessed 21 November 2015).

Parra-Quijano, M., Iriondo, J.M. and Torres, E. (2012a) Ecogeographical land characterization maps as a tool for assessing plant adaptation and their implications in agrobiodiversity studies. *Genetic Resources and Crop Evolution* 59, 205–217.
Parra-Quijano, M., Iriondo, J.M. and Torres, E. (2012b) Improving representativeness of genebank collections through species distribution models, gap analysis and ecogeographical maps. *Biodiversity Conservation* 21, 79–96.
Parra-Quijano, M., Torres Lamas, E., Iriondo Alegría, J.M. and López, F. (2014) *Capfitogen Tools. Programme to Strengthen National Plant Genetic Resource Capacities in Latin America*. FAO, Rome.
Peet, R.K. and Roberts, D.W. (2013) Classification of natural and semi-natural vegetation. In: van der Maarel, E. and Franklin, J. (eds) *Vegetation Ecology*, Second edn. Oxford University Press, Oxford, UK, pp. 27–70.
Pinheiro de Carvalho, M.Â.A., Nóbrega, H., Freitas, G., Fontinha, S. and Frese, L. (2012) Towards the establishment of a genetic reserve for *Beta patula* Aiton. In: Maxted, N., Dulloo, M.E., Ford-Lloyd, B.V., Frese, L., Iriondo, J.M., *et al.* (eds) *Agrobiodiversity Conservation: Securing the Diversity of Crop Wild Relatives and Landraces*. CAB International, Wallingford, UK, pp. 36–44.
Rebelo, A.G. (1994) Iterative selection procedures: centres of endemism and optimal placement of reserves. In: Huntley, B.J. (ed.) *Botanical Diversity in Southern Africa*. National Botanical Institute, Pretoria, pp. 231–257.
Rosenzweig, C. and Parry, M.L. (1994) Potential impact of climatic change on world food supply. *Nature* 367, 133–138.
Rubio, M.L., Iriondo, J.M., Parra, M. and Torres, E. (2013) National Strategy for the Conservation of Crop Wild Relatives of Spain. PGR Secure project. Available at: www.pgrsecure.bham.ac.uk/sites/default/files/documents/public/National_CWR_Conservation_Strategy_Spain.pdf (accessed 21 November 2015).
Scholten, M., Maxted, N. and Ford-Lloyd, B. (2011) UK national inventory of plant genetic resources for food and agriculture. Master thesis, School of Biosciences of the University of Birmingham, Birmingham, UK. Available at: www.brockwell-bake.org.uk/pics/UK_inventory.pdf (accessed 21 November 2015).
Vincent, H., Wiersema, J., Kell, S., Fielder, H., Dobbie, S., *et al.* (2013) A prioritized crop wild relative inventory to help underpin global food security. *Biological Conservation* 167, 265–275.
Wheeler, T. and Brown, J. (2013) Climate change impacts on global food security. *Science* 341, 508–513.
Williams, G.C. (1966) *Adaptation and Natural Selection*. Princeton University Press, Princeton, New Jersey.

# 18 Developing a Crop Wild Relative Conservation Strategy for Finland

H. Fitzgerald,[1]* H. Korpelainen[2] and M. Veteläinen[3]
[1]Finnish Museum of Natural History, University of Helsinki, Helsinki, Finland;
[2]Department of Agricultural Sciences, University of Helsinki, Helsinki, Finland;
[3]Boreal Plant Breeding Ltd, Jokioinen, Finland

## 18.1 Introduction

Crop wild relatives (CWR) are wild plant species that are related to cultivated plants. CWR species have the potential to enhance agricultural production by allowing crops to survive through plant breeding in the new environmental conditions resulting from climate change and help in providing future food security. Since CWR are valuable wild species, mostly not included in conservation programmes and themselves often threatened or growing in threatened habitats, they require urgent research and conservation.

This article discusses the process of preparing the Finnish crop wild relative conservation strategy, which was created as part of the EU Seventh Framework funded project: 'PGR Secure – Novel characterization of crop wild relative and landrace resources as a basis for improved crop breeding'. The article describes the methodology and results of the Finnish national crop wild relative checklist, prioritized list and inventory as well as the *in situ* and *ex situ* conservation gap analysis. When developing a CWR conservation strategy for Finland, the aim was to give practical suggestions for the conservation of the maximum taxonomic and genetic diversity of Finland's CWR through *in situ* and *ex situ* conservation.

## 18.2 Methodology

The methodology used in developing the Finnish CWR strategy was adapted from the method described by Maxted *et al.* (2011). This included the compilation of the CWR checklist and the collection of additional data on ecogeography, threats, uses and value of the species to enable the prioritization of the checklist. A conservation gap analysis was undertaken for the prioritized list. With the complementarity analysis, key areas were identified for *in situ* conservation. These were suggested to be established as genetic reserves in the future. The state of the present *ex situ* conservation of priority species was assessed and gaps were identified. Suggestions have been made for the collection of genetically representative samples of those CWR populations that are not sufficiently conserved at present.

### 18.2.1 CWR checklist and its prioritization

Finland's vascular plant diversity includes around 2500 taxa (Hämet-Ahti *et al.*, 1998; Lampinen and Lahti, 2010). To start with, a draft CWR checklist was obtained from PGR

*Corresponding author; e-mail: heli.fitzgerald@helsinki.fi

 *Enhancing Crop Genepool Use: Capturing Wild Relative and Landrace Diversity for Crop Improvement* (eds N. Maxted, M. Ehsan Dulloo and B.V. Ford-Lloyd)

Forum 'CWR Catalogue for Europe and the Mediterranean' (Kell *et al.*, 2005, 2008; Euro+Med, 2006), which included all the wild taxa that belong to the same genus as crop plants, such as food, fodder, forage, forestry, ornamental, medicinal and aromatic species. The draft checklist was then harmonized taxonomically with the 'Field Flora of Finland' (Hämet-Ahti *et al.*, 1998, 2005). As the checklist contained a large percentage of Finland's flora, it was prioritized to enable conservation planning. Data were collected on threat, use and value of the species.

The threat data consisted of the following categories: red listed species (Rassi *et al.*, 2010); species protected under national conservation decree (Anon., 1997); species listed in the EU Habitats Directive Annexes II, IV and V (EU, 1992); endemism in Europe (Bilz *et al.*, 2011); endemism in Finland (Hämet-Ahti *et al.*, 1998, 2005); and the list of species with a minimum of 20% of their European distribution in Finland, the 'OEK- species' (Rassi *et al.*, 2001). The value data consisted of the yield of CWR-related main crops in Finland in 2010 (Tike, 2010), yield of CWR-related herb crops in Finland 2010 (Galambosi, 2007) and the high priority human food crops in Europe (Kell *et al.*, 2012) based on the quantity (FAOSTAT, 2009) and economic value (Eurostat, 1995–2009) of their production. The use data included: Annex I species of the International Treaty on Plant Genetic Resources for Food and Agriculture (FAO, 2001); Nordic mandate species of root, oil and fibre plants, vegetables, fruit and berries, cereals and forage plants (NordGen thematic working groups); taxa listed as being used as medicinal, pharmaceutical, food, forage or fodder plants (Hanelt and IPK Gatersleben, 2001); and the species in genepools 1a, 1b and 2 (Harlan and de Wet, 1971) or taxon groups 1b, 2 and 3 (Maxted *et al.*, 2006) from The Harlan and de Wet Crop Wild Relative Inventory (Vincent *et al.*, 2013).

The prioritization process was started by selecting indigenous and archaeophyte taxa for prioritization from the checklist. Neophytes, which were introduced to Finland after the 17th century, were removed. However, it was recognized that neophyte species in Finland included many important crop wild relatives that might have formed adaptations to Nordic climate conditions and could therefore contain useful genetic material for breeding. The checklist was then prioritized on the three main criteria – use, value and threat – and 12 subcriteria (Fig. 18.1.) in the following stages: the vulnerable, endangered or critically endangered taxa were selected; taxa classified as near threatened were selected if they had a minimum of one point in the use criteria; and the non-threatened taxa having more than three points in the 12 subcriteria were selected into the priority list. Finally, the non-threatened subspecies were removed from the list.

### 18.2.2 Gap analysis

There were no conservation efforts identified particularly for CWR species in Finland, except that the threatened CWR taxa were included in threatened species conservation programmes. Several threatened CWR priority species have a conservation action plan by the Finnish Environment Institute (SYKE, 2010). Some threatened CWR taxa also have a special *in situ* conservation, management and monitoring plan under the extended national responsibility of the Natural Heritage Services of Metsähallitus (Metsähallitus, 2013). Additionally, the Finnish Forest Research Institute Metla has a programme to conserve the genetic resources of forest trees, and two of the priority CWR are conserved *in situ* in the Finnish Forest Research Institute's conservation areas (Metla, 2012).

The methodology of using gap analysis to evaluate CWR taxonomic and genetic diversity and to develop future strategies for their genetic conservation (Maxted *et al.*, 2008) was applied to Finnish CWR diversity. This included the following steps: circumscription of target taxon and target area; assessment of natural diversity (taxonomic, genetic, ecogeographic and threat assessment); assessment of current *in situ* and *ex situ* conservation; and establishing conservation priorities for the future.

The aim of the *in situ* gap analysis was to identify key locations of CWR taxa in Finland. First, the present conservation status of CWR taxa was analysed, followed by the identification of key areas containing CWR taxa by complementarity analysis. With the complementarity analysis, it was possible to find the minimum number of sites with the maximum number of CWR taxa, therefore conserving as many taxa as

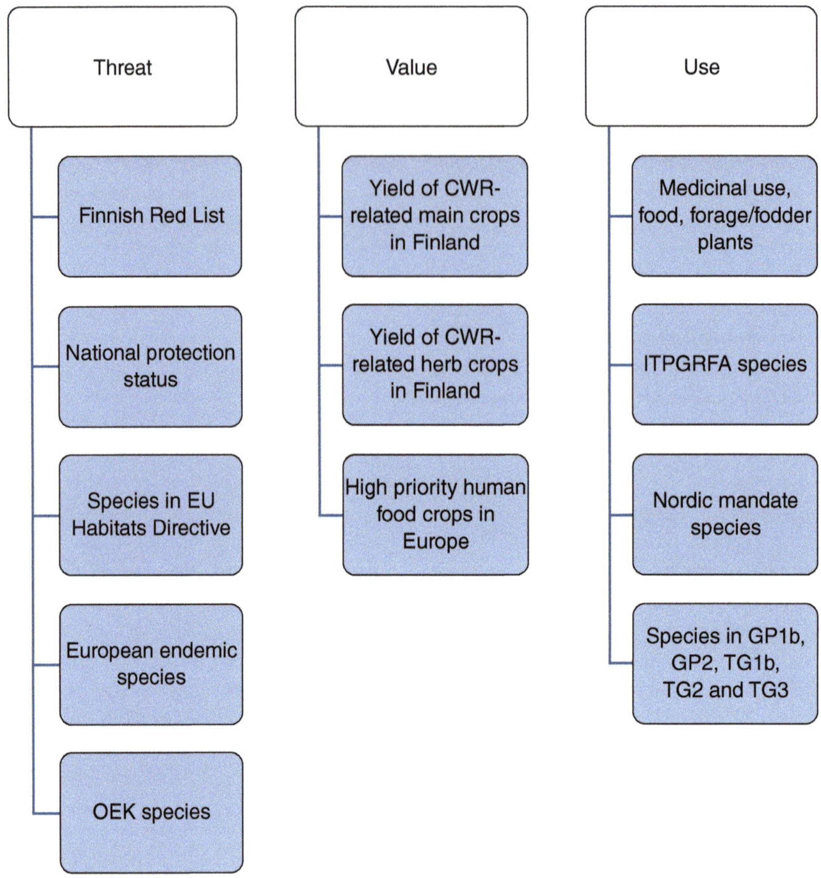


**Fig. 18.1.** Finnish CWR prioritization criteria and subcriteria. *Notes*: OEK species = Finnish 'responsibility species' that have a minimum of 20% of their European distribution within Finland; ITPGRFA = International Treaty on Plant Genetic Resources for Food and Agriculture; GP = genepool concept; TG = taxon group concept.

possible in a minimum number of locations. For the *in situ* gap analysis, the distribution data from the 'Atlas of the Distribution of Vascular Plants in Finland' (Lampinen and Lahti, 2010) were used on a 10 × 10 km grid level. Only the native and archaeophyte locations of threatened species were selected, as the neophyte locations were suspected to be of genetically alien material. Neophyte and extinct locations were therefore removed from the analysis. Threatened species' distribution data from 1990 onwards and non-threatened species' data from 1970 onwards were used. Additional mapping data for the *in situ* analysis were obtained from Maanmittauslaitos – open data (Maanmittauslaitos, 2013), OIVA – Ympäristö-ja paikkatietopalvelu asiantuntijoille (SYKE, 2012) and the Regional Government of Åland (Ahonen, Helsinki, 2013, personal communication).

The first 10 × 10 km grid site with the maximum number of CWR taxa was found by transferring the distribution data from Excel to ArcMap. Those taxa present on the first site were removed from the analysis, and the second site was found with the second highest concentration of CWR taxa in Finland. This process of complementarity analysis determined the CWR hotspots for Finland. The distribution maps were then overlaid with the protected area maps to find out whether CWR hotspots would fall in conservation areas. Additionally, conservation status data were looked into, giving information on what percentage of the CWR priority species populations were present in protected areas (Kemppainen and Eeronheimo, 2008, and Eeronheimo, Helsinki, 2013, personal communication).

The *ex situ* gap analysis was carried out by finding out which CWR taxa were already in conservation and which were not sufficiently conserved and therefore in need of conservation action. Current *ex situ* collections of Finnish native wild plant species were found to be mostly in botanic gardens, with some CWR taxa also conserved in the Nordic Genetic Resource Center, NordGen, collections. A newly established seedbank for threatened native vascular plant species in the Finnish Museum of Natural History includes several threatened CWR taxa in its collection lists. The data for the analysis were obtained directly from the institutions holding the *ex situ* collections, such as the Botanic Gardens of Helsinki, Turku and Oulu, MTT Agrifood Research Finland, Finnish Forest Research Institute and NordGen, with additional threatened species data from the results of the 'Vulnerability Assessment of Ecosystem Services for Climate Change Impacts and Adaptation' project (Hyvärinen *et al.*, 2011; Miranto *et al.*, 2012). Only the collections accompanied with a sufficiently accurate data level were considered to be in *ex situ* conservation following the description of origin data classes and genetic intactness classes by Miranto (2005). Additionally, only those taxa collected from five or more populations were accepted to be conserved in *ex situ* collections.

## 18.3 Results

### 18.3.1 Checklist and national inventory

Through the taxonomical harmonization of the draft checklist, the number of wild species in the same genus as cultivated crops in Finland was found to be 1905 taxa (Fitzgerald, *et al.*, 2013). Therefore, approximately 75% of the national flora consists of CWR taxa by a broad CWR definition (Maxted *et al.*, 2006). Most numbers of the priority species are found in the following families: Fabaceae, Rosaceae, Cyperaceae, Poaceae, Asteraceae, Brassicaceae and Caryophyllaceae. Important CWR genera for Finland include the ones identified as the wild relatives of '18 high priority human food crops/crop groups of Europe' (Kell *et al.*, 2012) and the ones listed as the 'priority human and animal food crops in the ITPGRFA' (FAO, 2001). These genera are therefore considered to be of particular importance: *Malus* (apples), *Fragaria* (strawberries), *Lactuca* (lettuce), *Lathyrus* (grass peas), *Elymus* (wheat complex), *Vicia* (faba bean/vetch), *Allium* (onions), *Agrostis*, *Festuca*, *Phleum*, *Poa* (grass forages), *Trifolium* (legume forages) and *Atriplex* and *Salsola* (other forages). In addition to those, the following CWR genera are important in Finland: *Ribes*, *Rubus*, *Hypericum* and *Urtica*, according to the yield of CWR-related main crops in Finland (Tike, 2010) and the yield of CWR-related herb crops in Finland (B. Galambosi, TEAR, Helsinki, 2011, personal communication).

The checklist prioritization resulted in a national inventory of 209 taxa (Fitzgerald, 2013). The ecogeographic data compiled from Rassi *et al.* (2010) and Hämet-Ahti *et al.* (1998, 2005) show the main habitats of priority species and the main threats to CWR diversity in Finland. Nearly one-quarter of the priority species are found growing on shoreline habitats such as the Baltic Sea shore, inland lake shores or riverbanks. Another quarter are found in rural biotopes and cultural habitats. The remainder of the CWR species grow in forest, mire, rocky, alpine and aquatic habitats. One-third of the species are threatened by the overgrowing of meadows and other open habitats. One-quarter are threatened by construction on land or of waterways. The remainder of the threats to CWR species includes peatland drainage and peat harvesting, forest management actions, chemical disturbances and random factors.

As a conclusion of the analysis of the ecogeographic data collected along the prioritization process, the Finnish CWR taxa requiring urgent conservation are those critically endangered taxa that are identified as having potential use and value – *Anthyllis vulneraria* ssp. *polyphylla*, *Artemisia campestris* ssp. *bottnica*, *Hypericum montanum*, *Pimpinella major*, *Rosa canina* and *Sorbus meinichii* – and those near threatened taxa identified as having potential use – *Malus sylvestris* and *Fragaria viridis*. It is also important to secure the genetic resources of non-threatened CWR taxa, particularly those identified as having potential use value, such as: *Alisma wahlenbergii*, *Allium schoenoprasum*, *Fragaria vesca*,

*Lathyrus sylvestris*, *Ribes nigrum*, *Ribes spicatum*, *Vicia cracca*, *Vicia hirsuta*, *Vicia sepium*, *Vicia tertrasperma* and *Vicia villosa*. Additionally, the wild relatives of the genera *Phleum* and *Fragaria* have value, due to their relevance for current plant breeding programmes in Finland.

### 18.2.2 Gap analysis

The *in situ* complementarity gap analysis results indicate that CWR species-rich areas in Finland are five sites that are distributed across the country. All five sites have a relatively high number of CWR taxa, but beyond these sites, the CWR diversity not found on the first five sites falls to less than five CWR taxa per 10 × 10 km grid square. The five hotspots are situated in: the Åland Islands, a large autonomous archipelago on the south-west coast of Finland; the Hanko Peninsula, the southernmost point of mainland Finland; the Oulanka area, in the middle-east region near the Russian border; and the Saana and Malla fells, in the far north-western corner of Finnish Lapland near the Swedish and Norwegian border (Fig. 18.2.). Over 60% of the Finnish CWR priority species are found in these five hotspots, making them ideal areas for genetic reserve conservation. Overall, half of the priority taxa analysed, 101 taxa out of 202, are found on the Åland Islands. This archipelago region is the most CWR species-rich area in Finland.

Due to mainland Finland and Åland having separate environmental legislation and administration, a second complementarity analysis was undertaken by treating mainland Finland and Åland data as two separate data sets to find out how the CWR hotspots would be distributed. The result was that the sites within Åland remained the same, but for mainland Finland two new sites were discovered in addition to the three sites found earlier (Table 18.1). These were found along the southern coastline, several islands in front of Helsinki and the Houtskari Islands in the Turku Archipelago. These sites hosted partially the same species as the sites in Åland. If the CWR diversity was looked at separately, the two hotspot sites in Åland would conserve 78% of the priority CWR found in Åland, and the five hotspots in mainland Finland would conserve approximately 60% of the CWR found in mainland Finland. However, some of the Finnish CWR species are only found in the Åland Islands, which makes it even more important to conserve sites as CWR genetic reserves both in mainland Finland and in Åland. If the genetic reserves were to be established, the first analysis that treats both regions together is preferable, since it would conserve the same amount of species in fewer areas.

The *ex situ* gap analysis found 56 out of 209 priority taxa to be in *ex situ* collections with sufficient origin data (Fig. 18.3). These collections are divided between the Finnish Botanic Gardens, NordGen and the Finnish Forest Research Institute. Most of the collections, however, are not genetically representative samples of the species distribution and would therefore need further collecting from diverse locations of their distribution. Only seven of the CWR priority taxa were collected from a minimum of five locations; *Allium schoenoprasum* var. *schoenoprasum*, *Elymus fibrosus*, *Festuca ovina*, *Festuca rubra*, *Phleum pratense* ssp. *pratense*, *Trifolium pratense* ssp. *pratense* and *Trifolium repens*. In all, 75% of the priority CWR taxa are not in any *ex situ* collection and should therefore be a future collection priority.

## 18.4 Discussion and Recommendations

The main suggestion for Finnish CWR conservation was that the valuable CWR taxa primarily should be conserved in their natural habitats, to allow populations to evolve and adapt to changing conditions. Ideally, CWR taxa should actively be conserved in genetic reserves, which could be established in the hotspot areas, and the target taxa should be included in the conservation area management and monitoring plan. Additionally, the taxa should be sampled for *ex situ* conservation for safety backup, their utilization in breeding programmes facilitated and, if needed, recovery or restoration programmes planned. There should be clarity in the conservation responsibility of these important plant species.

As a majority (76%) of the Finnish flora is CWR taxa by broad definition, many CWR already exist in protected areas by chance. However, the

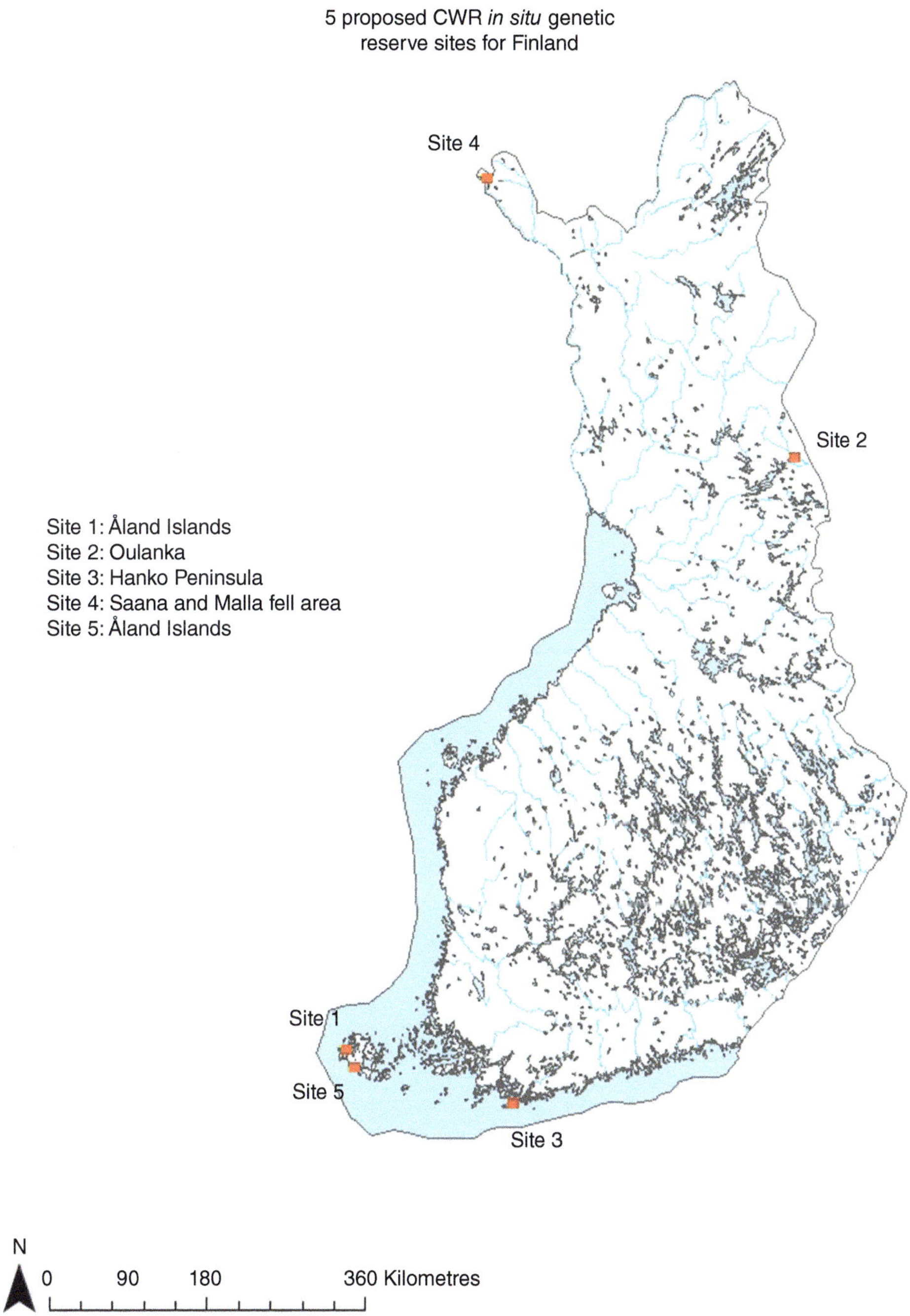


Fig. 18.2. Five CWR hotspots in Finland.

protection of most of these species was passive, as they were not taken into account when the conservation area management and monitoring plans were set up. The inclusion of CWR species, even the non-threatened species with a potential use value, in the protected area management plan would prevent a decline of the populations or other sudden adverse effects from

**Table 18.1.** Results of the two *in situ* complementarity analyses.

| Analysis 1: Mainland Finland and Åland together | | | | Analysis 2: Åland and mainland Finland separately | | | |
|---|---|---|---|---|---|---|---|
| 5 sites altogether | Location | CWR only in this site | Total CWR | 5 sites mainland Finland | Location | CWR only in this site | Total CWR |
| Site 1 | Åland Islands | 65 | 65 | Site 1 | Hanko Peninsula | 51 | 51 |
| Site 2 | Oulanka | 24 | 39 | Site 2 | Oulanka | 23 | 39 |
| Site 3 | Hanko Peninsula | 14 | 51 | Site 3 | Saana and Malla | 13 | 27 |
| Site 4 | Saana and Malla | 13 | 27 | Site 4 | Helsinki Islands | 10 | 39 |
| Site 5 | Åland Islands | 13 | 63 | Site 5 | Houtskari Islands | 7 | 34 |
| | | | | 2 sites Åland Islands | | | |
| | | | | Site 1 | Åland Islands | 65 | 65 |
| | | | | Site 2 | Åland Islands | 14 | 63 |

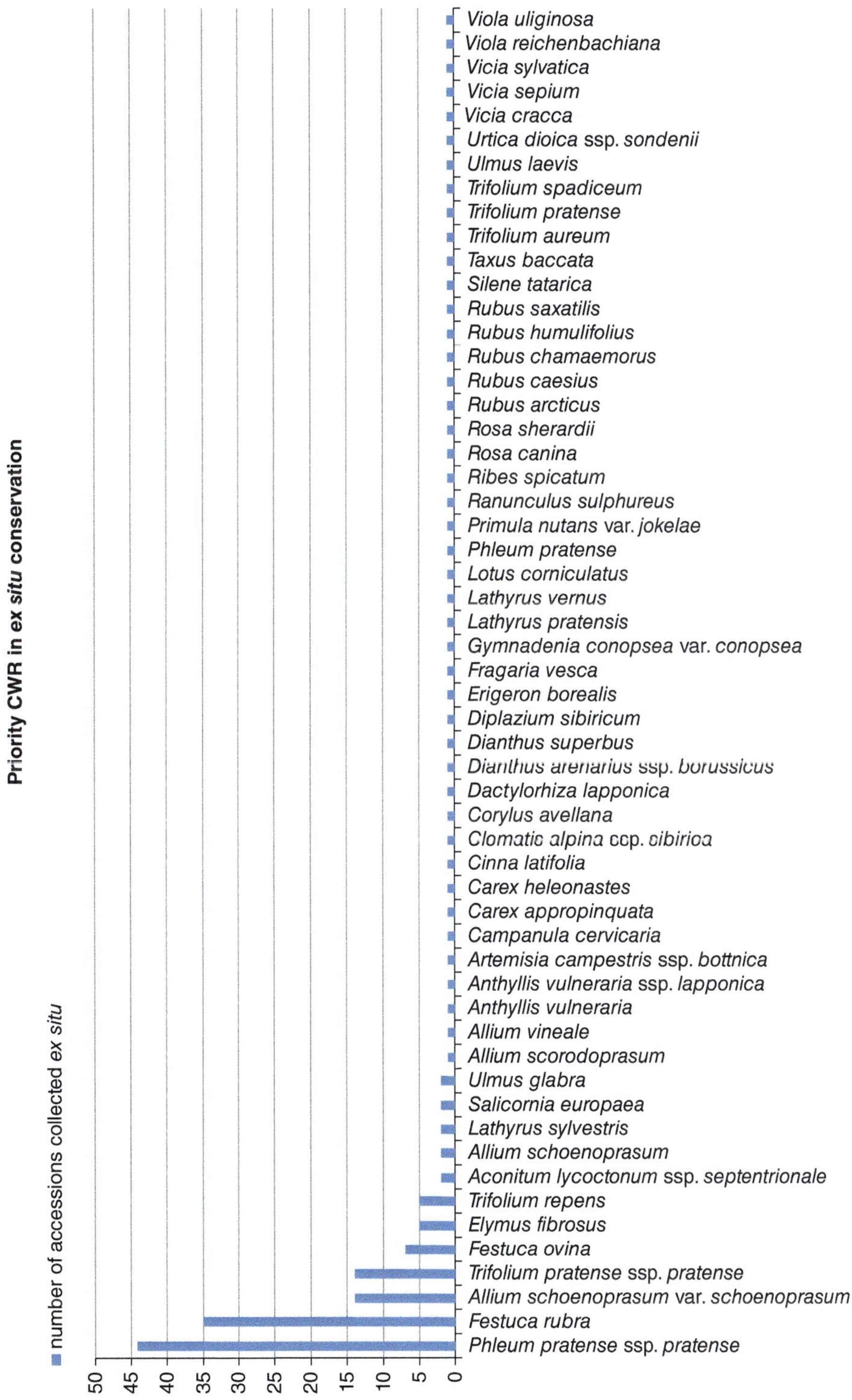


Fig. 18.3. Accessions of the priority CWR in *ex situ* conservation.

taking place. A monitoring plan of CWR population genetic diversity and natural dynamics would also be beneficial.

As *in situ* conservation protects species in their natural habitats, and many CWR species grow naturally in disturbed habitats such as wastelands, roadsides or field margins, methods to conserve them outside of protected areas would be necessary. Even weedy species may contain important genetic resources. As Maxted and Kell (2009) state, these habitats often contain large, thriving CWR populations that can provide dispersal and gene flow routes for other nearby CWR populations inside or outside of the protected areas. In practice, these sites would also need a monitoring and management agreement with the site owner. In Finland, the main threats to CWR species is the overgrowing of meadows and other open habitats, and as the main habitats are shorelines, lakesides and riverbanks, it appears that one of the main actions to conserve CWR taxa in Finland would be to prevent the overgrowing of waterside meadows, open cultural habitats and rural biotopes. The restoration of already changed habitats could be beneficial, as previous meadows and fields may hold a CWR seedbank in the soil.

Regarding *ex situ* conservation, the strategy recommends that the Finnish CWR priority species should be collected for *ex situ* backup conservation. The *ex situ* conservation of Finnish CWR taxa can be divided between NordGen and the threatened plant species seedbank in the Finnish Museum of Natural History. It would be beneficial if the seedbank supported potential users by providing data and access to CWR genetic resources.

The Finnish crop wild relative strategy gives practical guidelines for CWR *in situ* and *ex situ* conservation in Finland. However, addressing the following issues would bring the strategy forward: looking into the genetic reserve sites and populations in detail; creating a shorter priority list; incorporating breeders' and other users' views; creating conservation action plans and management and monitoring plans; undertaking genetic diversity analysis; and addressing the issue of climate change and the *in situ* conservation of CWR taxa.

## Acknowledgements

This research was funded by a grant from the EU Seventh Framework Programme for the PGR Secure project (Grant Agreement No 266394). The authors would like to acknowledge MTT Agrifood Research Finland, particularly Elina Kiviharju, and Maarit Heinonen, who was coordinating the Finnish participation in the PGR Secure project. We would also like to thank the following for their help during this project: Marko Hyvärinen, Marja Ruohonen-Lehto, Katileena Lohtander-Buckbee, Heikki Eeronheimo, Irmeli Ahonen, Terhi Ryttäri, Eija Kemppainen, Matti Yli-Rekola, Ritva Hiltunen, Bertalan Galambosi, Sampsa Lommi, Kirsi Valanne, Raino Lampinen and Sanna Laaka-Lindberg. The Finnish Museum of National History also kindly contributed to the funding of the strategy preparation.

## References

Anon. (1997) Nature Conservation Decree 160/1997 and its amendments. Helsinki, Finland. Available at: www.finlex.fi/en/laki/kaannokset/1997/19970160 (accessed 20 June 2015).

Bilz, M., Kell, S.P., Maxted, N. and Lansdown, R.V. (2011) *European Red List of Vascular Plants*. Publications Office of the European Union, Luxembourg.

EU (European Union) (1992) Council Directive 92/43/EEC on the conservation of natural habitats and of wild fauna and flora. The Council of the European Communities, Brussels, Belgium. Available at: eur-lex.europa.eu/legal-content/EN/TXT/?uri=celex:31992L0043 (accessed 20 June 2015).

Euro+Med (2006) Euro+Med PlantBase – the information resource for Euro-Mediterranean plant diversity. Available at: ww2.bgbm.org/EuroPlusMed/ (accessed 20 June 2015).

Eurostat (1995–2009) *Eurostat*, European Commission, Brussels, Belgium. Available at: ec.europa.eu/eurostat/web/agriculture/data/database (accessed 20 June 2015).

FAO (Food and Agriculture Organization) (2001) International Treaty on Plant Genetic Resources for Food and Agriculture. FAO, Rome. Available at: www.fao.org/AG/cgrfa/itpgr.htm (accessed 20 June 2015).

FAOSTAT (2009) *Agricultural Crop Production Data*. FAO, Rome. Available at: faostat.fao.org/ (accessed 20 June 2015).

Fitzgerald, H. (2013) *The National Crop Wild Relative Strategy Report for Finland*. MTT Report 121, MTT, Jokioinen, Finland, 96 pp. Available at: jukuri.mtt.fi/bitstream/handle/10024/481549/mttraportti121.pdf (accessed 20 June 2015).

Fitzgerald, H., Korpelainen, H. and Veteläinen, M. (2013) Prioritization of crop wild relatives in Finland. *Crop Wild Relative* 9, 10–12. Available at: www.pgrsecure.bham.ac.uk/sites/default/files/documents/newsletters/CWR_Issue_9.pdf (accessed 20 June 2015).

Galambosi, B. (2007) *Viljelyn merkitys luonnonyrttien raaka-ainehankinnassa Suomessa – Katsaus*. MTT-Mikkeli, Finland, pp. 17.

Hämet-Ahti, L., Suominen, J., Ulvinen, T. and Uotila, P. (1998) *Retkeilykasvio*, 4. uudistettu painos. Luonnontieteellisen keskusmuseon kasvimuseo, Helsinki.

Hämet-Ahti, L., Kurtto, A., Lampinen, R., Piirainen, M., Suominen, J., *et al.* (2005) Väre: Lisäyksiä ja korjauksia Retkeilykasvion neljänteen painokseen. *Lutukka* 21(2), 41–85.

Hanelt, P. and IPK Gatersleben (eds) (2001) *Mansfeld's World Database of Agricultural and Horticultural Crops*. Springer, Berlin/Heidelberg/New York. Available at: mansfeld.ipk-gatersleben.de (accessed 20 June 2015).

Harlan, J.R. and de Wet, J.M.J. (1971) Towards a rational classification of cultivated plants. *Taxon* 20, 509–517.

Hyvärinen, M., Miranto, M., Hiltunen, R. and Schulman, L. (2011) *Strategy and action plan for ex-situ conservation of threatened plants in Finland*. Available at: www.ymparisto.fi/download.asp?contentid=132157&lan=fi (accessed 20 June 2015).

Kell, S.P., Knüpffer, H., Jury, S.L., Maxted, N. and Ford-Lloyd, B.V. (2005) Catalogue of Crop Wild Relatives for Europe and the Mediterranean. University of Birmingham, Birmingham, UK. Available at: www.pgrforum.org/cwris/cwris.asp (accessed 20 June 2015).

Kell, S.P., Knüpffer, H., Jury, S.L., Ford-Lloyd, B.V. and Maxted, N. (2008) Crops and wild relatives of the Euro-Mediterranean region: making and using a conservation catalogue. In: Maxted, N., Ford-Lloyd, B.V., Kell, S.P., Iriondo, J., Dulloo, E. and Turok, J. (eds) *Crop Wild Relative Conservation and Use*. CAB International, Wallingford, UK, pp. 69–109.

Kell, S.P., Maxted, N. and Bilz, M. (2012) European crop wild relative threat assessment: knowledge gained and lessons learnt. In: Maxted, N., Dulloo, M.E., Ford-Lloyd, B.V., Frese, L., Iriondo, J.M. and Pinheiro de Carvalho, M.A.A. (eds) *Agrobiodiversity Conservation: Securing the Diversity of Crop Wild Relatives and Landraces*. CAB International, Wallingford, UK, pp. 218–242.

Kemppainen, E. and Eeronheimo, H. (2008) *Putkilokasvien suojelun edistäminen*. Suomen Ympäristökeskuksen raportteja, Helsinki, Finland. Available at: helda.helsinki.fi/bitstream/handle/10138/39676/SYKEra_34_2008.pdf?sequence=1 (accessed 20 June 2015).

Lampinen, R. and Lahti, T. (2010) Kasviatlas 2009. Helsingin Yliopisto, Luonnontieteellinen keskusmuseo, Kasvimuseo, Helsinki. Available at: www.luomus.fi/kasviatl/as (accessed 20 June 2015).

Maanmittauslaitos (2013) Avoimien aineistojen tiedostopalvelu. Available at: www.maanmittauslaitos.fi/aineistotpalvelut/latauspalvelut/avoimien-aineistojen-tiedostopalvelu (accessed 20 June 2015).

Maxted, N. and Kell, S.P. (2009) *Establishment of a Global Network for the In Situ Conservation of Crop Wild Relatives: Status and Needs*. FAO Commission on Genetic Resources for Food and Agriculture, Rome, 266 pp.

Maxted, N., Ford-Lloyd, B.V., Jury, S.L., Kell, S.P. and Scholten, M.A. (2006) Towards a definition of a crop wild relative. *Biodiversity and Conservation* 15(8), 2673–2685.

Maxted, N., Dulloo, E., Ford-Lloyd, B.V., Iriondo, J.M. and Jarvis, A. (2008) Gap analysis: a tool for complementary genetic conservation assessment. *Diversity and Distributions* 14, 1018–1030.

Maxted, N., Kell, S. and Magos Brehm, J. (2011) *Options to Promote Food Security: On-farm Management and In Situ Conservation of Plant Genetic Resources for Food and Agriculture*. FAO, Rome.

Metla (2012) Genetic diversity is conserved in gene reserve forests and collections. The Natural Resources Institute Finland. Available at: www.metla.fi (accessed 20 June 2015).

Metsähallitus (2013) Threatened species extended responsibility. Available at: www.metsa.fi/ (accessed 20 June 2015).

Miranto, M. (2005). Living collections of botanic gardens as a means of *ex situ* conservation – a case study of African violets (*Saintpaulia*) in Europe. MSc thesis, University of Helsinki. Available at: helda.helsinki.fi/bitstream/handle/10138/18935/livingco.pdf?sequence=1 (accessed 20 June 2015).

Miranto, M., Hyvärinen, M., Hiltunen, R. and Schulman, L. (2012) *Ex situ* conservation of threatened native plants in Finland: analysis of the current status. *Endangered Species Research* 17(3), 227–236.

Rassi, P., Hyvärinen, E., Juslén, A. and Mannerkoski, I. (eds) (2001) *The 2000 Red List of Finnish Species*. Ministry of the Environment and Finnish Environment Institute, Helsinki.
Rassi, P., Hyvärinen, E., Juslén, A. and Mannerkoski, I. (eds) (2010) *The 2010 Red List of Finnish Species*. Ympäristöministeriö and Suomen ympäristökeskus, Helsinki.
SYKE (2010) Species with action plan. Available at: www.ymparisto.fi/default.asp?node=1757&lan=sv (accessed 20 June 2015).
SYKE (2012) OIVA – Ympäristö-ja paikkatietopalvelu asiantuntijoille. Available at: wwwp2.ymparisto.fi/scripts/oiva.asp (accessed 20 June 2015).
Tike, Crop production statistics (2010) Yield of CWR related main crops 2010 Finland. Available at: stat.luke.fi/en/category/tilastot/satotilasto?page=2 (accessed 20 June 2015).
Vincent, H., Wiersema, J., Dobbie, S., Kell, S.P., Fielder, H., *et al.* (2013) Global checklist of priority crop wild relatives. Available at: www.cwrdiversity.org (accessed 20 June 2015).

# 19 Developing a National Crop Wild Relative *In Situ* Conservation Strategy for Lithuania: Creation of a National CWR Inventory and its Prioritization

J. Labokas,[1]* B. Karpavičienė,[1] V. Rašomavičius[1] and B. Gelvonauskis[2]
[1]*Institute of Botany, Nature Research Centre, Vilnius, Lithuania;*
[2]*Plant Gene Bank, Kėdainių r., Lithuania*

## 19.1 Introduction

*In situ* conservation, defined as 'the conservation of ecosystems and natural habitats and the maintenance and recovery of viable populations of species in their natural surroundings ... where they have developed their distinctive properties' (CBD, 1992), has received increasing attention from national governments (FAO, 2010) in view of the fact that it represents a dynamic conservation method that allows the maintenance of evolutionary processes. However, the most widespread means of conserving plant genetic resources has been through population sampling, seed collection and *ex situ* storage in seed banks. Yet crop wild relatives (CWR) continue to be underrepresented in genebanks (Maxted and Kell, 2009).

*In situ* CWR conservation efforts may be classified in one of three levels, at national, regional and global levels. At the global level, the conservation of CWR is highlighted within multiple international legislation policies. The Convention on Biological Diversity (CBD), at its tenth Conference of Parties in Nagoya, Japan, has endorsed the Strategic Plan on Biodiversity 2011–2020 and a set of 20 so-called Aichi Biodiversity Targets, of which target 13 directly relates to CWR protection (CBD, 2015). The *Second Global Plan of Action for Plant Genetic Resources for Food and Agriculture* (FAO, 2011) also highlights the importance of the *in situ* conservation of CWR in its priority actions on 'Survey and inventory of plant genetic resources for food and agriculture' (which includes CWR) and the priority action on the 'promotion of *in situ* conservation and management of CWR and wild food plants'. CWR have also been placed firmly on the international conservation agenda through Article 5 of the International Treaty on Plant Genetic Resources for Food and Agriculture (ITPGRFA) (FAO, 2001). It states that each contracting party shall promote the *in situ* conservation of wild crop relatives and wild plants for food production, including in protected areas, by supporting, *inter alia*, the efforts of indigenous and local communities.

The national level, which might include a subnational one, is a baseline, as successful systematic conservation largely depends on national legislation. It has been stressed that a national CWR conservation strategy should be viewed in the context of other policies and initiatives (Heywood and Iriondo, 2003) and not as a standalone approach to conservation.

*Corresponding author; e-mail: j.labokas@gmail.com

 

This chapter describes the process adopted in Lithuania in developing a list of priority CWR to serve as a basis for further systematic conservation planning and preparation of a national CWR *in situ* conservation strategy.

## 19.2 Country's Profile

Lithuania, with its area of 65,300 km$^2$, lies in northern Europe, on the eastern shores of the Baltic Sea, between the latitudes of 53°54′ N and 56°27′ N and longitudes of 20°57′ E and 26°50′ E. The climate is characterized as transitional between temperate maritime and continental. The mean annual temperature is 6.7–8.0°C and average annual precipitation amounts to 650–700 mm. There are four climatic subdivisions of the country based on the genetic factors of the climate in Lithuania (Bukantis, 2009).

The geographical position and climatic conditions predefine the vegetation zone, which is transitional between coniferous and broadleaved forests. The total forest area amounts to 33.3% of the country's territory, with *Pinus sylvestris* L. occupying 35.3% and *Picea abies* (L.) H. Karst. 20.8% of all forest stands (State Forest Service, 2013). An indicator of intensive forestry is the age structure of forests: about 50% of forest stands are no more than 50 years old (Kuliešis and Kulbokas, 2009).

Historically, Lithuania is an agricultural country. Despite some recession caused by political changes, currently the total area of utilized agricultural land has a trend of gradual increase from 2.689 million ha (Mha) in 2009 to 2.891 Mha in 2013 (Statistics Lithuania, 2013), i.e. from 41.2 to 44.3% of the country's territory.

There is a well-developed system of protected areas (PAs) in Lithuania, containing four categories of protection: (i) PAs of conservational priority; (ii) PAs of ecological protection priority; (iii) PAs of recuperational protection; and (iv) integrated PAs. Due to their particular importance, both PAs of conservational priority and integrated PAs are called special protected areas. They occupy 1023 thousand ha in total, or 15.7% of the country's territory (State Service for Protected Areas, 2014). There are 5 state strict reserves, 285 state reserves, 5 national parks, 30 regional parks, 1 biosphere reserve, 29 biosphere polygons, 3 recuperational sites and a number of municipal protected areas. The system of protected areas is based on the Law on Protected Areas (Ministry of Environment, 2001a). In Lithuania, the general provisions on the conservation of plant genetic resources are defined by the Law on National Plant Genetic Resources (Ministry of Environment, 2001b) and a series of related legal acts specifying its implementation.

The best situation regarding *in situ* conservation at the highest state level is with forest trees. In addition to the two state genetic reserves established in 2009, a draft decree was recently submitted to the government by the Ministry of Environment for the establishment of 112 more genetic reserves covering eight commercial tree species, including *Pinus sylvestris*, *Betula pendula* Roth, *Quercus robur* L., *Tilia cordata* Mill., *Picea abies*, *Alnus glutinosa* (L.) Gaertn., *Populus tremula* L. and *Fraxinus excelsior* L.

## 19.3 Methods and Sources of Information

### 19.3.1 National checklist

A floristic method as described by Maxted *et al.* (2013) was used to prepare the national checklist. The CWR checklist was developed by employing the CWR Catalogue for Europe and the Mediterranean (CWRIS) (Kell *et al.*, 2005, 2008) as an initial data source. The extracted checklist was harmonized with the national data sources, including *Flora of Lithuania* (Natkevičaitė-Ivanauskienė, 1959–1976; Natkevičaitė-Ivanauskienė *et al.*, 1980), *Dictionary of Plant Names* (Jankevičienė, 1998), *Vascular Plants of Lithuania* (Gudžinskas, 1999), *Flora of the Baltic Countries* (Laasimer *et al.*, 1993; Kuusk *et al.*, 1996, 2003), *Orchids (*Orchidaceae*) of Lithuania* (Gudžinskas and Ryla, 2006), *Red Data Book of Lithuania* (Rašomavičius, 2007), *European Red List of Vascular Plants* (Bilz *et al.*, 2011), the IUCN Red List (IUCN, 2014) and the Plant List (2010).

### 19.3.2 Prioritization of CWR

As active conservation cannot be planned for all the species, an attempt was made to develop prioritization criteria and to select the priority species to form the basis of an annotated inventory.

Following the Finnish model of species prioritization (Fitzgerald *et al.*, 2013), we have employed three main criteria and nine subcriteria (Table 19.1) for prioritizing CWR species. The three main criteria include:

**1.** Relative threat: in Lithuania, one of the major national policy instruments that apply to the conservation of the native wild flora is the Law on Protected Species of Fauna, Flora and Fungi (Ministry of Environment, 1997). In line with Article 4 of the law, the List of Strictly Protected Species of Fauna, Flora and Fungi of the Republic of Lithuania is being compiled based on the proposal of the Commission on the Lithuanian Red Data Book and approved by the Ministry of Environment (2010). Other sources of information included the databases of the Institute of Botany, Nature Research Centre, monographs (Gudžinskas and Ryla, 2006) and manuals (Gudžinskas and Balvočiūtė, 2007; Gudžinskas and Rašomavičienė, 2008). The relative threat was characterized by using the *Red Data Book of Lithuania* (Rašomavičius, 2007), with five categories of species (1 through 5), the *European Red List of Vascular Plants*, international policy instruments, including Habitats Directive, Bern Convention, CITES and EU Wildlife Trade Regulation (as referred to in Bilz *et al.*, 2011).
**2.** Value of CWR-related crops: Statistics Lithuania (2013) was consulted for the yields/acreages of agricultural and horticultural crops, as well as wild harvested plants; a list of high-priority human food crops in Europe, based on production quantity and economic value (Kell *et al.*, 2012), was checked.
**3.** Uses of species: sources of information included the criteria described by Labokas *et al.* (2012) and species listed in Annex I of the ITPGRFA (FAO, 2001), and those included in the national plant genetic resources lists (Plant Gene Bank, 2014) were considered; *Mansfeld's Encyclopedia of Agricultural and Horticultural Crops* (Hanelt and IPK Gatersleben, 2001) and the *European Pharmacopoeia* (EDQM, 2007) were consulted. In addition, The Harlan and de Wet CWR Inventory (www.cwrdiversity.org/checklist/), developed based on the genepool concept (Harlan and de Wet, 1971) and taxonomic group concept (Maxted *et al.*, 2006), was used to determine CWR relatedness to their crops (Vincent *et al.*, 2013).

The prioritization process was then carried out using the following steps:

**1.** All the species included in the *Red Data Book of Lithuania* were selected if they had at least one point in the four 'use' subcriteria.
**2.** The taxa not included in the *Red Data Book* but having at least three points in any of the remaining eight subcriteria and/or having several uses (food and medicinal, forage and medicinal, etc.) were included in the priority list.
**3.** Subspecies and varieties not threatened were removed from the list.

Table 19.1. The prioritization criteria for Lithuanian CWR. (From Fitzgerald *et al.*, 2013.)

1. Relative threat
   - Species of the *Red Data Book of Lithuania*
   - Species with narrowly specialized habitats and having relatively low resources, including those listed in the EU Wildlife Trade Regulation Annex D
2. Value
   - Yield/acreage of CWR-related crops in Lithuania
   - Yield/resources of wild harvested plants in Lithuania
   - High-priority human food crops in Europe based on production quantity and economic value
3. Use
   - Medicinal/pharmaceutical use, food, forage/fodder plants
   - Species listed in Annex I of the International Treaty on Plant Genetic Resources for Food and Agriculture (ITPGRFA)
   - Species included in the Lithuanian national plant genetic resources lists
   - Species in GP1b, GP2 and TG1b, TG2 and TG3

*Notes*: GP1b = genepool 1, species, easily mating with crops and producing strong, fertile progeny; GP2 = genepool 2, species, more distinct from crops, but still able to cross with them and produce at least some fertile offspring; TG1b = taxonomic group 1, the same species as crop; TG2 = taxonomic group 2, the same section/series as crop; TG3 = taxonomic group 3, the same subgenus as crop.

## 19.4 Results and Discussion

### 19.4.1 Creation of national CWR inventory

An initial list of the CWR for Lithuania, containing 1670 taxa, was extracted from the CWRIS catalogue. All alien species established in the country after 1500 AD were removed from the list. Only native species and archaeophytes were left, the latter amounting to around 100 species (Z. Gudžinskas, Vilnius, 2014, personal communication). Also, most hybrid species were excluded, as well as those with unconfirmed occurrence data. After some taxonomical harmonization, a national CWR checklist with a total of 1188 species, including 28 subspecific taxa, was produced. These make up 71% of the initial list. The checklist does not include microspecies of *Alchemilla* L. (25 currently known), *Euphrasia* L. (8), *Hieracium* L. (31), *Pilosella* Hill (35) and *Taraxacum* F.H. Wigg. (67), due to difficulties with their identification. Thus, the national CWR checklist contains 471 genera, divided into 117 families. The Asteraceae Dumort., Poaceae (R. Br.) Bernhart and Cyperaceae Juss. families contained the most number of species (Table 19.2).

As shown in Table 19.2, the top 15 families contain 795 species, making up 67% of the national CWR checklist, including 131 *Red Data Book* (RDB) species, comprising 60% of the whole RDB species list. The number of threatened RDB species, i.e. those of categories 1 (Endangered, E), 2 (Vulnerable, V) and 3 (Rare, R), amounts to 112. The most prominent family by the number of protected species is Orchidaceae, represented by 29 RDB species out of the total 36 occurring in the country. An analysis of the IUCN Red List of threatened species (IUCN, 2014) revealed that 366 species, or about 31% of the Lithuanian CWR national inventory, had been assessed for IUCN Red List categories for Europe, with the largest portion of 332 species being of Least Concern. These data were compared with the national *Red Data Book of Lithuania* and summarized in Table 19.3.

Table 19.3 shows that 218 species, i.e. around 18% of the national CWR inventory checklist, are included in the *Red Data Book of Lithuania*. Of these, 187 species, or around 16% of the CWR inventory, are nationally threatened (sum of the percentage in the first three categories, Table 19.3). Meanwhile, in Europe, only 18 species (less than 2%) of the checklist are assessed as threatened and near threatened. The regionally threatened species are *Neottianthe cucullata* (L.) Schltr. (EN, Orchidaceae), *Najas flexilis* (Willd.) Rostk. et W.L.E. Schmidt (VU, Najadaceae Rich.) and *Chimaphila umbellata* (L.)

**Table 19.2.** Top 15 families by number of species in the Lithuanian CWR checklist and numbers of the *Red Data Book of Lithuania* (RDB) species within them.

| Rank | Family | Number of genera | Number of species | Number of RDB species total | Number of RDB threatened species |
|---|---|---|---|---|---|
| 1 | Asteraceae | 42 | 132 | 12 | 11 |
| 2 | Poaceae | 49 | 107 | 17 | 11 |
| 3 | Cyperaceae | 13 | 91 | 18 | 15 |
| 4 | Fabaceae | 12 | 56 | 7 | 7 |
| 5 | Caryophyllaceae | 18 | 53 | 10 | 8 |
| 6 | Rosaceae | 16 | 47 | 2 | 2 |
| 7 | Scrophulariaceae | 13 | 43 | 6 | 6 |
| 8 | Brassicaceae | 26 | 42 | 4 | 3 |
| 9 | Lamiaceae | 20 | 42 | 9 | 8 |
| 10 | Ranunculaceae | 14 | 38 | 3 | 3 |
| 11 | Orchidaceae | 19 | 36 | 29 | 25 |
| 12 | Apiaceae | 27 | 34 | 9 | 8 |
| 13 | Rubiaceae | 4 | 29 | 4 | 4 |
| 14 | Polygonaceae | 5 | 25 | 0 | 0 |
| 15 | Ericaceae | 12 | 20 | 1 | 1 |
| Total | | 290 | 795 | 131 | 112 |

Table 19.3. Numbers of species in the Lithuanian CWR national inventory assessed by the IUCN Red List of Threatened Species for Europe (IUCN, 2014) and the *Red Data Book of Lithuania*.

| IUCN threat category, Europe | Number of species (% of CWR national inventory) | *Red Data Book of Lithuania* category | Number of species (% of CWR national inventory) |
|---|---|---|---|
| Endangered (EN) | 1 (0.1) | Endangered, 1 (E) | 57 (4.8) |
| Vulnerable (VU) | 2 (0.2) | Vulnerable, 2 (V) | 63 (5.3) |
| Near Threatened (NT) | 16 (1.3) | Rare, 3 (R) | 67 (5.6) |
| Data Deficient (DD) | 15 (1.3) | Indeterminate, 4 (I) | 21 (1.8) |
| Least Concern (LC) | 332 (27.9) | Restored, 5 (Rs) | 10 (0.8) |
| Total | 366 (30.8) | | 218 (18.3) |

W.P.C. Barton (VU, Ericaceae). The near threatened (NT) species include *Botrychium matricariifolium* A. Braun ex W.D.J. Koch, *Botrychium simplex* E. Hitchc. (both of Ophioglossaceae (R. Br.) C. Agardh), *Cypripedium calceolus* L., *Liparis loeselii* (L.) Rich., *Malaxis monophyllos* (L.) Sw. (all three of Orchidaceae), *Drosera anglica* Huds., *Drosera intermedia* Hayne (both of Droseraceae Salisb.), *Elatine alsinastrum* L. (Elatinaceae Dumort.), *Eriophorum gracile* W.D.J. Koch ex Roth (Cyperaceae), *Fraxinus excelsior* (Oleaceae Hoffmanns. et Link), *Helichrysum arenarium* (L.) Moench (Asteraceae), *Iris sibirica* L. (Iridaceae Juss.), *Potamogeton acutifolius* Link, *Potamogeton rutilus* Wolfg. (both of Potamogetonaceae Dumort.), *Sparganium natans* L. (Typhaceae Juss.) and *Linaria loeselii* Schweigg. (Scrophulariaceae), the only endemic in Lithuania.

Among the 15 regionally (Europe) Data Deficient species, there is a globally Endangered one, *Aldrovanda vesiculosa* L. (Droseraceae), as well as two globally Data Deficient species, i.e. *Malus sylvestris* Mill. and *Potentilla silesiaca* R. Uechtr. (both of Rosaceae). The latter is classified as regionally extinct (Poland) and possibly extinct (Lithuania, Latvia) (Bilz, 2013). Also, one regionally Least Concern species, *Alisma gramineum* Lej. (Alismataceae Vent.), is Data Deficient globally. In total, 150 species, or around 13% of the national CWR inventory, are assessed by the IUCN threat categories globally, with the largest portion of them, 145 species, being of the category Least Concern (IUCN, 2014).

### 19.4.2 Prioritization of national CWR inventory

Appendix 19.1 shows that in total 67 species, divided into 21 families, are covered by the major policy instruments, including national policy, in Lithuania. The summary list contains four former members of the genus *Lycopodium* L. that have had their names changed to *Diphasiastrum complanatum* (L.) Holub, *Diphasiastrum tristachyum* (Pursh) Holub, *Huperzia selago* (L.) Bernh. ex Schrank et Mart. and *Lycopodiella inundata* (L.) Holub. The total number of species covered by the European and international policy instruments in Lithuania is 59. The number of RDB species is 51 (76% of the total). It was observed that all of the 12 national policy species and most of the European and international policy species were RDB species of Lithuania. A minor exception involves two Rosaceae species, *Agrimonia pilosa* Ledeb. and *Potentilla silesiaca*, the latter being considered as possibly extinct in the country, four Lycopodiaceae P. Beauv. ex Mirb., seven Orchidaceae species, *Zostera marina* L., *Menyanthes trifoliata* L. and *Arctostaphylos uva-ursi* (L.) Spreng. Also, most of the 51 species assessed by the IUCN threat categories for Europe are Least Concern, except for one Endangered (*Neottianthe cucullata*), one Vulnerable (*Najas flexilis*), six Near Threatened (*Botrychium matricariifolium*, *Botrychium simplex*, *Cypripedium calceolus*, *Linaria loeselii*, *Liparis loeselii*, *Malaxis monophyllos*) and six Data Deficient (*Aldrovanda vesiculosa*, *Botrychium multifidum* (S.G. Gmel.) Rupr., *Herminium monorchis* (L.) R. Br., *Potentilla silesiaca*, *Pulsatilla patens* (L.) Mill., *Saxifraga hirculus* L.). Only seven species of the policy list are endemic to Europe, including two without confirmed occurrence and not included in the national RDB (*Epipactis purpurata* Sm. and *P. silesiaca*).

The process resulted in 180 prioritized species, representing 26 families and 78 genera (Table 19.4). The number of prioritized species

Table 19.4. Summary of the prioritized Lithuanian CWR inventory.

| Family | Genera | Species | Per cent of total species | Genera with numbers of species |
|---|---|---|---|---|
| Poaceae | 18 | 50 | 27.8 | *Agrostis* L. (4), *Alopecurus* L. (4), *Arrhenatherum* P. Beauv. (1), *Beckmannia* Host (1), *Bromopsis* Dumort. (Fourr.) (3), *Bromus* L. (1), *Dactylis* L. (2), *Elymus* L. (1), *Festuca* L. (11), *Glyceria* R. Br. (4), *Helictotrichon* Besser (1), *Hierochloe* R. Br. (2), *Hordelymus* (Jess.) Harz (1), *Lolium* L. (1), *Phalaroides* Wolf (1), *Phleum* L. (2), *Poa* L. (9), *Trisetum* Pers. (1) |
| Fabaceae | 11 | 49 | 27.2 | *Anthyllis* L. (2), *Astragalus* L. (2), *Lathyrus* L. (9), *Lotus* L. (3), *Medicago* L. (2), *Melilotus* Mill. (2), *Onobrychis* Mill. (1), *Ononis* L. (1), *Securigera* DC. (1), *Trifolium* L. (14), *Vicia* L. (12) |
| Rosaceae | 11 | 24 | 19.3 | *Fragaria* L. (2), *Malus* Mill. (1), *Prunus* L. (1), *Pyrus* L. (1), *Rosa* L. (4), *Rubus* L. (9), *Sorbus* L. (1), *Crataegus* L. (2), *Filipendula* Mill. (1), *Sanguisorba* L. (1), *Potentilla* L. (1) |
| Lamiaceae | 4 | 7 | 3.9 | *Leonurus* L. (2), *Mentha* L. (2), *Origanum* L. (1), *Thymus* L. (2) |
| Amaryllidaceae J. St.-Hil. | 1 | 6 | 3.3 | *Allium* L. (6) |
| Asteraceae | 5 | 6 | 3.3 | *Arnica* L. (1), *Arctium* L. (1), *Cichorium* L. (1), *Helichrysum* Mill. (1), *Taraxacum* F.H. Wigg. (2) |
| Ericaceae | 2 | 5 | 2.8 | *Arctostaphylos* Adans. (1), *Vaccinium* L. (4) |
| Hypericaceae Juss. | 1 | 4 | 2.2 | *Hypericum* L. (4) |
| Apiaceae | 4 | 4 | 2.2 | *Angelica* L. (1), *Carum* L. (1), *Daucus* L. (1), *Pastinaca* L. (1) |
| Chenopodiaceae Vent. | 2 | 4 | 2.2 | *Atriplex* L. (3), *Salsola* L. (1) |
| Boraginaceae Juss. | 3 | 3 | 1.7 | *Lithospermum* L. (1), *Pulmonaria* L. (1), *Symphytum* L. (1) |
| Grossulariaceae DC. | 1 | 3 | 1.7 | *Ribes* L. (3) |
| Polygonaceae | 2 | 2 | 1.1 | *Rumex* L. (1), *Bistorta* Adans. (1) |
| Corylaceae Mirb. | 1 | 1 | 0.6 | *Corylus* L. (1) |
| Gentianaceae Juss. | 1 | 1 | 0.6 | *Gentiana* L. (1) |
| Scrophulariaceae | 1 | 1 | 0.6 | *Gratiola* L. (1) |
| Araliaceae Juss. | 1 | 1 | 0.6 | *Hedera* L. (1) |
| Cannabaceae Endl. | 1 | 1 | 0.6 | *Humulus* L. (1) |
| Menyanthaceae (Dumort.) Dumort. | 1 | 1 | 0.6 | *Menyanthes* L. (1) |
| Myricaceae Blume | 1 | 1 | 0.6 | *Myrica* L. (1) |

| | | | | |
|---|---|---|---|---|
| Polemoniaceae Juss. | 1 | 1 | 0.6 | *Polemonium* L. (1) |
| Primulaceae Vent. | 1 | 1 | 0.6 | *Primula* L. (1) |
| Brassicaceae Burnett | 1 | 1 | 0.6 | *Thlaspi* L. (1) |
| Urticaceae Juss. | 1 | 1 | 0.6 | *Urtica* L. (1) |
| Valerianaceae Batsch | 1 | 1 | 0.6 | *Valeriana* L. (1) |
| Viburnaceae Raf. | 1 | 1 | 0.6 | *Viburnum* L. (1) |
| Totals: 26 | 78 | 180 | 106.5 | |

comprises around 15% of the national CWR inventory, the proportion close to those reported from the UK (Scholten *et al.*, 2008) and Finland (Fitzgerald, 2013), and those being developed in Norway (J. Phillips, Vilnius, 2014, personal communication).

Around 68% of the priority species are members of three families, Poaceae, Fabaceae and Rosaceae. Analysis of CWR priority species by their use categories shows that 55% of them are forage/fodder plants, with the remaining percentages shared nearly equally between fruit, vegetable and medicinal plant groups (Table 19.5).

A significant portion of species of the above grouping falls into more than one use category, which corroborates their prioritization. For instance, the species of food categories like *Angelica archangelica* L., *Carum carvi* L., *Ribes nigrum* L., *Rosa canina* L. and *Humulus lupulus* L. are officially recognized as medicinal plants by the *European Pharmacopoeia* (EDQM, 2007). Similarly, a forage/fodder species, *Melilotus officinalis* (L.) Pall., is an officially recognized medicinal plant, while a pharmacopoeian species, *Urtica dioica* L., is used in a breeding/cultivation programme aimed at fibre crop development.

With the growing interest in healthy foods and functional food additives, some comparatively new crops and their wild relatives are gaining more and more attention. This group primarily includes species such as *Sorbus aucuparia* L., *Viburnum opulus* L. and *Vaccinium* species (blueberries, cranberries and lingonberries), presently used in cultivation and partly in breeding, as well as *Corylus avellana* L., cultivated by many amateur growers. Many cultivars of these species are spreading countrywide.

The following genera of the '18 high-priority human food crops/crop groups of Europe' (Kell *et al.*, 2012) and of the 'priority human and animal food crops listed in the ITPGRFA' (FAO, 2001) are found in Lithuania and are therefore considered to be of particular importance: *Malus* (apple), *Daucus* (carrot), *Allium* (onion and garlic), *Cichorium* (chicory), *Pyrus* (pear), *Prunus* (plum and sloe), *Fragaria* (strawberry), *Lathyrus* (grass pea – GP1b, TG1b, GP2 and TG2 only), *Elymus* (wheat complex), *Vicia* (faba bean/vetch – mainly GP1b, TG1b, GP2 and TG2), *Agrostis*, *Alopecurus*, *Arrhenatherum*, *Dactylis*, *Festuca*, *Lolium*, *Phalaris* (syn. *Phalaroides*), *Phleum*, *Poa* (grass forages), *Astragalus*, *Coronilla* L. (syn. *Securigera* DC.), *Lotus*, *Medicago*, *Melilotus*, *Onobrychis*, *Trifolium* (legume forages), *Atriplex* and *Salsola* (other forages). In addition to these, the following CWR genera are important to Lithuania: *Ribes*, *Rubus*, *Carum* and a group of medicinal, aromatic and spice plants according to the acreages of CWR-related crops in Lithuania in 2013. Among wild harvested plants, *Vaccinium*, *Fragaria*, *Rubus*

**Table 19.5.** Use categories of Lithuanian CWR priority species.

| Use category | Families with numbers of genera/species | Species numbers | Per cent of total species |
|---|---|---|---|
| Forage/fodder | Poaceae (18/50), Fabaceae (10/48), Chenopodiaceae (1/1) | 99 | 55.0 |
| Fruit and berries | Rosaceae (7/19), Ericaceae (1/4), Grossulariaceae (1/3), Corylaceae (1/1), Viburnaceae (1/1) | 28 | 15.5 |
| Vegetables and spices | Amaryllidaceae (1/6), Apiaceae (4/4), Asteraceae (2/3), Brassicaceae (1/1), Cannabaceae (1/1), Chenopodiaceae (1/3), Lamiaceae (3/5), Myricaceae (1/1), Polygonaceae (1/1) | 25 | 14.0 |
| Medicinal/ pharmaceutical | Araliaceae (1/1), Asteraceae (3/3), Boraginaceae (3/3), Gentianaceae (1/1), Ericaceae (1/1), Fabaceae (1/1), Hypericaceae (1/4), Lamiaceae (1/2), Menyanthaceae (1/1), Polemoniaceae (1/1), Polygonaceae (1/1), Primulaceae (1/1), Rosaceae (4/5), Scrophulariaceae (1/1), Urticaceae (1/1), Valerianaceae (1/1) | 28 | 15.5 |
| Total | | 180 | 100 |

and medicinal plants are leading by size of yields harvested (Statistics Lithuania, 2013).

There is a group of CWR that, despite their immediate closeness to the crops, were not included in the priority list because of their weedy character. This is particularly the case with *Avena fatua* L. and *Raphanus raphanistrum* L., genepool primary (GP1) relatives to common oat and cultivated radish, respectively, the crops covered by the ITPGRFA. The other group, consisting of *Barbarea stricta* Andrz., *Barbarea vulgaris* W.T. Aiton, *Rorippa amphibia* (L.) Besser, *Rorippa palustris* (L.) Besser, *Rorippa sylvestris* (L.) Besser, and some others of the *Brassica* L. complex, also are characterized as weedy species, but they are remote relatives (TG4) to such minor crops like early yellowrocket, variableleaf yellowcress, etc. Therefore, all these species were not included in the prioritized CWR inventory.

## 19.5 Conclusion

The CWR priority list contains mainly native and archaeophyte species. The list is not definitive and is subject to change with relevant updated information on species occurrence, distribution and status. For instance, new entries may appear in the *Red Data Book of Lithuania* after some species-targeted surveys are performed. There are potentially valuable and well-adapted species among the neophytes in Lithuania, too; for instance, fruit/berry species, like *Hippophae rhamnoides* L., whose wild-collected yields are among the top five wild berries (Statistics Lithuania, 2013), and *Cerasus avium* (L.) Moench, field collections of which have been set up by horticultural scientists. Medicinal plants like *Sambucus nigra* L., *Acorus calamus* L. and *Oenothera biennis* L. also attract the attention of horticulturists and scientists. This group of species should be considered for *in situ* conservation in the future, as far as they are not invasive and not harmful to native plant communities and ecosystems.

Nevertheless, the aim of the current work has been achieved and the next step to be undertaken is ecogeographic diversity analysis – to collate the available baseline ecogeographic data for priority taxa and undertake threat assessment and gap analysis.

## References


Bilz, M. (2013) *Potentilla silesiaca*. The IUCN Red List of Threatened Species. Version 2014.2. Available at: www.iucnredlist.org (accessed 30 June 2015).

Bilz, M., Kell, S.P., Maxted, N. and Lansdown, R.V. (2011) *European Red List of Vascular Plants*. Publications Office of the European Union, Luxembourg.

Bukantis, A. (2009) *Climatic Subdivision. National Atlas of Lithuania.* National Land Service under the Ministry of Agriculture, Vilnius (in Lithuanian).

CBD (1992) Convention on Biological Diversity. CBD Secretariat, Ottawa, Canada. Available at: www.cbd.int/convention/text/ (accessed 30 June 2015).

CBD (2015) Aichi Biodiversity Targets. CBD Secretariat, Ottawa, Canada. Available at: www.cbd.int/sp/targets/default.shtml (accessed 30 June 2015).

EDQM (2007) *European Pharmacopoeia*. Sixth Edition. Vol 2. Council of Europe, Strasbourg. Available at: online6.edqm.eu/ep804/ (accessed 30 June 2015).

FAO (Food and Agriculture Organization) (2001) The International Treaty on Plant Genetic Resources for Food and Agriculture. FAO, Rome. Available at: www.planttreaty.org (accessed 30 June 2015).

FAO (2010) *The Second Report on the State of the World's Plant Genetic Resources for Food and Agriculture*. FAO, Rome.

FAO (2011) *Second Global Plan of Action for Plant Genetic Resources for Food and Agriculture*. Commission on Genetic Resources for Food and Agriculture. FAO, Rome. Available at: www.fao.org/docrep/015/i2624e/i2624e00.pdf (accessed 30 June 2015).

Fitzgerald, H. (2013) *The National Crop Wild Relative Strategy Report for Finland*. MTT Report 121, 95 pp. Available at: www.mtt.fi/mttraportti/pdf/mttraportti121.pdf (accessed 30 June 2015).

Fitzgerald, H., Korpelainen, H. and Veteläinen, M. (2013) Prioritization of crop wild relatives in Finland. *Crop Wild Relative* 9, 10–12. Available at: www.pgrsecure.org/publications (accessed 30 June 2015).

Gudžinskas, Z. (1999) *Vascular Plants of Lithuania*. Institute of Botany, Vilnius, 211 pp.

Gudžinskas, Z. and Balvočiūtė, J. (2007) *Medicinal Plants of Lithuania*. Šviesa, Kaunas, Lithuania, 160 pp. (in Lithuanian).
Gudžinskas, Z. and Rašomavičienė, B. (2008) *Spice Plants of Lithuania*. Šviesa, Kaunas, Lithuania, 152 pp. (in Lithuanian).
Gudžinskas, Z. and Ryla, M. (2006) *Orchids* (Orchidaceae) *of Lithuania* (in Lithuanian with English Preface and Summary). Institute of Botany, Vilnius, 104 pp.
Hanelt, P. and IPK Gatersleben (eds) (2001) *Mansfeld's Encyclopaedia of Agricultural and Horticultural Crops*. 6 vols. 1st English edn. Springer, Berlin/Heidelberg/New York, 3645 pp. Available at: mansfeld.ipk-gatersleben.de (accessed 30 June 2015).
Harlan, J.R. and de Wet, J.M.J. (1971) Towards a rational classification of cultivated plants. *Taxon* 20, 509–517.
Heywood, V. and Iriondo, J. (2003) Plant conservation: old problems, new perspectives RID B-3112-2008. *Biological Conservation* 113(3), 321–335.
IUCN (2014) The IUCN Red List of Threatened Species. Version 2014.2. Available at: www.iucnredlist.org (accessed 30 June 2015).
Jankevičienė, R. (Compiler) (1998) *Dictionary of Plant Names*. Institute of Botany, Vilnius, 523 pp.
Kell, S.P., Knüpffer, H., Jury, S.L., Maxted, N. and Ford-Lloyd, B.V. (2005) Catalogue of Crop Wild Relatives for Europe and the Mediterranean. University of Birmingham, UK. Available at: www.pgrforum.org/cwris/cwris.asp (accessed 30 June 2015).
Kell, S.P., Knüpffer, H., Jury, S.L., Ford-Lloyd, B.V. and Maxted, N. (2008) Crops and wild relatives of the Euro-Mediterranean region: making and using a conservation catalogue. In: Maxted, N., Ford-Lloyd, B.V., Kell, S.P., Iriondo, J., Dulloo, E. and Turok, J. (eds) *Crop Wild Relative Conservation and Use*. CAB International, Wallingford, UK, pp. 69–109.
Kell, S.P., Maxted, N. and Bilz, M. (2012) European crop wild relative threat assessment: knowledge gained and lessons learnt. In: Maxted, N., Dulloo, M.E., Ford-Lloyd, B.V., Frese, L., Iriondo, J.M. and Pinheiro de Carvalho, M.A.A. (eds) *Agrobiodiversity Conservation: Securing the Diversity of Crop Wild Relatives and Landraces*. CAB International, Wallingford, UK, pp. 218–242.
Kuliešis, A. and Kulbokas, G. (2009) *Lithuanian National Forest Inventory 2004–2008. Forest Resources and Their Dynamic*. Lututė, Kaunas, Lithuania, 88 pp.
Kuusk, V., Tabaka, L. and Jankevičienė, R. (eds) (1996) *Flora of the Baltic Countries. Compendium of Vascular Plants, Volume 2*. Tartu, Estonia, 372 pp.
Kuusk, V., Tabaka, L. and Jankevičienė, R. (eds) (2003) *Flora of the Baltic Countries. Compendium of Vascular Plants, Volume 3*. Tartu, Estonia, 408 pp.
Laasimer, L., Kuusk, V., Tabaka, L. and Lekavičius, A. (eds) (1993) *Flora of the Baltic Countries. Compendium of Vascular Plants, Volume 1*. Tartu, Estonia, 378 pp.
Labokas, J., Karpavičienė, B., Šveistytė, L., Radušienė, J. and Ložienė, K. (2012) Towards *in situ* conservation of crop wild relatives in Lithuania. In: Maxted, N., Dulloo, M.E., Ford-Lloyd, B.V., Frese, L., Iriondo, J.M. and Pinheiro de Carvalho, M.A.A. (eds) *Agrobiodiversity Conservation: Securing the Diversity of Crop Wild Relatives and Landraces*. CAB International, Wallingford, UK, pp. 91–95.
Maxted, N. and Kell, S.P. (2009) *Establishment of a Global Network for the In Situ Conservation of Crop Wild Relatives: Status and Needs*. FAO Commission on Genetic Resources for Food and Agriculture, Rome, 266 pp.
Maxted, N., Ford-Lloyd, B.V., Jury, S.L., Kell, S.P. and Scholten, M.A. (2006) Towards a definition of a crop wild relative. *Biodiversity and Conservation* 15(8), 2673–2685.
Maxted, N., Brehm, J.M. and Kell, S.P. (2013) *Resource Book for the Preparation of National Plans for Conservation of Crop Wild Relatives and Landraces.* FAO, Rome. Available from: www.fao.org/fileadmin/templates/agphome/documents/PGR/PubPGR/ResourceBook/TEXT_ALL_2511.pdf (accessed 30 June 2015).
Ministry of Environment, Republic of Lithuania (1997) Law on Protected Species of Fauna, Flora and Fungi. Ministry of Environment, Vilnius. Available at: www3.lrs.lt/pls/inter3/dokpaieska.showdoc_e?p_id=384575 (accessed 30 June 2015).
Ministry of Environment, Republic of Lithuania (2001a) Law on Protected Areas. Ministry of Environment, Vilnius. Available at: www3.lrs.lt/pls/inter3/dokpaieska.showdoc_e?p_id=300377&p_tr2=2 (accessed 30 June 2015).
Ministry of Environment, Republic of Lithuania (2001b) Law on National Plant Genetic Resources. Ministry of Environment, Vilnius. Available at: www3.lrs.lt/pls/inter3/dokpaieska.showdoc_e?p_id=230551&p_tr2=2 (accessed 30 June 2015).

Ministry of Environment, Republic of Lithuania (2010) Order on Approval of the List of Strictly Protected Species of Fauna, Flora and Fungi of the Republic of Lithuania. 1 April 2010, No D1-263 (in Lithuanian).
Natkevičaitė-Ivanauskienė, M. (ed.) (1959–1976) *Flora of Lithuania*, Volumes 1–5 (1 – 1959, 2 – 1963, 3 – 1961, 4 – 1971, 5 – 1976). Mintis and Mokslas, Vilnius, Lithuania.
Natkevičaitė-Ivanauskienė, M., Jankevičienė, R. and Lekavičius, A. (eds) (1980) *Flora of Lithuania*, Volume 6. Mokslas, Vilnius, Lithuania.
Plant Gene Bank (2014) Lists of national plant genetic resources with amendments. Available at: www.agb.lt/angi.htm (accessed 30 June 2015).
Plant List (2010) Version 1.1. (online). Available at: www.theplantlist.org (accessed 30 June 2015).
Rašomavičius, V. (editor-in-chief) (2007) *Red Data Book of Lithuania*. Ministry of Environment Vilnius. LUTUTė Publishing Company, Kaunas, Lithuania, 800 pp.
Scholten, M., Maxted, N., Kell, S.P. and Ford-Lloyd, B.V. (2008) Creation of a national crop wild relative strategy: a case study for the United Kingdom. In: Maxted, N., Ford-Lloyd, B.V., Kell, S.P., Iriondo, J., Dulloo, E. and Turok, J. (eds) *Crop Wild Relative Conservation and Use*. CAB International, Wallingford, UK, pp. 120–142.
State Forest Service (2013) *Lithuania's Forests*. Ministry of Environment, Vilnius.
State Service for Protected Areas under the Ministry of Environment (2014) Available at: www.vstt.lt/en (accessed 26 September 2014).
Statistics Lithuania (2013) Available at: osp.stat.gov.lt/en/web/guest/home (accessed 30 June 2015).
Vincent, H., Wiersema, J., Kell, S., Fielder, H., Dobbie, S., *et al*. (2013) A prioritized crop wild relative inventory to help underpin global food security. *Biological Conservation* 167, 265–275.

# Appendix

**Table A.19.1.** Species covered by national, European and international policy instruments and their threat status.

| Species | Family | Lithuanian RDB category | Strictly protected nationally | Habitats Directive Annexes | Bern Convention Annexes | CITES Annexes | EU Wildlife Trade Regulation | Endemic to Europe? | IUCN Red List Category (Europe) |
|---|---|---|---|---|---|---|---|---|---|
| *Agrimonia pilosa* | Rosaceae | | | II/IV | | | | | LC |
| *Aldrovanda vesiculosa* | Droseraceae | 1 (E) | | II/IV | I | | | | DD |
| *Arctostaphylos uva-ursi* | Ericaceae | | | | | | D | | LC |
| *Arnica montana* | Asteraceae | 2 (V) | | V | | | D | Yes | LC |
| *Botrychium matricariifolium* | Ophioglossaceae | 1 (E) | | | I | | | | NT |
| *Botrychium multifidum* | Ophioglossaceae | 3 (R) | | | I | | | | DD |
| *Botrychium simplex* | Ophioglossaceae | 1 (E) | | II/IV | I | | | | NT |
| *Carex davalliana* Sm. | Cyperaceae | 1 (E) | Yes | | | | | | LC |
| *Cephalanthera longifolia* (L.) Fritsch | Orchidaceae | 1 (E) | Yes | | | II | | | LC |
| *Cephalanthera rubra* (L.) Rich. | Orchidaceae | 1 (E) | | | | II | | | LC |
| *Coeloglossum viride* (L.) Hartm. | Orchidaceae | 2 (V) | | | | II | | | LC |
| *Corallorhiza trifida* Châtel. | Orchidaceae | 2 (V) | | | | II | | | LC |
| *Cypripedium calceolus* | Orchidaceae | 2 (V) | Yes | II/IV | I | II | A | | NT |
| *Dactylorhiza cruenta* (O.F. Müll.) Soó | Orchidaceae | 1 (E) | | | | II | | | |
| *Dactylorhiza fuchsii* (Druce) Soó | Orchidaceae | 4 (I) | | | | II | | | LC |
| *Dactylorhiza incarnata* (L.) Soó | Orchidaceae | 5 (Rs) | | | | II | | | LC |
| *Dactylorhiza longifolia* (Neuman) Aver. | Orchidaceae | 5 (Rs) | | | | II | | | |
| *Dactylorhiza maculata* (L.) Soó | Orchidaceae | 2 (V) | | | | II | | | LC |
| *Dactylorhiza majalis* (Rchb.) P.F. Hunt et Summerh. | Orchidaceae | 1 (E) | | | | II | | | LC |
| *Dactylorhiza ochroleuca* (Wüstnei ex Boll) Holub | Orchidaceae | 3 (R) | | | | II | | | |
| *Dactylorhiza traunsteineri* (Saut.) Soó | Orchidaceae | 2 (V) | | | | II | | | LC |
| *Dianthus arenarius* L. ssp. *arenarius* | Caryophyllaceae | 2 (V) | | II/IV | | | | Yes | LC |
| *Dianthus armeria* L. | Caryophyllaceae | 1 (E) | Yes | | | | | | |
| *Diphasiastrum complanatum* | Lycopodiaceae | | | V | | | | | |
| *Diphasiastrum tristachyum* | Lycopodiaceae | | | V | | | | | |
| *Dracocephalum ruyschiana* L. | Lamiaceae | 1 (E) | | | I | | | | LC |
| *Epipactis atrorubens* (Hoffm.) Besser | Orchidaceae | 2 (V) | | | | II | | | LC |

| | | | | | | | | | |
|---|---|---|---|---|---|---|---|---|---|
| *Epipactis helleborine* (L.) Crantz | Orchidaceae | | | | | II | | | LC |
| *Epipactis palustris* (L.) Crantz | Orchidaceae | | | | | II | | | LC |
| *Epipactis purpurata* Sm. | Orchidaceae | | | | | II | | Yes | LC |
| *Epipogium aphyllum* Sw. | Orchidaceae | 1 (E) | | | | II | | | LC |
| *Goodyera repens* (L.) R. Br. | Orchidaceae | | | | | II | | | LC |
| *Gymnadenia conopsea* (L.) R. Br. | Orchidaceae | 2 (V) | | | | II | | | LC |
| *Gymnadenia odoratissima* (L.) Rich. | Orchidaceae | 1 (E) | | | | II | | Yes | LC |
| *Hammarbya paludosa* (L.) Kuntze | Orchidaceae | 2 (V) | | | | II | | | LC |
| *Herminium monorchis* | Orchidaceae | 1 (E) | | | | II | | | DD |
| *Huperzia selago* | Lycopodiaceae | 5 (Rs) | | V | | | | | |
| *Hydrocotyle vulgaris* L. | Apiaceae | 1 (E) | Yes | | | | | | LC |
| *Isoetes lacustris* L. | Isoetaceae Rchb. | 1 (E) | Yes | | | | | | LC |
| *Linaria loeselii* | Scrophulariaceae | 2 (V) | | II/IV | I | | | Yes | NT |
| *Liparis loeselii* | Orchidaceae | 2 (V) | | II/IV | I | II | A | | NT |
| *Listera cordata* (L.) R. Br. | Orchidaceae | 2 (V) | | | | II | | | |
| *Listera ovata* (L.) R. Br. | Orchidaceae | | | | | II | | | |
| *Lobelia dortmanna* L. | Lobeliaceae | 1 (E) | Yes | | | | | | LC |
| *Lycopodiella inundata* R. Br. | Lycopodiaceae | 2 (V) | | V | | | | | |
| *Lycopodium annotinum* L. | Lycopodiaceae | | | V | | | | | |
| *Lycopodium clavatum* L. | Lycopodiaceae | | | V | | | D | | LC |
| *Malaxis monophyllos* | Orchidaceae | 2 (V) | | | | II | | | NT |
| *Melittis melissophyllum* L. | Lamiaceae | 1 (E) | Yes | | | | | | |
| *Menyanthes trifoliata* | Menyanthaceae | | | | | | D | | LC |
| *Najas flexilis* | Najadaceae | 1 (E) | | II/IV | I | | | | VU |
| *Neottia nidus-avis* (L.) Rich. | Orchidaceae | | | | | II | | | LC |
| *Neottianthe cucullata* (L.) Schltr. | Orchidaceae | 1 (E) | Yes | | | II | | | EN |
| *Ophrys insectifera* L. | Orchidaceae | 1 (E) | | | | II | | Yes | LC |
| *Orchis mascula* (L.) L. | Orchidaceae | 2 (V) | | | | II | | | LC |
| *Orchis militaris* L. | Orchidaceae | 2 (V) | | | | II | | | LC |
| *Orchis morio* L. | Orchidaceae | 2 (V) | | | | II | | | |
| *Orchis ustulata* L. | Orchidaceae | 1 (E) | Yes | | | II | | | |
| *Pedicularis sylvatica* L. | Scrophulariaceae | 1 (E) | Yes | | | | | | |
| *Platanthera bifolia* (L.) Rich. | Orchidaceae | | | | | II | | | LC |
| *Platanthera chlorantha* (Custer) Rchb. | Orchidaceae | 4 (I) | | | | II | | | LC |
| *Potentilla silesiaca* | Rosaceae | | | | I | | | Yes | DD |
| *Pulsatilla patens* | Ranunculaceae | 2 (V) | | II/IV | I | | | | DD |

*Continued*

Table A.19.1. Continued.

| Species | Family | Lithuanian RDB category | Strictly protected nationally | Habitats Directive Annexes | Bern Convention Annexes | CITES Annexes | EU Wildlife Trade Regulation | Endemic to Europe? | IUCN Red List Category (Europe) |
|---|---|---|---|---|---|---|---|---|---|
| *Saxifraga hirculus* | Saxifragaceae Juss. | 2 (V) | | II/IV | I | | | | DD |
| *Thesium ebracteatum* Hayne | Santalaceae R. Br. | 2 (V) | | II/IV | I | | | | LC |
| *Tofieldia calyculata* (L.) Wahlenb. | Melanthiaceae Batsch | 1 (E) | Yes | | | | | | |
| *Zostera marina* | Zosteraceae Dumort. | | | | I | | | | LC |
| Total: 67 | 21 | 51 | 12 | 18 | 14 | 36 | 6 | 7 | 51 |

*Notes*: E = endangered; V = vulnerable; R = rare; I = indeterminate; Rs = restored.

# 20 Priorities for the Conservation of Crop Wild Relatives at the Indian National Genebank

V. Gupta,* A. Kak and R.K. Tyagi
*Division of Germplasm Conservation, National Bureau of Plant Genetic Resources, New Delhi, India*

## 20.1 Introduction

The ultimate aim of conserving plant genetic resources (PGRs) in genebanks is to promote and facilitate the sustainable use of valuable resources by breeders/scientists/researchers in order to develop new varieties to ensure the food security of the increasing population by mitigating the adverse impact of climate change. Although intensive efforts have been made to conserve the genetic diversity of agriculturally important crops at the global level, very limited efforts have been made for the systematic conservation of crop wild relatives (CWR), which can offer a valuable genetic resource to breeders to meet the challenges of climate change and the rapidly changing demands of consumers. CWR are the reservoirs of many important traits by virtue of their adaptation to diverse habitats being wild, with no domestication. CWR, like any other form of biodiversity, are facing major threats to their existence. CWR have very restricted ecogeographic and specific distribution, and many of them are represented by inconspicuous population growth mixed with other wild species. The Second Report on the State of the World's Plant Genetic Resources for Food and Agriculture (FAO, 2010) predicts that as much as 16–22% of the wild relatives of important crops may disappear by 2055 because of changing climate. In spite of the increase in the number of genebanks, which now total 1750 worldwide, with about 130 holding 10,000 accessions of agricultural crops, the wild relatives are poorly represented in most *ex situ* collections. Only 2–10% of the collections of global genebanks comprise CWR accessions belonging to a very small proportion of the global CWR species (Maxted and Kell, 2009). In addition to their poor representation in genebanks globally, the diversity of CWR is also threatened in the wild, due either to habitat destruction, the introduction of alien species, urbanization, or the rapid shift from traditional farming to modern farming systems, or all of the above.

The requirement for the systematic conservation of CWR has been recognized by major organizations such as the Food and Agriculture Organization of the United Nations, the International Treaty for Plant Genetic Resources for Food and Agriculture (FAO, 2001), and in a number of other international treaties and policy documents. The Convention on Biological Diversity (CBD) recognizes conservation of CWR as a global priority (FAO 2001, 2011; CBD,

*Corresponding author; e-mail: vgupta1123@rediffmail.com

 *Enhancing Crop Genepool Use: Capturing Wild Relative and Landrace Diversity for Crop Improvement* (eds N. Maxted, M. Ehsan Dulloo and B.V. Ford-Lloyd) 

2010a,b). The global strategy for plant conservation 2011–2020 (CBD, 2010a) states in Target 9 that '70 per cent of the genetic diversity of crops including their wild relatives and other socio-economically valuable plant species should be conserved'.

The Indian subcontinent is a centre of domestication and diversification of many economically useful wild plant species that are distributed in different agroecological zones, such as the western Himalayas (125 species); the eastern Himalayas (82 species); the Gangetic plains (66 species); the north-eastern region (132 species); the Indus plains (45 species); Malabar/Western peninsula (145 species) and Deccan/Eastern peninsula (91 species) (Arora and Pandey, 2008). The wild genepool occurring in biologically distributed habitats is under threat of genetic erosion and requires not only timely measures for its collection and conservation but also compilation of all requisite information to make use of its wider adaptability/tolerance/resistance to diseases and insect pests, quality attributes and biotic and abiotic characters. Despite the fact that the wild genepool holds enormous promise for adapting agriculture to climate change, the exploitation of genes from wild relatives has been limited due to difficulty in making viable crosses between wild and domesticated species. The recent advances in recombinant DNA technology have raised the prospects of the transfer of genes controlling these useful traits to cultivated species, which was not possible earlier by conventional means. Not enough importance has been placed on the conservation and systematic utilization of CWR in India's crop improvement programmes, mainly because of the fact that it is very expensive and technically challenging to use CWR in breeding programmes. When CWR are used, this can lead to the identification of new sources of raw materials and can facilitate the utilization of CWR in the development of new varieties and hybrids particularly targeted for different environments including biotic and abiotic stresses. There is a need for a well-framed pre-breeding programme so as to deliver feasible intermediate products, which in turn can be used in the development of varieties with desired traits. Thus, there is an urgent need for the development of strategies and priorities for the collection and conservation of these species.

## 20.2 Methods

### 20.2.1 Preparation of a CWR inventory

The authors referred to the published literature of Arora and Nayar (1984), Pandey *et al.* (2005) and Singh *et al.* (2013) to generate a consolidated list of CWR distributed in the different agroclimatic zones of India. Further, this list was validated for authentic taxonomic status by referring to the online database of The Plant List (www.theplantlist.org). All synonyms were replaced by currently used validated taxonomic identity and authority. Finally, this authenticated list was then subdivided into crop groups, namely cereals, pseudocereals, millets, oilseeds, vegetables and grain legumes of globally identified food crops and fibres.

### 20.2.2 Gap analysis for CWR conservation

Gap analysis is a well-established conservation methodology used worldwide by ecological conservationists to identify areas in which selected elements of biodiversity are represented to prioritize these elements for conservation. It refers to a systematic method of analysing the degree of conservation of taxa in order to identify those locations, taxa and particular traits or adaptations that are either absent or undersecure in the conservation system. Although gap analysis is a fairly recent tool in the conservation of PGRs, it is proving to be an effective instrument in the planning of complementary conservation strategies that encompass both *in situ* and *ex situ* applications (Maxted *et al.*, 2011). The present analysis, based on the passport data associated with accessions conserved in the National Genebank at the National Bureau of Plant Genetic Resources (NBPGR) and their respective ecogeographic distribution data, yielded significant points that would help to enhance conservation efforts at both the generic and geographic levels. The methodology involves comparison between the natural diversity of CWR occurring *in situ* and its representation in *ex situ* collections, which in turn help in revising national strategies from time to time for the assessment of threat, as well as prioritizing conservation efforts.

A three-step methodology was adopted as follows, to find out the gaps in geographical representation and taxa representation in the genebank:

**1.** From the inventoried list prepared as above, the distribution data of all these CWR were compiled by referring to the work of Nayar *et al.* (2009) and Singh *et al.* (2013).
**2.** From the inventoried list, a list of CWR that were present in the National Genebank (NGB) was compiled to find out the gaps in the representation of taxa in the NGB. Further, information on the number of accessions of each taxa conserved in the NGB was also compiled.
**3.** Passport data of all the CWR present in the NGB, especially the names of places from where these collections were made, were compiled. This was further compared with the distribution data collated in step 1 above to ascertain the gaps in the collection and conservation of CWR in the NGB.

### 20.2.3 Prioritization of the taxa for conservation

After the gap analysis on the basis of availability in the NGB, the taxa were categorized into three categories:

**1.** High priority taxa, where a taxa is either absent or has less than 20 accessions.
**2.** Medium priority taxa, where a taxa is represented by more than 20 but less than 40 accessions.
**3.** Low priority taxa, where a taxa is represented by more than 40 accessions.

The targeted area of collection for these prioritized taxa for their conservation in the NGB is based on the 22 agrobiodiversity regions suggested by Nayar *et al.* (2009).

## 20.3 Results and Discussion

### 20.3.1 Inventory of CWR

Information on and the potential traits of CWR collected from different areas and conserved in the NGB are presented in Table 20.1. The CWR inventory prepared by comparing the passport data of collections conserved at the NGB and the respective ecogeographic distribution data revealed 238 CWR belonging to 54 taxa (Table 20.2). The family with highest number of CWR is the Poaceae, with 19 genera and 72 species, followed by the Fabaceae, with 10 genera and 58 species. The NGB conserves a total of 4121 accessions of CWR, which represents only 1.02% of the total collections (as on 30 September 2014, the total NGB holdings were 407,068, including safety duplicates and trial material). These CWR collections also include accessions with potential traits for utilization in crop improvement programmes (Table 20.1). There are relatively more CWR (18 species) with traits for biotic stress in comparison to those for abiotic stress (6 species) conserved in the NGB.

### 20.3.2 Prioritization for conservation

Based on the passport data, analysis of the information on the distribution range of the inventoried CWR occurring in India revealed that out of the 238 CWR species, 210 were categorized as high priority CWR for collection and conservation, as they were either not conserved or were represented by less than 20 accessions in the NGB, whereas medium and low priority species were 14 each. The *in situ* gap analysis, based on geographic distribution and conservation efforts by government authorities, identified five CWR-rich areas that could be designated as probable CWR reserves. These are the cold arid desert (10 species), western Himalayas (47 species), Deccan region (31 species), Konkan region (31 species) and the Malabar region (18 species). The cold arid desert has two endemic species of *Allium*, not yet collected along with CWR of the vegetable group. Among the species, *Allium chitralicum* F.T. Wang & Tang, *Allium gilgiticum* F.T. Wang & Tang, *Coix lacryma-jobi* var. *ma-yuen* (Rom. Caill.) Stapf., *Vigna hainiana* Babu, Gopin. & S.K. Sharma, *Vigna khandalensis* (Santapau) Sundararagh. and Wadhw and *Crotalaria madurensis* Wight & Arn. need to be collected as top priority, as they are endemic and are not represented in the NGB collection. Although the government of India has implemented an extensive *in situ* programme through establishing 69 National Parks and 410 Wildlife Sanctuaries, these are mainly for the animal genetic diversity occurring in the protected areas. The Western Ghats and Deccan Plateau

**Table 20.1.** CWR of potential value collected/conserved at the NGB. (From Pandey *et al.*, 2005; Singh *et al.*, 2013.)

| Botanical name of taxa (crop) | Traits of potential value | Area of collection |
|---|---|---|
| *Oryza nivara* (rice) | Resistance to *Grassy stunt virus* | Basti and Gonda, Uttar Pradesh |
| *Porteresia coarctata* (rice) | Hardiness trait for saline habitat | Sundarbans Delta, West Bengal |
| *Eleusine compressa* (finger millet) | Hardiness and drought tolerance | North-western arid tract |
| *Vigna mungo* var. *silvestris* (urdbean) | Tolerance to *Yellow mosaic virus* | Western and Eastern Ghats |
| *Vigna radiata* var. *sublobata* (mungbean) | Tolerance to *Yellow mosaic virus* | Western and Eastern Ghats |
| *Cicer soongaricum* syn. *Cicer microphyllum* (chickpea) | Cold hardiness and high number of seeds per pod | Lahaul and Spiti in cold arid region |
| *Sesamum laciniatum* (sesame) | Disease and pest resistance | Coastal Andhra Pradesh and Tamil Nadu |
| *Sesamum mulayanum* | Resistance to phyllody and powdery mildew | Western Ghats |
| *Sesamum prostratum* | Resistance to phyllody and powdery mildew | Western Ghats |
| *Linum perenne* (linseed) | Cold hardiness | Lahaul and Spiti, Himachal Pradesh and western Himalayas |
| *Citrus latipes* (lemon and limes) | Cold hardiness | Shillong and Khasi and Jaintia Hills |
| *Abelmoschus tuberculatus* (okra) | Tolerance to *Yellow mosaic virus* and fruit borer | Northern India |
| *Abelmoschus manihot* (okra) | Tolerance to *Yellow mosaic virus* and fruit borer | Northern India |
| *Cajanus scarabaeoides* (pigeonpea) | Resistance to pod fly damage and diseases | Western Peninsula, India Bihar, Deccan region |
| *Cajanus sericeus* (pigeonpea) | Resistance to pod fly damage and diseases | Western Peninsula, India Bihar, Deccan region |
| *Rhynchosia bracteata* (pigeonpea) | Resistance to pod fly damage and diseases | Western Peninsula, India Bihar, Deccan region |
| *Solanum incanum* (brinjal) | Resistance to disease and fruit borer | Shivalik ranges of western Himalayas |
| *Solanaum americanum* syn. *Solanum nigrum* (brinjal) | Resistance to bacterial wilt | Throughout India |
| *Solanum anguivi* syn. *Solanum indicum* (brinjal) | Resistance to Fusarium wilt | Tropical India |
| *Solanum khasianum* (brinjal) | Resistance to shoot and fruit borer | Western Himalayas, north-east hill region, Eastern Ghats |
| *Solanum torvum* (brinjal) | Resistance to Verticillium wilt and bacterial wilt | Tropical India |
| *Solanum sisymbriifolium* (brinjal) | Resistance to shoot and fruit borer | Eastern Himalayas |
| *Aegilops tauschii* (wheat) | Disease resistance | Stony alpine region of north-western Himalayas |
| *Glycine soja* (soybean) | Adaptability to colder region and shorter season | Eastern Himalayas |

alone have 15 National Parks and about 70 Wildlife Sanctuaries. These areas also protect the floristic diversity of the CWR present by virtue of its existence, but still a large number of high priority CWR present in other areas need to be brought under the network of protected areas such as the Lahaul-Spiti Valley in cold arid desert, where endemic cold-tolerant species like *A. chitralicum*, *A. gilgiticum*, *Cicer microphyllum* and other species occur.

The *ex situ* gap analysis of 22 agrobiodiversity zones of India based on the collections made

**Table 20.2.** Priority for collection and conservation of CWR in India (based on number of accessions conserved in the NGB).

| Agrobiodiversity hotspot region | High priority species | Medium priority species | Low priority species |
|---|---|---|---|
| Cold desert region | *Allium carolinianum* DC, *Allium chitralicum* F.T. Wang & Tang, *Allium gilgiticum* F.T. Wang & Tang, *Trigonella emodi* Benth., *Aegilops tauschii* Cosson, *Sorghum bicolor* Moench subsp. *drummondii* (Steud.) De Wet. ex Davidse, *Lathyrus humilis* (Ser.) Spreng., *Cicer macracanthum* Popov, *Cicer microphyllum* Royle, *Linum myscrense* B. Heyne ex Wall. | | |
| Western Himalayan region | *Allium altaicum* Pall, *Allium roylei* Stearn., *Allium macleanii* Baker, *Allium rubellum* M. Bieb, *Allium cernuum* Roth, *Allium chinense* G. Don., *Amaranthus blitum* L. , *Avena barbata* Pott., *Avena fatua* var. *fatua* L., *Avena fatua* var. *pilosa* Syme., *Avena ludoviciana* Dur., *Bassia scoparia* (L.) A.J. Scott, *Digitaria tomentosa* (J. Koenig ex Rottler) Henrard, *Eleusine indica* (L.), *Eremopyrum distans* (C. Koch), *Eremopyrum orientale* (L.), *Fagopyrum cymosum* Meissn., *Eremopyrum buonapartis* (Spreng.), *Elymus dasystachys* Trin., *Elymus dahuricus* Turcz., *Hordeum agriocrithon* Aberg., *Hordeum aegiceras* Nees ex Royle., *Hordeum leporinum* Link., *Hordeum murinum* subsp. *leporinum* (Link) Arcang., *Hordeum turkestanicum* Nevski., *Hordeum brevisubulatum* (Trin.) Link., *Hygroryza aristata* (Retz.), *Panicum trypheron* Schult., *Pennisetum purpureum* Schum., *Polytoca digitata* (L.) Druce., *Setaria pumila* (Foir.) Roem. & Schult., *Setaria verticellata* (L.) P. Beauv., *Sorghum bicolor* Moench subsp. *drummondii* (Steud.) De Wet. ex Davidse, *Sorghum cernuum* (Ard.) Host, *Sorghum controversum* (Steud.) Snowden, *Sorghum nitidum* (Vahl) Pers., *Macrotyloma sar-garhwalensis* R.D. Gaur & L.R. Dangwal, *Cajanus elongatus* (Benth.) Maesen, *Lathyrus humilis* (Ser.) Spreng., *Lathyrus aphaca* L., *Mucuna bracteata* DC., *Vigna radiata* var. *setosa*, *Brassica quadrivalvis* Hook. f. & Thomson, *Linum strictum* L., *Linum perenne* L., *Brassica tournefortii* Gouan, *Carthamus oxyacantha* M. Bieb. | *Abelmoschus esculentus* (L.) Moench, *Allium schoenoprasum* L., *Amaranthus gangeticus* L., *Carthamus lanatus* L. | *Abelmoschus moschatus* Medic., *Abelmoschus crinitus* Wall., *Pennisetum pedicellatum* Trin. |
| Eastern Himalayan region | *Allium sikkimense* Baker, *Allium pratti* Wright, *Allium rhabdotum* Stearn, *Cucumis hystrix* Chak, *Digitaria sanguinalis* (L.), *Hordeum glaucum* Steud. Syn., *Hordeum marinum* subsp. *glacum*, *Polytoca wallichiana* (Nees) Benth., *Porteresia coarctata* (Roxb.) Tateoka, *Glycine max* (L.) Merr. subsp. *soja* (Siebold & Zucc.) H. Ohashi | | |
| Brahmaputra valley region | *Cucumis hystrix* Chak., *Corchorus pseudo-olitorius* Islam & Zaid | | |
| Khasi–Jaintia–Garo Hills region | *Cucumis hystrix* Chak., *Momordica charantia* L., *Trichosanthes khasiana* Kundu, *Trichosanthes ovigera* Blume, *Trichosanthes wallichiana* (Ser.) Wight. *Avena fatua* var. *glabella* Malzev, *Coix lacryma-jobi var. ma-yuen*, *Digitaria cruciata* var. *esculenta* (Nees), *Cajanus elongatus* (Benth.) Maesen, *Cajanus sericeus* (Baker) Maesen, *Mucuna monosperma* Wight | | |

*Continued*

**Table 20.2.** Continued.

| Agrobiodiversity hotspot region | High priority species | Medium priority species | Low priority species |
|---|---|---|---|
| North-eastern hill region | *Cucumis hystrix* Chak., *Momordica subangulata* Blume, *Trichosanthes cordata* Roxb, *Trichosanthes ovigera* Blume syn. *Trichosanthes himalensis* C.B. Clarke, *Trichosanthes wallichiana* (Ser). Wight, *Crotalaria tetragona* Andrews, *Coix aquatic* Roxb., *Eleusine compressa* (Forssk.), *Oryza granulata* Nees et Arn. ex Steud, *Oryza meyeriana* (Zoll. & Moritzi) Baill., *Oryza minuta* J. Presl, *Panicum atrosanguinem* Hochst. ex A. Rich., *Polytoca wallichiana* (Nees) Benth, *Porteresia coarctata* (Roxb.) Tateoka, *Setaria sphacelata* (Sch.) Stapf. et C.E. Hubb. ex M.B. Moss., *Cajanus elongatus* (Benth.) Maesen, *Cajanus grandiflorus* (Baker) Maesen, *Cajanus mollis* (Benth.) Maesen, *Mucuna monosperma* Wight, *Vigna prainiana* Babu & S.K. Sharma, *Vigna clarkei* Prain, *Vigna trilobata* (L.) Verdc., *Cajanus goensis* Dalzell. | *Trichosanthes nervifolia* L. | *Oryza nivara* Sastry et Sharma, *Pennisetum pedicellatum* Trin. |
| Arid western region | *Cucumis prophetarum* L., *Urena lobata* L., *Crotalaria burhia* Benth., *Setaria intermedia* Roem. & Schult., *Cajanus albicans* (Wight & Arn.) Maesen, *Cajanus platycarpus* (Benth.) Maesen, *Cajanus sericeus* (Baker) Maesen, *Vigna dalzelliana* (Kuntze) Verdc., *Brassica rapa* L., *Linum mysorense* B. Heyne ex Wall., *Linum perenne* L. | | *Cucumis melo* var. *momordica* L. |
| Malwa Plateau and Central Highlands region | *Cucumis prophetarum* L., *Corchorus fascicularis* Lamk., *Hibiscus radiatus* Cav., *Pennisetum purpureum* Schum., *Polytoca digitata* (L.) Druce., *Setaria pumila* (Poir.) Roem. & Schult., *Setaria viridis* (L.) P. Beauv., *Sorghum bicolor* (L.) Moench, *Sorghum controversum* (Steud.) Snowden, *Sorghum nitidum* (Vahl) Pers., *Sorghum arundinaceum* (Desv.) Stapf, *Cajanus cajanifolius* (Haines) Maesen, *Mucuna pruriens* var. *utilis* (Wall. ex Wight) L.H. Bailey, *Vigna hainiana* Babu, Gopin. & S.K. Sharma | | |
| Kathiawar region | *Solanum purpurelineatum* Sabnis & Bhatt, *Cajanus platycarpus* (Benth.) Maesen, *Cajanus sericeus* (Baker) Maesen | | |
| Bundelkhand region | *Abelmoschus manihot* (L.) Medik. subsp. *tetraphyllus* (Roxb. ex Hornem.) Borss. var. *pungens* (Roxb.) Hochr., *Cucumis prophetarum* L., *Sorghum purpureosericeum* (A. Rich.) Schweinf. & Asch. | *Cucumis hardwickii* Royle, *Cucumis melo* L. | *Abelmoschus tuberculatus* Pal & Singh, *Abelmoschus crinitus* Wall, *Cucumis melo* L. |
| Upper Gangetic Plains | *Allium stracheyi* Baker, *Allium wallichii* Kunth., *Trigonella obcordata* Benth., *Trigonella occulta* Delile, *Boehmeria macrophylla* Hornem., *Amaranthus atropureus* Roxb., *Amaranthus blitum* L., *Panicum curviflorum* Hornem., *Panicum psilopodium* Trin., *Pennisetum purpureum* Schum., *Polytoca digitata* (L.) Druce., *Setaria pumila* (Poir.) Roem. & Schult., *Sorghum bicolor* (L.) Moench, *Sorghum nitidum* (Vahl) Pers., *Lathyrus aphaca* L., *Brassica rapa* L. | | *Abelmoschus tuberculatus* Pal & Singh, *Abelmoschus crinitus* Wall |

| | | | |
|---|---|---|---|
| Lower Gangetic Plains | *Momordica subangulata* Blume, *Trigonella occulta* Delile, *Hibiscus radiatus* Cav., *Amaranthus blitum* L. | | |
| Gangetic Delta region | *Coix lacryma-jobi* var. *ma-yuen* (Rom. Caill.) Stapf., *Echinochloa stagnina* (Retz.), *Leersia hackelii* Keng, *Pennisetum purpureum* Schum., *Polytoca digitata* (L.) Druce., *Setaria glauca* (L.) P. Beauv, *Setaria pumila* (Poir.) Roem. & Schult., *Sorghum bicolor* (L.) Moench, *Sorghum nitidum* (Vahl) Pers., *Vigna marina* (Burm.) Merr. | | *Abelmoschus manihot* (L) Medik subsp. *manihot* |
| Chota Nagpur region | *Sorghum controversum* (Steud.) Snowden | | |
| Bastar region | *Amaranthus spinosus* L. | | |
| Koraput region | *Trigonella occulta* Delile, *Amaranthus dubius* Mart. ex Thell, *Amaranthus tenuifolius* Willd, *Amaranthus viridis* L., *Oryza rufipogon* Griff. | *Oryza sativa* f. *spontanea* Roshev. | *Abelmoschus ficulneus* (L.) Wight & Arn., *Oryza officinalis* Wall. ex Watt |
| Southern Eastern Ghats region | *Solanum erianthum* D. Don, *Hibiscus eriocarpus* DC. , *Hibiscus tiliaceus* L., *Amaranthus tricolor* L., *Amaranthus paniculatus* L., *Amaranthus dubius* L., *Amaranthus tenuifolius* Willd., *Amaranthus viridis* L., *Echinochloa crusgalli* L., *Canavalia cathartica* Thouars | *Amaranthus dubius* Mart. ex Thell | *Solanum incanum* L., *Pennisetum pedicellatum* Trin. |
| Kaveri region | *Solanum nigrum* L., *Solanum virginianum* L. syn. *Solanum surattense* Burm. f., *Solanum xanthocarpum* Schard & H. Wendl., *Crotalaria madurensis* Wight & Arn., *Canavalia cathartica* Thouars, *Sesamum alatum* Thonn. | | *Solanum incanum* L. |
| Deccan region | *Solanum kurzii* Brace ex Prain, *Corchorus acutangulus* Lam., *Corchorus urticaefolius* Wight & Arn., *Corchorus tridens* L., *Hibiscus ficulneus* L., *Boehmeria platyphylla* D. Don, *Crotalaria rigida* B. Heyne, *Chionachne koenigii* Spreng, *Amaranthus blitum* L., *Amaranthus spinosus* L., *Bassia scoparia* (L.) A.J. Scott, *Chionachne semiteres* (Benth.) Henrard, *Coix lacryma-jobi* var. *ma-yuen* (Rom. Caill.) Stapf, *Echinochloa colona* (L.), *Oryza meyeriana* (Zoll. & Moritzi) Baill., *Panicum atrosanguineum* Hochst. ex A. Rich., *Panicum garadei* Raghavan and Karthick, *Panicum hippothrix* K. Schum, *Panicum notatum* Retz, *Panicum turgidum* Forssk, *Sorghum halepense* (L.) Pers., *Trilobachne cookei* (Stapf.) Schenck ex Henrard, *Cajanus volubilis* Blanco, *Cajanus albicans* (Wight & Arn.) Maesen, *Cajanus cajanifolius* (Haines) Maesen, *Neonotonia wightii* (Wight & Arn.) J.A. Lackey, *Canavalia cathartica* Thouars, *Vigna trilobata* (L.) Verdc. var. *pusilla*, *Cajanus heynei* (W. & A.) Maesen, *Cajanus rugosus* (Wight & Arn.) Maesen., *Cajanus trinervius* (DC.) Measen | *Amaranthus gangeticus* L., *Amaranthus hybridus* L., *Sesamum radiatum* Schumach. & Thonn. | *Amarathus caudatus* L., *Oryza rufipogon* Griff., *Pennisetum pedicellatum* Trin., *Sesamum mulayanum* N.C. Nair |

*Continued*

**Table 20.2.** Continued.

| Agrobiodiversity hotspot region | High priority species | Medium priority species | Low priority species |
|---|---|---|---|
| Konkan region | *Abelmoschus angulosus* Wall. ex Wight & Arn., *Abelmoschus ficulneus* (L.) Wight & Arn., *Abelmoschus manihot* (L.) Medik subsp. *manihot*, *Cucumis ritchiei* (Chkar) Jeffrey, *Cucumis setosus*, *Momordica tuberosa* (Roxb). Cogn, *Trichosanthes anamalaiensis* Bedd., *Hibiscus furcatus* Roxb., *Crotalaria nana* Burm. f., *Amaranthus tricolor* L., *Bassia scoparia* (L.) A.J. Scott, *Chionachne gigantea* (J. Koenig) Veldkamp, *Oryza minuta* Spreng., *Chionachne semiteres* (Benth.) Henrard, *Echinochloa crusgalli* L., *Nesphostylis bracteata* (Baker) D. Potter & J.J. Doyle, *Cajanus lineatus* (Wight & Arn.) Maesen, *Neonotonia wightii* (Wight & Arn.) J.A. Lackey, *Canavalia cathartica* Thouars, *Canavalia ensiformis* (L.) DC., *Canavalia rosea* (Sw.) DC., *Vigna vexillata* (L.) A. Rich., *Vigna grandis* (Dalzell & A. Gibson) Verdc., *Vigna radiata* var. *setulosa* (Dalzell) Ohwi & Ohashi, *Vigna vexillata* var. *wightii* (Bedd.) Babu & S.K. Sharma, *Vigna khandalensis* (Santapau) Sundararagh. & Wadhw, *Vigna minima* (Roxb.) Ohwi & H. Ohashi, *Vigna bourneae* Gamble, *Vigna mungo* var. *silvestris*, *Vigna dalzelliana* (Kuntze) Verdc., *Sesamum prostratum* Retz. | *Luffa acutangula* var. *amara*, *Amaranthus hybridus* L., *Carthamus oxyacantha* M. Bieb. | *Abelmoschus crinitus* Wall, *Cucumis melo* L., *Vigna radiata* var. *sublobata* (Roxb.) Verdc., *Vigna trilobata* (L.) Verdc. |
| Malabar region | *Momordica tuberosa* (Roxb). Cogn, *Solanum aculeatissimum* Jacq., *Solanum erianthum* D. Don., *Trichosanthes tricuspidata* Lour, *Solanum kurzii* Brace ex Prain, *Boehmeria glomerulifera* Miq., *Crotalaria clarkei* Gamble, *Crotalaria digitata* Hook., *Crotalaria grahamiana* Wight & Arn., *Crotalaria striata* DC., *Cajanus lineatus* (Wight & Arn.) Maesen, *Cajanus candollei*, *Neonotonia wightii* (Wight & Arn.) J.A. Lackey, *Canavalia cathartica* Thouars, *Vigna adenantha* (G. Mey.) Marechal & al., *Vigna pilosa* (Klein ex Willd.) Baker, *Vigna dalzelliana* (Kuntze) Verdc., *Sesamum laciniatum* Klein ex Willd. | *Abelmoschus manihot* (L.) Medik subsp. *manihot*, *Momordica charantia* L., *Solanum torvum* Sw., *Solanum nigrum* L., *Sesamum malabaricum* Burm. | *Trichosanthes cucumerina* L., *Solanum incanum* L. |
| Island region | *Crotalaria striata* DC., *Crotalaria retusa* L., *Oryza rufipogon* Griff., *Canavalia rosea* (Sw.) DC., *Mucuna monosperma* Wight, *Vigna marina* (Burm.) Merr. | | |

and their conservation at the NGB, north-eastern hill region (NEH), Malwa Plateau, Central Highlands and Gangetic Delta have been identified as diversity-rich areas for 69 species of vegetable and 27 of fibre crops. Since these areas are under direct human-mediated environmental changes, intensive collection efforts should be made to collect and conserve the maximum diversity in the NGB. CWR of taxa of *Cucumis*, *Trigonella*, *Amaranthus*, *Corchorus*, *Crotolaria*, *Momordica* and *Trichosanthes*, which are predominant in these areas, need immediate attention. Further analysis of various agrobiodiversity zones for the occurrence of these species is needed, and areas such as the Western Himalayan region, north-eastern hill region, Deccan Plateau and Konkan and Malabar regions are highly targeted for future collections and conservation thereof.

## 20.4 Future Needs and Conclusions

The design and implementation of national strategies are essential for the effective conservation of CWR. National government authorities and policy makers are required to develop protocols in view of national interests in the emerging intellectual property rights regimes. Since the aim is not simply to conserve the prioritized species but also to let their genetic make-up evolve under the changing natural conditions of biotic and abiotic stresses, conservation strategies should be based, therefore, on the assessment of the conservation status of prioritized CWR using both *in situ* and *ex situ* approaches. There is a an urgent need to develop a national CWR conservation strategy to guide national policy development programmes involving extensive trait-specific and area-specific explorations and collection missions to fill the gaps in the conservation of CWR in the NGB. Further, a national effort by a multidisciplinary team of scientists for characterization (including molecular characterization), evaluation and pre-breeding programmes, using biotechnological tools wherever required, should be made to advance the efficient utilization of conserved CWR in enhancing food and nutritional security.

## References

Arora, R.K. and Nayar, E.R. (1984) *Wild Relatives of Crop Plants in India.* Scientific Monograph 7. National Bureau of Plant Genetic Resources, New Delhi.

Arora, R.K. and Pandey, A. (2008) Conservation of crop wild relatives in India. *Crop Wild Relative Newsletter* 6, 16–18.

CBD (Convention on Biological Diversity) (2010a) *Global Strategy for Plant Conservation*. Secretariat of the Convention on Biological Diversity, Montreal, Canada.

CBD (2010b) *Strategic Plan for Biodiversity 2011–2020*. Secretariat of the Convention on Biological Diversity, Montreal, Canada.

FAO (Food and Agriculture Organization) (2001) *The Second Global Plan of Action for Plant Genetic Resources for Food and Agriculture*. FAO, Rome.

FAO (2010) Second report on *The State of the World's Plant Genetic Resources for Food and Agriculture.* FAO, Rome.

FAO (2011) *International Treaty on Plant Genetic Resources for Food and Agriculture*. FAO, Rome.

Maxted, N. and Kell, S. (2009) *Establishment of a Global Network for the In Situ Conservation of Crop Wild Relatives: Status and Needs*. Background study paper No 39. Commission on Genetic Resources for Food and Agriculture, FAO, Rome.

Maxted, N., Castañeda, N.P., Álvarez, H., Vincent, A. and Magos Brehm, J. (2011) *Gap Analysis: A Tool for Genetic Conservation in Collecting Plant Diversity: Technical Guidelines Update*. Bioversity International, Rome.

Nayar, M.P., Singh, A.K. and Nair, K.N. (2009) *Agrobiodiversity Hotspots in India: Conservation and Benefit Sharing*. Protection of Plant Varieties & Farmers' Rights Authority, New Delhi, 309 pp.

Pandey, A., Bhandari, D.C., Bhatt, K.C., Pareek, S.K., Tomar, A.K., *et al.* (2005) *Wild Relatives of Crop Plants in India: Collection and Conservation.* National Bureau of Plant Genetic Resources, New Delhi, 73 pp.

Singh, A.K., Rana, R.S., Mal, B., Singh, B. and Agrawal, R.C. (2013) *Cultivated Plants and Their Wild Relatives in India – An Inventory*. Protection of Plant Varieties and Farmers' Rights Authority, New Delhi, 215 pp.

# 21 Strategies for Detecting Climate Adaptations in Wild Pearl Millet for Future Breeding Use

C. Berthouly-Salazar,[1]* C. Mariac,[1] M. Couderc,[1] I.S. Ousseini,[1] S. Santoni,[2] M. Tenaillon[3] and Y. Vigouroux[1]
[1]*Institut de Recherche pour le Développement (IRD), UMR Diversité, Adaptation et Développement des Plantes (DIADE), Montpellier, France;* [2]*UMR AGAP 1334, INRA, Montpellier, France;* [3]*CNRS, UMR0320/UMR8120 Génétique Végétale, Gif-sur-Yvette, France*

## 21.1 Introduction

The current climatic change trajectory predicts major modifications of average temperature worldwide (Bindoff *et al.*, 2013). Regional temperature rise accompanied by rainfall scarcity will have major impacts on agriculture. It is expected that agriculture in sub-Saharan Africa will be particularly affected by these changes (Niang *et al.*, 2014). Sub-Saharan cereals agriculture relies mainly on native drylands cereals such as pearl millet and sorghum and non-native Asian rice and maize. In West Africa, where 16% of the population is undernourished (FAOSTAT, 2012), African native cereals adapted to arid and poor soil conditions play a crucial role in food security. Yet, with warmer conditions, it is expected that their production will be strongly affected. Lobell and Burke (2008) estimated that a reduction of one standard variation in rainfall during the growing season could damage millet production by as much as 10%. This emphasizes the critical need for both change in cropping systems and adapting local varieties to future harsher conditions. While improvement of local varieties could partially rely on the standing genetic variation available within the cultivated compartment, the unexplored yet greater reservoir of alleles contained in the wild relatives of crops offers a promising alternative.

Standing genetic variations in local crops can be an important source for adaptations. Vigouroux *et al.* (2011) showed that Nigerian pearl millet varieties had adapted to more arid conditions since the beginning of the drought episode in the 1970s. Hence, recent pearl millet varieties have a reduced flowering cycle, an effective adaptation to drought. Genetic analyses revealed that such a reduction was linked to an increased frequency of a specific allele from the PHYTOCHROME C gene, a gene link to photoperiodic response (Vigouroux *et al.*, 2011). This study demonstrated that within a few decades, farmers were able to mobilize standing genetic variation to adapt their crops to changing environmental conditions. Yet a common feature of domesticated genomes is the reduction of genetic diversity in crops compared with their wild progenitors (Glémin and Bataillon, 2009). In maize, genetic diversity in the cultivated compartment represents about

*Corresponding author; e-mail: cecile.berthouly@ird.fr

88% of the diversity contained in the wild compartment (Vigouroux *et al.*, 2005). Lentil domestication led to a loss of genetic diversity of approximately 40% (Alo *et al.*, 2011). For pearl millet, genetic loss between cultivated and wild types has been estimated at 20% (Mariac *et al.*, 2006; Oumar *et al.*, 2008). While the diversity of the cultivated compartment can be limiting, wild relatives' compartments encompass an interesting reservoir of diversity, and therefore of adaptations (Guarino and Lobell, 2011). Hence, wild relatives have encountered more extreme conditions and are therefore expected to carry greater resistance to biotic (pest, disease) or abiotic (drought, salinity, soil acidity) stresses, important traits that have been and could further be used for crop improvements (Hajjar and Hodgkin, 2007; Robert *et al.*, 2011). Yet characterizing, making use of and conserving such 'wild diversity' is challenging and ongoing efforts are currently being undertaken in this direction, as shown by publications mentioned in this book.

Until recently, heavy experimental designs with multiple generation backcrosses were needed to identify the genomic regions involved in adaptation. The number of markers, the availability of mapping populations and readily measurable phenotypes were all limiting factors for linkage mapping studies. The recent development of next generation sequencing (NGS) technologies has transformed genomic studies profoundly by opening the door to the acquisition of genome-wide marker data sets. By combining population genetics approaches and genome-wide sequencing, it has now become possible to reveal the genomic regions associated with adaptations. Such genome selection scans based on the pattern of diversity could be useful, particularly for traits lacking readily measurable phenotypes.

NGS produces millions of single nucleotide polymorphisms (SNPs) distributed across the whole genome, allowing for wide genome screening. At first, NGS was used entirely to sequence genomes of model organisms, including a few crops. Subsequently, those reference genomes were used to perform resequencing studies aiming at evaluating and investigating the patterns of genomic diversity within species. More recently, the development of genome reduction complexity techniques, such as 'genotyping-by-sequencing' (GBS; Elshire *et al.*, 2011), RADseq (Hohenlohe *et al.*, 2010) or transcriptome sequencing (RNAseq approach), that can be applied without the need of a reference genome (Lu *et al.*, 2013), have broadened the spectrum of species to non-model organisms. Selection scans can now therefore be applied to any organism.

Selection scans take advantage of the unusual polymorphism pattern of genomic regions targeted by selection, measured as deviations from a neutral model of evolution, to reveal it (Tenaillon and Tiffin, 2008). This unusual pattern includes significant local reduction – in the case of positive selection – or increase – in the case of balancing selection – of genetic diversity. In crops, such an approach was first undertaken on a small scale (few dozens of loci) to identify genes subject to selection during maize domestication (Vigouroux *et al.*, 2002; Tenaillon *et al.*, 2004; Yamasaki *et al.*, 2005). Later, genome scans in crops were coupled with NGS to reveal post-domestication selection of genes in maize landraces and cultivars (Hufford *et al.*, 2012, 2013) and of genes involved in the regulation of flowering time and phenology in wheat (Cavanagh *et al.*, 2013). Selection scans often identify a large set of 'candidate genes' compared to conventional quantitative trait loci (QTL) mapping, which interrogates only a few hundred loci at a time. A study based on 774 genes extrapolates that 2–4% of the maize genes have been subject to selective events during or after domestication (Wright *et al.*, 2005). Similar trends were observed for other studied plants. For example, a study found that 36 out of 492 (7%) sunflower genes exhibited evidence of selection (Chapman *et al.*, 2008). Similarly, about 4.38% of genes were affected significantly by selection during soybean domestication and genetic improvement (Li *et al.*, 2013).

However, recent studies have shown that selection scan approaches lead to a high number of false positives. Nielsen *et al.* (2007) compared a total of 803 genes identified as under selection by two different studies (Voight *et al.*, 2006; Wang *et al.*, 2006); they found that only seven genes were shared between them. The main reason for false positive discovery is that assumptions underlying the neutral model are in many cases unrealistic, thereby biasing statistical tests. The underlying neutral model can indeed be affected by many endogenous (Bierne *et al.*, 2011) or exogenous factors (Bierne *et al.*, 2013a,b). Briefly, two types of tests have been developed these past decades to perform selection scans. The first type are $F_{ST}$-based tests and rely on the assumption that if a locus is under selection, it will deviate from the neutral Wright distribution model (Beaumont and Nichols, 1996; Vitalis and Couvet, 2001; Foll and Gaggiotti, 2008; Bonhomme *et al.*, 2010). It allows comparisons of multiple populations, but $F_{ST}$-based tests are sensitive to the population structure and demographic histories. When spatial structure departs from Wright's island model of population subdivision, $F_{ST}$-based tests produce a high rate of false positives (Bierne *et al.*, 2013b). Besides, strong differentiation between populations and small selection coefficients considerably affect the efficiency of these methods (De Mita *et al.*, 2013). The second type of tests are based on correlation between genetic and environmental data (Joost *et al.*, 2007; Coop *et al.*, 2010; Poncet *et al.*, 2010). In addition to identifying targets of selection, such tests allow characterizing environmental factors governing selection, but more sampling points are needed. Correlation-based approaches are less, although still sensitive to the demographic model and strength of selection (De Mita *et al.*, 2013), but are more sensitive to patterns of isolation by distance than $F_{ST}$-based tests. A review of tests and their limitations can be found in Joost *et al.* (Joost *et al.*, 2013). Overall, genome scans tend to overestimate the number of loci under selection, and they are sensitive to current and past demographic histories including gene flow and underlying population structure. To overcome these limitations, one could consider investigating selection patterns along an environmental or domestication 'gradient' (i.e. wild–cultivated populations). For $F_{ST}$-based methods when a limited number of populations are sampled, the impact of population demographic history could be corrected by sampling at least two gradients and populations at the extremity of each gradient. The rationale behind the use of multiple gradients is to use them as replicates and identify common outlier loci among them to correct for false positives. When correlation-based methods are favoured, intensive population sampling along gradients is key (De Mita *et al.*, 2013). Regardless of the approach, multiple detection tests should be used as advocated by Pérez-Figueroa (Pérez-Figueroa *et al.*, 2010). Finally, coupling these selection scans with modelling approaches that incorporate past population histories into detection tests has been deemed an efficient method to limit false positive discovery (Clotault *et al.*, 2012; Roux *et al.*, 2013).

Pearl millet is a native sub-Saharan cereal that plays an important role in food security for the region. Genetic and archaeological studies identified a domestication centre in north Mali and north-west Niger around 4500 years ago (Oumar *et al.*, 2008; Manning *et al.*, 2011). The wild pearl millet distribution area ranges from Senegal to Sudan in the sub-Saharan region. The distribution area of wild pearl millet in Africa includes a maximum difference of 8°C in annual temperatures and 1300 mm in annual precipitation (www.worldclim.org/bioclim, data 2012), with altitudes ranging from 0 to above 1000 m. In this chapter, we present two strategies applied in two ongoing projects aimed at identifying the climatic adaptations in wild pearl millet (*Pennisetum glaucum* (L.) R. Br. ssp. *monodii*).

## 21.2 Methods and Results

The two projects presented below differ by the sampling and genotyping strategies. The first project includes a duplicate gradient and an RNAseq approach on four wild pearl millet populations, each distributed at the extremity of the two duplicated climatic gradients. The second project includes 48 populations distributed along a wider climatic and geographic gradient and uses a GBS approach (Table 21.1). It encompasses wider climatic conditions and more sampling points, as needed for the correlation-based detection tests.

The two genotyping methodologies employed include a genome reduction approach, and the main characteristics are presented in Table 21.2. The RNAseq approach is based on RNA transcripts and the availability of an assembled transcriptome. From a technical point of view, the main disadvantage of RNAseq is that it is a relatively costly method (around US$200 per sample) and good quality RNA extraction is less straightforward than DNA extraction. The GBS approach uses restriction enzymes (here *ApeKI*) to reduce genome

Table 21.1. Summary of characteristics and climatic gradients. Climatic variables are expressed in maximum difference.

| | Project 1 | | |
|---|---|---|---|
| | Gradient 1 | Gradient 2 | Project 2 |
| Sampling size: populations * samples | 2*18 | 2*18 | 48*2 |
| Genotyping approach | RNAseq | RNAseq | GBS |
| Altitude (m) | 317 | 295 | 903 |
| Annual mean temperature (°C) | 2.38 | 1.5 | 5.9 |
| Maximum temperature of warmest month | −1.1 | 0.1 | 9.6 |
| Minimum temperature of coldest month | 4.4 | 3.9 | 9.7 |
| Annual precipitation (mm) | 174 | 144 | 730 |
| Precipitation of wettest month | 65 | 50 | 226 |
| Precipitation of driest month | 0 | 0 | 0 |

Table 21.2. Summary of characteristics for the two genotyping approaches.

| | RNAseq | GBS + UNEAK pipeline |
|---|---|---|
| Approximate cost | 200 US$ | 50 US$ |
| RNA versus DNA extraction | – | + |
| Reference | Yes | No |
| Reads length after sequence cleaning | 35–95 bp | 64 bp |
| Number of SNPs before filtering | 505,487 | 262,928 |
| Number of SNPs after filtering | 87,218 | 21, 913 |
| Per cent of lost SNPs after filtering | 83 | 92 |
| Coverage for unfiltered SNPs (median) | 3.4 | 2.2 |
| Coverage for filtered SNPs (median) | 7.0 | 6.3 |
| Coverage for sample before filtering (median) | 5.0 | 3.6 |
| Coverage for sample after filtering (median) | 5.9 | 7.0 |
| Mean observed heterozygosity for filtered SNPs | 0.14 | 0.16 |
| Mean expected heterozygosity for filtered SNPs | 0.21 | 0.28 |

complexity. Compared with RNAseq, GBS is less expensive and costs about $50 per sample for 95-plex including bioinformatics procedures for SNP calling. A second, very strong advantage of genotyping by GBS is that it does not necessitate an assembled reference. Indeed, a specific and very easy bioinformatics pipeline has been developed to allow detecting single base polymorphism without the need of any reference, neither genome nor transcriptome (Lu *et al.*, 2013). Referred to as the UNEAK pipeline, it first performs a filtering based on reads and bases quality. Reads are then trimmed to 64 bp, identical reads are clustered into tags, and counts of these tags present in each bar-coded individual are stored. To be able to identify SNPs, the UNEAK pipeline performs a pairwise alignment of these tags, and only tag pairs with 1 bp mismatch are considered as candidate SNPs. With a certain error tolerance rate (here set to 0.03), only reciprocal pairs of tags are retained as SNPs. Briefly, in the RNAseq pipeline, we performed a first cleaning of raw sequences by removing bases of low quality and reads below 35 bp. We then mapped our clean reads to the reference; in this case, the transcriptome using BWA (Burrows-Wheeler Aligner) alignment software (Li and Durbin, 2009). To call SNPs and genotypes, we used the GATK (Genome Analysis Toolkit) and unified genotyper option (McKenna *et al.*, 2010).

According to the characteristics of both pipelines, we expect the UNEAK pipeline to be more conservative because the reads used are only 64 bp, while with our RNAseq analysis pipeline, they can range up to 95 bp. In addition, by allowing only one possible mismatch between 64 bp tags, the UNEAK pipeline will discard regions with very high diversity.

For both approaches, identified SNPs and genotypes are stored in a VCF (Variant Calling Format) file, where the position of SNPs, if available, SNP alleles, and coverage and quality descriptors information are stored. SNPs can be filtered according to coverage values or quality values. When a reference is available as for RNAseq, additional filtering can be performed according to mapping quality values. Both data sets were filtered for missing data, depth and quality according to their distribution values. Before filtering, we observed around 500,000 SNPs with a mean coverage of 6.4 ± 10.2 for the RNAseq approach on four populations (72 samples). Using the GBS approach with 48 populations (95 samples), we obtained approximately 260,000 SNPs with a mean coverage of 3.3 ± 3.2. Therefore, nearly twice as less SNPs were obtained with the GBS approach compared to the RNAseq approach. This difference was accentuated after filtering, since 92% of SNPs from GBS did not pass filters, compared to 83% for RNAseq. This is due partly to a higher rate of missing data with GBS. In the end, we obtained 87,218 SNPs for RNAseq and 22,380 SNPs for GBS. Coverage of both final data sets ranged from 6 to 7 for either SNPs or individuals, a sufficient and medium-high coverage for population genetic studies. Observed and expected heterozygosity values were slightly higher with SNPs from GBS (Table 21.2). This is not surprising since more populations, but also more genetically distant populations, were analysed in the GBS data set.

## 21.3 Conclusion

We present genotyping results from two projects aimed at identifying the genes linked to adaptation in wild pearl millet populations. The two projects differed by their sampling and genotyping strategies. We focus mainly on comparing two genotyping strategies, namely the RNAseq and the GBS approaches. Both approaches consist of reducing genome complexity. An inherent propriety of the RNAseq approach is that variants will be found in coding regions. To the contrary, the GBS approach uses restriction enzymes that are expected to cut randomly along the genome; variants should therefore, in theory, be found in coding and non-coding regions. However, the enzyme used was *ApeKI*, a methylation-sensitive enzyme that was recently found to be associated with sequencing bias towards coding regions (Schilling *et al.*, 2014). Overall, we found that both genotyping methods were able to identify tens of thousands of SNPs with a fourfold difference between methods after filtration, RNAseq being able to detect more SNPs than GBS. Undeniably, the major advantages of GBS are the relatively low cost and the ‘no reference’ UNEAK pipeline.

Wild relatives have long been used for the improvement of crops, either during domestication (Matsuoka *et al.*, 2002) or in breeding programmes (Hajjar and Hodgkin, 2007). They represent a reservoir of diversity and adaptations for current crops, a role that has become primordial in the actual climate change context. However, characterizing this diversity was very difficult in many aspects: cost, experimental design, small number of QTL markers, material with mapping populations and measure of phenotypes. The development of NGS approaches has revolutionized the characterization of species diversity. Selection scans allow the revealing of the genomic regions under selection with a higher precision than QTLs and without the need of a phenotype. Therefore, with these methods we will have access to genes with small effects and to more complex phenotypes such as roots and embryo development.

It is clear that in the near future, for each species, even non-model crops, we will have access to a list of candidate genomic regions potentially linked to adaptations. At that point, the challenge will be to retrieve these adapted phenotypes in a way that will be useable by breeders.

## Acknowledgements

This work was supported by the Agence National de la Recherche with funding grants No ANR-12-RPDOC-009-01 and No ANR-12-ADAP-0002. The authors declare no conflict of interest.

## References


Alo, F., Furman, B.J., Akhunov, E., Dvorak, J. and Gepts, P. (2011) Leveraging genomic resources of model species for the assessment of diversity and phylogeny in wild and domesticated lentil. *The Journal of Heredity* 102, 315–329.

Beaumont, M.A. and Nichols, R.A. (1996) Evaluation loci for use in the genetic analysis of population structure. *Proceedings of the Royal Society, London B* 263, 1619–1626.

Bierne, N., Welch, J., Loire, E., Bonhomme, F. and David, P. (2011) The coupling hypothesis: why genome scans may fail to map local adaptation genes. *Molecular Ecology* 20, 2044–2072.

Bierne, N., Gagnaire, P. and David, P. (2013a) The geography of introgression in a patchy environment and the thorn in the side of ecological speciation. *Current Zoology* 59, 72–86.

Bierne, N., Roze, D. and Welch, J.J. (2013b) Pervasive selection or is it...? Why are FST outliers sometimes so frequent? *Molecular Ecology* 22, 2061–2064.

Bindoff, N.L., Stott, P.A., AchutaRao, K.M., Allen, M.R., Gillett, N., *et al.* (2013) Detection and attribution of climate change: from global to regional. In: Stocker, T.F., Qin, D., Plattner, G.-K., Tignor, M., Allen, S.K., Boschung, J., Nauels, A., Xia, Y., Bex, V. and Midgley, P.M. (eds) *Climate Change 2013: The Physical Science Basis. Contribution of Working Group I to the Fifth Assessment Report of the Intergovernmental Panel on Climate Change*. Cambridge University Press, Cambridge, UK, and New York, pp. 867–952.

Bonhomme, M., Chevalet, C., Servin, B., Boitard, S., Abdallah, J., *et al.* (2010) Detecting selection in population trees: the Lewontin and Krakauer Test extended. *Genetics* 186, 241–262.

Cavanagh, C.R., Chao, S., Wang, S., Huang, B.E., Stephen, S., *et al.* (2013) Genome-wide comparative diversity uncovers multiple targets of selection for improvement in hexaploid wheat landraces and cultivars. *Proceedings of the National Academy of Sciences of the United States of America* 110, 8057–8062.

Chapman, M.A., Pashley, C.H., Wenzler, J., Hvala, J., Tang, S., *et al.* (2008) A genomic scan for selection reveals candidates for genes involved in the evolution of cultivated sunflower (*Helianthus annuus*). *The Plant Cell* 20, 2931–2945.

Clotault, J., Thuillet, A.-C., Buiron, M., De Mita, S., Couderc, M., *et al.* (2012) Evolutionary history of pearl millet (*Pennisetum glaucum [L.] R. Br.*) and selection on flowering genes since its domestication. *Molecular Biology and Evolution* 29, 1199–1212.

Coop, G., Witonsky, D., Di Rienzo, A. and Pritchard, J.K. (2010) Using environmental correlations to identify loci underlying local adaptation. *Genetics* 185, 1411–1423.

De Mita, S., Thuillet, A.-C., Gay, L., Ahmadi, N., Manel, S., *et al.* (2013) Detecting selection along environmental gradients: analysis of eight methods and their effectiveness for outbreeding and selfing populations. *Molecular Ecology* 22, 1383–1398.

Elshire, R.J., Glaubitz, J.C., Sun, Q., Poland, J.A., Kawamoto, K., *et al.* (2011) A robust, simple genotyping-by-sequencing (GBS) approach for high diversity species. *PloS ONE* 6, e19379.

FAOSTAT (Food and Agriculture Organization) (2012) FAOSTAT database. FAO, Rome. Available at: faostat3.fao.org/home/E (accessed 8 December 2015).

Foll, M. and Gaggiotti, O. (2008) A genome-scan method to identify selected loci appropriate for both dominant and codominant markers: a Bayesian perspective. *Genetics* 180, 977–993.

Glémin, S. and Bataillon, T. (2009) A comparative view of the evolution of grasses under domestication. *The New Phytologist* 183, 273–290.

Guarino, L. and Lobell, D.B. (2011) A walk on the wild side. *Nature Climate Change* 1, 374–375.

Hajjar, R. and Hodgkin, T. (2007) The use of wild relatives in crop improvement: a survey of developments over the last 20 years. *Euphytica* 156, 1–13.

Hohenlohe, P.A., Bassham, S., Etter, P.D., Stiffler, N., Johnson, E.A., *et al.* (2010) Population genomics of parallel adaptation in threespine stickleback using sequenced RAD tags. *PLoS Genetics* 6, e1000862.

Hufford, M.B., Xu, X., Van Heerwaarden, J., Pyhäjärvi, T., Chia, J. M., *et al.* (2012) Comparative population genomics of maize domestication and improvement. *Nature Genetics* 44, 808–811.

Hufford, M.B., Lubinksy, P., Pyhäjärvi, T., Devengenzo, M.T., Ellstrand, N.C., *et al.* (2013) The genomic signature of crop-wild introgression in maize. *PLoS Genetics* 9, e1003477.

Joost, S., Bonin, A., Bruford, M.W., Després, L., Conord, C., *et al.* (2007) A spatial analysis method (SAM) to detect candidate loci for selection: towards a landscape genomics approach to adaptation. *Molecular Ecology* 16, 3955–3969.

Joost, S., Vuilleumier, S., Jensen, J.D., Schoville, S., Leempoel, K., *et al.* (2013) Meeting review. Uncovering the genetic basis of adaptive change: on the intersection of landscape genomics and theoretical population genetics. *Molecular Ecology* 22, 3659–3665.

Li, H. and Durbin, R. (2009) Fast and accurate short read alignment with Burrows–Wheeler transform. *Bioinformatics* 25, 1754–1760.
Li, Y., Zhao, S., Ma, J.X., Li, D., Yan, L., *et al.* (2013) Molecular footprints of domestication and improvement in soybean revealed by whole genome re-sequencing. *BMC Genomics* 14, 579.
Lobell, D.B. and Burke, M.B. (2008) Why are agricultural impacts of climate change so uncertain? The importance of temperature relative to precipitation. *Environmental Research Letters* 3, 034007.
Lu, F., Lipka, A.E., Glaubitz, J., Elshire, R., Cherney, J.H., *et al.* (2013) Switchgrass genomic diversity, ploidy, and evolution: novel insights from a network-based SNP discovery protocol. *PLoS Genetics* 9, e1003215.
McKenna, A., Hanna, M., Banks, E., Sivachenko, A., Cibulskis, K., *et al.* (2010). The Genome Analysis Toolkit: a MapReduce framework for analyzing next-generation DNA sequencing data. *Genome Research* 20, 1297–1303.
Manning, K., Pelling, R., Higham, T., Schwenniger, J.-L. and Fuller, D.Q. (2011) 4500-Year old domesticated pearl millet (*Pennisetum glaucum*) from the Tilemsi Valley, Mali: new insights into an alternative cereal domestication pathway. *Journal of Archaeological Science* 38, 312–322.
Mariac, C., Robert, T., Allinne, C., Remigereau, M.S., Luxereau, A., *et al.* (2006) Genetic diversity and gene flow among pearl millet crop/weed complex: a case study. *TAG. Theoretical and Applied Genetics* 113, 1003–1014.
Matsuoka, Y., Vigouroux, Y., Goodman, M.M., Sanchez, J., Buckler, E., *et al.* (2002) A single domestication for maize shown by multilocus microsatellite genotyping. *Proceedings of the National Academy of Sciences of the United States of America* 99, 6080–6084.
Niang, I., Ruppel, O.C., Abdrabo, M.A., Essel, A., Lennard, C., *et al.* (eds) (2014) *Climate Change 2014: Impacts, Adaptation, and Vulnerability. Part B: Regional Aspects. Contribution of Working Group II to the Fifth Assessment Report of the Intergovernmental Panel on Climate Change*. Cambridge University Press, Cambridge, UK, and New York.
Nielsen, R., Hellmann, I., Hubisz, M., Bustamante, C. and Clark, A.G. (2007) Recent and ongoing selection in the human genome. *Nature Reviews. Genetics* 8, 857–868.
Oumar, I., Mariac, C., Pham, J.-L. and Vigouroux, Y. (2008) Phylogeny and origin of pearl millet (*Pennisetum glaucum* [L.] R. Br) as revealed by microsatellite loci. *Theoretical and Applied Genetics* 117, 489–497.
Pérez-Figueroa, A., García-Pereira, M.J., Saura, M., Rolán-Alvarez, E. and Caballero, A. (2010) Comparing three different methods to detect selective loci using dominant markers. *Journal of Evolutionary Biology* 23, 2267–2276.
Poncet, B.N., Herrmann, D., Gugerli, F., Taberlet, P., Holderegger, R., *et al.* (2010) Tracking genes of ecological relevance using a genome scan in two independent regional population samples of *Arabis alpina*. *Molecular Ecology* 19, 2896–2907.
Robert, T., Khalfallah, N., Martel, E., Lamy, F., Poncet, V., *et al.* (2011) Structure and evolutionary relationships within *Pennisetum* complex of species. In: Kole, C. (ed.) *Wild Crop Relatives: Genomic and Breeding Resources, Millets and Grasses*. Springer, New York, pp. 217–255.
Roux, C., Tsagkogeorga, G., Bierne, N. and Galtier, N. (2013) Crossing the species barrier: genomic hotspots of introgression between two highly divergent *Ciona intestinalis* species. *Molecular Biology and Evolution* 30, 1574–1587.
Schilling, M.P., Wolf, P.G., Duffy, A.M., Rai, H.S., Rowe, C.A., *et al.* (2014) Genotyping-by-sequencing for *Populus* population genomics: an assessment of genome sampling patterns and filtering approaches. *PloS ONE* 9, e95292.
Tenaillon, M.I. and Tiffin, P.L. (2008) The quest for adaptive evolution: a theoretical challenge in a maze of data. *Current Opinion in Plant Biology* 11(2), 110–115.
Tenaillon, M.I., U'Ren, J., Tenaillon, O. and Gaut, B.S. (2004) Selection versus demography: a multilocus investigation of the domestication process in maize. *Molecular Biology and Evolution* 21, 1214–1225.
Vigouroux, Y., McMullen, M., Hittinger, C.T., Houchins, K., Schulz, L., *et al.* (2002) Identifying genes of agronomic importance in maize by screening microsatellites for evidence of selection during domestication. *Proceedings of the National Academy of Sciences of the United States of America* 99, 9650–9655.
Vigouroux, Y., Mitchell, S., Matsuoka, Y., Hamblin, M., Kresovich, S., *et al.* (2005) An analysis of genetic diversity across the maize genome using microsatellites. *Genetics* 169, 1617–1630.
Vigouroux, Y., Mariac, C., De Mita, S., Pham, J.L., Gérard, B., *et al.* (2011) Selection for earlier flowering crop associated with climatic variations in the Sahel. *PloS ONE* 6, e19563.
Vitalis, R. and Couvet, D. (2001) Estimation of effective population size and migration rate from one- and two-locus identity measures. *Genetics* 157, 911–925.

Voight, B.F., Kudaravalli, S., Wen, X. and Pritchard, J.K. (2006) A map of recent positive selection in the human genome. *PLoS Biology* 4, e72.
Wang, E.T., Kodama, G., Baldi, P. and Moyzis, R.K. (2006) Global landscape of recent inferred Darwinian selection for Homo sapiens. *Proceedings of the National Academy of Sciences of the United States of America* 103, 135–140.
Wright, S.I., Bi, V., Schroeder, S.G., Yamasaki, M., Doebley, J.F., *et al.* (2005) The effects of artificial selection on the maize genome. *Science* 308, 1310–1314.
Yamasaki, M., Tenaillon, M.I., Bi, I.V., Schroeder, S.G., Sanchez-Villeda, H., *et al.* (2005) A large-scale screen for artificial selection in maize identifies candidate agronomic loci for domestication and crop improvement. *The Plant Cell* 17, 2859–2872.

# 22 Assessment of the Conservation Status of Mesoamerican Crop Species and Their Wild Relatives in Light of Climate Change

E. Thomas,[1]* M. Ramirez,[1] M. van Zonneveld,[2] J. van Etten,[2]
C. Alcázar,[1] M. Beltrán,[1] D. Libreros,[1] S. Pinzón,[1]
W. Solano[3] and G. Galluzzi[1]

[1]*Bioversity International, Cali, Colombia;* [2]*Bioversity International, Turrialba, Costa Rica;* [3]*Tropical Agricultural Research and Higher Education Centre (CATIE), Turrialba, Costa Rica*

## 22.1 Introduction

The landmass from Mexico to Panama, here referred to as Mesoamerica, is one of the most vulnerable areas susceptible to the negative impacts of climate change (BID and CEPAL, 2010). Almost all climate models predict a decrease in annual precipitation and increased drought in this area in the decades to come (Magrin *et al.*, 2014). It is thought that climate change will lead to an increase in the frequency or intensity of extreme meteorological events in the region, seriously affecting national infrastructure and production systems (ECLAC, 2014). The region has already shown that it is exceedingly vulnerable to extreme events. Hurricane Mitch in 1998 and Stan in 2005 destroyed livelihoods and set the economy back several years. Recent agricultural statistics show that variations in production in the region are due mostly to climate problems, in the form of excess or lack of rainfall, and the associated incidence of pests and diseases. It will be crucial to develop strategies and efficient and appropriate adaptation actions for the agricultural sector in order to improve food security in the region, reduce its vulnerability and increase the resilience of production systems (Lobell and Gourdji, 2012).

On the other hand, adapting to change is not exactly new in the history of agriculture. Domestication of plants and the origin of agriculture coincided precisely with the global warming that marked the end of the last ice age, about 10,000 years ago (Piperno, 2011). Since then, farmers around the world have modified the adaptive capacity of crops to thrive in a wide variety of climatic conditions, resulting in the rich diversity of plant genetic resources for food and agriculture (PGRFA) we have today (Pickersgill, 2007; Gremillion *et al.*, 2014). Maize, for example, is cultivated from sea level up to over 4000 m above sea level in the Bolivian Altiplano, a direct consequence of traditional agricultural innovation. Today, farmers face changes of a nature and scale never experienced in the history of agriculture. They need to increase their abilities

*Corresponding author; e-mail: evert.thomas@gmail.com

 

to use the diversity and wisdom accumulated over several thousands of years to improve cultivated species further, adapting them to new climatic conditions and meeting the challenge of feeding a growing population using fewer external inputs. The best use of regional genetic diversity will depend on our ability to maximize the synergy of traditional agricultural practices and the body of knowledge with modern, multidisciplinary scientific methods.

Mesoamerica is one of the primary centres of domestication and diversification of food crops in the world (Zizumbo-Villarreal and Colunga-GarcíaMarín, 2010). Some of the crops that sustain world food security, notably maize and beans, as well as some globally important income-generating crops, for example cotton, squash, avocado and peppers, were domesticated in Mesoamerica from wild plants and then spread to other regions (Brown *et al.*, 2013a,b,c, 2014; Kraft *et al.*, 2014; Colunga-GarcíaMarín and Zizumbo-Villareal, 2004; Galindo-Tovar *et al.*, 2008; Trucco and Trane, 2011; Hernández-López *et al.*, 2013; Fuentes and Santamaría, 2014). Mesoamerica is also a secondary centre of diversity for crops that play a more important role in food security outside of Mesoamerica, such as cassava (Schaal *et al.*, 2006) and sweet potato (Srisuwan *et al.*, 2006). The conservation and sustainable use of plant genetic resources of these and other Mesoamerican crops, including their wild relatives, are crucial for the adaptation of agriculture within the countries in the region, but also in other countries that depend on crops of Mesoamerican origin for their food security. In other words, countries are *interdependent* in achieving climate adaptation in agriculture. Climate adaptation and related actions in conservation and sustainable use of plant genetic resources are not the exclusive prerogative or responsibility of any one country. International coordination is needed to plan, finance and execute actions in these areas.

In response to these challenges and opportunities, the 'strategic action plan to strengthen the conservation and use of Mesoamerican plant genetic resources in adapting agriculture to climate change' (Ramírez *et al.*, 2013) was developed. The plan is a 10-year road map to strengthen the conservation, access and use of plant genetic resources in Mesoamerica, as a strategic element for food security and agricultural adaptation to climate change and other threats. It is organized across six thematic components: (i) conservation, including on-farm, *in situ* and *ex situ* conservation; (ii) sustainable use; (iii) institutions and policies; (iv) education and capacity building; (v) operationalization; and (vi) financial strategy.

The elaboration of the plan started with a diagnostic analysis of the scientific evidence on the current state of plant genetic resources for food and agriculture (PGRFA) in the region, and of climate change challenges and opportunities. In the next phase, the results of this baseline analysis were used as a basis for a broad participatory process involving regional stakeholders, who jointly identified priority actions to be included in the action plan. Results of the various analyses and documents used in the analyses are available to the public in Spanish on the ITZAMNÁ website (itzamna-mesoamerica.org). In this chapter, we present part of the diagnostic studies that were carried out to evaluate the on-farm, *in situ* and *ex situ* conservation status of PGRFA in Mesoamerica. The focus of this paper is on assessing the potential impact of anthropogenic climate change on the conservation of a set of Mesoamerican crop species and their wild relatives. We then use the results of this assessment to identify priority areas for on-farm and *in situ* conservation in Mesoamerica.

## 22.2 Methodology

### 22.2.1 Data collection

A set of ten genepools of crop species and their wild relatives were selected based on the following criteria (Table 22.1): (i) crops prioritized by regional and international bodies given their regional and global importance to food security and income generation; (ii) some of the main food crops widely grown in the region; and (iii) the representativeness of crop types: a combination of grain crops (maize, beans, amaranth), tubers (cassava, sweet potatoes), vegetables (cucurbits, amaranth, *Capsicum* peppers), fodder (*Tripsacum*) and perennial fruit species (papaya and avocado). Our assumption here was that an assessment of the conservation status of these genepools would provide a valid approximation

**Table 22.1.** The ten genepools considered in the present analysis and their importance.

| Genepool | Principal crops | Hemispheric conservation strategy score[a] | REMERFI collection priority[a] | ITPGRFA Annex 1[b] | Food security | Diets | Income | Fodder | Crop wild relatives |
|---|---|---|---|---|---|---|---|---|---|
| *Zea* | Maize | 57.5 | ✓ | ✓ | GR | | GR | GR | G |
| *Phaseolus* | Beans | 60 | ✓ | ✓ | GR | R | GR | | G |
| *Manihot* | Cassava | 50 | | ✓ | G | | GR | | G |
| *Ipomoea* | Sweet potato | 40 | | ✓ | G | G | | | G |
| *Cucurbita* | Cucurbits | 30 | ✓ | | | R | | | G |
| *Amaranthus* | Amaranth | 17.5 | | | R | R | | R | G |
| *Capsicum* | Hot and sweet peppers | 60 | ✓ | | | | GR | | G |
| *Carica* | Papaya | 17.5 | | | | GR | GR | | G |
| *Persea* | Avocado | 25 | | | | | GR | | G |
| *Tripsacum* | | – | | ✓ | | | | GR | G |

*Notes*: G = global importance (significant contribution to food security in areas outside of Mesoamerica); R = regional importance; REMERFI = Red Mesoamericana de Recursos Fitogenéticos; ITPGRFA = International Treaty on Plant Genetic Resources for Food and Agriculture. [a]CAPGERNet *et al.*, 2008; [b]FAO, 2009.

of the conservation status of Mesoamerican PGRFA in general. It is important to note here that, although we analysed *Tripsacum* as a separate genepool, consisting of both cultivated species and wild relatives, all species in the *Tripsacum* genepool are also considered wild relatives of cultivated maize. For example, Prischmann *et al.* (2009) have demonstrated experimentally the potential of *Tripsacum dactyloides* genes for introducing rootworm resistance in maize, as well as other potential traits related to drought tolerance and the ability to withstand aluminium toxicity in acidic soils.

We collected georeferenced presence data in Mesoamerica (herbarium records and genebank accessions) for the ten genepools, including 26 crop species (with eight additional subtaxa) and 358 of their wild relatives (with 52 additional subtaxa; Fig. 22.1). The main data sources were international genebanks (CIAT, CIMMYT, CIP, USDA, among others; using GENESYS as an entry point), scientific publications, herbaria data and GBIF (www.gbif.org). To complement this information and get a general idea of the composition and conservation status of collections in Mesoamerican genebanks, we circulated a survey, organized in Survey Monkey, to 24 operative genebanks in the region. The survey ran from 6 August to 19 October 2012. Replies came from 18 genebanks: 7 from Costa Rica, 5 from Mexico, 2 from Honduras, and 1 each from Nicaragua, Guatemala and Panama. Unfortunately, passport data from these genebanks were not available at that time for inclusion in the present analysis. Hence, the analyses presented here are based largely on accession data from international genebanks. Where relevant, we present some of the results from the genebank survey to complement the main trends observed.

The taxonomy of cultivated and wild species was verified with the Germplasm Resources Information Network (GRIN; www.ars-grin.gov) and The Plant List (www.theplantlist.org). Given taxonomic inconsistencies and incompleteness of data at subspecies level, we carried out analyses at the species level, by pooling all intraspecific taxa together. When sufficient data points were available, we analysed at the subspecies level. A distinction was made between cultivated and wild plant taxa. For example, the cultivated *Zea mays*, only contained subsp. *mays*, whereas the other subspecies were analysed separately as wild taxa. A similar approach was used for other crop species that still exist in wild forms (as is the case for *Phaseolus*); records were classified as wild or cultivated based on the relevant details mentioned in the passport data.

We carried out an initial data cleaning exercise to ensure the quality of data points. First, we eliminated duplicates, records or accessions not identified at species level or lacking coordinates, as well as presence points for which there was a mismatch between administrative area data mentioned in the passport and obtained after projection on a map. Subsequently, we excluded points that were environmental outliers. To this end, we used the coordinates of all presence

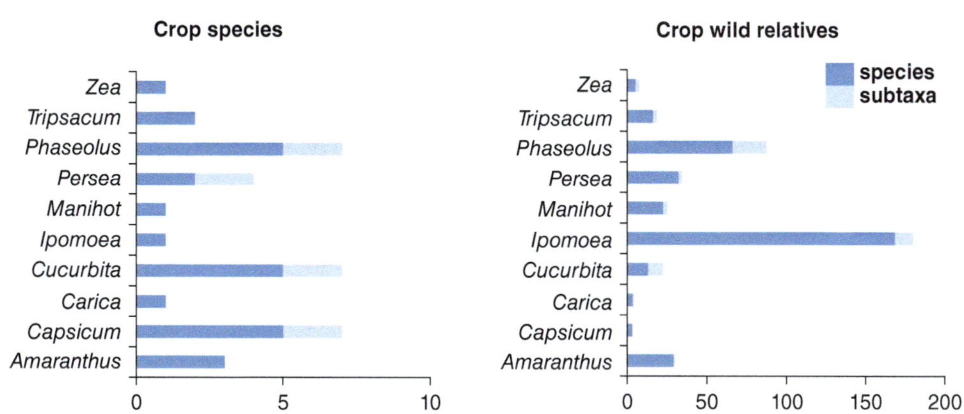


Fig. 22.1. Breakdown per genepool of numbers of species (dark blue) and additional subtaxa (light blue) considered in the present analysis.

points to extract for each point 19 bioclimatic variables obtained from the WorldClim data (Hijmans *et al.*, 2005). We excluded all points with outlier scores of 5 or more bioclimatic variables. To determine outlier scores, we calculated for each variable the first quartile (Q1) and third quartile (Q3) and the interquartile range (IQR) on the basis of the values for all points. We considered to be outlier scores all values that fell below Q1 – 2.5 * IQR or above Q3 + 2.5 * IQR. The corresponding presence points were excluded from the analyses.

### 22.2.2 Distribution modelling

To assess changes through time in the distribution of suitable habitats for each of the species considered here, we carried out a species distribution modelling exercise using Maxent, a robust modelling technique (Phillips *et al.*, 2006). Suitable areas identified by Maxent for crop species will not necessarily match the actual areas under cultivation. In some areas that are suitable, some crops may not actually be cultivated, and in some areas that are not suitable, some crops may actually be cultivated thanks to specific management practices; for example, through irrigation such as in vast areas in northern Mexico. However, putative habitat suitability through time as identified with Maxent can be expected to be a reasonable proxy for the potential impact of changing climate conditions on crop cultivation, that is, in the absence of changes in the adaptive potential of crops.

In line with Ramírez-Villegas and Jarvis (2010) and Ramírez-Villegas *et al.* (2014), we modelled only species with at least ten distinct locations (a combination of genebank accessions and herbarium records). Model calibration with Maxent for projection to current climate conditions was carried out based on data extracted from soil (FAO *et al.*, 2012), the ecoregion (FAO, 2001) and WorldClim climate maps (Hijmans *et al.*, 2005). For projections to past and future climate scenarios, models were calibrated based on current climate data only (averages of the period from 1960 to 1990). Modelling was performed at 2.5′ arc minutes spatial resolution, retaining only one presence point per grid cell. Background points (an overall maximum of 10,000 and maximum one per grid cell) were selected randomly from the area enclosed by a convex hull polygon constructed around all presence points and extended with a buffer corresponding to 10% of the polygon's largest axis.

Spatial autocorrelation in species presence points is an important concern in environmental niche modelling and can influence the quality of potential distribution maps strongly, and bias model evaluations based on cross-validation (Hijmans, 2012). We assessed the ability of Maxent to cope with this kind of spatial autocorrelation by calculating calibrated area under curve (cAUC) values and comparing these with a geographical null model, whereby habitat suitability decreased with increasing distance from presence points (see Hijmans, 2012). To this end, we (i) randomly partitioned both presence and background points in five groups and (ii) carried out five rounds of calibrating and testing each model (including the geographical null model), each time using four partitions for calibration and one partition for model testing, from which spatial sorting bias was removed following the procedure explained in Hijmans (2012). We then compared the resulting cAUCs obtained for Maxent and the geographical null model by means of Mann–Whitney tests. Average cAUC values for the null model were generally close to 0.5, which is equivalent to a random draw (Hijmans, 2012). In only one case (*Carica cnidoscoloides*, a wild relative of papaya) did the null model outperform Maxent, which was related to the restricted geographical range of known presence points.

The fitted models of both cultivated and wild species were projected to maps of current (WorldClim) and future climate. In line with others (Ramírez-Villegas *et al.*, 2014), for the future climate we used the average of 19 different climate models for the period 2040–2069 (2050s) and the SRES-A2 emission scenario, obtained from CMIP3 and downscaled by Ramírez-Villegas and Jarvis (2010). For wild species, we additionally projected the calibrated models to the average of two climate models (MIROC and CCSM) of the Last Glacial Maximum (LGM; ~21,000 years BP; Braconnot *et al.*, 2007). Suitability scores of individual projections theoretically range from 0 to 1. However, we restricted the modelled distributions visualized on maps to the maximum training sensitivity plus specificity threshold obtained from model calibration under current climate conditions. For projections to current climate conditions, we additionally restricted the modelled distribution to the same convex hull area used for the selection of background points.

### 22.2.3 Genetic data analysis

We mapped the distribution of the genetic diversity of *Z. mays* subsp. *mays* and *Phaseolus vulgaris* following the methodology described in van Zonneveld *et al.* (2012) and Thomas *et al.* (2012) at a resolution of 10 arc minute grid cells and circular neighbourhoods of 1 degree diameter. Molecular characterization data for maize and previously used by Vigouroux *et al.* (2008) were obtained from www.panzea.org, and for common bean from CIAT (International Center for Tropical Agriculture). The maize data set contained 449 individuals from Mesoamerica, genetically characterized by means of 92 nSSR (nuclear Simple Sequence Repeats) microsatellite markers, while the common bean data set comprised 274 individuals characterized with 36 nSSR markers.

### 22.2.4 On-farm and *in situ* conservation

In this chapter, we use the terms 'on-farm conservation' and '*in situ* conservation' to refer to the conservation of crop taxa in farmers' fields and of uncultivated crop wild relatives in natural or human-modified vegetation, respectively. Our reasoning for identifying priority sites for on-farm conservation was that, among other considerations, these should correspond to areas which are known currently to hold high crop species diversity and where environmental conditions are likely to remain favourable in the future, considering the expected impacts of climate change. To this end, we developed currently observed and modelled future (2050s) richness maps of crop species based on a combination of genebank accessions and herbarium records. To elaborate the currently observed richness map, we constructed circular areas around all known occurrences of each species with a radius of ~25 km, for which every observation was assumed to be representative, and then summed the species per grid cell with resulting maps obtained for all species. For a modelled future richness map, we summed all species with modelled distributions under climate conditions predicted for the 2050s per grid cell. Priority areas for on-farm conservation were obtained by intersecting both maps after restricting them to the areas covered by cells with richness values above the third quantile (75th percentile) of richness scores in all grid cells of each respective map.

A similar procedure was used for identifying priority areas for the *in situ* conservation of crop wild relatives, but an additional modelled richness map for the LGM was used to detect areas where climate conditions were likely to have permitted the persistence of a high diversity of wild relatives since the last glacial period and will continue to do so under climate change. Areas that have acted as glacial refugia are likely to contain higher levels of genetic diversity compared to expansion areas within species' ranges (cf. Thomas *et al.*, 2012, 2015). Therefore, we expect priority areas identified for crop wild relatives to contain not only high species diversity but also high intraspecific genetic diversity.

The above methodology is focused on identifying putative hotspots of crop species and their wild relatives, but does not consider the complementarity of different hotspots in efficiently conserving all species considered here. We therefore carried out an additional reserve selection exercise, after Rebelo and Sigfried (1992), to find the lowest number of geographical units to conserve the highest possible species diversity. For this, we used maps of protected areas (UNEP-WCMC, 2010) and the distribution of linguistic groups (WLMS, 2005; considered a proxy for the presence of indigenous and local communities) as possible geographical units.

## 22.3 Results

### 22.3.1 Documentation status of accessions

A comparison of the accessions conserved in international genebanks and Mesoamerican genebanks points to a certain degree of complementarity (Table 22.2). Where international genebanks specialize in ensuring the conservation of their mandate crops that tend to be staples with global distribution and significance (maize and relatives, beans, cassava, sweet potato), Mesoamerican genebanks have invested more in the conservation of more regionally or locally distributed PGRFA (amaranth, cucurbits, chilli peppers). Regional genebanks are also important for the conservation of tree crops, whose recalcitrant seeds require their conservation as living collections.

Although, based on Table 22.2, the number of accessions held by international and Mesoamerican genebanks appears impressive, the large majority of these are not yet documented adequately. Figure 22.2 shows that major constraints to an efficient assessment of the conservation status of PGRFA in Mesoamerican genebanks relate to the lack of georeferences, incomplete or non-existent taxonomic identification data and the unavailability of passport data in digital format. Accessions conserved in international genebanks appeared to suffer from similar flaws, given that, after data cleaning, we retained a total of 30,402 records, of which only about one-fifth (~6000) corresponded to genebank accessions, which was far less than the numbers indicated in Table 22.2.

Table 22.2. Comparison of accessions from Mesoamerica conserved in regional and international genebanks. Numbers are based on survey data from 18 Mesoamerican genebanks and data extracted from international genebanks at the time of this research (2012), respectively, and are to be considered as rough estimates (duplicates were removed to the extent possible). International genebanks are likely to conserve more accessions, but information was not available online in 2012.

| | Number of accessions in Mesoamerican genebanks | Number of accessions in international genebanks |
|---|---|---|
| *Zea* | >28,000 | >15,000 |
| *Phaseolus* | >9,000 | >25,000 |
| *Manihot* | <250 | >1,350 |
| *Ipomoea* | <150 | >1,050 |
| *Cucurbita* | >3,100 | >300 |
| *Capsicum* | >1,900 | >1,000 |
| *Carica* | <100 | >10 |
| *Persea* | >650 | 0 |
| *Tripsacum* | <50 | >100 |
| *Amaranthus* | >550 | >100 |

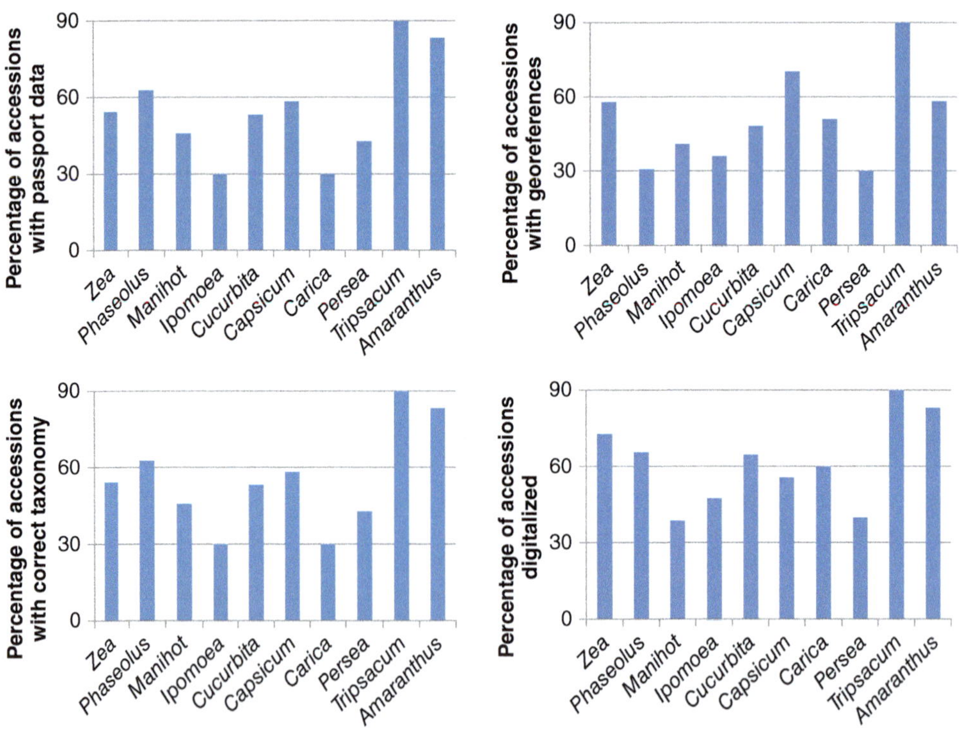


Fig. 22.2. Documentation status of accessions conserved by Mesoamerican genebanks (average values across 18 genebanks).

While not all species on our list were represented in genebank collections, a fourth of the accessions were crop wild relatives, which seemed to reflect the interest of international genebanks in conserving these genetic resources (Table 22.3). However, these accessions represent but a fraction of the actual diversity and geographic coverage of the crop wild relatives considered here, as illustrated by Fig. 22.3, which shows a comparison between a crop wild relative richness map and the number of species collected, and the geographic area covered, by genebank collection missions. It is clear that with the limited data available to us at the time of these analyses (2012), it was not possible to carry out a robust gap analysis. An excellent resource for assessing collection gaps of the wild relatives of a number of important crops is the 'Crop Wild Relatives and Climate Change' (2013) webpage.

### 22.3.2 Climate change impact

Based on the results obtained from distribution modelling under the current and future climate, under a worst-case scenario (i.e. in the absence of technological and agricultural innovations, and improved management practices), climate change is expected to cause substantial changes in the range and distribution of suitable areas for most crop species and their wild relatives. Table 22.4 shows that by the 2050s, all crop taxa considered here are expected to lose between 10 and 81% of the areas identified as suitable under current climate conditions, implying that in the absence of human intervention the consequences for food production could be severe. For all taxa, new suitable areas are likely to appear as a consequence of climate change, but these are highly variable (0.3–938.9% of current suitable areas) and for most species do not compensate for the losses expected, leading to net decreases in suitable areas. Furthermore, Table 22.4 suggests that a reduction in habitat suitability may be more severe for the most important crops for human food consumption.

Table 22.3. Herbarium records and genebank accessions retained after data cleaning.

| | Cultivated species | Crop wild relatives |
|---|---|---|
| Herbarium records | 6,831 | 17,541 |
| Genebank accessions | 4,832 (accounting for 21 species and 5 additional subtaxa) | 1,168 (accounting for 85 species and 10 additional subtaxa) |
| Total | 11,693 | 18,709 |

An interesting observation from Table 22.4 is that modelling at the subspecies level often shows less pessimistic results. Good examples are *Ipomoea batatas* (sweet potato) and *Capsicum annuum* (pepper), for which net suitability losses are predicted, while for *I. batatas* var. *batatas* and *C. annuum* var. *annuum* net suitability gains are expected. For other crops like *Phaseolus vulgaris* and *Persea americana*, suitability losses are less severe when modelled at subspecies level. Although these discrepancies can relate partly to differences in ecological niche construction by Maxent, given the variable numbers of presence points available at species and subspecies level, these observations suggest that taxa at subspecies level have different growth requirements and different degrees of adaptive potential.

Figure 22.4 shows the potential impact of climate changes on environmentally suitable areas for maize and common bean from the present to the 2050s. Consistent with the results presented in Table 22.4, suitability losses could be very severe under a worst-case scenario. Based on the mapped distributions of genetic diversity for both crops, the implications for the on-farm conservation of their genetic resources seem to be less dramatic (Fig. 22.4). The areas where the highest levels of diversity have been observed, and which correspond to the putative areas where the domestication of both crops may have been initiated (Matsuoka *et al.*, 2002; Bitocchi *et al.*, 2012), are likely to remain suitable in the future. Nonetheless, Fig. 22.4 suggests that negative impacts may be expected along the margins of these diversity hotspots, and a more detailed study is necessary to evaluate fully the eventual need for targeted interventions to ensure efficient long-term on-farm conservation.

Figure 22.5 presents a summary of the expected impact of climate change on the suitability of all crop species and their wild relatives

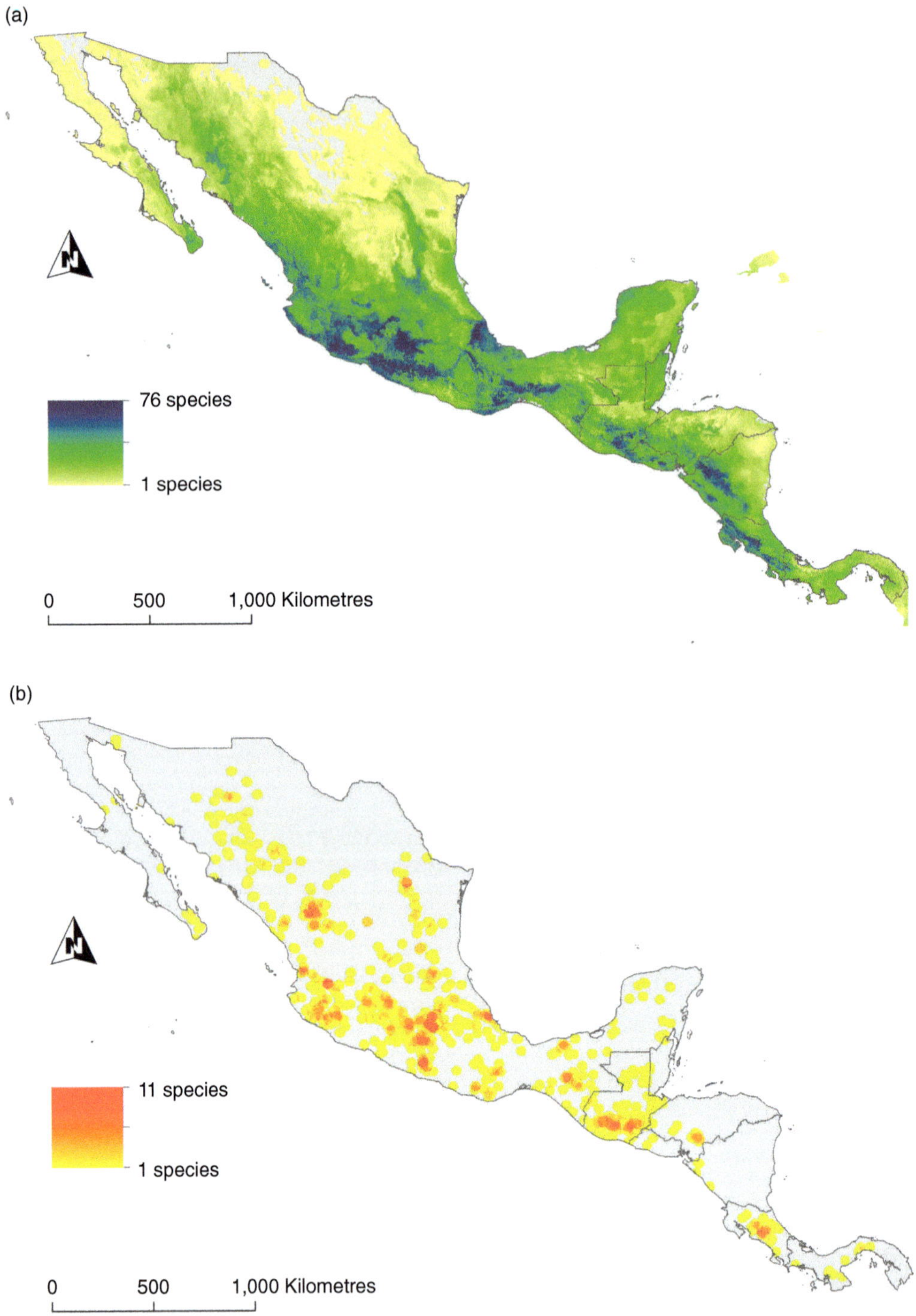


**Fig. 22.3.** (a) Species richness map of crop wild relatives, constructed by overlaying modelled distributions of individual species; (b) map showing the number of species collected, and geographic area covered, by genebank collection missions considered in our data set. Each genebank accession was considered to be representative for a circular area with a radius of ~25 km.

**Table 22.4.** Predicted gains, losses and net surface changes in suitability areas of crop taxa from the present to the 2050s, based on environmental niche modelling. Taxa at subspecies level are subsets of corresponding taxa at species level. Numbers are intended to give an impression of the scale of potential suitability changes and should not be taken as absolute estimates.

| Crop species | Current suitable area gained (%) | Current suitable area lost (%) | Net change in suitable area (per cent of current suitable area) | Net change in suitable area (millions of hectares) |
|---|---|---|---|---|
| *Capsicum chinense* | 206.7 | 9.6 | +197.1 | +56.7 |
| *Cucurbita argyrosperma* | 113.1 | 17.7 | +95.4 | +41.4 |
| *Cucurbita argyrosperma* subsp. *argyrosperma* | 128.0 | 15.6 | +112.4 | +37.7 |
| *Capsicum pubescens* | 938.9 | 36.5 | +902.4 | +34.4 |
| *Phaseolus acutifolius* | 55.0 | 13.2 | +41.9 | +27.9 |
| *Capsicum baccatum* | 92.2 | 18.1 | +74.1 | +21.7 |
| *Phaseolus acutifolius* var. *acutifolius* | 52.1 | 16.0 | +36.1 | +21.0 |
| *Amaranthus cruentus* | 125.2 | 29.5 | +95.7 | +17.4 |
| *Ipomoea batatas* var. *batatas* | 211.5 | 60.5 | +151.0 | +9.3 |
| *Capsicum annuum* var. *annuum* | 55.3 | 45.0 | +10.3 | +2.5 |
| *Tripsacum dactyloides* | 36.3 | 32.5 | +3.8 | +1.3 |
| *Persea americana* var. *americana* | 55.3 | 67.0 | –11.8 | –1.2 |
| *Cucurbita moschata* | 33.6 | 35.7 | –2.0 | –1.3 |
| *Amaranthus hypochondriacus* | 34.3 | 53.1 | –18.8 | –3.1 |
| *Persea schiedeana* | 10.2 | 51.6 | –41.4 | –3.5 |
| *Phaseolus coccineus* | 12.3 | 20.4 | –8.1 | –3.6 |
| *Cucurbita ficifolia* | 14.2 | 33.2 | –19.0 | –4.0 |
| *Persea americana* var. *drymifolia* | 40.1 | 61.7 | –21.7 | –6.4 |
| *Phaseolus dumosus* | 0.3 | 62.6 | –62.4 | –12.0 |
| *Phaseolus vulgaris* var. *vulgaris* | 17.5 | 39.9 | –22.4 | –15.1 |
| *Tripsacum andersonii* | 1.1 | 80.9 | –79.8 | –23.9 |
| *Phaseolus lunatus* | 5.2 | 43.6 | –38.4 | –27.8 |
| *Carica papaya* | 12.5 | 70.0 | –57.5 | –33.2 |
| *Ipomoea batatas* | 3.5 | 65.0 | –61.5 | –33.2 |
| *Manihot esculenta* | 7.1 | 66.9 | –59.8 | –33.9 |
| *Cucurbita pepo* | 13.1 | 76.9 | –63.8 | –35.3 |
| *Capsicum frutescens* | 5.7 | 80.5 | –74.8 | –37.0 |
| *Phaseolus vulgaris* | 5.8 | 42.5 | –36.7 | –37.2 |
| *Persea americana* | 5.1 | 63.5 | –58.4 | –40.4 |
| *Zea mays* subsp. *mays* | 1.4 | 38.6 | –37.1 | –44.9 |
| *Capsicum annuum* | 6.9 | 61.8 | –54.9 | –50.7 |

(a)

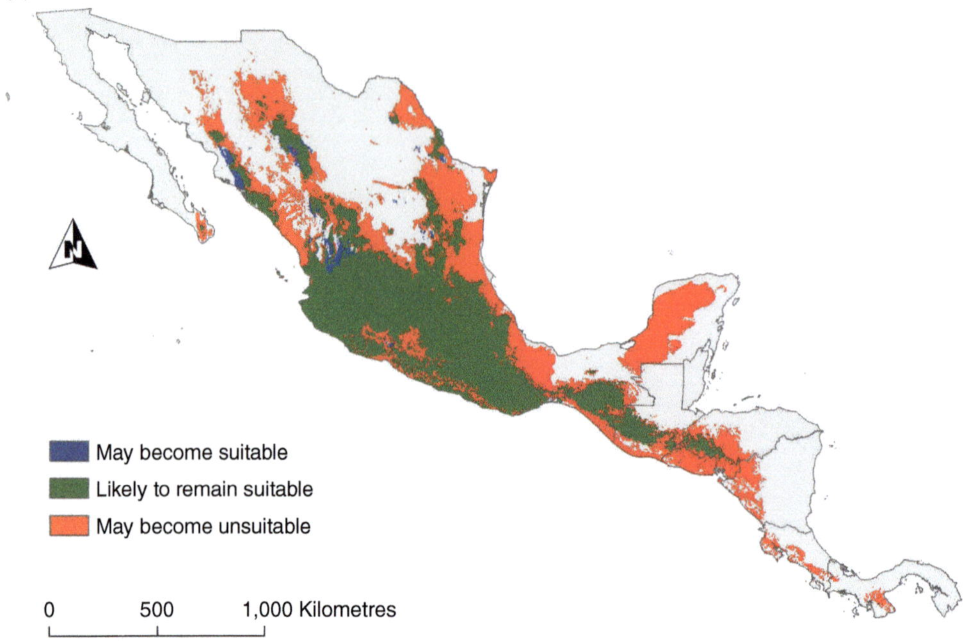


(b)

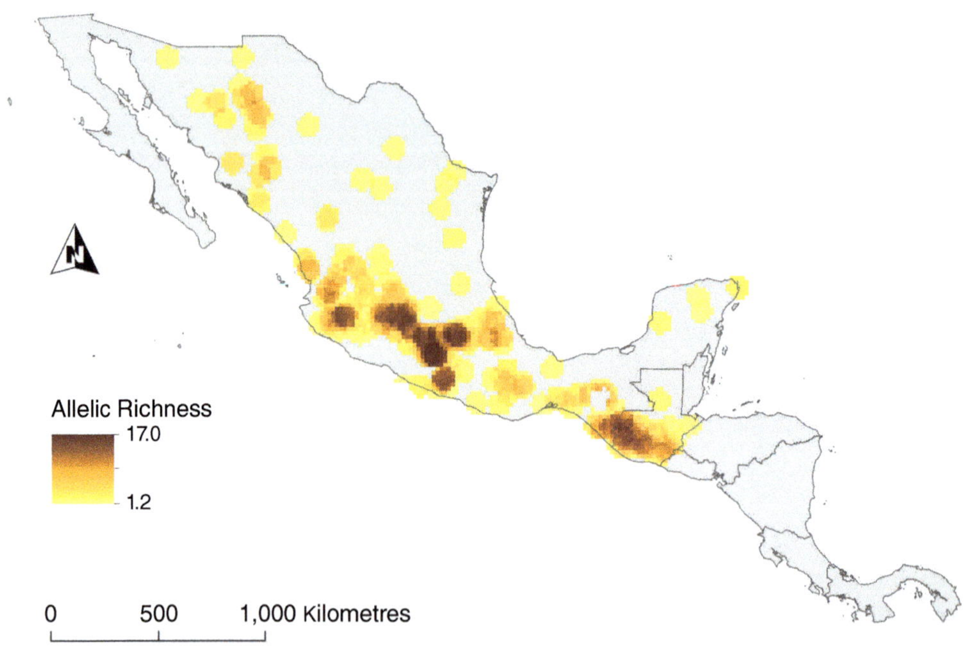


Fig. 22.4. (a and c) Expected changes in habitat suitability from the present to the 2050s as a consequence of climate changes for maize (Zea mays L. subsp. mays) and common bean (Phaseolus vulgaris L.); (b and d) distribution of genetic diversity (allelic richness) of maize and common bean.

(c)

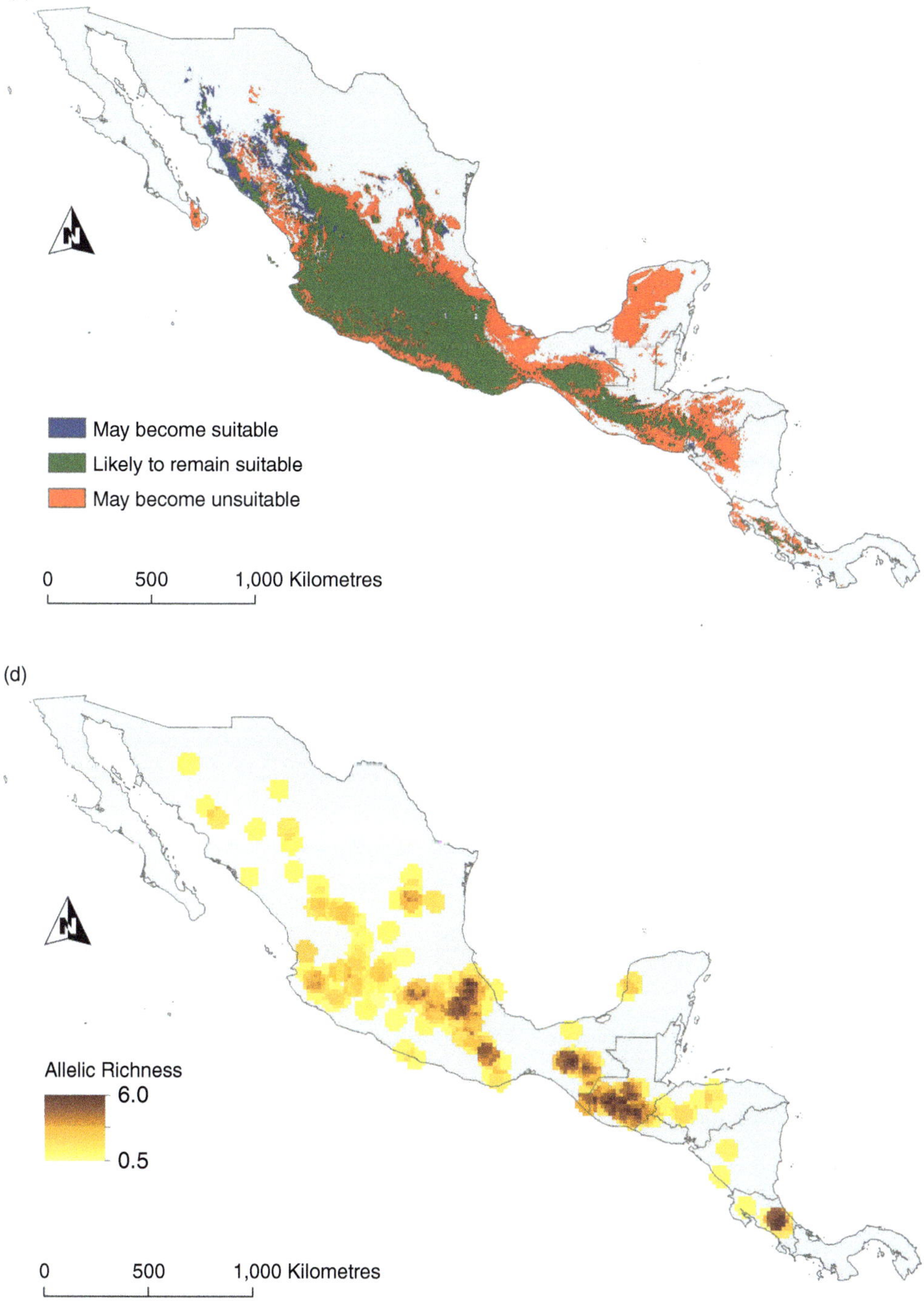


Fig. 22.4. Continued.

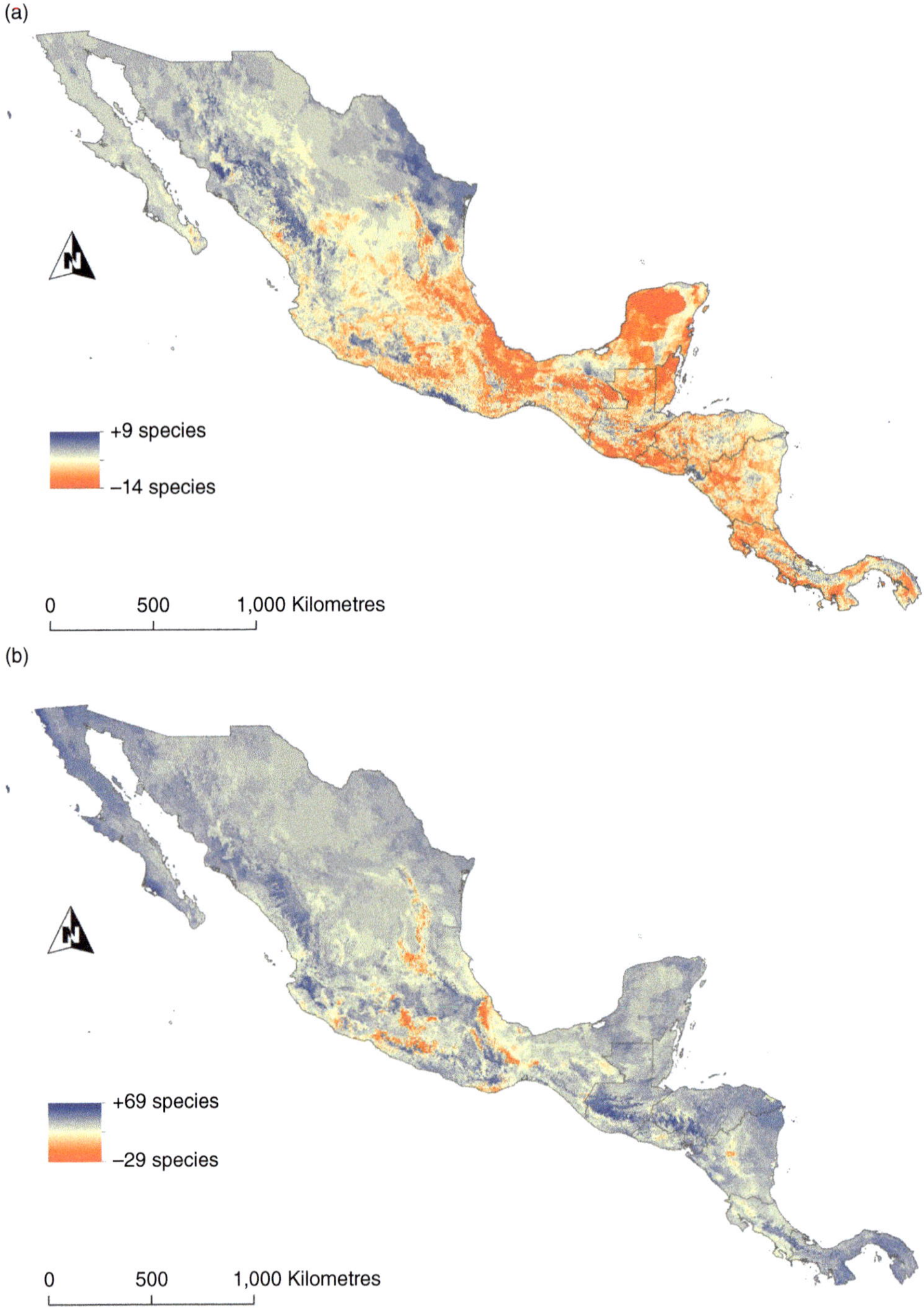


Fig. 22.5. Summary of changes in habitat suitability from the present to the 2050s as a consequence of climate change for the ensemble of the crop species (a) and their wild relatives (b) considered here. By the 2050s, red areas may become unsuitable for a higher number of species than presently, while blue areas may become suitable for a higher number of species than presently. Maps of individual species can be consulted at: itzamna-mesoamerica.org.

considered here across Mesoamerica. The general trend observed is that under a 'business as usual' scenario, most parts of the subcontinent are expected to suffer suitability losses for crop species; only in northern Mexico is there a predominance of suitability gains. Almost an opposite pattern is observed for crop wild relatives, for which, by the 2050s, climate change is predicted to result in net suitability gains in most areas, that is, under a scenario of unrestricted dispersal, although clearly some areas are expected to suffer suitability loss for wild species, too (Table 22.5). For most genepools, these losses are likely to be widely compensated by suitability gains, under a scenario of unlimited dispersal. However, this trends needs to be interpreted carefully, since there are large differences in the numbers of wild relatives considered per genepool (Fig. 22.1). There seems to be a positive association between the number of crop wild relatives in a genepool and the expected net suitability changes. For example, for *Carica* and *Capsicum*, only two and three wild relatives, respectively, were considered here; and for these genepools precisely, the most pessimistic suitability changes were observed.

Also, by pooling the results of crops and wild relatives, as in Table 22.5 and Fig. 22.5, some more dramatic forecasts may be obscured. A good example relates to the wild relatives of cultivated maize, which attain the highest levels of species richness in the very same area where the genetic diversity of maize is highest (Figs 22.4b and 22.6a). Although the environmental conditions in this area are expected in the future to remain suitable for the cultivation of maize (Fig. 22.4a), the local persistence of its wild relatives may be more problematic under climate change (Fig. 22.6b). At the same time, there are vast areas on the subcontinent that are expected to gain suitability in the future and overcompensate for the losses expected. The new areas could thus help to ensure the *in situ* conservation of wild relatives, but this will likely require human-assisted migration.

### 22.3.3 On-farm and *in situ* conservation

Figure 22.7 shows the distribution of priority areas for on-farm and *in situ* conservation of the crop species and their wild relatives considered here. Although only 31% of the priority areas for crop species coincide with the location of different linguistic groups (taken as a proxy for the areas inhabited by indigenous and local communities), all crop species can be conserved easily at least once in a few areas, the most important ones being the areas inhabited by the Maya from Yucatán, Mexico, the Q'eqchi from Guatemala and the Mixe from Totontepec, Mexico. It is clear, however, that to conserve effectively the intra-specific diversity existing in each of these crop species, it will be necessary to target a series of hotspots across Mesoamerica that conserve complementary diversity. For crop wild relatives,

**Table 22.5.** Predicted gains, losses and net changes in suitability areas (in millions of hectares), from the present to the 2050s, of crop species and their wild relatives pooled together per genepool, as well as for all genepools together, based on environmental niche modelling.

| | Crop species | | | | Crop wild relatives | | | |
|---|---|---|---|---|---|---|---|---|
| Genepool | Gain | Loss | Net change | | Gain | Loss | Net change | |
| *Cucurbita* | 123.4 | 65.9 | +57.5 | ↑ | 158.5 | 12.5 | +146.0 | ↑ |
| *Amaranthus* | 24.7 | 12.7 | +12.0 | ↑ | 108.0 | 63.9 | +44.1 | ↑ |
| *Capsicum* | 67.1 | 43.6 | +23.5 | ↑ | 5.7 | 12.8 | −7.1 | ↓ |
| *Carica* | 7.2 | 40.4 | −33.2 | ↓ | 1.9 | 13.9 | −11.9 | ↓ |
| *Tripsacum* | 12.5 | 34.9 | −22.4 | ↓ | 89.6 | 29.7 | +59.9 | ↑ |
| *Ipomoea* | 1.9 | 35.1 | −33.2 | ↓ | 223.5 | 18.9 | +204.6 | ↑ |
| *Phaseolus* | 40.4 | 72.7 | −32.3 | ↓ | 198.6 | 26.9 | +171.7 | ↑ |
| *Manihot* | 4.0 | 37.9 | −33.9 | ↓ | 242.0 | 1.0 | +241.0 | ↑ |
| *Persea* | 4.2 | 46.4 | −42.3 | ↓ | 207.0 | 5.4 | +201.6 | ↑ |
| *Zea* | 1.7 | 46.7 | −44.9 | ↓ | 89.3 | 22.2 | +67.1 | ↑ |
| Mesoamerica | 138.3 | 141.3 | −3.1 | | 241.1 | 10.5 | +230.5 | |

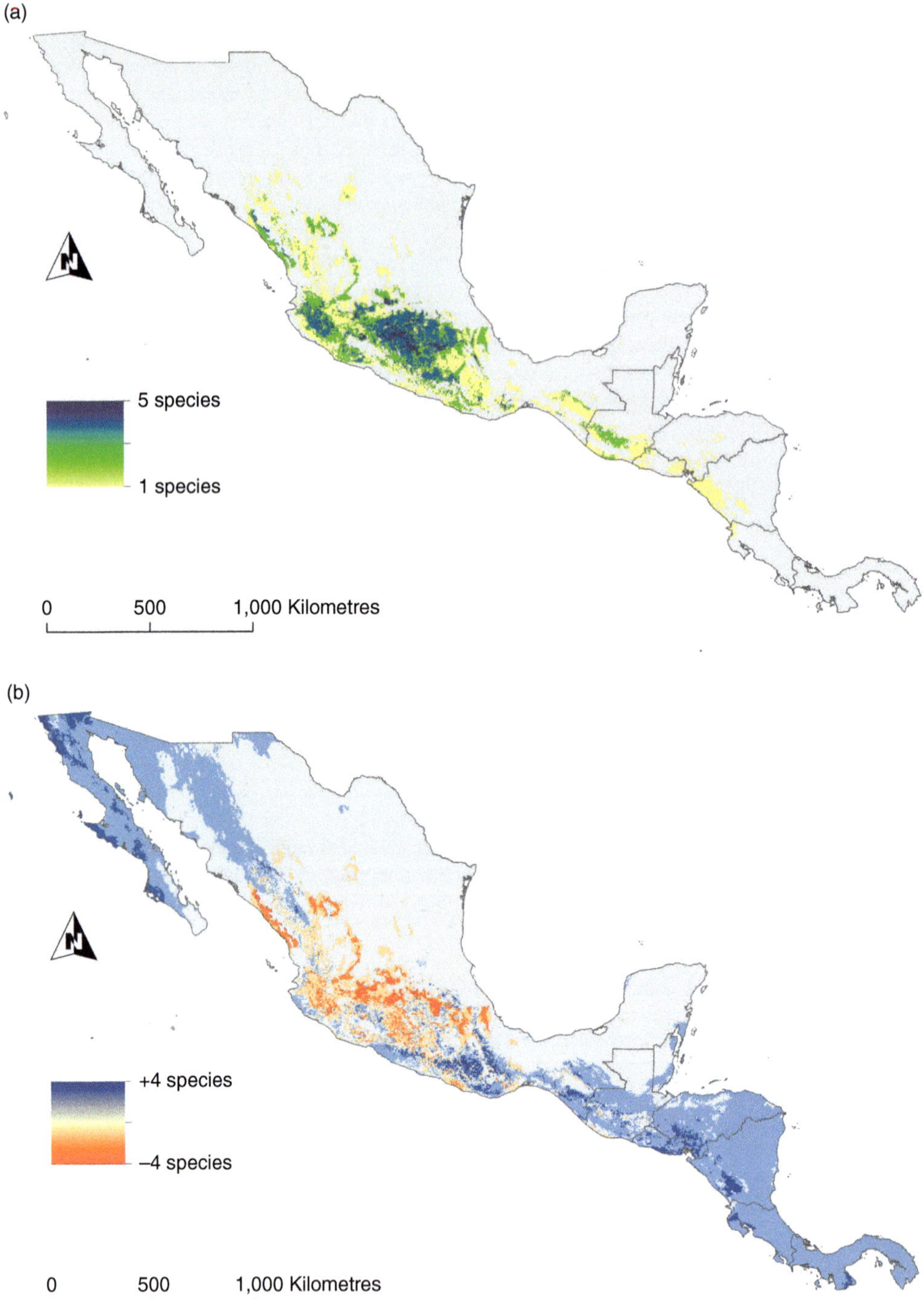


**Fig. 22.6.** (a) Richness distribution of maize's crop wild relatives (*Zea* genus) constructed by overlaying modelled distributions of individual species; (b) expected changes in habitat suitability from the present to the 2050s as a consequence of climate change for the ensemble of maize's wild relatives. Red areas are expected to become unsuitable for one or more species by the 2050s, while blue areas may become suitable for a higher number of species than presently.

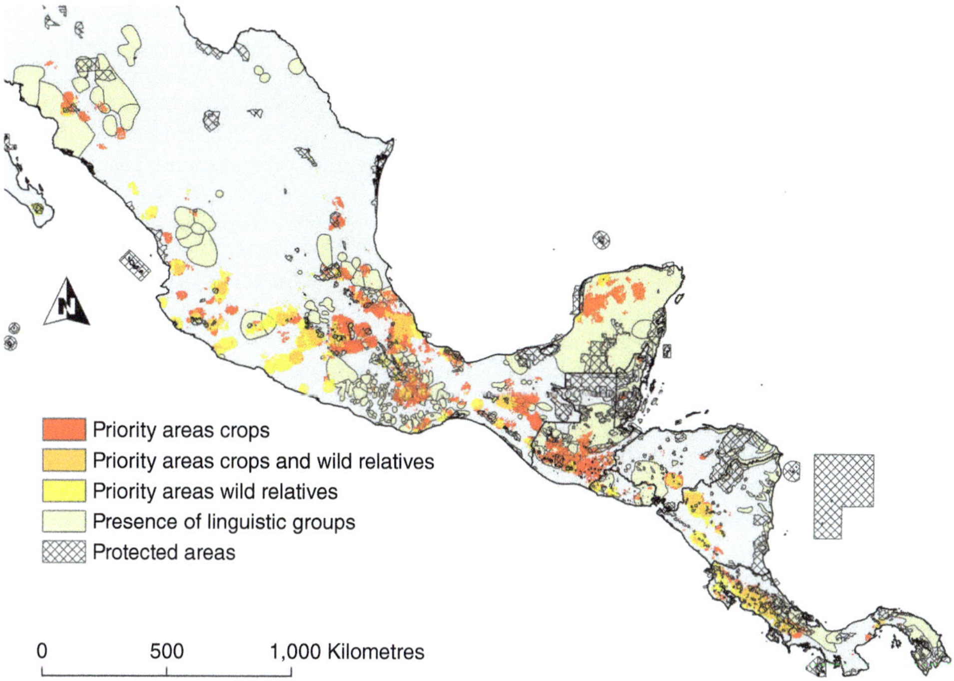


Fig. 22.7. Priority areas for the on-farm and *in situ* conservation of crop species and their wild relatives, respectively, compared with the distribution of protected areas and the presence of linguistic groups that serve as a proxy for occupation by indigenous and local communities.

less encouraging results were obtained: only 12% of the priority areas overlap with protected areas, and based on the data available to use, only 61% of all species are known to occur in at least one of the current set of protected areas (the most important ones being the Sierra Gorda, the Los Tuxtlas and the Sierra La Laguna areas, all located in Mexico). Hence, the currently existing protected areas seem inadequate to achieve the effective conservation of crop wild relatives. Golicher *et al.* (2012) came to comparable conclusions in a similar modelling exercise for Mesoamerican tree species.

Interestingly, Fig. 22.7 shows the strong overlap between priority areas for crop species (35% of priority areas) and their wild relatives (54%), suggesting that at least part of on-farm (crops) and *in situ* conservation (wild relatives) efforts could target the same areas. Intuitively, priority areas for on-farm conservation would be those coinciding with, or located close to, areas inhabited by indigenous and traditional farmer communities, who in many cases are the creators and the de facto custodians of cultivated PGR. As it is unlikely that new protected areas will be created for the specific purpose of conserving crop wild relatives, it may make sense to develop participatory conservation strategies for both crop species and their wild relatives, together with indigenous and traditional farmer communities. A combination of Mesoamerican protected areas and areas inhabited by ILCs (Indigenous and Local Communities) would allow the effective conservation of all crop taxa and at least 77% of all wild relatives considered here.

## 22.4 Discussion

### 22.4.1 Documentation state of accessions in genebanks

Although our analysis was developed around the genepools of ten priority crops, the general trends observed can be expected to be representative of the conservation status of other cultivated and wild species of importance to Mesoamerican agriculture. It is clear that significant efforts have been made in Mesoamerica over time to conserve agrobiodiversity *ex situ*, *in situ* and on farm, in complementary ways to the strategies implemented by international genebanks, in line with similar evolutions elsewhere such as the Andean region (Galluzzi and López Noriega, 2014). But our results also highlight important limitations affecting the capacity to tap fully the potential that these resources hold for responding to present and future climate change challenges.

A limitation that Mesoamerican national genebanks share with *ex situ* institutions in many developing countries (FAO, 2010) is the lack of adequate financial and technical support to the conservation, characterization and documentation (e.g. through common quality standards for passport data) of stored material. Detailed and reliable knowledge of what is already there and in what state is, nonetheless, essential to enable a more efficient use and conservation of *ex situ* collections. Although the data available to us preclude a proper gap analysis,[1] it is clear that significant gaps remain in the germplasm currently conserved in Mesoamerican and international genebanks. Filling the remaining gaps in a strategic manner will require serious efforts to improve the documentation and characterization status of existing *ex situ* collections. This is important, not only to safeguard genetic resources that otherwise will be lost forever as a consequence of the ongoing processes of environmental degradation and loss of biocultural diversity but also to allow their use in responding to current and future challenges, such as climate change. A better documentation and characterization of existing collections will also permit their more efficient use in climate change adaptation and breeding programmes, among others. At the same time, it is crucial to carry out documentation and characterization initiatives in a coordinated, synergistic and standardized manner.

In response to these needs, the recently adopted 'strategic action plan to strengthen the conservation and use of Mesoamerican plant genetic resources in adapting agriculture to climate change' (referred to as SAP in what follows; Ramírez *et al.*, 2013) calls for a more efficient use of existing strengths in terms of human resources and genebank facilities in the different Mesoamerican countries, improving cooperation, synergies and complementarities between regional and international entities, as well as with community organizations. The SAP also calls for more initiatives to identify and use promising materials focused on adaptive traits and resistance to biotic and abiotic disturbances. This will not only require a more efficient sharing of good quality passport and characterization data in the region but also the design and institutionalization of agile mechanisms for distributing traditional and improved seeds of good quality.

### 22.4.2 Climate change impact

In line with previous findings, our assessment of the impact of climate on Mesoamerican PGRFA suggests that under a 'business as usual' scenario, climate change can be expected to cause substantial changes in the range and distribution of suitable areas for most cultivated crop species and their wild relatives. Responses to changes in climate are species specific; some crops are likely to experience net suitability gains, while others are expected to suffer net losses. However, uncertainties remain about the direction of expected responses. For example, Lobell *et al.* (2008) predicted gains in the average yield of maize and cassava in Central America and the Caribbean, contradicting the negative trends found here, and by others before us (Bellon *et al.*, 2011; Monterroso Rivas *et al.*, 2011). Regardless, from a precautionary perspective, it is of course best to prepare for a worst-case scenario.

One constant in forecasts is that nearly all species are likely to experience displacements of suitable areas, independent of whether net gains or losses are predicted. This means that for agriculture in Mesoamerica to keep up productivity in the future, human-assisted migration of crops to non-traditional cultivation areas and the use of seed sources of the same crops that are better matched to future climate will be among the adaptation strategies that need to be considered (see Bellon *et al.*, 2011, for maize). Taking full advantage of the intraspecific diversity in crop species will be part of the solution. Our results suggest that less pessimistic distribution projections are generally obtained when modelling crops at subspecies level, in line with similar observations made elsewhere (Pearman *et al.*, 2010; Oney *et al.*, 2013). Our observations, nonetheless, show only the top of the iceberg, and a more rigorous assessment of all different taxa at the intraspecific level is necessary to allow for more accurate estimates of the potential suitability impacts of climate change. As argued before, this will first require further standardization of taxonomic classification systems and their horizontal implementation by genebanks in Mesoamerica and beyond. The results obtained from other studies have shown the vast potential of employing intraspecific crop diversity for adapting to climate change. For example, future projections of each of the 59 described maize races in Mexico show contrasting responses to climate change (Ureta *et al.*, 2012), implying that a reshuffling of races within and beyond traditional maize growing areas might already represent an important step towards upholding productivity (Bellon *et al.*, 2011). This of course may work well on paper, but in practice will require implementation strategies that are fully embedded in and aligned with the sociocultural realities and culinary preferences of local farmers (Hellin *et al.*, 2014).

The improvement of the adaptive capacity of existing intraspecific crop taxa, as well as the development of new ones through breeding, is another important adaptation strategy to cope with the negative impacts of climate change (Lobell *et al.*, 2008). As mentioned before, landraces that have adapted to different ecological zones as a consequence of human and natural selection are a treasure trove for the discovery of new traits of interest. Also, crop wild relatives hold great potential to contribute genes that increase the adaptive capacity of crop species under climate change. Their genetic resources are largely unexplored compared to crop species and hold promise to contain many types of adaptive traits like resistance to diseases (Tanksley and McCouch, 1997). Over the past decades, there has been a stable increase in the rate of release of cultivars containing genes from crop wild relatives (Hajjar and Hodgkin, 2007). Our finding that crop wild relatives overall are expected to suffer much less from the effects of climate change, or might even benefit from it (Fig. 22.5; Table 22.5), could suggest that at least part of these species possess genes permitting this adaptive capacity, which may in turn have potential for introgression in crop species that are less well prepared to adapt to changing climatic conditions. Similar observations have been made for rice wild relatives in Colombia (Thomas *et al.*, 2016). On the other hand, not all crop wild relatives we have considered here may have the same potential for use in crop improvement programmes. For example, many of the 169 *Ipomoea* wild relatives are unlikely to be used for this purpose in the near future, either because they do not have traits of interest to breeders or because of technological or biological constraints to the introgression of genes, related to incompatibility issues, among others.

### 22.4.3 *In situ* and on-farm conservation

The *in situ* and on-farm conservation of wild relatives and crop species, respectively, are essential components in the adaptation of plant genetic resources to climate change. In wild species, the process of natural selection depends on population size, the amount of genetic variation among individuals, selection pressure and gene flow (the movement of either pollen or reproductive material) from neighbouring populations (Frankel, 1974; Hoyt, 1998). The *in situ* conservation of crop wild relatives allows species' populations to be exposed continually to natural selection, allowing their gradual adaptation to changing environmental conditions (Meilleur and Hodgkin, 2004; Heywood *et al.*, 2007; Jarvis *et al.*, 2008), thus securing the maintenance of a potential source of adapted genes for breeding. The results from our modelling exercise suggest that most crop wild relatives are expected to experience larger suitability gains than losses in the future under a scenario of unlimited dispersal (Fig. 22.5; Table 22.5). Although this is a hopeful forecast, it is important to realize that predicted range expansions will require effective ecological connectivity and gene flow between source populations of crop wild relatives and the areas they are expected to be able to colonize. Ensuring ecological connectivity will be more important for some species than for others, depending on their reproductive biology (notably pollination and dispersal mechanism). For some species, more drastic interventions may be needed, such as for maize's wild relatives (*Zea* genus), which attain the highest levels of species richness in the very same area where the genetic diversity of maize is highest (Figs 22.4b and 22.6a), (also see Kato Yamakake *et al.*, 2009; CONABIO, 2011). It is precisely in this area where our analysis predicts that the negative impact of climate change will be strongest (Fig. 22.6b), in line with Ureta *et al.* (2012). At the same time, significant suitability gains are predicted, but most areas are located large distances from current presence sites. Given the importance of maize's crop wild relatives for sourcing genes of interest for breeding programmes, at least two strategies to ensure the effective *in situ* conservation of these species' genetic resources may be appropriate: one targeted on promoting local adaptation and another on translocation to new suitable areas; for example, through assisted migration.

On-farm conservation is the interplay of processes of human and natural selection in farmers' fields (Brush, 1991, 2000; Wood and Lenné, 1997). Maintaining crop diversity on farm allows farmers continuously to select crops displaying traits of interest, including those related to adaptive capacity under climate change; it also allows for gene exchange to occur with surrounding wild or weedy relatives (Harlan, 1965; de Wet and Harlan, 1975), continually providing farmers with new variants for experimentation and selection. The success of on-farm conservation is highly dependent on the maintenance of traditional farming practices (Bellon and van Etten, 2014). In traditional systems, cultural and biological diversity are interlinked and interdependent (Harmon, 1996; Brown, 2009; Brown *et al.*, 2013a), in an interplay that is the heart of 'biocultural' approaches to conservation (Maffi, 2005). Based on these considerations, we believe that successful on-farm conservation of Mesoamerican crops requires a biocultural approach involving indigenous and traditional farming communities across the region's cultural and environmental gradients. Here, we used the distribution of linguistic groups as a proxy for the distribution of cultural diversity in Mesoamerica, but it is likely that it can also be indicative of the distribution of inter- and intraspecific crop diversity (Leclerc and Coppens d'Eeckenbrugge, 2012; Brown *et al.*, 2013a; Westengen *et al.*, 2014).

Our results suggest there are vast opportunities to integrate *in situ* (crop wild relatives) and on-farm (crop species) conservation objectives. The observation that the current network of protected areas in Mesoamerica is insufficient to ensure the effective conservation of crop wild relatives, together with the fact that there is a significant overlap between priority areas for on-farm and *in situ* conservation (Fig. 22.7), suggests that at least part of *in situ* conservation efforts for crop wild relatives could focus on the same areas as for the on-farm conservation of Mesoamerican crop diversity. This concept is in line with similar proposals made previously (Harvey *et al.*, 2008) and is also fully recognized by the SAP. Since all countries in the region harbour priority areas for the conservation of both crops and their wild relatives (see Fig. 22.7), shared and well-coordinated regional efforts are the only way forward to achieve adequate coverage of these key areas. To achieve this, the SAP proposes on-farm and *in situ* management to focus on indigenous territories as key areas for on-farm conservation, while integrating these with existing wild biodiversity conservation plans. In particular, the plan puts forward an approach centred on the establishment and management of biocultural territories defined as 'the set of biological resources, comprising PGRFA diversity and the landscapes and ecosystems which host them, as well as to the traditions and good agricultural practices of rural populations, including indigenous and local communities' (Ramirez *et al.*, 2013). As an important support measure, the plan calls for efforts to provide incentives to farmers and indigenous people to strengthen their role as custodians, given the ongoing dynamics of rural development and cultural change. This includes institutional recognition of such a role, taking stock of international agreements in this field and in particular from the concept of farmers' rights stated in article 9 of the International Treaty on Plant Genetic Resources for Food and Agriculture. As emphasized by the SAP, it is of utmost importance to integrate the biocultural territories in existing conservation initiatives in the Mesoamerican region, such as the network of protected areas and the Mesoamerican biological corridor, so as to promote ecological connectivity throughout agricultural landscapes and protected areas to facilitate the migration of wild relatives, as well as important pollinators indispensable for sustaining agricultural production.

As a final remark, we believe it is encouraging to observe how, in addition to the broad and active participation of regional stakeholders during the process of elaboration and discussion of the SAP, the policy community responded to the final document, leading to its endorsement at the level of the CAC (Consejo Agropecuario Centro Americano), which brings together the Ministries of Agriculture of all countries in the region. Furthermore, the plan is being used by PGRFA institutions as a roadmap to guide them in their activity planning and priority setting, and in their applications for funding.

## Acknowledgements

The authors thank the Benefit-sharing Fund of the International Treaty on Plant Genetic Resources for Food and Agriculture, as well as the CGIAR Research Program on Climate Change, Agriculture and Food Security (CCAFS) and Forest, Trees and Agroforestry (FTA) for providing project funding. We are grateful to Nora Castañeda for useful comments on a previous version of this chapter.

## Endnote

[1] The results of the gap analyses we did perform based on the data available to us can be consulted at itzamna-mesoamerica.org.

## References

Bellon, M.R. and van Etten, J. (2014) Climate change and on-farm conservation of crop landraces in centres of diversity. In: Jackson, M., Ford-Lloyd, B. and Parry, M. (eds) *Plant Genetic Resources and Climate Change*. CAB International, Wallingford, UK, pp. 137–150.

Bellon, M.R., Hodson, D. and Hellin, J. (2011) Assessing the vulnerability of traditional maize seed systems in Mexico to climate change. *Proceedings of the National Academy of Sciences of the United States of America* 108, 13432–13437.

BID and CEPAL (2010) *Cambio Climático: Una Perspectiva Regional*. Comisión Económica para América Latina y el Caribe (CEPAL), Banco Interamericano de Desarrollo (BID), Santiago.

Bitocchi, E., Nanni, L., Bellucci, E., Rossi, M., Giardini, A., *et al.* (2012) Mesoamerican origin of the common bean (*Phaseolus vulgaris* L.) is revealed by sequence data. *Proceedings of the National Academy of Sciences of the United States of America* 109, 5148–5149.

Braconnot, P., Otto-bliesner, B., Harrison, S., Joussaume, S., Peterchmitt, J.Y., *et al.* (2007) Results of PMIP2 coupled simulations of the Mid-Holocene and Last Glacial Maximum – Part 1: experiments and large-scale features. *Climate of the Past* 3, 261–277.

Brown, C.H. (2009) Development of agriculture in prehistoric Mesoamerica: the linguistic evidence. In: Staller, J.E. and Carrasco, M. (eds) *Pre-Columbian Foodways*. Springer, Berlin, pp. 71–107.

Brown, C.H., Luedeling, E., Wichmann, S. and Epps, P. (2013a) The paleolinguistics of domesticated squash. In: Quinlan, M. and Lepofsky, D. (eds) *Explorations in Ethnobiology: The Legacy of Amadeo Rea*. Society of Ethnobiology, Vancouver, BC, Canada, pp. 132–161.

Brown, C.H., Clement, C.R., Epps, P., Luedeling, E. and Wichmann, S. (2013b) The paleolinguistics of domesticated chili pepper (*Capsicum* spp.). *Ethnobiology Letters* 4, 1–11.

Brown, C.H., Clement, C.R., Epps, P., Luedeling, E. and Wichmann, S. (2013c) The paleolinguistics of domesticated manioc (*Manihot esculenta*). *Ethnobiology Letters* 4, 61–70.

Brown, C.H., Clement, C.R., Epps, P., Luedeling, E. and Wichmann, S. (2014) The paleobiolinguistics of maize (*Zea mays* L.). *Ethnobiology Letters* 5, 52–64.

Brush, S.B. (1991) A farmer-based approach to conserving crop germplasm. *Economic Botany* 45, 153–165.

Brush, S.B. (2000) *Genes in the Field. On-Farm Conservation of Crop Diversity*. Lewis Publishers, Boca Raton; IDRC, Ottawa; IPGRI, Rome.
CAPGERNet, NORGEN, REDARFIT, PROCISUR's, REGENSUR, REMERFI and TROPIGEN (2008) Hemispheric strategy for conservation and use of plant genetic resources for the Americas. Available at: infoagro.net/archivos_Infoagro/Infotec/biblioteca/FORAGRO2008/CDavidson.pdf (accessed 9 December 2015).
Colunga-GarcíaMarín, P. and Zizumbo-Villarreal, D. (2004) Domestication of plants in Maya lowlands. *Economic Botany* 58, S101–S110.
CONABIO (2011) *Contribuciones al Conocimiento de los Maíces Nativos y sus Parientes Silvestres en México.* Availableat: www.biodiversidad.gob.mx/genes/pdf/proyecto/Mapa_I_Maices_CONABIO_2011_bajax.pdf (accessed 18 September 2014).
Crop Wild Relatives and Climate Change (2013) Online resource. Available at: www.cwrdiversity.org (accessed 18 September 2014).
de Wet, J.M.J. and Harlan, J.R. (1975) Weeds and domesticates: evolution in the man-made habitat. *Economic Botany* 29, 99–107.
ECLAC (2014) *The Economics of Climate Change in Latin America and the Caribbean: Paradoxes and Challenges*. CEPAL, Santiago.
FAO (Food and Agriculture Organization) (2001) GeoNetwork. FAO, Rome. Available at: www.fao.org/geonetwork/srv/en/metadata.show?id=1255 (accessed 26 September 2014).
FAO (2009) International Treaty on Plant Genetic Resources for Food and Agriculture. FAO, Rome. Available at: www.fao.org/docrep/011/i0510e/i0510e00.htm (accessed 26 September 2014).
FAO (2010) *The Second Report on the State of the World's Plant Genetic Resources for Food and Agriculture*. FAO, Rome.
FAO, IIASA, ISRIC, ISS-CAS and JRC (2012) *Harmonized World Soil Database* (version 1.2). Food and Agriculture Organization of the United Nations, Rome/IIASA, Luxemburg.
Frankel, O.H. (1974) Genetic conservation: our evolutionary responsibility. *Genetics* 78, 53–65.
Fuentes, G. and Santamaría, J.M. (2014) Papaya (*Carica papaya* L.): origin, domestication, and production. *Plant Genetics and Genomics: Crops and Models* 10, 3–15.
Galindo-Tovar, M.E., Ogata-Aguilar, N. and Arzate-Fernandez, A.M. (2008) Some aspects of avocado (*Persea americana* Mill.) diversity and domestication in Mesoamerica. *Genetic Resources and Crop Evolution* 55, 441–150.
Galluzzi, G. and López Noriega, I. (2014) Conservation and use of genetic resources of underutilized crops in the Americas – a continental analysis. *Sustainability* 6, 980–1017.
Golicher, D.J., Cayuela, L. and Newton, A.C. (2012) Effects of climate change on the potential species richness of Mesoamerican forests. *Biotropica* 44, 284–293.
Gremillion, K.J., Barton, L. and Piperno, D. (2014) Particularism and the retreat from theory in the archaeology of agricultural origins. *Proceedings of the National Academy of Sciences of the United States of America* 11, 6171–6177.
Hajjar, R. and Hodgkin, T. (2007) The use of wild relatives in crop improvement: a survey of developments over the last 20 years. *Euphytica* 156, 1–13.
Harlan, J.R. (1965) The possible role of weed races in the evolution of cultivated plants. *Euphytica* 14, 173–176.
Harmon, D. (1996) Losing species, losing languages: connections between biological and linguistic diversity. *Southwest Journal of Linguistics* 15, 89–108.
Harvey, C., Komar, O., Chazdon, R., Ferguson, B.G., Finegan, B., *et al.* (2008) Integrating agricultural landscapes with biodiversity conservation in the Mesoamerican hotspot. *Conservation Biology* 22, 8–15.
Hellin, J., Bellon, M.R. and Hearne, S.J. (2014) Maize landraces and adaptation to climate change in Mexico. *Journal of Crop Improvement* 28, 484–501.
Hernández-López, V.M., Vargas-Vázquez, M.L.P., Muruaga-Martínez, J.S., Hernández-Delgado, S. and Mayek-Pérez, N. (2013) Origen, domesticación y diversificación del frijol común. Avances y perspectivas. *Revista Fitotecnia Mexicana* 36, 95–104.
Heywood, V., Casas, A., Ford-Lloyd, B., Kell, S. and Maxted, N. (2007) Conservation and sustainable use of crop wild relatives. *Agriculture, Ecosystems and Environment* 121, 245–255.
Hijmans, R.J. (2012) Cross-validation of species distribution models: removing spatial sorting bias and calibration with a null model. *Ecology* 93, 679–88.
Hijmans, R.J., Cameron, S.E., Parra, J.L., Jones, P.G. and Jarvis, A. (2005) Very high resolution interpolated climate surfaces for global land areas. *International Journal of Climatology* 25, 1965–1978.

Hoyt, E. (1988) *Conserving the Wild Relatives of Crops*. International Board for Plant Genetic Resources, International Union for Conservation of Nature and Natural Resources and World Wide Fund for Nature, Rome.

Jarvis, A., Lane, A. and Hijmans, R.J. (2008) The effect of climate change on crop wild relatives. *Agriculture, Ecosystems and Environment* 126, 13–23.

Kato Yamakake, T.A., Mapes Sánchez, C., Mera Ovando, L.M., Serratos Hernández, J.A. and Bye Boettler, R.A. (2009) *Origen y Diversificacion del Maíz: Una Revisión Analítica*. Universidad Nacional Autónoma de México, Comisión Nacional para el Conocimiento y Uso de la Biodiversidad, Mexico D.F.

Kraft, K.H., Brown, C.H., Nabhan, G.P., Luedeling, E., Luna Ruiz, J.D.J., *et al.* (2014) Multiple lines of evidence for the origin of domesticated chili pepper, *Capsicum annuum*, in Mexico. *Proceedings of the National Academy of Sciences of the United States of America*, 111, 6165–6170.

Leclerc, C. and Coppens d'Eeckenbrugge, G. (2012) Social organization of crop genetic diversity. The G × E × S interaction model. *Diversity* 4, 1–32.

Lobell, D.B. and Gourdji, S.M. (2012) The influence of climate change on global crop productivity. *Plant Physiology* 160, 1686–1697.

Lobell, D.B., Burke, M.B., Tebaldi, C., Mastrandrea, M.D., Falcon, W.P., *et al.* (2008) Prioritizing climate change adaptation needs for food security in 2030. *Science* 319, 607–610.

Maffi, L. (2005) Linguistic, cultural, and biological diversity. *Annual Review of Anthropology* 34, 599–617.

Magrin, G.O., Marengo, J.A., Boulanger, J.-P., Buckeridge, M.S., Castellanos, E., *et al.* (eds) (2014) Climate Change 2014: Impacts, Adaptation, and Vulnerability. Part B: *Regional Aspects.* Contribution of Working Group II to the Fifth Assessment Report of the Intergovernmental Panel on Climate Change. Cambridge University Press, Cambridge, UK, and New York. Available at: ipcc-wg2.gov/AR5/report/final-drafts/ (accessed 26 September 2014).

Matsuoka, Y., Vigouroux, Y., Goodman, M.M., Sanchez, G.J., Buckler, E., *et al.* (2002) A single domestication for maize shown by multilocus microsatellite genotyping. *Proceedings of the National Academy of Sciences of the United States of America* 99, 6080–6084.

Meilleur, B.A. and Hodgkin, T. (2004) *In situ* conservation of crop wild relatives: status and trends. *Biodiversity and Conservation* 13, 663–684.

Monterroso Rivas, A., Conde Álvarez, C., Rosales Dorantes, G., Gómez Díaz, J.D. and Gay García, C. (2011) Assessing current and potential rainfed maize suitability under climate change scenarios in México. *Atmósfera* 24, 53–67.

Oney, B., Reineking, B., O'Neill, G. and Kreyling, J. (2013) Intraspecific variation buffers projected climate change impacts on *Pinus contorta*. *Ecology and Evolution* 3, 437–449.

Pearman, P.B., D'Amen, M., Graham, C.H., Thuiller, W. and Zimmermann, N.E. (2010) Within-taxon niche structure: niche conservatism, divergence and predicted effects of climate change. *Ecography* 33, 990–1003.

Phillips, S.J., Anderson, R.P. and Schapire, R.E. (2006) Maximum entropy modeling of species geographic distributions. *Ecological Modelling* 190, 231–259.

Pickersgill, B. (2007) Domestication of plants in the Americas: insights from Mendelian and molecular genetics. *Annals of Botany* 100, 925–940.

Piperno, D.R. (2011) The origins of plant cultivation and domestication in the New World tropics: patterns, process, and new developments. *Current Anthropology* 52, S453–S470.

Prischmann, D.A., Dashiell, K.E., Schneider, D.J. and Eubanks, M.W. (2009) Evaluating *Tripsacum*–introgressed maize germplasm after infestation with western corn rootworms (Coleoptera: Chrysomelidae). *Journal of Applied Entomology* 133, 10–20.

Ramírez, M., Galluzzi, G., van Zonneveld, M., Thomas, E., van Etten, J., *et al.* (2013) *Strategic Action Plan to Strengthen Conservation and Use of Mesoamerican Plant Genetic Resources in Adapting Agriculture to Climate Change*. Bioversity International, Cali, Colombia.

Ramírez-Villegas, J. and Jarvis, A. (2010) *Downscaling Global Circulation Model Outputs: The Delta Method*. Decision and Policy Analysis Working Paper No 1. Centro Internacional de Agricultura Tropical (CIAT), Cali, Colombia.

Ramírez-Villegas, J., Cuesta, F., Devenish, C., Peralvo, M., Jarvis, A., *et al.* (2014) Using species distributions models for designing conservation strategies of tropical Andean biodiversity under climate change. *Journal for Nature Conservation* 22, 391–404.

Rebelo A.G. and Sigfried W.G. (1992) Where should nature reserves be located in the Cape Floristic Region, South Africa? Models for the spatial configuration of a reserve network aimed at maximizing the protection of diversity. *Conservation Biology* 6, 243–252.

Schaal, B.A., Olsen, K.M. and Carvalho, L.J.C.B. (2006) Evolution, domestication, and agrobiodiversity in the tropical crop cassava. In: Motley, T.J., Zerega, N. and Cross, H. (eds) *Darwin's Harvest: New Approaches to the Origins, Evolution, and Conservation of Crops*. Columbia University Press, New York, pp. 269–284.

Srisuwan, S., Sihachakr, D. and Siljak-Yakovlev, S. (2006) The origin and evolution of sweet potato (*Ipomoea batatas* Lam.) and its wild relatives through the cytogenetic approaches. *Plant Science* 171, 424–433.

Tanksley, S.D. and McCouch, S.R. (1997) Seed banks and molecular maps: unlocking genetic potential from the wild. *Science* 227, 1063–1066.

Thomas, E., van Zonneveld, M., Loo, J., Hodgkin, T., Galluzi, G., *et al.* (2012) Present spatial diversity patterns of *Theobroma cacao* L. in the Neotropics reflect genetic differentiation in Pleistocene refugia followed by human-influenced dispersal. *PLoS ONE* 7, e47676.

Thomas, E., Alcazar Caicedo, C., McMichael, C.H., Corvera, R. and Loo, J. (2015) Uncovering spatial patterns in the natural and human history of Brazil nut (*Bertholletia excelsa*) across the Amazon basin. *Journal of Biogeography* 42, 1367–1382, doi:10.1111/jbi.12540.

Thomas, E., Tovar, E., Villafañe, C., Bocanegra, J.L. and Moreno, R. (2016) An assessment of the distribution, genetic diversity and potential spatiotemporal scale of alien gene flow in rice crop wild relatives (Oryza spp.) in Colombia. PLoS ONE (forthcoming).

Trucco, F. and Trane, P.J. (2011) Amaranthus. In: Kole, C. (ed.) *Wild Crop Relatives: Genomic and Breeding Resources, Vegetables Department of Crop Sciences*. University of Illinois, Urbana, pp. 11–21.

UNEP-WCMC (2010) Data standards for the World Database on Protected Areas (WDPA). Available at: old.unep-wcmc.org/world-database-on-protected-areas-wdpa-data-standards_966.html (accessed 9 December 2015).

Ureta, C., Martínez-Meyer, E., Perales, H.R. and Alvarez-Buylla, E.R. (2012) Projecting the effects of climate change on the distribution of maize races and their wild relatives in Mexico. *Global Change Biology* 18, 1073–1082.

van Zonneveld, M., Scheldeman, X., Escribano, P., Viruel, M.A., Van Damme, P., *et al.* (2012) Mapping genetic diversity of cherimoya (*Annona cherimola* Mill.): application of spatial analysis for conservation and use of plant genetic resources. *PLoS ONE* 7, e29845.

Vigouroux, Y., Glaubitz, J.C., Matsuoka, Y., Goodman, M.M., Sánchez, G.J., *et al.* (2008) Population structure and genetic diversity of New World maize races assessed by DNA microsatellites. *American Journal of Botany* 95, 1240–1253.

Westengen, O.T., Okongo, M.A., Onek, L., Berg, T., Upadhyaya, H., *et al.* (2014) Ethnolinguistic structuring of sorghum genetic diversity in Africa and the role of local seed systems. *Proceedings of the National Academy of Sciences of the United States of America* 111, 14100–14105.

WLMS (2005) World language mapping system. Available at: www.worldgeodatasets.com/language/ (accessed 26 September 2014).

Wood, D. and Lenné, J.M. (1997) The conservation of agrobiodiversity on-farm: questioning the emerging paradigm. *Biodiversity and Conservation* 6, 109–129.

Zizumbo-Villarreal, D. and Colunga-GarcíaMarín, P. (2010) Origin of agriculture and plant domestication in West Mesoamerica. *Genetic Resources and Crop Evolution* 57, 813–825.

# 23 Landrace Conservation of Maize in Mexico: an Evolutionary Breeding Interpretation

**H. Perales***
*El Colegio de la Frontera Sur, Carretera Panamericana y Periferico Sur s/n, San Cristobal, Mexico*

## 23.1 Introduction

At the inception of the crop genetic resources conservation movement, *ex situ* methods were the preferred way to proceed (Frankel, 1970). This contrasts to the preferences for wild species where *ex situ* methods are generally viewed as remedial, a welcomed addition but no way forward. Frankel noted (1970) that 'we conclude that the prospects of long-term conservation of "primitive" populations in their natural habitat, with anything like their current population dynamics, is generally much more tenuous than that of wild plants'. The indispensable nature of crop genetic resources for human survival and the uncertain nature of the conservation of crop variants in farmers' fields were strong arguments for this position.

The view of the relationship between *in situ* and *ex situ* conservation for crop genetic resources has changed since the 1990s. Case studies documented that the expected replacement of traditional landraces with commercial seeds was not as simple and straightforward as expected previously (Brush, 1995). The Convention on Biological Diversity (UN, 1993) highlighted the need for the *in situ* conservation of crops and instituted the political milieu for our present view of the complementarity of *ex situ* and *in situ* conservation. Nonetheless, the preference for the *ex situ* method and its influence for on-farm conservation has remained.

Justifications for plant genetic resources conservation – both on farm and *ex situ* – have emphasized the need for available genes for geneticists and breeders. Notable advantages for on-farm conservation include the impossibility for germplasm banks to capture all the complexities of the variations in the field, that on-farm conservation is a natural laboratory, that evolutionary processes are creating new germplasm and thereby being a backup for germplasm banks, which permits resampling. Still, even when recognizing that increasing and maintaining private benefits for farmers is crucial for on-farm conservation (Brush, 2000; Bellon, 2004), the availability of genes for geneticists is the central justification. Several factors weaken this proposition. In particular, germplasm banks are still very much underused, despite them being much easier and cheaper to access than on-farm varieties; therefore, it is difficult to envision that under normal conditions breeders will seek on-farm resources to carry out their work. In addition, biotechnology has broken down species barriers and most work is done with genes from bacteria and other microorganisms. The main objective of on-farm conservation may be not

*E-mail: hperales@ecosur.mx

 *Enhancing Crop Genepool Use: Capturing Wild Relative and Landrace Diversity for Crop Improvement* (eds N. Maxted, M. Ehsan Dulloo and B.V. Ford-Lloyd) 

genes but rather people's livelihoods, as these varieties are the way millions are solving their seed requirements. On-farm conservation is for household livelihoods, gene conservation is the by-product of farmers generating landraces. This chapter describes some aspects that structure maize diversity in the landscape and considers whether an evolutionary breeding interpretation for the case is possible.

## 23.2 Maize in Mexico

The economic environment for maize production in Mexico has changed dramatically in the last 20 years; however, the on-farm conservation of maize has not been affected greatly. Mexico signed the North American Free Trade Agreement (NAFTA) with the USA and Canada, which came into law in 1994. The treaty has provisions for maintaining maize tariffs, since it was considered that local production could not compete with that of the USA and Canada (Nadal, 1999). It was widely expected that maize production in Mexico would decline and, at the same time, maize biodiversity would be affected (Nadal, 2000). There were changes in regional production (Sweeney *et al.*, 2013), but the total area sown with maize was relatively stable at around 8 Mha (million hectares) between 1994 and 2014, yet total production even increased from 18.1 Mt in 1993 to 22.7 Mt in 2013 (SIAP, 2015). The number of farmers planting maize was similar in the two agricultural censuses available, 2.75 and 2.83 million households for 1991 and 2007, respectively (INEGI, 2015a,b). About 85% of the farmers planting maize do so in fields of less than 5 ha and under rainfed conditions. Governmental policy has favoured large-scale producers in the north with irrigation and high-quality lands (Fox and Haight, 2010), while the official policy for small-scale producers has been to induce them to abandon maize and switch to other agricultural activities or to sell labour. Nonetheless, at the national level, farmers have not abandoned maize production, as described above or in recent published case studies (Eakin *et al.*, 2014).

We have only one national estimate of the adoption of commercial seeds for Mexico. Aquino *et al.* (2001) calculated that commercial varieties comprised 22% of total seed requirements in Mexico (hybrids 15% and open pollinated varieties 7%); thus, 78% of the seed used each year in Mexico is farmer-saved seed. Not all farmer-saved seeds are traditional landrace seeds (Aguirre *et al.*, 2000; Bellon and Risopoulos, 2001; Bellon *et al.*, 2006); some are recycled (advanced generations of commercial seeds) or creolized (advanced generations of commercial seeds introgressed with traditional landraces). We do not have a national estimation of the quantity of recycled and creolized seeds; local estimates have been 10.8% recycled and 27.0% creolized for the seed lots of communities in Frailesca, Chiapas (Bellon *et al.*, 2006), 4.4 and 11%, respectively, for communities in the coast of Oaxaca (Bellon *et al.*, 2006), or only 3.5% creolized in Guanajuato (Aguirre *et al.*, 2000). Thus, it is safe to assume that at least 50% of the seed planted each year in Mexico is of extant traditional landraces; that is, more than 2 million farmers with 4 Mha or more. A governmental programme to support on-farm conservation of these landraces started in 2009 and is limited to communities that are at or near natural reserves (SEMARNAT, 2015). The programme consists mainly of a subsidy for planting traditional landraces and presently reaches barely 0.05% of farmers that use these seeds. The on-farm conservation of maize landraces in Mexico is a robust de facto process after more than 60 years of plant breeding efforts.

Maize agriculture in Mexico has had a bimodal 'modern/peasant' structure that has persisted to the present (Hewitt, 1976; Eakin *et al.*, 2014). Land with irrigation, mostly in the northern states, accounts for 18% of annual maize area and 43% of production (SIAP, 2015). These lands are managed with high inputs and can attain yields as large as those in any other region in the world. However, in Mexico, most of the maize is planted under rainfed conditions, commonly in lands that are too steep or stony to be farmed with machinery. Nonetheless, this bimodal structure is not all straightforward; even in the high-input agriculture of the north, maize is a family activity and the small-scale 'traditional' famers use chemical inputs extensively (Eakin *et al.*, 2014). For the peasant sector, maize is primarily food security with very low or nil net income, and it seems that highest yield is not the objective. It is not uncommon to hear farmers state that 'maize is not a business' (Perales *et al.*, 1998); nonetheless, abandonment of maize production seems to be very low, even after NAFTA (Eakin *et al.*, 2014).

## 23.3 Factors that Structure Maize Diversity in the Landscape

In Mexico, maize diversity is classified as races, a proposal of Anderson and Cutler (1942) formalized by Wellhausen *et al.* (1952). Races were purposefully defined 'as loosely as possible... [as] a group of related individuals with enough characteristics in common to permit their recognition as a group' and sought to be a natural form of classification that represented phylogenetic relationships (Anderson and Cutler, 1942). Sánchez *et al.* (2000) have listed 59 races for Mexico, and within each race there is variation in grain colour and texture, some plant characteristics and phenology. These races are planted in Mexico between 0 and 2900 metres above sea level (masl), and their distribution correlates with climate variables. There is a large amount of variation within and between races for climate adaptation, with annual mean temperatures ranging from 11.3 to 26.6°C, and 426–4245 mm annual rainfall (Ruiz *et al.*, 2008). Four groups of races have been characterized for environmental conditions: temperate to semi-hot (annual mean temperature between 14 and 21°C), semi-hot to hot (20–25°C), very hot and wet (24.5–27.5°C) and a special group that comprises Jala and two other races. It is interesting to note that the climatic ranges of maize are greater than those of the closest teosinte, suggesting that it evolved adaptability after domestication (Ruiz *et al.*, 2008).

Many races or varieties have been reported in case studies, though the common pattern is that few dominate almost all environments. For example, 21 races are reported in the germplasm banks for Chiapas, but only 11 comprise at least 1% of all accessions (Perales and Hernández, 2005). In a systematic sample for Chiapas taken between 1999 and 2004 (119 communities and 2073 household), there were eight common races, but only three represented more than 80% of the seed lots (Brush and Perales, 2007). The distribution of these three races correlated directly with the environmental classification of Ruiz *et al.* (2008), with Tuxpeño in the very hot, Comiteco in the semi-hot and Oloton in the temperate environment. This pattern of few dominant and several minor varieties is also common at the community level, with one to three varieties sown in 80% or more of the maize area (Perales *et al.*, 2003a). The average number of varieties per farmer in a community has been reported at between 1.1 and 2.7, though there are two extreme cases with 5.0 and 5.8 seed lots/farmer, with no intermediate reports between 2.7 and 5.0 (Dyer *et al.*, 2014).

Altitude and irrigation are major influences that correlate strongly with the presence of landraces and commercial cultivars. In a transect from the highlands near Mexico City (2400 masl) to intermediate elevations in the state of Morelos (1200 masl), Perales *et al.* (2003b) reported that the highlands were completely dominated with traditional landraces (100%), while in the lower elevations, hybrids and their advanced generations were common (34 and 37%, respectively). This is significant, because the community near Mexico City has a much better infrastructure, access to markets and education, and has four major research institutions with more than 50 years of breeding programmes. Also, production of the landraces in the highlands was mostly for the market (66% of production), and where hybrids were present, only 11–29% was sold. This also contradicts simplistic expectations that commercial varieties are destined for markets and landraces for consumption. A similar situation is present in Chiapas (Brush and Perales, 2007), where above 1400 masl landraces completely dominate the environment. Arslan and Taylor (2009) applied an econometric model to a national level survey and found that irrigation was positively associated with and highly significant as a factor to explain commercial cultivar use, but not so for traditional cultivars. This agrees well with anecdotal observations. Most of the area of maize with irrigation is in the northern states of Mexico, although there are some significant areas in the centre of the country. Some areas with good rainfed conditions are also planted with commercial cultivars. Arslan and Taylor (2009) also found other variables that were associated significantly with landraces or commercial cultivars. Land area with maize had a negative coefficient for landraces, implying that these were used mostly in smaller fields. The use of machinery was associated positively with commercial cultivars, and slope, drought and altitude had negative coefficients. There is also evidence that economic well-being affects the use of commercial seed. In lowland Chiapas, non-poor households use more commercial hybrids than poor farmers, who tend to use more open pollinated cultivars and landraces, though both groups use all types of varieties (Bellon *et al.*, 2006). It is also of interest that fertilizers and herbicides are prevalent in farming systems using traditional landraces. In Chiapas, more than 70% of the households use fertilizers and herbicides, though 95% of those with commercial hybrids do so (Perales, unpublished results). Partial adoption of commercial seeds is common for small-scale and semi-commercial farmers (Bellon, 1991; Bellon and Brush, 1994); many farmers use these seeds without discarding landraces, commonly because of contrasting traits of interest. For example, hybrids are recognized to be better for lodging resistance and for sale of grain, whereas landraces resist ear rot and storage pests better (Bellon *et al.*, 2006).

Although there is large variation between races in morphological characters, it has been found that the genetic distance between them is relatively small. Working with several races and populations from communities of Oaxaca, Pressoir and Berthaud (2004a,b) found no isolation by distance and among village differentiation ($F_{st}$ = 0.003) lower than within village differentiation ($F_{st}$ = 0.011). Also, they found significant homozygote excess than that expected under the Hardy–Weinberg equilibrium ($F_{is}$ = 0.138) and that this was related to variation in the mean anthesis-silking interval and flowering range of a given population. For central Chiapas, van Heerwaarden *et al.* (2009) also found larger morphologic than genetic distances. On a continental scale for America, Vigouroux *et al.* (2008) analysed 310 races and also found greater within-race variation (24%) than among races (7.6%). In this case, genetic distance correlated best with geographic distance (0.410) and weakly with altitude (0.044) and landrace name (0.065). Even though there are private alleles in maize races (Reif *et al.*, 2006), most of the genes are shared between them.

Pressoir and Berthaud (2004a) propose that high levels of seed exchange among farmers prevent population differentiation at community and regional levels. Notwithstanding, most seed exchange seems to be within communities (Badstue *et al.*, 2006; Brush and Perales, 2007). In a 3-year study of six communities of Oaxaca with almost only traditional landraces, Badstue *et al.* (2006) found that seed transactions were between 0.23 and 0.31 acquisitions/farmer/year and between 0.24 and 0.39 distributions/farmer/year. In the sample of Chiapas referred to above (Brush and Perales, 2007), the original acquisition of the seed lots was between 64 and 95% within the community, with the lowest values in lower elevations. Badstue *et al.* (2006) did not record the origin of the seed providers, but it can be inferred that about 95% were from within the communities themselves. It is significant that the most important reason for acquiring maize seed in Oaxaca (Badstue *et al.*, 2007) is for experimentation purposes (32.5%), followed by commencing to farm (27.9%) and lack of sufficient seed (18.2%). This is to illustrate that even in communities where landraces are dominant, farmers are not traditional in a static sense, they are not only willing to try alternative seed types but actively seek them.

A common assumption is that commercial cultivars are superior to landraces (Brush, 1989), an idea related in particular to yield. For the case of maize in Mexico, this is not a straightforward assumption that can be generalized, in particular for temperate and marginal environments. In Table 23.1, a series of experiments in the central

Fig. 23.1. Mexican farmer with maize landraces. (Photograph courtesy of C. Dietz.)

highlands of Mexico are summarized; these were comparisons of a group of landraces and improved varieties. In most cases, the best landrace was as good as or better than the best-improved variety. It has been noted that the superiority of improved types under an experimental station environment is commonly lost when planting under farmers' conditions (Muñoz *et al.*, 1976a). In Table 23.2, further experiments under marginal conditions are presented; it is not rare that landraces are competitive in these environments. This is not to say that commercial hybrids are no good, but the environments where they are superior are commonly limited to hot climates with irrigation or good rainfall, as noted above. Perales *et al.* (1998) proposed that where landraces were being cultivated, these were competitive or superior from the agronomic or economic point of view.

More than 60 years ago, Anderson (1947) noted that 'maize is a sensitive mirror of the people who grow it', and other scholars agreed (Hernandez, 1985), even though demonstrations were not formalized. Perales *et al.* (2005) presented evidence that two ethnolinguistic groups in Chiapas had maize diversity that was not all explained by environmental adaptation, even though there were no apparent differences in the classification and management of maize. In an additional study comparing maize used by mestizos and indigenous people throughout Chiapas, Brush and Perales (2007) showed that both groups preferred the same type of maize in different environments. However, distinct patterns of each group were also apparent; mestizos had a higher preference for the Tuxpeño race and commercial seed, and indigenous people for the Oloton and Comiteco races. Indigenous people also kept their seed lots for longer periods, had more variation in seed colour and had a higher preference for sources of seed within the community.

Maize quality is also a factor that seems to determine the conservation of some landraces; anecdotal descriptions are common, though no specific study has tested the relationship between particular uses and maize types. Some speciality dishes require a particular type of maize, such as the case of pozole, a soup of boiled and burst grains with meat (similar to hominy of the USA). Pozole is only made with maize from the races Ancho or Cacahuacintle, because they have large

**Table 23.1.** Average yield (t/ha) for the best landrace and improved cultivar in several experiments in the central highlands of Mexico.

| | Source | | | | | |
|---|---|---|---|---|---|---|
| Cultivar and other | CIMMYT (1969) | CIMMYT (1973) | Muñoz *et al.* (1976b) | Tut *et al.* (1985) | Herrera *et al.* (1991) | Herrera *et al.* (1991) |
| State of experiments | Puebla | Puebla | Tlaxcala, Mexico, Hidalgo | Puebla | Puebla high yield | Puebla low yield |
| Number of sites | 6 | 7 | 13 | 4 | 5 | 6 |
| Landrace | 5.4 | 5.5 | 3.1 | 5.1 | 3.7 | 1.3 |
| Improved | 5.3 | 5.6 | 2.8 | 4.5 | 4.4 | 0.6 |

**Table 23.2.** Average yield (t/ha) for the best landrace and improved cultivar in several experiments in marginal environments in the south of Mexico. (From Muñoz, 2005.)

| | State and region | | | | | | | |
|---|---|---|---|---|---|---|---|---|
| | Michoacan | | | Chiapas | Oaxaca | | | |
| Cultivar | Sierra | Centre | Coast | Meseta Comiteca | High Mixteca | Low Mixteca | Central valleys | Isthmus |
| Landrace | 4.6 | 5.6 | 2.7 | 4.9 | 2.7 | 2.8 | 4.1 | 2.8 |
| Improved | 2.1 | 5.2 | 2.5 | 4.1 | 2.3 | 2.5 | 3.3 | 2.0 |

grains that burst suitably for the dish; these maize races have premium prices of up to two or three times those of generic grain. Other dishes or forms of preparing maize have reputed requirements, such as the Bolita race for clayuda (a special large, semi-rigid tortilla) or the Zapalote Grande race for totopo (a toasted tortilla speciality), but no details are available as to its veracity. Markets based on speciality characteristics linked to dishes seem to be behind the increased importance of Ancho-Pozolero and blue grain types in Mexico State (Perales *et al.*, 2003a; Keleman and Hellin, 2009), and this might also be the case for other varieties. Nevertheless, tortillas are the most important use of maize in Mexico, and almost all varieties are good for this purpose. So, even if quality is important, the majority of maize types can only be marketed as generic grain with no premium price.

There are also several minor qualities of the varieties that affect the decisions whether to cultivate commercial seeds or landraces. For example, the husk of hybrids available in Mexico is typically very short and does not cover the tip of the ear, facilitating the entrance of storage insects that become a pest at the end of the season. In contrast, landraces have 5 cm or more of husk over the ears, limiting insect entrance. When farmers are interested in consuming their own maize, hybrids are therefore considered unsuited for year-round consumption. Also, in Mexico, husks are used for tamales, and some maize types have been selected for wide husks that have premium value at markets.

A further important aspect is that landraces are not all static. Some farmers purposefully attempt to cross their landraces with commercial varieties (Perales *et al.*, 2003b), and seed lots of this type are not uncommon (Bellon *et al.*, 2006). This same process might also be happening between landraces; it is very common that different maize types coexist in adjoining fields and they frequently cross between them (Louette, 2000). For example, Perales *et al.* (2003b) described how a common important variety in the semi-hot environment of Morelos, Ancho-Pozolero, was not recognized as having been inherited from their forefathers. They stated unequivocally that their fathers gave them a variety named Ancho, and that after an introduction of a Pozolero that had better market price they found that their Ancho could also pass in the market for premium price. It was not agreed if they only started selecting their Ancho for wider grains or if the Pozolero crossed with Ancho, creating Ancho-Pozolero, but it was considered a new not an inherited form.

Even though there is large morphologic variation between maize races, and the effect of human agency in the process appears self-evident, we do not have adequate details of the effect of artificial selection. Louette and Smale (2000) described farmers' seed selection practices in Jalisco and found that farmers maintained variety ideotype selecting against off-types, even while being subject to significant gene flow through cross pollination. Nonetheless, since selection practices are basically done on harvested ears, mainly large, well-developed and healthy, improvement appears to be limited. Based on two cycles of farmer and random selection from an initial population, few characteristics exhibited differences in the progeny. Also, it is of interest that in farmers' opinions they could modify characteristics of a variety based on changing planting date, applying fertilizer or planting next to a different variety, but not by seed selection (Louette and Smale, 2000). Soleri and Cleveland (2001) drew similar conclusions working in Oaxaca; they found that farmers thought characters like ear length and seed size were related to the growing environment and were not inherited; thus, they had low expectations for a response to selection. A comparison between random and farmer selected ears for three generations showed few differences between the first random selection and the third farmer selection, but not so between the third random and farmer selection; actual response to farmer selection was approximately zero (Soleri *et al.*, 2000).

Collections of maize diversity in Mexico have been carried out since 1943, and after a new broad sample that ended 5 years ago there is a georeferenced database with more than 18,000 accessions classified by race. An analysis of this database reveals that the largest diversity, measured as race richness, occurs in the west and in central Mexico; the north-east and eastern side is typically less diverse (Perales and Golicher, 2014). Some other important aspects are that, of the 59 reputed races, only 47 had enough samples to be considered in the analysis, 5 races were very common and half of the 47 were rare or very rare (Wellhausen *et al.*, 1952, had noted the same 60 years ago), and abundance correlated well (Spearman rho was between 0.63 and 0.72, all significant) between three periods into which the database was segmented (1943–1954, 1968–1979 and 1997–2009). Also, all races sampled in the 1950s were resampled in the last sample and, based on modelled distributions, nationwide richness had not decreased (Perales and Golicher, 2014). Eleven biogeographic regions based on race composition were proposed, with six of these containing 90% of all collections for 38 of the 47 races (Perales and Golicher, 2014).

## 23.4 An Evolutionary Breeding Interpretation of Maize in Mexico

More than 50 years ago, Suneson (1956) and Bal *et al.* (1959) reported the results of an experiment that had been almost dismissed until recently. Suneson extended previous work undertaken by H.V. Harlan and M.L. Martini and called the method evolutionary breeding, which consisted of 'natural selection in concert with time and population diversity' (Suneson and Wiebe, 1962; Suneson, 1969). In Suneson's experiment, there were several composite cross populations of barley (large number of hybrids bulked), originated by H.V. Harlan, that were permitted to evolve for up to 29 generations; compared with a common cultivar, these were fairly competitive and sometimes slightly superior in yield (though the generality of this has been contested, see Hockett *et al.*, 1983; Soliman and Allard, 1991). In recent years, the method has again been put to the test for organic and low-input systems (see, for example, Murphy *et al.*, 2004; Phillips and Wolfe, 2005).

An evolutionary breeding scenario, within a subsistence food security focus, could explain the widespread and persistent success of maize landraces in Mexico. Extant maize landraces have never been formally started as broad composites, but it can be proposed that these populations do have broad genetic bases that are continuously refreshed by seed exchanges and pollen flow. These populations are changing, in the evolutionary breeding sense, mostly under natural selection, with varying degrees of artificial selection. Not all these efforts need to be successful, though the populations that perform well are surely distributed within communities. The process need not be the most efficient or quick, but for these farmers it could be adequate and as good as other alternatives available. Producing for subsistence does not disengage farmers completely from markets (Dyer *et al.*, 2006). This being the case, most small-scale farmers do not seem focused on the highest yields; they are interested in obtaining a sensible production adequate for household food security at the minimum cost possible. Even if evolutionary breeding is not as good as conventional breeding, in this scenario evolutionary breeding might be working satisfactorily for these farmers and their families.

A related issue is how crop evolution will unfold in the future. Currently, after several millennia where farmers were the only force in plant evolution (apart from natural selection), plant breeders have become the foremost driving force. This idea is evident in Frankel *et al.*'s (1995, p. 84) statement that 'it may well be the case that ... it is not the superiority of landraces which is the cause of their retention, but the lack of local plant breeders who could produce improved varieties in tune with the agro-ecosystem'. That is, if we had more breeders (and resources), no landraces would remain; thus, no on-farm conservation. However, could it be that there will never be enough breeders for all environments? Could it be that the social and environmental complexity is such that institutional breeding will never be the only process? It is possible that the future will entail complex solutions where farmer-mediated crop evolution is one part of the solutions. On-farm conservation is a by-product of farmers' participation in crop evolution; as long as they continue to produce viable varieties for themselves, *in situ* conservation will endure. If farmers fade off as evolutionary agents, on-farm conservation will be an artificial construct not much different from *ex situ* conservation. On-farm conservation has still been viable for many farmers of maize in Mexico; can this continue to be so?

## 23.5 What Should Conservation Efforts Focus On?

On-farm conservation efforts for maize landraces in Mexico have focused on subsidies. These have been small awards to farmers near or at nature reserves that already cultivate landraces (in 2009, there were 25,826 beneficiaries; by 2013, only 10,258 were supported), or to farmers that were chosen as 'custodians' of the particular races (in 2014, there were 214 farmers paid to keep the 59 reputed races). What can we expect from this approach and what do we want or need for the on-farm conservation of maize in Mexico?

The first issue that we need to confront is the scale; de facto on farm conservation involves more than 2 million farmers and more than 4 Mha. If the objectives for the on-farm conservation of maize in Mexico are to focus on a representation of diversity available, as is presently the case, what would it imply for the rest of de facto conservation? The profusion of diversity rests, in part, on the large-scale nature of the phenomena; with only a handful of farmers involved, maybe *ex situ* methods are better suited. Vigouroux *et al.* (2008) calculated that with 172 accessions of the 964 samples (17.8%) studied, they could represent 80% of the 3752 different alleles they observed. If specific populations are our main objective, on-farm conservation could only be symbolic and the genes conserved would not be much different from those of a germplasm bank.

A next issue is richness, commonness and rarity. From the analysis of germplasm collections and modelling of race distributions, it is clear that the west and centre of the country have larger richness; the east and north have less races present. However, even lower richness regions have particular races that are not distributed in high richness areas, which implies that concentrating only in high richness areas will not be enough to conserve maize diversity. Rare races are usually considered in conservation efforts, but what of common races? In Mexico, common races are extremely important for the livelihoods of those households that plant them; these races are the foundation of food security for these households and others in rural areas. These races should not be disregarded because they are not under risk; efforts should be focused on supporting the continuation of their competitiveness.

Under an evolutionary breeding scenario, common races would have a special place in the process. The importance of continuing evolution for on-farm conservation has been mentioned in the literature, though what part of the evolutionary process is central has not been defined. For example, it has been suggested that resistance to diseases or insects is important for this continuing evolution, though it can be proposed that it is farmers' participation in crop evolution that should be the central concern. On-farm conservation is the product of farmers as actors in crop evolution; they are still producing viable varieties for their own needs. Farmers' involvement in crop evolution would not guarantee the permanence of the same varieties nor of their integrity purely as historic forms. Expecting unchanging conditions and believing that farmers will participate for the long run with this kind of restriction is unsustainable. As Bennett noted some 50 years ago, 'the purpose of conservation is not to capture the present moment of evolutionary time, in which there is no special virtue, but to conserve the material so that it will continue to evolve' (cited in Frankel, 1970, p. 476). On-farm conservation is for farmers' livelihoods; the by-product can be genes for breeders, but this is a by-product not the objective. We need to change our mentality from a germplasm bank focus to an evolutionary perspective and imagine a complex interplay of actors, where not only breeders have a place in the future of crop evolution. There is no true impediment for farmers to maintain an effective place in the process of crop evolution: they have had a place historically and they still have a place at present. As Sir Joseph Hutchinson (1965) noted 'Our limitations are the limitations of our scientific insight and imagination, rather than of the biological material with which we work'.

## References

Aguirre Gomez, J.A., Bellon, M.R. and Smale, M. (2000) A regional analysis of maize biological diversity in southeastern Guanajuato, Mexico. *Economic Botany* 54, 60–72.

Anderson, E. (1947) Field studies of Guatemalan Maize. *Annals of the Missouri Botanical Garden* 34, 433–467.

Anderson, E. and Cutler, H.C. (1942) Races of *Zea mays*: I, Their recognition and classification. *Annals of the Missouri Botanical Garden* 29, 69–88.

Aquino, P., Carrión, F., Calvo, R. and Flores, D. (2001) Selected maize statistics. In: Pingali, P.L. (ed.) *CIMMYT 1999–2000 World Maize Facts and Trends: Meeting World Maize Needs*. CIMMYT, Mexico D.F., pp. 45–57.
Arslan, A. and Taylor, J.E. (2009) Farmer's subjective valuation of subsistence crops: the case of traditional maize in Mexico. *American Journal of Agricultural Economics* 91, 956–972.
Badstue, L.B., Bellon, M.R., Berthaud, J., Juárez, X., Rosas, I.M., *et al*. (2006) Examining the role of collective action in an informal seed system: a case study from the central valleys of Oaxaca, Mexico. *Human Ecology* 34, 249–273.
Badstue, L.B., Bellon, M.R., Berthaud, J., Ramírez, A., Flores, D., *et al*. (2007) The dynamics of farmers' maize seed supply practices in the Central Valleys of Oaxaca, Mexico. *World Development* 35, 1579–1593.
Bal, B.S., Suneson, C.A. and Ramage, R.T. (1959) Genetic shift during 30 generations of natural selection in barley. *Agronomy Journal* 51, 555–557.
Bellon, M.R. (1991) The ethnoecology of maize variety management: a case study from Mexico. *Human Ecology* 19, 389–418.
Bellon, M.R. (2004) Conceptualizing interventions to support on-farm genetic resource conservation. *World Development* 32, 159–172.
Bellon, M.R. and Brush, S.B. (1994) Keepers of maize in Mexico. *Economic Botany* 48, 196–209.
Bellon, M.R. and Risopoulos, J. (2001) Small-scale farmers expand the benefits of improved maize germplasm: a case study from Chiapas, Mexico. *World Development* 29, 799–811.
Bellon, M.R., Adato, M., Becerril, J. and Mindek, D. (2006) Poor farmers' perceived benefits from different types of maize germplasm: the case of creolization in Lowland Tropical Mexico. *World Development* 34, 113–129.
Brush, S.B. (1989) Rethinking crop genetic resource conservation. *Conservation Biology* 3, 19–29.
Brush, S.B. (1995) *In situ* conservation of landraces in centers of crop diversity. *Crop Science* 35, 346–354.
Brush, S.B. (2000) The issues of *in situ* conservation of crop genetic resources. In: Brush, S.B. (ed.) *Genes in the Field, On-farm Conservation of Crop Diversity*. IPGRI-IDRC-Lewis, Boca Raton, Florida, pp. 3–26.
Brush, S.B. and Perales, H.R. (2007) A maize landscape: ethnicity and agro-biodiversity in Chiapas Mexico. *Agriculture, Ecosystems and Environment* 121, 211–221.
CIMMYT (1969) *The Puebla Project 1967–1969*. CIMMYT, El Batán, Texcoco, Mexico, pp. 57–65.
CIMMYT (1973) *El Plan Puebla, siete años de experiencia: 1967–1973*. CIMMYT, El Batán, Texcoco, Mexico, pp. 41–45.
Dyer, G.A., Boucher, S. and Taylor, J.E. (2006) Subsistence response to market shocks. *American Journal of Agricultural Economics* 88, 279–291.
Dyer, G.A., López, A., Yúnez, A. and Taylor, J.E. (2014) Genetic erosion in maize's center of origin. *Proceedings of the National Academy of Sciences of the United States of America* 111, 14094–14099.
Eakin, H., Perales, H., Appendini, K. and Sweeney, S. (2014) Selling maize in Mexico: the persistence of peasant farming in an era of global markets. *Development and Change* 45, 133–155.
Fox, J. and Haight, L. (2010) *Subsidios para la desigualdad: las políticas públicas del maíz en México a partir del libre comercio*. Woodrow Wilson International Center for Scholars. México D.F.
Frankel, O.H. (1970) Genetic conservation in perspective. In: Frankel, O.H. and Bennett, E. (eds) *Genetic Resources in Plants: Their Exploration and Conservation*. IBP Handbook No 11. International Biological Program and Blackwell, Oxford, UK, pp. 469–489.
Frankel, O.H., Brown, A.D.H. and Burdon, J.J. (1995) *The Conservation of Plant Biodiversity*. Cambridge University Press, Cambridge, UK.
Hernandez, E. (1985) Maize and man in the greater Southwest. *Economic Botany* 39, 416–430.
Herrera, B.E., Lopez, H. and Lopez, P.A. (1991) Estrategia de mejoramiento genético para cultivos básicos de temporal en el Plan Puebla. In: Anon., *Sintesis del informe anual 1991*. Colegio de Postgraduados, Puebla, Mexico, pp. 25–28.
Hewitt, C. (1976) *Modernizing Mexican Agriculture: Socioeconomic Implications of Technological Change 1940–1970*. United Nations Institute for Research on Society and Development, Geneva, Switzerland.
Hockett, E.A., Eslick, R.F., Qualset, C.O., Dubbs, A.L. and Stewart, V.R. (1983) Effects of natural selection in advanced generations of barley Composite Cross II. *Crop Science* 23, 752–756.
Hutchinson, J.B. (1965) Crop plant evolution: a general discussion. In: Hutchinson, J.B. (ed.) *Essays on Crop Plant Evolution*. Cambridge University Press, London, pp. 167–181.
INEGI (Instituto Nacional de Estadística y Geografía) (2015a) Censo Agrícola y Ganadero 1991. Available at: www.inegi.org.mx/est/contenidos/proyectos/Agro/ca1991/Resultados_Agricola/default.aspx (accessed 31 July 2015).

INEGI (2015b) Censo Agrícola, Ganadero y Forestal 2007. Available at: www.inegi.org.mx/est/contenidos/proyectos/Agro/ca2007/Resultados_Agricola/default.aspx (accessed 31 July 2015).

Keleman, A. and Hellin, J. (2009) Specialty maize varieties in Mexico: a case study in market-driven agro-biodiversity conservation. *Journal of Latin American Geography* 8, 147–174.

Louette, D. (2000) Traditional management of seed and genetic diversity: what is a landrace? In: Brush, S.B. (ed) *Genes in the Field, On-farm Conservation of Crop Diversity*. IPGRI-IDRC-Lewis, Boca Raton, Florida, pp. 109–142.

Louette, D. and Smale, M. (2000) Farmers' seed selection practices and traditional maize varieties in Cuzalapa, Mexico. *Euphytica* 113, 25–41.

Muñoz, A. (2005) *Centli-Maiz*. Colegio de Postgraduados, Montecillo, Mexico, 211 pp.

Muñoz, A., Carballo, A. and Gonzalez, V.A. (1976a) Mejoramiento del maíz en el CIAMEC. I. Analisis critico y reenfoque del programa. In: Anon., *Memoria VI Congreso Nacional de Fitogenetica*. Sociedad Mexicana de Fitogenetica, Monterrey, Mexico, pp. 124–130.

Muñoz, A., Gonzalez, V.A., Livera, M., Lopez, A. and Ron, J. (1976b) Mejoramiento del maíz en el CIAMEC. II. Ampliacion de la base germoplasmica y su aprovechamiento considerando caracteres agronómicos y rendimiento. In: Anon., *Memoria VI Congreso Nacional de Fitogenetica*. Sociedad Mexicana de Fitogenetica, Monterrey, Mexico, pp. 113–123.

Murphy, K., Lammer, D., Lyoin, S., Carter, B. and Jones, S.S. (2004) Breeding for organic and low-input farming systems: an evolutionary–participatory breeding method for inbred cereal grains. *Renewable Agriculture and Food Systems* 20, 48–55.

Nadal, A. (1999) Issue Study 1. Maize in Mexico: some environmental implications of the North American Free Trade Agreement (NAFTA). In: Anon., *Assessing Environmental Effects of the North American Free Trade Agreement (NAFTA): An Analytic Framework (Phase II) and Issue Studies*. Commission for Environmental Cooperation, Montreal, Canada, pp. 65–182.

Nadal, A. (2000) *Mexican Corn: Genetic Variability and Trade Liberalization*. PROCIENTEC Documento de Trabajo1-06. El Colegio de Mexico, Mexico, 13 pp.

Perales, H. and Golicher, D. (2014) Mapping the diversity of maize in Mexico. *PLoS ONE* 9(12), e114657.

Perales, H. and Hernández, J.M. (2005) Diversidad del maíz en Chiapas. In: González, M., Ramírez, N. and Ruiz, L. (eds) *Diversidad biológica de Chiapas*. Plaza y Valdés/ECOSUR/COCYTECH, México, pp. 337–355.

Perales, H.R. (1998) Conservation and evolution of maize in the Amecameca and Cuautla valleys of Mexico. PhD dissertation, University of California, Davis.

Perales H.R., Benz, B.F. and Brush, S.B. (2005) Maize diversity and ethnolinguistic diversity in Chiapas, Mexico. *Proceedings of the National Academy of Sciences of the United States of America* 102, 949–954.

Perales, R.H., Brush, S.B. and Qualset, C.O. (1998) Economic and agronomic competitiveness of maize landraces and *in situ* conservation in Mexico. In: Smale, M. (ed.) *Farmers, Gene Banks and Crop Breeding: Economic Analyses of Diversity in Wheat, Maize, and Rice*. CIMMYT and Kluwer Academic Publishers, Norwell, Massachusetts, pp. 109–125.

Perales, R.H., Brush, S.B. and Qualset, C.O. (2003a) Dynamic management of maize landraces in Central Mexico. *Economic Botany* 57, 21–34.

Perales, R.H., Brush, S.B. and Qualset, C.O. (2003b) Landraces of maize in Central Mexico: an altitudinal transect. *Economic Botany* 57, 7–20.

Phillips, S.L. and Wolfe, M.S. (2005) Evolutionary breeding for low input systems. *Journal of Agricultural Science* 143, 245–254.

Pressoir, G. and Berthaud, J. (2004a) Patterns of population structure in maize landraces from the Central Valleys of Oaxaca in Mexico. *Heredity* 92, 88–94.

Pressoir, G. and Berthaud, J. (2004b) Population structure and strong divergent selection shape phenotypic diversification in maize landraces. *Heredity* 92, 95–101.

Reif, J.C., Warburton, M.L., Xia, X.C., Hoisington, D.A., Crossa, J. *et al.* (2006) Grouping of accessions of Mexican races of maize revisited with SSR markers. *Theoretical Applied Genetics* 113, 177–185.

Ruiz, J., Durán Puga, N., Sánchez González, J., Ron Parra, J., González Eguiarte, D., *et al.* (2008) Climatic adaptation and ecological descriptors of 42 Mexican maize races. *Crop Science* 48, 1502–1512.

Sánchez G., J.J., Goodman, M.M. and Stuber, C.W. (2000) Isozymatic and morphological diversity in the races of maize of Mexico. *Economic Botany* 54, 43–59.

SEMARNAT (Secretaría de Medio Ambiente y Recursos Naturales) (2015) Programa de Conservación de Maíz Criollo. Available at: www.conanp.gob.mx/maiz_criollo/ (accessed 31 July 2015).

SIAP (Servicio de Información Agroalimentaria y Pesquera) (2015) Producción anual por cultivo, maíz. Available at: www.siap.gob.mx/cierre-de-la-produccion-agricola-por-cultivo/ (accessed 31 July 2015).

Soleri, D. and Cleveland, D.A. (2001) Farmers' genetic perceptions regarding their crop populations: an example with maize in the Central Valleys of Oaxaca, Mexico. *Economic Botany* 55, 106–128.

Soleri, D., Smith, S.E. and Cleveland, D.A. (2000) Evaluating the potential for farmer and plant breeder collaboration: a case study of farmer maize selection in Oaxaca, Mexico. *Euphytica* 116, 41–57.

Soliman, K.M. and Allard, R.W. (1991) Grain yield of composite cross populations of barley: effects of natural selection. *Crop Science* 31, 705–708.

Suneson, C.A. (1956) An evolutionary plant breeding method. *Agronomy Journal* 48, 188–191.

Suneson, C.A. (1969) Evolutionary plant breeding. *Crop Science* 9, 119–121.

Suneson, C.A. and Wiebe, G.A. (1962) A "Paul Bunyan" plant breeding Enterprise with barley. *Crop Science* 2, 347–348.

Sweeney, S., Steigerwald, D.G., Davenport, F. and Eakin, H. (2013) Mexican maize production: evolving organizational and spatial structures since 1980. *Applied Geography* 39, 78–92.

Tut, C., Gaona, R., Rangel, J.L., Garcia, R., Aguirre, D., *et al.* (1985) *Analisis de los diferentes programas de investigación relacionados con el cultivo de maíz en el área de influencia del Campo Agricola Experimental Tecamachalco, Puebla.* Instituto Nacional de Investigaciones Agricolas, Tecamachalco, Puebla, Mexico, 75 pp.

United Nations (1993) Convention on Biological Diversity. Rio de Janeiro.

van Heerwaarden, J., Hellin, J., Visser, R.F. and van Eeuwijk, F.A. (2009) Estimating maize genetic erosion in modernized smallholder agriculture. *Theoretical Applied Genetics* 119, 875–888.

Vigouroux, Y., Glaubiktz, J., Matsuoka, Y., Goodman, M.M., Sanchez, J., *et al.* (2008) Population structure and genetic diversity of New World maize races assessed by DNA microsatellites. *American Journal of Botany* 95, 1240–1253.

Wellhausen, E.J., Roberts, L.M., Hernandez, E. and Mangelsdorf, P.C. (1952) *Races of Maize in Mexico.* Bussey Institution of Harvard University, Cambridge, Massachusetts, 223 pp.

# 24 Use of Spontaneous Sexually Produced New Landraces of a Vegetatively Propagated Crop of the Andes (*Oxalis tuberosa* Mol.) to Enhance *In Situ* Conservation

M. Bonnave,[1]* T. Bleeckx,[1] F. Terrazas[2] and P. Bertin[1]
[1]*Earth and Life Institute, Université catholique de Louvain, Louvain-la-Neuve, Belgium;* [2]*Fundación PROINPA (Promoción e Investigación de Productos Andinos), Cochabamba, Bolivia*

## 24.1 Introduction

Many cultivated species are propagated vegetatively. The main advantage of vegetative propagation for farmers is that it allows agronomically valuable genotypes, which generally display hybrid vigour due to their heterozygosity, to be fixed and propagated easily. In cases of strict vegetative propagation, the only sources of variation inside a lineage are mutations, which can be accumulated over generations and fixed easily by farmers. Strictly vegetatively propagated crops are very sensitive to genetic erosion due to genetic drift and selection. In selecting the propagules for the next crop generation, farmers choose not to propagate all lineages at the same rate, and in the long term, some genotypes are inevitably lost. However, numerous vegetatively propagated crops still have a functional sexual reproduction system. Allogamous sexual reproduction permits the recombination of parental genotypes into new genotypes, thereby enhancing the diversity available to farmers. Recombination allows the combination of many favourable mutations in the same lineage, rather than their competition among lineages (Jarvis *et al.*, 2000; McKey *et al.*, 2010).

Vegetatively propagated crops that are also capable of allogamous sexual reproduction may combine the advantages of both reproduction systems. This is the case for economically important crops such as potato (*Solanum* spp.), sweet potato (*Ipomea batatas* Lam.) or cassava (*Manihot esculenta* C.). For these crops, plant breeding programmes have permitted spectacular progress to be made in terms of productivity, resistance to pathogens, etc. These programmes are usually conducted by using a crossing step in order to produce botanical seed and genetic variation (Ceccarelli *et al.*, 2009; Bradshaw, 2010). Nevertheless, modern plant breeding science is not the only way to use sexual reproduction of vegetatively propagated crops. Studies show that farmers take advantage of the sexual reproduction of these crops in traditional agricultural systems. For instance, cassava farmers in Guyana actively seek volunteer plants produced by sexual reproduction and take care of these plants until maturity. If they are found satisfactory, these new genotypes will be multiplied

*Corresponding author; e-mail: maxime.bonnave@gmail.com

 *Enhancing Crop Genepool Use: Capturing Wild Relative and Landrace Diversity for Crop Improvement* (eds N. Maxted, M. Ehsan Dulloo and B.V. Ford-Lloyd)

as a new landrace or assigned to an existing landrace (Elias *et al.*, 2001a). This practice considerably increases the genotypic and intra-varietal diversity that is managed by these farmers (Elias *et al.*, 2001b). Similar practices were described for sweet potato in New Guinea (Yen, 1960; Bulmer, 1965). In the case of potato, traditional farmers of a community in Peru would undertake active botanical seed propagation when they observed yield decreases due to tuber 'tiring', which was most likely due to virus disease accumulation in the tubers. They would then collect botanical seeds, sow them in beds and carry out a selection of the best new landraces based on the size, number and quality of the tubers. According to Quiros *et al.* (1992), this would explain the extensive morphologic and varietal variability observed in the region.

They are many underused and neglected crop species that are mainly propagated vegetatively but which also display a functional sexual reproduction system. There is an important lack of knowledge on their reproductive systems and on the effect that sexual reproduction could have on the genetic diversity of these crops (McKey *et al.*, 2010). In the Andes only, eight of these crops are grown (Hermann and Heller, 1997). Oca (*Oxalis tuberosa* Mol.) is one of these crops. It is traditionally cultivated in the Andean highlands for the consumption of its tubers and forms an integral part of the local staple diet, showing interesting nutritional value (Campos *et al.*, 2006; Chirinos *et al.*, 2009). Potato is the main crop in this region. By following it in the rotation, oca plays an important role in interfering with the life cycle of potato pests (Terrazas and Valdivia, 1998). Oca presents a fully functional, mainly allogamous, sexual reproduction characterized by a heteromorphic self-incompatibility system (Trognitz *et al.*, 1998). Although the global genetic diversity of oca (measured with ISSR (Inter Simple Sequence Repeat) molecular markers) appeared to be low when compared to other crops, a high number of morphological variation and of distinct genotypes was observed (Pissard *et al.*, 2006). Genetic erosion of oca has been observed in several regions of Peru and Bolivia (Terrazas and Valdivia, 1998; Tapia and Estrella, 2001; FAO, 2010). One of the main reasons for this erosion is that farmers are discouraged from growing oca due to low market prices and to the competition with potato, which benefits from higher market prices and the availability of modern, high-yielding varieties. An important intra-landrace genetic diversity was observed, with both ISSR and SSR molecular markers (Pissard *et al.*, 2008; Bleeckx, 2012). According to Gibbs (1976), sexual reproduction would have been dominant before domestication, leading to the creation of highly heterozygous genotypes and a rich variety of morphotypes, which would then have been culled and fixed by the farmers by preferential selection and vegetative propagation. As farmers traditionally identify landraces by a small number of morphological criteria such as colour, shape and the taste of the tuber, it is likely that confusion between plants sharing the same or very similar morphotypes and different genotypes occurred, and that these genotypes were considered as one variety by the farmers, explaining the intra-landrace diversity observed (Bianco and Sachs, 1997; Bradbury and Emshwiller, 2011). However, Pissard *et al.* (2006; 2008) hypothesized that the large number of genotypes and the intra-varietal diversity could be explained by the incorporation of new genotypes created in recent sexual reproduction events.

## 24.2 *In situ* Sexual Reproduction of Oca and the Creation of New Genotypes

In a recently published study, we showed that sexual reproduction was still functional in natural conditions (Bonnave *et al.*, 2014). Many plantlets issued from sexual reproduction were found in fields that previously had been cultivated with oca (Fig. 24.1). These plantlets could be differentiated from the numerous plants growing from remnant tubers from the previous harvest because of the presence of cotyledons. Molecular studies conducted with SSR markers showed that all the collected samples of sexually produced plantlets showed unique genetic profiles that clearly differentiated them from the potential parental landraces (i.e. the varieties that were previously grown in the field where the plantlets were found). Sexually produced plantlets exhibited much less structure and much more genetic variability than the potential parental varieties. The samples were clearly linked to both of their potential parental landraces, confirming the allogamy of oca in field conditions.

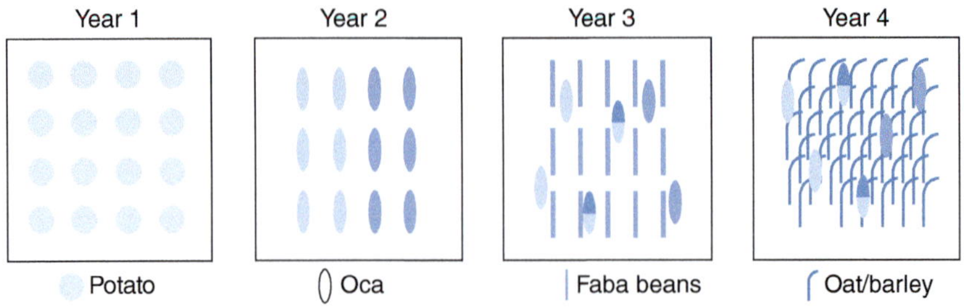


**Fig. 24.1.** Typical rotation pattern in the Andes highlands. The oca plants (ovals) in the faba bean and oat/barley fields are *k'ipas*. The plain blue scales are plants issued from vegetative propagation and the mixed blue scales are ones issued from sexual reproduction, illustrating the combination of the two parental genotypes.

## 24.3 Farmers' Management of *K'ipas* and Knowledge of Sexual Reproduction

The question still remained about the possible incorporation of sexually produced genotypes into germplasm. A few preliminary interviews with farmers were made at the same time as the collection of plantlets (Bonnave *et al.*, 2014). In Quechua, all plants of a crop species found in a field where this crop was cultivated the previous year were called *k'ipas* by the farmers, without exception (Academia mayor de la lengua Quechua, 2005). It can be translated as 'volunteer plant'. The interviews showed that farmers sometimes used the tubers of *k'ipas* as seeds. However, they all thought that these tubers were issued exclusively from the vegetative propagation of the varieties that were cultivated in the previous years and assigned these tuber seeds to these varieties. However, at harvest time, it is difficult to differentiate morphologically the plants issued from vegetative propagation from those issued from sexual reproduction (as opposed to cassava). Therefore, tubers of plantlets originated from sexual reproduction and sharing similar morphotypes to cultivated landraces could be incorporated unknowingly into these varieties.

These preliminary interviews paved the way to another study being conducted on farmers' practices regarding *k'ipas* and their knowledge of sexual reproduction. This was necessary in order to better assess the probability and the context of the incorporation of sexually produced genotypes into the cultivated germplasm of farmers. More information was needed on the different uses of the *k'ipas* that were made by farmers, on the reason for the selection of *k'ipas* for planting material, on the management of the *k'ipa* planting material and on the farmer's awareness of the sexual reproduction of oca. In order to achieve these objectives, 38 farmers from four communities in the Candelaria district (Colomi municipality, Cochabamba department, Bolivia) were interviewed. This region, considered as a microcentre of diversity for Andean tuber crops, was chosen because many plantlets issued from sexual reproduction were observed in *k'ipa* fields, and because of the good relationship maintained between the PROINPA (Promoción e Investigación de Productos Andinos) foundation and the local farmers. These particular communities were selected because they were located in different agroecological zones, between which agricultural practices vary, according to Terrazas and Valdivia (1998). All the interviews were realized in Quechua. General information such as gender and age was gathered first. Farmers were then addressed with open questions on the use of *k'ipas* of oca and on their knowledge of the sexual reproduction of oca and potato (in order to compare between the two crops).

The results showed that farmers had various uses for *k'ipas*. Two-thirds of the farmers used *k'ipas* for something other than pig feeding. Among these, 100% used the tubers of *k'ipas* for self-consumption, 75% sold the tubers at the market and 75% used them as seeds, in different frequencies (Fig. 24.2). The use of *k'ipas* as seeds was correlated to the age of the farmers: only 28% of the younger farmers used *k'ipas* as seeds, while 73% of the older farmers used them for this

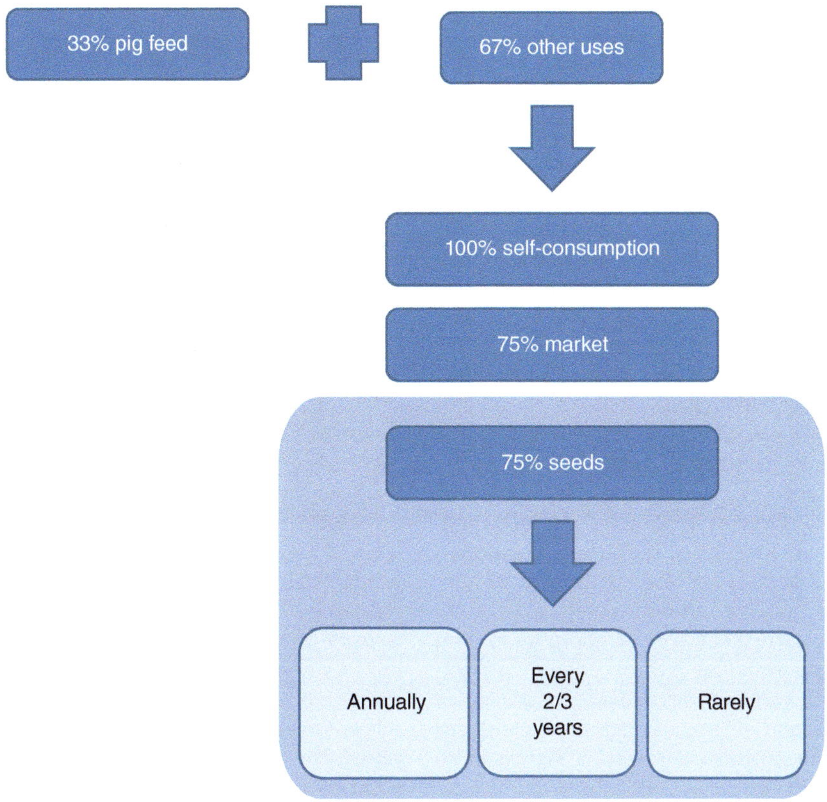


Fig. 24.2. Proportions of the different uses of *k'ipas* and the frequency of use of *k'ipas* as seeds (in percentage of interviewed farmers).

purpose. The reasons given by the farmers for not using *k'ipas* as seeds are: (i) they had not considered it; (ii) it leads to a low yield; (iii) it transmits diseases; and (iv) tubers of *k'ipas* sometimes differ morphologically from the landraces they grow. To the contrary, the reasons farmers use *k'ipas* as seeds are: (i) a lack of seeds at sowing time due to poor calculation of the needed quantity; (ii) a lack of seeds due to frost damage; and (iii) the fact that the tubers of *k'ipas* are directly mixed with the tubers of the cultivated landraces at the moment of harvest and therefore cannot be distinguished at the sowing period. According to the farmers, the selection of the tubers of *k'ipas* for planting material is based on morphological criteria, generally the principal and secondary colours of the tubers. The selected seeds are then assigned to known landraces (presently or previously cultivated). New names are never given to the collected planting material. The *k'ipa* seeds are then planted mixed with normal seeds of the assigned landrace (80% of the farmers) or planted separately to assess their quality in the first year (20% of the farmers).

Knowledge concerning the sexual reproduction of oca is very poor, as only 18% of the farmers interviewed were aware that fruit set of oca could occur, while 90% of farmers were aware that it could occur for potato (Table 24.1). This can be explained by the fact that potato fruits are much larger than oca fruits.

Farmers even used a specific Quechua word to designate potato fruit: '*mak'unkura*'. Moreover, it seems that most of the farmers who had knowledge of true potato seeds acquired it from other farmers or from technicians of the PROINPA foundation. While 42% of the farmers knew about the possibility of obtaining plants from true potato seeds, only one farmer knew about this possibility for oca. Finally, 16% of the farmers knew that new landraces could be obtained with these true potato seeds. Knowledge of the

Table 24.1. Farmers' knowledge of the sexual reproduction of potato and oca.

| | Potato | Oca |
|---|---|---|
| Farmers who observed fruit set | 90% | 18% |
| Farmers knowing that plants can grow from botanical seeds | 42% | 3% |
| Farmers knowing that new varieties can be obtained | 16% | 3% |

sexual reproduction of oca was correlated with age: only 7.1% of the younger farmers knew that oca could set fruit, while 40% of the older farmers knew this.

After all the interviews were conducted, a group feedback phase was realized at the request of the farmers. The results of the interviews were summarized to the farmers and then a class on the sexual reproduction of potato and oca was given. Farmers expressed great interest in sexual reproduction and the possible creation of new landraces. Some of them knew that a potato participatory plant breeding programme had been carried out in a neighbouring district, Morochata, resulting in the creation of interesting potato varieties that had also been adopted in the Colomi region. To read more about this programme, consult the work of Gabriel *et al.* (2000, 2004).

## 24.4 Conclusions and Perspectives

These results clearly show that *k'ipa* planting material is incorporated into the landraces cultivated by farmers. Farmers seem to use *k'ipas* as planting material when they face an unexpected lack of this material. These farmers do think that this planting material only originates from vegetative propagation and they always collect the tubers of *k'ipas* that are similar to the landraces they grow. However, as the selection of the planting material is based on very few morphological criteria, the incorporation of sexually produced new genotypes, which was hypothesized after the preliminary work of this study, now seems highly probable. Although apparently unintentional, such incorporations may have been, and still could be, a key asset in the adaptation of the oca crop to the evolution of ecological conditions. As they present similar agricultural management, it is probable that this phenomenon also occurs for other neglected Andean tuber crops such as mashua (*Tropaeolum tuberosum* R. & P.) or ulluco (*Ullucus tuberosus* C.). Moreover, this unintentional incorporation of new genotypes, even if practised at low frequencies, could play an important role in the evolution and the structure of the genetic diversity of other important, vegetatively propagated crops such as potato, cassava, sweet potato, yam (*Dioscorea* spp.), fruit trees, etc., in traditional systems.

However, the results also showed that the use of *k'ipas* as seeds seemed to decline in the young generation, and that this generation had very little knowledge of sexual reproduction. If farmers were to abandon completely the resort to *k'ipas* when they lack planting material, the window of opportunity for new genotype incorporation into cultivated germplasm would be closed. The germplasm would then only rely on point mutations of fixed genotypes to adapt itself to the changing environmental, cultural and economic conditions. Nevertheless, farmers showed great interest when informed about the possibility of creating new landraces by means of sexual reproduction. Given proper training, farmers could intentionally propagate genotypes issued from sexual reproduction in natural conditions. The natural selection that occurs in field conditions should lead to the creation of a genepool of plantlets that are adapted to local environmental conditions. Farmers could then select from these the plantlets that are of interest to them. This invariably would lead to a rise of the genotypic diversity they manage. Moreover, these practices could be the first step to the creation of participatory plant breeding (PPB) programmes. Such programmes would allow the creation of locally adapted modern varieties by using local landraces as genitors (Sthapit *et al.*, 1996). Therefore, the new varieties would include the genetic diversity that was conserved before in the local landraces. This could allow a balance between the goal of maintaining genetic diversity *in situ* and improving varieties according to the needs of farmers. A simplified way to produce new genotypes of interest would be to plant two or more compatible landraces that have been selected (by farmers and scientists) for the traits that are needed in an improved variety. Open pollinations between the selected varieties, germination and selection for adaptation to the processes of the local environmental conditions should occur in natural conditions. Scientists and farmers could then jointly evaluate the clonal lineages of these spontaneous new genotypes for their agronomic performances and morphological and culinary characteristics. An important side effect of the PPB programmes is the empowerment of participant farmers. Farmers obtain new knowledge and skills by participating in such programmes. Moreover, they are the first ones to access the new varieties and distribute them to other farmers by the informal seed system. In the Andes, conserving and distributing seeds is usually correlated with wealth and social prestige (Zimmerer, 1996; Bianco and Sachs, 1997). Therefore, participating farmers could become nodal farmers, implicated in the conservation of genetic resources and the dissemination of the new varieties created.

## Acknowledgements

The authors would like to thank all the farmers who contributed to this study and Eddy Lopez for the hard work realized during the interviews. MB acknowledges the support of fellowships from Fonds pour la formation à la Recherche dans l'Industrie et dans l'Agriculture (Fund for Research Training in Industry and Agriculture), Belgium.

## References

Academia mayor de la lengua Quechua, de, A.M. (2005) *Diccionario Quechua-Español-Quechua*. Gobierno regional de Cusco, Cusco, Peru.

Bianco, M. and Sachs, C. (1997) Growing Oca, Ulluco, and Mashua in the Andes: socioeconomic differences in cropping practices. *Agriculture and Human Values* 15, 267–280.

Bleeckx, G. (2012) Relation entre l'hétérostylie, la reproduction sexuée et la diversité génétique de la oca (*Oxalis tuberosa* Mol.). MSc thesis. Université catholique de Louvain, Louvain la Neuve, Belgium.

Bonnave, M., Bleeckx, G., Rojas Beltrán, J., Maughan, P., Flamand, M.-C., *et al.* (2014) Farmers' unconscious incorporation of sexually-produced genotypes into the germplasm of a vegetatively-propagated crop (*Oxalis tuberosa* Mol.). *Genetic Resources and Crop Evolution* 61, 721–740.

Bradbury, E.J. and Emshwiller, E. (2011) The role of organic acids in the domestication of *Oxalis tuberosa*: a new model for studying domestication resulting in opposing crop phenotypes. *Economic Botany* 65, 76–84.

Bradshaw, J.E. (2010) *Root and Tuber Crops (Handbook of Plant Breeding)*. 1st edn. Springer, New York.

Bulmer, R. (1965) Beliefs concerning the propagation of new varieties of sweet potato in two New Guinea Highlands societies. *Journal of Polynesian Society* 74, 237–239.

Campos, D., Noratto, G., Chirinos, R., Arbizu, C., Roca, W., *et al.* (2006) Antioxidant capacity and secondary metabolites in four species of Andean tuber crops: native potato (*Solanum sp.*), mashua (*Tropaeolum tuberosum* Ruiz & Pavón), oca (*Oxalis tuberosa* Molina) and ulluco (*Ullucus tuberosus* Caldas). *Journal of the Science of Food and Agriculture* 86, 1481–1488.

Ceccarelli, S., Guimarães, E.P. and Weltzien, E. (2009) *Plant Breeding and Farmer Participation*. FAO, Rome.

Chirinos, R., Betalleluz-Pallardel, I., Huamán, A., Arbizu, C., Pedreschi, R., *et al.* (2009) HPLC-DAD characterisation of phenolic compounds from Andean oca (*Oxalis tuberosa* Mol.) tubers and their contribution to the antioxidant capacity. *Food Chemistry* 113, 1243–1251.

Elias, M., McKey, D., Panaud, O., Anstett, M.C. and Robert, T. (2001a) Traditional management of cassava morphological and genetic diversity by the Makushi Amerindians (Guyana, South America): perspectives for on-farm conservation. *Euphytica* 120, 143–157.

Elias, M., Penet, L., Vindry, P., McKey, D., Panaud, O., *et al.* (2001b) Unmanaged sexual reproduction and the dynamics of genetic diversity of a vegetatively propagated crop plant, cassava (*Manihot esculenta* Crantz), in a traditional farming system. *Molecular Ecology* 10, 1895–1907.

FAO (Food and Agriculture Organization) (2010) State of diversity of major and minor crops. Appendix 4 in: *Report on the State of the World's Plant Genetic Resources for Food and Agriculture*. FAO, Rome.

Gabriel, J., Torrez, R. and Thiele, G. (2000) Participatory approaches in potato improvement: experiences of PROINPA in Bolivia. In: Almekinders, C.J. and de Boef, W.S. (eds) *Encouraging Diversity: The Conservation and Development of Plant Genetic Resources.* Intermediate Technology Publications, London, pp. 194–199.

Gabriel, J., Herbas, J., Salazar, M., Ruiz, J., López, J., *et al.* (2004) *Participatory Plant Breeding: A New Challenge in the Generation and Appropriation of Potato Varieties by Farmers in Bolivia.* PGRA, CGIAR and Fundación PROINPA, Cali, Columbia.

Gibbs, P. (1976) Studies on the breeding system of *Oxalis tuberosa* Mol. *Flora* 165, 129–138.

Hermann, M. and Heller, J. (1997) *Andean Roots and Tubers: Ahipa, Arracacha, Maca and Yacon.* International Plant Genetics Research Institute, Rome.

Jarvis, D., Myer, L., Klemick, H., Guarino, L. and Smale, M. (2000) *A Training Guide for In Situ Conservation On-farm.* IPGRI, Rome.

McKey, D., Elias, M., Pujol, B. and Duputié, A. (2010) The evolutionary ecology of clonally propagated domesticated plants. *New Phytologist* 186, 318–332.

Pissard, A., Ghislain, M. and Bertin, P. (2006) Genetic diversity of the Andean tuber-bearing species, oca (*Oxalis tuberosa* Mol.), investigated by inter-simple sequence repeats. *Genome* 49, 8–16.

Pissard, A., Rojas-Beltran, J. and Faux, A. (2008) Evidence of intra-varietal genetic variability in the vegetatively propagated crop oca (*Oxalis tuberosa* Mol.) in the Andean traditional farming system. *Plant Systematics and Evolution* 270, 59–74.

Quiros, C., Ortega, R., Raamsdonk, L., Herrera-Montoya, M., Cisneros, P., *et al.* (1992) Increase of potato genetic resources in their center of diversity: the role of natural outcrossing and selection by the Andean farmer. *Genetic Resources and Crop Evolution* 39, 107–113.

Sthapit, B.R., Joshi, K.D. and Witcombe, J.R. (1996) Farmer participatory crop improvement. III. Participatory plant breeding, a case study for rice in Nepal. *Experimental Agriculture* 32, 479–496.

Tapia, C. (2001) Quantification of genetic erosion in ullucus (*Ullucus tuberosus* Caldas), oca (*Oxalis tuberosa* Mol.) and mashua (*Tropaeolum tuberosum* R. & P.) in agroecosystesm of the provinces on Canar, Chimborazo and Tugurahua, Ecuador. Paper presented for the workshop, "*In Situ* Conservation of Agrobiodiversity: Scientific and Institutional Experiences and Implications for National Policies." International Potato Center (CIP), La Molina Peru, August 14–17, 2001.

Terrazas, F. and Valdivia, G. (1998) Spatial dynamics of in situ conservation: handling the genetic diversity of Andean tubers in mosaic systems. *Plant Genetic Resource Newsletter* 114, 9–15.

Trognitz, B., Hermann, M. and Carrión, S. (1998) Germplasm conservation of oca (*Oxalis tuberosa* Mol.) through botanical seed. Seed formation under a system of polymorphic incompatibility. *Euphytica* 101, 133–141.

Yen, D.E. (1960) The sweet potato in the Pacific: the propagation of the plant in relation to its distribution. *Journal of Polynesian Society* 69, 368–375.

Zimmerer, K.S. (1996) *Changing Fortunes: Biodiversity and Peasant Livelihood in the Peruvian Andes.* University of California Press, Berkeley, California.

# 25 A Long-term Systematic Monitoring Framework for On-farm Conserved Potato Landrace Diversity

S. de Haan,[1]* S. Polreich,[1] F. Rodriguez,[1] H. Juarez,[1]
F. Plasencia,[1] R. Ccanto,[2] C. Alvarez,[3] A. Otondo,[4] H. Sainz,[5]
C. Venegas[6] and J. Kalazich[7]

[1]*International Potato Center (CIP), Lima, Peru;* [2]*Grupo Yanapai, Huancayo, Peru;* [3]*Centro Andino de Educacion y Promociona (CADEP), Cusco, Peru;* [4]*Instituto Nacional de Innovación Agropecuaria y Forestal (INIAF), La Paz, Bolivia;* [5]*Universidad Pública de El Alto (UPEA), La Paz, Bolivia;* [6]*Centro de Educacion y Tecnología (CET), Chiloé, Chile;* [7]*Instituto de Investigaciones Agropecuarios (INIA), Osorno, Chile*

## 25.1 Introduction

Global environmental change has stimulated different scientific communities to develop frameworks for the systematic monitoring of the conservation status of ecosystems, habitats or specific species, ranging all the way from coral reefs, tropical glaciers to endangered flora. Examples of such monitoring initiatives include the Global Observation Research Initiative in Alpine Environments (GLORIA), NOAA's Integrated Coral Observing Network (ICON), World Glacier Monitoring Service (WGMS) and IUCN Red List of Threatened Species, among many others.

The crop genetic resources conservation community has long expressed concerns about landrace loss and genetic erosion (Ochoa, 1975; Hawkes, 1983; FAO, 2010). Yet hard evidence of loss of alleles, landraces or shifts in their frequencies is commonly lacking as a consequence of the non-existence of baseline or time series data, fixed monitoring sites, standardized methods or solid partnerships, among other prerequisites needed for systematic monitoring of the conservation status of crop genetic diversity in centres of origin.

There are several challenges for the advancement of *in situ* conservation science. One of them relates to the need to develop standard procedures and metrics to measure temporal changes in the conservation status of landraces and wild relatives. Examples of crop genetic *in situ* monitoring, particularly of cultivated landraces, are rare. Some exceptions exist (Salick, 2012), but these are generally non-systematic in the sense that they do not cover multiple benchmark sites or countries within the centre of origin of a particular crop species, apply standard procedures that are concerted and easily accessible or replicable across species, and foresee regular time intervals with local partnerships for sustained monitoring.

Systematic monitoring is only possible if standard procedures for baseline research, time

*Corresponding author; e-mail: s.dehaan@cgiar.org

 

series and spatial comparisons are developed and agreed upon by different stakeholders. Since the start-up of the CGIAR Research Program on Roots, Tubers and Bananas (CRP-RTB), the International Potato Center (CIP) has gradually advanced the implementation of a hotspot-based *in situ* network for potato landraces in the crop's centre of origin. This process started in 2012, and the network is called the *Chirapaq Ñan* Initiative (Rainbow Route in the Quechua language). The development and implementation of a systematic monitoring approach has involved the following components: (i) hotspot identification, followed by standard procedures to measure; (ii) total diversity; (iii) relative diversity; (iv) spatial diversity; and (v) collective knowledge. Outputs include different information systems about the landrace conservation status. Networking and attention to country interests, farmer participation and benefit sharing are essential elements of the framework.

## 25.2 Hotspot Identification

The term 'hotspot' is frequently used in agrobiodiversity science, but may refer to different levels of resolution ranging from country to plot level (Galluzi *et al.*, 2010; Zimmerer, 2013, 2014) and species to landrace level (Meilleur and Hodgkin, 2004; Das *et al.*, 2013). Here, hotspots refer to geographies with a high concentration of distinct landraces. Hotspots serve as representative sites to monitor systematically the conservation status at the intraspecific level, in this case of potato landraces.

Different complementary criteria and methods were used to identify hotspots of contemporary potato landrace diversity. Criteria included: (i) the biogeography of cultivated species and cultivar groups, specifically *Solanum tuberosum* (Andigenum, Chilotanum, Phujera, Goniocalyx and Stenotomum cultivar groups), *Solanum ajanhuiri*, *Solanum curtilobum* and *Solanum juzepczukii* (Hawkes, 1990; Ovchinnikova *et al.*, 2011); (ii) expert opinion and consensus ($n$ = 410 experts); (iii) ethnicity and linguistic diversity (Steward, 1946; Adelaar, 2007); (iv) the distribution and presence of wild relatives from the primary genepool (*Solanum acaule*, *Solanum berthaultii*, *Solanum brevicaule*, *Solanum candolleanum*, *Solanum okadae*, *Solanum infundibuliforme* and *Solanum vernei*); (v) perceived threat level to conservation; and (vi) partnership strength and interest (Horton *et al.*, 2009).

Geographical information system (GIS) spatial overlays were used to map the landrace diversity hotspots. Frequencies of consensus were used to give higher weight to hotspots that had been highlighted by several experts. As a result, the following hotpots were identified for different countries: Argentina (Jujuy province), Bolivia (department of La Paz), Chile (Chiloe province), Colombia (department of Nariño, Cauca province), Ecuador (Chimborazo province) and Peru (departments of Cusco, Apurimac, Huancavelica and Huánuco). Within each department or province, specific benchmark sites were identified at the district and community level. The surface area of the specific hotspots identified varied from 225 to 1200 km$^2$.

## 25.3 Baseline Research

Detailed baseline research is conducted at the hotspot level involving multiple dimensions. This process started under the umbrella of the *Chirapaq Ñan* Initiative and has initially involved two hotspots in Peru, followed by an additional hotspot in Bolivia. In-depth baseline research takes 36–48 months and will result in open access databases that make the information available to a wide public and includes catalogues, red lists and maps.

### 25.3.1 Total diversity

A detailed inventory of all landraces present in the hotspot documents total landrace diversity. During three to four consecutive seasons, on-farm trials are installed with individual custodian farmers and farmer groups. Standard procedures are followed for: (i) separation of seed tubers; (ii) commencement of characterization trials; (iii) morphological characterization (Gómez, 2000); (iv) DNA fingerprinting (Ghislain *et al.*, 2009); (v) plant photography; (vi) flow cytometry to determine ploidy; and (vii) data management and storage. Main variables include total landrace and allele diversity, but additional (comparative) variables that can be determined include heterozygosity, allele frequency and population differentiation, among other complementary variables (vernacular nomenclature, cultivar group distribution, etc.). The main output involves an (online) hotspot-based catalogue with detailed information about each individual landrace present in the hotspot (see CIP, 2006; Monteros *et al.*, 2010; Scurrah *et al.*, 2013).

### 25.3.2 Relative diversity

A quantitative assessment is made of the relative abundance of landraces. Standard procedures involve field sampling of all potato fields grown by a randomly selected sample of households living within the communities covering the hotspot ($n$ = 150–200 households). A minimum of 200 random tuber samples per field are collected and the identity of each sample is determined to the landrace level using vernacular nomenclature. Based on local expert knowledge, involving elder farmers (especially women) and an experienced agronomist from the region, a master list of landrace names, synonyms and homonyms is established. A database is constructed and the following indicators are calculated: (i) overall cultivar frequency (OCF) and (ii) relative cultivar frequency (RCF). The OCF is a measure of evenness based on the number of households conserving a specific landrace, while the RCF is a measure of relative abundance based on extensive field sampling (Table 25.1).

The final output is a red list of the conservation status of landraces. Such red lists do exist at the species level for flora and fauna and for crop wild relatives (Butchart *et al.*, 2004; MMAyA, 2009; IUCN, 2012). Furthermore, a qualitative approach exists for landraces (Padulosi and Dulloo, 2012). However, the *Chirapaq Ñan* Initiative is the first landrace-level monitoring platform to apply a robust quantitative approach. The red list offers a snapshot of the conservation status of all the individual landraces present in a pre-defined hotspot at any given moment.

### 25.3.3 Spatial diversity

Beyond genetic and species diversity, the landscape-level spatial distribution of landraces is important to monitor. Potato landraces in the Andes are concentrated in a tight altitudinal belt at the upper altitudinal limits of where agriculture is possible (De Haan and Juarez, 2010). Standard procedures basically involve participatory GIS with exactly the same households ($n$ = 150–200) involved in the field-level sampling described in the previous paragraph (Juarez *et al.*, 2011). The GPS coordinates of each field are taken (corners, centre) and linked with other key variables for the same unit, such as property regime (communal, private usufruct), cultivar category (floury landraces, bitter landraces or bred varieties) and total number of landraces, among other relevant data (tillage system, previous crops grown, etc.).

Involvement of local authorities and youth is essential and will allow for detailed mapping in social contexts that may otherwise be opposed to

**Table 25.1.** Two practical indicators specifying the relative abundance of individual landraces.

| Indicator | Equation | Specification | Variables |
|---|---|---|---|
| OCF | OCF = (CCF community 1 + CCF community 2 + ...)/*n* communities sampled | Community cultivar frequency (CCF) = (number of households conserving a specific cultivar/total household sample size of the community) * 100% | Very few households (OCF <1%); few households (OCF <5%); many households (OCF <25%); most households (OCF >25%) |
| RCF | RCF = (HCF household 1 + HCF household 2 + ...)/*n* households sampled | Household cultivar frequency (HCF) = (sample size of a specific cultivar/total tuber sample size of the household) * 100% | Very scarce (RCF <0.05); scarce (RCF <0.10); uncommon (RCF <0.25); common (RCF <1.00); abundant (RCF >1.00) |

*Notes*: OCF = overall cultivar frequency; RCF = relative cultivar frequency.

defining field boundaries. The geospatial database allows for the calculation of a range of baseline indicators per individual landrace, or for each of three cultivar categories, including: altitudinal distribution and plasticity (minimum, maximum average in metres above sea level), total and field-size area ($m^2$), concentration (number of landraces by altitudinal belt), etc. (Figs 25.1 and 25.2). Outputs include distribution maps of landrace diversity. These maps are shared with community authorities and may include additional layers of information that are useful for land-use planning.

### 25.3.4 Collective knowledge

The collective knowledge that is intrinsically linked to landraces constitutes a unique property. It exists while landraces are managed and maintained by farmers. Collective knowledge may include folk taxonomy and nomenclature, farmer cuisine, management practices, medicinal use, rituals or cultural expressions and, importantly, mechanisms of practice-based intergenerational transfer of know-how. Collective knowledge is dynamic and its monitoring as a non-biological property poses a serious challenge. Collective knowledge may not always be strictly collective and indeed also refer to specialized practices that are only known by a few.

Partners of the *Chirapaq Ñan* Initiative decided early on that databases were not the best platform for baseline documentation of collective knowledge. It was decided that baseline documentation should be linked to youth engagement, intergenerational dialogue and the involvement of schools to develop adapted formal curricula based on local knowledge (Fig. 25.3). Traditionally, formal education in the Andes has ignored the local languages (Quechua and Aymara) and culture (Stobart and Howard, 2002; Regalsky and Laurie, 2007), but during the past decade, the implementation of adapted intercultural bilingual education has increased in Bolivia, Ecuador and Peru (Hornberger, 2000). The main method employed involved teachers and students applying participatory action research, including videography, drawings, open interviews and sound recordings (Nazarea, 2006). The raw content was then used by teachers and specialists in education to develop a multimedia output titled *las papitas de mi comunidad* (the potatoes from my community). This education tool offers a partial baseline of localized potato knowledge for three hotspots. The aim is to expand the pilot and thereby the documentation of collective knowledge of landrace diversity. Baseline indicators

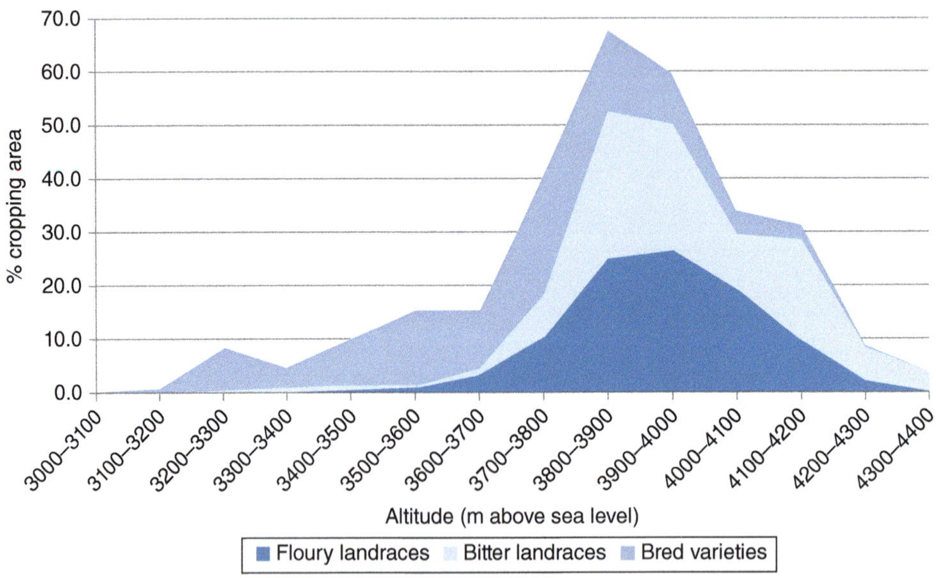


Fig. 25.1. Contemporary spatial distribution map of the cropping area of three landrace categories by altitudinal belts for central and southern Peru based on participatory GIS (regions of Pasco, Junín, Huancavelica, Apurimac and Cusco; $n$ = 666 households/2226 fields).

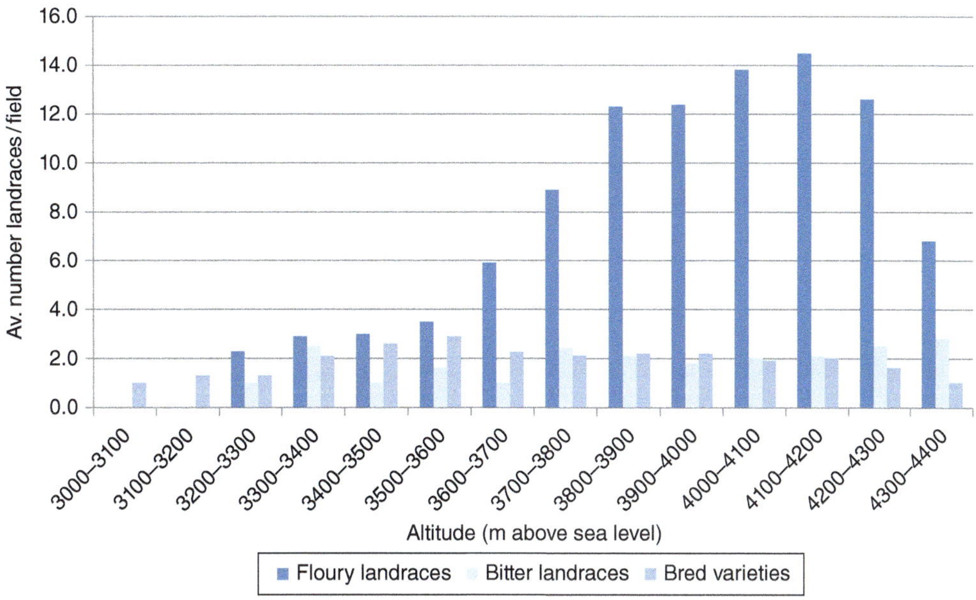


Fig. 25.2. Contemporary spatial distribution map of the average number of landraces per field for three landrace categories by altitudinal belts for central and southern Peru based on participatory GIS (regions of Pasco, Junín, Huancavelica, Apurimac and Cusco; $n$ = 666 households/2226 fields).

Fig. 25.3. Two students from a secondary school in the Apurimac hotspot in southern Peru review participatory videography material capturing collective knowledge about potato diversity.

are of a qualitative nature and include: type of knowledge, number of practices documented, attributes by landrace, among others.

### 25.3.5 Threats to conservation

Given that landraces are managed by smallholder farmers, it is important to understand local perceptions about future conservation, including incentives, disincentives and threats. A baseline of drivers underlying change dynamics, either loss or enrichment, is as important as a baseline of the actual crop genetic diversity itself. To date, the *Chirapaq Ñan* Initiative has not yet engaged in the baseline documentation of incentives, disincentives or threats. Climate stations have been installed in all hotspots. However, besides climate change (Vuille *et al.*, 2003; Perez *et al.*, 2010) there may be many other factors that potentially can affect the fate of on-farm conserved landraces, including migration (Rudel and Richards, 1990), mining (Bury, 2004) and pests and diseases (Giraldo *et al.*, 2010; Kroschel *et al.*, 2013), among others. As yet, no standard procedures have been developed for the baseline documentation of threats to landrace diversity. However, this will be a key priority for future research at each hotspot.

## 25.4 A Network Approach

### 25.4.1 Institutional partnerships

Solid partnerships and local leadership at the hotspot level are essential for the success of the long-term and systematic monitoring of landrace diversity. An international centre such as CIP can facilitate coordination, communication and the use of standard procedures among actors and hotspots. However, the grassroots monitoring work and field-level coordination at specific hotspots has to be led by national and local institutions.

The *Chirapaq Ñan* Initiative has adopted a consortium approach to monitoring. This implies that several institutions with complementary interest and expertise are stimulated to join forces at the hotspot level. A typical consortium can include a national agricultural research institute, NGO, university and municipalities. At the hotspot level, the consortium is led by one organization. However, ideally, each member of the consortium should fulfil a complementary role for the action research to be effective.

### 25.4.2 Farmers' alliances and participation

Farmers, both women and men, are the primary actors realizing *in situ* conservation through the active management of landraces, from planting, harvesting, storage and sales to processing. Basically, conservation is the outcome of diverse utility rationales. Yet a characteristic of monitoring is that it is passive and potentially extractive and does not necessarily provide direct utility to a farmer. Therefore, monitoring ideally should be accompanied by processes that respond to local needs, and demands that do not directly influence landrace diversity as such but rather improve farm management, family health and nutrition and/or smallholder organization. In practice, demands raised by farmers and communities participating in the *Chirapaq Ñan* Initiative have ranged from capacity building in pest and disease control, exchange of knowledge during field days and support for self-organization. Also, a pilot experience linking corporate social responsibility with benefit sharing has been initiated. Specifically, funds made available by a breeding company are used by a recently established association of custodian farmers to support the cultivation of landraces, family health and/or schooling of children.

## 25.5 Information Management

It is important that baselines are published during the research process, or at the latest once the documentation process is completed. Baseline data should be openly accessible and include clear information about the procedures applied and a data dictionary. The use of different platforms, including digital web-based and printed media, is ideal. The *Chirapaq Ñan* Initiative stores the biological data of each individual landrace in an open access BioMart platform. This includes Simple Sequence Repeats (SSR) markers, morphological descriptor and ploidy data and overviews of vernacular nomenclature, among other information. Furthermore, each fully characterized landrace will be documented in a digital hotspot-based catalogue that is searchable. The catalogue will include photographs and red lists. An advanced pilot already exists and the first online version will be launched in 2016. Spatial diversity data will be accessible as raw data and processed in high-resolution maps. Information about collective knowledge will mostly be available locally, particularly in schools. Once a hotspot-level baseline is complete, it will also be published in printed media.

## 25.6 Perspectives and Next Steps

Systematic long-term monitoring is anticipated to provide robust intelligence about the landrace conservation status of potato in its centre of origin. Baseline research will allow for future time series comparison, genetic gap analysis and spatial distribution mapping. An important outcome will be that decision makers will have adequate information to design policies and prioritize interventions based on baseline documentation and trend monitoring. The participation of custodian farmers, NGOs, governments and indigenous organizations is essential for the success of long-term monitoring. Each hotspot needs a representative organization to provide leadership and facilitate the grassroots participation of farmers and their communities.

Standard procedures and metrics have been developed for potato. These are currently being proof-tested and applied in three hotspots in Peru and Bolivia. The monitoring approach will gradually be expanded to other landrace diversity hotpots that have been identified in Argentina, Chile, Colombia and Ecuador. The framework for potato can be expanded to other crops and regions. Eventually, a global network for systematic monitoring can be established. The CGIAR Research Program on Roots, Tubers and Bananas (CRP-RTB) is already pursuing such an approach for cassava, bananas and plantains, yam, taro, sweet potato and other minor roots and tuber crops such as oca, mashua, olluco and yam bean.

## References


Adelaar, W. (2007) *The Languages of the Andes*. Cambridge University Press, Cambridge, UK.

Bury, J. (2004) Livelihoods in transition: transnational gold mining operations and local change in Cajamarca, Peru. *Geographical Journal* 170(1), 78–91.

Butchart, S.H.M., Stattersfield, A.J., Bennum, L.A., Shutes, S.M., Akcakaya, H.R., *et al.* (2004) Measuring global trends in the status of biodiversity: red list indices for birds. *Plos Biology* 2(12), 2294–2304.

CIP (2006) *Catálogo de Variedades de Papa Nativa de Huancavelica – Perú.* Federación Departamental de Comunidades Campesinas de Huancavelica (FEDECCH), International Potato Center (CIP), Lima.

Das, B., Sengupta, S., Parida, S.K., Roy, B., Ghosh, M., *et al.* (2013) Genetic diversity and population structure of rice landraces from Eastern and North Eastern States of India. *BMC Genetics* 14, 71, doi: 10.1186/1471-2156-14-71.

De Haan, S. and Juarez, H. (2010) Land use and potato genetic resources in Huancavelica, Central Peru. *Journal of Land Use Science* 5(3), 179–195.

FAO (Food and Agriculture Organization) (2010) *The Second Report on the State of the World's Plant Genetic Resources for Food and Agriculture*. FAO, Rome.

Galluzi, G., Eyzaguirre, P. and Negri, V. (2010) Home gardens: neglected hotspots of agro-biodiversity and cultural diversity. *Biodiversity and Conservation* 19, 3635–3654.

Ghislain, M., Núñez, J., Del Rosario Herrera, M., Pignataro, J., Guzman, F., *et al.* (2009) Robust and highly informative microsatellite-based genetic identity kit for potato. *Molecular Breeding* 23, 377–388.

Giraldo, D., Juarez, H., Pérez, W., Trebejo, I., Yzarra, W., *et al.* (2010) Severity of potato late blight (*Phytophthora infestans*) in agricultural areas of Peru associated with climate change. *Revista Peruana Geo-Atmosférica* (2), 56–67.

Gómez, R. (2000) *Guía para las Caracterizaciones Morfológicas Básicas en Colecciones de Papa*. International Potato Center (CIP), Lima.

Hawkes J.G. (1983) *The Diversity of Crop Plants*. Harvard University Press, Cambridge, Massachusetts.

Hawkes, J.G. (1990) *The Potato: Evolution, Biodiversity & Genetic Resources.* Smithsonian Institution Press, Washington.

Hornberger, N.H. (2000) Bilingual education policy and practice in the Andes: ideological paradox and intercultural possibility. *Anthropology & Education Quarterly* 31(2), 173–201.

Horton, D., Prain, G. and Thiele, G. (2009) *Perspectives on Partnership: A Literature Review*. Social Sciences Working Paper 2009-3. International Potato Center (CIP), Lima.

IUCN (2012) The IUCN Red List of Threatened Species. International Union for Conservation of Nature (IUCN), Gland, Switzerland. Available at: cmsdocs.s3.amazonaws.com/IUCN_Red_List_Brochure_2014_LOW.PDF (accessed 5 February 2015).

Juarez, H., Plasencia, F. and De Haan, S. (2011) Zooming in on the secret life of potato genetic resources: high technology meets old-fashioned footwork. In: ESRI (ed.) *Conservation Mapbook*. ESRI, New York, pp. 64–67.

Kroschel, J., Sporleder, M., Tonnang, H.E.Z., Juarez, H., Carhuapoma, P., *et al.* (2013) Predicting climate change caused changes in global temperature on potato tuber moth *Phthorimaea operculella* (Zeller) distribution and abundance using phenology modeling and GIS mapping. *Agricultural and Forest Meteorology* 170, 228–241.

Meilleur, B.A. and Hodgkin, T. (2004) *In situ* conservation of crop wild relatives: status and trends. *Biodiversity and Conservation* 13, 663–684.

MMAyA (2009) *Red Book of Crop Wild Relatives in Bolivia*. Bioversity International, Ministerio de Medio Ambiente y Agua (MMAyA), La Paz, Bolivia.

Monteros, C., Yumisaca, F., Andrade, J. and Reinoso, I. (2010) *Catálogo Cultivares de Papas Nativas Sierra Centro Norte del Ecuador: etnobotánico, morfológico, agronómico y calidad*. Instituto Nacional de Investigaciones Agropecuarias (INIAP), International Potato Center (CIP), Quito.
Nazarea, V.D. (2006) Local knowledge and memory in biodiversity conservation. *Annual Review of Anthropology* 35, 317–335.
Ochoa, C.M. (1975) Potato collecting expeditions in Chile, Bolivia and Peru, and genetic erosion of indigenous cultivars. In: Frankel, O.H. and Hawkes, J.G. (eds) *Crop Genetic Resources for Today and Tomorrow*. Cambridge University Press, Cambridge, UK, pp. 167–173.
Ovchinnikova, A., Krylova, E., Gravrilenko, T., Smekalova, T., Zhuk, M., *et al.* (2011) Taxonomy of cultivated potatoes (*Solanum* section *Petota*: Solanaceae). *Botanical Journal of the Linnean Society* 165, 107–155.
Padulosi, S. and Dulloo, E. (2012) Towards a viable system for monitoring agrobiodiversity on farm: a proposed new approach for red listing of cultivated plant species. In: Padulosi, S., Bergamini, N. and Lawrence, T. (eds) *On Farm Conservation of Neglected and Underutilized Species: Trends and Novel Approaches to Cope with Climate Change*. Proceedings of an International Conference, Frankfurt, 14–16 June 2011. Bioversity International, Rome, pp. 171–200.
Perez, C., Nicklin, C., Dangles, O., Vanek, S., Sherwood, S., *et al.* (2010) Climate change in the high Andes: implications and adaptation strategies for small scale farmers. *International Journal of Environmental, Cultural, Economic and Social Sustainability* 6, 71–88.
Regalsky, P. and Laurie, N. (2007) 'The school, whose place is this'? The deep structures of the hidden curriculum in indigenous education in Bolivia. *Comparative Education* 43(2), 231–251.
Rudel, T.K. and Richards, S. (1990) Urbanization, roads, and rural population change in the Ecuadorian Andes. *Studies in Comparative International Development* 25(3), 73–89.
Salick, J. (2012) Indigenous peoples conserving, managing, and creating biodiversity. In: Gepts, P., Famula, T.R., Bettinger, R.L., Brush, S.B., Damania, A.B., McGuire, P.E. and Qualset, C.O. (eds) *Biodiversity in Agriculture: Domestication, Evolution and Sustainability*. Cambridge University Press, Cambridge, UK, pp. 426–444.
Scurrah, M., De Haan, S. and Winge, T. (2013) Cataloguing potato varieties and traditional knowledge from the Andean highlands of Huancavelica, Peru. In: Anderson, R. and Winge, T. (eds) *Realising Farmers' Rights to Crop Genetic Resources: Success Stories and Best Practices*. Routledge, New York, pp. 65–79.
Steward, J.H. (1946) *Handbook of South American Indians*. Volume 2. The Andean Civilizations. Smithsonian Institution, Washington.
Stobart, H. and Howard, R. (eds) (2002) *Knowledge and Learning in the Andes: Ethnographic Perspectives*. Liverpool University Press, Liverpool, UK.
Vuille, M., Bradley, R.S., Werner, M. and Keimig, F. (2003) 20th century climate change in the tropical Andes. *Climatic Change* 59, 75–99.
Zimmerer, K.S. (2013) The compatibility of agricultural intensification in a global hotspot of smallholder agrobiodiversity (Bolivia). *Proceedings of the National Academy of Sciences* 111(8), 2769–2774.
Zimmerer, K.S. (2014) Conserving agrobiodiversity amid global change, migration, and nontraditional livelihood networks: the dynamic uses of cultural landscape knowledge. *Ecology and Society* 19(2), 1, dx.doi.org/10.5751/ES-06316-190201.

# 26 A European *In Situ* (On-farm) Conservation and Management Strategy for Landraces

V. Negri,[1]* P. Freudenthaler,[2] F. Gasi,[3] N. Maxted,[4] P. Mendes Moreira,[5] S. Strãjeru,[6] A. Tan,[7] M. Veteläinen,[8] R. Vogel[9] and J. Weibull[10]

[1]*Department of Agricultural, Nutritional and Environmental Sciences, University of Perugia, Perugia, Italy;* [2]*Austrian Agency for Health and Food Safety (AGES), Department for Plant Genetic Resources, Linz, Austria;* [3]*Faculty of Agriculture and Food Sciences, University of Sarajevo, Sarajevo, Bosnia and Herzegovina;* [4]*School of Biosciences, University of Birmingham, Birmingham, UK;* [5]*Department of Agronomy, Coimbra College of Agriculture, Coimbra, Portugal;* [6]*Vegetal Genetic Resources Bank, Suceava, Romania;* [7]*Soyak Mavisehir A. 14/3, Izmir, Turkey;* [8]*Boreal Plant Breeding Ltd, Jokioinen, Finland;* [9]*Landesamt für Umwelt, Gesundheit und Verbraucherschutz Brandenburg (LUGV), Eberswalde, Germany;* [10]*Swedish Board of Agriculture, Plant and Environment Department, Jönköping, Sweden*

## 26.1 Why We Need a European Strategy

Landraces (LR), and similar genetic material like forage ecotypes, are important components of plant genetic resources for food and agriculture (PGRFA). As Esquinas-Alcazar (1993) writes: 'The heterogeneous varieties of the past [i.e. LR] have been and still are the plant breeder's raw material. They have been a fruitful, sometimes the sole, source of genes for pest and disease resistance, adaptation to difficult environments, and other agricultural traits like the dwarf-type in grains that have contributed to the green revolution in many parts of the world'. Their critical importance for breeding remains, particularly given we are living through a period of climate changes and increased agroenvironmental unpredictability.

Under *in situ* (on-farm) conservation:

- LR continue to evolve in response to biotic and abiotic pressures and to adapt to their environment (Tiranti and Negri, 2007; Rhoné *et al.*, 2008, 2010; Pusadeea *et al.*, 2009; Negri and Tiranti, 2010; Vigouroux *et al.*, 2011). They represent, therefore, a dynamic, always updated source of alleles for the diverse aims of crop improvement, particularly for crops that are growing in adverse environmental conditions resulting from climate change.
- Local communities of farmers that continue LR cultivation
  - have the opportunity to rely for (at least part of) their income on typical, high-value, unique products;

*Corresponding author; e-mail: valeria.negri@unipg.it

 

- are able to develop their skills and capacities in farming systems based on higher inter- and intraspecific diversity (as such, answering the consumer demand for sustainable production systems) and in selection (for yield and quality) and conservation methods;
- maintain genetic diversity in agroecosystems;
- maintain different traditions and uses of a crop while extending crop and varietal uses (Martin *et al.*, 2009; Torricelli *et al.*, 2009; Rijal, 2010; Moudrý *et al.*, 2011);
- maintain peoples' cultural, social and ritual values linked to a certain crop (Brush, 2000); and
- maintain sustainable agroecosystems and crucial agroecosystem services.

Due to the modernization of agriculture and breeding for uniformity (Negri, 2005; Negri *et al.*, 2009; Maxim *et al.*, 2010), the loss of LR has been, and continues to be, extensive, and therefore there is a pressing need to conserve extant LR actively *ex situ* and *in situ*, as well as to develop strategies that enable us to achieve this goal.

In the European LR context, *in situ* (on-farm) conservation means promoting their continued cultivation by farmers in the adaptation areas, or, *sensu lato*, to reintroduce into cultivation from genebank holdings LR that were once cultivated in a certain area, or even to introduce LR into new cultivation areas. Specifically for horticultural crops in the latter context, this may include the introduction into urban and peri-urban areas, where the aim may be as much social welfare as crop production. In all mentioned cases, the grower/maintainer/farmers or farmer communities/associations are central in the process of *in situ* (on-farm) conservation: their economic benefit, or other personal motivations, from LR cultivation should always be considered in a strategy.

## 26.2 The International Context and Policy Background Requires Actions for LR Conservation

The development of a firm systematic basis for European *in situ* (on-farm) conservation is required by the international context in which each European country and the European Union (EU) exist, in terms of the wealth of the indigenous LR resource, the threat it faces, the desire to conserve the resource actively, the potential for its greater exploitation and the policy obligations inherent in the ratification of the various global and regional conventions and treaties. The planning and policy tool to achieve such systematic LR conservation and use is the European regional strategy for LR conservation and use.

Twenty-three years after the landmark 1992 Earth Summit in Rio and the Convention on Biological Diversity (CBD, 1992), the Rio+20 declaration on 'The future we want' reaffirmed the need to improve food security, based on sustainable agricultural practices that preserve natural resources, including genetic diversity, by building on enhanced agricultural research and stronger international cooperation. As party to the Convention on Biological Diversity (CBD, 1992), the EU agreed that by 2020 the genetic diversity of cultivated plants, farmed and domesticated animals and of wild relatives, including other socio-economically as well as culturally valuable species, was to be maintained, and strategies had to be developed and implemented for minimizing genetic erosion and safeguarding their genetic diversity.

In the tenth meeting, the Conference of the Parties to the CBD (October 2010, in Japan) adopted a revised and updated Strategic Plan for Biodiversity for the 2011–2020 period (which the United Nations General Assembly declared the United Nations Decade on Biodiversity), and specific targets (the Aichi Targets) to reduce biodiversity loss by 2020 (see: www.cbd.int/sp/targets). The *Reduction of the direct pressures on biodiversity and the promotion of sustainable use* target foresees that agriculture and forestry are managed sustainably, ensuring the conservation of biodiversity. In particular, the target concerning the improvement of the status of biodiversity by safeguarding ecosystems, species and genetic diversity, which includes the genetic diversity of cultivated plants and farmed and domesticated animals and of wild relatives, is maintained. Further, that strategies are developed and implemented for minimizing genetic erosion, safeguarding genetic diversity and promoting sustainable use of the conserved resource.

During the past 15 years, there have been major policy developments with impact on the conservation, use and exchange of PGRFA. Undoubtedly, the International Treaty on PGRFA

(ITPGRFA; FAO, 2001) and the Second Global Plan of Action (GPA; FAO, 2011) are the most important, as they reflect wide consensus among states and have a binding nature for many European states and the EU. The ITPGRFA (FAO, 2001) clearly: recognizes the role and rights of farmers in conserving, using and improving agricultural genetic resources; allows the fair and equitable sharing of benefits arising from their use; pursues the fair agricultural policies that promote the development and maintenance of diverse farming systems and enhance the sustainable use of agricultural biological diversity and other natural resources; promotes the survey, inventory and collection of PGRFA and supports farmers and local communities' efforts to manage and conserve them on farm.

The Second GPA (FAO, 2011), prepared under the aegis of the Commission on Genetic Resources for Food and Agriculture, updates the Global Plan of Action for the Conservation and Sustainable Use of PGRFA, adopted in 1996, at the Fourth International Technical Conference on PGRFA. The Second GPA responds to the needs and priorities identified in the Second State of the World's PGRFA (FAO, 2011). It specifically calls for:

- 'surveying and inventorying PGRFA';
- 'supporting on-farm management and improvement of plant genetic resources for food and agriculture';
- 'expanding characterization, evaluation and further development of specific subsets of collections to facilitate use';
- 'supporting plant breeding, genetic enhancement and base-broadening efforts';
- 'promoting diversification of crop production and broadening crop diversity for sustainable agriculture'; and
- 'promoting development and commercialization of all varieties, primarily farmers' varieties/LR and underutilized species'.

while the new EU Biodiversity Strategy, 'Our life insurance, our natural capital: an EU biodiversity strategy to 2020' (European Parliament Resolution, 2012) lays down the framework for EU action during this decade, in order to meet the commitments made by EU leaders in March 2010. The strategy is also the EU's means of implementing the CBD Strategic Plan for Biodiversity into EU policies and actions, a 'National Biodiversity Strategy and Action Plan' (NBSAP) in the CBD terminology. In addition to the EU Biodiversity Strategy, nearly all EU Member States have also developed their own NBSAPs (see: biodiversity.europa.eu), further adding to the implementation of the CBD and related international agreements at national level through a wide range of national and subnational policies and measures. Target 3 of the EU biodiversity strategy calls for 'increasing the contribution of agriculture and forestry to maintaining and enhancing biodiversity' and Target 6 for 'helping avert global biodiversity loss'. The resolution of the European Parliament on the EU 2020 Biodiversity Strategy also indicates that the key to the EU 2020 Biodiversity Strategy is the reform of the Common Agricultural Policy (CAP), which is 'designed to support farming that ensures food safety (in a context of climate change) and promote sustainable and balanced development across all Europe's rural areas, including those where production conditions are difficult'.

The June 2013 reform of the CAP focused on three priorities: (i) viable food production; (ii) sustainable management of natural resources; and (iii) balanced development of rural areas throughout the EU. CAP specifically mentions measures to 'help farming meet the challenges of soil and water quality, biodiversity and climate change'. It also specifically mentions the need to 'favour crop diversification, ... and conserving areas of ecological interest'. For rural development, it foresees, among its six priorities, 'restoring, preserving and enhancing ecosystems and the possibility for the Member States/regions to design thematic sub-programmes to pay especially detailed attention to [among others] the climate change mitigation/adaptation and biodiversity issues'.

The European Commission Report to the European Parliament, the Council and the European Economic and Social Committee – Agricultural Genetic Resources – From Conservation to Sustainable Use (2013) recalls 'the need to conserve and sustainably use genetic resources' and ensures that 'this objective is properly catered for in the ongoing development of relevant policies and programmes, notably:

- the Rural Development Policy, via its agri-environmental measures to target the level of practical farming and via the European Innovation Partnership to bridge practice needs with research activities and foster interaction between relevant actors;

- the Research & Innovation Policy with its Framework Programme Horizon 2020 to build up the knowledge base on genetic diversity in agriculture'.

However, Member States have not received inputs from the Commission on how or where to address exactly measures or programmes in favour of agrobiodiversity conservation. Finally, specifically regarding LR conservation, the Commission Directives 2008/62/EC 20 June 2008, 2009/145/EC 26 November 2009 and 2010/60/EU 30 August 2010 on seed production and marketing opened a new way for their conservation because they are aimed 'to ensure *in situ* conservation and the sustainable use of PGR', as their premise states, although they focus on seed production and marketing instead of genetic resource conservation *per se*. Previous European seed regulation made it impossible to market LR seed because the registration to the Common Catalogue also required, beside distinctness and stability, uniformity, which is a trait that LR do not express. However, at present, the application of the above-mentioned Directives has only partially favoured the registration of LR maintained on farm or preserved in *ex situ* collections (see Spataro and Negri, 2013, for a discussion). Another step forward could be taken if LR diversity was incorporated as a core element of all EU co-financed rural development programmes and agri-environment schemes, or even if LR diversity was recognized as a key need for rural development programmes and agri-environment schemes and semi-natural parks or other types of protected areas.

## 26.3 What To Conserve?

In drafting any conservation strategy, the conservation focus should be set clearly. There are two possible foci for *in situ* (on-farm) conservation (Negri *et al.*, 2014b):

- on a particular cultivated LR *per se*; and
- on a particular agroecosystem where a/several LR is/are cultivated; that is, using a more holistic approach.

In the first case, the aim is to maintain a particular LR without reference to the (*sensu lato*) environmental context in which it is grown. This is the most common approach, and many illustrations of this approach exist (see, for example, in Negri *et al.*, 2012, 2013a; Veteläinen *et al.*, 2009b), showing that it allows for developing and directly using the intrinsic diversity of LR for farmers' economic profit or other needs.

In the second case, the aim takes into account the environment of adaptation; that is, the abiotic and biotic context where the LR evolved its distinctive characteristics. This second approach reflects the definition of *in situ* conservation of the CBD (CBD, 1992; Definition Article 2: Use of Terms) and the ITPGRFA (FAO, 2001; Article 2: Use of Terms); that is, '*In situ* conservation means the conservation of ecosystems and natural habitats and the maintenance and recovery of viable populations of species in their natural surroundings and, in the case of domesticated or cultivated species, in the surroundings where they have developed their distinctive properties'. Here, the environmental services other than direct profit or other personal motivations of a farmer are also considered important and valuable for agriculture. Agriculture does take advantage of the services provided by the environment (pollination, biological pest control, maintenance of soil structure and fertility, nutrient cycling and hydrological services (Fisher *et al.*, 2007; Power, 2010)), and in some cases is based on 'naturality' (e.g. pastoral activities on natural grasslands). Such an approach is particularly relevant given the European perspective of halting the loss of biodiversity and the degradation of ecosystem services in the EU by 2020, and of protecting, valuing and appropriately restoring biodiversity by 2050 (European Parliament Resolution, 2012) and the recommendations (priority activities 2, 10 and 11) of the Second GPA (FAO, 2011).

In contrast to biodiversity conservation initiatives, there has been little attention paid to identify the agrobiodiversity richest environments (e.g. most appropriate areas, MAPAs) where to set LR conservation actions is priority, although the topic is worthy of investigation. The only reference available (Negri *et al.*, 2012) suggests dividing a territory into areas where (i) the LR diversity, (ii) agroecosystem ecological diversity and (iii) the presence of protected areas are assessed.

## 26.4 Constraints and Opportunities for Wider Cultivation of LR in Europe

It is obvious that an effective and efficient conservation strategy should rely principally on overcoming constraints and reinforcing opportunities for LR cultivation. This is particularly important since common methodologies for *in situ* (on-farm) conservation have not yet been established. A review of the present constraints and opportunities for wider LR cultivation on farms and in gardens in Europe was carried out within the frame of the EC funded PGR Secure project (Negri *et al.*, 2013a, 2014a). In summary, the *main constraints* are identified as follows:

- *Lack of complete LR data* (crops cultivated, number, location, traits, socio-economic environment, etc.) for all European countries, which may in turn contribute to a European strategy for LR *in situ* (on-farm) conservation.
- *Lack of policies concerning LR* in situ *(on-farm) conservation at the European level*, although recognizing that such policies do exist in a few countries.
- *Insufficient* ex situ *safety duplication of LR*, which both hampers conservation and makes difficult the re/introduction of LR into cultivation.
- *Insufficient market availability of sufficient LR seed for farmers*, which makes LR direct use difficult.
- *Lack of knowledge on the level of LR diversity that is maintained on farm and on how this diversity evolves over time*, which inhibits the development of an effective management strategy for the maintenance of diversity on farms.

Meanwhile, there are many *successful examples of LR use across Europe* that illustrate their conservation *in situ* (on farm) does exist, and which are listed below:

- The development of *typical local products* (also those awarded European quality marks like Controlled Designation of Origin (DOC), Controlled Designation of Origin Guaranteed (DOCG), for wines, and Protected Designation of Origin (PDO), Protected Geographical Indication (PGI), Traditional Specialty Guaranteed (TSG), for other products), for which many examples are available across Europe, that come from LR (for EU countries see: ec.europa.eu/agriculture/quality). Typical products awarded quality marks are lucrative on the market because of the added value that consumers attribute to their superior quality and their link with the history, culture and traditions of the area. Strengthening the relationships between the agrifood system and the community-based management of plant genetic resources could lead to more effective *in situ* (on-farm) conservation. If LR exist and consumers are willing to pay a premium price for them, a self-sustainable system could be triggered. In this way, the cultivation of LR would become advantageous for local farmers, and effective *in situ* (on-farm) conservation could become a reality.
- Development of *new (e.g. environmentally friendly/organic) farming systems based on 'diverse' varieties*. For example, the German biosphere reserve of Schorfheide-Chorin successfully marketed organic bread and beer obtained from LR. Subsequently, old cultivars were developed and successful networking of farmers, bakers and brewers was achieved. A similar approach has been taken for the marketing of Scottish whisky (Martin *et al.*, 2009). The use of 'diverse' materials in organic farming, such as LR, is increasing as well, and consequently so is the seed request of such materials.
- Development of *local food supply systems*, including community and home gardens. For example, a very successful campaign of 'Coldiretti' (the main Italian farmer organization) promotes in the local markets 'food from 0 kilometre' that is sometimes obtained from LR. In Portugal, urban agriculture has been able to improve the quality of life from both food and social aspects (Mendes Moreira, Coimbra, 2015, personal communication). To link this sort of activity to a wider use of LR could help greatly in achieving their *in situ* (on-farm) conservation.
- Development of *green care and social farming activities*. The focus point here is the client, and cultivated plants are the tool to improve human mental and physical health and the quality of life, which makes LR the most

appropriate germplasm to use (see: www.cost.eu/COST_Actions/fa/Actions/866). In horticultural therapy, plant biodiversity is used to help disabled people by exposing them to multiple sensory stimuli and soliciting their care towards the living being, the 'plant'. Linking this activity to LR diversity maintenance also helps in *in situ* (on-farm) conservation (Scartazza *et al.*, 2015).

- Development of *companies purposely dedicated to produce and sell seed/propagation material of LR*. For example, the 'Programme for Diversity of Cultivated Plants: keeping the cultivated heritage alive' in Sweden, through public–private sector (including seed industries) cooperation, offers small seed samples/plants from the national genebank to be grown and propagated further. Such activities could be taken as exemplar activities of LR reintroduction from genebanks. Similar activities, based on the efforts of the public sector, are also in progress in other countries (e.g. Hungary: Baktay and Simon, Chapter 28, this volume, 2016; Italy: Negri, Perugia, 2014, personal communication; Romania: Străjeru and Constantinovici, 2014).
- The *use of LR in conventional breeding and unconventional (participatory) breeding* for obtaining varieties that: (i) improve adaptation to local conditions (which is particularly relevant under the changing climate) and disease resistance (Newton *et al.*, 2011); (ii) give products that are appreciated by consumers; or (iii) satisfy special needs, which is also a reason for *in situ* (on-farm) conservation. The 'Vaso' project in Portugal developed new broad-based cultivars of maize (Mendes Moreira, 2006; Vaz Patto *et al.*, 2013), as well as the EC FPVII 'Solibam' project for maize, wheat, barley, broccoli and tomato (Chable *et al.*, 2014).
- *Successful LR on-farm conservation policies in some European countries* (e.g. Italy, Finland) exist that can be followed in order to promote *in situ* (on–farm) conservation. In Italy, regional legislations exist that protect agrobiodiversity and offer subsidies to 'custodian farmers' (Negri, 2012; Negri and Torricelli, 2014).
- Finally, *favourable seed legislation policies* (Commission Directives 2008/62/EC 20 June 2008 and 2009/145/EC 26 November 2009 on 'conservation varieties'), allowing the seed of variable populations like LR to be marketed, under certain restrictions, are still in force in the EU, which provides options for a wider use of LR in agriculture, for developing niche seed markets and consequently also new seed companies for these markets (Spataro and Negri, 2013). Temporary derogations are also provided for marketing the seed of experimental populations (Commission Implementing Decision 2014/150/EU of 18 March 2014).

It should also be noted that the use of 'diverse' varieties, such as LR, on farms fits perfectly within the new CAP (see above). So there is a push from farmers to maintain LR diversity but there is also a pull from the consumer, with numerous examples of consumer-based organizations or initiatives that, like the Slow Food movement, demand greater food diversity.

## 26.5 A Generic European LR *In Situ* (On-farm) Conservation Strategy: Recommendations

The need for a European LR conservation strategy is justified by the requirement for a more coherent strategy across Europe that can promote enhanced consumer choice and the cultivation of greater diversity while also promoting the systematic maintenance of that diversity and raising awareness of the value of LR diversity to wealth, health and well-being. To develop the strategy, we took into account the main recommendations from (i) the European continental perspective and (ii) the national perspectives of selected countries.

As for the former, the recommendations included in a document prepared for European Cooperative Programme for Plant Genetic Resources (ECPGR) concerning scope and methods to favour *in situ* (on-farm) conservation in Europe (Negri *et al.*, 2014b) were considered. As for the latter, the recommendations from countries that represent very different situations across Europe in terms of agroclimatic conditions, LR on-farm diversity and socio-economic context (Finland, Italy and the UK; Heinonen and collaborators, 2014; Maxted *et al.*, 2014; Negri and Torricelli, 2014) were considered.

In spite of the differences that exist among European countries, common needs and elements for a strategic and cooperative approach to *in situ* (on-farm) conservation do exist. This includes both conservation and enhancement of use actions, as recommended and detailed below. Conservation actions:

- compile an LR inventory and ensure it is regularly updated
- educate and raise public awareness of local LR diversity
- safe backup of on-farm LR diversity in *ex situ* conservation
- promote LR reintroduction from genebanks to on-farm
- increase European coordination in developing and implementing measures for LR conservation
- make available adequate funds for LR *ex situ* and *in situ* (on-farm) conservation actions
- make available adequate funds for carrying out research into LR diversity in the context of climate change and unpredictability.

Enhancement of LR use:

- promote the use of home garden LR in community and home gardens
- promote the registration of LR as 'conservation varieties' and other related designations
- promote the awarding of quality marks for products coming from LR
- promote typical local products from LR
- carry out campaigns aimed to promote local economies based on nearby products from LR
- develop workable value chain systems for LR
- stimulate the use of LR in plant breeding and participatory plant breeding programmes, especially those aimed at creating varieties suitable to environmentally friendly agronomic systems and for future climatic conditions
- stimulate the use of LR in 'green care' and 'social farming' activities aimed at increasing people's well-being or services to the community.

Many of these recommended activities can take advantage of the policy and legislative European opportunities for supporting LR cultivation mentioned above. The conservation and promotion of use actions in favour of LR could be carried out by initially concentrating efforts on the most threatened LR and on MAPAs. Among the conservation actions identified above, the need for European LR inventories has already been stressed, and has been reinforced for a long time by international policies and strategies for a sustainable use of PGRFA, and more recently by many papers referring specifically to Europe (Maxted *et al.*, 2009; Veteläinen *et al.*, 2009a, 2009b, 2012; Negri *et al.*, 2014a).

Although many activities to increase awareness and capacities have been carried out (see: www.pgrsecure.org, LR help desk), lack of funding has hampered the compilation of LR inventories across the whole of Europe. We note that no action for the development of European plans for *in situ* (on-farm) conservation and their implementation can be taken if there is no informative base (and, of course, the same considerations hold true at the national level). The main drawback to developing a European *in situ* (on-farm) conservation strategy of LR continues to be the lack of information on their existence, location and use.

Following the CBD (1992), the Second GPA (FAO, 2011) policy and strategy for *in situ* conservation and management stresses that 'The surveying and inventorying of PGRFA should be considered as the first step in the process of conservation and reducing the rate of biodiversity loss'. In addition, the ITPGRFA (FAO, 2001), to which many European countries and the EU are contracting parties, recommends that: 'Each contracting party should promote or support appropriate farmers and local communities with efforts to manage and conserve on-farm their plant genetic resources for food and agriculture' (article 5c).

The compilation of plant genetic resource inventories should also contribute towards meeting the targets of the EU 2020 Biodiversity Strategy (European Parliament Resolution, 2012) since, at present, there is no systematic *in situ* (on-farm) conservation of LR in Europe. For this urgent need, we propose the methods described below.

## 26.6 Compilation of National and European Inventories of LR That Are Maintained *In Situ* (On-farm)

Although there may be many specific approaches to establishing a European LR inventory, within the strategic context these may be characterized as both a 'bottom-up additive' and a 'top-down strategy'.

### 26.6.1 A bottom-up additive strategy

While recognizing the sovereignty of each country over its own genetic resources, a European inventory should be based on national inventories and each country's decision to contribute the data that are needed to compile the European inventory. The procedures and actors of this strategy are presented graphically in Fig. 26.1 and can be summarized in the following steps:

**1.** Creation of official national inventories of (extant, reintroduced and introduced) LR.
**2.** Merging of national inventories into a single European inventory following a national process regarding the data to be included.
**3.** Construction of a solid, search-efficient database that contains the European *in situ* (on-farm) inventory information.

Following the example of EURISCO (eurisco.ipk-gatersleben.de), there could be a role for the members of the European Cooperative Program Genetic Resources (ECPGR) On-farm Working Group (WG). They should initially

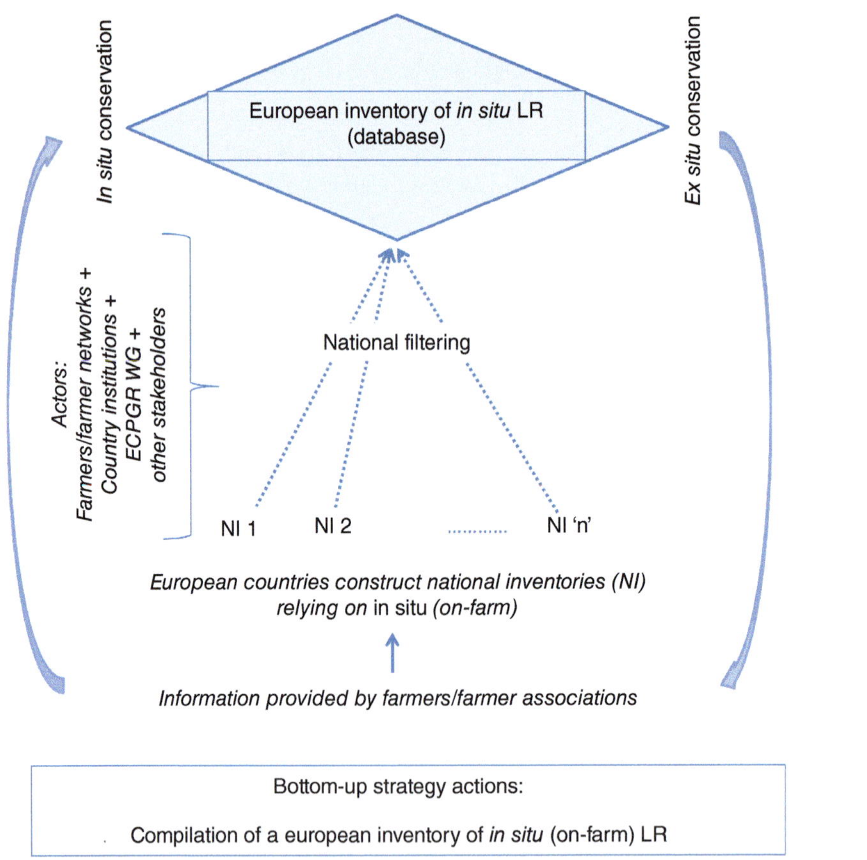


**Fig. 26.1.** The compilation of a European inventory of *in situ* (on-farm) LR through an approach aimed at conserving a certain PGR population *per se*: bottom-up strategy actions needed. Arrows show periodic reworking.

encourage each country's government to construct a national inventory of on-farm maintained LR. Gathering data for inventories can rely on information provided directly by farmers/farmer associations maintaining LR, ECPGR WG members and other stakeholders. The establishing of inventories can take advantage of information recording tools already purposely developed (the 'Descriptors for web-enabled national *in situ* LR inventories' and the 'Tool for recording national *in situ* landrace inventory data', both available from the www.pgrsecure.org LR help desk).

Inventories of this type have already been developed in some countries (i.e. Italy and Finland; Negri *et al.*, 2013b; Heinonen and collaborators, 2014). It is noteworthy that these inventories are already used in Italian regions to fund, through the European Agricultural Fund for Rural Development (EAFRD) activities aimed to protect, monitor and enhance the utility of *in situ* (on-farm) PGRFA. Once these national inventories have been carried out, and after a filtering process scrutinized by the national inventory focal points in agreement with the national coordinators, their data can be merged into a unique European inventory.

European data should be stored in a database that is linked to other formal information sources (EURISCO, Central Crop Databases, genebank databases, PGR Diversity Gateway) and to farmer/farmer organization network databases (Fig. 26.1 diamond).

### 26.6.2 A top-down strategy

However, considering that national inventories cannot be carried out immediately in some European countries, and that certain areas of Europe have different levels of on-farm maintained diversity, a 'top-down' approach could also be taken (Fig. 26.2). Such an approach would require an initial informative data set on diversity maintained *in situ* (on-farm). This could be developed by retrieving available bibliographic information and database data through the ECPGR WG members and other stakeholders, or querying EURISCO, GENESYS (www.genesys-pgr.org/welcome) or other international databases holding information on variable material and subsequently by checking actual existence in the field.

From the initial informative data set, LR maintained *in situ* (on-farm) that are of priority importance for Europe and worldwide would be subsequently identified. However, criteria to identify materials of priority importance for Europe, and worldwide, need to be discussed and agreed upon first.

Inclusion in the European inventory of national data would ultimately be the decision of the appropriate national agency. Of course, the development of a European inventory would require appropriate funds to be allocated by the European Commission after it eventually adopts our European LR *in situ* (on-farm) conservation strategy to fulfil the targets of the EU 2020 Biodiversity Strategy (European Parliament Resolution, 2012).

It must be noted finally that both the bottom-up and the top-down strategies would require periodic updating because *in situ* (on-farm) inventories deal with very dynamic situations (growers come and go, so situations can change rapidly) and research is continuously collecting new data on LR diversity maintained in the field and its utility. The consolidated European inventory will serve as a basis for promoting not only *in situ* (on-farm) but also *ex situ* conservation actions.

As mentioned above, these plans should rely on different elements (depending on the country and country area), but the information collected will assure the possibility of:

- collecting LR not already present in *ex situ* collections. A recent gap analysis carried out for Italy shows that most *in situ* (on-farm) LR (97.4%) have no matching record in EURISCO and are possibly not conserved in the main genebanks (Negri and Torricelli, 2014);
- promoting the use of LR in agriculture and, as such, also achieving their *in situ* (on-farm) conservation as mentioned above;
- promoting research on variable materials regarding:
    - within- and among-genetic diversity level, for traits conferring
        - resistance/tolerance to biotic stresses
        - resistance/tolerance to abiotic stresses
        - quality
    - evolution of *in situ* (on-farm) genetic diversity under changing climatic conditions;

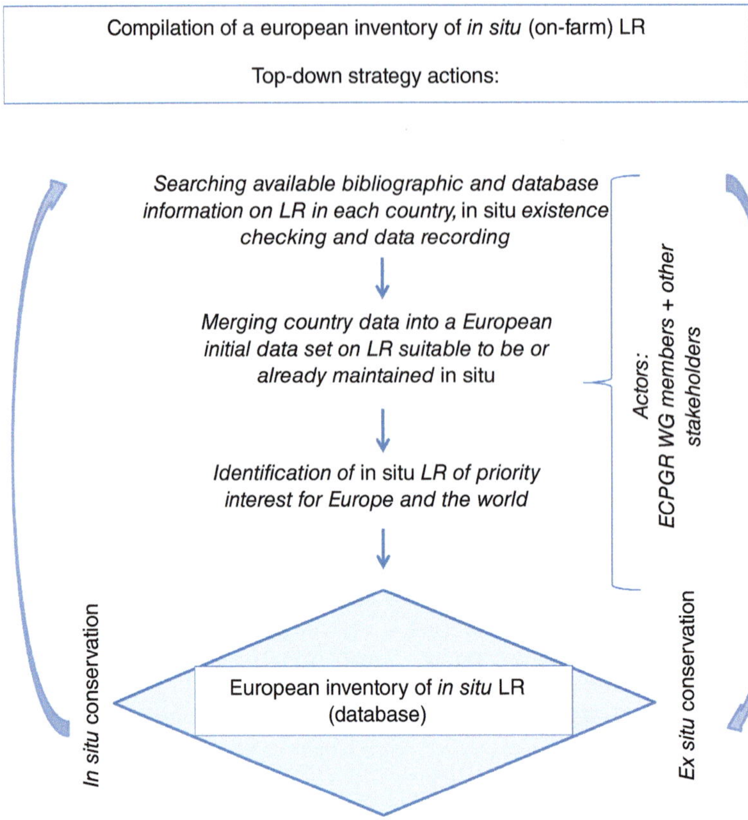


Fig. 26.2. The compilation of a European inventory of LR that are maintained *in situ* (on-farm) through an approach aimed at conserving a certain PGR population *per se*: top-down strategy actions needed. Arrows show periodic reworking.

   - a level of genetic diversity that can be maintained in a LR under different agroecosystems; that is, ecosystems that differ for their physical (i.e. climatic/pedological) and biological (i.e. number of species) traits and for management applied to them (e.g. 'organic/low' input versus 'conventional' agronomic systems);
   - socio-economic factors that drive conservation, in such answering unsolved research questions (Veteläinen *et al.*, 2009a,b); and
- developing the research needed to identify agrobiodiversity hotspots (MAPAs) and compiling their European inventory, (see below).

The development of a European inventory also:

- allows for the assessment of the overall implementation progress and related follow-up processes of the Second GPA (FAO, 2011) following the criteria and indicators set by the Commission on Genetic Resources for Food and Agriculture (CGRFA-15/15/Inf.23, www.fao.org/3/a-mm564e.pdf) (i.e. most of the indicators and questions mentioned are included in the 'Descriptors for web-enabled national *in situ* LR inventories' from the www.pgrsecure.org LR help desk);
- enables the development of effective biodiversity and agrobiodiversity indicators to permit more precise monitoring of LR diversity;
- facilitates cooperation among ECPGR WGs and supports their activities within ECPGR; and
- facilitates cooperation among the formal sector and the networks of farmers and farmer organizations.

Finally, it will be a useful example to develop *in situ* conservation actions further at the global level.

## 26.7 The Compilation of Country and European MAPA Inventories

Considering the holistic approach that has been described above, the identification of MAPAs would promote the planning and implementation of *in situ* conservation activities at national, European and global levels. In fact, MAPAs could be proposed to national authorities as the sites where *in situ* (on-farm) conservation activities should be initiated or enhanced with priority because of their high value for agrobiodiversity conservation. On the basis of national inventories of variable materials maintained on-farm, and the information contained in them, hotspots of diversity (MAPAs) could be identified following a bottom-up or a top-down strategy similar to that described above for inventorying variable material.

From national inventories of MAPAs, a European MAPA inventory could then be developed that would serve as the basis of an integrated system of high-value areas for agrobiodiversity conservation in Europe (Fig. 26.3).

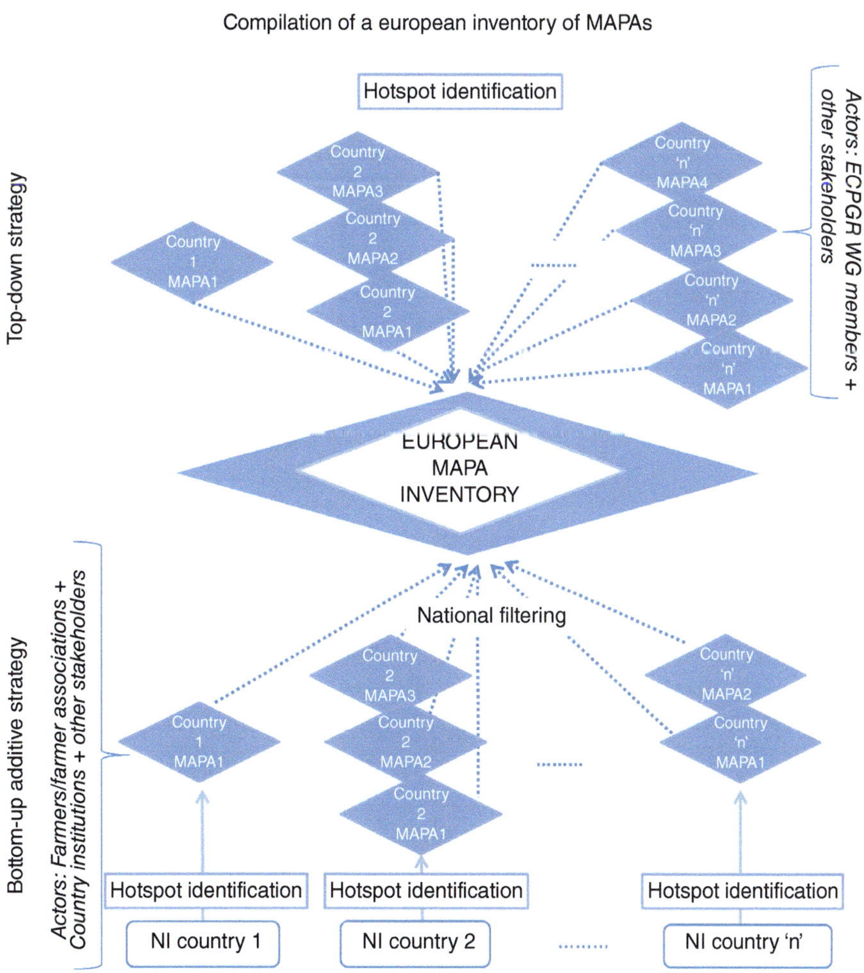


Fig. 26.3. The European MAPA identification following a bottom-up (bottom) and a top-down (upper) strategy.

In particular, MAPAs could be usefully integrated into the Natura 2000 site network, which currently addresses only general nature conservation of priority habitats and threatened species. In addition, at the global level, MAPAs might be usefully integrated into the FAO GIAHS – Globally Important Agricultural Heritage Systems (www.fao.org/giahs/giahs-home/en) – that promotes the dynamic conservation and adaptive management of unique traditional agricultural heritage systems around the world. This addresses the actions recommended by the Second GPA (FAO, 2011) and the present EU policy requirements, as well as foreseen achievements in terms of conservation of biodiversity and agricultural policies[1] mentioned above.

As a consequence, the compilation of a European inventory of MAPAs would:

- facilitate the development of plans for *in situ* biodiversity conservation and their implementation at national and European level, in general; and
- facilitate the application of EU policies in favour of biodiversity conservation and agriculture.

Finally, it will be useful to develop *in situ* conservation actions at the global level.

## 26.8 A European Network of Unique Materials and Sites for Coordinated and Integrated *In Situ* (On-farm) and *Ex Situ* Conservation Activities

It is imperative for agrobiodiversity-based industry and conservation to maintain unique variable materials, like extant LR that are cultivated in just one farm and/or that are characterized by unique alleles, as well as unique MAPAs, like those including relic/irreplaceable habitats and a high number of unique variable materials. Unique variable materials should be sampled immediately and conserved in genebanks as backup collections, so as to make them available for re/introduction and research and breeding.

From the European inventories of MAPAs and LR, sites and materials that have distinctive and unique characteristics should be selected and proposed by ECPGR *in situ* (on-farm) WG members to European countries (and through them to the EU) as sites and materials that deserve the highest conservation priorities. For them, specific national and European funding for conservation, monitoring and managing should be foreseen that allows for efficient and effective conservation across years.

The unique *in situ* (on-farm) maintained LR would be the source of unique *ex situ* collections, while at the same time offering an opportunity to monitor how diversity evolves in response to climate changes and farmer management. All the same, unique MAPAs will offer the possibility of monitoring over time the changes of the three components of agrobiodiversity: habitats, number of species and level of intraspecific diversity. In the end, creating a European network of unique materials and sites would lead to a fully coordinated and integrated *in situ* (and, via the population safety backup, *ex situ*) conservation (Fig. 26.4) of the salient agrobiodiversity hotspots and PGRs for the benefit of future European generations.

## 26.9 Conclusions

To promote efficiently and effectively continued LR cultivation by farmers in Europe, we propose a series of coordinated actions. The main drawback to implementation remains a lack of specific funding for such a programme. In the opinion of the authors, the European Union should meet this commitment, as the continued availability of LR diversity is a clear European public good that underpins the future of food security in Europe.

## Acknowledgements

The contribution of Isabelle Goldringer (INRA, France) and Maarit Heinonen (LUKE, Finland) to this paper is acknowledged. The research leading to these results has received partial funding from the European Community's Seventh Framework Programme (FP7/ 2007-2013) under Grant Agreement No GA266394 PGR Secure.

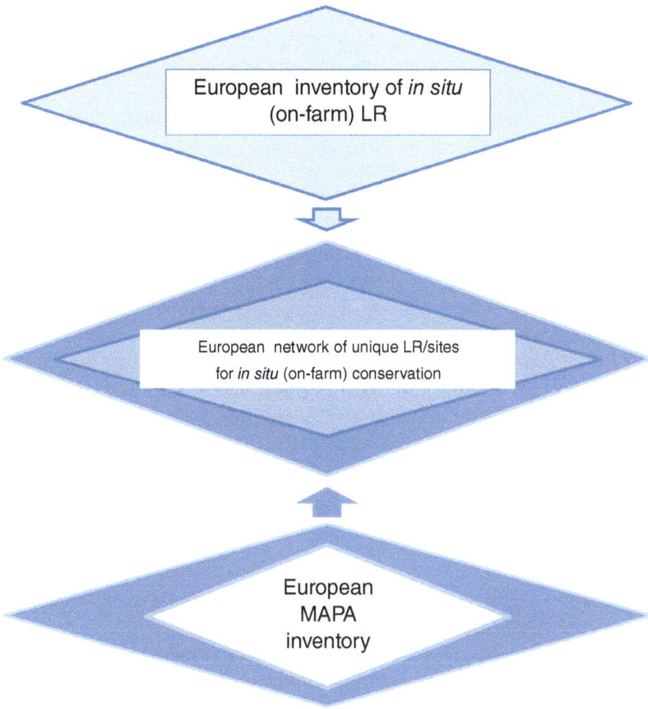


Fig. 26.4. A European network of unique LR and sites for coordinated and integrated *in situ* conservation.

## Endnote

[1] SEC (2011) 540 Final; European Parliament Resolution, 2012.

## References


Brush, S.B. (2000) The issues of *in situ* conservation of crop genetic resources. In: Brush, S.B. (ed.) *Genes in the Field*. IPGRI, Rome, IDRC, Ottawa, Lewis Publisher, Boca Raton, pp. 3–26.

CBD (Convention on Biological Diversity) (1992) Text and Annexes. Secretariat of the Convention on Biological Diversity, Montreal. Available at: www.cbd.int/convention (accessed February 2013).

Chable, V., Goldringer, I., Howlett, S.A., Barberi, P., Miko, P., *et al*. (2014) Diversity strategies for organic and low input agricultures and their food systems. Book of abstracts of Solibam final congress, Nantes, 7–9 July 2014. Available at: orgprints.org/27582/1/Solibam+final+congress+-+book+of+abstracts.pdf (accessed 28 December 2015).

Commission Directive 2008/62/EC of 20 June (2008) *Providing For Certain Derogations For Acceptance Of Agricultural LRs And Varieties Which Are Naturally Adapted To The Local And Regional Conditions And Threatened By Genetic Erosion And For Marketing Of Seed And Seed Potatoes Of Those LRs And Varieties*. OJ L 162, 21.6.2008. European Commission, Brussels, pp. 13–19.

Commission Directive 2009/145/EC of 26 November (2009) *Providing For Certain Derogations, For Acceptance Of Vegetable LRs And Varieties Which Have Been Traditionally Grown In Particular Localities And Regions And Are Threatened By Genetic Erosion And Of Vegetable Varieties With No Intrinsic Value For Commercial Crop Production But Developed For Growing Under Particular Conditions And For Marketing Of Seed Of Those LRs And Varieties*. OJ L 312, 27.11.2009. European Commission, Brussels, pp. 44–54.

Commission Directive 2010/60/EU of 30 August (2010) *Providing For Certain Derogations For Marketing Of Fodder Plant Seed Mixtures Intended For Use In The Preservation Of The Natural Environment Text with EEA Relevance*. OJ L 228, 31.8.2010. European Commission, Brussels, pp. 10-14.
Commission Implementing Decision 2014/150/EU of 18 March (2014) *On The Organisation Of A Temporary Experiment Providing For Certain Derogations For The Marketing Of Populations Of The Plant Species Wheat, Barley, Oats And Maize Pursuant To Council Directive 66/402/EEC*. OJ L 82, 20.3. 2014. European Commission, Brussels, pp. 29–36.
Esquinas-Alcazar, J.T. (1993) Plant genetic resources. In: Hayward, M.D., Bosemark, N.O. and Romagosa, I. (eds) *Plant Breeding: Principles and Prospects*. Chapman & Hall, London, pp. 33–51.
European Commission (2013) Agricultural Genetic Resources – from conservation to sustainable use. Report to the European Parliament, the Council and the European Economic and Social Committee. European Commission, Brussels. Available at: ec.europa.eu/agriculture/genetic-resources/pdf/swd-2013-486_en.pdf (accessed February 2013).
European Parliament Resolution (2012) Our life insurance, our natural capital: an EU biodiversity strategy to 2020 (2011/2307(INI)). European Parliament Resolution of 20 April 2012. European Parliament, Brussels, Belgium. Available at: ec.europa.eu/environment/nature/biodiversity/comm2006/pdf/EP_resolution_april2012.pdf (accessed 28 December 2015).
FAO (Food and Agriculture Organization) (2001) International Treaty on Plant Genetic Resources for Food and Agriculture. FAO, Rome. Available at: www.planttreaty.org (accessed February 2014).
FAO (2011) Second Global Plan of Action for Plant Genetic Resources for Food and Agriculture. FAO, Rome. Available at: www.fao.org/agriculture/crops/core-themes/theme/seeds-pgr/gpa/en (accessed February 2014).
Fisher, B., Costanza, R., Kerry Turner, R. and Morling, P. (2007) Defining and classifying ecosystem services for decision making. *CSERGE Working Paper EDM*, No 07-04.
Heinonen, M. and contributors (2014) Landrace *in situ* Conservation Strategy for Finland. Available at: www.mtt.fi/mttraportti/pdf/mttraportti163.pdf (accessed February 2014).
Martin, P., Wishart, J., Cromarty, A. and Chang, X. (2009) New markets and supply chains for Scottish bere barley. In: Vetelainen, M., Negri, V. and Maxted, N. (eds) *European Landraces: On-farm Conservation, Management and Use*. Bioversity Technical Bulletin No 15, Bioversity International, Rome, pp. 251–263. Available at: www.bioversityinternational.org/index.php?id=19&user_bioversitypublications_pi1[showUid]=3252 (accessed February 2014).
Maxim, A., Sandor, M., Străjeru, S., Jidavu, M., Sima, R., *et al.* (2010) *Agrobiodiversitate si Bioconservare*. Risoprint, Cluj-Napoca, Romania, pp. 173–177, 206–581.
Maxted, N., Negri, V. and Vetelainen, M. (2009) Landrace inventories: needs and methodologies. In: Vetelainen, M., Negri, V. and Maxted, N. (eds) *European LRs: On-farm Conservation, Management and Use*. Bioversity Technical Bulletin No 15, Bioversity International, Rome, pp. 45–52. ISBN 978-92-9043-805-2. Available at: www.ecpgr.cgiar.org/fileadmin/bioversity/publications/pdfs/1347_European_landraces_on-farm_conservation_management_use.pdf (accessed 28 December 2015).
Maxted, N., Scholten, M., Ford-Lloyd, B., Allender, C., Astley, D., *et al.* (2014) Landrace conservation strategy for the United Kingdom. Available at: www.pgrsecure.bham.ac.uk/sites/default/files/documents/deliverables/D4.3_UK_landrace_conservation_strategy.pdf (accessed 28 December 2015).
Mendes Moreira, P.M.R. (2006) Participatory maize breeding in Portugal. A Case Study. *Acta Agronomica Hungarica* 54(4), 431–439.
Moudrý, J., Konvalina, P., Stehno, Z., Capouchová, I. and Moudrý, J. Jr (2011) Ancient wheat species can extend biodiversity of cultivated crops. *Scientific Research and Essays* 6(20), 4273–4280.
Negri, V. (2005) Agro-biodiversity conservation in Europe: ethical issues. *Journal of Agricultural and Environmental Ethics* 18(1), 3–25.
Negri, V. (2012) Policies supportive of on-farm conservation and their impact on custodians farmers in Italy. In: Padulosi, S., Bergamini, N. and Lawrence, T. (eds) *On Farm Conservation of Neglected and Underutilized Species: Status, Trends and Novel Approaches to Cope with Climate Change*. Proceedings of an International Conference, Frankfurt, 14–16 June 2011. Bioversity International, Rome, pp. 211–217.
Negri, V. and Tiranti, B. (2010) Effectiveness of *in situ* and *ex situ* conservation of crop diversity. What a *Phaseolus vulgaris* L. landrace case study can tell us. *Genetica* 138, 985–999.
Negri, V. and Torricelli, R. (2014) The Italian landrace conservation strategy. Available at: www.pgrsecure.bham.ac.uk/sites/default/files/documents/deliverables/D4.2_IT_landrace_conservation_strategy.pdf (accessed 20 February 2015).
Negri, V., Maxted, N. and Veteläinen, M. (2009) European LR conservation: an introduction. In: Veteläinen, M., Negri, V. and Maxted, N. (eds) European landraces: on-farm conservation, management and use.

Bioversity Technical Bulletin No 15. Bioversity International, Rome. Available at: www.bioversityinternational.org/index.php?id=19&user_bioversitypublications_pi1[showUid]=3252 (accessed February 2015).

Negri, V., Barocco, R., Pacicco, L., Veronesi, F. and Venanzoni, R. (2012) An approach towards prioritising LR rich areas as a priority for protection in Europe. In: Maxted, N., Dulloo, M.E., Ford-Lloyd, B.V., Frese, L., Iriondo, J.M. and Pinheiro de Carvalho, M.A.A. (eds) *Agrobiodiversity Conservation: Securing the Diversity of Crop Wild Relatives and Landraces*. CAB International, Wallingford, UK, pp. 118–124.

Negri, V., Heinonen, M. and Maxted, N. (2013a) On-farm landrace conservation. Presentations of the Workshop 'On the conservation and sustainable use of plant genetic resources in Europe: a stakeholder analysis', Wageningen, 25–29 November 2013. Available at: www.nordgen.org/ngdoc/plants/Samarbeten_och_natverk/PGR_secure_workshop2013/5_LR_Conservation.pdf (accessed February 2015).

Negri, V., Pacicco, L., Bodesmo, M. and Torricelli, R. (2013b) *The first Italian inventory of in situ maintained landraces*. On CD ROM. ISBN 978-88-6074-279-7. Morlacchi Editrice, Perugia, Italy. Available at: vnr.unipg.it/PGRSecure/html/project.html (accessed February 2015).

Negri, V., Torricelli, R., Maxted, N. and Heinonen, M. (2014a) European landrace conservation strategy. Available at: www.pgrsecure.bham.ac.uk/sites/default/files/documents/deliverables/D4.5_European_generic_LR_conservation_strategy.pdf (accessed February 2015).

Negri, V., Freudenthaler, P., Gasi, F., Goldringer, I., Mendes Moreira, P., *et al*. (2014b) The ECPGR concept for *in situ* (on-farm) conservation in Europe. Available at: www.pgrsecure.org, LR helpdesk (accessed February 2015).

Newton, A.C., Akar, T., Baresel, J.P., Bebeli, P.J., Bettencourt, E., *et al*. (2011) Cereal landraces for sustainable agriculture. In: Lichtfouse, E., Hamelin, M., Debaeke, P. and Navarrete, M. (eds) *Sustainable Agriculture*, Volume 2. Springer Science & Business Media, New York, pp. 147–186.

Power, A.G. (2010) Ecosystem services and agriculture: tradeoffs and synergies. *Philosophical Transactions of the Royal Society B* 365, 2959–2971.

Pusadeea, T., Jamjoda, S., Chiangb, Y.-C., Rerkasema, B. and Schaalc, B.A. (2009) Genetic structure and isolation by distance in a landrace of Thai rice. *PNAS* 106, 13880–13885.

Rhoné, B., Remoué, C., Galic, N., Goldringer, I. and Bonnin, I. (2008) Insight into the genetic bases of climatic adaptation in experimentally evolving wheat populations. *Molecular Ecology* 17, 930–943.

Rhoné, B., Vitalis, R., Goldringer, I. and Bonnin, I. (2010) Evolution of flowering time in experimental wheat populations: a comprehensive approach to detect genetic signatures of natural selection. *Evolution* 64(7), 2110–2125.

Rijal, D. (2010) Role of food tradition in conserving crop landraces on-farm. *Journal of Agriculture and Environment* 11, 107–119.

Scartazza, A., Mancini, M.L., Negri, V., Di Baccio, D., Mattioni, C., *et al*. (2015) Promoting human well-being by caring the local biodiversity in Enabling Gardens. *Landraces* 3, 6. Available at: www.pgrsecure.bham.ac.uk/sites/default/files/documents/newsletters/Landraces_Issue_3.pdf (accessed February 2015).

SEC (2011) Our life insurance, our natural capital: an EU biodiversity strategy to 2020. Final Commission Staff Working Paper Impact Assessment. Accompanying the document Communication from the Commission to the European Parliament, the Council, the European Economic and Social Committee and the Committee of the Regions. Available at: ec.europa.eu/environment/nature/biodiversity/comm2006/pdf/2020/1_EN_impact_assesment_part1_v4.pdf (accessed February 2015).

Spataro, G. and Negri, V. (2013) The European seed legislation on conservation varieties: focus, implementation, present and future impact on landrace on-farm conservation. *Genetic Resources and Crop Evolution* 60, 2421–2430.

Străjeru, S. and Constantinovici, D. (2014) Use of plant genetic resources held in the collection of Suceava Genebank, Romania. In: Meglic V. (ed.) *SEEDNet the Way Ahead – Crop Sustaln Workshop,* Ljubljana, Slovenia, 5–6 November 2014. Book of abstracts, Kmetijski Institut Slovenije, Ljublijana, pp. 51.

Tiranti, B. and Negri, V. (2007) Selective micro-environmental effects play a role in shaping genetic diversity and structure in a *Phaseolus vulgaris* L. landrace. *Molecular Ecology* 16, 4942–4955.

Torricelli, R., Quintaliani, L. and Falcinelli, M. (2009) The 'Farro' (*Triticum dicoccon* Schrank) from Monteleone di Spoleto (Valnerina Valley, Umbria). In: Veteläinen, M., Negri, V. and Maxted, N. (eds) European landraces: on-farm conservation, management and use. Bioversity Technical Bulletin No 15. Bioversity International, Rome, pp. 183–186. Available at: www.bioversityinternational.org/index.php?id=19&user_bioversitypublications_pi1[showUid]=3252 (accessed February 2015).

Vaz Patto, M.C., Mendes-Moreira, P., Alves, M.L., Mecha, E., Brites, C., *et al.* (2013) Participatory plant quality breeding: an ancient art revisited by knowledge sharing. The Portuguese experience. In: Andersen, S.B. (ed) *Plant Breeding from Laboratories to Fields*. InTech, Zagreb, pp. 255–288.

Veteläinen, M., Maxted, N. and Negri, V. (2009a) A European strategic approach to conserving crop landraces. In: Veteläinen, M., Negri, V. and Maxted, N. (eds) *European Landraces: On-farm Conservation, Management and Use*. Bioversity Technical Bulletin No 15. Bioversity International, Rome, pp. 305–325 ISBN 978-92-9043-805-2. Available at: www.bioversityinternational.org/index.php?id=19&user_bioversitypublications_pi1[showUid]=3252 (accessed 28 December 2015).

Veteläinen, M., Maxted, N. and Negri, V. (2009b) *European Landraces: On-farm Conservation, Management and Use.* Bioversity Technical Bulletin No 15. Bioversity International, Rome. ISBN 978-92-9043-805-2. Availableat:www.bioversityinternational.org/index.php?id=19&user_bioversitypublications_pi1[showUid]=3252 (accessed 28 December 2015).

Veteläinen, M., Negri, V., and Maxted, N. (2012) A second look at the European Strategic Approach to Conserving Crop Landraces. In: Maxted, N., Dulloo, M.E., Ford-Lloyd, B.V., Frese, L., Iriondo, J.M. and Pinheiro de Carvalho, M.A.A. (eds), *Agrobiodiversity Conservation: Securing the Diversity of Crop Wild Relatives and Landraces*. CAB International, Wallingford, UK, pp 181-185.

Vigouroux, Y., Barnaud, A., Scarcelli, N. and Thuillet, A.-C. (2011) Biodiversity, evolution and adaptation of cultivated crops. *Comptes Rendus Biologies* 334, 450–457.

# 27 Using Landraces in Agriculture, Food and Cooking: Experiences Around a Large City in Southern Europe

A. Lázaro,* I. Fernández and C. de Lorenzo
*IMIDRA – Finca el Encín, Madrid, Spain*

## 27.1 Introduction

Many network strategies and programmes have been established since the necessity arose to preserve plant biodiversity. The Community of Madrid – central Spain – set up a germplasm bank in 1995. It is located at the IMIDRA, which is the Institute for Food and Agriculture Research supported by the Regional Administration of Madrid. This collection of plant genetic resources, which currently contains 203 accessions, consists mainly of landraces cultivated in the Madrid region, within 60 km around the large city of Madrid (5,457,561 inhabitants), for at least three generations. This is to say, these accessions were cultivated before the development of modern agriculture, which occurred in Spain around the 1970s. Of these conserved accessions, 91% are horticultural crops: melons, tomatoes, beans, peppers, garlic and others (Escribano and Lázaro, 2012; Escribano *et al.*, 2012; Lázaro *et al.*, 2013; Lázaro and de Lorenzo, 2015). Such an amount of agro/biodiversity, collected in a small area (8,028.5 km$^2$), and mainly not dedicated to agriculture, is evidence of the biodiversity richness of southern Europe.

Accessions arrived by farmer donations, collecting missions, or through local association donations (cooperatives, local groups supported by the European LEADER initiative, civic environmental education centres, etc.). As is the case with many other collections, the germplasm bank in Madrid has provided plant material to those users who have requested it. What type of user is demanding plant genetic resources? To answer this question, plant receivers were always asked to fill in a short questionnaire before the material was sent to them. The information required focused on the type of intended use, the purpose of the petition and whether the material was going to be grown under organic agricultural practices.

With the collaboration of the Complutense University of Madrid, the preference and commercial facts underlying landrace demand were explored. A study of the perception of traditional local products has been conducted, both in consumers at home and at restaurants using Madrid landraces. An online questionnaire was proposed and consumers answered, marking their level of agreement or disagreement with the following potential characteristics of landraces: they are of better quality, have better odour and more flavour, are available only in season, are more expensive, fresher, healthier, less beautiful, or tastier. The analysis of such information provides a picture of how civil society perceives the value of plant genetic resources, which means the actual support by public budgets for such collections.

*Corresponding author; e-mail: almudena.lazaro@madrid.org

 *Enhancing Crop Genepool Use: Capturing Wild Relative and Landrace Diversity for Crop Improvement* (eds N. Maxted, M. Ehsan Dulloo and B.V. Ford-Lloyd) 

## 27.2 Results

Figure 27.1 shows the number of accessions (both seeds and explants) released to users from the IMIDRA germplasm collection in the past 5 years. It is clearly shown that the number of donations supplied increases year on year, demonstrating the growing interest of today's society in local plant varieties. Figure 27.2 shows the different types of germplasm users, and Fig. 27.3 the intended purpose for the requested material. The seed bank was able to supply a growing number of seed requests: in 2013, more than 200 users requested 440 accessions. Only 23.3% of the delivered germplasm was grown under an organic agricultural system; 66.2% of users in the last decade were farmers or hobby cultivators and 22.9% were public institutions. There were also many other users such as co-operatives, foundations, associations and, notably, chefs. The main destination of the plant material was for cultivation in the home garden for domestic consumption (62.3%), followed by cultivation for research tasks (11.8%). Other confirmed uses were production on farms, education, cultivation at restaurants and finally breeding. Therefore, the main destination of the conserved landraces was recorded as home consumption and use in gastronomy. Experience demonstrates that most of the users were private citizens who requested old varieties to grow in their garden for self-consumption. The second group of users were public institutions, mainly research centres, technological or teaching centres.

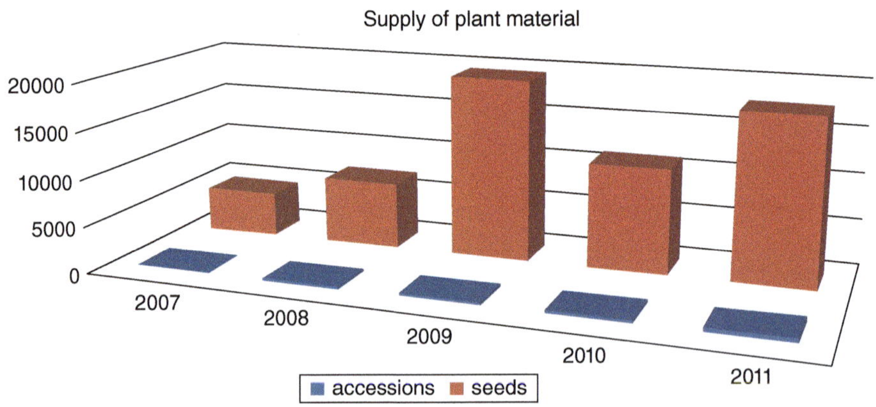


Fig. 27.1. Number of accessions (both seeds and explants) donated by germplasm collection for 5 years.

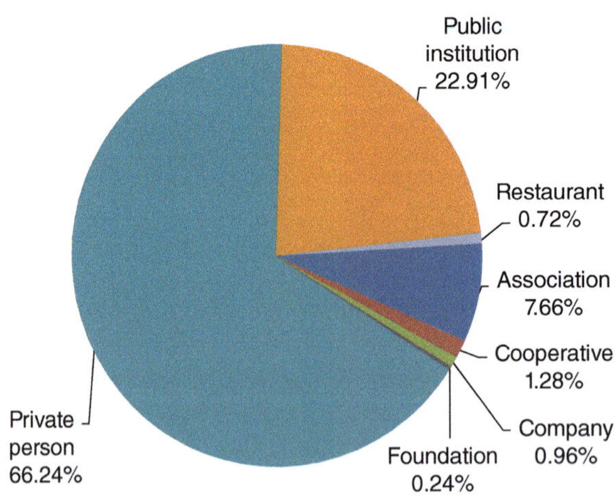


Fig. 27.2. Germplasm user type.

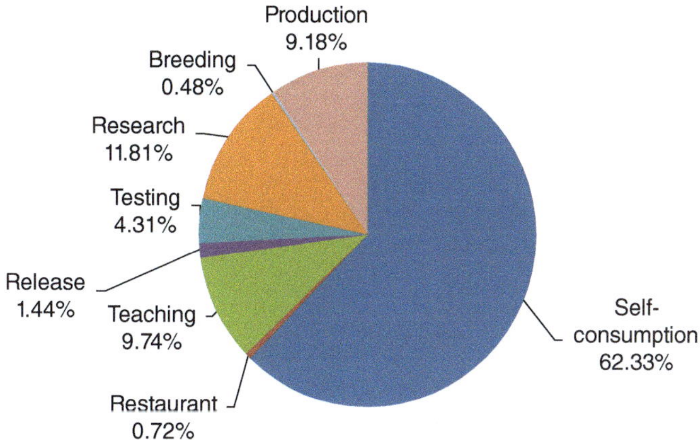


Fig. 27.3. Reported use of requested germplasm.

Breeders, who had been considered one of the main users of seed banks (Engels *et al.*, 2002), were in this case a minority. Restaurants were included among the minor users.

The number of annual seed requests has grown so substantially that the process of disseminating the material needed to be changed to avoid the distribution of entire collections. In 2012, IMIDRA stopped supplying seed requests and started producing seedlings for sale. The result was that the amount of material supplied to users doubled (Fig. 27.4). One of the reasons is the increasing use of landrace accessions in restaurants. Spanish cooks have become social and media stars. Spanish cooking is known worldwide. Ideas such as '0 kilometre' cooking have boosted the request for these landraces, which has been extenuated by celebrity chefs using landrace materials and promoting their use in social networks, TV programmes and other media. This movement chimes with the desire for 'safe traditional crops', mainly because of their sensorial features, and has got deeply into citizens' preferences and perceptions. Figure 27.5 shows some products made with landraces supplied by the IMIDRA germplasm collection.

Not only celebrity chefs but also large food distribution chains have shown interest in traditional varieties. Therefore, a cooperation agreement between IMIDRA, a large supermarket chain and farmers has been signed, in order to sell traditional varieties to the public. This has marked the reconnection of the importance of the local 'terroir' or 'land connection' with large urban markets. In supermarkets, local varieties are located and signalled at a particular corner for products of proximity. Following a survey of consumers, 1238 answers (Fig. 27.6) indicated that they valued old varieties mainly for their taste (32.5%) and their better odour and flavour (29.4%). The substantive issue was: why do you pay more for a traditional crop? The consumer answer was the perceived better quality of landrace varieties.

## 27.3 Conclusions

The preservation of plant genetic resources is based on the central premise of their value to future generations, to be able to breed crop varieties and face new challenges. This proposal is still valid. But, in addition, our experience shows that today's consumers, especially those living in European cities, are demanding food quality, flavours and textures, and they are accessing these from landrace collections. European citizens value such vintage accessions due to their sensorial characteristics. A social current towards concepts such as proximity, organic, seasonal products, local diversity, identity, history and one's roots with the land is growing in large cities. Flavours, quality and diversity are still in folk memory. These landraces, linked with the tradition of farming and culinary folk memory,

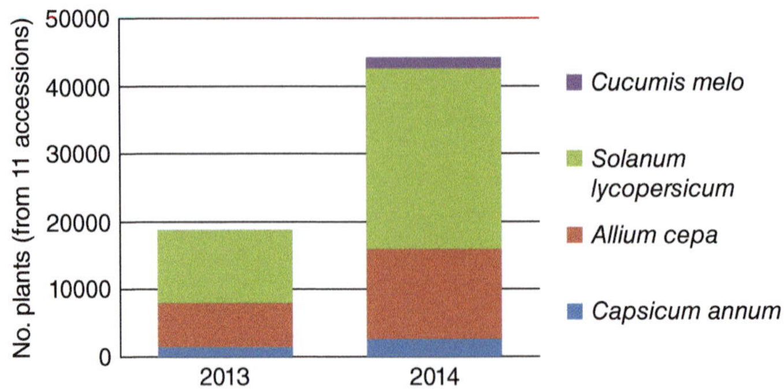


**Fig. 27.4.** Number of seedlings obtained from conserved accessions (Madrid landraces of four crops) provided to farmers and other users over 2 years.

**Fig. 27.5.** Three different recipes, prepared with traditional or old tomato varieties. (a) Three textures, three flavours (sweet tomato flesh in puree, acid gelatine from tomato seeds and aromatic tomato water). (b) Cold tomato soup (tomato, onion and Mediterranean spices thoroughly cooked in extra virgin olive oil at mild temperature, pureed and left to go cold). (c) Marinated sardine fillet on tomato tartar (tomato, egg, onion, caper, olive, all dressed with an extra virgin olive oil-based emulsion with spices).

are characterized by a highly remarkable morphological and sensorial diversity (Escribano and Lázaro, 2012; Escribano *et al.*, 2012; Lázaro *et al.*, 2013; Lázaro and de Lorenzo, 2015). These are some of the strengths of landraces for present and future generations; they are currently providing an important economic and sustainable activity for urban communities.

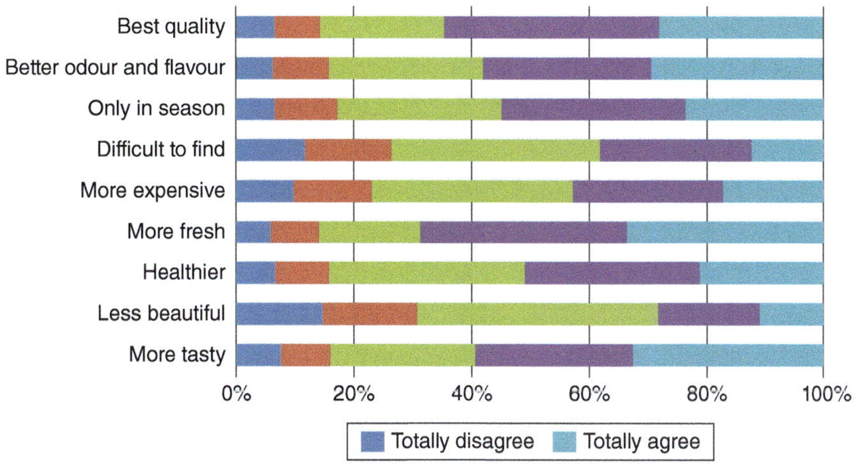


Fig. 27.6. Consumers' perception of the strengths and weaknesses of landrace varieties.

## References

Engels, J.M.M., Ramantha-Rao, V., Brown, A.H.D. and Jackson, M.T. (2002) *Managing Plant Genetic Diversity*. CAB International, Wallingford, UK.

Escribano, S. and Lázaro, A. (2012) Sensorial characteristics of Spanish traditional melon genotypes: has the flavor of melon changed in the last century? *European Food Research and Technology* 234(4), 581–592.

Escribano, S., Lázaro, A., Cuevas, H.E., López-Sesé, A.L. and Staub, J.E. (2012) Spanish melons (*Cucumis melo* L.) of the Madrid provenance: a unique germplasm reservoir. *Genetic Resources and Crop Evolution* 59(3), 359–373.

Lázaro, A. and de Lorenzo, C. (2015) Texture analysis in melon landraces through instrumental and sensory approaches. *International Journal of Food Properties* 18(7), 1575–1583.

Lázaro, A., Villar, B., Aceituno-Mata, L., Tardío, J. and De la Rosa, L. (2013) The Sierra Norte of Madrid: an agrobiodiversity refuge for common bean landraces. *Genetic Resources and Crop Evolution* 60(5), 1641–1654.

# 28 Hungarian Strategies for the Conservation of Crop Wild Relative and Landrace Diversity

**B. Baktay* and A. Simon**
*Research Centre for Agrobiodiversity, Tápiószele, Hungary*

## 28.1 Introduction

The Hungarian National Strategy for the Conservation of Plant Genetic Resources for Food and Agriculture 2014–2020 was endorsed by the Hungarian Minister of Rural Development in 2013. The document was prepared by the Ministry of Rural Development, the National Genebank Council and by different experts from relevant institutions. The vision and targets have been developed in line with the Convention on Biological Diversity (CBD) and the International Treaty on Plant Genetic Resources for Food and Agriculture (ITPGRFA). Target 13 of the CBD Strategic Plan 2011–2020 invites parties to maintain the genetic diversity of cultivated plants and of wild relatives, including other socio-economically as well as culturally valuable species and to develop and implement strategies for minimizing genetic erosion and safeguarding their genetic diversity by 2020 (CBD, 2010). The Hungarian strategy sets a vision for 2020 backed up by 14 targets, which provide a comprehensive background for the conservation of plant genetic resources for food and agriculture in the next 7 years in Hungary.

The Centre for Plant Diversity (CPD, Tápiószele, Hungary) is the largest seed bank in Hungary. The institute is specialized for the *ex situ* conservation of crops and vegetables and systematically has collected their landraces (LR) and crop wild relatives (CWR) in the past 50 years. In addition, the 5-year-long Pannon Seed Bank Life+ Project started in 2010. The establishment of a joint seed bank for agricultural and wild flora is a good and demonstrative example in Europe in line with the objectives of the CBD.

## 28.2 Centre for Plant Diversity

The predecessor of the institute was established in Tápiószele (Hungary) by Lajos Szelényi's Agricultural and Educational Foundation in the second half of the 19th century. The first centralized collection of genetic resources for various crops had been established in 1950, and the Institute for Agrobotany was founded in 1958. Later, cooled storage rooms were built and a programme was initiated aimed at updating collections to meet the International Standards for Genebanks. Genetic resource activities have been extended and updated, and a National Centre for Agrobiodiversity was organized. Developments included the establishment of a National Base Collection for seed-propagated crops and their wild relatives, as well as the organization of a National Database for PGRFA (Plant Genetic Resources for Agriculture) (L. Holly, 1992, Hungary, personal communication). The Centre

*Corresponding author; e-mail: baktay@mail.nodik.hu

for Plant Diversity was founded in 2010, and this new institute also acts as a technical coordination centre and provides support for the Hungarian Genebank Council.

The institute has a nationwide responsibility for the technical coordination of all plant genetic resources collections. The Institute is also responsible for the development and maintenance of field crop and vegetable genetic resource collections, in addition to coordinating plant genetic resource activities in Hungary. These collections can supply initial material for breeding programmes, as well as for botanical, biochemical, cytological and genetic research. Seed samples are distributed, together with the relevant information, to users in Hungary and worldwide.

### 28.2.1 Genebank maintenance

Two collections have been set for the medium- and long-term conservation of seed-propagated crops, and meristem cultures are used in the case of vegetatively propagated crops. In the case of the active collection, the storage temperature is 0°C, which provides shorter viability but facilitates easier and more frequent access to the samples. In the chambers of the base collection, the temperature is 20°C below zero, and the estimated lifespan exceeds 100 years for most of the species. At the end of the 1970s, *in vitro* methods were introduced for the maintenance of vegetatively propagated crops. The techniques applied for preserving potato collections are based on growing sterile meristem cultures under suboptimal conditions. The total number of accessions is 51,354 in the genebank of the Centre for Plant Diversity (Table 28.1 and Fig. 28.1).

The harvesting of raw seeds is followed by cleaning, considering the nature and special requirements of genebank activity. During the preparation of seeds for storage, cleaned seeds are sampled and divided in order to assess the thousand seed mass (TSM), to prepare samples for germination tests and facilitate medium- and long-term conservation. Changes in the viability of seeds kept in the storage chambers are checked by germination tests every 10 years, then every 5 years. When viability declines to a lower limit, the sample is rejuvenated.

**Table 28.1.** Storage methods and numbers of accessions in the Centre for Plant Diversity genebank.

| Storing method | Accessions | Number of samples |
|---|---|---|
| Active | 48,884 | 86,942 |
| Base | 8,335 | 11,754 |
| *In vitro* | 556 | 556 |
| Vegetative | 209 | 209 |
| Total | 57,984 | 99,461 |

### 28.2.2 Multiplication and regeneration of accessions

In order to obtain sufficient amounts of high-quality seed for medium- and long-term conservation, evaluation and distribution, the institute has nearly 300 ha of land, of which 20–30 ha is used for genebank nurseries and other field trials each year. During multiplication, special attention is paid to spatial isolation. The species of which the cross-pollination rate is expected to be less than 2–3% are regenerated without isolation on the multiplication plots of the institute. Large collections of the same cross-pollinated species are multiplied by using artificial isolation methods (houses for isolation, isolation cages, paper and cheese-cloth bags, etc.).

### 28.2.3 The characterization and evaluation of PGRs

The characterization and evaluation of PGR collections is carried out using internationally accepted (International Plant Genetic Resources Institute) descriptor lists. Biochemical, resistance and morphometric analyses ensure proper assessment for the genetic composition of collected accessions.

### 28.2.4 Documentation

The documentation of the passports and the evaluation and genebank management data of the PGR are computerized to suit the development of a national PGRFA database. The new database system is under development.

### 28.2.5 International participation

Gene preservation and other activities relating to PGRFA, in accordance with international

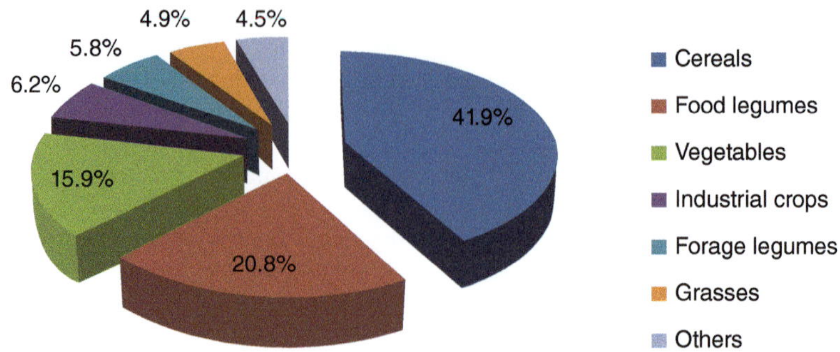


Fig. 28.1. Agrobiodiversity collection held in the Centre for Plant Diversity, Tápiószele, Hungary.

conventions and agreements such as the CBD and the ITPGRFA, have high priority in Hungary. Due to the significant global decline of agrobiodiversity in the last century, collaboration on PGRFA, especially at the regional level, is essential. Thus, the CPD participates in the ECPGR (European Cooperative Programme for Plant Genetic Resources) and other international and European programmes, including EU working groups and FAO (Food and Agriculture Organization) work on PGRFA.

Hungary, as the Presidency of the Visegrad Group in the first half of 2014, hosted a workshop in order to provide a consultation platform for experts dealing with national activities and strategies regarding the preservation of PGRFA in seed banks. The aim of the workshop was to contribute to developing and improving an effective and successful cooperation within the region. Experts from countries from the Visegrad Group and Austria, Bulgaria, Croatia, Romania, Serbia, Slovenia and Ukraine were present at the workshop. The workshop provided a good opportunity for participants to discuss ongoing processes based on international agreements, programmes and institutions (e.g. ECPGR, AEGIS, ITPGRFA, CBD, etc.), as well as to define potential synergy in these areas or a concept for common collaboration.

## 28.3 National Strategy for the Conservation of PGRFA 2014–2020

The Hungarian National Strategy for the Conservation of Plant Genetic Resources for Food and Agriculture 2014–2020 was endorsed by the Hungarian Minister of Rural Development in April 2013. The document was prepared by experts from the Ministry of Rural Development and from other institutions, and by the National Genebank Council. The vision and targets of the strategy have been developed in line with the CBD, the ITPGRFA and the EU Biodiversity Strategy to 2020.

The vision for the future is the long-term conservation of the existing diversity of our national treasures, Hungary's genetic resources of food crops and microorganisms free of genetic damage, as far as is possible the realization of their actual economic value and the spread of their sustainable use in their natural environment, and the promotion of their use in research, education and Hungarian breeding efforts. The main goal of the strategy is to identify the major objectives needed for the conservation of Hungary's plant genetic resources for food – that is, genetic materials of crops, grape varieties, fruit, vegetables, herbs, spices and essential oil plants, and microorganisms important in agriculture and the food industry – in line with the vision for the future, and also to identify the measures that ensure the achievement of these for the period up to 2020.

The strategy includes 14 targets:

- Institutional structure and legislative background:
  Target 1: prevent the fragmentation of the institutional system and the coordination of the work of institutions involved in genetic conservation.
  Target 2: ensure the continuous operation of the Plant Genebank Council and ask for their expertise in all matters related to PGR conservation.

Target 3: ensure the proper implementation of the FAO treaty by 2014.
Target 4: ensure the continuous availability of EU and national funding for the conservation of national PGR collections.
Target 5: encourage the implementation of ministerial regulations on landraces.

- *Ex situ* conservation in genebanks:
Target 6: create catalogues and proper documentation of PGRs in all institutions involved in genetic conservation activities by 2014, the continuous updating of these, and extending and continuously updating the national database. Compile detailed documentation on collected local varieties and landraces, and the traditional knowledge related to these.
- *In situ* on-farm conservation:
Target 7: involve conserved PGRs in breeding programmes and make genetic materials available to Hungarian farmers and gardeners for on-farm use. Exploit the landscape ecology potential of landraces in the development of the agricultural cultivated landscape.
Target 8: maintain and develop fruit field collections. In accordance with the regulation on fruit landraces, conserve the registered landraces in national field collections.
Target 9: promote community action on PGR conservation.
- Research, education and awareness raising:
Target 10: identify, collect and document the genetic resources within Hungary.
Target 11: encourage and support agricultural research for the use of genebank materials in Hungary's research institutes.
Target 12: launch information experiments and participatory breeding programmes based on Hungarian varieties, landraces, local varieties and traditional varieties, in cooperation with farmers and researchers/scientists.
Target 13: the awareness of society and decision makers in relation to the importance and societal and environmental use of the conservation of plant genetic resources needs to be increased, and the use of landraces and other Hungarian varieties needs to be revived.
- International cooperation:
Target 14: meet the adopted international objectives, strengthen current international cooperation and extend international relations in the field of PGR conservation.

## 28.4 Collecting Programme 2013–2014

Target 10 of the National Strategy for the Conservation of Plant Genetic Resources for Food and Agriculture 2014–2020 emphasizes the need for further collection missions in Hungary in order to ensure the collection and conservation of traditional varieties and landraces. In 2013, the CPD developed a 2-year collecting programme. The main goal of this project was to explore regions in Hungary where collecting activity had been very low in the past 40 years and to collect seed samples of cultivated and wild plant species. In the past 10 years, the number of collected accessions has been very diverse in each year (Fig. 28.2). In 2013, two collecting trips were organized in the north-east and western part of Hungary (5 days in Zala, and also in Zemplén). The collection mission conducted in the Zemplén mountains has revealed a range of diversity of locally grown traditional varieties. These varieties and landraces are cultivated by elderly home gardeners, and there is an urgent need to conserve these varieties in the genebank. The traditional knowledge that still exists on landrace maintenance may also form a basis for local NGO's on-farm conservation activities.

Without the establishment of the Pannon Seed Bank, some wild species and the richness of the Pannonian flora may be threatened by extinction. In order to make the Pannon Seed Bank operational, seed samples of Hungarian wild plant species need to be collected. One of the main aims of collecting missions is to collect the valuable wild species of the region. During the mission in Zemplén, 32 accessions of 30 species were collected: species of nature conservation importance were the following: *Gymnadenia conopsea* (L.) R. Br., *Linum flavum* L. and *Chamaecytisus albus* (Hacq.) Rothm. (Gyovai, 2013).

## 28.5 Multiplication and Regeneration

One of the main tasks of the CPD is the multiplication and regeneration of accessions in order to obtain sufficient quantities of high-quality seeds for medium- and long-term conservation, evaluation and distribution. The number of multiplied and regenerated seed samples has increased over latter years (Fig. 28.3). The reasons for this increase are:

**1.** Lack of capacity and financial resources meant routine regeneration had been postponed, but these seed samples were regenerated in 2013 to avoid loss of the accession.
**2.** In 2011, a 5-year project was started in several Hungarian institutes, including the CPD. The project requires a minimum of 2000 seeds in each accession that are important as part of the national collection. This project ensures the safety conservation of each accession, and the safety duplication as well. The National Safety Duplication Storage was launched in 2014.

In 2013 in the CPD, more than 7000 accessions were planted in the fields for multiplication and

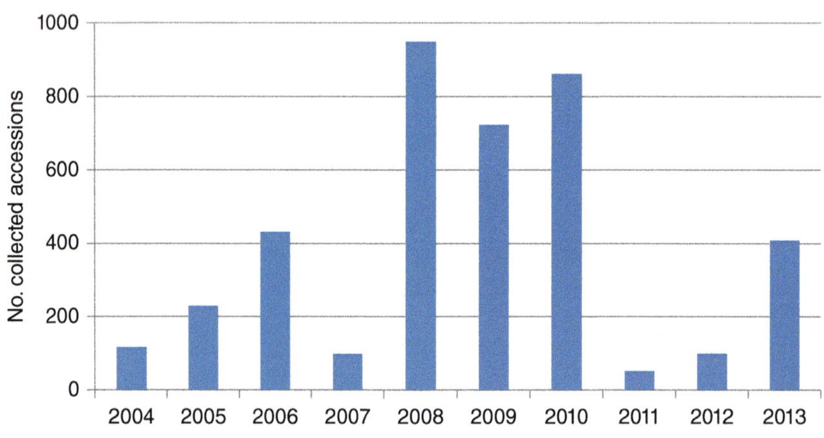


Fig. 28.2. Germplasm accessions collected from 2004 to 2013.

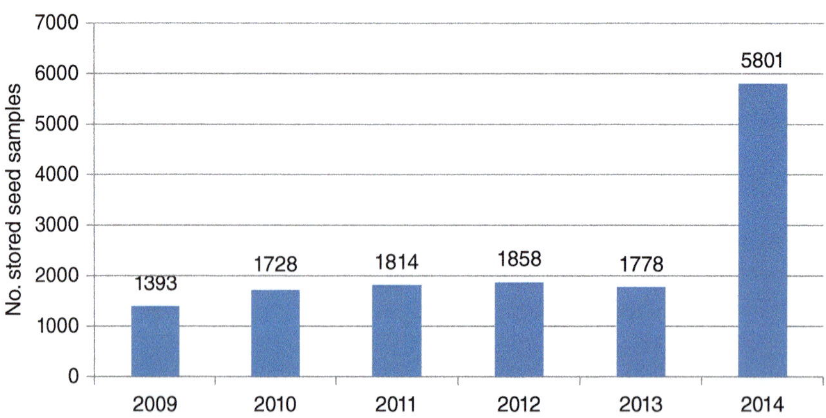


Fig. 28.3. Seed samples 2009–2014 stored in CPD.

regeneration, mostly LR of cereals and vegetables, but also many CWR and medicinal plants.

## 28.6 Public Delivery

Extra seeds from the multiplied and regenerated seed samples are offered for public delivery. In the past 10 years, the average number of delivered seed samples was from 2000 to 4000 (Fig. 28.4). In 2014, the number of seed samples sent by the CPD to users was more than 12,000. The accessions including Multilateral System (MLS) are available for all kinds of purposes. CPD provides genebank material if the recipient accepts the conditions of Standard Material Transfer Agreement (SMTA). In the case of hobby gardeners and farmers who request accessions for their own consumption and/or for crop production (90% of requests), a special transfer agreement is used that is based on the original SMTA.

Following discussion with gardeners and farmers, the main requirements are as follows:

- The *recipient* multiplies the samples in the year of the transfer.
- In the interest of obtaining seeds with sufficient quantity and quality, the *recipient* does his or her best in the process of multiplication.
- To avoid genetic erosion, in the case of cross-pollinated species, the *recipient* pays special attention to the application of proper isolation methods during the multiplication process.
- The *recipient* accepts and undertakes that he or she fills in and returns the questionnaire no later than 31 December of a given year, which the *provider* sends together with this agreement. The purpose of this is for the *provider* to collect information on the multiplication process and the experiences of the *recipient* regarding the usage of the accessions handed over in the frame of the agreement.

Furthermore, the *recipient* accepts that:

- The *provider* reserves the right to check the process and backgrounds/circumstances of the multiplication.
- The *provider*, in an emergency and in case of successful multiplication, may demand and claim a certain amount of seeds from the accession be handed over.
- The *provider* may exclude the *recipient* from the additional public delivery and seed request, if the *recipient* does not meet the engagements.

In January 2013, a special seed request system was established by the CPD using the webpage of

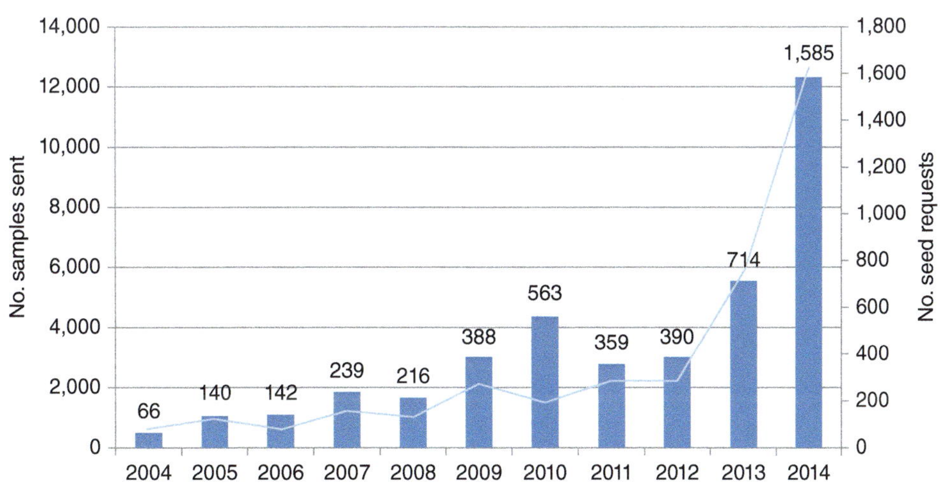


**Fig. 28.4.** Public delivery of germplasm accessions 2004–2010 from CPD.

the institute. More than 1500 seed requests were handled using the new system. The list of accessions includes mostly landraces, in order to promote the sustainable use and on-farm conservation of LR in Hungary. Anyone can apply for accessions listed on the website, but the maximum order is ten accessions/person. Seeds are sent by post with the signed SMTA. In return, the institute requests a completed information sheet on the experiences regarding the accessions ordered, which must be sent back to the institute at the end of the year. The experience of using LR is very positive. In Hungary, there is a great demand for LR use, and the number of seed requests, the number of LR-related events in Hungary and the activity of NGOs underpin the necessity of the conservation and multiplication of LR in genebanks.

Another good example of the on-farm conservation of old varieties in Hungary is 'Agreement on the cooperation of fruits varieties adopted to regional characteristics and in cultivation for a long time in the Carpathian Basin', which from 1 January 2013 has been coordinated by the CPD. From that time, more than 30 gardens have been established all over Hungary and more than 1000 grafts planted in the gardens. The main goal of this agreement is LR conservation, but all gardens have importance in the awareness-raising tools.

## 28.7 Awareness-raising and Education

From 2012, the CPD has paid special attention to awareness-raising and education. In order to present the genebank's work and activities, and to highlight the importance of conserving PGRs, the institute can be visited by anyone who applies for guidance. The CPD has had more than 1500 visitors in each year for the past 2 years, mostly groups of schoolchildren and students, but also groups of experts, scientists, decision makers and local inhabitants. In 2012, an exhibition of the Pannon Seed Bank Project titled 'From seed to seed' was developed at the CPD that presented the whole life cycle of plants (Fig. 28.5). More than 150 native wild species are presented, emphasizing the importance and richness of the plant diversity of Hungary. The exhibition is

Fig. 28.5. 'From seed to seed' exhibition at CPD.

interactive; thus, visitors can test and touch all the interactive components of the exhibition. Using special games for children – and adults, too – the fruit and seeds of protected, strictly protected and invasive plant species, the utilization of herbs, and poisonous plants, are all presented.

## 28.8 Pannon Seed Bank Life+ Project

The Pannon Seed Bank Life+ Project, 'The establishment of the Pannon Seed Bank for the long-term *ex situ* conservation of Hungarian vascular wild plants', ran from 1 January 2010 to 31 December 2014. The main aim of the project was the collection and long-term conservation of seed samples of approximately 50% of the Pannon flora – at least 800 species, including CWR of the Hungarian native wild vascular flora. The project was supported by the LIFE financial instrument of the European Community and the Hungarian Ministry of Agriculture. During collection, great emphasis has to be placed on species of nature conservation importance, especially on rare species or populations. In the case of species that are of ecological importance, the focus is on the collection of samples in diverse environmental conditions, characteristic populations of species found in special and rare plant associations, and typical species of wetland habitats endangered by climate change. CWR, forage species, herbs, drought-tolerant species, dominant species of forest associations and rare and threatened native weed species are considered species of ecologic importance.

## References

CBD (2010) *Strategic Plan for Biodiversity 2011–2020*. Secretariat of the Convention on Biological Diversity. Montreal, Canada. Available at: www.cbd.int/decision/cop/?id=12268 (accessed 29 October 2010).

Gyovai, Á.P. (2013) Landraces in Zemplén-mountains the first target area of collection mission program 2013–2014. *Tájökológiai Lapok* 11, 367–371.

# 29 Assessment of Italian Landrace Density and Species Richness: Useful Criteria for Developing *In Situ* Conservation Strategies

**R. Torricelli,* L. Pacicco, M. Bodesmo, L. Raggi and V. Negri**
*Department of Agricultural, Nutritional and Environmental Sciences, University of Perugia, Perugia, Italy*

## 29.1 Introduction

Landraces (LR) are a key component of agrobiodiversity; being increasingly threatened, they deserve immediate conservation actions (Vetelainen *et al.*, 2009). There is an imperative to maintain LR diversity *in situ* (on-farm) and increase their use. In fact, in the actual climate change and climate unpredictability scenario, LR are key sources of genetic variation for sustaining agricultural diversity and for secure food availability for the future. In this regard, the 'PGR Secure project' aims to develop conservation strategies for European crop wild relative and LR diversity and to enhance their use as a means of underpinning European food security in the face of climate change.

In Italy, the risks related to plant genetic resource erosion and to the loss of agrobiodiversity were providentially perceived earlier than in other countries, and various studies on an LR census have been conducted (Negri, 2003; Concezzi *et al.*, 2008; Falcinelli and Lorenzetti, 2008; Negri *et al.*, 2012). Furthermore, some Italian regions (i.e. Emilia-Romagna, Friuli Venezia-Giulia, Lazio, Marche, Toscana and Umbria) have approved and implemented laws to safeguard and promote biodiversity and local genetic resources conservation and utilization (Negri, 2012; Negri and Torricelli, 2013), and recently the Ministry of Agriculture has developed specific Guidelines for the Conservation of Genetic Resources for Food and Agriculture (Marino, 2013; www.politicheagricole.it/flex/cm/pages/ServeBLOB.php/L/IT/IDPagina/305).

In Italy, LR are still maintained *in situ* because they are widely used to make products that are awarded quality labels like Controlled Designation of Origin (DOC), Protected Designation of Origin (PDO), Protected Geographical Indication (PGI), Traditional Specialty Guaranteed (TSG) or Certified Product from Organic Agriculture. These products are commercialized across Italy and abroad, taking advantage of commercial chains specialized in typical, high-quality food. In other cases, LR products are relegated to home production and consumption because of the appreciation and traditional use of that particular family (Negri, 2003). Also, some LR are used and maintained *in situ* (on-farm) because they guarantee specific adaptation to particularly adverse pedoclimatic conditions. For example, adaptability and good productivity under difficult or harsh conditions, together with better

*Corresponding author; e-mail: renzo.torricelli@unipg.it

resilience in comparison with modern varieties, are the main reasons for the utilization of most forage LR (Negri, 2003). Finally, LR are often preferred to commercial varieties in those types of agriculture where environmental health is strongly aimed for. Many organic farmers already use LR, and an increasing number of them are looking for LR seed because in organic agriculture local adaptation and tolerance to biotic/abiotic stresses are particularly relevant traits. There are also examples of varieties developed from LR that have been commercialized successfully in both *Brassica* spp. (Branca *et al.*, 2002, 2007) and *Medicago sativa* L. (Falcinelli and Torricelli, 2001).

Within the PGR Secure project, we focused on developing tools, inventories and strategies for European and Italian LR conservation (Negri *et al.*, 2013, 2014; Negri and Torricelli, 2014). In this context, the aim of the present work was to summarize inventory data recorded for each Italian region and to assess the LR density index and species richness level through the Italian regions. This information can be used to develop or enhance political and economic actions for LR conservation by Italian regions (Negri *et al.*, 2012; Marino, 2013; Negri and Torricelli, 2014). In fact, in Italy, the responsibility of maintaining genetic resources lies with the regions.

## 29.2 Materials and Methods

The first step carried out was the preparation of an 'official' inventory of LR. To achieve this objective, official data on LR maintained *in situ* were requested from the Ministry of Agriculture, the Consultative Body of the regions and each single region officer in charge of genetic resources. On the basis of the data received, the First Italian Inventory Of *In Situ* Maintained LR was compiled that contained detailed information on the species, common names, geographic references and other information relative to the LR (Negri *et al.*, 2013).

Subsequently, the whole Italian territory was subdivided into quadrants (2 × 2 km size) according to Gargano (2011). Quadrants were then merged to obtain 20 × 20 km grids, each cell being georeferenced using the WGS84-UTM32N system (EPSG: 32632). By optimizing data in this way, 995 quadrants were obtained.

Finally, the Landrace Density Index (LDI), which provides information on LR density, and the Shannon Index H′ (Shannon and Weaver, 1949; Spellerberg and Fedor, 2003), which gives information on species richness, were calculated for the entire Italian territory. In particular: (i) LDI is the LR number per quadrant; (ii) H′ is an estimate of the species richness and is calculated, as:

$$H' = -\sum_{i=1}^{s_i} p_i * \ln p_i$$

where $s_i$ is the number of LR species and $p_i$ is the LR frequency (i.e. number of LR belonging to each species/LR total number). H′ was calculated only in those quadrants where LDI ≥ 1.

## 29.3 Results and Discussion

A total of 4806 accessions representing 2365 LR were inventoried across Italy (Negri *et al.*, 2014). The recorded LR, mostly found in home gardens, belong to 329 different species, including fruit trees, vegetables, grain legumes, forage crops, cereals, ornamental plants and other species. The highest LR number was recorded in Umbria (378), Calabria (288), Sicily (251), Basilicata (212) and Campania (203), which accounted for more than 50% of the total recorded LR (Fig. 29.1). According to our data, LR are mostly used for human food; in fact, fruit trees (apple, pear, cherry, plum, grape and olive trees) are 73% of recorded LR, followed by herbaceous plants (25%, grain legumes, vegetables and cereals). Only 2% of LR maintained on-farm are forage crops. The species most frequently cultivated as LR are *Prunus avium* L. (18%), *Malus domestica* Borkh. (17%), *Pyrus communis* L. (15%), *Vitis vinifera* L. (9%), *Phaseolus vulgaris* L. (7%), *Olea europaea* L. (4.8%) and, to a lesser extent, *Ficus carica* L., *Solanum tuberosum* L. and *Capsicum annuum* L. Regarding geographical distribution, the 4806 recorded accessions are scattered from sea level to mountainous areas, but they are found most frequently inland in hill sites (i.e. those from 350 to 513 m above sea level). Most LR are found at low latitudes, between 37°N and 41°N. Differently, *C. annuum*, *V. vinifera* and *P. avium* LR occur frequently in a more northerly

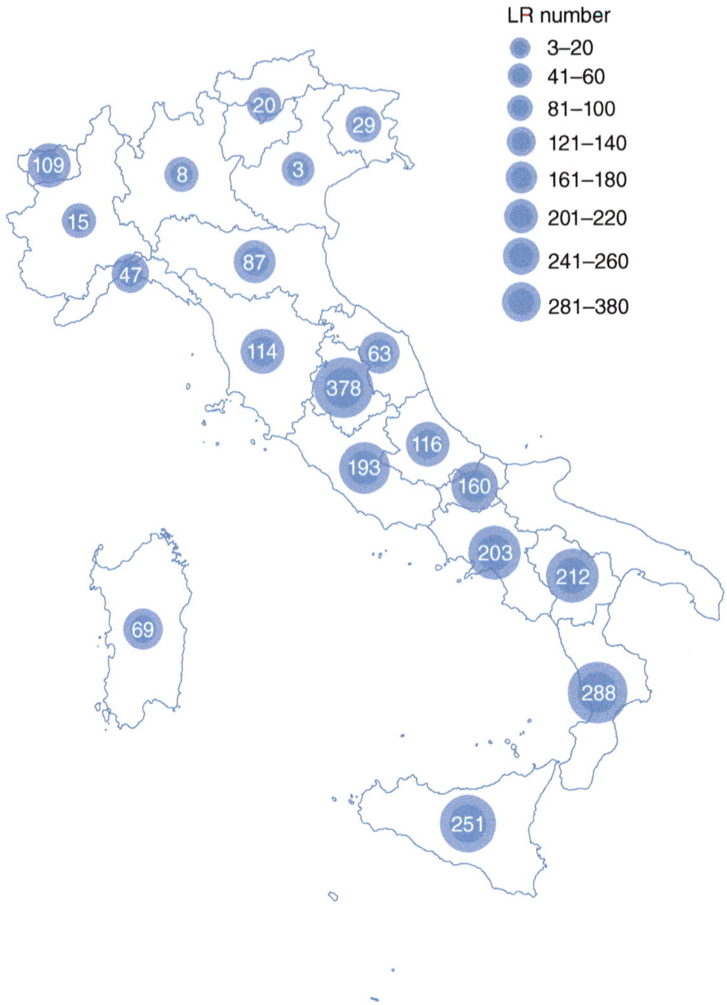


Fig. 29.1. LR distribution by Italian region.

position (mean latitude = 44.24°, 42.89° and 41.81°, respectively), while *M. domestica* (41.79°) and *P. communis* (40.96°) LR are found most frequently in southern areas. As regards longitude, *C. annuum* LR show concentrated, circumscribed values, while *P. avium*, *M. domestica* and *P. communis* are more widespread.

The analysis of LR diversity estimated through LDI and H′ shows that the highest values of LDI were found in Umbria, Basilicata and Lazio (Fig. 29.2a), while in Basilicata, Molise and Umbria, which have the highest number of species cultivated as LR, the highest values of H′ were detected (Figs 29.2b and c). However, fairly good values of diversity were detected also in other regions of northern (Valle D'Aosta), central (Toscana and Abruzzo) and southern Italy (Calabria). As shown in Fig. 29.1, the majority of LR are distributed in the Italian regions located on Apennine areas that are characterized by high ecological diversity. The Apennines is a land system that preserves, and in some cases enhances, the biological diversity threatened by changes of land use and by diffuse abandonment (Farina, 1995). In fact, the heterogeneity of the landscape has favoured the development and conservation of different farming systems. Over thousands of years, people living in these

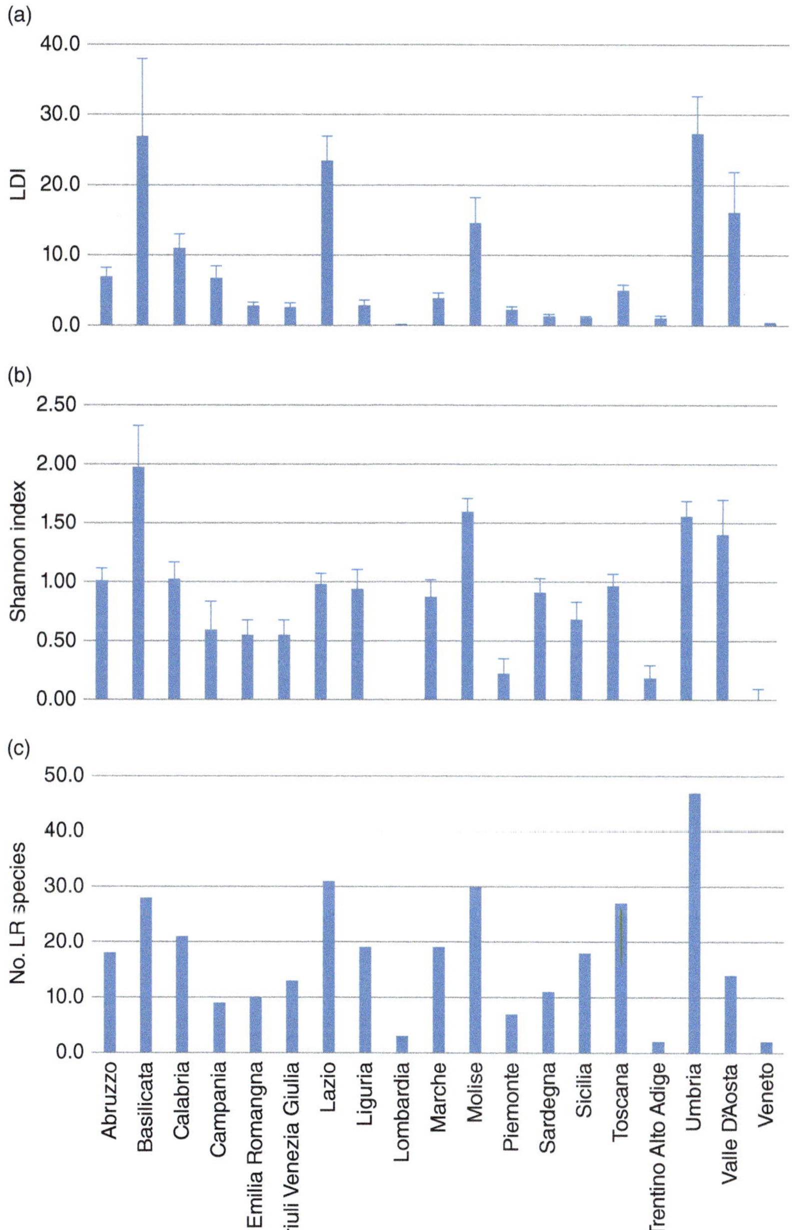


Fig. 29.2. Mean and SE of: (a) LDI and (b) Shannon index; (c) total number of species cultivated as LR in the Italian regions.

areas have understood how to use their lands to their best according to soil type and climatic conditions. Over the years, this process shaped the landscape and led to the establishment of small fields, each different from the other in terms of crop production and quality (Ambrosi, 1981; Farina, 1995). A peculiar combination of mountain, hill and plain, characterized by specific environment conditions, allowed the formation of strongly different ecological niches (Pavia *et al.*, 2007), where the LR were developed and maintained over time. Accordingly, it is in the Apennine regions that LR diversity is highest (Fig. 29.2). As mentioned previously, some of these regions have already approved laws in favour of LR conservation, which will favour LR maintenance.

## 29.4 Conclusions

The compilation of an inventory is the starting point for any conservation action, addressed either to some agrobiodiversity components or globally to them all. However, having inventoried LR does not mean *per se* they are safeguarded on behalf of future generations. Besides increasing awareness at every level of the importance of landraces, policy actions that are aimed both at conservation and use should be implemented, as suggested in the framework of PGR Secure (Negri and Torricelli, 2014).

Among LR conservation actions, we can list the following: continuously update the inventory; promote safe backup of LR in public *ex situ* collections; increase coordination in developing and implementing measures for LR conservation; provide adequate funds for *ex situ* and *in situ* conservation and research; promote the registration of LR as 'conservation varieties' (Spataro and Negri, 2013); promote the use of home garden LR in community and home gardens; promote the awarding of quality marks for products derived from LR; promote typical local products derived from LR to the widest extent; and promote the use of LR in plant breeding and participatory plant breeding programmes, especially those aimed at creating varieties suitable to environmentally friendly agronomic systems.

It should be noted that almost all of these actions have already been foreseen in the Italian Guidelines for the Conservation of Genetic Resources for Food and Agriculture (Marino, 2013; www.politicheagricole.it/flex/cm/pages/ServeBLOB.php/L/IT/IDPagina/305) and by the Italian National Plan for Agrobiodiversity applied in part by some of the Italian regions.

They are also favoured by the Common Agricultural Policy (CAP) 2014–2020. The 2014–2020 CAP is, in fact, oriented strongly to diversify crops and conserve and restore diverse agricultural systems. It specifically mentions measures to help farming meet the challenges of soil and water quality, biodiversity and climate change. It also specifically mentions the need to favour crop diversification and conserving areas of ecological interest.

Through the 2014–2020 CAP, which provides large funding for the above-mentioned activities, the Italian regions, which have the responsibility of maintaining genetic resources, can have the economic means to apply actions in favour of agrobiodiversity, in general, and LR, in particular, conservation. These actions could be applied initially to those areas within regions that are richest in agrobiodiversity, previously identified as the Most Appropriate Areas (MAPAs) in earlier studies (Negri *et al.*, 2012). The MAPAs are areas richest in all of the agrobiodiversity components (inter- and intraspecific diversity for plants, and microbes), which also includes protected wild forms of life (Negri *et al.*, 2012). The identification of these areas (Pacicco *et al.*, unpublished results) will help the purpose.

## Acknowledgements

The research received funding from the European Community's Seventh Framework Programme (FP7 2007–2013) under Grant Agreement No 266394 'PGR Secure'. Thanks are due to all the Italian regions and autonomous provinces and to all the officers who contributed to the compilation of the inventory.

## References

Ambrosi, A.C. (1981) *Lunigiana: La preistoria e la romanizzazione, I –La Preistoria*. Itinerari Educativi, Centro Aullese di Ricerche e di Studi Lunigianesi, Aulla, Italy.

Branca, F., Li, G., Goyal, S. and Quiros, C.F. (2002) Survey of aliphatic glucosinolates in Sicilian wild and cultivated Brassicaceae. *Phytochemistry* 59(7), 717–724.

Branca, F., Chiarenza, G.L., Palmigiano, S., Ruggeri, A. and Castellano, A. (2007) Diversity of Sicilian broccoli landraces with regard to the main biometric characteristics and the antioxidant compounds of the head. *Agroindustria* 6, 101.

Concezzi, L., Gramaccia, M. and Mauceri, S. (2008) Realizzazione di un Catalogo Nazionale delle Varietà Locali. In: *Rete Interregionale per la Ricerca Agraria, Forestale, acquacoltura e pesca, Azioni di innovazione e ricerca a supporto del piano nazionale sementiero*. PRIS2. Litograf Editor, Italy, pp. 531–540.

Falcinelli, M. and Lorenzetti, S. (2008) Le varietà locali presenti in Italia e la loro salvaguardia per la difesa della biodiversità nazionale. In: *Rete Interregionale per la Ricerca Agraria, Forestale, acquacoltura e pesca, Azioni di innovazione e ricerca a supporto del piano nazionale sementiero*. PRIS2. Litograf Editor, Italy, pp. 503–508.

Falcinelli, M. and Torricelli, R. (2001) Costituzione di una varietà di erba medica per l'agricoltura biologica. *Sementi Elette* 47, 36–39.

Farina, A. (1995) Upland farming systems of the Northern Apennines. In: Halladay, P. and Gilmour, D.A. (eds) *Conserving Biodiversity Outside Protected Areas: The Role of Traditional Agro-ecosystems*. IUCN, Gland, Switzerland, and Cambridge, UK, pp. 123–135.

Gargano, D. (2011) Verso la redazione di nuove Liste Rosse della flora d'Italia: una griglia standard per la misura dell'*Area of Occupancy (AOO)*. *Informatore Botanico Italiano* 43, 455–458.

Marino, M. (2013) The new Italian guidelines for conservation and characterization of animal, microbial and plant biodiversity of agricultural interest. *Landraces* 2. Available at: pgrsecure.org/helpdesk_lr ISSN: 2281-2067 (accessed 22 December 2014).

Negri, V. (2003) Landraces in central Italy: Where and why they are conserved and perspectives for their on-farm conservation. *Genetic Resources and Crop Evolution* 50, 871–885.

Negri, V. (2012) Policies supportive of on-farm conservation and their impact on custodian farmers in Italy. In: Padulosi, S., Bergamini, N. and Lawrence, T. (eds) *On-farm Conservation of Neglected and Underutilized Species: Status, Trends and Novel Approaches to Cope with Climate Change*. Proceedings of an International Conference, Frankfurt, 14–16 June 2011. Bioversity International, Rome, pp. 211–216.

Negri V. and Torricelli R. (2013) Regional laws protect genetic resources in Italy. *Landraces* 2. Available at: pgrsecure.org/ helpdesk_lr ISSN: 2281-2067 (accessed 22 December 2014).

Negri, V. and Torricelli, R. (2014) The Italian landrace conservation strategy. Available at: www.pgrsecure.org (accessed 22 December 2014).

Negri, V., Barocco, R., Pacicco, L., Veronesi, F. and Venanzoni, R. (2012) Landraces in Europe: an approach towards identifying landrace-rich areas as a priority for protection. In: Maxted, N., Dulloo, M.E., Ford-Lloyd, B.V., Frese, L., Iriondo, J.M. and Pinheiro de Carvalho, M.A.A. (eds) *Agrobiodiversity Conservation: Securing the Diversity of Crop Wild Relatives and Landraces*. CAB International, Wallingford, UK, pp. 118–124.

Negri, V., Pacicco, L., Bodesmo, M. and Torricelli, R. (2013) The first Italian inventory of *in situ* maintained landraces. Morlacchi Editore, CD-ROM multimediale. Available at: vnr.unipg.it/PGRSecure/start.html (accessed 22 December 2014).

Negri, V., Torricelli, R., Maxted, N. and Heinonen, M. (2014) European landrace conservation strategy. Available at: www.pgrsecure.org/helpdesk_lr (accessed 22 December 2014).

Pavia, R., Barbagiovanni, M.I., Della Strada, G., Piazza, M.G., Engel, P., *et al.* (2007) Autochthonous fruit tree germplasm at risk of genetic erosion found in home gardens in the region of Latium (Italy). In: Bailey, A., Eyzaguirre, P. and Maggioni, L. (eds) *Crop Genetic Resources in European Home Gardens*. Proceedings of a Workshop, 3–4 October 2007, Ljubljana, Slovenia. Bioversity International, Rome, pp. 21–25.

Shannon, C.E. and Weaver, W. (1949) *The Mathematical Theory of Communication*. University of Illinois Press, Urbana, Illinois.

Spataro, G. and Negri, V. (2013) The European seed legislation on conservation varieties: focus, implementation, present and future impact on landrace on-farm conservation. *Genetic Resources and Crop Evolution* 60, 2421–2430.

Spellerberg, I.F. and Fedor, P.J. (2003) A tribute to Claude Shannon (1916–2001) and a plea for more rigorous use of species richness, species diversity and the 'Shannon-Wiener' index. *Global Ecology and Biogeography* 12, 177–179.

Vetelainen, M., Negri, V. and Maxted, N. (2009) A European strategic approach to conserving crop landraces. In: Vetelainen, M., Negri, V. and Maxted, N. (eds) *European Landraces: On-farm Conservation, Management and Use*. Bioversity Technical Bulletin 15, Bioversity International, Rome, pp. 305–325.

# 30 Chickpea Wild Relatives and Landraces of Georgia

A. Korakhashvili*
*National Academy of Sciences of Georgia, Tbilisi, Georgia*

## 30.1 Introduction

In Georgia, the chickpea (*Cicer arietinum* L.) is a traditional grain legume crop influencing the food supply chain, especially in warmer and drier regions. This crop has been a dominant legume species in Georgian agriculture during the past 30 centuries, but no longer is so, being replaced by other grain legumes such as bean and soybean. Nevertheless, demand for chickpea is growing here at present, because of the good nutritive value of the seeds and the recently changing climate creating favourable conditions for chickpea cultivation. Food legumes in Georgia, though secondary to cereals in terms of production and consumption, have an important role in both sustainable crop production systems and human nutrition. They form an important component of the population's diet in the developing countries of Transcaucasia. Chickpea is not only a source of human food but also is a powerful tool for restructuring the arable lands of Georgia and heightening the fertility of soils. Our forefathers knew well the high nutritive value of this crop and its positive effect, including its medicinal characteristics, on humans and their domestic livestock, who consumed the crop green or as straw.

## 30.2 Active Conservation of Georgian Plant Genetic Resources

Georgia's genetic resource programme of grain legumes is committed to collecting, evaluating, documenting and renewing their biodiversity, to use for the genetic enhancement of legumes (Bedoshvili, 2010). This diversity occurs in two forms – wild relatives of legume species and farmer-selected crop landraces. The latter are the result of thousands of years of risk-averse selection by the farmers themselves, but the wild relatives are equally important, for they are constantly adapting in order to survive changes in climate, soil, pollution levels, exposition and other aspects of their environment. It is not only chickpea and other legumes species that are diverse. Much of the land area of Georgia was once range and shrub land. Even now, there is a vast array of species and ecotypes of annual legumes capable of surviving some of the most arid (east Georgia) and humid (west Georgia) conditions.

Each country has an obligation to retain its plant genetic resource diversity as part of its natural wealth *in situ* and under *ex situ* conditions. This is particularly important in Georgia, because it is home to the primary and secondary

*E-mail: akorakhashvili@yahoo.com

hearths of cultural plant origin. Georgia is a centre of origin for barley, millet, flax, vetch, pea, chickpea, lucerne, clover and fruit, being first collected extensively in the 1920s and 1930s. Significant *in situ* diversity still remains in Georgia, as evidenced by more recent field surveys (Jorjadze and Berishvile, 2009). By the resolutions of the Academy of Agricultural Sciences and the Ministry of Agriculture of Georgia, the Georgia Agrarian University (GAU) has been carrying out re-collection and *ex situ* storage, including surveys of local legumes' germplasm, collection, conservation and enhancement, particularly of chickpea. These works are implemented mainly through the financial support of international institutions such as ICRISAT, GTZ, ACDI-VOCA, ICARDA, WB, etc. As a result of such activity, we have collected 692 samples of food and feed legumes over the past 12 years: soybean, 68; faba bean, 24; chickpea, 46; lentil, 36; pea, 7; haricot beans, 83; as well as *Trifolium* spp., 89; *Astragalus*, 22; *Medicago*, 88; *Onobrychis*, 46; *Melilotus*, 19; *Lupinus*, 16; *Galega*, 19; *Vicia*, 112; and *Lathyrus*, 17.

## 30.3 Chickpea Conservation

The chickpea genetic resources collection is maintained in the genebank of the Agrarian University of Georgia. This institution is mandatorily responsible for legumes in Georgia. A set of 46 chickpea accessions is maintained in the genebank. In 2011–2012, these accessions were regenerated (as a winter and spring crop) using advanced growing technology of seed pilling in accordance with a special plan for regeneration, enhancement and protection. The level of genetic diversity and the effectiveness of the new germplasm acquired to enrich the collection were evaluated. All accessions needed regeneration and duplication abroad. This process resulted in the release of three new chickpea varieties: Zedazeni, Aragvi and Eliksiri.

The GAU has been engaged in food legume collection and breeding for over 80 years, with programmes for their improvement. The national agricultural research systems of Georgia have made efforts to increase the production of food legumes both vertically (through high inputs, including irrigation) and horizontally (increasing land area). Recently, however, national programme scientists and research administrators in Georgia have realized that diversification is needed in a legume-based, sustainable cropping system. In 2012, regenerated material was obtained from all chickpea accessions. Recently acquired new germplasm contributed the entire new entire collection; that is, almost 100% of the diversity collected originated from the recently acquired accessions. Field and laboratory analysis revealed the importance of collecting missions to the enrichment of collections.

The overall objective of the food legume germplasm regeneration programme is for the future improvement programmes of Georgia (chickpea, lentil, bean and soybean) to increase the productivity and sustainability of the farming system in a partnership between the AUG and Georgian farmers. This objective is pursued through: improving the methodologies for food legume breeding; training of AUG staff in the renewal of germplasm; the improvement of production practices and genetic stocks with resistance to various biotic and abiotic stresses; sharing the improved germplasm between the AUG and farmers; and its evaluation under targeted environments for identifying improved lines for future use. During the same year, the 46 chickpea accessions were evaluated for various biotic and abiotic stresses in varying agronomic backgrounds. The improved genetic materials were shared in the form of three different sets of trials in three different nurseries. The decentralization of breeding activities with ICARDA in the South Caucasus and Central Asia resulted in the identification of promising lines of chickpea. In accordance with International Plant Genetic Resources Institute (IPGRI) descriptors, safety cleaned and packaged new germplasm was sent to the genebanks of the AUG (Tbilisi, Georgia), ICARDA (Aleppo, Syria) and Svalbard Global Seed Vault (Spitsbergen, Norway) for the benefit of present and future generations.

## 30.4 Conclusion

A successful germplasm enhancement programme requires the development of an international research infrastructure with the participation of European countries, including Georgia. The opportunity for interaction afforded by international networks will be invaluable in developing a long-term plan for applying classical and contemporary techniques effectively to the germplasm needs of chickpea producers. The availability of appropriate information on the chickpea germplasm of Georgia is a critical success factor in breeding and the biodiversity protection factor in the agri-food sector's global responsibility for the production of such food and the preservation of the quality of natural environments. We understand that free exchange of chickpea germplasm on a worldwide basis between plant scientists is vital, as genetic diversity is the foundation of all plant improvement programmes. The integration of European countries in The European Cooperative Programme for Plant Genetic Resources (ECPGR) Working Groups provides a good opportunity for European scientists and professionals who are engaged in the collection, systematization, development and use of modern chickpea diversity, which could support successful wild relative and landrace breeding work and agrobiodiversity protection for the development of sustainable agriculture.

## References

Bedoshvili, D. (2010) National Report on the State of Plant Genetic Resources for Food and Agriculture in Georgia. Food and Agriculutre Organization of the United Nations, Rome.

Jorjadze, M. and Berishvile, T. (2009) On-farm conservation of crop landraces in Georgia. In: Veteläinen, M., Negri, V. and Maxted, N. (eds) *European Landraces: On-farm Conservation, Management and Use*. Bioversity International Technical Bulletin 15. Bioversity International, Rome, pp. 244–250.

# 31 Landrace Inventories and Recommendations for *In Situ* Conservation in Finland

**M. Heinonen***

*Natural Resources Institute Finland, Genetic Diversity, Jokioinen, Finland*

## 31.1 Need for Landrace Inventory

In Finland, landraces and local strains are still cultivated to some extent, especially landraces of cereals, forages, fruit, berries and some vegetables. However, there are no comprehensive statistics on landrace cultivation. Since 2000, a decree for conservation varieties, their seed production, approval and marketing has been applied to cereals, forages, pulses and some other arable land crops in Finland. This was the first European support system for the on-farm cultivation of arable landraces and old cultivars (Paavilainen, 2009). Today, there are 29 registered conservation varieties. Furthermore, eight local strains of forages and one landrace potato have been accepted to the National List of Plant Varieties. (Evira, 2014). Many papers specifically referring to Europe (e.g. Veteläinen *et al.*, 2009) have stressed the need for a landrace *in situ* inventory. In addition, international policies and strategies for a sustainable use of plant genetic resources for food and agriculture (PGRFA) have also emphasized this need.

In Finland, the first collecting missions were organized by professional plant breeders, who collected and studied an extensive amount of landrace samples for breeding material during the early 1900s. The Nordic Center for Plant Genetic Resources (NordGen) conducted its inaugural landrace collecting missions of cereals and forages during the late 1970s and early 1980s in Finland. Later, in the mid-1990s and 2000s, cereal landrace inventories were conducted by the Seed Testing Department and MTT Agrifood Research Finland (Heinonen and Veteläinen, 2009).

The aims of the current Finnish landrace inventories were to complete the national *ex situ* collections with variety proved new accessions, to improve the quality of reference material (both genotype and phenotype identifiers) for the future use of variety verification, to collect farmer knowledge and to promote the continuity of landrace growing. This chapter provides a description of the Finnish landrace inventory process applied. We present the general inventory results, as well as discuss recommendations for national landrace *in situ* conservation drawn from the inventory experience.

## 31.2 Landrace Inventory Methods

During 2011–2014, landrace *in situ* inventories were carried out by MTT Agrifood Research

*E-mail: maarit.heinonen@luke.fi

Finland. The target taxa were apple (*Malus domestica*), potato onion (*Allium cepa* Aggregatum group) and cereals (winter and spring rye (*Secale cereale*); spring barley (*Hordeum vulgare*); winter and spring wheat (*Triticum aestivum*); and oats (*Avena sativa*)). The data of conservation varieties and landraces accepted in the National List of Plant Varieties were also collated from the Finnish Food Safety Authority register (Evira, 2014).

Different inventory methods were developed depending on the target taxa and the facilities for variety identification. The shared method with all inventories was to prepare and release public calls or announcements to find farmers and gardeners with potential landraces to obtain plant material for further evaluation. This was an essential starting point for all target taxa because there was only one limited register of landrace growers: the register of conservation varieties. Landrace call materials were prepared for both nationwide and local or territorial search missions. The latter approach of restricted geographical location was especially workable because it attracted the media, which broadcast the calls for plant materials, and which proved to be time-efficient and cost-effective. The media was especially active in spreading information about local apple varieties, and according to the media follow-up in MTT Agrifood Research Finland, there have been well over 100 news items and articles in the Finnish media. We had the unique possibility to collaborate also with national audiovisual media, especially with calls for potato onions and cereal landraces, and to plan and execute two short (10–30 min) episodes in one magazine programme and one cooking programme. Call materials consisted of press releases to the media, short articles for publication in local print and digital media and trade magazines, and posters offered to public events and websites.

Call materials contained landrace descriptions, as detailed as possible to limit the amount of public response. We succeeded in the requests for potato onions and were contacted about plant material of potato onions only, no other types of onions. On the other hand, calls for local apple varieties produced a massive amount of contacts (over 500) about old apple trees, the informants of which had very little or no knowledge of their origin (of seed planted, foreign variety or local variety). Neither MTT Agrifood Research Finland (currently Natural Resources Institute Finland) nor any other expert organization in Finland had the resources nor the methods available to evaluate them all.

The most important reference for the landrace inventory process, both for call materials and for variety verification, has been phenotypic and other written data dating from the early 20th century, usually collected by the plant breeders of that time. Local history data (especially family, village), printed in newspaper articles or books, and old photographs were also very useful, if available. The problem with using the latter source of material is that collecting and analysing it is especially time-consuming. During the inventory projects, a web-based announcement form for all PGRFA important taxa was designed and released on the public website of the Finnish National PGR Programme for Agricultural and Horticultural Plants (MTT, 2014a).

All types of landrace calls requested farmer or gardener knowledge on the origin, cultivation history and description of their potential landraces as well as farmer background data.

These preliminary data formed the basis for the detailed studies. The landraces with most potential were selected by evaluating the material provided against the old literature and other landrace knowledge available and, most importantly, against farmer information about his or her landrace (see the descriptors for *in situ* inventories; Negri *et al.*, 2012). All inventories aimed to identify a landrace, distinguishing it from other landrace strains, as well as bred and foreign cultivars.

After releasing the calls for potato onions known to have been grown for at least 50 years, we received 45 contacts from potato onion growers from throughout Finland. We asked these growers to send a couple of bulbs as a sample for DNA analysis and 41 samples were received. The bulbs were planted in pots and grown in a glasshouse to obtain a couple of centimeters of green shoot for DNA extraction.

Twenty-nine accessions of potato onions and six shallots (*A. cepa* Ascalonium group) were accepted for long-term storage in national field collections. No identification of accession genotypes had previously been carried out. Consequently, the potato onions and shallots in the national field collection, as well as the potato onions still in cultivation, were analysed with nine microsatellite DNA markers developed for the *A. cepa* group (Fischer and Bachmann, 2000). Two bulb onions (*A. cepa* Cepa group) (one yellow and one red) were used as a standard. Altogether, 72 samples were analysed with nine microsatellite DNA markers. The DNA analysis results of the cultivated samples were compared with the results of the accessions in the national collection in order to estimate the genetic variation within the material and to see if any clones had been distributed to various locations.

While we searched for potato onions on the taxa level, we needed to approach the landrace *in situ* inventory of apples differently: on the local variety level. Fifty-one accessions of local apple varieties were accepted for long-term storage in the national field collection, and the identification of accession genotypes was carried out earlier. About 30 local apple varieties described in the pomological literature in the early 20th century were still missing from the collection. Furthermore, it is unclear whether all the varieties in the apple collection are genuine ones since name confusions are known to be common. The data of origin and morphological descriptions of the collection apples were, in many cases, limited and scattered.

Because the collection apple material had originally been collected mainly from nurseries, the collection holder had very limited grower contacts. Hence, the call materials needed to be based on the list of local apple varieties gathered from a range of the Finnish pomological literature of the late 19th and early 20th centuries. We listed 80 local apple varieties with their given names and their origin municipality, and we searched for original mother trees and/or old trees crafted from the mother tree of those varieties. During the inventory, we became aware of about 20 additional local apple varieties, especially originated from northern Finland, which had been not described in the old pomological literature but which met the prerequisites of the local variety (spread to local cultivation). We included those local varieties in the inventory.

In total, leaf samples from 268 separate apple trees were selected for DNA genotype analysis. On average, leaf and apple samples were needed from two to five separate trees per one variety studied, to compare and to find out which one was the genuine one if the mother tree was already dead. For the most part, we received leaf samples for DNA analysis and apple samples for phenotype analysis via mail posted by the growers, who had been given very detailed instructions in order to avoid unsuitable samples. We visited 75 old gardens from where the local apple variety was known to originate or where an old tree of that potential variety existed in order to interview the growers as well as to gather leaf and apple samples and visual material (photographs; observation). Leaf samples were used for DNA extraction and were analysed with eight microsatellite DNA markers (Liebhard *et al.*, 2002). The DNA analysis results of *in situ* samples were compared with the results of the accessions in the national collection and with morphological descriptors.

For the cereal landrace *in situ* inventory, we updated the inventory made during 2006–2008, when we received 46 notifications of landraces and old bred cultivars (Heinonen and Veteläinen, 2009). The seed material received from farmers was evaluated for phenotype traits by specialists on cereal genetic resources. No DNA analysis was available. During the re-inventory, farmer descriptor data (Negri *et al.*, 2012) were collected. To find new farmer contacts, as a part of a 30 minute TV cooking program, we made a request with media professionals for cereal landraces. We received ten new contacts, three of which were potential landraces (rye) and the remainder were old cultivars. The phenotype evaluation awaits further resources.

## 31.3 Major Inventory Results

In total, 22 different genotypes (clones) of potato onions were found by DNA analysis. Sixteen of these were already in the national field collection. About half of the collection accessions were overlapping clones (duplicates). Among the cultivated samples, there were six genotypes that did not exist in the national collection but which have now been included. By tradition, potato onion clones were handed out to other home gardeners. *In situ* data include two particularly widespread potato onion landraces: the northern and the eastern types. Additionally, four other landrace potato onion types have been cultivated in at least three different localities (municipalities). The DNA results offer a proper basis for drawing up morphological descriptions of potato onion varieties for use by the national collection, and especially for the needs of potential commercial use in the future. Noteworthy was also the fact that the onion material we received from growers was all from potato onions and not, for example, from shallots. Among the target taxa, the potato onion call and landrace verification was the most straightforward, since no plant breeding programme of potato onions has taken place in Finland, and farmer knowledge verified that all the clones had been cultivated for over 50 years in the same garden or farm.

The opposite was the case with local apple varieties, since there has been a centuries long tradition of cultivating foreign varieties (especially Russian and Swedish) and also of seed planting in Finland. In the landrace apple inventory, the first task was to identify the difference between a seed-planted mother tree in one garden and those that had been spread by local cultivation as clones. After doing this, we were ready to collect diverse written and oral material and to test it against DNA analysis. Due to this laborious process, the apple inventory has produced the results of variety verification for 70 local apple varieties, but about 20 candidates were not able to be verified with available data. We have been unable to locate nine of the local apple varieties listed because we have received no grower information on them.

The local apple variety inventory revealed several missing accessions in the national field collection, and also some accessions that were not variety verified or were the wrong ones (name confusion). The inventory also yielded phenotype and genotype identifiers of every local apple variety for the purposes of conservation and sustainable use. Farmer knowledge and historic origin data have also been added as essential background data to these identifiers. The list of home gardeners (over 100) and the locations of the variety verified old apple trees will also serve as a backup resource for the national collection. The completed DNA reference data of local apple varieties will enable future inventories to count them *in situ*. The inventory showed the considerable interest of citizens towards old apple trees but highlighted their weak and limited knowledge of their origin.

The current cereal inventory has added only a few new farmers to the contact list of the earlier inventory in 2008. The call material may not have reached enough farmers, but the number of landrace cereal farmers seems to be minor and it may not be possible to find many more extant landraces.

Those landraces registered as conservation varieties or accepted on the Finnish Plant Variety List along with bred varieties have been identified as landraces, but not much other *in situ* descriptor information has been gathered; for example, farmer knowledge on the use of the yield.

The inventories produced results of 154 landraces, of which 106 have been variety verified by the inventory projects (Table 31.1). Some of them, especially local apple variety clones and landrace strains of grass plant crops, are rather commonly cultivated throughout Finland, in particular in central and eastern parts.

## 31.4 Recommendations for Finnish *In Situ* Landrace Conservation

On the basis of the information gathered in the inventory projects, recommendations for the landrace *in situ* conservation strategy for Finland were drawn up.

**1.** Complete national *in situ* inventory for all crops, including ornamental plants: an extensive view of landraces in cultivation is essential for the implementation of the Finnish National PGR Programme for Agricultural and Horticultural Plants; to enable support and enhance *in situ* management and use. Landrace *in situ* inventories need to be extended and more crops involved. More effort needs to be put into collecting farmer- and/or gardener-based knowledge on their landraces, cultivation practices and crop use to complete the landrace knowledge base. There is also a requirement for the DNA identification of genotypes to help in variety verification, especially of vegetatively propagated crops. DNA marker profiles and reference data need to be broadened on two levels: (i) among crops already studied (potato onion, fruit, berries, potato) by incorporating new samples of genotypes, and (ii) by introducing new

**Table 31.1.** Number of identified and variety verified landraces on farms/in gardens in Finland. (From Heinonen, 2014.)

| Crop vernacular name | Scientific name | Total number of *in situ* accessions | Registered as conservation variety | Accepted to the Finnish Plant Variety List |
|---|---|---|---|---|
| Potato onion | *Allium cepa* Aggregatum group | 41[a] | 0 | 0 |
| Swede | *Brassica rapa* var. *napobrassica* | 2 | 0 | 1 |
| Turnip (in slash-and-burn cultivation) | *Brassica rapa* subsp. *rapa* | 2 | 2 | 0 |
| Oat | *Avena sativa* | 2 | 1 | 0 |
| Red and white clover | *Trifolium* spp. | 10 | 7 | 3 |
| Barley | *Hordeum vulgare* | 1 | 1 | 0 |
| Timothy | *Phleum pratense* | 2 | 0 | 2 |
| Rye | *Secale cereale* | 22[b] | 8 | 1 |
| Apple | *Malus domestica* | 70[c] | 0 | 0 |
| Potato | *Solanum tuberosum* | 2 | 0 | 1[d] |

*Notes*: [a]Includes several of the same clones. In total, 22 different clones identified. [b]Includes some duplicates (i.e. same landrace is cultivated on two or three farms). [c]These are mother trees or old clones; one accession per local variety. In total, about 100 landrace apple varieties were identified; about 20 samples are still under variety verification (morphological and DNA fingerprinting). [d]Landrace 'Puikula' cultivated and packed in Lappland has been awarded the EU quality mark 'Protected destination of origin (PDO)' since 1997.
Note that except for some potato onions and rye *in situ*, all inventoried landraces are extant in their original area (see Negri *et al.*, 2014) and no duplicated accessions have been included in the table.

crops (especially vegetables, ornamental plants) for DNA identification. Local and territorial NGOs (e.g. associations of local heritage, rural and home economics, youth and museums) are potential collaborators in collecting local landraces for evaluation and also in preserving them locally in living plant collections and on farms and in gardens.

**2.** Promote safe landrace conservation: more efforts are needed to persuade farmers, especially young farmers, in active landrace *in situ* management. On the one hand, the support programme for conserving landrace varieties needs to be less bureaucratic and costly in order to attract more farmers, and on the other hand there needs to be more informal support tools (farmer networks, etc.). Some support tools for horticultural in-garden management also need to be established. An in-garden network will also serve as an essential backup reserve for field collections of horticultural crops that are increasingly threatened by predisposition of plant pests, viruses and bacteria (e.g. apple proliferation phytoplasma and *Erwinia amylovora*).

**3.** Promote sustainable and active use of landraces: in order to promote landrace use, more characterization and evaluation information is required. In addition to research on resistance and biotic stresses, more emphasis on nutrition and other quality properties of landraces is needed. In order to develop market opportunities for landrace niche products, promoting local food chains is required. We are also lacking knowledge on desirable consumer landrace products. Actions are needed to broaden the landrace user community to bring on board breeders and farmers/gardeners, and also, for example, schools, local museums, owners of public and private historic gardens, local heritage associations, rural and home economics associations and youth associations.

**4.** Promote landrace reintroduction from genebanks: the reintroduction of landraces back to on-farm management will diversify the genetic base of landraces held on farms. This will also benefit the readaptation of landraces to local environments after some time.

**5.** Continue promoting landrace awareness among stakeholders and the public: raising further awareness of the significance and value of landraces is needed. Both local and nationwide promotion events, for example seminars and public events ('ask and show landraces'), provide a forum for exchanging knowledge between scientists and local people, serving the aims of both finding new landrace accessions and of promoting their cultivation and use. Teaching and lecturing activities on landraces at universities and agricultural/horticultural schools educate the new generation to actively value and use them. The Finnish National PGR Programme for Agricultural and Horticultural Plants with experts is currently preparing learning materials to elementary, high and technical school levels of landraces aiming to get landrace knowledge into the official syllabuses of the Finnish education system. The Finnish landrace information website also needs to be improved by providing more taxa and detailed information (MTT, 2014b). Active awareness building also aims to gain more national support for research on landraces and for development work on their sustainable use.

## 31.5 Conclusion

The more verified and diverse scientific, farmer and consumer knowledge we have on national landraces maintained *in situ* and *ex situ*, the higher will be their value, and this also promotes their use. Furthermore, it assists in meeting the current EU genetic resources policy, which is shifting from conservation to sustainable use of agricultural genetic resources (European Commission, 2013).

Currently, landraces are grown mainly for home consumption and only a few landrace-based products are available for consumers, with the exception of one potato landrace registered under the terms of Protected Designation of Origin (PDO) and the traditional turnip for slash-and-burn cultivation, which are retail traded nationwide. Although 29 seed-propagated landraces (cereals, grass plant crops, turnip type of slash-and-burn cultivation) are registered as conservation varieties, mainly these are cultivated on one single farm. Eight seed-propagated landrace varieties (mostly grass plant crops) have been accepted for the National List of Plant Varieties, and these have a larger cultivating area with many farmers. Twelve of the trademark FinE© (Finnish Elite) horticultural plant varieties, which have proved to be valuable in cultivation, originate from the old local varieties. They, as well as one-third of the landrace apple varieties, are widely marketed by nurseries, and are therefore common in home gardens. The total cultivation area of each landrace apple variety ranges from a handful of gardens to hundreds, or in some cases thousands, of gardens. A handful of horticultural farms have commercial apple production with landrace varieties.

Mainly landrace products are occasionally sold at local markets, events, small bakeries and agritourism farms as typical or niche products. However, Finns are more and more interested in landrace products. According to a recent national study (Tienhaara *et al.*, 2013), 70% of consumers are willing to buy, regularly or occasionally, landrace plant products: apples, bread and potatoes.

## Acknowledgements

The research has involved many experts providing practical work, advice and contacts. Particular gratitude is due to Merja Veteläinen (Boreal Plant Breeding Ltd), Kaarina Paavilainen (Finnish Food Safety Authority Evira), Elina Kiviharju (Finnish National Plant Genetic Research Programme for Agricultural and Horticultural Plants), and colleagues at MTT Agrifood Research Finland for practical work: Hilma Kinnanen, Kristiina Antonius, Pirjo Tanhuanpää, Ritva Valo, Jaana Ala-Kaarre, Leena Lohermaa, Sirpa Moisander, Merja Hartikainen and Terhi Suojala-Ahlfors. Very special thanks go to the numerous private home gardeners and farmers growing landraces, because without their knowledge and plant samples, this research would not have been possible.

The landrace *in situ* inventory and variety verifications were funded by a grant from the EU Seventh Framework Programme for the PGR Secure project (Grant Agreement No 266394), Finnish Cultural Foundation, Uusimaa Regional Fund, MTT Agrifood Research Finland and the Finnish National Plant Genetic Resources Programme for Agricultural and Horticultural Plants.

## References

European Commission (2013) Agricultural Genetic Resources – from conservation to sustainable use. Report from the Commission to the European Parliament, the Council and the European Economic and Social Committee. Brussels, 28.11.2013 COM (2013) 838 final. Available at: ec.europa.eu/agriculture/genetic-resources/pdf/com-2013-838_en.pdf (accessed 1 October 2014).

Evira, Finnish Food Safety Authority (2014) National List of Plant Varieties. *Finnish Plant Variety Journal* 2, 1–58.

Fisher, D. and Bachmann, K. (2000) Onion microsatellites for germplasm analysis and their use in assessing intra- and interspecific relatedness within the subgenus *Rhizirideum*. *Theoretical and Applied Genetics* 101, 153–164.

Heinonen, M. and Veteläinen, M. (2009) Cereal landrace inventories in Finland. In: Veteläinen, M., Negri, V. and Maxted, N. (eds) European landraces: on-farm conservation, management and use. Bioversity Technical Bulletin 15, pp. 70–78. Available at: www.bioversityinternational.org/index.php?id=19&user_bioversitypublications_pi1[showUid]=3252 (accessed 29 July 2015).

Heinonen, M. (2014) Landrace *in situ* Conservation Strategy for Finland. MTT Report 163, MTT Helsinki. Available at: urn.fi/URN:ISBN:978-952-487-570-7 (accessed 11 March 2016).

Liebhard, R., Gianfrancecshi, L., Koller, B., Ryder, C.D., Tarchini, R., *et al.* (2002) Development and characterisation of 140 new microsatellites in apple (*Malus* × *domestica* Borkh.). *Molecular Breeding* 10, 217–241.

MTT (2014a) Kasvigeenivarakuulutus (Call for plant genetic resources in situ. An electronic form. In Finnish.) Available at: www.luke.fi/kasvigeenivarat (accessed 29 July 2015).

MTT (2014b) *MaatiaisTietoPankki* (Web-based Finnish landrace information site. In Finnish.) Available at: www.luke.fi/kasvigeenivarat-->Maatiaistietopankki (accessed 29 July 2015).

Negri, V., Maxted, N., Torricelli, R., Heinonen, M., Veteläinen, M., *et al.* (2012) Descriptors for web-enabled national in situ landrace inventories. 18 pp. Available at: www.pgrsecure.bham.ac.uk/sites/default/files/documents/helpdesk/LRDESCRIPTORS_PGRSECURE.pdf (accessed 29 July 2015).

Negri, V., Torricelli, R., Maxted, N. and Heinonen, M. (2014) *Generic European LR diversity in situ (on-farm) conservation strategy.* PGR Secure D.4.5: 30 p. Available at: www.pgrsecure.bham.ac.uk/sites/default/files/documents/helpdesk/D4.4_European_specific_LR_conservation_strategy.pdf (accessed 11 March 2016).

Paavilainen, K. (2009) National policies and support systems for landrace cultivation in Finland. In: Veteläinen, M., Negri, V. and Maxted, N. (eds) European landraces: on-farm conservation, management and use. Bioversity technical bulletin 15, pp. 296–299. Available at: www.bioversityinternational.org/index.php?id=19&user_bioversitypublications_pi1[showUid]=3252 (accessed 29 July 2015).

Tienhaara, A., Ahtiainen, H. and Pouta, E. (2013) Consumer and citizen roles and motives in the valuation of agricultural genetic resources in Finland. Paper presented at European Association of Environmental and Resources Economists, 26–29 June 2013, Toulouse, France. Available at: www.webmeets.com/EAERE/2013/m/viewpaper.asp?pid=870 (accessed 29 July 2015).

Veteläinen, M., Negri, V. and Maxted, N. (eds) (2009) European landraces: on-farm conservation, management and use. Bioversity Technical Bulletin No. 15, Bioversity International, Rome. Available at: www.bioversityinternational.org/index.php?id=19&user_bioversitypublications_pi1[showUid]=3252 (accessed 29 July 2015).

# 32 Community Biodiversity Management (CBM): A Participatory Methodology that Integrates Empowerment, Livelihoods and On-farm Management of Agrobiodiversity

A. Subedi*
*Centre for Development Innovation, Wageningen University and Research Centre, Wageningen, the Netherlands*

## 32.1 Introduction

In the past two decades, a number of initiatives in various parts of the (developing) world focused on generating evidence of the successful implementation of *in situ* conservation of agrobiodiversity on-farm and translating the conservation practices that fit in the context of the livelihoods of small-scale, and often poor, farmers (Jarvis *et al.*, 2011). Over the years, it became clear that the dynamic nature of the management and utilization of plant genetic resources (PGRs) by farmers needed to support dynamic processes and team up with farming communities that had their own ideas and priorities in relation to agricultural development and the use of their genetic resources (de Boef *et al.*, 2012a).

Based on those experiences, development and conservation organizations now approach PGR conservation in the context of sustainable livelihood development, and food sovereignty. The initiatives used a multitude of crops, in dissimilar locations and situations. Each individual experience, while alone modest, contributes to a comprehensive body of knowledge and experiences on how to associate PGR conservation with farmers' livelihood development, which is commonly referred to as community biodiversity management (CBM) as the overarching methodology (Subedi *et al.*, 2005a; de Boef *et al.*, 2013).

CBM focuses on increasing the decision-making power of communities and securing access to and control over their biological and genetic resources for sustainable livelihood management. CBM methodology integrates local knowledge and practices and is driven by local social systems, local rules and institutions (Subedi *et al.*, 2005a). As part of the process, a local organization, either a non-governmental organization (NGO) or civil society organization (CSO), or government agency for rural development and extension services, engages itself in strengthening capabilities, supporting and mentoring decision-making processes and reinforcing local institutions, until they are fully autonomous in decision making at the community level, while considering agrobiodiversity as one of their important livelihood assets. Such autonomous decision making contributes to the empowerment of farming communities, which ultimately results in a situation where the community

*E-mail: abishkar.subedi@wur.nl

purposely and collectively manages its biodiversity in a sustainable manner. Within community biodiversity management, communities are approached as open systems, simply because of the multitude of scalar interactions with the dynamics and constantly changing patchwork of relations that define agrobiodiversity (de Boef *et al.*, 2013).

Despite a diverse array of specific historical, political, cultural, biological and agroecological contexts, it is possible to identify common conditions in which the CBM methodology contributes effectively to the empowerment of farming communities, and leads to *in situ* conservation or on-farm management of agrobiodiversity. Organizations involved in CBM may use dissimilar terminologies, but they often share the same goals.

However, many development and conservation organizations employ CBM as a means of contributing to sustainable livelihood development, and the *in situ* conservation of agrobiodiversity on-farm, its proven practices and the practical sharing of experiences concerning CBM, is limited. This gap is identified in the *Second Report on the State of the World's Plant Genetic Resources for Food and Agriculture* (FAO, 2010). Drawing on in-depth experiences from Brazil, Ethiopia, India, France, Nepal, and reference cases from several other countries, this paper provides an overview of CBM methodology and synthesizes the key role of the most commonly used participatory practices that integrate empowerment, livelihoods and on-farm management of agrobiodiversity.

## 32.2 Community Biodiversity Management (CBM)

CBM methodology consists of three major elements: (i) components or key steps; (ii) community-based practices; and (iii) participatory tools (Table 32.1). CBM components or key steps include the way that facilitating organizations contribute to the empowerment and autonomous decision-making process of farming communities. The second element is community-based practices, which are a basket of choices; based on effective implementation, the community can secure their livelihood by ensuring the on-farm management of agrobiodiversity. The third element is participatory tools, which are very short term and are used to implement community practices.

Table 32.1. Components, practices and tools associated with community biodiversity management. (From Sthapit *et al.*, 2008; de Boef and Thijssen, 2007.)

| Components/key steps | Community/participatory practices | Participatory tools |
|---|---|---|
| • Enhancing community awareness on conservation and diversity<br>• Increasing the community's understanding of diversity, social networks and organization<br>• Establishing institutional working modalities<br>• Strengthening the capabilities of community organizations<br>• Consolidating the community's role in planning and implementation<br>• Establishing a community biodiversity management fund<br>• Social learning and scaling-up for collective community action | • Diversity/seed fair<br>• Diversity block<br>• Diversity kit<br>• Diversity field flora<br>• Rural radio programme<br>• Community biodiversity register<br>• Participatory plant breeding (grassroots breeding, participatory variety selection, participatory genetic enhancement, participatory domestication)<br>• Community seed bank, community genebank, community seed reserve<br>• Value addition and marketing local agrobiodiversity products, local enterprises, local food cafe | • Four-cell analysis<br>• Diversity wheel<br>• Social seed network analysis<br>• Matrix ranking<br>• Preference ranking<br>• Resource map<br>• Rainbow diagram<br>• Venn diagram<br>• Stakeholder analysis and so on |

## 32.3 Community Practices that Enhance CBM

### 32.3.1 Diversity fair

The diversity fair is a popular practice employed by various organizations to sensitize communities to the value and importance of biodiversity and traditional knowledge (Rijal *et al.*, 2000). The diversity fair has its origin in the Andean countries (Tapia and Rosa, 1993), and since the early 1990s has spread globally as a practice used by organizations engaged in facilitating CBM for conservation and development. The latest global survey shows that the diversity fair is promoted largely by NGOs and UN agencies (de Boef *et al.*, 2012b).

Extensive experiences can be found in Nepal, where the diversity fair is a collective process and a competitive event during which farmers collect and display plant parts, seed, fruit samples and traditional food items. The organization of the fair begins with the formation or identification of participating farmers. Farming communities set the date and venue, and decide on what activities should be carried out to organize the fair. The farmers share biodiversity-related information and associated traditional knowledge (ATK), either orally, through illustrations, or through displays on their stalls. This practice fulfils multiple objectives, such as creating awareness, locating diversity and custodians, documenting information on ATK and exchanging seed/planting materials and knowledge. The diversity fair offers an easy way for PGR professionals to collect the seeds of local varieties for genebanks, which may serve for *ex situ* conservation or for further crop improvement (Shrestha *et al.*, 2013a).

### 32.3.2 Diversity kit

The 'diversity kit' is an assemblage of small quantities of seed that is selected by farmers, containing both local and improved varieties but without hybrid or transgenic seed. The diversity kit has been implemented in Nepal since 1998, to deploy the diversity, increase the local availability of the seed of neglected and underutilized crops, and increase the farmer's participation in the on-farm management project (Sthapit *et al.*, 2006). In a municipality, in the Santa Catarina State of Brazil, the diversity kit evolved in 2004–2009 as a collective practice of the community to regain local seed security, restoring traditional knowledge and culture and enhancing self-reliance on household food production, thereby restoring self-esteem and food sovereignty (Canci *et al.*, 2013). In Guraciaba, the diversity kit is made up of a box consisting of several small packets, containing at least four to eight different crop species and up to five varieties of each crop. Detailed information on all crop varieties was included, such as local name, use values and cultivation practices. Each kit box was labelled with pictures of local rice and bean crops, illustrating local food culture.

A survey conducted in Guraciaba showed that a total of 300 farming households were able to benefit from the diversity kit, obtaining access to more than 16 food crops and 53 varieties. The survey showed that before the kits were distributed, only 25% of farming households were growing rice and 50% beans. On distribution of the kit in some of the communities, these numbers increased to 58% for rice and to approximately 90% for beans. In addition to this, more than 35% of households interviewed considered they were able to produce healthy food again at the household level (Vicente *et al.*, 2010).

### 32.3.3 Community Biodiversity Register

A Community Biodiversity Register (CBR) refers to a record maintained by community members of the genetic resources (GRs) in a community, including information on the custodians, passport data, agroecology and cultural and use values. It is a community-level information database recorded in a register for keeping an inventory of the biodiversity and traditional knowledge, and for monitoring the local crop diversity (Subedi *et al.*, 2005b). The concept started as a People's Biodiversity Register (PBR) in India in early 1995 (Utkarsh, 1999). After this, diverse types of institutions used the CBR method for different purposes, and as a consequence, the methodologies for CBRs have evolved in varying ways. A critical review of the existing practices in implementing a CBR has revealed two distinct types: (i) documentation of biodiversity at the local political unit level, assisted by government, universities or NGOs, with objectives to reaffirm the sovereign rights and as evidence of prior knowledge of the country of its biological resources; and (ii) a CBR as a participatory practice to empower local communities to document their important genetic resources and traditional knowledge for developing community-based conservation and development plans that enhance their livelihoods (Subedi *et al.*, 2005b). In the first type of CBR, we found a successful case example in India. The Plant Biodiversity Register has been recognized by national policy under the Biological Diversity Act, 2002, and the Biological Diversity Rules, 2004. Until August 2014, India had documented successfully 1863 PBRs from 13 different states (NBA, 2014). In the second type of CBR, a successful example was found in Nepal, where CBR implementation began in 1998 under the on-farm management of an agrobiodiversity project where community-based organizations (CBOs) played a key role in developing local conservation with a plan to conserve local crop varieties, the wild relatives of cultivated crops and wetland biodiversity (Subedi *et al.*, 2005b).

### 32.3.4 Participatory plant breeding

Participatory plant breeding (PPB) is the overarching terminology that covers a diversity of methods in which farmers and scientists collaborate in plant breeding or become involved in the identification of superior varieties, with the basic objective of addressing more effectively the needs of farmers in the marginal areas of developing countries (Witcombe and Joshi, 1995; Ceccarelli *et al.*, 1996; Morris and Bellon, 2004; Almekinders *et al.*, 2006). Based on the degree of collaboration between farmers and scientists, various methods of PPB are being used to meet the diverse objectives of farmers and the suitability of the reproduction system of the targeted crops, which is illustrated in Table 32.2. The most common PPB methods are described briefly below.

**Table 32.2.** Participatory plant breeding (PPB) methods categorized by the degree of collaboration between farmers and scientists. (From Morris and Bellon, 2004; de Boef *et al.*, 2013.)

| Key steps | GB | PGE | PVS | PD | PPB |
|---|---|---|---|---|---|
| Diversity assessment | F, S | F, S | F, S | F, S | F, S |
| Setting breeding goals | F, S | F, S | F, S | F | F, S |
| Selection of parents | | | | S | F, S |
| Making crosses | | | | F, S | S |
| Early selection | | | | S | F, S |
| Advance selection/ trials | F, S | F, S | F, S | F, S | F, S |
| Variety release | F, S | F, S | | | F, S |
| Seed production | F | F | F | F | F |
| Dissemination | F | F | F | F | F |

*Notes*: F = farmers; S = scientists; GB = grassroots breeding; PGE = participatory genetic enhancement; PVS = participatory variety selection; PD = participatory domestication; PPB = participatory plant breeding.

#### *Grassroots breeding (GB)*

A method in which breeders strengthen the skills of farmers in the selection, seed production and marketing of neglected and underutilized crops; that is, those crops and species for which no or very limited capacity exists within formal breeding programmes. GB is applied in situations where the production systems are rich in genetic diversity but many individual farmers have problems in terms of its access, availability and quality. Sthapit *et al.* (2013) reported a detailed experience of GB from Nepal in a number of crops; for example, sponge gourd, taro, cauliflower, rice bean, chilli and rice. The implementation of GB consisted of seven steps, namely: the participatory rural appraisal of local PGRs; the assessment of farmer-preferred traits to determine available variability; the selection of traits preferred by farmers and scientists; the selection of varieties or species from the target environment; community-based seed multiplication of selected varieties/species; distribution of seed/planting material; and monitoring effectiveness at the household and community level.

#### *Participatory genetic enhancement (PGE)*

The aim of PGE is to select the farmer- or market-preferred variety within the existing diversity of a specific crop. PGE includes the process of participatory survey and selection up to the identification of more stable and uniform varieties. Examples of PGE can be found in maize (Brazil, France and Guatemala), rice (India, Nepal) and beans (Nicaragua) (de Boef *et al.*, 2013). PGE-led varieties are released formally and seed production may have a more formal set up, as in the case of rice in Nepal (Gyawali *et al.*, 2010), or the varieties may return to more informal seed systems and follow paths of seed dissemination similar to those of local varieties or the products of grassroots breeding (Lassaigne and Kendall, 2013).

#### *Participatory variety selection (PVS)*

PVS is a rapid and cost-effective way of identifying farmer-preferred varieties if a suitable choice of varieties exists. A successful PVS programme has four steps, namely identifying farmers' needs, searching for suitable material to test with farmers, experimenting on its acceptability in farmers' fields, and a wider dissemination of farmer-preferred varieties (Witcombe *et al.*, 1996). The purpose of PVS can be varied, such as the selection of superior varieties among the existing local varieties, or comparing improved varieties with local varieties or the selection of final products from the breeding programmes. Increasingly, PVS is becoming one of the final steps of a PPB programme to select farmer- or client-preferred stable breeding lines before the variety release application.

#### *Participatory domestication (PD)*

For perennial and often semi-domesticated species, such as tropical fruit, collaboration between scientists and farmers may result in a process that is referred to as participatory domestication (PD). Dos Santos *et al.* (2013) reported a detailed methodology of PD in pineapple guava (*Acca sellowiana*) from Brazil. PD implementation involves seven steps, namely: the development of a collaborative framework; the formation of a farmer working group; the assessment of production systems; the participatory assessment of market-oriented traits; developing a breeding strategy (such as grafting); mass selection; and on-farm evaluation.

#### *Complete participatory plant breeding (PPB)*

In PPB, farmers and breeders collaborate in most of the steps. Witcombe *et al.* (1996) suggested that PPB was more resource consuming. PPB needs to be used when the possibilities of PVS have been exhausted, or when the search process fails to identify any suitable varieties for testing. However, in recent years within PPB, a more effective division of responsibility has been sought to define what the farmer and breeder should contribute to increase breeding efficiency and minimize cost. A PPB-led rice variety was released in early 1996 in Nepal (Joshi *et al.*, 1997).

The use of participatory methods in plant breeding or selection of 'farmer-' or 'client-' preferred varieties is now becoming popular. National Agriculture Research Institutes (NARIs), universities, International Agriculture Research Centres (IARCs) and NGOs are the major implementers of PPB (de Boef *et al.*, 2012b). The available data show that at least 50 countries currently use various methods of PPB (Table 32.3).

### 32.3.5 Community seed bank

The concept of a community seed bank (CSB), begun in the early 1980s by civil society organizations' work in India to address the local conservation of PGRs, provides evidence that smallholder farmers are able to back up conventional genebanks, ensure local seed supply and promote food sovereignty (Mooney, 1992). These critical assumptions of the early 1980s are now proven as the key role of CSBs in many communities in the world (Shrestha *et al.*, 2013b). The latest case experiences show that CSBs are also found to be a reliable adaptation option to the local effect of climate change, such as in the case of Nicaragua, Honduras and Guatemala (hurricane and droughts) and Bangladesh (river floods). The involvement of various stakeholders in CSBs has resulted in different terminology in the course of CSB advancement to fulfil the different objectives of farmers or gardeners around the world. In many countries of the 'south', they are usually known as community gene or seed banks, where the seed users are smallholder farmers of the same village or same district, while in the 'north', they are often known as

Table 32.3. List of countries and crops where various methods of PPB are being used. (From FAO, 2010; data collected by the author.)

| Continents | Key crops | Countries |
|---|---|---|
| Africa | Maize, rice, sorghum, millet, barley, pulses, beans, Bambara groundnut | Algeria, Angola, Benin, Burkina Faso, Egypt, Eritrea, Ethiopia, Libya, Malawi, Mali, Morocco, Namibia, Senegal, Tunisia, Uganda, Zambia, Zimbabwe (17) |
| Asia | Barley, date palm, wheat, rice, maize, melon, grape, chickpea, lentil, sesame, cocoa, finger millet, vegetables, root crops | Azerbaijan, Bangladesh, Bhutan, China, India, Indonesia, Iran, Iraq, Jordan, Laos, Lebanon, Malaysia, Nepal, the Philippines, Sri Lanka, South Korea, Thailand, Vietnam, Yemen (19) |
| Europe | Potato, maize | Netherlands, Portugal, Turkey (3) |
| South America and Caribbean | Bean, maize, rice, sorghum, potato, cocoa, banana, coffee, pumpkin, pigeon pea, underutilized crops | Brazil, Bolivia, Colombia, Costa Rica, Cuba, Dominican Republic, Ecuador, Guatemala, Nicaragua, Venezuela, Peru (11) |

seed saver groups, associations or networks, where seed users are (hobby) farmers or gardeners who might be in different countries. The various types of CSBs we find include a community genebank, a community seed reserve, community seed huts, seed wealth centres, and so on.

CSBs in developing countries are largely community driven, capacitated in seasonal seed storage facilities, maintain both local and improved seed varieties and the seed transactions occur in non-monetary forms in the majority of cases. Most CSBs are established with donor funding support and are facilitated by CSOs and NGOs. Early exceptions are Ethiopia and Nepal, where national plant genetic resource programmes (NPGRPs) are directly involved in establishing CSBs. The latest survey showed that besides the dominant role of CSOs and NGOs, other regional/international organizations and UN agencies are increasingly spreading their support (de Boef *et al.*, 2012b). Currently, CSBs are practised in more than 26 countries in the world, representing Asia (Bangladesh, Bhutan, China, India, Malaysia, Philippines, Sri Lanka), Africa (Burkina Faso, Ethiopia, Malawi, Rwanda, South Africa, Uganda, Zimbabwe), Australia, Europe (France, Greece, Spain) and North and South America (Bolivia, Brazil, Canada, Guatemala, Honduras, Mexico, Nicaragua). Data collected from nine countries showed the existence of over 470 community seed banks, contributing to meet the local needs of farmers or gardeners and covering the majority of food security crops (Table 32.4). The semi-arid region of Paraíba in Brazil, which has the largest network with 205 CSBs, provides a source of seed to 7145 farming households (Table 32.5). In Nepal, the majority of CSB users are the economically poor category farmers of the village. The sustainability of CSBs is found to be a critical challenge in many parts of the world as donor funds come to an end. This is the case when large investment has been made in physical infrastructures set up, while little effort is made in building the capacity of local institutions. In Nepal, combined strategies found to be successful in developing sustainable CSBs include: developing capacitated local institutions to manage CSBs; establishing a community fund; selling improved varieties seed in cash to the community outside the village; and building strong partnerships with local government agencies, agriculture development programmes and NPGRPs.

### 32.3.6 Adding value and creating markets for local agrobiodiversity products

Farming communities are the custodians of agrobiodiversity, and when they engage in value addition and marketing of agrobiodiversity products, that is those based on the local crops and varieties that farmers cultivate and on the non-timber forest products they gather from their agricultural landscape, they obtain direct

Table 32.4. Status of community seed banks (CSB) in nine different countries. (From Ramprasad, 2007; Shrestha *et al.*, 2013b.)

| Country | Name of facilitating organization | Major crops included | Number of CSBs | Average number of members per CSB |
|---|---|---|---|---|
| Bangladesh | UBINIG | Rice, pulses, oilseeds, spices, leafy vegetables, fruit, timber plants, medicinal plants | 29 | 1300 |
| Brazil | ASA-PB | Maize, cowpea, lima bean, common bean | 205 | 20 |
| Canada | Seeds of Diversity | Vegetables, fruit, grains, flowers, herbs | 1 | 1400 |
| Ethiopia | IBC/EOSA | Wheat, chickpea, grass pea | 14 | 680 |
| France | AgroBio Périgord | Maize | 8 | 280 |
| India | MSSRF, Green Foundation and network, Deccan Development Society | Rice, millets, sorghum, pulses | 83 | 40 |
| Mali | USC | Sorghum, millet, cowpea, Bambara groundnut | 7 | 280 |
| Mexico | Instituto Simone Weil | Maize, vegetables, medicinal plants | 10 | 200 |
| Nepal | LI-BIRD, NARC, DOA, Oxfam Nepal, Parivartan Nepal | Rice, legumes, gourds, millets, vegetables | 115 | 95–360 |

Table 32.5. Role of community seed banks (CSBs) in Paraíba, Brazil. (From Londres and Almeida, 2009.)

| Season | Number of CSBs | Number of households | Total seed stock (t) | Number of municipalities |
|---|---|---|---|---|
| 1998–1999 | 62 | 1860 | 15 | N/A |
| 1999–2000 | 129 | 3838 | 66 | N/A |
| 2000–2001 | 220 | 6920 | 100 | 55 |
| 2002–2003 | 175 | 7441 | 78 | 51 |
| 2003–2004 | 205 | 7170 | 128 | 60 |
| 2004–2005 | 207 | 7145 | 161 | 56 |
| 2005–2006 | 228 | 6560 | N/A | 61 |
| 2007–2008 | 205 | 3730 | 56 | 60 |

*Note*: N/A = data not available.

economic benefits from maintaining diversity. A common feature among conservation and development organizations supporting community biodiversity management is that they strengthen the capabilities of farming communities in developing their institutional capacity, structures and linkages with the value chain of those potential agrobiodiversity products. The latest global survey shows that NGOs and CSOs are the key stakeholders that support the community in value addition and market development. The same survey also shows that private sector participation is noteworthy (de Boef *et al.*, 2012b).

Markets may be accessed for specific species, enhancing their value through emphasizing their association with a unique landscape or production system. An example can be seen with the marketing of the 'tea' leaves of native yerba mate plants (*Ilex paraguariensis*), which are harvested in collectively managed forests of the faxinais in Brazil and which obtain higher prices in nearby urban markets when compared with the cultivated product (Sedrez dos Reis *et al.*, 2013), or where coffee beans gathered from natural coffee populations in the Afromontane forests in Ethiopia target global premium markets (Feyissa *et al.*, 2013). Products may be specific varieties such as local varieties of rice, for example Tam Xoan aromatic rice in the Red River Delta in Vietnam (Thomas and Dao The Anh, 2013) and Pokhareli Jethobudho aromatic rice in the Seti river valley in Nepal (Gyawali *et al.*, 2010). In the Jeypore Tract of India, market linkages of Kalajeera rice contributed to increases in three dimensions – the area of production, the number of farmers growing and the grain price, which led to secure conservation of this specific local rice (Table 32.6).

Other examples address the processing of unique species and varieties, by farming groups or associated artisan food producers, into special products such as: flour or bread made from the Meunier d'Apt traditional wheat variety from Lubéron, France (Kastler and Moÿ, 2013); Andean blackberry marmalade and spicy pepper pastes produced by members of the conservation cooperative in Cotacachi and marketed at tourist markets in nearby Otavalo and Quito in Ecuador (Tapia and Carrera, 1993); umbuzada, a delicacy based on the fruits of the umbuzeiro tree and prepared with milk and sugar, marketed in nearby rural towns in the semi-arid region of north-eastern Brazil but also reaching people with north-eastern backgrounds as far afield as São Paulo (Peroni *et al.*, 2013).

All these products are produced by smallholder farmer associations, often groups of poor, women and/or tribal farmers, and are based on unique and often rare agrobiodiversity. Creating alternatives to the global agro-industrial food market system by stimulating local products for local markets, with perhaps traditional and even new products, and by targeting consumers who recognize and appreciate those products and their provenance, has created incentives for farming communities to continue to manage local agrobiodiversity (Jarvis *et al.*, 2011; de Boef *et al.*, 2013).

Table 32.6. Status and marketing of Kalajeera rice in the Jeypore Tract, India. (From MSSRF, 2009.)

| | Year | | | |
|---|---|---|---|---|
| Information | 2004 | 2006 | 2008 | 2010 |
| Cultivated area (hectares) | 12 | 41 | 50 | 71 |
| Number of households | 49 | 126 | 159 | 212 |
| Price of grain (US$/kg) | 0.20 | 0.30 | 0.30 | 0.34 |

### 32.3.7 Community biodiversity management fund

The main purpose of the community biodiversity management (CBM) fund is to promote collective actions for agrobiodiversity management and to enhance rural development through creating a local mechanism that provides easy access to financial resources (Shrestha *et al.*, 2013c). The CBM fund was piloted successfully in Nepal in 2002 to overcome the challenges in creating incentives for farming communities who are involved in an agrobiodiversity management project. The basic working modality of a CBM fund is similar to rural microcredit schemes. However, it has its own salient features, which are analysed briefly in Table 32.7. One of the conditions of accessing a CBM fund is that a community member who accesses the loan is required to conserve at least one of the most threatened crop diversities in the village. A CBM fund has two funding windows: (i) a revolving fund that provides loans to community members for small enterprise development, and this is the major portion of a CBM fund; and (ii) a seed fund that is non-refundable, awarded mainly for long-term conservation work such as supporting participatory plant breeding work in the village or rewarding custodian farmers.

Experiences from Nepal show that a CBM fund is a mechanism from which local institutions can generate regular income in the form of interest, with which they are able to continue their basic conservation work and livelihood development work when externally financed projects have been phased out. For instance, in Nepal, *Jaivikshrot Sanrachhan Abhiyan* in Begnas village and the Agriculture Development and Conservation Society in Kachorwa village have been able to mobilize more than US$12,000 and US$8000, respectively. Both organizations have their own staff and run programmes for

Table 32.7. Key features of CBM funds and rural microcredit schemes.

| Key features | CBM fund | Microcredit schemes |
|---|---|---|
| Key purpose | Support local income-generating activities and contribute to conserving local biodiversity | Rural development |
| Ownership of fund/scheme | Community members or community-based organization | Implementing financial institutions, government, NGOs/projects |
| Rules/guidelines | Each village or community group design their own rules/ guidelines | Same set of rules applies to a large area as developed by implementers |
| Priority | Rural poor, women or other deprived community members and custodians who maintain the rich agrobiodiversity | Rural poor, women or other deprived community members |

income-generating activities. They do not need external financial support to run their regular CBM activities (Shrestha *et al.*, 2012).

## 32.4 Conclusion

The importance of agrobiodiversity management and the sustainable use of agrobiodiversity are increasingly being recognized at the international level, and key policy developments to promote and regulate this area have taken place, including the Convention on Biological Diversity (CBD) and the International Treaty on PGRFA (ITPGRFA). The *Second Report on the State of the World's Plant Genetic Resources for Food and Agriculture* (FAO, 2010) highlighted an increase in the use of participatory approaches in implementing on-farm management projects. Scientific understanding of the on-farm management of genetic diversity has increased.

The latest global survey shows that the on-farm management of agrobiodiversity is currently not well reflected within many NPGRPs, and suggests that many development organization programmes are working outside the direct sphere of influence and control of national programmes (Linn *et al.*, 2014). Out of a surveyed 877 respondents, 18% of on-farm management practitioners are either unaware of or do not consider their work to be any part of (21%) the NPGRP. This is supported by NPGRP managers and policy makers, the majority of whom consider on-farm management to be only moderately (37%) or to a limited extent (36%) part of the NPGRP. On the other hand, 60% of NPGRP managers and policy makers claim to have a good or comprehensive knowledge of the work done by on-farm management practitioners. In contrast, over half of the on-farm management practitioners (56%) reported that they were unfamiliar with the NPGRPs in their country. This indicates that direct linkage between the groups is very limited (Linn *et al.*, 2014).

This analysis supports the hypothesis that CBM practices that contribute to on-farm management (OFM) are, to a large extent, supported by stakeholders who are not engaged directly in the conservation and use of PGRFA, and therefore are not associated with NPGRPs. This leads to the recognition of roles for new types of PGR stakeholders while developing NPGRP programmes and strategies.

The FAO (2010) report has indicated gaps in linkages with education institutes. This is mainly because higher education and professional development institutions that include the conservation and use of PGRs and biodiversity in their curricula depend primarily on the existing body of knowledge, which emphasizes *ex situ* conservation and scarcely incorporates approaches that address conservation and use within a livelihood and sustainable development setting. This bias needs to be addressed in the formation of future conservation and development professionals. At present, no specific methodology or practices have been put in place for preparing future conservation and development professionals with the right attitudes, skills and capabilities. Universities have a role to play in ensuring that future generations of these professionals are better equipped to take on the facilitation roles required to make

CBM work, while also contributing through research to the design of community-based strategies. The universities need to address the complex aspects of development studies, such as empowerment, sovereignty and resilience in their curricula (de Boef *et al.*, 2013).

Despite the enormous efforts from development organizations and international agricultural research organizations, wider impact has not yet been achieved. The compartmentalized approach needs to be broadened to integrate multi-stakeholder partnerships that promote new PGR governance structures between farming communities, development organizations, NPGRPs, agriculture universities, the national private sector and agriculture development programmes, which will promote food and nutrition security objectives that can lead to potential sea changes to mainstream CBM methodology that will enhance the livelihoods of farming communities, thereby contributing to the on-farm management of agrobiodiversity in the longer run.

## References


Almekinders, C., Hardon, J., Christinck, A., Humphries, S., Pelegrina, D., *et al.* (eds) (2006) *Bringing Farmers Back into Breeding: Experiences with Participatory Plant Breeding and Challenges for Institutionalization.* AgroSpecial 5. Agromisa, Wageningen, the Netherlands.

Canci, A., Guadanin, C.A., Henke, J.P. and Lazzari, L. (2013) The diversity kit: restoring farmers sovereignty over food, seed and genetic resources in Guaraciaba, Brazil. In: de Boef, W.S., Peroni, N., Subedi, A., Thijssen, M.H. and O'Keeffe, E. (eds) *Community Biodiversity Management: Promoting Resilience and the Conservation of Plant Genetic Resources.* Routledge, Abingdon, UK, pp. 32–36.

Ceccarelli, S., Grando, S. and Booth, R.H. (1996) International breeding programmes and resource-poor farmers: crop improvement in difficult environments. In: Eyzaguirre, P. and Iwanaga, M. (eds) *Proceedings of a Workshop on Participatory Plant Breeding,* 26–29 July 1995, Wageningen, the Netherlands. International Plant Genetic Resources Institute, Rome, pp. 99–116.

de Boef, W.S. and Thijssen, M.H. (2007) *Participatory Tools Working with Crops, Varieties and Seeds. A Guide for Professionals Applying Participatory Approaches.* Wageningen International, the Netherlands.

de Boef, W.S., Thijssen, M.H., Shrestha, P., Subedi, A., Feyissa, R., *et al.* (2012a) Moving beyond the dilemma: practices that contribute to the on-farm management of agrobiodiversity. *Journal of Sustainable Agriculture* 36, 788–809.

de Boef, W.S., Dulloo, E., Subedi, A. and Nilsen, L.B. (2012b) A preliminary analysis of the global survey: on-farm management of plant genetic resources for food and agriculture. Input to the technical workshop 'Towards the establishment of a global network for in situ conservation and on-farm management of plant genetic resources for food and agriculture (PGRFA)'. Available at: www.fao.org/fileadmin/templates/agphome/documents/PGR/ITWG/ITWG6/workshop/Preliminary_analysis-OFMSurvey.pdf (accessed 15 April 2015).

de Boef, W.S., Subedi, A., Peroni, N., Thijssen, M. and O'Keeffe, E. (2013) *Community Biodiversity Management: Promoting Resilience and the Conservation of Plant Genetic Resources.* Routledge, Abingdon, UK.

Dos Santos, K., Peroni, N., Guries, R.P., Donazzolo, J. and Nodari, R.O. (2013) Participatory domestication of the fruit tree species, feijoa (*Acca sellowiana*) in Brazil. In: de Boef, W.S., Peroni, N., Subedi, A., Thijssen, M.H. and O'Keeffe, E. (eds) *Community Biodiversity Management: Promoting Resilience and the Conservation of Plant Genetic Resources.* Routledge, Abingdon, UK, pp. 240–245.

FAO (Food and Agriculture Organization) (2010) *Second Report on the State of the World's Plant Genetic Resources for Food and Agriculture.* FAO, Rome.

Feyissa, R., Gezu, G., Tsegaye, B. and Dessalegn, T. (2013) On-farm management of plant genetic resources through community seed banks in Ethiopia. In: de Boef, W.S., Peroni, N., Subedi, A., Thijssen, M.H. and O'Keeffe, E. (eds) *Community Biodiversity Management: Promoting Resilience and the Conservation of Plant Genetic Resources.* Routledge, Abingdon, UK, pp. 26–31.

Gyawali, S., Sthapit, B.R., Bhandari, B., Bajracharya, J., Shrestha, P.K., *et al.* (2010) Participatory crop improvement and formal release of Jethobudho rice landrace in Nepal. *Euphytica* 176, 59–78.

Jarvis, D.I., Hodgkin, T., Sthapit, B.R., Fadda, C. and Lopez-Noriega, I. (2011) An heuristic framework for identifying multiple ways of supporting the conservation and use of traditional crop varieties within the agricultural production system. *Critical Reviews in Plant Science* 30, 125–176.

Joshi, K.D., Sthapit, B.R., Gurung, R.B., Gurung, M.P. and Witcombe, J.R. (1997) Machhapuchhre-3 the first rice variety developed through participatory plant breeding approach released for mid to high-altitudes of Nepal. *International Rice Research Notes* 22(2), 12.

Kastler, G. and Moÿ, A.C. (2013) The dynamic management and regional marketing of a local wheat variety by farmers in the Lubéron, France. In: de Boef, W.S., Peroni, N., Subedi, A., Thijssen, M.H. and O'Keeffe, E. (eds) *Community Biodiversity Management: Promoting Resilience and the Conservation of Plant Genetic Resources.* Routledge, Abingdon, UK, pp. 213–216.

Lassaigne, B. and Kendall, J. (2013) Farmer and participatory maize breeding: increasing farmers' autonomy and promoting the use of diversity in France. In: de Boef, W.S., Peroni, N., Subedi, A., Thijssen, M.H. and O'Keeffe, E. (eds) *Community Biodiversity Management: Promoting Resilience and the Conservation of Plant Genetic Resources*. Routledge, Abingdon, UK, pp. 246–252.

Linn, B.N., Subedi, A., Dullo, M.E., Ghosh, K., Chavez-Tafur, J., *et al.* (2014) The relationship between national plant genetic resources programmes and practitioners promoting on-farm management: results from a global survey. *Plant Genetic Resources* 12, 143–146.

Londres, F. and Almeida, P. (2009) *Impacto do controle corporativo no setor de sementes sobre agricultores familiares e sistemas alternativos de distribuição: estudo de caso do Brasil.* AS-PTA, Rio de Janeiro-RJ, Brazil. Available at: aspta.org.br/wp-content/uploads/2011/05/Estudo-Sementes-ASPTA-WoW-AA-2009-FINAL.pdf (accessed 21 December 2015).

M.S. Swaminathan Research Foundation (MSSSRF) (2009) *'Kalajeera' A Local Variety to Strengthen the Economy of Tribal Communities*. M.S. Swaminathan Research Foundation, Chennai, India.

Mooney, P.R. (1992) Toward a folk revolution. In: Cooper, D., Vellvé, R. and Hobbelink, H. (eds) *Growing Diversity: Genetic Resources and Local Food Security*. Intermediate Technology Publications, London, pp. 125–145.

Morris, M.L. and Bellon, M.R. (2004) Participatory plant breeding research: opportunities and challenges for the international crop improvement system. *Euphytica* 136, 21–35.

NBA (2014) *People's Biodiversity Register – Revised PBR Guidelines*. National Biodiversity Authority, India. Available at: nbaindia.org/content/105/30/1/pbr.html (accessed 21 August 2014).

Peroni, N., de Albuquerque, P., Luiza de Assis, A., Machado de Freitas, E. and Neto, L. (2013) The domestication of landscapes and cultural keystone species in a context of community biodiversity management, in Brazil. In: de Boef, W.S., Peroni, N., Subedi, A., Thijssen, M.H. and O'Keeffe, E. (eds) *Community Biodiversity Management: Promoting Resilience and the Conservation of Plant Genetic Resources*. Routledge, Abingdon, UK, pp. 145–150.

Ramprasad, V. (2007) Community seed banks for maintaining genetic diversity. *LEISA Magazine*, June 2007, p. 23.

Rijal, D.K., Rana, R.B., Subedi, A. and Sthapit, B.R. (2000) Adding value to landraces: community-based approaches for *in situ* conservation of plant genetic resources in Nepal. In: Friis-Hansen, E. and Sthapit, B.R. (eds) *Participatory Approaches to the Conservation and Use of Plant Genetic Resources*. IPGRI, Rome, pp. 166–172.

Sedrez dos Reis, M., da Silva, C.V., Mattos, A.G., Zechini, A., Mantovani, A., *et al.* (2013) Caívas and their contribution to the conservation of Atlantic forest landscapes in Brazil. In: de Boef, W.S., Peroni, N., Subedi, A., Thijssen, M.H. and O'Keeffe, E. (eds) *Community Biodiversity Management: Promoting Resilience and the Conservation of Plant Genetic Resources*. Routledge, Abingdon, UK, pp. 32–36.

Shrestha, P., Sthapit, S., Paudel, I., Subedi, S., Subedi, A., *et al.* (2012) *A Guide to Establishing a Community Biodiversity Management Fund for Enhancing Agricultural Biodiversity Conservation and Rural Livelihoods*. LI-BIRD, Pokhara, Nepal.

Shrestha, P., Vernooy, R. and Chaudhary, P. (2013a) *Community Seed Banks in Nepal: Past, Present, Future*. Proceedings of a National Workshop, LI-BIRD/USC Canada Asia/Oxfam/The Development Fund/IFAD/Bioversity International, 14–15 June 2012, Pokhara, Nepal.

Shrestha, P., Gezu, G., Swain, S., Lassaigne, B., Subedi, A., *et al.* (2013b) The community seed bank: a common driver for community biodiversity management. In: de Boef, W.S., Peroni, N., Subedi, A., Thijssen, M.H. and O'Keeffe, E. (eds) *Community Biodiversity Management: Promoting Resilience and the Conservation of Plant Genetic Resources*. Routledge, Abingdon, UK, pp. 109–117.

Shrestha, P.K., Subedi, A., Peroni, N. and de Boef, W.S. (2013c) Community biodiversity management: defined and contextualized. In: de Boef, W.S., Peroni, N., Subedi, A., Thijssen, M.H. and O'Keeffe, E. (eds) *Community Biodiversity Management: Promoting Resilience and the Conservation of Plant Genetic Resources*. Routledge, Abingdon, UK, pp. 19–25.

Sthapit, B.R., Shrestha, P.K. and Upadhyay, M.P. (2006) *Good Practices: On-farm Management of Agricultural Biodiversity in Nepal.* NARC, LI-BIRD, IPGRI and IDRC, Kathmandu.

Sthapit, B.R., Khadka, K., Shrestha, P., Subedi, S. and Poudel, I. (2013) Grassroots breeding of local crops and varieties in support of community biodiversity management and resilience in Nepal. In: de Boef, W.S., Peroni, N., Subedi, A., Thijssen, M.H. and O'Keeffe, E. (eds) *Community Biodiversity Management: Promoting Resilience and the Conservation of Plant Genetic Resources*. Routledge, Abingdon, UK, pp. 233–239.

Subedi, A., Shrestha, P., Sthapit, B.R., Rijal, D.K., Rana, R.B., *et al.* (2005a) Community biodiversity management (CBM): lessons learned from the in situ conservation project, In: Sthapit, B.R., Upadhyay, M.P., Shrestha, P.K. and Jarvis, D.I. (eds) *On-farm Conservation of Agricultural Biodiversity in Nepal. Vol II: Managing Diversity and Promoting its Benefits*. Proceedings of the Second National Workshop, 25–27 August 2004, Nagarkot, Nepal. IPGRI, Rome, pp. 56–70.

Subedi, A., Sthapit, B.R., Upadhyay, M.P. and Gauchan, D. (eds) (2005b) *Learning from Community Biodiversity Register in Nepal*. Proceedings of the National Workshop, 27–28 October 2005, Khumaltar, Nepal. NARC, Kathmandu.

Tapia, M.E. and Rosa, A. (1993) Seed fairs in the Andes: a strategy for local conservation of plant genetic resources. In: de Boef, W., Amanor, K. and Wellard, K. with Bebbington, A. (eds) *Cultivating Knowledge.* Intermediate Technology Publications, London, pp. 111–118.

Thomas, F. and Dao The Anh (2013) Marketing local rice varieties in Vietnam, supported by their participatory genetic enhancement and intellectual property rights. In: de Boef, W.S., Peroni, N., Subedi, A., Thijssen, M.H. and O'Keeffe, E. (eds) *Community Biodiversity Management: Promoting Resilience and the Conservation of Plant Genetic Resources*. Routledge, Abingdon, UK, pp. 207–212.

Utkarsh, G. (1999) People's Biodiversity Register. *ILEIA Newsletter*, December 1999, p. 28.

Vicente, N.R., Fantini, A.C., Alves, A.C. and Canci, A. (2010) Avaliação da efetividade do kit de diversidade. In: Canci, A., Alves, A.C. and Guadagnin, C.A. (eds) *Kit diversidade: estratégias para a segurança alimentar e valorização das sementes locais*. Instituto de Agrobiodiversidade e Desenvolvimento Socioambiental, Guaraciaba, SC, Brazil, pp. 171–185.

Witcombe, J. and Joshi, A. (1995) Farmer participatory approaches for varietal breeding and selection and linkages to the formal seed sector. In: Eyzaguirre, P. and Iwanaga, M. (ed.) *Proceedings of a Workshop on Participatory Plant Breeding,* 26–29 July 1995, Wageningen, the Netherlands. International Plant Genetic Resources Institute, Rome, pp. 57–65.

Witcombe, J.R., Joshi, A., Joshi, K.D. and Sthapit, B.R. (1996) Farmers participatory crop improvement. I. Varietal selection and breeding methods and their impacts on biodiversity. *Experimental Agriculture* 32, 445–460.

# 33 Evolutionary Plant Breeding: A Method for Rapidly Increasing On-farm Biodiversity to Support Sustainable Livelihoods in an Era of Climate Change

M. Rahmanian,[1*] K. Razavi,[1] R. Haghparast,[2] M. Salimi[1] and S. Ceccarelli[3]
[1]*Centre for Sustainable Development (CENESTA), Tehran, Iran;*
[2]*Dryland Agriculture Research Institute of Iran, Maragheh, Iran;*
[3]*Via delle Begonie, Ascoli Piceno, Italy*

## 33.1 Introduction

Agricultural biodiversity is widely recognized to be essential for adapting successfully to climate change – particularly the conservation and use of landraces and crop wild relatives (FAO, 2010; Cardinale *et al.*, 2012; Hooper *et al.*, 2012; HLPE, 2012). Yet despite wide recognition for the importance of agricultural biodiversity to adaptation, the policies (and budgets) of most governments, research institutes and international agencies continue to favour breeding and marketing of single and uniform varieties. Single varieties can yield well under ideal conditions, but this means the manufacturing of 'ideal' environments through the use of chemical inputs and significant amounts of water. In Iran, for example, while there is a high-profile national policy for self-sufficiency in wheat, the main approach has been to rely on high-yielding uniform varieties that are dependent on irrigation and therefore increase stress on water supplies in this arid country. Rapid increase of on-farm biodiversity is therefore a matter of urgency in an era of climate change.

One major challenge is that farmers' access to diversity can be limited. The programme on evolutionary plant breeding in Iran illustrates one possible approach for the rapid increase of on-farm biodiversity. This initiative grew out of a programme of participatory plant breeding (PPB) implemented by an Iranian NGO (CENESTA) working in collaboration with farmers and breeders (Ceccarelli, 2012). Though initially conceived as a strategy to make farmers more independent from institutions and to adapt crops to climate change, evolutionary plant breeding has proved to be a promising model of providing a large number of farmers access to a great amount of biodiversity in a relatively short time. It has also been suggested as a way to develop varieties specifically adapted to organic agriculture (Ceccarelli, 2014).

## 33.2 From Single Lines to Mixtures

The first PPB trials started in Iran in 2006 with the cooperation of CENESTA, local farmers, ICARDA, the Department of Agriculture of

*Corresponding author; e-mail: maryam.rahm@gmail.com

Kermanshah province and the Dryland Agriculture Research Sub-Institute at Sararood (DARSI). The trials of wheat and barley were conducted on several farms in the provinces of Kermanshah and Semnan. As with most PPB trials, the trials started with a large number of lines. Four years of selection by farmers successively narrowed these down to a handful of locally adapted varieties, and farmers cultivated these as single varieties (e.g. Zeyd-Ali and Saman – two varieties of wheat selected by farmers).

During the second year, a parallel line of work started to develop. Salvatore Ceccarelli (ICARDA), who helped start the PPB programme in Iran, shared the experiences of farmers from other parts of the world who were experimenting with mixtures (Box 33.1). Many farmers he had met insisted that mixtures of several different varieties (and sometimes even several crops) led to good yield stability, and in some cases to a higher-quality product. These farmers believed that the enhanced genetic diversity of the mixtures brought greater stability to their yields over time. In fact, the benefits of crop diversity have been well documented in the scientific literature. Di Falco and Chavas (2006), for example,

**Box 33.1.** Making the best of 'leftover' experimental seed.

Speaking with a group of farmers in Iran, Salvatore Ceccarelli shared the story of Syrian farmers engaged in a PPB project (Fig. 33.1). Their experiment had started with 160 different lines of barley in 200 plots. Each year the breeders would harvest only a small representative sample from each plot, leaving the rest of the plot for the farmers to harvest and use as they wished. During one visit, he asked the farmers if they were happy with the lines they had selected. They responded that they were, but that nothing was better than the mixture. 'Which mixture?', asked Salvatore; the farmers explained that after the breeders had taken their samples, they had harvested the entire field – all 200 plots – with a combine harvester and instead of using it as animal feed (as the breeders had assumed they would), they had decided to plant it as a mixture. The result was a mixture that was better than any of the individual lines – and the commercial varieties – on their own.

**Fig. 33.1.** Syrian farmer, Ali Turkye, with ICARDA's Michael Michael in a field of 'Yana' – the name Ali gave to the mixture of the 'leftovers'. (Photograph courtesy of Salvatore Ceccarelli.)

review the literature that shows that cultivating several varieties that respond differently to weather and temperature randomness contributes to resilience, including to drought, pests and diseases. Both pest populations and weeds are more likely to grow and spread when they face a genetically uniform crop. These exchanges led the Iranian farmers to sow mixtures of the selected lines from the PPB trials. Some knew from experience that the reason why their landraces were still so popular (mostly in rainfed agriculture) was that they were not genetically uniform; roughly 70% of rainfed wheat in Iran uses a single landrace: 'Sardari'.

At harvest time, they were pleased with the results, and many farmers have continued experimenting with making their own mixtures, including commercially available varieties in some cases. They developed so many different mixtures of wheat and barley that they decided to start trials to compare their mixtures, which include 16 distinct mixtures of wheat and 21 distinct mixtures of barley. Some of these mixtures are composed of only a handful of varieties, while others have several hundred; some contain only landraces, while others are composed of new lines and commercial varieties, or both. After 3 years, farmers have observed that over time a number of the varieties tend to become dominant in the mixtures, suggesting that mixtures must be actively managed by farmers through a selection process in order to maintain the original diversity.

## 33.3 Our Ideas Evolve, As Do Our Seeds

Since creating mixtures, the team have developed an idea to bring even greater biodiversity into farmers' fields, and this has been termed 'evolutionary plant breeding' (EPB). EPB represents a dynamic and inexpensive strategy that will enhance quickly the adaptation of crops to climate change (Ceccarelli *et al.*, 2010, 2013). The idea of EPB is not new. Harlan and Martini (1929) proposed the composite cross method of breeding barley. The method relies on the bulking of $F_1$ progenies, obtained by the hybridization of varieties of diverse origin and genetic make-up, to generate a heterogeneous population of recombinant genotypes. The population is then multiplied in the specific environments where locally adapted varieties are needed. The procedure should guarantee that the frequencies of adapted genotypes increase without creating a uniform population. Suneson (1956) further developed the idea into a formal 'evolutionary plant breeding method'. Using this method, he registered nine barley composite crosses, meant to be used for the generation of new varieties, as parent lines and material for the study of population genetics. The success of such breeding methods depends on recombination and segregation and the extent to which survival over many generations in mass-propagated populations is correlated positively with agricultural value (Allard and Hansche, 1964). The first participatory evolutionary breeding programme was started in 2008 by planting a mixture of nearly 1600 barley $F_2$. This represented ICARDA's entire barley crossing programme of that year and included a wide range of germplasm: from the wild progenitor, *Hordeum spontaneum*, to landraces from several countries and to modern breeding materials. Besides Iran, this barley population was planted in Syria, Jordan, Algeria and Eritrea, and 2 years later in Italy.

In a programme of EPB, farmers are asked to plant and harvest a small (4–5 kg) sample of seed in the same place for several successive years. These populations are currently evolving in a multitude of environments, chosen by the farmers and characterized by single abiotic or biotic stresses, or combinations of stresses, and under different types of agronomic management. The expectation is that the frequency of genotypes with adaptation to local conditions will increase gradually. It is also possible, and actually desirable, to plant samples in other locations affected by different stresses, or different combinations of stresses, by sharing the population with other farmers. For example, in Iran, the barley population that was planted by five farmers in two provinces in 2008, spread to 50 farmers in four provinces in the cropping season 2010/11. Figure 33.2 shows the location of these PPB and EPB trials in Iran. A similar approach was taken in Italy, where the evolutionary populations of wheat and barley were grown in 14 regions in 2013.

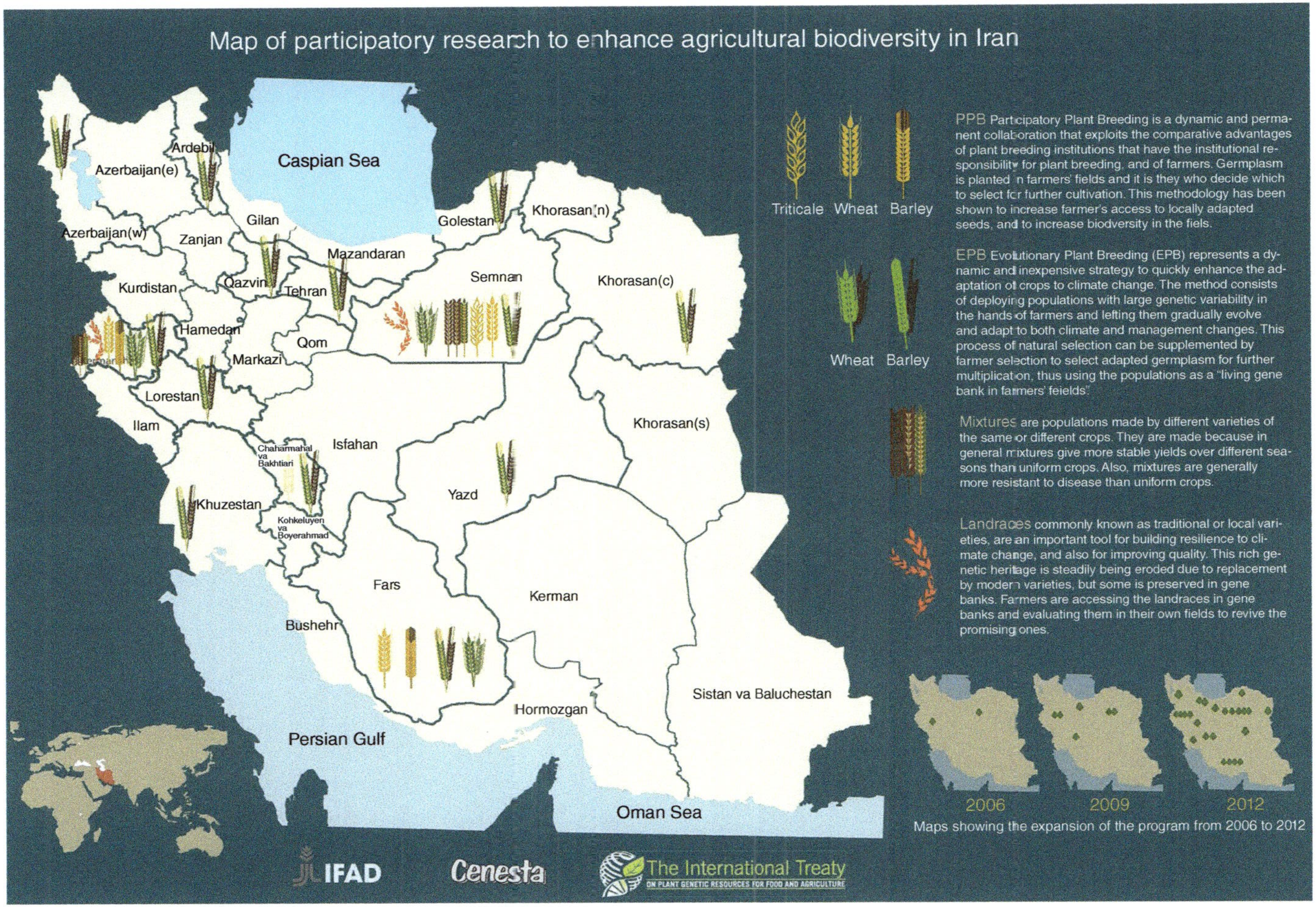


**Fig. 33.2.** Maps of PPB and EPB trials in Iran in 2010/11.

## 33.4 Evolutionary Populations: Living Genebanks in Farmers' Fields

EPB represents an attempt to make the process of *in situ* conservation, or rather management, more dynamic by combining participation and evolution in participatory evolutionary breeding programmes. One way of exploiting the progressive adaptation of the evolutionary populations is to consider them as evolving sources of new cultivars, progressively better adapted to the evolving agronomic and climatic conditions – in other words, a living genebank in farmers' fields. To do this, farmers (by themselves or jointly with scientists) can select the most desirable spikes and use them in participatory breeding programmes. For farmers who prefer to sow mixtures rather than single varieties, the evolutionary populations could serve as a source of germplasm for creating new mixtures. The importance of having secure access to such a collection of seeds became apparent in Jordan and Ethiopia, for instance, where farmers and scientists turned to evolutionary populations once the civil war in Syria disrupted their regular source of breeding materials. Farmers became the owners of their future, with the best varieties evolving in their fields, and there was less or no further need to buy commercial seed.

Another approach to dynamic *in situ* conservation would be to create new evolutionary populations by making mixtures of landraces obtained from genebanks. While genebanks perform an important role in the conservation of landraces, it is recognized that one of their limitations is that they freeze not only seeds but also their evolution at the time of collection. Landraces and wild relatives must therefore also be conserved *in situ*; that is, in their own native environment. Farmers in Garmsar who have evaluated a collection of landraces of wheat and barley (160 accessions of each) have made their own evolutionary populations by creating mixtures of these landraces.

## 33.5 Unexpected Findings: Outscaling of Evolutionary Populations

In the first year of the EPB trials, each farmer was given 4 kg of seed for a small experimental plot. The expectation was that they would continue to sow only 4 kg each year, enough to allow the population to evolve and to act as a source of locally adapted germplasm. The principle in all participatory research is that plot sizes should be kept to a minimum in order to avoid a large burden on farmers who dedicate land for research. One of the most unexpected outcomes of the evolutionary population trials was that some farmers decided to sow *all* the seed they had harvested because of the promising yields they achieved (Box 33.2). In other words, they immediately started multiplying the seed and cultivating it on a wide scale as their *main crop*. It had not been expected that the population would be so well adapted (this was especially the case in rainfed regions) that farmers would be inspired to cultivate it on a wider scale. Diversity is what makes landraces so popular: 'Sardari' has consistently outperformed 'improved varieties' in Iran's rainfed regions. However, 'Sardari' is susceptible to yellow rust and lodging in wet years.

## 33.6 Evolutionary Populations and the Question of the Consumer

Many people who hear about EPB are concerned about the quality of the final product. Is the final product of suitable quality, or even usable? In Iran, the trials have so far included only wheat and barley, and there have been no reasons for concern. In the case of barley, which is mostly used as an animal feed in Iran, there are no concerns about cooking quality. In terms of quality as animal feed, farmers in Garmsar (some of whom also raise livestock) requested a protein analysis of several varieties of barley in their PPB trials. The results showed that the evolutionary population of barley had more protein than the local barley variety (an improved variety from the national research system).

In the case of wheat, farmers and bakers in the provinces of Seman and Kermanshah have tried making bread from the evolutionary populations and have been pleased with the results. Some are even marketing the bread in local artisanal bakeries. Exchanges with farmers in both France and Italy (associated with the Reseau Semences Paysannes and the Rete Semi Rurali,

**Box 33.2.** Early and promising results

In 2008/9, the team started an evolutionary plant breeding (EPB) programme for rainfed barley using a mixture of 1600 $F_2$ populations received from ICARDA. The seed was planted in a farmer's field in the Dalahoo region, Kermanshah, west Iran. In spite of late sowing, undesirable in that cold region, the farmer noted its good performance, and in 2009/10, he planted the harvested seed on 0.5 ha. The evolutionary population out-yielded the local barley and performed almost as well as the improved barley cultivar, 'Sararood-1'. The farmer was satisfied with the population's performance, even in a dry season, and decided to cultivate it on a wider scale and to distribute the seed to other farmers.

Based on the success of EPB with barley, in 2009/10, Reza Haghparast, a wheat breeder at DARSI, established a similar programme for bread wheat. He used the backup bread wheat seeds conserved in DARSI's seed store, together with seeds from segregating populations from its wheat breeding programme. The evolutionary population was planted in a farmer's field in the Sahne region in Kermanshah province in 2009/10, a year that saw higher than average rainfall in Kermanshah. The population yielded 3500 kg/ha under rainfed conditions, while the very popular local variety, 'Sardari', suffering from lodging and rust, produced only 1500 kg/ha. In the following year, which was relatively dry and characterized by many stresses, the evolutionary population outperformed 'Sardari' again.

Results were also promising in very hot regions. In 2012/13, the evolutionary population of wheat produced 1700 kg/ha in a hot region of Kermanshah province (Sar-e Pol-e Zahab), while the local variety produced only 800 kg/ha. Also in Kermanshah province, in Ravansar, the yield was on average 16% greater than local varieties, and in Sarab-e Niloofar, yields were 51% greater than 'Sardari'. That same year, the Department of Agriculture of Kermanshah province approved and funded a multi-year project on EPB throughout the province.

Outscaling has been less widespread in irrigated regions, but there are nevertheless promising results. In 2011/12, an evolutionary population of wheat, also provided by DARSI, was cultivated in Garmsar (Semnan province) under irrigated conditions and produced 9.2% more than 'Pishtaz', an improved variety used as the local check. The farmer who cultivated this field claimed that the evolutionary population had less weeds and pests than 'Pishtaz', and while the former was not treated with pesticides and herbicides, the latter had to be heavily treated. This suggests that evolutionary populations could be both very useful in organic agriculture and cheaper to grow.

respectively) confirmed that creating mixtures not only brings greater yield stability but also greater aroma and quality to the bread, though of course the composition of the mixture is important.

Obviously, the suitability of evolutionary populations as the main crop depends on the use of the crop and the cultural preferences of local farmers and consumers. What is clear is that in the case that the crop does not lend itself to being consumed as a mixture (such as in the case of many vegetable varieties), the evolutionary populations can still serve as living genebanks and the source of new adapted germplasm that can be multiplied as single lines. For example, the use of EPB with vegetables is currently under way in Italy with tomato, beans and zucchini (courgette) (Fig. 33.3). Here, the evolutionary population was made by allowing 11 commercial hybrids to cross-pollinate naturally in the field (a project of AIAB, the Italian Association for Organic Agriculture). Researchers have found fruit with colour and shape absent in the original hybrids. One farmer has selected five plants out of the 200, each with a different fruit shape, and is planning to market them as soon as he harvests. He is keeping the population by harvesting one fruit from each plant, including the five selected.

## 33.7 Biodiversity, Evolutionary Populations and Seed Laws

Farmers make up an important portion of the Iranian population (25%) and economy (20%). The vast majority of Iranian farmers are smallholders (75% of farmers have less than 5 ha of land) and rely mainly on farm-saved seed for the vast majority of cereals, legumes, paddy rice and forage crops. These crops make up the major part of the crops grown in the country: 86% of the total area under cultivation.

Fig. 33.3. A field of 200 plants of zucchini in Italy.

Iran's seed law (SPCRI, 2003) states that all registered seed must uphold the criteria of being new, distinct, uniform and stable. It is clear that evolutionary populations could never meet these criteria: it is precisely their diversity and dynamic evolution that are their main advantages. However, this law addresses only seed that is marketed, and since the evolutionary populations are not marketed, there is currently no legal impediment to this work. According to the Seed and Plant Policy Document of the Islamic Republic of Iran (SPCRI), 'Farmers who produce farm-saved seed and plants have the right to store, use, exchange, and distribute their own plant materials' (SPCRI, 2015).

The negative impacts of many seed laws on biodiversity have become the cause of concern from the perspective of farmers' rights. According to the Special Rapporteur on the Right to Food, 'The expansion of IP rights can constitute an obstacle to the adoption of policies that encourage the maintenance of agrobiodiversity and reliance on farmers' varieties' (De Schutter, 2009). If Iran were to join the International Union for the Protection of New Varieties of Plants (UPOV) (and there is interest among some breeders in Iran to do so), then there would be major limitations on evolutionary populations. First, while the 1961 and 1978 versions of UPOV automatically allow the use of farm-saved seed by farmers, the 1991 version (which Iran would have to accept if it decided to join UPOV) would allow the use of farm-saved seed only as an optional exception and only if royalty were paid on seed (GRAIN, 2007). Second, in many countries that have developed seed laws based on the UPOV system, the marketing of seeds is based on the prerequisite of registration and certification. Intellectual Property Rights (IPR) regimes such as UPOV require that all certified seed adhere to the criteria of distinctness, uniformity and stability. As already mentioned, the evolutionary populations are useful precisely because they do not adhere to these criteria that favour uniformity and outlaw diversity.

## 33.8 Where Next?

The first national workshop on EPB was organized in Shiraz in January 2013, in which farmers from several provinces shared their experiences with each other. Regular local, regional and national workshops and field visits are needed to strengthen farmers' knowledge about how to use these populations. At the same time, it is important to develop awareness of the potential impacts of different seed laws and policies on farmers' rights to save, exchange, develop and use their seeds sustainably.

The evolutionary populations of wheat and barley continue to be distributed throughout Iran, both through farmer-to-farmer exchanges and through exchanges organized through DARSI and CENESTA. The main challenge is to keep track of the distribution of the populations, as well as gathering data on their performance and giving support to farmers (explaining what these populations are and how they can be used), etc. Part of the challenge of tracking their distribution is related to naming the populations: what to call the different populations and when to give the same population a different name when it has evolved in different environments? Another challenge is having enough of the original seed to distribute to farmers in very different environments (rather than giving farmers seed that has already evolved in a region which is quite different from theirs). Finally, access to the seed to make new evolutionary populations requires the cooperation of breeders and genebanks, which is a challenge for some farmer and civil society organizations.

## References


Allard, R.W. and Hansche, P.E. (1964) Some parameters of population variability and their implications in plant breeding. *Advances in Agronomy* 16, 281–325.

Cardinale, B.J., Duffy, J.E., Gonzalez, A., Hooper, D.U., Perrings, C., *et al.* (2012) Biodiversity loss and its impact on humanity. *Nature* 486, 59–67.

Ceccarelli, S. (2012) Living seed – breeding as co-evolution. In: *Seed Freedom: A Global Citizens Report*, Navdanya, New Delhi, pp. 39–46.

Ceccarelli, S. (2014) GMO, organic agriculture and breeding for sustainability. *Sustainability* 6, 4273–4286.

Ceccarelli, S., Grando, S., Maatougui, M., Michael, M., Slash, M., *et al.* (2010) Plant breeding and climate changes. *Journal of Agricultural Science* 148, 627–638.

Ceccarelli, S., Galié, A. and Grando, S. (2013) Participatory breeding for climate change-related traits. In: Kole, C. (ed.) *Genomics and Breeding for Climate-Resilient Crops*, Vol 1. Springer-Verlag, Berlin, Heidelberg, pp. 331–376.

De Schutter, O. (2009) UN Special Rapporteur on the Right to Food. *Seed policies and the right to food: enhancing agrobiodiversity and encouraging innovation.* UN Doc A/64/170, 23 July 2009.

Di Falco, S. and Chavas, J.P. (2006) Crop genetic diversity, farm productivity and the management of environmental risk in rainfed agriculture. *European Review of Agricultural Economics* 33(3), 289–314.

FAO (Food and Agriculture Organization) (2010) *Second Report on the State of the World's Plant Genetic Resources for Food and Agriculture.* FAO, Rome.

GRAIN (2007) The end of farm-saved seed? Industry's wish list for the next revision of UPOV. Available at: www.grain.org/fr/article/entries/58-the-end-of-farm-saved-seed-industry-s-wish-list-for-the-next-revision-of-upov (accessed 31 July 2015).

Harlan, H. and Martini, M. (1929) A composite hybrid mixture. *Journal American Society of Agronomy* 21, 487–490.

HLPE (High Level Panel of Experts on Food Security and Nutrition) (2012) *Food Security and Climate Change*. FAO, Rome.

Hooper, D.U., Adair, E.C., Cardinale, B.J., Byrnes, J.E., Hungate, B.A., *et al.* (2012) A global synthesis reveals biodiversity loss as a major driver of ecosystem change. *Nature* 486, 105–108.

SPCRI (2003) Act of Plant Varieties Registration, Control and Certification of Seeds and Plant Materials. Available at: www.spcri.ir/en/Pages/ActofPlantVarietiesRegistration,Control.aspx (accessed 31 July 2015).

SPCRI (2015) Seed and Plant Policy Document of the Islamic Republic of Iran. Available at: www.spcri.ir/en/Pages/SeedandPlantPolicy.aspx (accessed 31 July 2015).

Suneson, C.A. (1956) An evolutionary plant breeding method. *Agronomy Journal* 48, 188–191.

# 34 Value Chain Development: A Silver Bullet for Agrobiodiversity Conservation and Use?

A.G. Drucker,[1]* and J. Appels[2]
[1]*Bioversity International, Rome, Italy;* [2]*Wageningen University, Development Economics Group, Wageningen, the Netherlands*

## 34.1 Introduction

The diversity of cultivated crop species/varieties and livestock breeds has been and continues to be the basis of our food supply and good nutrition, and contributes importantly to the capacity of agricultural and food systems to cope with global change. Nevertheless, as agrobiodiversity provides a mixture of private and public benefits, with markets capturing only a part of their value and thus underestimating their true worth, crop species/variety and livestock breed choices made by farmers may lead to less genetic diversity being conserved and used than is considered socially desirable (Pearce and Moran, 1994; Heal *et al.*, 2004; Drucker, 2007). Private benefits include direct use values (e.g. crop, plant and livestock products for food, clothing, shelter, medicinal and sociocultural purposes). Public benefits include indirect use values (such as the maintenance of agroecosystem resilience, gene flow and evolutionary processes, as well as traditional knowledge and cultural practices) and non-use values (e.g. option and existence values, where the former is a kind of insurance value given risk aversion and uncertainty about the future – such as that arising from climate change or new disease challenges). The *Second Report on the State of the World's Plant Genetic Resources for Food and Agriculture* (FAO, 2009) and *The State of the World's Animal Genetic Resources for Food and Agriculture* (FAO, 2007) in fact depict an unprecedented loss occurring across the globe of crop species/varieties, livestock breeds and associated traditional knowledge. An important part of this loss is the unintended effect of increased agricultural intensification. While this intensification creates enormous benefits for humanity, there is increasing evidence that this loss can have important negative consequences (Heal *et al.*, 2004; MEA, 2005), including the loss of resources with high productivity (and market potential) in the type of marginal environments farmed by the poor.

Although concerns about genetic erosion in crops have led to efforts to 'insure' against losses by sampling and storing large numbers of landraces and wild relatives of cultivated plants *ex situ* in collections or genebanks, *in situ*/on-farm approaches increasingly have been recognized as a necessary complement.

Yet given that the costs of such conservation tend to be local (i.e. at the farm or community level), while the benefits tend to be regional, national or even global, poor farmers cannot be

*Corresponding author; e-mail: a.drucker@cgiar.org

expected to be able to afford to conserve genetic resources purely for the benefit of wider society without adequate incentives to do so. Such spatial differentiation of the relative costs and benefits of the maintenance of biodiversity in general has led the Convention on Biological Diversity (CBD), which views *in situ* conservation as its preferred approach, to call for the development of positive incentives (COP 10 Decision X/2) for the conservation of biodiversity as part of its Aichi Target 3, which forms part of its strategic plan for 2011–2020 (CBD, 2010).

One such incentive mechanism, often seen as a positive and self-sustaining one, is associated with support for increased orientation towards expanding agricultural markets. It is argued that such support can potentially generate enhanced private benefits for farmers through, *inter alia*, access to improved species/varieties, increased choices of input suppliers and product outlets, increased accessibility to credit, better management capacity, improved employment opportunities and associated income generation (Eaton and Shepherd, 2001; Winters *et al.*, 2009). The potential financial sustainability over the longer term of such market-driven approaches makes them highly attractive. Consequently, value chain development (VCD) has gained attention increasingly in recent years as a tool for harnessing the potential of existing agricultural market channels to promote the use of threatened genetic resources, with consumers ultimately paying for the *in situ*/on-farm maintenance of locally adapted genetic resources through such mechanisms as eco-labelling, certification or denomination of origin schemes.

Nevertheless, although it has been shown that such VCD can facilitate conservation through the use of specific livestock breeds (among others, see: LPP *et al.*, 2010; Mathias *et al.*, 2010; Lauvie *et al.*, 2011; FAO, 2013a; Tienhaara *et al.*, 2013) and neglected and underutilized crop species/varieties[1] (NUS) (see references in Table 34.1), the potential to use such an approach as a cornerstone of a wide-ranging, cost-effective, diversity-maximizing national agrobiodiversity conservation and use strategy remains to be explored. In particular, relying solely on market development might be a dangerous strategy for the conservation of a diverse genetic resource pool, especially as market conditions can change rapidly and generally consumers and agribusiness tend to favour a narrow suite of crop species/varieties or animal breeds. Market chain approaches may also require relatively high initial investments to generate appropriate product volumes, with such volumes being far in excess of those required to achieve modest conservation goals and, where overly successful, may even displace other threatened plant and animal genetic resources (Narloch *et al.*, 2011a).

This paper thus seeks to contribute to an assessment of the role that value chain projects have played in agrobiodiversity conservation, specifically for crop genetic resources. In particular, we explore: (i) the degree to which value chain projects have focused on genetic resources that are actually threatened and form part of a strategic conservation portfolio; (ii) the degree to which conservation goals are, in practice, embedded in the broader set of the goals of value chain projects; and (iii) the effectiveness of value chain projects in reaching their stated conservation goals.

## 34.2 Conceptual Background to Neglected and Underutilized Species Value Chain Development

Biological diversity is increasingly recognized as a global asset of tremendous value to present and future generations. However, the full value of such diversity is often not fully reflected in the marketplace. VCD approaches seek to harness the underdeveloped potential of such resources, thereby generating social benefits, such as poverty reduction, income and employment generation, and environmental conservation. VCD approaches place at their centre the interrelatedness of actors in the value chain and seek to facilitate a positive or desirable change in the chain to extend or improve productive operations. Interventions can focus on improving business operations at the level of producers, processors and other actors in the chain and/or the (contractual) relationships among them; as well as improve the flow of knowledge, information and innovation. VCD can also foster overall coordination in the chain; the participation of selected beneficiaries in local, national or global value chains, the reduction of entry barriers and a higher share of value addition for certain actors (UNIDO, 2011).

NUS-VCD refers to a specialized application of the regular value chain concept. Related projects specifically aim to enhance the use of NUS by combining conservation goals with a contribution to improving rural livelihoods by using the market as a tool to achieve economic, environmental and social impact (Will, 2008; Gotor *et al.*, 2013).

NUS-VCD approaches are typical examples of what Ferraro and Kiss (2002) refer to as indirect strategies for investing in agrobiodiversity conservation. Unlike direct payment approaches, VCD approaches are seen as a preferred strategy by many, given that they can be potentially self-sustaining in the long run (Will, 2008). NUS-VCD projects can focus on different market scales (local, national and international), depending on the type of NUS and the type of goals that are set. Dawson *et al.* (2007) argue that local markets may be the most effective for promoting relatively modest increases in the use of a wide range of species/varieties, and thereby avoiding domination/displacement by any single one. While the majority of NUS-VCD projects do indeed focus on local markets as a starting point (as will be seen below), a shift towards national markets may provide additional opportunities for value addition (e.g. through processing for transport and increased longevity/shelf-life) for a relatively small number of species/varieties. The shift from national markets to international markets is less common for NUS-VCD projects, given the need for increased uniformity and the existence of non-tariff commercial barriers such as the European Union Regulation on Novel Food (Padulosi *et al.*, 2011a). Nevertheless, examples of such market-scale shifts for NUS do exist, such as that of quinoa, which is now widely exported from Andean countries, as well as cultivated in an increasing number of countries (Giuliani *et al.*, 2012).

### 34.2.1 NUS-VCD limitations and complementarity of direct approaches

With a view to evaluating the role of NUS-VCD in achieving conservation outcomes for threatened crop genetic resources, it is useful to place the above NUS value chain within a broader conceptual framework.

The unprecedented scale of genetic erosion taking place and the limited conservation resources available suggest that interventions need to be prioritized. In terms of the implementation of such a 'strategic' conservation plan, a key input relates to the identification of a priority conservation portfolio (Krishna *et al.*, 2013). Weitzman (1993) shows how a cost-effective, diversity-maximizing conservation portfolio can be identified by combining diversity, threat and conservation cost data. Simianer *et al.* (2003), Reist-Marti (2003) and Zander *et al.* (2009) have demonstrated how the Weitzman approach can be applied to animal genetic resources; while more recently, a plant genetic resource application has also emerged (Samuel *et al.*, 2013). The approach has strong appeal due to its rigorous mathematical justification and the possibility to derive optimum conservation decisions with well-defined properties. It addresses the 'Noah's Ark' question of 'what should be conserved?', with the identification of such a priority conservation portfolio aiming to ensure that for any given level of resource allocation, the widest range of diversity possible can be targeted for conservation.

Having identified such a portfolio, a process of triage can then be applied to determine what proportion of the portfolio may have high market potential, what proportion has unknown market potential and what proportion has no current market potential. A range of market/value chain analysis tools could be applied for the purpose of this triage; for example, growth and competitiveness assessments may be realized using measures of unmet market demand (quantities or product range), unused competitive advantages (e.g. unique product, cost advantage, proximity to markets) and by accounting for the potential for geographical expansion. Such portfolio identification and analysis would help to ensure that those crop species/varieties selected for value addition and marketing do indeed form part of the priority conservation portfolio, rather than just having high market potential that is currently unexploited as a result of market frictions and externalities (e.g. lack of information regarding nutritional and health benefits, (organic) production practices or origin).

For those genetic resources within the portfolio that have unknown market potential, more direct intervention strategies (such as payments/ rewards for agrobiodiversity conservation services (PACS) – discussed below in further detail) are likely to be required. Such strategies would permit farmers to experiment in a relatively risk-free environment, in order to help them and market specialists determine if there is indeed a potential for VCD. For those that are shown ultimately to have no current market potential, longer-term PACS schemes to reward farmers for maintaining such diversity *in situ* on-farm for the public good would need to be designed, along with consideration of *ex situ* conservation strategies.

## 34.3 Methods

A literature review of NUS-VCD projects in developing countries in Africa, Asia and the Americas realized between 1997–2014 was carried out. In addition to published papers and grey literature, project information was gathered through expert interviews and an e-mail survey in English, French and Spanish. A total of 36 projects were identified, of which 28 (see Table 34.1) were selected for further analysis. The remainder were excluded as either they were already closely related to the other projects under review, they did not fully meet this study's definition of being focused on a NUS and/or the literature on the project was too minimal to conduct an analysis. A summary of the 28 studies is provided in Table 34.1, while full details can be found in Appels (2014).

Projects may be divided into two major groups: those that carried out research only (R) and those that realized both research and development interventions (R&D). The former have no intervention component, instead focusing largely on market and value chain analysis and concluding with recommendations regarding potential future interventions. By contrast, the R&D projects have involved actual interventions and their impact may be assessed. Projects were analysed in terms of the frequency of mention of specific: (i) project goals; (ii) analytical tools; (iii) genetic resource selection criteria; (iv) market constraints; (v) recommendations; and (vi) stated outcomes/impacts. These were rated according to the following Likert scale: rarely, occasionally and frequently.

## 34.4 Findings

Both the R and R&D NUS-VCD projects do occasionally state that they have an explicit conservation goal, aiming to create a bridge between market potential and conservation. Yet market goals are generally found to be given greater weight than the conservation ones; and an examination of the actual selection criteria used to determine the crop species/varieties focus of such projects suggests that there is a focus on individual NUS with market potential. This raises questions regarding the degree to which such projects are in fact being applied strategically to a diverse portfolio of threatened crop species/varieties.

Of the 18 genetic resource selection criteria identified, which covered economic factors, plant characteristics, value chain actor preferences and a range of strategic/developmental considerations, none relate explicitly to importance for conservation. Although the potentially conservation-related criterion 'wide genetic variability' is mentioned occasionally for both types of project, such criteria as 'occurrence in natural habitat' and, perhaps the most relevant, 'extent of anthropogenic pressure' are mentioned occasionally or rarely from a research perspective only (i.e. under the R projects). Neither of the latter two is mentioned in projects with actual interventions (i.e. the R&D projects). Furthermore, even where threat/anthropogenic pressure is being accounted for, not every crop species or variety that is considered threatened is necessarily dissimilar enough to be considered to form part of a given priority conservation portfolio.

Despite this lack of a strategic orientation with regard to the potential conservation focus of NUS-VCD projects, a number of the recommendations made and interventions carried out could be considered to be contributing to conservation outcomes, particularly those related to germplasm access (including the improvement of *ex situ* conservation practices), introduction and characterization. Furthermore, a number of specific conservation outcomes arising from actual project interventions are reported; the most important of which is a positive impact on variety richness, which is mentioned occasionally. This is an interesting finding, as it suggests that NUS-VCD projects may in some contexts indeed be associated with genetic resource conservation outcomes at the infraspecific level. Although an

Table 34.1. Overview of NUS-VCD project type, location and crop focus.

| Project number (type[a]) | Crop(s) focus | Country | Reference |
|---|---|---|---|
| 1 (R) | African garden egg | Ghana | Horna and Gruère *et al.* (2007) |
| 2 (R) | Amaranthus, African nightshades, African eggplant, spider plant, cowpea, okra | Kenya and Tanzania | Muhanji *et al.* (2011) |
| 3 (RD) | African leafy vegetables | Kenya | Gotor and Irungu (2010); FAO (2013b) |
| 4 (RD) | Indigenous leaf vegetables | Kenya | Ngugi *et al.* (2007) |
| 5 (RD) | Soybeans | Kenya | Chianu *et al.* (2009) |
| 6 (R) | Bambara groundnut | Swaziland and Zimbabwe | Hampson *et al.* (2000); Azam-Ali *et al.* (2001) |
| 7 (R) | Cape gooseberry | Uganda | Barirega *et al.* (2014) |
| 8 (R) | Bambara groundnut | West Africa (focus on Nigeria and Zimbabwe) | Hillocks *et al.* (2012) |
| 9 (RD) | Henna, nigella, coriander, cumin | Yemen | Gotor *et al.* (2013) |
| 10 (R) | Apple, apricot, mulberry | Afghanistan and Tajikistan | Giuliani *et al.* (2011) |
| 11 (R) | Minor millets | India | Gruère *et al.* (2007); Vijayalakshmi *et al.* (2010) |
| 12 (R) | Amla, tamarind, kokum | India | Daniel and Dudhade (2006) |
| 13 (R) | Malabar nut tree, redstem wormwood, cleavers, green amaranth, ivy, common nettle | Pakistan | Ahmad and Javad (2007) |
| 14 (R) | Mango rambutan, mangosteen, longkong, sala | Thailand | Author survey; and Kruijssen, Netherlands, 2014, personal communication |
| 15 (R) | Fig, jujube, laurel, caper, purslane, mallow | Syria | Giuliani (2005, 2007); Guiliani and Buerli (2006); Kruijssen *et al.* (2007) |
| 16 (R) | Emmer | Turkey | Giuliani *et al.* (2009) |
| 17 (RD) | Minor millets and Andean grains | Bolivia, India, Nepal and Peru | Author survey; and Padulosi, Italy, 2014, personal communication |
| 18 (RD) | Potatoes | Bolivia, Ecuador and Peru | Devaux *et al.* (2007) |
| 19 (RD) | Ahipa, arracacha, maca, yacon | Bolivia, Ecuador and Peru | Author survey; and Hermann and Heller (1997) |
| 20 (RD) | Quinoa, canahua, amaranth | Bolivia and Peru | Padulosi *et al.* (2014) |
| 21 (R) | Quinoa | Bolivia and Peru | Author survey; and Hellin, Mexico, 2014 personal communication |
| 22 (R) | Cherimoya | Bolivia, Ecuador and Peru | Vanhove and Van Damme (2013) |
| 23 (R) | Peach palm | Brazil, Colombia, Costa Rica and Peru | Clement *et al.* (2004) |
| 24 (R) | Maize | Guatemala | Isakson (2011) |
| 25 (R) | Maize | Mexico | Keleman and Hellin (2009) |
| 26 (RD) | Capsicum | Peru and Bolivia | Jäger *et al.* (2014) |
| 27 (RD) | Capsicum | Peru | Garcia-Yi (2014) |
| 28 (R) | Copoazu, seje, peach palm, Amazon grape, manaca, temare, caimito | Venezuela | Looy *et al.* (2008) |

*Notes*: [a]R = research project; RD = research and development project.

increased NUS area under cultivation is mentioned only rarely as a project outcome, this is nonetheless important as, while there remains a lack of consensus as to which precise factors and thresholds should be part of a crop threat assessment methodology (for example, for livestock/animal genetic resources (AnGR) the 'at risk' population threshold has been defined by FAO (2013) as between 2000 and 6000 breeding females; a similarly defined threshold does not exist for crops), clearly the area under cultivation is likely to form a key measure and does indeed feature in emerging risk assessments. Additional outcomes related to the 'changed image of species' and the 'improved availability of seeds and planting material' (both mentioned only rarely) may also be considered to be contributing to conservation outcomes. Nevertheless, conservation outcomes considered to be associated with enhanced farmer use did not seem to be mentioned as prominently as might be implied by the stated project conservation goals, and in any case, not all the projects were found to be associated with specific conservation outcomes.

Furthermore, a clear description of how those conservation outcomes that did exist were actually determined was not always apparent. Interventions related to conservation are mostly poorly described, perhaps because of the tendency to focus on the market interventions. Furthermore, the project conservation goals are often formulated very broadly (e.g. aiming 'to promote biodiversity' or 'to create a link between conservation and markets'), making them difficult to measure and evaluate. Overall, the evaluation tools used to determine conservation impact are also poorly described, making it challenging to determine the actual contribution that the NUS-VCD projects had on the conservation of threatened species/varieties, and on the feasibility of achieving conservation goals through these projects.

Such challenges are highlighted further by the fact that no examples of impact assessment activities over longer periods (beyond the end-of-project evaluation) were reported. Although NUS-VCD projects are often promoted as being self-sustaining in the long term, we were unable to identify any such evidence of long-run self-sustainability in the project literature. While an insufficient demonstration of the self-financing conservation dimension of NUS-VCD projects has also been found in the context of wild biodiversity studies (Wells *et al.*, 1998; Hughes and Flintan, 2001; Ferraro and Kiss, 2002), we acknowledge that this does not necessarily mean that none of the projects reviewed will be sustainable over the longer term.

This lack of long-term monitoring/impact assessment also means that potential leakage effects (i.e. unintended negative socio-economic and/or environmental effects arising from the NUS-VCD intervention process) were almost never mentioned in the project literature. In particular, only monitoring/impact assessment during the project life and its immediate aftermath is unlikely to be able to address concerns regarding potential displacement that may be caused by successful NUS-VCD. Such concerns arise from the fact that even where adequate market potential currently exists for some species/varieties, their development may turn out to be so successful that the displacement of other priority crop species/varieties considered to be threatened occurs. Examples of such displacement include minor millets in India (Padulosi *et al.*, 2011b) and Andean grains in Latin America (Rojas *et al.*, 2009). Market development opportunities can thus change into threats in the absence of sustainability constraints.

These findings do not necessarily imply that NUS-VCD projects have had a negative impact on biodiversity conservation, nor that there are no benefits from implementing NUS-VCD projects. A range of significant positive NUS-VCD outcomes were identified that have impacted the livelihoods of smallholder farmers positively. Nevertheless, the case studies have revealed that care needs to be taken when making the assumption that NUS-VCD projects have a potentially positive impact on conservation. Even in the case studies identified as having the best conservation outcomes (such as Devaux *et al.*, 2007; Jäger *et al.*, 2014 [unpublished results]; Padulosi *et al.*, 2014), there remains a lack of strong scientific evidence that such an assumption holds. In addition to the lack of long-term impact assessment in order to reveal whether NUS-VCD projects are in fact self-sustaining over the longer term, the strength of the market potential selection criteria in NUS-VCD projects makes it a challenging stand-alone tool with which to conserve a broad range of threatened species/varieties. NUS-VCD may raise the financial profitability only of those local genetic resources that closely match consumers' current tastes and preferences, while continuing to neglect a large share of the conservation portfolio and their non-use values. For example, of 3000 quinoa varieties in Peru, only 15–20 have national/international market potential (Rabines, Lima, 2014, personal communication); while of the thousands of mango varieties in India, only five have international market potential (H. Lamers, Delhi, 2014, personal communication). Even where NUS-VCD would be so successful as to double or triple these numbers, a large percentage of the overall portfolio of genetic resource diversity with important public good values would remain unaffected. Such NUS-VCD mechanisms alone are thus unlikely to be a silver bullet, capable of correcting for market failures associated with the public goods dimensions of crop genetic resource diversity maintenance.

### 34.4.1 Exploring the need for a complementary incentive mechanism

The role of emerging and innovative agrobiodiversity-focused payments for ecosystems services – so-called payments (or rewards) for agrobiodiversity conservation services (PACS)[2] – incentive schemes in helping to overcome some of these challenges appear to be well worth exploring. Whereas NUS-VCD aims to enhance agricultural market returns so as to allow poor smallholder farmers (both women and men) to capture more of the private benefits of NUS production, the conditional payments/rewards made under PACS schemes aims to allow them to capture the public good values generated by their maintenance of such genetic resources. In this context, it is also possible to appreciate how the two different approaches contribute to the internalization of different kinds of ecosystem service externality; for example, with NUS-VCD projects seeking to internalize nutrition, production practices (e.g. organic) or origin externalities, while PACS is focused more directly on the genetic resource loss externality *per se*.

PACS incentive schemes have been shown (Krishna *et al.*, 2013; Narloch *et al.*, 2011a,b, 2013, and in press) to be a potentially effective instrument for promoting the cost-effective maintenance of safe minimum populations of threatened NUS crop varieties while accounting for pro-poor/social equity considerations. Where interventions are not required on an annual basis, as was found to be the case following the 2010–2011 PACS scheme in Bolivia and Peru (Drucker *et al.*, 2015), conservation outcomes may even be as financially sustainable as NUS-VCD projects, once initial VCD investment costs are accounted for and the fact that conservation goals may be expected to be significantly smaller than the minimum volumes required for VCD.

In particular, a PACS scheme and its underlying strategic portfolio would ensure that: (i) agricultural market development efforts are focused on priority threatened species/varieties; (ii) that by encouraging the maintenance of safe minimum populations of specific threatened crop genetic resources (through the establishment of a minimum population sustainability constraint beyond which the population is not allowed to fall), any displacement would have only a limited impact on other priority species/varieties within the conservation portfolio; and (iii) that specific crop genetic resources that are no longer widely used can be adequately evaluated for their agricultural market development potential through farmer experimentation. Where such market potential turns out to exist, VCD support could then be used as a longer-term source of funding for those crop genetic resources that at the beginning of the project had unknown market potential but which, after initial PACS support, were shown to have such potential. The overall goal is that through such time-limited PACS support, the genetic resources that constitute a strategic asset of the poor can be increasingly marketed locally and in more distant markets. For those that continue to have no current market potential but can be associated with the provision of public good ecosystem services (e.g. agroecosystem resilience, including climate change adaptation, maintenance of evolutionary processes, traditional knowledge and future option values), alternative sources of sustainable financing will need to be found.

NUS-VCD and PACS should thus be viewed as complementary approaches (Narloch *et al.*, 2011a), including due to the fact that a number of the in-kind rewards used may also directly support NUS-VCD interventions. Such in-kind rewards, many of which could be provided by existing government agricultural development and education programmes, include, *inter alia*: (i) information (e.g. school materials, extension literature); (ii) technology transfer: hardware, software and know-how (e.g. access to quality seed and diversity, agricultural machinery, farm inputs, other infrastructure, including associated know-how); (iii) training (e.g. improved agronomic practices, participatory plant breeding); (iv) joint research and development (e.g. farmer experimentation and participatory evaluation); (v) institutional capacity building (e.g. strengthening of institutions of collective action, development of value chain partnerships, establishment of biodiversity fairs/awards, community biodiversity registers); and (vi) local income generation and employment (e.g. opportunities to participate in genetic resources (GR) participatory monitoring and verification activities). Furthermore, where there is a spatial (community-level) overlap between those priority crop genetic resources with market potential and those without, the provision of VCD-related support may be made conditional on the maintenance of small areas of those threatened crop genetic resources without market potential. A conceptual basis for PACS financing strategies, through private and public sector service beneficiary and purchaser identification/mapping and dialogue, has also been developed recently (Drucker *et al.*, 2013; Martinez-de la Cruz *et al.*, 2015).

## 34.5 Conclusions

An extensive literature review of NUS-VCD projects that contain a conservation element was undertaken, together with expert interviews and a case study review of 28 projects from 1997 to 2014. The resulting analysis reveals the complex nature of NUS-VCD projects when it comes to defining, contributing to and evaluating conservation goals. Care thus needs to be taken when recommending such approaches as a tool for generating positive conservation outcomes. To date, such projects have lacked a strategic approach, instead being realized on an ad hoc individual basis, focusing on individual NUS with market potential. The degree to which NUS-VCD projects can contribute strategically to the maintenance of a diverse portfolio of threatened genetic resources thus remains to be demonstrated.

With a view to supporting the conservation components of future NUS-VCD projects and developing improved strategies and methods, it is apparent that such projects need to be implemented within a more strategic framework that ensures that such work is carried out on genetic resources that are indeed within a priority conservation portfolio. Even when applied within such a strategic context, project conservation goals will need to be articulated more clearly and evaluated, if evidence of conservation outcomes is to be generated. The use of emerging agrobiodiversity project assessment protocols (e.g. Bellon *et al.*, 2015a,b) and agrobiodiversity-relevant ecosystem service indicator development – related to the maintenance of landscape-level resilience, traditional knowledge and future option values (Willaarts and Drucker, 2015 [unpublished results]) – may be useful in this context.

## Acknowledgements

This work was supported by the Consortium Research Programme on Policies, Institutions and Markets.

## Endnotes

[1] NUS may also be referred to as orphan, abandoned, new, underutilized, neglected, lost, underused, local, minor, traditional, forgotten, alternative, niche, promising or underdeveloped (Padulosi and Hoeschle-Zeledon, 2004). For the purposes of this study, we use Gruère *et al.*'s (2007) definition, which defines an NUS as any agricultural or non-timber forest species, collected, managed or cultivated that simultaneously captures the following three characteristics: (i) the species is locally abundant in developing countries, but globally rare; (ii) there is little scientific knowledge about the (use of the) species, but local users possess practical knowledge about it; and (iii) the use of the species is limited relative to its economic potential.

[2] PACS may be defined as a subcategory of agriculture-related payment for ecosystem services (PES) that focuses on socially valuable and threatened local plant and animal genetic resources. For more information regarding PACS, including scientific publications, factsheets and video, see: www.bioversityinternational.org/pacs.

## References


Appels, J. (2014) Value chain development for conservation? Concepts, evidence and challenges. Mimeo. Bioversity International, Rome.

Ahmad, S.S. and Javed, S. (2007) Exploring the Economic Value of Underutilized Plant Species in Ayuba National Park. *Pakistan Journal of Botany* 39(5), 1435–1442.

Azam-Ali, S.N., Sesay, A., Karikari, S.K., Massawe, F.J., Aguilar-Manjarrez, J., Bannayan, M. and Hampson, K.J. (2001) Assessing the potential of an underutilized crop–a case study using bambara groundnut. *Experimental Agriculture* 37(4), 433–472.

Barirega, A. (2014) Potential for Value Chain Improvement and Commercialization of Cape Gooseberry (Physalis peruviana L.) for Livelihood Improvement in Uganda. *Ethnobotany Research & Applications* 12, 131–140.

Bellon, M.R., Gotor, E. and Caracciolo, F. (2015a) Conserving landraces, improving livelihoods: how to assess the success of on-farm conservation projects? *International Journal of Agricultural Sustainability* 13(2), 167–182.

Bellon, M.R., Gotor, E. and Caracciolo, F. (2015b) Assessing the effectiveness of projects supporting on-farm conservation of native crops: evidence from the High Andes of South America. *World Development* 70, 162–176.

CBD (Convention on Biological Diversity) (2010) Strategic Plan for Biodiversity 2011–2020. Secretariat of the Convention on Biological Diversity, Montreal, Canada. Available at: www.cbd.int/decision/cop/?id=12268 (accessed 30 June 2015).

Chianu, J.N., Ohiokpehai, O., Vanlauwe, B., Adesina, A., De Groote, H. and Sanginga, N. (2009) Promoting a Versatile but yet Minor Crop: Soybean in the Farming Systems of Kenya. *Journal of Sustainable Development in Africa* 10(4), 324–344.

Clement, C.R., Weber, J.C., Leeuwen, J. Van, Domian, C.A., Cole, D.M. and Ar, L.A. (2004) Why extensive research and development did not promote use of peach palm fruit in Latin America. *Agroforestry Systems* 61(1-3), 195–206.

Daniel, J.N. and Dudhade, P.A. (2006) *Analysis of Economic Characteristics of Value Chains of Three Underutilised Fruits of India.* Research Report No. 3. International Centre for Underutilised Crops, Colombo, Sri Lanka.

Dawson, I.K., Guarino, L. and Jaenicke, H. (2007) Underutilised plant species: impacts of promotion on biodiversity. Position Paper No 2, International Centre for Underutilised Crops, Colombo, Sri Lanka.

Devaux, A., Velasco, C., López, G., Bernet, T., Thiele, G., *et al.* (2007) *Collective Action for Innovation and Small Farmer Market Access: The Papa Andina Experience*. CAPRi Working Paper No 68. CGIAR Systemwide Program on Collective Action and Property Rights (CAPRi), c/o International Food Policy Research Institute, Washington, DC.

Drucker, A.G. (2007) Methods for economic valuation. In: Food and Agriculture Organization (eds) *State of the Art in the Management of Animal Genetic Resources. State of the World's Animal Genetic Resources*. Section E, Part 4. FAO, Rome, pp. 429–442.

Drucker, A.G., Padulosi, S. and Jager, M. (2013) No free lunches: PES and the funding of agricultural biodiversity conservation – insights from a competitive tender for quinoa-related conservation services in Bolivia and Peru. Multi-stakeholder dialogue on Case studies on Remuneration of Positive Externalities (RPE)/Payments for Environmental Services (PES), 12–13 September. FAO, Rome.

Drucker, A.G., Narloch, U., Pascual, U., Soto, J.L., Pinto, M., Midler, E., Valdivia, E. and Rojas, W. (2015) *Domestication of Payments for Ecosystem Services: New Evidence from the Andes*. CAPRi Working Paper No. 118, International Food Policy Research Institute, Washington, DC.

Eaton, C.S. and Shepherd, A.W. (2001) *Contract Farming – Partnerships for Growth*. AGS Bulletin 145. FAO, Rome.

FAO (Food and Agriculture Organization) (2007) *The State of the World's Animal Genetic Resources for Food and Agriculture*. FAO, Rome.

FAO (2009) *Second Report on the State of the World's Plant Genetic Resources for Food and Agriculture*. FAO, Rome.

FAO (2013a) *In vivo* conservation of animal genetic resources. *FAO Animal Production and Health Guidelines*. No 14. FAO, Rome. Available at: www.fao.org/docrep/018/i3327e/i3327e.pdf (accessed 30 June 2015).

FAO (2013b) Local biodiversity in production, market and consumption. African Leafy Vegetables in Kenya. Agriculture and nutrition: lessons from the field and good practices. FAO Nutrition Division, Rome. Available at: www.fao.org/fileadmin/user_upload/wa_workshop/Factsheet/2.FSNL_Fact_sheet_African LeafyVegetables_Kenya.pdf (accessed 15 December 2015).

Ferraro, P.J. and Kiss, A. (2002) Direct payments to conserve biodiversity. *Science* 298(5599), 1718–1719.

Garcia-Yi, J. (2014) Market Participation and Agro-Biodiversity Loss: The Case of Native Chili Varieties in the Amazon Rainforest of Peru. *Sustainability* 6(2), 615–630, doi:10.3390/su6020615.

Giuliani, A. (2005) *The Caper in Syria: Nature's Gift to Syria's Resource Poor.* Bioversity International, Rome.

Giuliani, A. (2007) *Developing Markets For Agrobiodiversity. Securing Livelihoods in Dryland Areas.* Earthscan Publications, London.

Giuliani, B.A., and Buerli, M. (2006). Reorganizing the market chain of capers in Syria : a multi-stakeholder process. *Rural Development News* 1, 21–25.

Giuliani, A., Karagöz, A., and Zencirci, N. (2009) Emmer (*Triticum dicoccon*) Production and Market Potential in Marginal Mountainous Areas of Turkey. *Mountain Research and Development* 29(3), 220–229, doi:10.1659/mrd.00016.

Giuliani, A., van Oudenhoven, F., and Mubalieva, S. (2011) Agricultural Biodiversity in the Tajik Pamirs. *Mountain Research and Development* 31(1), 16–26, doi:10.1659/MRD-JOURNAL-D-10-00109.1.

Giuliani, A.F., Hintermann, F., Rojas, W. and Padulosi, S. (2012) *Biodiversity of Andean Grains: Balancing Market Potential and Sustainable Livelihood*. Bioversity International, Rome.

Gotor, E., and Irungu, C. (2010). The impact of Bioversity International's African Leafy Vegetables programme in Kenya. *Impact Assessment and Project Appraisal* 28(1), 41–55, doi:10.3152/14615511 0X488817.

Gotor, E., Caracciolo, F., Canto, G.M.B. and Al Nusairi, M. (2013) Improving rural livelihoods through the conservation and use of underutilized species: evidence from a community research project in Yemen. *International Journal of Agricultural Sustainability* 11(4), 347–362.

Gruère, G.P., Nagarajan, L. and King, E.I.O. (2007) *Collective Action and Marketing of Underutilized Plant Species: The Case of Minor Millets in Kolli Hills, Tamil Nadu, India*. CGIAR Systemwide Program on Collective Action and Property Rights, International Food Policy Research Institute, Washington, DC.

Hampson, K.J., Azam-Ali, S.H., Sesay, A., Mukwaya, S. and Azam-Ali, S.N. (2000) *Assessing opportunities for increased utilisation of bambara groundnut in Southern Africa*. Internal Report. Tropical Crops Research Unit, University of Nottingham. 52 pp.

Heal, G., Walker, B., Levin, S., Arrow, K., Dasgupta, P., *et al.* (2004) Genetic diversity and interdependent crop choices in agriculture. *Resource and Energy Economics* 26, 175–184.

Hermann M. and Heller, J. (eds) (1997) *Andean roots and tubers: Ahipa, arracacha, maca and yacon. Promoting the conservation and use of underutilized and neglected crops*. Institute of Plant Genetics and Crop Plant Research, Gatersleben, Germany and International Plant Genetic Resources Institute, Rome.

Hillocks, R.J., Bennett, C. and Mponda, O.M. (2012) Bambara nut: A review of utilisation, market potential and crop improvement. *African Crop Science Journal* 20(1), 1–16.

Horna, D. and Gruère, G. (2006) Marketing underutilized crops for biodiversity: The case of African garden egg (*Solanum aethiopicum*) in Ghana. In: *8th Bioecon Conference*, Kings College, Cambridge, UK, pp. 1–20.

Hughes, R. and Flintan, F. (2001) *Integrating Conservation and Development Experience: A Review and Bibliography of the ICDP Literature.* International Institute for Environment and Development, London.

Isakson, S.R. (2011) Market provisioning and the conservation of crop biodiversity: An analysis of peasant livelihoods and maize diversity in the Guatemalan highlands. *World Development* 39(8), 1444–1459. doi:10.1016/j.worlddev.2010.12.015.

Keleman, A. and Hellin, J. (2009) Specialty Maize Varieties in Mexico: A Case Study in Market-Driven Agro-Biodiversity Conservation. *Journal of Latin American Geography* 8(2), 147–174.

Krishna, V., Drucker, A.G., Pascual, U., Raghu, P.T. and King, E.D.I.O. (2013) Estimating compensation payments for on-farm conservation of agricultural biodiversity in developing countries. *Ecological Economics* 87, 110–123.

Kruijssen, F., Keizer, M., and Giuliani, A. (2007) *Collective Action for Small-Scale Producers of Agricultural Biodiversity Products.* CAPRi Working Paper No. 71, IFPRI, Washington, DC.

Lauvie, A., Audiot, A., Couix, N., Casabianca, F., Brives, H., *et al.* (2011) Diversity of rare breed management programmes: between conservation and development. *Livestock Science* 140, 161–170.

Looy, T., Carrero, G.O., Mathijs, E., and Tollens, E. (2008) Underutilized agroforestry food products in Amazonas (Venezuela): a market chain analysis. *Agroforestry Systems* 74(2), 127–141, doi:10.1007/s10457-008-9110-0.

LPP, LIFE Network, IUCN–WISP and FAO (2010) Adding value to livestock diversity – marketing to promote local breeds and improve livelihoods. FAO Animal Production and Health Paper No 168. FAO, Rome. Available at: www.fao.org/docrep/012/i1283e/i1283e00.htm (accessed 30 June 2015).

Martinez-de la Cruz, D., Drucker, A.G., Cadima, X., Neves, B. and Gatto, P. (2015) *Assessing Public and Private Sector Current and Potential Involvement in the Conservation and Use of Agrobiodiversity: A Bolivian Case Study*. Bioversity International, Rome.

Mathias, E., Mundy, P. and Köhler-Rollefson, I. (2010) Marketing products from local livestock breeds: an analysis of eight cases. *Animal Genetic Resources* 47, 59–71. Available at: www.fao.org/docrep/013/i1823t/i1823t00.pdf (accessed 30 June 2015).

MEA (Millennium Ecosystem Assessment) (MEA) (2005) *Ecosystems and Human Well-being*. Island Press, Washington, DC.

Muhanji, G., Roothaert, R.L., Webo, C., and Stanley, M. (2011) African indigenous vegetable enterprises and market access for small-scale farmers in East Africa. *International Journal of Agricultural Sustainability* 9(1), 194–202, doi:10.3763/ijas.2010.0561.

Narloch, U., Drucker, A.G. and Pascual, U. (2011a) Payments for agrobiodiversity conservation services for sustained on-farm utilization of plant and animal genetic resources. *Ecological Economics* 70(11), 1837–1845, doi:10.1016/j.ecolecon.2011.05.018.

Narloch, U., Pascual, U. and Drucker, A.G. (2011b) Cost-effectiveness targeting under multiple conservation goals and equity considerations in the Andes. *Environmental Conservation* 38(4), 417–425.

Narloch, U., Pascual, U. and Drucker, A.G. (2013) How to achieve fairness in payments for ecosystem services? Insights from agrobiodiversity conservation auctions. *Land Use Policy* 35, 107–118.

Narloch, U., Drucker, A.G. and Pascual, U. (*in press*) What role for cooperation in conservation tenders? Paying farmer groups in the High Andes. *Land Use Policy* (Special Issue on Competitive Tenders). doi:10.1016/j.landusepol.2015.09.017.

Ngugi, I.K., Gitau, R., and Nyoro, J.K. (2006) *Access to High Value Markets by Smallholder Farmers of African Indigenous Vegetables*. International Institute for Environment and Development, London.

Padulosi, S. and Hoeschle-Zeledon, I. (2004) Underutilized plant species: what are they? *Leisa Magazine* 20(1), 5–6.

Padulosi, S., Bergamini, N. and Lawrence, T. (2011a) *On-Farm Conservation of Neglected and Underutilized Species: Status, Trends and Novel Approaches to Cope with Climate Change.* Proceedings of an International Conference, 14–16 June 2011, Frankfurt. Bioversity International, Rome.

Padulosi, S., Heywood, V., Hunter, D. and Jarvis, A. (2011b) Underutilized species and climate change: current status and outlook. In: Yadav, S.S., Redden, R.J., Hatfield, J.R., Lotze-Campen, H. and Hall, A.E. (eds) *Crop Adaptation to Climate Change*. John Wiley & Sons and Blackwell Publishing Ltd, Chichester, UK, pp. 507–521.

Padulosi, S., Amaya, K., Jager, M., Gotor, E., Rojas, W., *et al.* (2014) A holistic approach to enhance the use of neglected and underutilized species: the case of Andean grains in Bolivia and Peru. *Sustainability* 6(3), 1283–1312.

Pearce, D. and Moran, D. (1994) *The Economic Value of Biodiversity*. Earthscan, IUCN The World Conservation Union, p. 104. Available at: www.cbd.int/doc/external/iucn/iucn-biodiversity-value-1994-en.pdf (accessed 30 June 2015).

Reist-Marti, S.B., Simianer, H., Gibson, J., Hanotte, O. and Rege, J.E.O. (2003) Weitzman's approach and conservation of breed diversity: an application to African cattle breeds. *Conservation Biology* 17, 1299–1311.

Rojas, W., Valdivia, R., Padulosi, S., Pinto, M., Soto, J.L., *et al.* (2009) From neglect to limelight: issues, methods and approaches in enhancing sustainable conservation and use of Andean grains in Peru and Bolivia. *JARTS Supplement* 92, 1–32.

Samuel, A., Drucker, A.G., Andersen, S.B., Simianer, H. and van Zonneveld, M. (2013) Development of a cost-effective diversity-maximising decision-support tool for *in situ* crop genetic resources conservation: the case of cacao. *Ecological Economics* 96, 155–164.

Simianer, H., Marti, S.B., Gibson, J., Hanotte, O. and Rege, J.E.O. (2003) An approach to the optimal allocation of conservation funds to minimize loss of genetic diversity between livestock breeds. *Ecological Economics* 45, 377–392.

Tienhaara, A., Ahtiainen, H. and Pouta, E. (2013) Consumers as conservers – could consumers' interest in a specialty product help to preserve endangered Finncattle? *Agroecology and Sustainable Food Systems* 37, 1017–1039.

UNIDO (United Nations Industrial Development Organization) (2011) *Pro-poor Value Chain Development: 25 Guiding Questions for Designing and Implementing Agroindustry Projects*. UNIDO, Vienna.

Vanhove, W. and Van Damme, P. (2013) Value chains of cherimoya (*Annona cherimola* Mill.) in a centre of diversity and its on-farm conservation implications. *Tropical Conservation Science* 6(2), 158–180.

Vijayalakshmi, D., Geetha, K., Gowda, J., Ravi, S.B., Padulosi, S. and Mal, B. (2010) Empowerment of women farmers through value addition on minor millets genetic resources: a case study in Karnataka. *Indian Journal of Plant Genetic Resources* 23(1), 132–135.

Weitzman, M.L. (1993) What to preserve? An application of diversity theory to crane conservation. *The Quarterly Journal of Economics* 108(1), 157–183.

Wells, M., Guggenheim, S., Khan, A., Wardojo, W. and Jepson, P. (1998) *Investing in Biodiversity. A Review of Indonesia's Integrated Conservation and Development Projects*. World Bank Publications, Washington, DC.

Will, M. (2008) *Promoting Value Chains of Neglected and Underutilized Species for Pro-Poor Growth and Biodiversity Conservation. Guidelines and Good Practices*. Global Facilitation Unit for Underutilized Species, Rome.

Winters, P., Davis, B., Carletto, G., Covarrubias, K., Quinones, E.J., *et al.* (2009) Assets, activities and rural income generation: evidence from a multicountry analysis. *World Development* 37(9), 1435–1452.

Zander, K.K., Drucker, A.G., Holm-Mueller, K. and Simianer, H. (2009) Choosing the 'cargo' for Noah's Ark – applying Weitzman's approach to Borana cattle in East Africa. *Ecological Economics* 68(7), 2051–2057.

# 35 Moving Slowly Towards the Light: A Review of Efforts to Create a Global System for PGRFA Over the Last Half-century

M. Halewood*
*Bioversity International, Rome, Italy*

## 35.1 Introduction

For 50 years, under the auspices of the United Nations (UN), the international community has been working to develop a globally coordinated system to conserve, make available and sustainably use plant genetic resources for food and agriculture (PGRFA). As part of that effort, countries have worked together successfully to develop a range of policy instruments and programmes that are part of the system. But challenges remain – some rooted in tensions still unresolved after half a century. These challenges need to be addressed before all of the requisite elements of a functional, efficient and equitable global system are in place.

It is beyond the scope of this chapter to analyse all the programmes, policies and practices developed so far as part of the evolving global system of PGRFA conservation and use. Such reviews exist already (FAO, 1998; Hodgkin *et al.*, 2013). Furthermore, one has only to look at the recent reports from a number of UN bodies to appreciate that the international community continues to be engaged in developing new instruments, policies and programmes to promote internationally enhanced, coordinated and 'linked up' PGR conservation and sustainable use.[1] Instead, for the most part, this chapter focuses on what it identifies as the historical 'core components' of the global system. And it analyses the progress that has been made, and challenges that have been encountered, in the realization of these core components.

In summary, this chapter argues that for almost 50 years there has been a high level of agreement and commitment within the international community concerning the basic vision and core components of a global system of conservation and use: (i) an internationally linked network of *ex situ* collections providing facilitated access to PGRFA for agricultural research and breeding; (ii) a global information system related to those collections; and (iii) financial support to strengthen the capacity of developing countries to participate in the global system, and incentives or rewards for farmers as diversity stewards. However, getting these components into place and operational has been beset by uncertainties and disagreements about the conditions under which all stakeholders, and countries in particular, are willing to commit themselves to participate. In retrospect, it

*E-mail: m.halewood@cgiar.org

appears most of the uncertainties and disagreements have concerned:

- How much PGRFA should be included and from what sources?
- Who should be required to share monetary benefits, based on what formula?
- What is the legal status of PGRFA and who has the 'deepest' rights of control over them?

This chapter will analyse how the framing and treatment of these issues has been addressed in the context of international efforts to promote and operationalize the global system. As part of this exercise, it is useful to divide the past 50 years of policy development concerning this global system under the UN into three epochs, as set out in Fig. 35.1.

The chapter is divided into sections, each corresponding to one of these epochs. By the end of the chapter, readers should have a clear understanding of the significant progress that the international community has made in the direction of operationalizing these three interlinked components, as well as the outstanding challenges that need to be addressed.

Before proceeding, it is important to make one disclaimer. By focusing on these particular components over the life of the global system, I am not suggesting that other components are unimportant, or that there are not gaps in the overall system that should be addressed. For example, it has been argued that considerably more can and should be done under the aegis of the global system to support *in situ* conservation and sustainable use, in the context of both plant breeding and on-farm and agricultural landscape biodiversity management (Hodgkin *et al.*, 2013). I concur with those arguments. However, such an analysis is beyond the current scope of this chapter.

## 35.2 Epoch 1: Consolidation of Basic Vision with Core Components: 1965–1983

The first epoch starts with the creation of the Food and Agriculture Organization (FAO) Panel of Experts on Plant Exploration in 1965, and

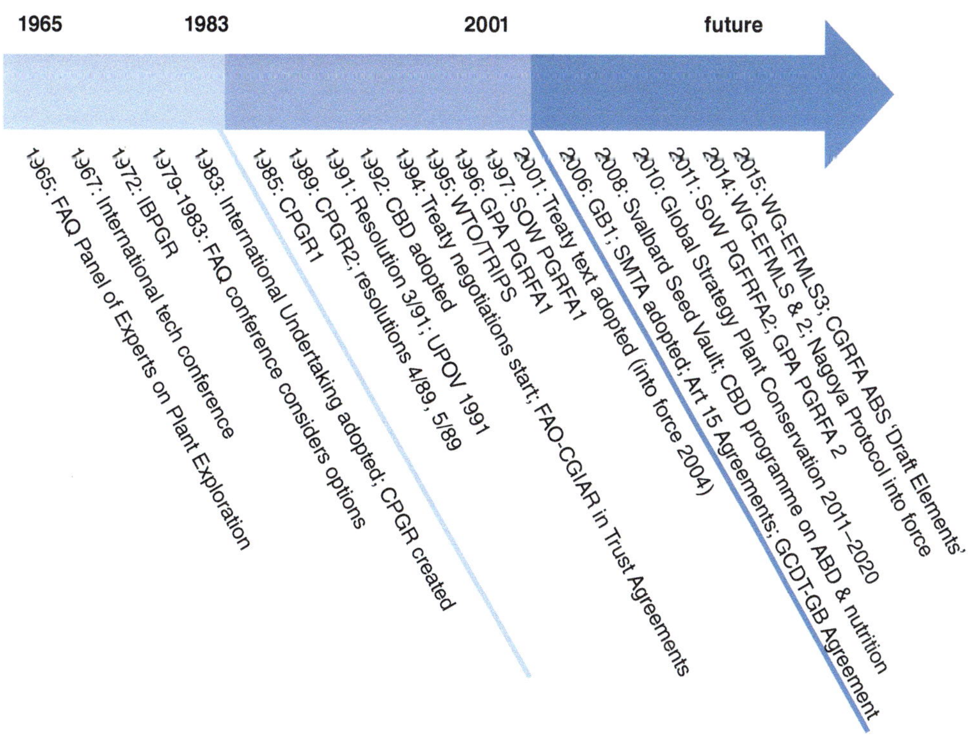


Fig. 35.1. Three epochs of development of a global PGRFA system.
*Note*: ABD = agricultural biological diversity.

concludes in 1983 with the FAO Council's creation of the Commission on Plant Genetic Resources (CPGR) and adoption of the International Undertaking on PGRFA (International Undertaking). In this period, there was a gradual consolidation of the basic vision of the core components of the global system.

Pistorius (1997) and Esquinas-Alcazar *et al.* (2013) provide thorough accounts of the years leading up to the adoption of the CPGR and the International Undertaking in 1983. Other contemporary commentators on events at the time, including Mooney (1983) and Kloppenberg (1988), described concerns about international PGRFA governance and (mis)use; concerns that contributed to the adoption of both the CPGR and International Undertaking (Halewood, 2010). The International Undertaking called for the establishment of a 'global system' building on the 'international arrangements, being carried out under the auspices of FAO and other organizations in the United Nations System, by national and regional institutions and institutions supported by the CGIAR'. The rationale for the creation of the global system, as stated in Article 7 of the International Undertaking, was to ensure the development of: 'an internationally coordinated network of national, regional and international centres, [...] that have assumed responsibility to hold, for the benefit of the international community and on the principle of unrestricted exchange, base or active collections of the plant genetic resources of particular plant species', and 'a global information system, [...] relating to plant genetic resources maintained in the aforementioned collections'. Furthermore, the global system was also meant to ensure that 'institutional capability within developing countries' to participate in the conservation and use-related activities under the International Undertaking were 'adequately funded'. The terms of reference for the newly minted CPGR, as set out in FAO Council Resolution 9/83, included recommending measures to ensure the comprehensiveness of the global system and the efficiency of its operation in line with 'the Undertaking'.

Among the various UN bodies that have been involved in the development of the global system since its creation, the FAO CPGR (later renamed the Commission on Genetic Resources for Food and Agriculture (CGRFA))[2] has been the historical centre.

## 35.3 Epoch 2: Conceptual Revolution/Resolution: 1984–2001

The International Undertaking anticipated that 'governments or institutions' would notify FAO of their willingness to include their collections in the international network 'under the auspices or the jurisdiction of FAO' and that they would, 'whenever requested by FAO, make material in the base collection available to participants in the Undertaking, for purposes of scientific research, plant breeding or genetic resource conservation, free of charge, on the basis of mutual exchange or on mutually agreed terms' (Article 7.2). The International Undertaking also affirmed the 'universally accepted principle that plant genetic resources are a heritage of mankind and consequently should be available without restriction' (Article 1).

It was due largely to uncertainties and tensions around these two points that countries did not proactively commit themselves, or their collections, to the global system during this era. Eight countries abstained from adopting the International Undertaking on the grounds that, among other things, the principle of common heritage (as far as PGRFA was concerned) was inconsistent with their commitments to promoting and respecting plant variety protection (PVP) rights (Mekouar, 2002; Esquinas-Alcazar *et al.*, 2013). Efforts to appease these countries, and to bring them into the International Undertaking fold, culminated 6 years later in FAO Council Resolution 4/89, which offered the interpretation that PVP rights as set out in the International Convention for the Protection of New Varieties of Plants (UPOV) Convention (1978) were consistent with the principle of common heritage.

Furthermore, to appease the appeasers, a companion resolution adopted the same day (Resolution 5/89) set out for the first time in an international agreement, the principle of Farmers' Rights.[3] Resolution 4/89 – also adopted at the same time – stated that 'the best way to implement the concept of Farmers' Rights is to ensure the conservation, management and use of plant genetic resources, for the benefit of present and future generations of farmers', and that this could be achieved through the International Fund for Plant Genetic Resources, already established by FAO. To reflect the responsibility of those countries that have benefitted most from the use of germplasm, the Fund would benefit from being supplemented by further contributions from adhering governments, on a basis to be agreed upon, in order to ensure for the Fund a sound and recurring basis. The International Fund should be used to support plant genetic conservation, management and utilization programmes, particularly within developing countries and those that are important sources of plant genetic material.

Meanwhile, it was noted during the first session of the CPGR, in 1985, that despite their commitments pursuant to the International Undertaking, some countries were, de facto, restricting access to PGRFA within their borders, and that the system 'did not provide all legal guarantees many considered necessary to ensure unrestricted exchange of materials from base collections' (CPGR, 1985, paragraph 29). It was decided that governments and organizations should sign agreements with FAO, formalizing their commitments under the international network, and a number of model agreements were developed over the following years. Countries baulked at models that involved transferring their PGRFA collections 'unconditionally' to FAO and 'renouncing the right to subject the designated germplasm to national legislation' and allowing FAO to 'determine all policies in respect of activities related to the designated germplasm' (CPGR, 1991, Annex J). Instead, almost without exception, CPGR members expressed a preference for model agreements whereby they would place 'designated germplasm' 'under the auspices of FAO within the international network of base collections ...' but 'retain ownership of the resources' and make it 'available when necessary for the purpose of scientific research, plant breeding or genetic resource conservation, without restriction, either directly to users or through FAO, either on mutually agreed terms or free of cost' (CPGR, 1991, Annex J; Halewood, 2010).

Over the same period, dissatisfaction concerning low levels of financial commitments from developed countries for developing country capacity-building activities, lingering resentment about the resolution 5/89 compromise and increasing pressure to extend Intellectual property (IP) protection through World Trade Organization/Agreement on Trade Related Aspects of Intellectual Property Rights (WTO/TRIPs) and UPOV Convention renegotiations reinforced many countries' sense that continued commitment to the principle of common heritage was potentially undermining their ability to negotiate more preferential benefit-sharing terms in return for proactively engaging in the international network (Halewood, 2014). Ultimately, in 1991, the FAO Council adopted Resolution 3/91, which offered the interpretation that the principle of common heritage was, in addition to being subject to the primacy of PVP rights, also subject to countries' sovereign rights of control over PGRFA within their boundaries. The 'universally accepted principle' was now basically overturned. One year later, in 1992, the text of the Convention on Biological Diversity (CBD) was adopted, underscoring the same principle, to wit recognizing 'the sovereign rights of States over their natural resources', and that the 'authority to determine access to genetic resources rests with the national governments and is subject to national legislation' (Article 15.1).

One might imagine that these developments would have signalled the end of countries' efforts to provide a firm legal/policy foundation for the international network of base collections. However, that was not the case. Indeed, while the underlying legal conceptual status of PGRFA had been completely reversed over the period of 1983–1991/92, efforts to develop firm legal/policy footing for the international network (and PGRFA generally) increased in the years that followed (Halewood, 2014). Under the aegis of the CPGR, countries continued their work to develop agreements whereby organizations and countries would include their materials in the international network. Ultimately, no countries ended up signing such agreements, as their efforts to develop acceptable agreements were eventually overtaken, and rendered redundant, by the negotiations of the International Treaty on Plant Genetic Resources for Food and Agriculture (ITPGRFA).[4] The CGIAR Centres signed variants of these agreements with FAO in 1994 to place their collections 'in trust for the benefit of the international community', with the undertaking that the centres would make them available 'without restriction' for 'scientific research, plant breeding, or genetic resources conservation', along with any other available 'related information'. The CGIAR Centres would not claim ownership over the germplasm; nor would they seek intellectual property rights on designated germplasm or related information (SGRP, 2003).

Another development that reflected countries' (both north and south) desire to establish and implement a global system was the adoption, in 1996, by 150 countries, of the Global Plan of Action (GPA) for the Conservation and Sustainable Utilization of PGRFA; it was endorsed by the FAO Conference in 2007 along with *The State of the World's Plant Genetic Resources for Food and Agriculture* (SOW). As stated in the latter, 'The GPA provides the overall framework or blueprint for the Global System and the periodic SOW reports provide a mechanism for monitoring progress and evaluating the system' (FAO, 2010, at 151).

The United Nations Conference on Environment and Development, through Agenda 21, called for a further strengthening of the Global system (UNCED, 1992, paragraph 14.59). In 1994, the Second Conference of the Parties (COP) to the CBD recognized the unique features of agricultural biological diversity and the need for distinctive solutions. In 1996, COP 3 decided to develop a programme of activities on agricultural biological diversity.[5] The CBD deepened its work on agrobiodiversity over the following years, adopting a multi-year programme of work at COP 5 in 2005.[6]

The strongest evidence of unextinguished momentum for advancing realization of the core components of the global system was the launch of renegotiations of the International Undertaking to be in conformity with the CBD (based on the principle of national sovereign rights to regulate access to PGRFA). The Nairobi Final Act, 1992, which adopted the text of the CBD, noted the outstanding issue of *ex situ* collections of PGRFA collected before 1993, which were explicitly excluded from the scope of the CBD.[7] The FAO Council responded by calling on the CPGR to provide a forum for the renegotiation of the International Undertaking, in harmony with the CBD. These negotiations started in 1994 and ended in 2001 with the adoption of the text of the ITPGRFA. The ITPGRFA creates the multilateral system of access and benefit sharing that provides a legally binding foundation for countries to pool and share PGRFA.

The ITPGRFA is based explicitly on the recognition of countries' sovereign rights to control access to genetic resources. Indeed, that principle is one of the basic building blocks of the ITPGRFA's multilateral system of access and benefit sharing. As stated in ITPGRFA Article 10.2: 'In the exercise of their sovereign rights, the Contracting Parties agree to establish a multilateral system, which is efficient, effective, and transparent, both to facilitate access to plant genetic resources for food and agriculture, and to share, in a fair and equitable way, the benefits arising from the utilization of these resources, on a complementary and mutually reinforcing basis.'[8] On the basis of their importance for global food security, and countries' interdependence on them, the Treaty negotiators selected a list of 64 crops and forages to be included in the multilateral system. They are listed in Annex 1 of the Treaty. The PGRFA of the 64 crops and forages that are included in the multilateral system come from three basic sources. The first source is PGRFA that are automatically included in the multilateral system by virtue of a Contracting Party ratifying or acceding to the ITPGRFA. This consists of Annex 1 PGRFA that are 'under the management and control of contracting parties and in the public domain' (Article 11.2). The scope of materials to which this formula extends depends in part on the political and legal systems of the countries concerned, an issue that is discussed at length elsewhere (Nijar, 2012; TAC-MLS/SMTA, 2015). That said, in most countries in the multilateral system, most of the materials in national genebanks (except those subject to intellectual property rights (IPRs) or held in black box conditions) would likely be automatically included, and materials in farmers' fields or held by companies would not (since they are under the management or control of the farmers and companies concerned). It is important to note that the adoption of this formula obviates the need for the kinds of agreements with countries that were being developed under the framework of the International Undertaking.

The second source of PGRFA in the multilateral system is that which is voluntarily included by natural and legal persons (Article 11.2). Member states agree under the Treaty to provide incentives for such voluntary inclusions (Article 11.3). The third source of PGRFA in the multilateral system is international institutions that sign agreements with the governing body to place their collections under the auspices of the ITPGRFA.

ITPGRFA member states agree to provide facilitated access to materials in the multilateral system for the purposes of 'utilization and conservation for research, breeding and training for food and agriculture, provided that such purpose does not include chemical, pharmaceutical and/or other non-food/feed industrial uses' (Article 12.3(a)). Such access is to be provided for free, or for minimal administrative fees (Article 12.3 (b)). In the event that a recipient incorporates material accessed from the multilateral system into a new PGRFA product, commercializes it and restricts others from using that product for further research and breeding, that recipient is obliged to pay an equitable share of benefits to an international benefit-sharing fund (BFS) established under the ITPGRFA. The ITPGRFA text did not include all of the necessary details for the monetary benefit-sharing clauses to be immediately operational. Instead, it committed the first meeting of the governing body (after the ITPGRFA came into force) to 'determine the level, form and manner of the payment, in line with commercial practice' (Article 13.2.d(ii)). Negotiations to resolve these issues took place during the third epoch described below.

The ITPGRFA created a financial mechanism or 'Trust Account' (as it is referred to under Article 19.3(f)) – later named the International Benefit-sharing Fund – for the collection and disbursement of money. The funds are meant to be used to support priority activities under the GPA and to support farmers in developing countries (and economies in transition) who conserve and use plant genetic resources sustainably.

The Treaty also calls for the creation of a global information system concerning scientific, technical and environmental issues related to PGRFA.

The ITPGRFA is an international, legally binding agreement regarding core components of the global system. In this regard, it represents a significant advance beyond the never-finalized efforts to formalize conditions for countries' participation in the international network of collections under the framework of the International Undertaking, during a period when PGRFA were considered common heritage. In retrospect, it seems reasonable to conclude that the gradual shift from conceiving of PGRFA as common heritage to seeing them as subject to IPRs and national sovereignty contributed to this progress. It appears that once countries' deeper sovereign rights of control over plant genetic resources had been recognized, they were more comfortable moving forward with developing rules for the international sharing of those same resources. In the end, countries proactively exploited tools developed to control access to genetic resources to construct a system to share them: the multilateral system of access and benefit-sharing (Halewood, 2014).

## 35.4 Epoch 3: The Plant Treaty Era: 2001–?

### 35.4.1 Progress in the implementation of the ITPGRFA

In the 15 years since the ITPGRFA was adopted, there has been significant progress in its institutionalization and implementation. It came into force in 2004. A special intergovernmental group was formed to negotiate the standard material transfer agreement (SMTA) to use for all transfers of PGRFA in the multilateral system. That committee met over the course of 2 years, leading up to the first meeting of the governing body in 2006. The committee's negotiations were extremely important, since the treaty text adopted in 2001 left a number of access and benefit-sharing issues still unresolved related to the form and amount of necessary payments, as noted above. In 2006, the first meeting of the Governing Body of the Parties adopted the SMTA. It includes a default benefit-sharing obligation (the 'Article 6.7 option'), which is triggered if a recipient commercializes a new PGRFA product that incorporates any material from the multilateral system, and refuses to allow others to use the new product for research and plant breeding. In such cases, the commercializer is obliged to send 1.1% of gross sales of that product to the BSF. Alternatively, recipients can elect to pay 0.5% of all its gross sales of all PGRFA products of the same crop over a period of 10 years, regardless of whether they incorporate materials from the multilateral system of access and benefit-sharing (MLS) and regardless of whether they are available for others to use for further research and breeding. In return, the recipient enjoys facilitated access to all PGRFA of that particular crop for the same period of time, from all potential providers in the multilateral system. The recipient can renew this arrangement or allow it to lapse, after which time, if they incorporate the accessed material in a new PGRFA product, they will be obliged to make a payment on all sales of that particular new PGRFA product (from hereon, this shall be referred to as the Article 6.11 option).

The governing body adopted a Funding Strategy in 2006. In 2009, the governing body adopted the Strategy for the Implementation of the Benefit-sharing Fund, which set a financial target of US$116 million for the BSF over the course of 5 years, ending in 2014. It was anticipated that a relatively small proportion of this total would come from commercial users through the operation of the mandatory financial benefit-sharing formula. While the BSF has fallen considerably short of its target, enough money was made available from voluntary contributions from governments to support three rounds of disbursements to date for projects in developing countries. The first round of proposals was approved in 2009 (total value US$580,000), the second round was approved in 2011 (total value US$5.5 million) and the third round was approved in 2015 (total value over US$10 million). The governing body also signed an agreement with the Global Crop Diversity Trust (GCDT), recognizing it as an essential element of the Funding Strategy. The GCDT has, among other things, provided considerable financial and technical support to a number of genebanks around the world, coordinated a global programme of regeneration and safety backup (and availability through the multilateral system) of unique, threatened accessions of Annex 1 crops and forages, and provided support for the operation of the Svalbard Global Seed Vault.[9] Seventeen international organizations have signed agreements with the ITPGRFA governing body, placing their collections under the Treaty's framework. Out of the 17, 11 are International Agricultural Research Centres of the CGIAR, which have included approximately 700,000 accessions under the Treaty framework.

Each year, the CGIAR Centres distribute approximately 500,000 samples of PGRFA to recipients around the world under the SMTA: with approximately 86% sent to developing countries and countries with economies in transition. Approximately 75% of that material has been distributed by the Centres' breeding programmes; and 25% by the genebanks. Details about the Centres' distributions and acquisitions of materials, and the challenges they have faced, and practices they have developed to operate under their agreements with the ITPGRFA's governing body, have been reported elsewhere.[10] The ITPGRFA Secretariat estimates that approximately 1.6 million accessions are included in the MLS, with a significant representation of the genepool of food security crops and their wild relatives. Approximately 600 accessions are transferred every day through the MLS.[11] As of April 2015, 132 countries were members of the ITPGRFA.

Despite this positive progress, the ITPGRFA has encountered a number of challenges. Over the period of 2009–2014, the BSF received only approximately 14% of the projected funds needed. Almost all of the contributions to the BSF have been voluntary contributions from governments or international agencies. No money has been directed to the fund as a result of the operation of the mandatory benefit-sharing conditions included in the SMTA. Many of the commercial users who would be most likely to trigger the mandatory benefit-sharing clauses of the SMTA are not taking materials from the multilateral system. Either they have sufficient genetic diversity to meet their needs for the time being, or they are able to access materials from other sources. They complain that both of the benefit-sharing formulae include clauses and obligations that are too far removed from standard commercial practices for them to choose freely to take materials from the multilateral system and expose their companies to legal benefit-sharing conditions. One of their main complaints is that the default benefit-sharing formula is triggered by *any* incorporation of multilateral system PGRFA into new products, rather than an easier to track and trace minimum percentage threshold (e.g. 10%). A second, closely related complaint is that the benefit-sharing formula extends benefit-sharing obligations indefinitely to all uses of accessed material, no matter how far into the future. The ultimate end result being that the benefit-sharing obligation will have a viral effect, eventually spreading to a wider range of materials in seed companies' portfolios (ISF, 2007). To date, there have been no published reports of recipients opting to receive material under the Article 6.11 option.

Furthermore, the amount of PGRFA that is functionally included in the multilateral system is also lower than might reasonably have been expected when the Treaty came into force and the SMTA was adopted. This outcome is partly a practical consequence of the formula regarding materials that are automatically included in the multilateral system. Unless information about such materials is actually published, it is difficult, if not impossible, for potential users to know it exists and to ask for it. Partly in recognition of this fact, the ITPGRFA Secretary sent messages to all contracting parties asking them to confirm what collections or materials were included in the multilateral system, with the idea that such information could be published on the Treaty website. Governing body resolutions regarding the multilateral system have reiterated similar requests. Responses from parties clarifying what materials are included have been returned relatively slowly. As of April 2015, only 32 countries out of the 132 state parties (24%) had sent responses.[12] Furthermore, apart from some details included in country reports to the governing body, there is very little evidence of natural and legal persons voluntarily including PGRFA in the multilateral system. It is fairly clear that the issues of lower than targeted benefit-sharing funds and lower than reasonably expected levels of inclusion are interlinked. Some country representatives have stated in governing body and intersessional meetings that the lack of funds directed to the BSF creates a disincentive for them to proactively identify materials already in the multilateral system and/or to encourage natural and legal persons to include materials voluntarily.

In light of the observed behaviour of potential commercial users, and potential providers, of PGRFA through the multilateral system, it has become evident that some readjustments to the access and benefit-sharing conditions that were struck by the negotiators of the ITPGRFA in 2001, and the SMTA in 2006, are necessary. To this end, at its fifth session, in September 2013, the ITPGRFA governing body created the Ad Hoc Open Ended Working Group to Enhance the Functioning of the MLS (WG-EFMLS), with the mandate to develop a range of optional measures to: '(a) increase user-based payments and contributions to the Benefit-sharing Fund in a sustainable and predictable long-term manner, and (b) enhance the functioning of the Multilateral System by additional measures'.[13] The first two meetings of the WG-EFMLS were held in May and December 2014 in Geneva.[14] A third meeting will be held before the sixth session of the governing body in October 2015.

So far, the WG-EFMLS has focused on measures to increase user-based payments. In this context, most attention is being focused on options to simultaneously increase payments to the multilateral system and create a flatter, lower-transaction cost system with minimal tracking and tracing requirements to respond to some of the complaints from commercial users. In principle, the most direct way to increase contributions to the BSF, to lower costs/concerns about tracking and tracing, and increase PGRFA available through the MLS would be for member states to agree to make annual payments to the BSF based on a percentage of all seed sales within their borders. They could then recoup those amounts from companies (or public research organizations, if they are selling seed) according to their own national systems. However, a number of contracting parties have indicated they are not willing to consider this option; instead, they insist that the mandate of the WG-EFMLS is to look for options that apply directly to commercial users and not to national governments as intermediaries.

One reform on which most regions appear to agree is that some level of payment should be required from commercializers who make their products available for research and breeding. This should be a lower rate than that paid by commercializers who insist on restrictions.

So far, the most animated discussions within the WG-EFMLS have focused on developing a revised form of Article 6.11, to make it more attractive, and further lower the transaction costs associated with its operation. Among other things, this could include increasing the difference in payment rates between Articles 6.11 and 6.7. Some regions (and observers) are suggesting that the existing default Article 6.7 option should be eradicated entirely, and that the SMTA should only include the revised Article 6.11 formula. According to a revised Article 6.11 option, the SMTA would state that the recipient would have facilitated access to all the PGRFA of a particular crop, or to all crops, in the multilateral system, and would, in return, make an annual payment to the BSF based on sales of PGRFA products of that crop, or all crops combined. The obligation would last a number of years, after which it could be renewed. This much is consistent with the current Article 6.11 option. The innovation currently being considered by the WG-EMLS is to introduce an exhaustion date, after which time it would not be necessary to share monetary benefits based on commercialization. There are a number of exhaustion formulae under discussion; perhaps the simplest is that the benefit-sharing formula would not be applicable 10 years (or some other number of years) after the last receipt of materials from the multilateral system. This represents an important change from the current system, whereby after the expiry of the second option, payments would revert to a hybridized form of the default option, which reintroduces, de facto, the need to track and trace all along, thereby vitiating the purported track and trace free benefits of the second option. And the benefit-sharing requirement continues in perpetuity.

It is too early to predict what options the WG-EFMLS will ultimately forward to the governing body for consideration at its next meeting in October 2015, or what the governing body will ultimately decide. It is quite possible that the governing body will not be able to make a final decision at that time, and will need to prolong the process.

The fact that the governing body decided to launch this process of review and (hopefully) renewal reflects positively on the member states of the ITPGRFA as a collective community. The ITPGRA was not, and is not, a failed international agreement. While not living up to many people's expectations, it is, nonetheless, performing an important function. Launching such a process comes with risks; parties may not agree to any revisions, and a failure to agree on a new formula could lead to unprecedented levels of dissatisfaction with the status quo. Alternatively, a new formula that misjudges the willingness of users to subject themselves to a new, revised monetary benefit-sharing formula could lead to still more avoidance of the multilateral system as a whole. That said, the fact that approximately 130 member states collectively decided (as the governing body) to take the risk and bet on a favourable outcome is evidence of the widely shared appreciation of the importance of PGRFA for local, national and global food security, and our international interdependence on those resources. Furthermore, with the adoption of the ITPGRFA in 2001 and the 9 years of experience operating under the MLS (2006–2015), the global community may reasonably feel, looking backwards, that the extra distance it needs to cover now is not nearly as hazardous or long as the terrain it has already covered during the previous epochs of consolidation and conceptual revolution. If commitment to operationalizing the core elements of the global system could weather the principle of universal heritage being superseded by intellectual property rights and national sovereignty, and the International Undertaking and the International Network of *Ex situ* Collections being superseded by the ITPGRFA (FAO, 2010, at 151), and 30 years of fighting over funding, then there are good reasons to believe it will survive – and guide – the revision of the access and benefit-sharing provisions of the SMTA. If successful, it is hard to say how long the Plant Treaty epoch will last; it could be a very long time.

### 35.4.2 Progress in other fora: CBD, CGRFA

Up until this point, I have focused entirely on issues related to the ITPGRFA in this description of the third epoch. However, there have been a number of other very important developments during this period. Some of these developments represent organic growth and maturation of the global system, building on previous progress. For example, 111 countries submitted national reports to form the basis for the *Second Report on the State of the World's Plant Genetic Resources for Food and Agriculture*, which was endorsed by the CGRFA in 2009. In 2011, the CGRFA adopted the Second Global Plan of Action for Plant Genetic Resources for Food and Agriculture. The Svalbard Global Seed Vault opened its doors for the first deposit of reproductive materials in 2010. It currently holds 864,000 seed samples, of over 5000 species, collected from 232 countries.[15] While the deposited seeds are not considered global public goods (future access is restricted to the actual depositors),[16] the creation of this ultimate backup collection represents an important step in the evolution of the global system.

Other developments have taken place more or less in parallel, in other fora, beyond the FAO, CGRFA and the governing body of the ITPGRFA. The Eighth Conference of the Parties to the CBD (COP 8) in 2008 adopted a cross-cutting programme on agricultural biological diversity and nutrition. COP 10, in 2010, adopted the Global Strategy for Plant Conservation 2011–2020. It also adopted the text of the Nagoya Protocol on Access to Genetic Resources and the Fair and Equitable Sharing of Benefits Arising from their Utilization at the Convention on Biological Diversity.[17] It came into force in November 2014. The COP 10/11 decision adopting the text of the Nagoya Protocol makes specific (positive) references to the ITPGRFA, as does the protocol's preamble. While the ITPGRFA is not mentioned explicitly in Article 4 'Relationship to other international agreements', that article clearly operates to recognize both the ITPGRFA and the agreements that international organizations signed with the ITPGRFA governing body, placing their materials under the ITPGRFA's framework. As such, it is clear that the ITPGRFA and the CBD and Nagoya Protocol do not actually overlap, and they do cover different materials and/or uses of materials. They can, and should be, implemented in mutually supportive ways (Cabrera *et al.*, 2013; Halewood *et al.*, 2013; Halewood, 2015). The Nagoya Protocol Article 8 on 'special considerations' states that in the development and implementation of its access and benefit-sharing legislation or regulatory requirements, each Party shall 'consider the importance of genetic resources for food and agriculture and their special role for food security'. While this can be interpreted and implemented in a number of ways, it makes sense that countries should consider implementing this article by developing procedures for the expedited consideration of requests for access to genetic resources for food and agriculture that are not included in the multilateral system of the ITPGRFA. In January 2015, the CGRFA adopted the Draft Elements to Facilitate Domestic Implementation of Access and Benefit-Sharing for Different Subsectors of Genetic Resources for Food and Agriculture.[18] The Draft Elements underscore the distinctive nature of GRFA (including PGRFA) and suggest principles that countries can follow when developing mechanisms to regulate access to GRFA not covered by the multilateral system (or Article 15 agreements) in the spirit of Nagoya Protocol Article 8. While not traditionally listed among the instruments that are part of the global system, it is important that countries implement the Nagoya Protocol in ways that complement, and support, the conservation and sustainable use of plant genetic resources for food and agriculture that are not included in the multilateral system of access and benefit sharing, particularly for food security crops.

## 35.5 A Brief Review of Progress to Date on the Key Components

As far as what I have described as the core components are concerned, where do things stand in 2015? Harkening back to the vision set out in the International Undertaking, the multilateral system constitutes 'an internationally linked network of *ex situ* collections providing facilitated access to PGRFA for agricultural research and breeding'. Significantly, it also includes, at least in principle, *in situ* materials that are either 'under the management and control of contracting parties and in the public domain' (Moore and Tymowski, 2005) or that are included voluntarily. In this respect, the multilateral system exceeds the expectations of the International Undertaking.

It seems likely that many of the *ex situ* collections that negotiators had in mind when the International Undertaking was adopted, and when they were subsequently developing model agreements for including collections in the network, are actually currently included in the multilateral system by virtue of the combination of: (i) the country concerned ratifying the ITPGRFA and (ii) the automatic inclusion of materials that are under the management and control of the contracting party and in the public domain (Thormann *et al.*, 2015). That said, more research is necessary to know the proportion of those collections actually available under the multilateral system.

A range of would-be providers are holding back from proactive engagement with the multilateral system, partly because they are unhappy with the benefit sharing that has been possible under the multilateral system. In this regard, the multilateral system has not managed to overcome some of the historical tensions that characterized the unsuccessful process to develop agreements for countries to place collections in the international network (under the International Undertaking).

The range of crops and forages included in the multilateral system is narrower than what was considered as potentially part of the global system in the first epoch, where no crop lists or limits were established under the International Undertaking. As a result, there is a risk that a range of important food security crops that are left out of Annex 1 will receive less financial support for collection maintenance or research through the BSF or GCDT (Hodgkin *et al.*, 2013). And of course, since they are not included in the multilateral system, there is no guarantee of facilitated access to the PGRFA of those crops for the purposes of agricultural research and breeding. What is not available through the multilateral system needs to be accessed subject to whatever other national access laws exist.

It is not necessary in this context to repeat the previous discussion about the state of the BSF. Most, perhaps all, developing countries and developing country farmers would say that the vision of financial support generated through an international benefit-sharing fund for their capacity strengthening (as conceived of in the early days of the CPGR, resolutions 4/89 and 5/89 and the ITPGRFA itself) is far from adequately realized.

On the other hand, the international community is proactively engaging in a process that could re-set the benefit-sharing formulae to increase the flow of resources to the BSF, while making the multilateral system more attractive and useful for commercial users, and creating incentives for providers to proactively make materials available through the system. It is too early to predict whether these efforts will be successful, but it seems that, under the framework of the ITPGRFA, solutions to long-standing challenges are within grasp.

The global information system envisaged in the International Undertaking and called for again in the ITPGRFA does not yet exist, *per se*. In the interim, between the fifth and sixth sessions of the governing body, the ITPGRFA Secretariat has coordinated a background study concerning elements that could be included in the global information system. Quite likely, the system will consist of a linked set of existing (and future) accession-level information systems, with the capacity to link in characterization and evaluation data from a range of sources. It could also include information about *in situ* PGRFA. One of the biggest current information nodes that could play an important role in the evolution of the global information system is Genesys, which aggregates accession-level information about materials held in the international collections hosted by the CGIAR Centres, European National Inventories (whose records are recorded in the European Plant Genetic Resources Search Catologue (EURISCO, 2015)) and the US Department of Agriculture. Efforts to support national genebanks to adopt and use standardized information systems, and eventually link them into a global information system network, will be very important. The scene is set for this kind of information sharing; the SMTA includes an undertaking by recipients to share non-confidential information derived from use of the multilateral system through the global information system. As soon as some system (or network of systems) is officially recognized as 'the Article 17 global information system', it will be possible for researchers to get into the habit of sharing information through that system. Again, it appears that the official creation, and longer-term evolution, of a global information system is within reach.

If the international community (supported by non-governmental groups) is successful in these most recent efforts, and these components of the global system are finally, after 50 years, operationalized, the Plant Treaty epoch will be set to continue for a long time to come.

## Endnotes

[1] For example, during its last session in 2015, the Commission on Genetic Resources for Food and Agriculture, of the Food and Agriculture Organization of the UN, considered targets and indicators for PGRFA, endorsed voluntary guidelines for national PGRFA strategies and national seed policies, Voluntary Guidelines for Mainstreaming Biodiversity into Policies, Programmes and National and Regional Plans of Action on Nutrition, and Voluntary Guidelines to Support the Integration of Genetic Diversity into National Climate Change Adaptation Planning.

[2] From its creation in 1983, it was called the Commission on Plant Genetic Resources. It was renamed the Commission on Genetic Resources for Food and Agriculture to reflect a broadened mandate in 1995.

[3] Resolution 5/89 states that 'Farmers' Rights mean rights arising from the past, present and future contributions of farmers in conserving, improving, and making available plant genetic resources, particularly those in the International Community, as trustee for present and future generations of farmers, for the purpose of ensuring full benefits to farmers, and supporting the continuation of their contributions, as well as the attainment of the overall purposes of the International Undertaking' (both Resolutions 4/89 and 5/89 are available at: www.fao.org/docrep/x5588E/x5588e06.htm#e.%20commission%20on%20plant%20genetic%20resources%20and%20the%20international%20undertaking:%20prog).

[4] A CGRFA report from 1997 stated that continued work developing draft agreements with countries was 'to some extent on hold, pending the outcome of the negotiations for the revision of the International Undertaking', as they would 'have a bearing on the final form and content of future agreements' (CGRFA, 1997).

[5] www.cbd.int/convention/results/?id=7107&l0=AGR&l2=<NONE>&l4=<NONE>&l5=<NONE>&l6=<NONE>&l7=<NONE>&l8=<NONE> (accessed 13 April 2015).

[6] www.cbd.int/decision/cop/default.shtml?id=7147 (accessed 13 April 2015).

[7] Paragraph 4 of Resolution 3 ('The Interrelationship Between the Convention on Biological Diversity and the Promotion of Sustainable Agriculture') of the Nairobi Final Act states that the conference recognized 'the need to seek solutions to outstanding matters concerning plant genetic resources within the Global System for the Conservation and Sustainable Use of Plant Genetic Resources for Food and Sustainable Agriculture, in particular: (a) Access to *ex situ* collections not acquired in accordance with this Convention; and (b) The question of farmers' rights' (available at: www.cbd.int/doc/handbook/cbd-hb-09-en.pdf).

[8] The following recital is also included in paragraph 13 of the Preamble of the ITPGRFA: 'in the exercise of their sovereign rights over their plant genetic resources for food and agriculture, states may mutually benefit from the creation of an effective multilateral system for facilitated access to a negotiated selection of these resources and for the fair and equitable sharing of the benefits arising from their use'.

[9] For more about the Svalbard Global Seed Vault see: www.regjeringen.no/en/topics/food-fisheries-and-agriculture/agriculture/svalbard-global-seed-vault/id462220/.

[10] The CGIAR reports to the Second, Third and Fourth Sessions of the Governing body are available at: www.planttreaty.org/sites/default/files/gb2i11e.pdf, www.planttreaty.org/sites/default/files/gb3i15e.pdf, and www.planttreaty.org/sites/default/files/gb4i05e.pdf, respectively.

[11] See the ITPGRFA website at: www.planttreaty.org/content/recent-progress.

[12] The list of country notifications is posted on the ITPGRFA website at: www.planttreaty.org/inclusions.

[13] See governing body resolution 2/2013 at: www.planttreaty.org/sites/default/files/RES2_2013_en.pdf.

[14] The official reports of these meetings are available at: www.planttreaty.org/sites/default/files/OWG-EFMLS-1_RE.pdf and www.planttreaty.org/sites/default/files/OWG-EFMLS-2_en.pdf.

[15] www.nordgen.org/sgsv/index.php? page=welcome (accessed 13 April 2015).

[16] www.nordgen.org/sgsv/scope/sgsv/files/SGSV_Deposit_Agreement.pdf (accessed 13 April 2015).

[17] www.cbd.int/decisions/cop/? m=cop-10 (accessed 13 April 2015).

[18] 'Elements to Facilitate Domestic Implementation of Access and Benefit-Sharing for Different Subsectors of Genetic Resources for Food and Agriculture', Appendix 2 of the Report of the Fifteenth Regular Session of the Commission on Genetic Resources for Food and Agriculture, CGRFA-15/15/Report, available at: www.fao.org/3/a-mm660e.pdf.

## References

Cabrera Medaglia, J., WalløeTvedt, M., Perron-Welch, F., Jørem, A. and Phllips, F.-K. (2013) The interface between the Nagoya Protocol on ABS and the ITPGRFA at the international level: potential issues for consideration in supporting mutually supportive implementation at the national level. FNI Report 1/2013. Fridtjof Nansen Institute, Lysaker, Norway. Available at: www.abs-initiative.info/629.html (accessed 28 February 2015).

CGRFA (1997) *Report of the Eighth Regular Session of the Commission on Genetic Resources for Food and Agriculture*. CGRFA-8/99/REP Commission on Genetic Resources for Food and Agriculture, Food and Agriculture Organization of the United Nations (FAO), Rome.

CPGR (1985) *Report of the First Session of the Commission on Plant Genetic Resources*. Rome, 11–15 March 1985, Doc. CPGR/85/REP. Commission on Genetic Resources for Food and Agriculture, FAO, Rome.

CPGR (1991) *Report of the Fourth Session of the Commission on Plant Genetic Resources to the Ninety-Ninth Session of the UNFAO Council*. Doc. CL 99/16. Commission on Genetic Resources for Food and Agriculture, FAO, Rome.

Esquinas-Alcazar, J., Hilmi, A. and Lopez Noriega, I. (2013) A brief history of the negotiations of the International Treaty on Plant Genetic Resources for Food and Agriculture. In: Halewood, M., Lopez Noriega, I. and Louafi, S. (eds) *Crop Genetic Resources as a Global Commons: Challenges in International Governance and Law*. Routledge, Oxford, UK, pp. 35–49.

EURISCO (2015) EURISCO database. Available at: eurisco.ipk-gatersleben.de/apex/f?p=103:1:20535553703486 (accessed 17 August 2015).

FAO (Food and Agriculture Organization) (1998) *The State of the World's Plant Genetic Resources for Food and Agriculture*. FAO, Rome.

FAO (2010) *The Second Report on the State of the World's Plant Genetic Resources for Food and Agriculture*. FAO, Rome.

Halewood, M. (2010) Governing the management and use of pooled microbial genetic resources: lessons from the global crop commons. *International Journal of the Commons* 4(1), 404–436. Available at: www.thecommonsjournal.org/index.php/ijc/article/view/152/114 (accessed 17 August 2015).

Halewood, M. (2013) What kind of goods are plant genetic resources for food and agriculture? Towards the identification and development of a new global commons. *International Journal of the Commons* 7(2), 278–312.

Halewood, M. (2014) International efforts to pool and conserve crop genetic resources in times of radical legal change. In: Cimoli, M., Dosi, G., Maskus, K.E., Okediji, R.L. and Reichman, J.L. (eds) *Intellectual Property Rights: Legal and Economic Challenges for Development*. Oxford University Press, Oxford, UK, pp. 288–322.

Halewood, M. (ed.) (2015) *Mutually Supportive Implementation of the Plant Treaty and the Nagoya Protocol: A Primer for National Focal Points and Other Stakeholders*. Discussion Paper. Bioversity International, Rome.

Halewood, M., Andrieux, E., Crisson, L., Gapusi, J.R., Wasswa Mulumba, J., *et al.* (2013) Implementing 'mutually supportive' access and benefit sharing mechanisms under the Plant Treaty, Convention on Biological Diversity, and Nagoya Protocol. *Law, Environment and Development Journal* 9(2), 68–96.

Hodgkin, T., Demers, N. and Frison, E. (2013) The evolving global system of conservation and use of plant genetic resources for food and agriculture: what is it and where does the Treaty fit in? In: Halewood, M., Lopez Noriega, I. and Louafi, S. (eds) *Crop Genetic Resources as a Global Commons: Challenges in International Governance and Law*. Routledge, Oxford, UK, pp. 344–367.

ISF (International Seed Federation) (2007) *ISF Position Paper on Access to Plant Genetic Resources for Research and Breeding.* Available at: www.worldseed.org/isf/on_sustainable_agriculture.html (accessed 5 January 2016).

Kloppenburg, J. (1988) *First the Seed. The Political Economy of Plant Biotechnology. 1492–2000.* Cambridge University Press, Cambridge, UK.

Mekouar, A. (2002) A global instrument on agrobiodiversity: the International Treaty on Plant Genetic Resources for Food and Agriculture. FAO Legal Papers Online. Available at: www.fao.org/fileadmin/user_upload/legal/docs/lpo24.pdf (accessed 5 January 2016).

Mooney, P. (1983) The law of the seed. Another development and plant genetic resources. *Development Dialogue* 1(2), 1–172.

Moore, G. and Tymowski, W. (2005) *Explanatory Guide to the International Treaty on Plant Genetic Resources for Food and Agriculture.* International Union for the Conservation of Nature, Gland, Switzerland.

Nijar, G.S. (2012) *Malaysia's Implementation of the Multilateral System of Access and Benefit Sharing.* Bioversity International, Rome, and Malaysian Agricultural Research Development Institute, Kuala Lumpar.

Pistorius R. (1997) *Scientists, Plant and Politics. A History of the Plant Genetic Resources Movement.* International Plant Genetic Resources Institute, Rome.

System-wide Genetic Resources Programme (SGRP) (2003) Booklet of CGIAR Centre Policy Instruments, Guidelines and Statements on Genetic Resources, Biotechnology and Intellectual Property Rights. Version II. Rome: Bioversity International, Rome. Available at: www.bioversityinternational.org/e-library/publications/detail/booklet-of-cgiar-centre-policynbsp-instruments-guidelines-and-statements-on-genetic-resources-biotechnology-and-intellectual-property-rights-version-ii-rome-july-2003/ (accessed 5 January 2015).

TAC-MLS/SMTA (2015) *Opinions and Advice of the Ad Hoc Technical Advisory Committee on the Multilateral System and the Standard Material Transfer Agreement*. FAO, Rome.

Thormann, I., Fiorino, E., Halewood, M. and Engels, J.M.M. (2015) Plant genetic resources collections and associated information as a baseline resource for genetic diversity studies: an assessment of the IBPGR-supported collections. *Genetic Resources Crop Evolution* 62(3), doi:10.1007/s10722-015-0231-9.

UNCED (1992) *Agenda 21*. UNCED, Rio de Janeiro, Brazil. Available at: sustainabledevelopment.un.org/content/documents/Agenda21.pdf (accessed 17 August 2015).

# 36 On the Conservation and Sustainable Use of Plant Genetic Resources in Europe: A Stakeholder Analysis

L. Frese,[1*] A. Palmé,[2] G. Neuhaus,[1] L. Bülow,[1] N. Maxted,[3] G. Poulsen[2] and C. Kik[4]

[1]*Julius Kühn Institute, Federal Research Centre for Cultivated Plants, Institute for Breeding Research on Agricultural Crops, Quedlinburg, Germany;* [2]*Nordic Genetic Resource Center, Alnarp, Sweden;* [3]*School of Biosciences, University of Birmingham, Birmingham, UK;* [4]*Centre for Genetic Resources, the Netherlands (CGN), Wageningen University and Research Centre, Wageningen, the Netherlands*

## 36.1 Introduction

Europe will in the future be facing increasing challenges for food security, largely caused by a growing human population and a changing climate (Underwood *et al.*, 2013). In addition, other factors such as changing consumer demands and environmental regulations are likely to have a considerable impact on food production. Actions on several different levels are needed to address these challenges (Underwood *et al.*, 2013). A central issue is the genetic enhancement of crops to give large, high-quality harvests in sustainable agricultural production systems.

Adapting crops to new conditions and demands is possible, and is something that humans have done since they invented agriculture. However, adaptation is not possible without genetic variation, and genetic variation within modern crops is limited. On the other hand, there is a wealth of genetic variation available in former breeder's varieties, landraces (LR) and crop wild relatives (CWR), and traits that are predicted to be important in breeding in the future, such as disease and pest resistance, drought/salt tolerance and nutrient-use efficiency, can be found among these important plant genetic resources (PGRs). It is therefore essential that these resources are conserved safely and that their use by the breeding and research sectors, as well as farmers working in the conventional and ecological farming sector, is facilitated (Underwood *et al.*, 2013).

Here, we present a study on the current status of conservation and use of plant genetic resources in Europe (current chapter) and on how this could be improved and developed towards a European Plant Germplasm System (EU-PGS) (see Frese *et al.*, Chapter 37, this volume, 2016). In the current chapter, we focus on the methods we used, our vision of what an improved system for conservation and use of plant genetic resources should look like and also the main strengths, weaknesses, opportunities and threats in the current system.

*Corresponding author; e-mail: lothar.frese@jki.bund.de

## 36.2 Approach

### 36.2.1 Data collection

The main aim of our study was to gain a better understanding of the plant germplasm system (PGS) in Europe, and based on that knowledge, develop a vision for an improved PGS and recommend actions to attain this vision. Our first step towards this goal was to identify the main PGS stakeholders, these being: genebanks, research organizations, breeding companies, agro-NGOs (non-governmental organizations) and government. We realize that there are also other stakeholders affecting the system; for example, farmers and consumers that affect the demand for new varieties and food products. However, we have chosen to focus on what we see as the central stakeholder group.

At the start of the PGR Secure project in 2011, key persons were identified in each of the countries and they were asked to recommend stakeholder group representatives familiar with the PGS in their country. Subsets of these were subsequently interviewed using a semi-structured interview approach. In total, 133 interviews were conducted in 21 European countries. From the interviews, we gained qualitative data and opinions, which were used to conduct analyses of the status of conservation and use at the country or regional level (Fig. 36.1). Reports were written in cooperation with the country's key persons and published for the following countries/regions: Austria, the Baltic Countries (Estonia, Latvia and Lithuania), Bulgaria, Czech Republic, France, Germany, Greece, Italy, the Netherlands, the Nordic Countries (Iceland, Norway, Sweden, Finland and Denmark), Poland, Romania, Slovenia, Spain and the UK (see www.nordgen.org/index.php/en/content/view/full/2481).

In the next phase of the project, information from the interviews was used to develop a set of questions for an online questionnaire, which was sent to a total of 1160 e-mail addresses from the five stakeholder groups. The list of questionnaire recipients was compiled with the help of the country's key persons and other information sources such as the World Information and Early Warning System (WIEWS) established by FAO, and published participant lists from conferences. A total of 226 people responded to the online questionnaire (19.5%), but only a subset of these, 131, completed all the questions. Both the partial and complete answers were used in the analysis, which was conducted separately for each stakeholder but including all geographic regions. By a combination of interviews and online questionnaires, a good overview of the status of PGR conservation and use was gained across geographic regions of the EU and Europe as a whole (Fig. 36.2).

### 36.2.2 SWOT analysis

We have chosen to analyse the data using a SWOT analysis. At the heart of this analysis is the identification of the Strengths, Weaknesses, Opportunities and Threats for an organization. The SWOT analysis is useful when trying to understand complex systems by identifying both internal (strengths and weaknesses) and external (opportunities and threats) factors (Fig. 36.1) affecting the system. The main aim is to use the increased knowledge and understanding of the system to develop strategies for its improvement.

On the European level, formal SWOT analyses were conducted in three separate steps for each stakeholder group. First, a target state was identified. Here, we ask ourselves: what kind of system do we aim for? This step is not always performed explicitly in SWOT analyses, but what is considered to be the ideal state of a system does affect what is considered a strength or weakness, opportunity or threat. Therefore, to describe the target state explicitly makes the process more transparent. The second step is to identify the internal and external factors that affect the system and categorize them as strengths, weaknesses, opportunities or threats. The third, and last, step of the analysis is to develop strategies on how to improve the system. Here, we used the information from step two to identify four different types of strategies: strength and opportunity strategy (SO), strength and threat strategy (ST), weakness and opportunity strategy (WO) and weakness and threat strategy (WT), see Table 36.1. The general approach is to take advantage of the strengths and opportunities of the system, while minimizing the weaknesses and threats. In the SWOT analyses on the country/regional level, a less formal approach was used, where step two

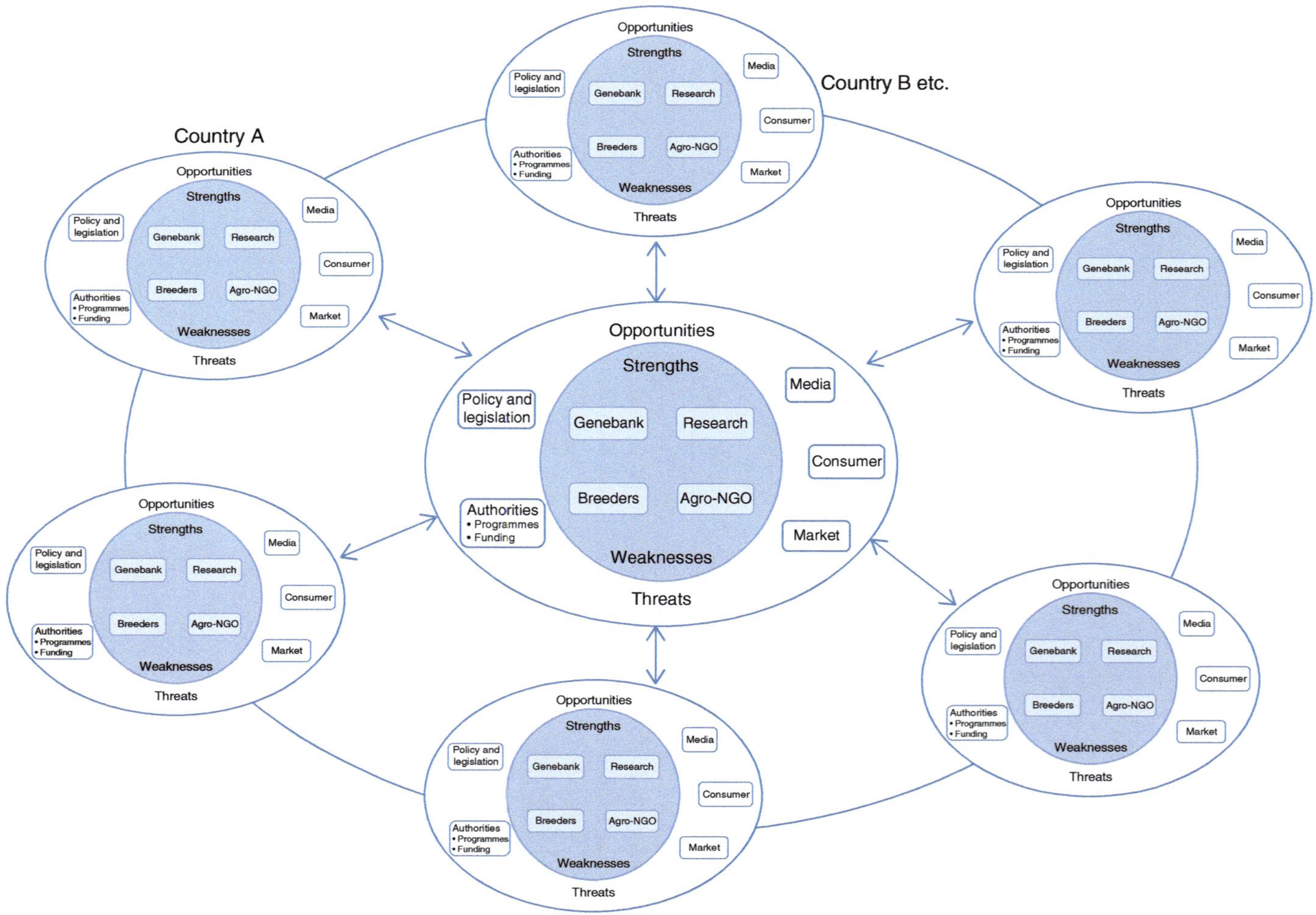


**Fig. 36.1.** Separate analyses were conducted to understand the status of PGR conservation and use within countries/regions and on the European level. Semi-structured interview data were used mainly for the former analysis (outer circles) and the online questionnaire data for the latter (inner circle). The white part of a circle indicates external factors, i.e. those that cannot be controlled by the analysed organization, while the blue part shows internal factors that are under the control of the organization.

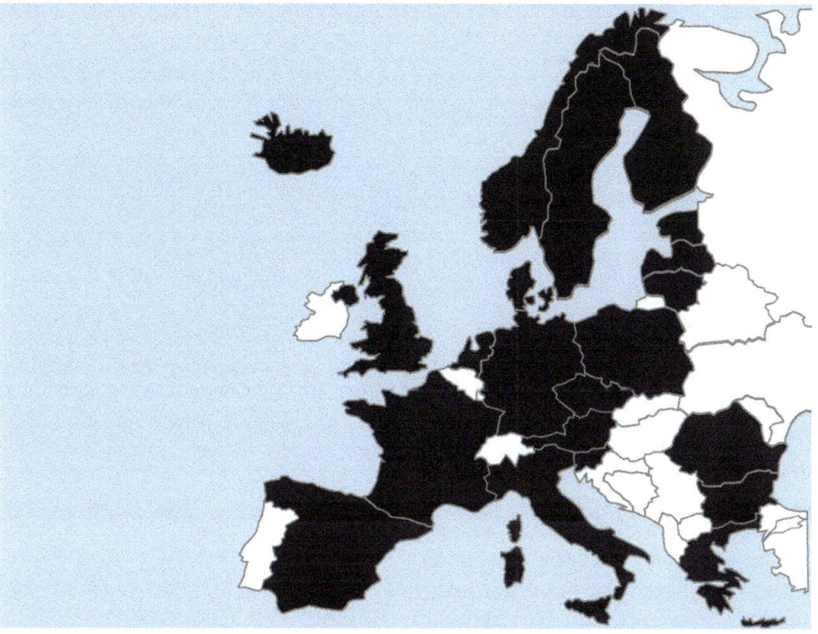

Fig. 36.2. Data on the conservation and use of plant genetic resources were collected by semi-structured interviews and/or online questionnaire from stakeholders in the countries marked in black.

Table 36.1. Descriptions of the different types of strategies that can be developed within a SWOT analysis to improve the analysed system.

| Strategy | Abbreviation | Explanation |
|---|---|---|
| Strength and opportunity strategy | SO | Use strengths to take advantage of opportunities |
| Strength and threat strategy | ST | Use strengths to reduce the likelihood and impact of threats |
| Weakness and opportunity strategy | WO | Make use of opportunities to overcome weaknesses |
| Weakness and threat strategy | WT | Minimize weaknesses and threats |

described above was followed by recommendations on how to improve the system.

Preliminary versions of the SWOT analyses, both on the European level and the country/regional level, were used as input for a stakeholder workshop held in November 2013 in Wageningen (the Netherlands). Over 80 people attended the workshop, representing the PGS community and including genebanks, breeding companies, public research, agro-NGOs and government representatives. The SWOT results were discussed at length in smaller stakeholder workshops and in plenary sessions that included all stakeholders. In this way, substantial feedback was received and incorporated into the report, which was then circulated among the rapporteurs and chairs of the workshop before the final report versions were produced.

## 36.3 Conservation of PGR in Europe: Targets and SWOT

As described above, the analysis of conservation and use of PGRs in Europe was made at both the country/regional level and the European level. Below, we present a summary of the results on the European level and ask the reader to consult the country/regional reports for local analyses and the online questionnaire report for more details on the European-level analysis (see www.nordgen.org/index.php/en/content/view/full/2481).

### 36.3.1 Genebanks

Genebanks are a central stakeholder for the *ex situ* conservation of PGRs. In addition, they also have a central but underutilized role in *in situ* and on-farm conservation. Another key issue for genebanks is the responsibility to facilitate the use of their collections by making information and PGRs easily accessible, promoting its evaluation and taking part in pre-breeding projects.

#### *Target states*

What is regarded as the ideal state of the genebanks sector in Europe, or of any other system, is, of course, subjective. Here, we have taken the basic idea that PGRs should be conserved safely and used efficiently due to their agricultural and cultural importance. With this basic assumption in mind, and the information from the interviews and questionnaire, as well as the feedback from the workshop, the following ideal European genebank sector was described:

**1.** Target state 1: European countries interacting with the EU-PGS do so on the basis of a national biodiversity/agrobiodiversity strategy and a national PGR programme.
**2.** Target state 2: the national PGS develop and implement a comprehensive conservation and use strategy for PGRs, including stable funding, which combines the complementary advantages of the *ex situ*, *in situ* and on-farm management techniques.
**3.** Target state 3: the national genebanks, as the core element of the national plant germplasm system, are juridically independent or have earmarked funding within their host institution and are adequately funded by their governments.
**4.** Target state 4: national PGS operations meet a mutually agreed minimal set of quality management system criteria. These criteria include: (i) national inventories of genebank accessions, and CWR and landraces held *in situ* or on-farm are regularly updated and uploaded to EURISCO; (ii) genebank accessions are conserved according to a jointly agreed guideline such as the AEGIS Quality System (AQUAS) (Engels *et al.*, 2009); (iii) characterization and evaluation data are recorded and made available to potential users; (iv) national plant germplasm systems apply germplasm access and exchange rules detailed in Regulation (EU) No 511/2014 of the European Parliament and of the Council of 16 April 2014 on compliance measures for users from the Nagoya Protocol on Access to Genetic Resources and the Fair and Equitable Sharing of Benefits Arising from their Utilization in the Union (EU, 2014); and (v) germplasm collecting is based on gap analysis and guided by a European-wide review system for collecting mission proposals.
**5.** Target state 5: EU approves a regulation required to provide long-term co-funding for the establishment, improvement and operation of EU-PGS infrastructures.

#### *SWOT*

Based on the information gathered from the online questionnaire and the Wageningen workshop, the strengths, weaknesses, opportunities and threats for the genebank sector in Europe were identified. The top three factors in each category are presented in Table 36.2.

### 36.3.2 Public research organizations

Public research has an important role to play in the conservation and use of PGRs in Europe, since they provide much of the knowledge base for conservation, plant breeding and policy decisions (see, for example, EASAC, 2011). To conduct efficient and reliable conservation, scientifically sound information is needed in all parts of the conservation process, such as sampling, documentation, regeneration, seed storage, conservation *in situ* site selection and establishment, and monitoring of *in situ* sites. At the root of conservation is the understanding of the material to be conserved; that is, the primary, secondary and tertiary genepool of a crop, as well as the structure of genetic diversity within each species. Studies on different levels and different parts of each genepool are therefore needed, such as, for example, those performed by Kilian *et al.* (2007), Hagenblad *et al.* (2012), Solberg *et al.* (2014) and Wedelsbäck Bladh *et al.* (2014) and summarized by Vincent *et al.* (2013). Public research also provides plant breeders with new tools to make breeding more efficient, such as

Table 36.2. The three most important strengths, weaknesses, opportunities and threats impacting the genebank sector in Europe.[a]

| Strengths | Weaknesses |
|---|---|
| • In general, the crop group composition of *ex situ* holdings managed within the 17 interviewed countries matches the interests and needs of the user community (public research, plant breeders, farmers and agro-NGOs).<br>• All of the interviewed genebanks have established computerized genebank information systems. Some of them operate web-based information systems, allowing users to search for passport, characterization and evaluation data and to order accessions.<br>• All countries provide some funds for conservation activities through national agrobiodiversity and other environmental protection strategies. | • Most genebanks are not independent with respect to their programme, funds or staff. Crop experts are thus forced to prioritize research work at the expense of conservation work.<br>• *Ex situ* management of PGRs of many EU genebanks is not based on defined, written quality management standards such as AQUAS. Limited funding exists for basic tasks of maintaining and regenerating PGRs.<br>• Genebanks rarely lead the development of national *in situ* management and on-farm management strategies. Many diverse governmental authorities are responsible for *in situ*/on-farm management, and national *in situ*/on-farm strategies are missing. |
| **Opportunities** | **Threats** |
| • PGR conservation activities are backed up by the policy through national agrobiodiversity and other nature protection strategies in most countries, and in some countries conservation measures are further detailed in national PGR expert programmes.<br>• EU programmes can support national *ex situ*, *in situ* and on-farm conservation activities.<br>• EURISCO could potentially act as the information backbone of an EU-PGS and the goal should be one integrated European information system. ECCDBs (European Central Crop Databases) exist and demonstrate how characterization and evaluation data can be recorded according to the single observation concept and made available online. | • Insufficient support of germplasm conservation activities at all governmental levels.<br>• The ECPGR (European Cooperative Programme for Plant Genetic Resources) is currently being reformulated; however, without an increased budget. If the programme continues to rely on input in kind by the stakeholder community, it may no longer be sustained by the community. This would result in the loss of the only body in Europe experienced in the coordination of PGR activities.<br>• The native plant species and habitat conservation sector and the agricultural PGR conservation sector do not routinely cooperate in the field of *in situ* conservation. |

*Note*: [a]A full report of all the strengths, weaknesses, opportunities and threats identified in the study can be found at: www.nordgen.org/index.php/en/content/view/full/2481, stakeholder workshop 2013.

high-throughput molecular tools, predictive characterization and analysis tools and phenotyping approaches. In addition, some researchers carry out pre-breeding and provide germplasm to plant breeders. Last, but not least, researchers have the possibility to provide policy makers with scientifically based evidence and knowledge upon which they can base decisions.

### Target states

The following ideal public research sector was defined within this study:

**1.** Target state 1: public research provides all the basic knowledge needed for efficient, evidence and scientifically based conservation strategies that integrate *in situ*, *ex situ* and on-farm management and encompass the primary, secondary and tertiary genepool of crops. Public research provides a research foundation for the integration of *ex situ*, *in situ* and on-farm conservation techniques, and the ways in which they are best combined and applied for genepools of specific crops.

**2.** Target state 2: the effects and efficiency of complementary (*in situ*, on-farm, *ex situ*) conservation strategies is monitored and adaptive management is applied. Indicators for monitoring diversity at the habitat level, the interspecific level and intraspecific level are improved or developed, tested and applied. Research develops indicators and their metrics allowing the monitoring of genetic diversity within the native European genepools of crops to prevent genetic erosion (Bonneuil *et al.*, 2012; FAO, 2015) and to ensure complementary conservation of crops and wild relatives.

**3.** Target state 3: data are made publicly available in web-based information systems, accessible through a single entry point. Development and further improvement of plant genetic resources information systems at the European and national levels includes the active participation of research institutes, allowing recording of conservation, characterization and evaluation data following the single observation concept (primary data, as opposed to summary statistics). Research institutes, as holders and one of the main users of germplasm, accept co-responsibility for the establishment of a comprehensive EU information infrastructure for plant genetic resources.

**4.** Target state 4: framework conditions (policy, funding and infrastructure) facilitate exploitation of a wide range of PGRs, including CWR and LR, in breeding programmes. Public research institutes maintain and use genetic diversity in pre-breeding programmes. Breeding research institutes maintain, enhance and promote the use of breeding pools of prioritized crops, including neglected and underutilized species, as well as those of potential value for the adaptation of European agriculture to climate change.

#### *SWOT*

Based on the information gathered from the online questionnaire and the Wageningen workshop, the strengths, weaknesses, opportunities and threats for the public research sector in Europe were identified. The top three factors in each category are presented in Table 36.3.

### 36.3.3 Breeders

The role of plant breeders consists of adapting crops and varieties to sustainable production systems and in supplying farmers with seeds and planting materials. An ideal European variety breeding sector should be able to meet the challenges of a changing climate, growing population, demands for decreased environmental impacts and evolving consumer demands. Past achievements show that there is a great potential in breeding for meeting farmer- and consumer-based needs. In the past decades, plant breeding contributed about 50% of the improved productivity, while the combination of improved fertilization, agricultural techniques and plant protection account for the remaining 50% of the improvement (von Witzke and Noleppa, 2013).

#### *Target states*

In this study, the ideal European breeding sector was defined with the following four target states:

**1.** Target state 1: plant breeders continue to interact with the other stakeholder groups of the EU-PGS, not only on the basis of market requirements but also on national and EU-wide PGR strategies and PGR action plans.

**2.** Target state 2: together with the public breeding research sector, commercial plant breeders maintain genetic diversity in breeding pools of a range of crops and produce varieties allowing farmers to optimize cultivation systems and adapt them to changing climatic conditions. There is good access to the characteristics and information on released varieties.

**3.** Target state 3: plant breeding activities of smaller companies are strengthened through Public Private Partnership (PPP) programmes. PPP projects develop new tools for breeding and enhance both diversity within crops as well as crop species diversity via pre-breeding.

**4.** Target state 4: breeders can access germplasm and information easily. Two especially important issues are: (i) easy access to germplasm and related information is made possible via a European plant genetic resources information infrastructure; and (ii) easy access to germplasm is facilitated by an extended species list of Annex I of the ITPGRFA (the International Treaty on Plant Genetic Resources for Food and Agriculture).

#### *SWOT*

Based on the information gathered from the online questionnaire and the Wageningen workshop,

Table 36.3. The three most important strengths, weaknesses, opportunities and threats that are affecting public research organizations in Europe.[a]

| Strengths | Weaknesses |
|---|---|
| • The EU plant genetic research community holds a wide range of crop-specific expertise and methodological expertise such as trait-specific evaluation methods, predictive characterization, genomics, biodiversity and informatics.<br>• PGR collections are linked to public research institutes that publish significant characterization and evaluation data.<br>• Research institutes cooperate with all stakeholder groups at the national and European level. | • Financial means for research are increasingly restricted, short-term funding prevails and financial resources are distributed unevenly between the EU member states, leading to a loss of expertise and long-term breeding research programmes.<br>• Instability of research institute policies concerning PGR and pre-breeding research programmes.<br>• Several institutes operate their own collections. This lack of rationalization may hinder accessibility of germplasm unnecessarily. |
| **Opportunities** | **Threats** |
| • The majority of research institutes base their work on national PGRFA expert programmes.<br>• Significant amounts of genetic diversity for pre-breeding research programmes are available at genebanks. The interest in improving the conservation and utilization of CWR and LR is increasing worldwide.<br>• Increasing availability of efficient new predictive characterization, genotyping and phenotyping tools open up for potentially cheaper and more efficient identification and in-depth evaluation of genetic resources. | • Short-term project funding for breeding research and instabilities of policies (i.e. target crops/products) in reaction to the development of global, regional and national markets.<br>• Insufficient visibility of PGR collections (passport, characterization and evaluation data) on the Internet is hindering the selection of the most appropriate accessions, which in turn limits conserved materials use severely.<br>• Lack of implementation of PGR conservation and use policies in Europe are limiting access and potential use to PGR resources. As a consequence, CWR are not well represented in genebanks and LR conservation is not performed systematically. |

*Note*: [a]A full report of all the strengths, weaknesses, opportunities and threats identified in the study can be found at: www.nordgen.org/index.php/en/content/view/full/2481, stakeholder workshop 2013.

the strengths, weaknesses, opportunities and threats for the plant breeding sector in Europe were identified. The top three factors in each category are presented in Table 36.4.

### 36.3.4 Agro-NGOs

This stakeholder group has a rather complex composition if all groups and activities across the EU member states are viewed together. Also, a corresponding diversity of objectives is associated with the diverse target groups that exist. The position of agro-NGOs in the national PGS and EU-PGS is more difficult to determine than is the case for genebanks, public research institutes and plant breeders. The interests range from members of seed-saver organizations wishing to contribute to the conservation of LR as hobby gardeners, to extension services like organizations aiming at the promotion of products based on LR and varieties valued by the local people and grown in a specific region by conventional agricultural or horticultural enterprises to those companies involved in organic or biodynamic production. Independently of the individual interest and expertise of agro-NGO members, these associations, organizations and farm enterprises have an important role to play with regards to the establishment of complementary PGRFA conservation programmes and general awareness raising of the value of PGRFA diversity. They are the agents on the ground (farmer, gardener, members of seed-saver organizations) as concerns the maintenance and improvement of locally adapted forms of crop species.

Table 36.4. The three most important strengths, weaknesses, opportunities and threats that are affecting the plant breeding sector in Europe.[a]

| Strengths | Weaknesses |
|---|---|
| • Breeders as a group develop many varieties for a wide range of crop species and agroenvironments.<br>• The knowledge and professional experiences available among breeders allows forecasting of future needs in plant breeding.<br>• Breeders cooperate in pre-competitive research projects with other stakeholders, creating increased opportunities for innovation and increased efficiency. | • Smaller companies have limited capacities for scientific cooperation and for conducting their own pre-breeding programmes or applying breeding tools.<br>• Limited breeding in some fruit trees, edible fungi, grain legumes, forages and neglected crops, as well as crops for marginal areas in Europe.<br>• Limited knowledge of international ABS (access and benefit sharing) arrangements within smaller commercial companies. |
| **Opportunities** | **Threats** |
| • Public funds for research allow plant breeders to cooperate with the breeding research sector and other stakeholders.<br>• Breeding research is a highly innovative sector (genomics, phenomics, bioinformatics, new tools and procedures to facilitate selection) that transfers knowledge and pre-bred material to the plant breeding sector.<br>• Plant breeders have access to conserved PGRFA diversity, most commonly via public genebanks. | • International ABS and MLS (multilateral system) arrangements should facilitate the exchange of PGRs, but it is not currently always working efficiently and therefore limits the exchange.<br>• Lack of characterization and evaluation data in (inter)national databases limits targeted choice of accessions from genebank holdings.<br>• Limited public funds for breeding research and maintenance of genebanks is restricting potential collaboration and causing considerable problems. |

*Note*: [a]A full report of all the strengths, weaknesses, opportunities and threats identified in the study can be found at: www.nordgen.org/index.php/en/content/view/full/2481, stakeholder workshop 2013.

### *Target states*

The ideal agro-NGO sector is defined as follows:

**1.** Target state 1: it is a pre-condition that the interests of agro-NGOs are considered in national expert programmes and a representative appointed as a member of the board supervising the implementation and revision of the national PGR programme. *Ex situ* conservation actions complementing on-farm conservation actions are established and operating.

**2.** Target state 2: LR and other types of traditionally grown cultivars of crops are conserved on the farm and sold to consumers. The conservation cost is largely supplemented by sales revenues. Existing markets for local or regional traditional food or goods are promoted and new food systems including appropriate markets for diversity are developed.

**3.** Target state 3: the consumers' interest in products produced from LR and traditional crops will change over time. These changes can be associated with loss of a farmer's knowledge. *Ex situ* backup samples of on-farm conserved germplasm must be taken and stored in genebanks, along with the farmer's knowledge in the genebank's information system. The establishment of a systematic backup system is required to guarantee the complementarity of *ex situ* and on-farm conservation actions for a longer period; that is, at least 30 years. The target would be to have backup systems established and operating, both for genetic material and associated information.

**4.** Target state 4: seed supply systems operated by agro-NGOs are sustained, promoted and strengthened at the national and European level. The agro-NGO (informal) sector and the formal sector (genebank, public research, plant breeding) exchange seeds and information on the basis of a regulated, transparent germplasm exchange system. Both sectors interact on the basis of reciprocal respect. New EU regulations are reviewed for perverse unintentional impacts on PGRFA and its maintenance by NGO communities and ensure protection of the livelihoods of agro-initiatives and farmers' rights.

### *SWOT*

Based on the information gathered from the online questionnaire and the Wageningen workshop, the strengths, weaknesses, opportunities and threats for the agro-NGOs in Europe were identified. The top three factors in each category are presented in Table 36.5.

## 36.3.5 Governments

The conservation and sustainable use of PGRFA is part of the governmental public service mission. The EU member states, as well as the EU, are contracting parties to the CBD (Convention of Biological Diversity) and the ITPGRFA and governmental departments and experts are involved in follow-up negotiations in the context of these two important international agreements. Despite these facts, the support of PGRFA activities is decreasing within the member states and is also limited at the EU Commission DG AGRI (the Directorate-General for Agriculture and Rural Development). The investments of the EU agricultural sector in the safeguarding of its own most important resources, the inter- and intraspecific diversity in agricultural genepools, is relatively small compared to investments of the DG research into plant genetics and plant breeding research.

### *Target states*

The ideal governmental sector is defined by the following three target states:

**1.** Target state 1: improved coordination of policies on PGRFA conservation and use, including their implementation and the removal of perverse incentives, in harmony

Table 36.5. The three most important strengths, weaknesses, opportunities and threats that are affecting European agro-NGOs.[a]

| Strengths | Weaknesses |
|---|---|
| • On-farm conservation of LR promotes local adaptation to different environmental conditions. This is a unique selling point of the on-farm conservation technique.<br>• Strong links with societal groups such as farmers, gardeners, consumers and the educational sector. Agro-NGOs lobby for, conserve, develop and communicate know-how on cultivation, processing and utilization of PGRs.<br>• Significant influence on biodiversity policy issues through lobbying policy makers and public awareness raising. This strength builds on local, regional and national level bottom-up approaches. | • Lack of or insufficient national LR inventories in some countries.<br>• Limited involvement, training and technology transfer with the national PGR programme and with other stakeholders.<br>• Limited (employed) staff and annual budgets of agro-NGOs. |
| **Opportunities** | **Threats** |
| • The need for on-farm conservation actions is addressed by the provisions of the International Treaty and addressed in EU Council regulations.<br>• The EU Biodiversity Strategy 2020 advocates the need to promote the use of traditional agricultural varieties specific to certain regions.<br>• On-farm conservation can benefit from societal trends such as ecological farming and regional food consumption attitudes and preferences. | • Future EU seed directives may limit the modus operandi and livelihood of agro-NGOs.<br>• Strengthening of intellectual property rights (International Union for the Protection of New Varieties of Plants (UPOV) 91, patents, copyrights) may cause conflicts of interest between the governmental and non-governmental sector.<br>• The insufficient information systems operated by the governmental sector do not allow recording, documentation and dissemination of farmers' knowledge. |

*Note*: [a]A full report of all the strengths, weaknesses, opportunities and threats identified in the study can be found at: www.nordgen.org/index.php/en/content/view/full/2481, stakeholder workshop 2013.

with international commitments and frameworks to be implemented at both national and European level.

**2.** Target state 2: in consultation with the EU Commission, a European PGRFA strategy and corresponding action plan for *ex situ* and *in situ* and on-farm conservation and use, possibly within the framework of ECPGR (European Cooperative Programme for Plant Genetic Resources), and in harmony with the FAO Second Global Plan of Action for Plant Genetic Resources for Food and Agriculture (GPA), is established.

**3.** Target state 3: existing infrastructures of the ECPGR and its members are strengthened, including AEGIS and EURISCO, to arrive at a sufficiently funded organizational and technical European infrastructure for PGRFA *ex situ, in situ* and on-farm conservation, as well as use that connects all relevant stakeholder groups, integrates existing national components and assists in the enhancement of these national components.

### *SWOT*

Based on the information gathered from the online questionnaire and the Wageningen workshop, the strengths, weaknesses, opportunities and threats for the government sector in Europe were identified. The top three factors in each category are presented in Table 36.6.

## 36.4 Conclusions

There are many notable strengths and competences within the plant germplasm system in Europe today, and there are also several promising opportunities coming from outside the system. Perhaps most importantly, there is today a general consensus that food diversity based on genetic resources is important and should be protected by conservation measures. In the same vein, member states have accepted the international legal obligations for the conservation of PGRs for food and agriculture, which puts legal and ethical pressure on policy makers to act to conserve PGRFA diversity. Within the current system, the conservation sector does conserve a substantial amount of genetic resources *ex situ*, and the crop diversity conserved does largely mirror the needs of the users. There are also public funds available for conservation, evaluation, PGR research and pre-breeding, and there is an interest among the stakeholders to cooperate within such integrated projects. A large knowledge base and high competence in combination with innovative thinking among stakeholders leads to successful projects and the development of efficient new tools that can be used in conservation and breeding.

However, there are several major problems that need to be addressed if we are to secure reliable conservation of essential PGRs and make it possible to use these resources efficiently in future efforts to assure food security. A central problem is that PGR issues often have a low priority, both at the European and national governmental levels, which leads to insufficient support of conservation activities and a lack of implementation of conservation and use policies.

*Ex situ* conservation is the most well-established conservation approach for PGRFA, but most of the European genebanks are reliant on governmental support with regards to direct funding or research programmes, and underfunding frequently puts PGRs at risk. In addition, most genebanks do not follow defined standards to assure transparency and a minimum quality of the work. Visibility and access to *ex situ* collections are often limited, and there is a lack of relevant evaluation data available in the databases.

Another issue affecting several stakeholders is the prevalence of short-term funding and instabilities of policies. Both conservation and breeding are long-term efforts that demand long-term commitments. For example, pre-breeding projects are crucial to bridge the gap between genetic resources and conventional breeding but need funding over a long period to be successful.

A range of difficulties arise when implementing on-farm and *in situ* conservation. At the heart of the problem is perhaps uncertainty regarding responsibility; do ministries of agriculture or environment take the lead, or better still, can they work together? At the governmental level, the responsibility for *in situ* conservation is often shared between authorities, and the terms of cooperation and responsibility are not always clear. Genebanks have traditionally worked with *ex situ* conservation and have not

Table 36.6. The three most important strengths, weaknesses, opportunities and threats that are affecting the government sector.[a]

| Strengths | Weaknesses[b] |
|---|---|
| • Member state governments have accepted legal obligations for PGRFA conservation and support policy processes and measures at the national, international and European levels, such as the International Treaty, national biodiversity, CBD/agrobiodiversity and the EU Biodiversity Strategies to 2020.<br>• Governments coordinate biodiversity PGRFA programmes by establishing National Focal Points and Competent National Authorities and fund national conservation and use activities on PGRFA.<br>• Most member states provide long-term funding to maintain PGR conservation. | • In some countries, PGRFA issues have low priority within the ministry(ies) responsible and the complexities of GR issues are not well perceived by people in a position of responsibility at the national or European level.<br>• If ministries, such as agriculture and environment, share responsibilities for conservation, they do not always cooperate in a systematic and structured manner on PGRFA issues, which is a strong disadvantage, in particular in the case of *in situ* conservation approaches. |
| **Opportunities** | **Threats** |
| • In the past, stakeholders involved in PGR issues expressed deviating views on the value of PGRFA and the risks arising from genetic erosion. Today, all communities support the need for comprehensive complementary conservation measures.<br>• Ongoing European and national programmes on biofuels, ecosystem services, etc., already promote agrobiodiversity and genetic resources.<br>• The EU Preparatory Action will feed the process on GR conservation and sustainable use. | • The PGR community and the commercial plant breeding sector in Europe as a whole have so far failed to link the issue of biodiversity/PGRFA protection to climate change mitigation, food security and wealth creation.<br>• The food processing sector is the largest industry in the EU. Despite this, the economic importance and the number of enterprises and employees in the agricultural sector, it is declining in many European countries. In addition, the critical view of the public on agricultural policies and its consequences does not create a favourable condition for a strong governmental support of agrobiodiversity issues. Taken together, this could lead to an insufficient budget for agricultural research, including that related to breeding and PGRFA conservation.<br>• Fear of increased administrative burdens and distrust of existing legal frameworks result in industries/breeders establishing private genebanks that might undermine public funding for existing public genebanks. |

*Note*: [a]A full report of all the strengths, weaknesses, opportunities and threats identified in the study can be found at: www.nordgen.org/index.php/en/content/view/full/2481, stakeholder workshop 2013. [b]The workshop members identified only two weaknesses in total.

risen to the challenge to take a leading role in the development of *in situ* and on-farm conservation strategies. In many countries, inventories of LR and CWR are still missing, and so are conservation strategies targeted at these important genetic resources.

In this study, we have identified a long list of weaknesses and threats. However, our main message is that these can be overcome, but actions are needed at both the national and the European level. To this end, we have come up with recommendations for future actions, which are discussed in the next chapter in this book (see Frese *et al.*, Chapter 37, this volume).

## Acknowledgements

This research was funded by a grant from the EU Seventh Framework Programme for the PGR

Secure project (Grant Agreement No 266394). We would also like to thank the country key persons and consultants. Without their strong support we would not have been able to conduct the interviews in their countries. We are also very grateful to the stakeholder group representatives who have participated in interviews and have answered the online questionnaire, as well as the participants of the workshop in Wageningen 2013. These were the main input sources for our study.

## References

Bonneuil, C., Goffaux, R., Bonnin, I., Montalent, P., Hamon, C., *et al.* (2012) A new integrative indicator to assess crop genetic diversity. *Ecological Indicators* 23, 280–289.

EASAC (2011) Plant genetic resources for food and agriculture: roles and priorities in the European Union. *EASAC policy report* 17, Halle (Saale), Germany.

Engels, J., Maggioni, L. and van Hintum, T. (2009) Quality system for AEGIS Version 15. Available at: www.ecpgr.cgiar.org/fileadmin/templates/ecpgr.org/upload/AEGIS/FOR_WEB_FINAL/QualitySystemfinal.pdf (accessed 3 December 2015).

EU (2014) Regulation (EU) No 511/2014 of the European Parliament and of the Council of 16 April 2014 on compliance measures for users from the Nagoya Protocol on Access to Genetic Resources and the Fair and Equitable Sharing of Benefits Arising from their Utilization in the Union. *Journal of the European Union L* 150/59.

FAO (2015) Reporting format for monitoring the implementation of the Second Global Plan of Action for Plant and Genetic Resources for Food and Agriculture. Food and Agriculture Organization of the United Nations, Rome. Available at: www.fao.org/3/a-ml478e.pdf (accessed 29 July 2015).

Hagenblad, J., Asplund, L., Balfourier, F., Ravel, C. and Leino, M.W. (2012) Strong presence of the high grain protein content allele of NAM-B1 in Fennoscandian wheat. *Theoretical and Applied Genetics* 125, 1677–1686.

Kilian, B., Özkan, H., Kohl, J., von Haeseler, A., Barale, F., *et al.* (2007) Haplotype structure at seven barley genes: relevance to gene pool bottlenecks, phylogeny of ear type and site of barley domestication. *Molecular Biology and Evolution* 24, 2657–2668.

Solberg, S.Ø., Brantestam, A.K., Kylin, M., Bjørn, G.K. and Thomsen, M.G. (2014) Genetic variation in Danish and Norwegian germplasm collections of hops. *Biochemical Systematics and Ecology* 52, 53–59.

Underwood, E., Baldock, D., Aiking, H., Buckwell, A., Dooley, E., *et al.* (2013) Options for sustainable food and agriculture in the EU. Synthesis report of the STOA Project. 'Technology Options for Feeding 10 Billion People'. Institute for European Environmental Policy, London/Brussels. Available at: www.europarl.europa.eu/RegData/etudes/etudes/join/2013/513539/IPOL-JOIN_ET%282013%29513539_EN.pdf (accessed 17 December 2015).

Vincent, H., Wiersema, J., Kell, S.P., Dobbie, S., Fielder, H., *et al.* (2013) A prioritised crop wild relative inventory as a first step to help underpin global food security. *Biological Conservation* 167, 265–275.

Wedelsbäck Bladh, K., Liljeroth, E., Poulsen, G., Yndgaard, F. and Kolodinska Brantestam, A. (2014) Genetic diversity in Nordic horseradish, *Armoracia rusticana*, as revealed by AFLP markers. *Genetic Resources and Crop Evolution* 61, 383–394.

Witzke, von H. and Noleppa, S. (2013) Studie zur gesellschaftlichen Bedeutung der Pflanzenzüchtung in Deutschland. *Agrar-Europe* 25(17), 1–26.

# 37 Towards an Improved European Plant Germplasm System

L. Frese,[1]* A. Palmé,[2] L. Bülow[1] and C. Kik[3]
[1]Julius Kühn Institute, Federal Research Centre for Cultivated Plants, Institute for Breeding Research on Agricultural Crops, Quedlinburg, Germany; [2]Nordic Genetic Resource Center, Alnarp, Sweden; [3]Centre for Genetic Resources, the Netherlands (CGN), Wageningen University and Research Centre, Wageningen, the Netherlands

## 37.1 Introduction

Europe is the cradle of modern plant breeding. One of the most famous European scientists, N.I. Vavilov, founded the first genebank worldwide to maintain crop diversity and to service crop plant research and plant breeding. The first ontologies for genebank information systems were developed in Europe, and the first computerized genebank information system became operational in Europe in the 1970s (Seidewitz, 1974, 1979). At this time, the development of the later Genetic Resources Information Network (GRIN) of the US National Plant Germplasm System was lagging behind (April, 1979). The European Cooperative Programme for Plant Genetic Resources (ECPGR) became operational in 1980 (IBPGR, 1984). It was established primarily to complement and support the national *ex situ* conservation, documentation and utilization activities in European countries.

The International Biological Programme emphasized the importance of wild relatives of economic species and called for the organized protection of these species in their natural habitats *in situ* (Frankel and Hawkes, 1975) in the 1970s. The term 'genetic reserve' had already been coined by the Australian geneticist and doctorate of the Agricultural University of Berlin, Sir Otto Frankel (Frankel, 1970; Evans, 1999). In the 1990s, European scientists elaborated the genetic reserve conservation technique (Maxted *et al.*, 1997). The ECPGR established the *In situ* and On-farm Conservation Network in 2000 (Laliberté *et al.*, 2000), which has developed and improved systematic approaches for the *in situ* management of crop wild relatives (CWR; Kell *et al.*, 2012) and landraces (LR; Negri *et al.*, 2014) over the years. Despite the obviously high scientific and technical innovation potential, Europe has thus far failed to establish a fully integrated plant germplasm system that halts genetic erosion in domesticated species and their wild relatives and at the same time services research and plant breeding in a way that satisfies users.

As no elaborate study has been carried out to analyse the plant genetic resources conservation and use problems in Europe to date, a study was undertaken within the framework of PGR Secure (see Frese *et al.*, Chapter 36, this volume, 2016). The rationale behind this study is the presence of substantial conservation and use constraints of PGRs in Europe. These constraints are influencing innovation negatively and our potential to meet efficiently the food security

*Corresponding author; e-mail: lothar.frese@jki.bund.de

 

challenges caused by, for example, a growing human population and a changing climate. Targets and strategies were presented in a report and reviewed by a large number of representatives from the genebank, public research, plant breeding, agro-NGO and policy sector during a European workshop held in Wageningen, the Netherlands, in November 2013. The suggestions of the participants have been taken into consideration and are reflected in 12 recommendations presented in the discussion paper 'Towards a European plant germplasm system – the third way' (see www.nordgen.org/index.php/en/content/view/full/2481) and summarized in Table 37.1. This chapter focuses on recommendations addressing the main problems of managing *ex situ* PGRs in Europe.

## 37.2 Development of Infrastructures for PGR Conservation

The absence of a coherent European infrastructure for the conservation of PGRs is the major constraint of the PGR system in Europe. This view was shared by the workshop participants. We propose, therefore, the development of a *technical*, *organizational* and *information* infrastructure that supports the implementation of conservation measures and that adequately services the user community (Recommendation 1, 2 and 3, Table 37.1). We called it the European Plant Germplasm System (EU-PGS). The remit of an EU-PGS would be much larger than a European network of genebanks. The system would rather integrate the stakeholder groups responsible for PGR conservation, for plant genetic and plant breeding research, for plant breeding, as well as agro-NGOs. The EU-PGS would organize their relationships and the flow of information and material so as to achieve the best possible conservation and use of PGRs in Europe, thus avoiding overlap and maximizing cooperative benefits. As no such system currently exists, task-sharing and coordinated actions for the conservation of PGRs *ex situ*, *in situ* and on-farm have only sporadically been conducted on the basis of project funding (EC, 2013).

The EU member states and the European Commission have not succeeded in halting the loss of biodiversity in Europe (EC, 2011). It is therefore very important to establish a system that at least contributes effectively to the conservation and sustainable use of PGRs for food and agriculture that are native to Europe or are being held in European genebanks. An EU-PGS (Fig. 37.1)

**Table 37.1.** A summary of the 12 recommendations on how the status of *ex situ* conservation and use of PGRs could be improved. For further details on the recommendations see 'Towards a European plant germplasm system – the third way' (www.nordgen.org/index.php/en/content/view/full/2481).

| Number | Recommendation |
|---|---|
| 1 | Establishment of a European plant germplasm system |
| 2 | Establishment of a technical EU infrastructure for the organization of conservation of plant genetic resources for food and agriculture measures |
| 3 | Establishment of an EU information infrastructure for conservation of PGRFA |
| 4 | Disentangle genebank tasks from plant breeding research and plant breeding tasks at the national level |
| 5 | Establish a legal basis for conservation of PGRFA in the EU |
| 6 | Inventory of financial means available to genebanks and estimation of financial means needed for a fully functioning European network of genebanks |
| 7 | Increase the visibility of genebanks on the Internet |
| 8 | Clear uncertainties concerning access and benefit-sharing (ABS) rules so that breeding companies can take economic decisions on a safe legal basis |
| 9 | Research should be strengthened to understand better the amount and geographic distribution of genetic diversity present in priority crop genepools |
| 10 | Develop a European infrastructure for long-term crop-specific pre-breeding programmes |
| 11 | The European agro-NGOs and their influence should be strengthened |
| 12 | Establishment of a European network of private–public partnership programmes for evaluation of PGRFA in Europe |

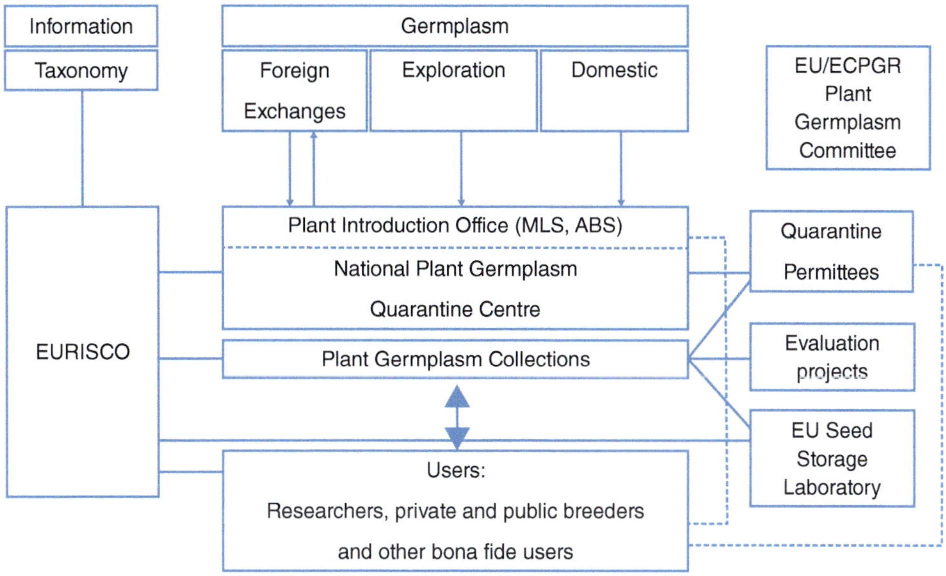


Fig. 37.1. Schema of an EU *ex situ* plant germplasm system.

would have elements corresponding to the National Plant Germplasm System (NPGS) of the USA. The NPGS has been described by White *et al.* (1989). The USDA/ARS is coordinating this decentralized network of crop-specific collections managed by partner institutions located in the federal states. Cooperation between partners is supported by the National Center for Genetic Resources Preservation (NCGRP), which is the central germplasm repository and is responsible for the long-term preservation of all accessions maintained by the NPGS. All partners are linked through the Genetic Resources Information Network (GRIN) (Williams, 2005). The reason why we used the schema of the NPGS published by White *et al.* (1989) to draw Fig. 37.1 simply is to demonstrate that we Europeans have already created several of the elements of an EU-PGS; they are not, however, well connected and are often far from sufficiently funded.

A central taxonomy unit advises and supports collectors and collection and database managers. Information and germplasm is exchanged with European counterparts and partners outside of Europe. A central Plant Introduction Office supervises germplasm exchange, advises on issues in the context of the Multilateral System (MLS) of the International Treaty on Plant Genetic Resources for Food and Agriculture, the Access and Benefit Sharing (ABS) system of the Convention on Biological Diversity, plant explorations and collecting missions outside of Europe, and is kept informed on domestic explorations and collecting missions. The EU Plant Germplasm System is supervised by a Plant Germplasm Committee. National Plant Germplasm Systems run quarantine centres and issue quarantine permits according to EU rules. EURISCO is imbedded in a European PGR Information Infrastructure. It functions as a central access point within a network of national information systems that keep and manage data on PGRs. Data from evaluation projects are recorded and managed by these information infrastructure components and are made available to the user community via EURISCO. The central seed storage laboratory stores backup samples of germplasm collections held within the EU Plant Germplasm System.

European countries have already decided to establish a European genebank collection that will be composed of accessions maintained within a decentralized network of national holdings. Unlike the NPGS, the task sharing will be organized collection-wise and not crop-wise. The ECPGR member countries supported this policy decision and started implementing A European Genebank Integrated System (AEGIS), called the virtual European genebank. By signing the AEGIS memorandum of understanding, governments commit themselves to facilitate access to AEGIS accessions held by their Associate Member Institution(s) (AMs) and to distribute this germplasm at their responsibility and expense (Engels and Maggioni, 2013). The implementation of the AEGIS quality system (AQUAS) by AMs is part of this process. The management of germplasm collections at a high-quality level is a very important step towards the organization of an EU-PGS, as it harmonizes management procedures across genebanks in Europe. Comparable management procedures create trust, which is a precondition for the organization and implementation of task sharing within any institutional network. Once fully developed, AEGIS would have features similar to the NPGS, that is, a set of easily accessible accessions, but would lack a central backup collection comparable to the NCGRP of the USDA/ARS. The access to AEGIS accessions will be facilitated through the European PGR Search Catalogue (EURISCO), which potentially can be developed towards the European equivalent of the Genetic Resources Information Network (GRIN) managed by the Database Management Unit of the USDA/ARS.

Two independent bodies guide PGR activities in Europe. The committee on the conservation, characterization, collection and utilization of genetic resources in agriculture was established to assist the European Commission in the implementation of Council Regulation (EC) No 870/2004, and is called the 'EU genetic resources committee'. The second body is the ECPGR programme, which is guided by the steering committee representing the member countries. Some of the experts are involved in both the ECPGR and the EU genetic resources committee. A Plant Germplasm Committee could be composed of experts from both groups and supported by the ECPGR Secretariat.

The NPGS today organizes 42 Crop Germplasm Committees that are composed of scientist users with specific expertise. They provide advice to the NPGS and organize evaluation and pre-breeding projects, which over the years have generated large quantities of descriptive data and a steady flow of pre-bred material (for example, Panella *et al.*, 2015) registered as germplasm releases. The data and germplasm releases are accessible to any user in the world (Williams, 2005; GRIN, 2014). In the EU, countries with a strong seed industry provide support to crop-specific networks (Lefort *et al.*, 1997) or associations composed of experts from genebanks, the public research sector and breeding companies that screen, evaluate, document and utilize genetic resources (BMELV, 2012). These national expert groups are basically open to any institution or person in Europe who is willing to contribute to the joint activities and to share the benefits, but this seldom happens as there is no legal or institutional framework to encourage cooperation across countries. Though slow, the AEGIS process is valuable as it may lead, if adequately funded, to a technical and organization infrastructure for PGR conservation and use, which in future could be developed into an EU-PGS. The AEGIS process is based on political agreements, and progress very much depends on the extent of input in kind that can be provided by the associated member institutions. To date, 34 European countries have signed the AEGIS Memorandum of Understanding, four of them have nominated altogether 12,500 AEGIS accessions, the majority of which are managed by genebanks in Germany and the Netherlands (ECPGR, 2014). This is a small percentage of the approximately 1.1 million accessions registered by EURISCO.

The ECPGR programme is, for financial and pragmatic reasons, still concentrated on the implementation of the *ex situ* conservation strategy in Europe. Actions of the new ECPGR programme focus on four main outputs; that is, AEGIS, EURISCO, the agreement on a European *in situ* and on-farm concept and the strengthening of relations between genebanks and users of germplasm (ECPGR, 2014). When ECPGR member countries reach an agreement on *in situ* and on-farm conservation and management concepts, the expert community can start elaborating a European strategy for PGRs that integrate *ex situ*, *in situ* and on-farm conservation techniques.

The ECPGR community is currently debating and reviewing two concept papers: the *in situ* concept 'preserving diversity: a concept for *in situ* conservation of crop wild relatives in Europe' and the on-farm concept 'the ECPGR concept for *in situ* (on-farm) conservation in Europe'. The first concept foresees the establishment of a network of genetic reserves for *in situ* conservation of CWR in Europe (Maxted *et al.*, 2015), as part of a global network (Maxted *et al.*, 1997; Maxted and Kell, 2009). The latter proposes the building of individual national and European networks of on-farm sites for coordinated and integrated *in situ* (on-farm) conservation of unique LR materials (Negri *et al.*, 2014). While the debate is going on, species continue to fall through the PGR conservation security net, as illustrated by Fig. 37.2. As an example, *Beta lomatogona* is a CWR of beet, but it is hard-seeded and adapted to high altitude, so is difficult to maintain *ex situ* and is certainly not actively conserved *in situ*, although it may be passively conserved in some existing protected areas. Attempts to introgress its genes into sugarbeet failed in the 1980s, and subsequently it has not been used as a crossing partner again, even though it does have some traits of interest to breeders.

Each of the three conservation strategies implies a wide array of biological, ecological, technical, organizational and scientific challenges. It is not very realistic to assume that a complementary European conservation strategy can be undertaken by institutions of the individual European countries without a supranational programme, joint planning, effective guidance and control. This is another argument in favour of an EU-PGS.

## 37.3 Legal Aspects

The organization and implementation of a truly complementarily PGR conservation strategy requires an EU-PGS having the mandate and means to operate a complex programme. We suggest, therefore, the establishment of a legal basis for the conservation of PGRs for food and agriculture in the EU. In our vision, an EU-PGS is composed of sovereign, national units, coordinated by a European plant germplasm conservation agency (possibly the ECPGR), with EURISCO as the single entry point for users, and co-funded by the EU. Within this EU-PGS, *ex situ*, *in situ* and on-farm conservation should be integrated and the work should follow a set of common standards. The national units should be legally independent organizations with separate powers. Genebanks are mostly attached to research institutes and are seldom independent in terms of mandate, programme, staff and budget. The tasks of genebanks are often intermingled with breeding research, pre-breeding and breeding activities. Under such conditions, staff are often forced to assume the roles of a collection manager and a pre-breeder as well. If capacities are limited, resources are concentrated on breeding aspects at the disadvantage of conservation of PGRs.

Fig. 37.2. *Beta lomatogona*, a CWR of potential use for beet breeding.

Interviewees from the genebank stakeholder group mentioned declining budgets for genebanking. Twenty out of these 24 respondents to the online questionnaire described the shift from permanent funding to funding of temporary projects as a major constraint in genebanking. When we asked interview partners which genebanks worked most effectively, interviewees representing genebanks, breeders and agro-NGO all mentioned NordGen, Centre for Genetic Resources, The Netherlands (CGN), Leibniz Institute of Plant Genetics and Crop Plant Research (IPK), and the NPGS of the USDA/ARS. Genebank representatives in 14 countries were asked about the juridical status of the genebank in their country. Genebanks in 11 countries are dependent. Only one genebank, the NordGen, is fully independent and two, CGN and IPK, dispose of an earmarked budget (PGR Secure, 2014). Likewise, the high-performing NPGS has an earmarked budget on hand to finance routine genebank operations. We concluded that independent genebanks functioned better than those that could not decide freely on the work programme, deployment of staff and budget. National units of an EU-PGS should have the same legal status. We suggest, therefore, disentangling genebank tasks from plant breeding research and plant breeding tasks at the national level (Recommendation 4 and 5, Table 37.1).

PGRs are considered one of the most important assets for adapting crop species to new agricultural production conditions caused by climate change. It is therefore of utmost importance to establish a technical European infrastructure that supports *ex situ, in situ* and on-farm conservation and allows straightforward access to information and germplasm. A regulation on PGRFA would initiate a process leading to an EU conservation strategy, an EU action plan and corresponding national strategies and plans. Costing of the work related to strategy implementation would be an important task, since strategy papers remain vague as long as they do not define a set of realistic targets and explain the financial means needed to achieve these targets. In this context, Council Regulation (EEC) No 1210/90 of 7 May 1990 on the establishment of the European Environment Agency and the European Environment Information and Observation Network is of interest. Regulation (EC) No 401/2009 is the amended version.

Although this is an EU regulation, it also encourages pan-European cooperation. Paragraph 12 of the regulation's preamble stresses that it is desirable '[...] to be open to other countries which share the concern of the Community and the Member States for the objectives [...]'. Regulation (EC) No 401/2009 enabled the foundation of the European Environment Agency (EEA), the establishment and operation of the European Union Nature Information System (EUNIS) and the network of Natura 2000 protected areas, as well as the creation of management structures. The EEA is managed by the Executive Director, who disposes of an independent budget controlled by the European Union. The agency is guided by the EEA Management Board. Article 4, 3, of the regulation foresees National Focal Points to facilitate communication between national partner agencies and the EEA.

This regulation could serve as a starting point for the elaboration of a regulation on an EU-PGS. The ECPGR Secretariat could be developed towards a European Plant Germplasm Conservation Agency, vested to manage a budget that is required to develop the infrastructure and to operate a European plant germplasm conservation programme that integrates the best components of *in situ*, on-farm and *ex situ* conservation techniques. We are convinced that most of the technical and organizational constraints today limiting the conservation and sustainable use of CWR and LR are implications of the legal gap for PGR conservation measures in the EU.

## 37.4 Financial Aspects

The current weaknesses in the European way of managing and utilizing PGRs can be overcome through improved task sharing and cooperation, in combination with increased support on both the national and EU level. We doubt, however, that task sharing and cooperation, discussed by ECPGR working groups for decades, can be organized effectively and efficiently without a legal basis, activities focused on selected crop genepools, stronger guidance and coordination and a funding mechanism that strengthens national partners instead of taking away capacities from national systems through the input in kind approach.

There is vivid debate on access and benefit sharing and the legal mechanisms required to comply with the Nagoya Protocol, while there is no debate on the costs of conservation actions in Europe. Virchow (2003) in the chapter 'Efficient Conservation of Crop Genetic Diversity: The Costs and Implications' wrote that 'benefit sharing should evolve towards cost sharing' and '... without detailed knowledge about the costs of conservation the cost sharing cannot be implemented in a way that secures the sustainable conservation of crop genetic resources'. Prior to demanding more financial support, the European PGR conservation community should therefore inventory the financial means available to genebanks and estimate the financial means needed for the establishment of a fully operative and functioning European PGR conservation programme (Recommendation 6, Table 37.1). This work should be performed as part of the process initiated by Engels *et al.* (2012) with the paper 'Strategy Paper on the ECPGR Relationship with the European Union/European Commission'.

Today, European institutions pursue two ways of funding PGR conservation and use activities. First, the ECPGR relies on input in kind of institutions located in the member countries. Members provide the ECPGR with funds for the secretariat, the AEGIS process, the operation of EURISCO and for funding small actions of working groups and meetings of the steering committee. The member states de facto paid €1,950,286 for phase VIII (2009–2013) and are currently signing for phase IX. As in the past, the progress to be achieved by working groups relies on the interest of individuals in specific subjects and the ability of working groups to attract project funds from sources outside the ECPGR programme.

Many of the working groups succeeded in attracting project funds for PGR conservation and use actions through the European Community Programmes on the Conservation, Characterization, Collection and Utilization of Genetic Resources in Agriculture (Council Regulation (EC) No 1467/1994 and Council Regulation (EC) No 870/2004). Both programmes proved beneficial to working groups in that additional financial means became available for the collection, evaluation, documentation and exploitation of genetic resources. The actual amount of EU co-funding of PGR projects implemented in the second programme amounted to €6.4 million for arable crops, fruit, vegetables and spices (Council Regulation (EC) No 870/2004) (EC, 2013). In the field of PGRs, neither of the programmes helped in the building of a European infrastructure for PGR conservation.

The ECPGR stands on a vulnerable base. It has to balance the overriding long-term interest of the steering committee against the interests of individuals who invest work time into the ECPGR and who expect to derive some short-term benefit from participating in working group activities. The appeal of the steering committee to invest work time in the establishment of AEGIS exemplifies this conflict between overriding and individual interests. AEGIS does not create short-term individual benefit. On the contrary, individuals are called upon by the steering committee to provide input in kind to this process at a time when the staff and financial resources available to institutions at the national level are being decreased. Individuals are in conflict with personal/institutional targets and targets of the steering committee promising the emergence of a fully operative European programme, a promise that has been made since the establishment of the ECPGR and which so far has not materialized in 2015; that is, about 30 years after the establishment of the ECPGR. With this contribution to conference proceedings, we do not question the aims of the ECPGR programme but the funding concept applied to reach its targets.

The infrastructure of a European Plant Germplasm System should meet the requirements of *ex situ*, *in situ* and on-farm conservation techniques; namely, it should be based on a complementary PGR conservation strategy and work plans. The European Plant Germplasm System should be established and be fully operational in sufficient time to be able to mitigate climate change-induced damages to agricultural production systems through plant germplasm utilization programmes. The forecasted timing and impacts of climate change on agriculture (Jarvis *et al.*, 2008) will not allow us to try for another 30 years to establish an efficient and effective European Plant Germplasm System.

Without doubt, this ambitious goal cannot be achieved with an approved ECPGR budget of €2.8 million for a programme phase of 5 years (ECPGR, 2014) and an unknown amount invested by institutions into the ECPGR on an input in kind basis.

A funding mechanism exists in the EU that is completely independent from ECPGR activities. It is called agri-environmental measures (AEMs) and provides payments to farmers who subscribe, on a voluntary basis, to environmental commitments related to the preservation of the environment. AEMs are designed to encourage farmers to protect and enhance the environment on their farmland by paying them for the provision of environmental services. Farmers commit, for a minimum period of at least 5 years, to adopt environmentally friendly farming techniques that go beyond legal obligations. In return, farmers receive payments that provide compensation for the additional costs and income foregone resulting from applying those environmentally friendly farming practices in line with the stipulations of agri-environment contracts. Agri-environment payments encourage farmers to adopt agricultural activities or levels of production intensity that deliver positive environmental outcomes. The conservation of high-value habitats and their associated biodiversity is an example of commitments covered by national/regional agri-environmental schemes. The costs are part-financed from the EU budget. Mathews (2011) speaks of €22 billion payments for agri-environment measures during the 2007–2013 research and development (RD) programming period.

In September 2014, the European Court of Auditors repeated its criticism and wrote that the European Commission and member states must ensure better use of the European Regional Development Fund (ERDF) for halting biodiversity loss. It is strange to see that **€42.7 billion** was foreseen for agri-environment payments in the 2007–2013 programming period of the Rural Development Programme of the EU-28 Member States and that this amount was spent without much measurable effect (Whitefield, 2006; ECA, 2014), while the ECPGR, once founded by European countries with the objective of conserving one of the most important resources of agriculture, the PGR, is supported by up to 43 ECPGR member states with **€2.8 million** for a similar programme period.

We think, therefore, that it is time to discuss a third way towards the conservation and sustainable use of PGRs in Europe and suggest investing in establishing an EU-PGS similar to the NPGS of the USA. We are convinced that such a system can only function as efficiently and effectively as the NPGS if it is reasonably funded. As a rule of thumb, *ex situ* conservation, including related research costs, amounts to approximately €60/accession and year: 1,725,315 accessions are kept in European genebanks (FAO, 2010), resulting in an assumed total annual cost for *ex situ* conservation of €103,518,900/year for the whole of Europe. In view of the several billion euros spent on agri-environmental measures within the EU-28, a budget of €100 million/year is not unreasonable. The question rather is whether the stakeholder groups and policy makers feel that having an EU-PGS is worth this amount.

## 37.5 Genepools as Management Units

Research should be strengthened to understand better the amount and geographic distribution of genetic diversity present in priority crop genepools (Recommendation 9, Table 37.1). As funds for any biodiversity conservation programme are always limited, the proponents of Council Directive 92/43/EEC on the Conservation of Natural Habitats and of Wild Fauna and Flora, known as the Habitats Directive, suggested focusing conservation activities on a defined set of species. To this end, nature conservation experts elaborated a list of so-called policy species, Annex II 'Animal and Plant Species of Community Interest whose Conservation Requires the Designation of Special Areas of Conservation' of the Directive. The main aim of the Habitat Directive is to promote the maintenance of biodiversity by requiring member states to take measures to maintain or restore natural habitats and wild species listed in the Annexes to the Directive at a favourable conservation status, introducing robust protection for those habitats and species of European importance (EC, 1992).

With regard to political enforceability, it would be a strategically good decision to streamline a European plant germplasm conservation programme in a similar way by prioritizing crop genepools. Kell *et al.* (see Chapter 11, this volume) identified 25 crop genepools of high importance for European agriculture. Species that occur in Europe and are part of these genepools could be defined as PGRFA policy species for which conservation measures must be given high funding priority.

Our technical capabilities and capacities to measure genetic diversity with molecular genetic markers are increasing steadily. The fast development of molecular and bioinformatic tools already contribute to a better understanding of the processes and selective forces causing spatial and temporal genetic differentiation within a species (Gregorius *et al.*, 2003; van Zonneveld *et al.*, 2014). Recent progress in conservation genetics and genomics will increasingly allow us to devise and organize complementary conservation strategies and programmes (van Zonneveld *et al.*, 2014), the effects of which can be monitored with molecular markers and assessed according to the S.M.A.R.T.E.R. (goals that are specific, measurable, achievable, relevant and time-bound, and can be evaluated and reviewed) criteria. Unlike AEMs, an EU-PGS could demonstrate that investments in conservation actions cause measurable effects that can be assessed.

Complementary PGR conservation strategies and programmes rather concern genetic diversity within crop genepools than genetic diversity distributed across different crops' genepools. Breeding is a crop-specific activity that relies on novel genetic variation contained in the crop species and related wild species. The natural and logical management entity for conservation and use is thus the crop-specific genepool.

## 37.6 Conclusions

ECPGR is at a tipping point. The renewed ECPGR programme, the growing interest and involvement of the EC in genetic resources issues and vast progress in conservation science and plant breeding research offer chances for opening a new chapter in PGR conservation and use in Europe. While in the 1980s plant germplasm activities in Europe and the USA were comparable, European countries are clearly lagging behind today. The deviating development is attributable to the weak financial support of the ECPGR by the member countries and gaps in the EU legislation.

The conservation of genetic diversity native to or existing in European countries calls for coordinated, concerted and adequately funded actions, which cannot be organized by the member states themselves in a satisfactory manner. Hence, there is a significant role to play for the EU according to Article 5, 3 of the Treaty on European Union (EU, 2010) which details that 'under the principle of subsidiarity, in areas which do not fall within its exclusive competence, the Union shall act only if and in so far as the objectives of the proposed action cannot be sufficiently achieved by the Member States, either at central level or at regional and local level, but can rather, by reason of the scale or effects of the proposed action, be better achieved at Union level'.

Each nation can, of course, try to improve work related to the conservation and sustainable use of PGRs individually. However, a joint European infrastructure, strategy and programme that allow task sharing and promote cooperation is a more rational approach as it avoids unnecessary expenditures. Our SWOT analysis has shown that the objectives of the ECPGR cannot be sufficiently achieved by the member countries, either at central level or at regional and local level, but rather can be better achieved at Union level. There are, therefore, good reasons for the EU to take action in favour of our most important natural resources for food, the genetic diversity of crops and related species.

## Acknowledgements

This research was funded by a grant from the EU Seventh Framework Programme for the PGR Secure project (Grant Agreement No 266394).

## References

April, J. (1979) An information system to service the US National Plant Germplasm System. In: *Technical Aspects of Information Management and Means of Communication in Plant Genetic Resources Work for a Future Utilization of Genetic Material in Plant Breeding. Proc. of a Colloquium held in Braunschweig on 7th and 8th November 1978*. Institut für Pflanzenbau und Saatgutforschung, Bundesforschungsanstalt für Landwirtschaft, Braunschweig-Völkenrode (FAL), Germany, pp. 73–78.

BMELV (2012) Nationales Fachprogramm zur Erhaltung und nachhaltigen Nutzung pflanzengenetischer Ressourcen landwirtschaftlicher und gartenbaulicher Kulturpflanzen. Bundesministerium für Ernährung, Landwirtschaft und Verbraucherschutz, Berlin.
EC (European Commission) (1992) Council Directive 92/43/EEC of 21 May 1992 on the conservation of natural habitats and of wild fauna and flora. *Official Journal* L 206, 22/07/1992 P. 0007–0050.
EC (2011) Communication from the Commission to the European Parliament, the Council, the Economic and Social Committee and the Committee of the Regions. Our life insurance, our natural capital: an EU biodiversity strategy to 2020. 3.5.2011, COM(2011) 244 final, Brussels.
EC (2013) Preserving genetic resources in agriculture. Achievements of the 17 projects of the Community Programme 2006–2011. European Commission, Belgium.
ECA (European Court of Auditors) (2014) Special Report, No 12. Is the ERDF effectively funding projects that directly promote biodiversity under the EU biodiversity strategy to 2020? European Union, Luxembourg.
ECPGR (European Cooperative Programme for Plant Genetic Resources) (2014) A European Genebank Integrated System. Available at: www.ecpgr.cgiar.org/aegis/about-aegis/ (accessed 8 December 2015).
Engels, J. and Maggioni, L. (2013) AEGIS Distribution Policy for European Collection Material. ECPGR Secretariat, Maccarese, Rome.
Engels, J., Freudenthaler, P., Landbo, L. and Latorre, F. (2012) Strategy Paper on the ECPGR Relationship with the European Union/European Commission. ECPGR, Rome.
EU (European Union) (2010) Consolidated version of the Treaty on European Union. *Official Journal of the European Union* C83/18.
Evans, L.T. (1999) Sir Otto Herzberg Frankel. 4 November 1900–21 November 1998: Elected F.R.S. 1953 *Biographic Memorials of Fellows of the Royal Society* 45, 165–181, 1748–8494.
FAO (Food and Agriculture Organization) (2010) *The Second Report on the State of the World's Plant Genetic Resources for Food and Agriculture*. Commission on Genetic Resources for Food and Agriculture, FAO, Rome.
Frankel, O.H. (1970) Genetic conservation of plants useful to man. *Biological Conservation* 2, 162–169.
Frankel, O.H. and Hawkes, J.G. (eds) (1975) *Crop Genetic Resources for Today and Tomorrow*. International Biological Programme 2, Cambridge University Press, Cambridge, UK.
Gregorius, H.-R., Gillet, E.M. and Ziehe, M. (2003) Measuring differences of trait distributions between populations. *Biometrical Journal* 45(8), 959–973.
GRIN (Germplasm Resources Information Network) (2014) USDA/ARS NPGS Germplasm Resources Information Network. Available at: www.ars-grin.gov/npgs/ (accessed 10 October 2014).
IBPGR (International Board for Plant Genetic Resources) (1984) *The ECP/GR. Introduction to the European Cooperative Programme for Conservation and Exchange of Crop Genetic Resources*. AGPG/IBPGR 84/148, October 1984, Rome.
Jarvis, A., Lane, A. and Hijmans, R.J. (2008) The effect of climate change on crop wild relatives. *Agriculture, Ecosystems and Environment* doi:10.1016/j.agee.2008.01.013.
Kell, S., Maxted, N., Frese, L. and Iriondo, J.M. (2012) *In situ* conservation of crop wild relatives: a strategy for identifying priority genetic reserve sites. In: Maxted, N., Dulloo, M.E., Ford-Lloyd, B.V. Frese, L., Iriondo, J.M. and M.Â.A. Pinheiro de Carvalho (eds) *Agrobiodiversity Conservation: Securing the Diversity of Crop Wild Relatives and Landraces.* CAB International, Wallingford, UK, pp. 7–19.
Laliberté, B., Maggioni, L., Maxted, N. and Negri, V. (compilers) (2000) *ECP/GR In situ and On-farm Conservation Network. Report of a joint meeting of a Task Force on Wild Species Conservation in Genetic Reserves and a Task Force on On-farm Conservation and Management, 18–20 May 2000, Isola Polvese, Italy.* International Plant Genetic Resources Institute, Rome.
Lefort, M., Arbez, M., Chauvet, M., Datteé, Y., Guiard, J., *et al.* (1997) The French strategy for the management of plant genetic resources. *Plant Varieties and Seeds* 10, 153–162.
Mathews, A. (2011) Court of Auditors launches broadside against deficiencies in agri-environment schemes. Available at: capreform.eu/court-of-auditors-launches-broadside-against-deficiencies-in-agri-environment-schemes/ (accessed 10 October 2014).
Maxted, N. and Kell, S.P. (2009) *Establishment of a Global Network for the In Situ Conservation of Crop Wild Relatives: Status and Needs*. FAO Commission on Genetic Resources for Food and Agriculture, Rome, 266 pp.
Maxted, N., Hawkes, J.G., Ford-Lloyd, B.V. and Williams, J.T. (1997) A practical model for *in situ* genetic conservation. In: Maxted, N., Ford-Lloyd, B.V. and Hawkes, J.G. (eds) *Plant Genetic Conservation: The* in situ *Approach.* Chapman and Hall, London, pp. 339–367.

Maxted, N., Avagyan, A., Frese, L., Iriondo, J.M., Magos Brehm, J., *et al*. (2015) Preserving diversity: a concept for *in situ* conservation of crop wild relatives in Europe: version 2. *In Situ* and On-farm Conservation Network, European Cooperative Programme for Plant Genetic Resources, Rome. Available at: www.pgrsecure.org/documents/Concept_v2.pdf (accessed 29 July 2015).

Negri, V., Freudenthaler, P., Gasi, F., Goldringer, I., Moreira, P.M., *et al*. (2014) *The ECPGR Concept for* In Situ *(on-farm) Conservation in Europe*. *In Situ* and On-farm Conservation Network, European Cooperative Programme for Plant Genetic Resources, Rome.

Panella, L., Hanson, L.E., McGrath, J.M., Fenwick, A.L., Stevanato, P., *et al*. (2015) Registration of FC305 Multigerm Sugarbeet Germplasm Selected from a Cross to a Crop Wild Relative. *Journal of Plant Registration* 9, 115–120.

PGR Secure (2014) PGR Secure project. Available at: www.pgrsecure.org (accessed 10 October 2014).

Seidewitz, L. (1974) *Thesaurus for the International Standardisation of Gene Bank Documentation. Part I. Cereals (incl. Maize, Rice and Sorghum).* Genbank, Institut für Pflanzenbau und Saatgutforschung der Forschungsanstalt für Landwirtschaft, Braunschweig-Völkenrode, Germany.

Seidewitz, L. (1979) Data base concepts for a rational handling of information in relation to plant genetically valuable material. In: *Technical Aspects of Information Management and Means of Communication in Plant Genetic Resources Work for a Future Utilization of Genetic Material in Plant Breeding. Proc. of a Colloquium held in Braunschweig on 7th and 8th November 1978.* Institut für Pflanzenbau und Saatgutforschung, Bundesforschungsanstalt für Landwirtschaft, Braunschweig-Völkenrode (FAL), Germany, pp. 31–71.

van Zonneveld, M., Dawson, I., Thomas, E., Scheldeman, X., Etten, van J., *et al*. (2014) Application of molecular markers in spatial analysis to optimize *in situ* conservation of plant genetic resources. In: Tuberosa, R., Graner, A. and Frison, E. (eds) *Genomics of Plant Genetic Resources, Volume 1. Managing, Sequencing and Mining Genetic Resources*. Springer, Heidelberg, Germany, pp. 67–91.

Virchow, D. (2003) Efficient conservation of crop genetic diversity: the costs and implications. In: Virchow, D. (ed.) *Efficient Conservation of Crop Genetic Diversity. Theoretical Approaches and Empirical Studies.* Springer, Berlin, pp. 3–13.

White, G.A., Shands, H.L. and Lovell, G.R. (1989) History and operation of the National Plant Germplasm System. In: Janick, J. (ed.) *Plant Breeding Reviews: The National Plant Germplasm System of the United States, Volume 7.* John Wiley & Sons, Inc, Hoboken, New Jersey, pp. 5–56.

Whitefield, J. (2006) Agriculture and environment: how green was my subsidy? *Nature* 439, 908–909, doi:10.1038/439908a.

Williams, K.A. (2005) An overview of the US National Plant Germplasm System's exploration program. *HortScience* 40(2), 297–301.

# 38 Impact of the Genetic Resources Policy Landscape on Food Security: An Assessment of the Genetic Resources and Intellectual Property Rights Programme

E. Thörn,[1]* C.-G. Thornström[2] and I. Virgin[3]
[1]*Swedish University of Agricultural Sciences, Alnarp, Sweden;*
[2]*Swedish University of Agricultural Sciences, Uppsala, Sweden;*
[3]*Stockholm Environment Institute, Stockholm, Sweden*

## 38.1 Introduction

The focus of the United Nations Millennium Development Goals is to reduce poverty and hunger while at the same time improving health and education and ensuring environmental sustainability (United Nations Millennium Declaration, 2000). The challenge of securing an ample supply of food is enormous, and will remain a major challenge for the foreseeable future. A sustainable increase in global agricultural production based on increased productivity with reduced resource consumption is, therefore, urgently needed. Improved crop and livestock breeding programmes based on old and new technologies and continuous innovations through strengthened scientific and technological developments are key in this regard. In addition to resource scarcity (including water, nutrients and arable land), climate change is one of the most serious threats facing the human race (FAO, 2011). Climate change will most likely cause dramatically changed conditions for agriculture worldwide, and this is expected to be particularly severe in the southern hemisphere. Drought will become more frequent and severe in many parts of the world, and heavy rainfalls and flooding will increase in other parts. An additional threat to food production and food security is the emergence and changed distribution of new and altered crop diseases (Edame *et al.*, 2011).

In recent times, there has been a growing appreciation of the role of intellectual property rights (IPRs) in inspiring the human capital formation, knowledge diffusion, benefit sharing and technological innovation that is necessary for national development. Intellectual property (IP) plays a very important role in economic development, and this new environment has created both threats and opportunities for countries, especially in the south, which must develop a clear understanding of developments in IP and their implications. To be able to protect knowledge and genetic resources (GRs), and to obtain the benefits of these resources, countries must formulate efficient national and institutional policies and regulations and must engage in the international debate.

International conventions and treaties, together with rapid advances in the biosciences, have led to new conditions for access to GRs and knowledge. The Convention of Biological Diversity (CBD) and the requirements under WTO/

*Corresponding author; e-mail: eva.thorn@slu.se

TRIPS (World Trade Organization/Agreement on Trade-Related Aspects of Intellectual Property Rights) have led to stricter mechanisms for access to and the control of genetic resources, as well as new protocols for the protection of biological innovations. Through the CBD, access to GRs is under the sovereign control of states, and this forces institutions and corporations to negotiate with the gene-rich developing countries directly for access to biological resources. At the same time, the use of genetic engineering in the production of novel organisms (e.g. GMO crops) connects IPRs more tightly to the development of genetic resources. As national capabilities in IPR-intensive areas such as biotechnology increase, the need for harmonization with the global efforts for stronger IPRs (e.g. TRIPS) has become more important. The Food and Agriculture Organization (FAO) International Treaty for Plant Genetic Resources for Food and Agriculture (ITPGRFA) has contributed to an easier and more efficient flow of genetic material across national borders, through the multilateral system (MLS) that aims to facilitate the exchange of germplasm across borders. Breeders' exemptions in UPOV (International Union for the Protection of New Varieties of Plants (Union Internationale pour la Protection des Obtentions Végétales)) provide mechanisms for facilitated access to germplasm in consistency with the ITPGRFA and CBD. Requirements under WTO/TRIPS, however, have caused stricter controls and new protocols for the protection of bioinnovations.

The CBD, UPOV, ITPGRFA, and WTO/TRIPS regulate different dimensions of more or less the same biological matter and related information. The concepts agreed on in the different treaties focus on three overriding outcomes: new rules for access, innovation and knowledge. For example, the CBD focuses on bilateral access subject to national legislation, while the ITPGRFA is based on multilateral access to specific crop genera and on specific regulations that restrict products from received material to further use in plant breeding. UPOV is a *sui generis* system outside the patent/IP system in the TRIPS agreement. To complicate matters further, the CBD refers to the protection of traditional/indigenous knowledge, while the ITPGRFA refers to the protection of farming-systems knowledge (Thornström, 2012).

Challenges for future food security are greatest in the developing world. The list of problems and needs is extensive and includes weak markets, lack of capital, poor infrastructure, diseases, droughts, nutrient-deficient soils, land tenure challenges and degraded natural resource bases. With the advent of genetic engineering and novel multilateral IPR agreements, a new chapter in the history of crop breeding has begun. Modern biosciences, such as bioinformatics and genetic engineering, have had a major impact on crop breeding worldwide. Countries without the competence to use modern bioscience techniques are missing the opportunity to use these powerful technologies to develop their own productive, resource efficient and sustainable crop and animal production systems for food, feed and agroindustrial products.

The crop breeding systems in developing countries have to become more efficient in serving farmers with new local varieties. It is, however, unlikely that the private sector will play a leading role in the development of the high-quality varieties that are most relevant to the needs of resource-poor and vulnerable farmers. Thus, for many key important food crops in the south, the dissemination of new varieties will rely heavily on the public sector for both formal and informal seed sector programmes. Strong public research efforts will be essential for harnessing and adapting new breeding tools to the needs of small-scale farmers in developing countries. In this context, ownership of patents and the restrictions on access to plant genetic resources (PGRs), technology and know-how is critical, and this ownership restricts the freedom of public institutions to operate. Therefore, close collaboration is needed between public sector research and development institutions and the private sector.

Several developing countries have a *sui generis* system for plant variety protection (PVP). The problem is that these PVP systems are usually not effective and not even well known to local breeders, let alone implemented. As a consequence, these PVP systems are not very supportive for research, competition, or the improvement of the local crops grown by farmers. It should also be noted here that only four African countries are members of UPOV, and none of these belong to the least developed countries (LDCs) (Merso, 2013).

IP regulations exist in many developing countries; however, these laws are mostly very old and are often limited to covering only patents and trademarks. In order to make them compatible with TRIPS standards, there is an urgent need for the transformation of these legal systems. It is, however, obvious that the LDCs in particular will need support during this process if they are to be accepted by the WTO members and be ready to access the WTO. Capacity building is, therefore, a very important issue for these countries, especially when it comes to the implementation of regulations (Merso, 2013).

During the transition period to update IP practices, guidance and training will be essential for the LDCs. They will need to develop and improve negotiation skills in international environments so as to share in the benefits of the commercialization of their knowledge or genetic resources when these are patented in developed countries. They will need the capacity to analyse and assess the risks and benefits of harmonization of their current legal systems with international patent laws and to assess the impacts of IP standards and how these can stimulate innovation and technology transfer (Commission on Intellectual Property Rights, 2002).

The IPR and legal regimes governing the issues of the use and exchange of genetic resources have become more complicated and stricter. Thus, policy makers, scientists and other practitioners – especially those in resource-poor countries – face considerable challenges in formulating IPR policies and in negotiating appropriate agreements. This new environment has created both threats and opportunities for countries in the south, and these countries must develop a clear understanding of recent developments and their implications. In light of this, the Swedish International Development Cooperation Agency (Sida) decided to stimulate and facilitate the integration of the developing countries into the international IP system by supporting the Genetic Resources and Intellectual Property (GRIP) programme that was developed and implemented by the Swedish University of Agricultural Sciences (SLU) together with the Stockholm Environment Institute (SEI).

## 38.2 The Genetic Resources and Intellectual Property Rights Programme (GRIP)

GRIP was a Sida-funded international training programme that was implemented from 2003 to 2013. The primary aim of the GRIP programme was to assist policy makers, researchers and practitioners from developing countries to understand better the background and implications of the regulatory and policy landscape related to IPRs and the exchange of genetic resources. The programme also sought to enhance the practitioners' managerial and technical skills in the field of IP, to contribute to processes of change and development in the participants' organizations and to provide guidance in the formation of policies concerning GR/IP and innovation systems. The overall development objective of the GRIP programme was to improve the target countries' capacity to serve their informed national interests in GR/IP matters and to develop further the infrastructure needed in terms of professionals and institutions/functions and thus contribute to long-term development and poverty alleviation.

The GRIP programme was realized by means of advanced training for senior professionals from the government sector, national research and development institutions, and private sector organizations in charge of managing and defining regulatory and policy frameworks related to IP and the access and benefit sharing of genetic resources and related knowledge. The GRIP programme dealt with matters embracing many disciplines such as genetics, technology, law and policies, as well as the links between these and issues such as the exchange of genetic resources, international agreements, access and benefit sharing, farmers' rights, IPRs, regional and international policies, and the impact of these respective policies on the management of genetic resources at different national and institutional levels. The training courses offered through the GRIP programme also facilitated a better understanding of how to develop appropriate IPR instruments for beneficial partnerships between various actors in national, regional and international settings such as universities, public–private partnerships, etc. One focus of the GRIP programme was on IP in relation to PGRs and crop breeding.

The GRIP programme trained 330 professionals from more than 50 countries, with a focus on countries in sub-Saharan Africa, South-east Asia and the Andean community. During the period of 2003 to 2013, a total of 15 training programmes were held. In 2008, an expanded format of the GRIP programme was introduced that consisted of three phases: one training course in Sweden; an individual work period on a personal project of change (PoC) in one's home country during which the participants were supposed to contribute to the strengthening of IPR/genetic resources regimes in their respective home countries; and a regional follow-up seminar. During the last 2 years of the programme, the number of courses offered doubled annually.

The positive outcome of the first phase of the GRIP programme was affirmed in an evaluation seminar held in Bangkok in 2007 during which GRIP alumni shared their experiences and described developments within the GR/IP field in their respective home countries (Thornström, 2007; Thornström and Virgin, 2009).

## 38.3 Scope and Methodology of the GRIP Impact Assessment

An impact assessment of the GRIP programme was carried out in 2013 after the last group of participants had completed the course. The focus of the assessment was on the implementation of international, national and institutional GR/IP regimes in the participant's home countries and on how the participants had influenced policy development at the institutional and national level. In particular, it was important to elucidate the implementation and impact of the participants' individual PoCs and to clarify if and how knowledge gained through the GRIP training contributed to the personal careers of the participants.

The specific objectives of the evaluation were to assess, identify and demonstrate changes since 2009, to evaluate remaining gaps and needs regarding national and institutional development of IP/GR regulatory regimes and to appraise the impact of the GRIP training. It was also expected that the evaluation would reveal gaps and further needs for training and capacity building within GR/IP and associated fields. The main stakeholders involved in the evaluation were all GRIP alumni from 2009 to 2013, and the assessment was focused at the national, regional and institutional levels. The questions addressed during the evaluation dealt with the outcomes and impacts from capacity building with regard to GR/IP policies, legislations, regulations and new national and institutional GR/IP frameworks/regimes, and to what extent the GRIP programme had contributed to these developments.

The assessment was developed with the Netigate programme (www.netigate.se). An electronic questionnaire with a total of 41 general and specific questions was prepared and sent out by e-mail to the GRIP alumni. The respondents were given a period of 3 months to send in their responses. Data were collected, compiled and analysed from the answers received.

## 38.4 Results

A total of 166 people participated in the GRIP training from 2009 to 2013. Table 38.1 shows the number of participants per region and country. A few countries, such as Kenya, South Africa, DPR Korea and Colombia, had significantly more participants than other countries. The high number of participants from DPR Korea was due to a special agreement for capacity building between Sweden and DPR Korea.

The questionnaire was sent out to 164 of the participants. Sixteen of the e-mail addresses bounced back, and it was also impossible to reach the participants from DPR Korea by e-mail. A total of 88 persons took part in the survey, and this was 54% of the total number of the GRIP participants and 60% of those who had received the e-mail with the questionnaire. Because the number of responses from each country differed significantly, the results can only be considered indicative of the current situation in the respective countries.

Several questions focused on the participants' PoCs, and 66% of the respondents reported that their PoCs had been implemented successfully. Some of the respondents stated that their PoCs were either in the process of being implemented or were at the discussion level. Reasons given for PoCs not being implemented were mostly funding problems and, in a few cases, political problems. Because the majority of the respondents had only recently finished the training, it is obvious that they would need more

Table 38.1. Number of GRIP participants according to region and country.

| Africa | | Asia | | Latin America and the Caribbean | | Europe | |
|---|---|---|---|---|---|---|---|
| Country | Number of participants | Country | Number of participants | Country | Number of participants | Country | Number of participants |
| Benin | 1 | Bangladesh | 2 | Brazil | 1 | Bosnia and Herzegovina | 3 |
| Botswana | 2 | Cambodia | 1 | Chile | 2 | Georgia | 1 |
| Cameroon | 1 | China | 8 | Colombia | 10 | Kosovo | 1 |
| Egypt | 2 | India | 4 | Ecuador | 3 | Montenegro | 1 |
| Ethiopia | 1 | Indonesia | 1 | Mexico | 1 | Serbia | 3 |
| Ghana | 1 | Iraq | 1 | Peru | 5 | Turkey | 1 |
| Kenya | 16 | Jordan | 2 | Suriname | 1 | | |
| Malawi | 1 | DPR Korea | 10 | Trinidad and Tobago | 1 | | |
| Morocco | 2 | Kyrgyzstan | 1 | | | | |
| Mozambique | 2 | Laos | 3 | | | | |
| Nigeria | 2 | Malaysia | 1 | | | | |
| Rwanda | 4 | Nepal | 2 | | | | |
| South Africa | 10 | Pakistan | 1 | | | | |
| Sudan | 1 | Palestine | 1 | | | | |
| Tanzania | 8 | Philippines | 4 | | | | |
| Tunisia | 2 | Sri Lanka | 4 | | | | |
| Uganda | 8 | Tajikistan | 1 | | | | |
| Zambia | 5 | Thailand | 7 | | | | |
| Zimbabwe | 4 | Vietnam | 5 | | | | |
| Total | 73 | | 59 | | 24 | | 10 |

time for the implementation process, and some were even still working on the finalization of their PoCs. Due to the fairly short time since many of the respondents had finalized their PoCs, the impact of the projects has so far been limited. However, reports on the PoCs still showed a variety of outcomes. On a personal level, many of the respondents described increased knowledge and understanding of IP matters and increased negotiation skills. On the institutional level, the PoCs were reported to have contributed to capacity building and awareness raising, as proudly reported by one of the respondents: 'I have educated more than 5,000 lawyers on PVP in Kenya – and I am proud of that – and I have also educated breeders on the procedures they should follow in case their intellectual property is infringed upon' (Aprinah Shikoli, Kenya, 2013, personal communication). Specific institutional outcomes included the establishment of technology transfer offices, the development of IP and access and benefit-sharing policies, and licensing agreements. Increased attention to traditional knowledge (TK) was a recurrent issue in the responses, and one person said: 'Successful commercialization of indigenous crops is still new to the organization, and the model is seen as an example of commercialization and the way to manage indigenous crops' (Cecilia Bester, South Africa, 2013, personal communication). Similar outcomes were reported at the national level, including the implementation of international conventions and treaties, the enactment of legislation related to GRs, contributions to national IP policies, increased conservation activities, enhanced private investment in seed production and variety protection, and the establishment of national PGR programmes. As stated in one of the responses: 'The country gained its first document that precisely described the importance of conservation and the utilization of plant genetic resources and the importance of international legislation regarding this specific matter' (Jasmin Grahic, Bosnia and Herzegovina, 2013, personal communication).

In general, the respondents reported that the national capacity related to the adoption and implementation of international instruments had increased to a very large extent or significantly (62%) during the last 5 years prior to the survey. It was also reported by 75% of the respondents that their countries had adopted and begun to implement regulations and policies related to the exchange of genetic resources and to the sharing of benefits from such resources. Examples of international instruments that were implemented included the CBD, UPOV, Cartagena Protocol, ITPGRFA and Nagoya Protocol. Likewise, much had happened on the national level, and laws had been enacted in the areas of environmental protection, agricultural production, food safety, variety protection, farmers' rights, plant health, protection of TK, technology transfer, biosafety and biodiversity. On the question of institutional GR/IP development, the degree of the adoption/implementation of regulations and policies related to IP and genetic resource utilization was lower (46%) compared to the national level. The most frequent answer to this question was the development of institutional IP policies, followed by technology transfer policies and technology transfer offices. A large number of the respondents (63%) reported institutional improvements regarding capacities related to the handling of GR/IP matters compared to 5 years prior to the survey. Only 7% reported that their institutions had not made any improvements.

Although quite a large number of the countries had ratified the CBD and the ITPGRFA, only 50% of the respondents reported that their institutions had developed or implemented policies/regulations on the access and exchange of GRs. The most frequently mentioned were standard/material transfer agreements (MTAs or SMTAs), followed by regulations for prior informed consent (PIC) and mutually agreed terms (MATs).

The capacity-building component of the GRIP training resulted in increased involvement in training activities carried out by the course participants. A total of 66% of the respondents reported that they had been engaged in the training of a large number of students, colleagues and farmers through seminars, regular university courses and workshops. In addition, it was stated by 48% of the respondents that the participation of women in training activities had increased from 2009 to 2013. This applied to all levels of training, from farmers to graduate students. The respondents were asked to estimate the number of new or additional IPRs their organization had applied for compared to 5 years ago (Fig. 38.1). The answers indicate that there

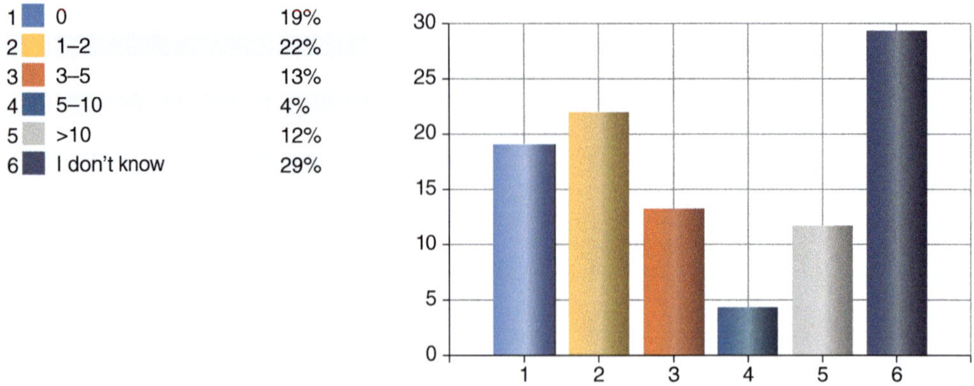


**Fig. 38.1.** Estimate the number of new or additional IPRs your organization has applied for and been granted compared to 5 years ago.

was an increase in the protection of innovations. The picture is similar for the number of new or additional technology transfer/license agreements that had been entered into by the respondent's organization compared to 5 years ago (Fig. 38.2). Awareness raising is an important issue in most of the countries in order to educate all levels of society about GR/IP matters, and 57% of the respondents stated that these kinds of activities had given positive results, including among local communities.

Finally, the respondents were asked to give their opinion on what improvements were needed in order to make current regulations such as the CBD, protection of TK, etc., work better in their countries. The most frequent answer to this question was, 'increased awareness of GR/IP issues'. The following are quotations from some of the respondents in this regard:

> The lawmakers must enact a comprehensive law on TK that shall embody the various areas of TK, including traditional medicine knowledge. At the moment, the country only has a national policy that touches on the subject matter and that obviously does not have the full force of law. It is imperative for lawmakers to see the urgency in the subject matter so as to curb bio-piracy and other concerns associated with TK.
>
> (Raya Salim, Kenya, 2013, personal communication)

> The basic needs of farmers should be addressed in our country. Land holding is very much limited to poor farmers, therefore they do not cooperate with the implementation of the CBD and relevant acts. Protection of TK and awareness are important. Training is one of the important tools, and there should be continuous education on this matter.
>
> (Banshi Sharma, Nepal, 2013, personal communication)

> There is a need for more flexibility to make many of the regulations work better. For example, in the CBD the issue of 'Disclosure of Source' should be removed or modified.
>
> (Steve Taiwo, Nigeria, 2013, personal communication)

> TK is under-appreciated in my country due to a history of colonization. My country is a signatory to the CBD and has a national law in place based on it; nevertheless, the understanding of the CBD and its application at a national level is far from being clear and as a consequence access to genetic resources is undermined, and this in turn negatively impacts national scientific research.
>
> (Maria Virginia Jaramillo Navarro, Colombia, 2013, personal communication)

> We need to have more people who can generate proposals for improvement within the state agencies. There must be greater technical and scientific components and only minor political components in the state agencies charged with regulating this issue.
>
> (Giovanny Garavito, Colombia, 2013, personal communication)

GR/IP legislation seems to be in place in most of the countries, but the bottleneck is most often the implementation of the legislation due to a

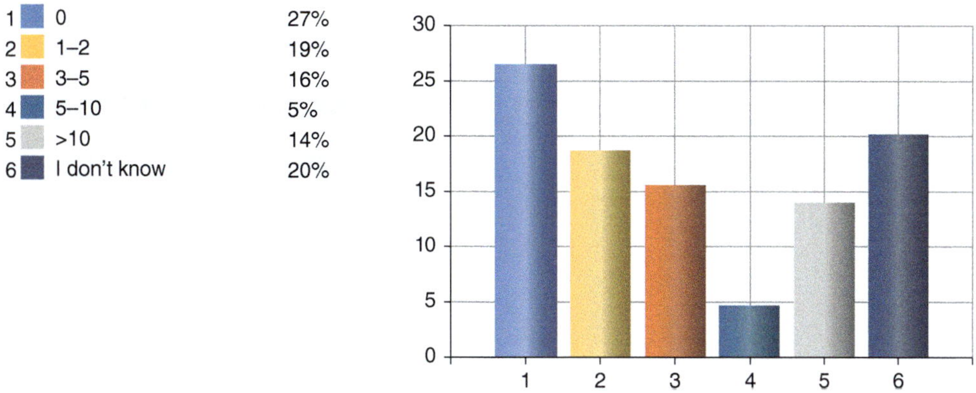


Fig. 38.2. Estimate the number of new or additional technology transfer/license agreements your organization has entered into compared to 5 years ago.

lack of established procedures for managing and monitoring the legislation, as well as a lack of human resources. The respondents were asked to give their views on what measures could be taken in order to break this stalemate. Many of the respondents proposed the establishment of specialized authorities or agencies or moving the executive responsibility from the ministries to public or private organizations, or even to NGOs. The following are some of the comments from the respondents:

> Fall-back positions would be any public institutions or facilities that have the expertise, will, and understanding to work on such important issues.
>
> (Sreten Terzic, Serbia, 2013, personal communication)

> A private-sector organization can be trained to perform such functions or the work can be outsourced by the government to independent organizations.
>
> (Russell Pili, Philippines, 2013, personal communication)

> In my opinion there is a need to establish an international agency that will oversee how government agencies monitor and manage the existing national legislation.
>
> (Noel Mbonde, Tanzania, 2013, personal communication)

> Public awareness of various legal instruments and their benefits can have a long-lasting effect on breaking the stalemate, and this would help in implementing these regulations. To some extent, NGOs can also help in implementing these legislations.
>
> (Sanjay Singh, India, 2013, personal communication)

> Pertinent NGOs could be educated and encouraged to manage and/or monitor the regulations.
>
> (Julius Siwale, Zambia, 2013, personal communication)

## 38.5 Conclusions

It can be firmly concluded that the GRIP programme significantly improved the course participants' knowledge and understanding of GR and IP matters. This has led to increased skills in their daily work, which in turn has led to a positive influence on their careers, such as promotions, more advanced tasks and even new positions. These recently acquired skills and experiences have also been of use for their organizations and their professional environments through training events, discussions and advisory tasks.

The PoCs of the participants have, in most cases, been received positively by their employers, and many have even already been implemented. This is, of course, expected because, before taking on the assignment, the projects were, in most cases, endorsed by their superiors.

It is clear that a large number of developing countries have signed or ratified international instruments related to GRs and IP. However, much remains to be done before functional structures can be put in place and clear processes for their implementation can be developed. This has an influence on the daily operations of organizations dealing with these matters, and often results in complicated situations or even serious restraints on the handling and trading of GRs. Therefore, much still needs to be done, but the GRIP participants have definitely contributed to an increased awareness of current obstacles and needs at all levels.

Research and innovations are important for developing countries, and it is evident that there is great potential for their competitiveness in the near future. Nevertheless, a successful outcome will need to have supportive policies and structures in place. The good news is that this is under way in most countries and has been implemented successfully in some. The GRIP participants have certainly provided constructive contributions to this progress.

From the reports and statements given by the GRIP participants, it can be established that the capacity-building objective of the GRIP training has been fulfilled. Furthermore, changes in the form of improved structures, new and more efficient routines and supportive functions have all strengthened the prospects for progressive economic development for many of the target countries. According to the respondents to the impact assessment questionnaire, the GRIP programme has, to a large extent, been instrumental in this ongoing progress.

When taking an overall look at the GRIP experience for more than a decade, it is evident that one needs to examine critically the nature of the legal concepts agreed on in different treaties; that is, the scope of protection and rights conferred. It is necessary, therefore, to translate what is politically desirable into what is technically and scientifically possible in view of available financial, political and administrative resources, and in view of several dozen other international treaties that require national implementation. Adding to these challenges are, of course, the concept of *intangible genetic information*; that is, digitalized genetic information (bioinformatics/synthetic biology) and the concept of national sovereignty.

## References


Commission on Intellectual Property Rights (2002) *Integrating Intellectual Property Rights and Development Policy – Report of the Commission on Intellectual Property Rights, September 2002*. Commission on Intellectual Property Rights, London.

Edame, G.E., Anam, B.E., Fonta, W.M. and Duru, E.J.C. (2011) Climate change, food security and agricultural productivity in Africa: issues and policy directions. *International Journal of Humanities and Social Science* 1(21), 205–223.

FAO (Food and Agriculture Organization) and Earthscan (2011) *The State of the World's Land and Water Resources for Food and Agriculture: Managing Systems at Risk*. FAO, Rome, and Earthscan, London.

Merso, F. (2013) *IP trends in African LDCs and the LDC TRIPS Transition Extension*. Policy Brief No 26, June 2013. International Centre for Trade and Sustainable Development (ICTSD), Geneva.

Thornström, C.-G. (ed.) (2007) *Exchanging Biodiversity – National Regulatory Strategies and Capacity Building*. Swedish University of Agricultural Sciences, Alnarp, Sweden.

Thornström, C.-G. (2012) International conventions and agreements – consequences for international trade and utilization of biological matter, including microorganisms. In: Sundh, I., Wilcks, A. and Goettel, M. (eds) *Beneficial Microorganisms in Agriculture, Food and the Environment – Safety Assessment and Regulations*. CAB International, Wallingford, UK, pp. 293–307.

Thornström, C.-G. and Virgin, I. (eds) (2009) *GRIP 2003–2008 – Genetic Resources and Intellectual Property Rights. Advanced International Training Programme Final Report*. Swedish University of Agricultural Sciences, Alnarp, Sweden.

United Nations Millennium Declaration (2000) UN General Assembly resolution A/RES/55/2. Available at: www.un.org/millennium/declaration/ares552e.htm (accessed 14 March 2016).

# 39 What Do We Have To Lose? Monitoring Crop Genetic Diversity

**M.E. Dulloo,* I. Thormann and A.G. Drucker**
*Bioversity International, Rome, Italy*

You can't manage what you don't measure
(Peter Drucker)

## 39.1 Introduction

Crop genetic resources for food and agriculture are the basis for sustainable agriculture and food and nutrition security, as well for maintaining essential ecosystem services (Hajjar *et al.*, 2008; FAO, 2015). While they play an essential role in securing basic human food and nutritional needs (Dulloo *et al.*, 2014), they are also critical in maintaining healthy evosystems (services provided by evolutionary services) (Faith *et al.*, 2010). These help support landscape-level agroecosystem resilience, as well as allowing plants and animals to undergo natural evolutionary processes (Frankham, 2010), which in turn generate broad genetic variation that is essential for crops, trees and animals to adapt to change (Bellon, 2009). Genetic diversity thus represents the essential raw material for species to evolve and to adapt. Maintaining crop diversity is a key strategy for farmers around the world to guarantee their sustenance. Increasing the sustainable use of agricultural biodiversity in production and consumption systems – both in landscapes and in diets – will be an important part of the solution to the challenge of meeting future food and nutrition security.

The global challenges associated with the conservation of genetic diversity are manifold (Dulloo, 2013). Policy makers are sometimes unaware of the role that these resources play in ensuring sustainable global food and nutrition security, and so policies can inadvertently threaten this diversity. The long-term public benefits of crop, tree and animal diversity to society at large are poorly understood compared to their short-term private benefits. The loss of crop genetic resources is happening at an unprecedented rate, and may lead to a loss of options for improving agriculture. A key challenge is to understand the rate at which these resources are being lost and what the drivers are. Climate change is one of the main drivers, but not the only one. Some other examples of drivers are habitat change, invasive alien species, overexploitation and pollution (Millennium Ecosystem Assessment, 2005). FAO (2010) also recognizes the introduction of new varieties as a significant factor in the loss of traditional crop diversity. How to monitor the changes in diversity, especially *in situ* and on farm, is itself a major challenge.

*Corresponding author; e-mail: e.dulloo@cgiar.org

 

### 39.1.1 Evidence of genetic erosion

Genetic erosion is considered a major threat to the sustainable production of food worldwide. Fowler and Mooney (1990) described it as 'the biggest single environmental catastrophe in history' that threatens humanity's food supplies. Global concern about biodiversity loss, and in particular the genetic diversity of agricultural crops, has led to major international conventions, treaties and global action plans on the conservation and use of agricultural diversity. In the preamble of the International Treaty on Plant Genetic Resources for Food and Agriculture (ITPGRFA), it is mentioned that the global community is alarmed by the continuing genetic erosion of plant genetic resources for food and agriculture (PGRFA) (FAO, 2002a). The *Second Global Plan of Action on Conservation and Sustainable Use of Plant Genetic Resources for Food and Agriculture* by the UN Food and Agriculture Organization (FAO) also reports that 'genetic erosion is continuing unabated in many regions of the world and that the genetic vulnerability of crops has further increased' (FAO, 2012). Also the *Strategic Plan for Biodiversity 2011–2020*, with its Aichi Targets adopted at the Tenth Conference of Parties of the Convention on Biological Diversity (CBD), recognized the importance of minimizing genetic erosion and safeguarding genetic diversity. Aichi Target 13 is devoted to the safeguarding of genetic diversity of cultivated plants and farmed and domesticated animals and of wild relatives, including other socio-economically as well as culturally valuable species (CBD, 2010a). The importance of conserving the genetic diversity of cultivated crops is also explicit in Target 9 of the CBD *Global Strategy for Plant Conservation* 2011–2020 (CBD, 2010b): '70 per cent of the genetic diversity of crops including their wild relatives and other socio-economically valuable plant species conserved, while respecting, preserving and maintaining associated indigenous and cultural knowledge'.

### 39.1.2 Genetic erosion on-farm

With regard to crop genetic diversity on-farm, the replacement of locally distributed and traditionally diverse landraces by modern varieties is generally considered to be the main threat to diversity and the main reason for genetic erosion. This has been substantiated by many reports and studies (Baur, 1914; Bennett, 1968; Frankel and Bennett, 1970; Wilkes and Wilkes 1972; Hawkes, 1983; Arunachalam, 1999; Hutchinson and Weiss, 1999; Frankel, 1970; Thrupp, 1998; Tapia and Estrella, 2001; Brush, 2004; Chaudhuri, 2005; Megersa, 2014). For example, in Ecuador, genetic erosion in three Andean tuber species varied from 25% to 46.5%, based on a socio-economic study, as well as morphological and genetic comparison of tubers collected for the study from *ex situ* collections (Tapia and Estrella, 2001). Megersa (2014) assessed erosion in barley in the North Shewa zone of Ethiopia, recording a loss of 65% of landraces between 1994 and 2010. This was attributed to the introduction of improved varieties and other crops, recurrent drought, changed land-use patterns and a lack of policy support. In two districts in the highlands of West Shewa, of 14 barley landraces described by farmers in 1994, only 4 were still found in cultivation in 2010. In Italy, early surveys from the 1920s to the 1950s observed high genetic erosion (13.2% per annum) on wheat, while from the 1950s until 1980, erosion rates of 0.48 and 4% per annum were estimated (Hammer and Laghetti, 2005). On the island of Favignana, there was an erosion rate of 12.2% per annum, leading to the extinction of the last wheat landraces of this island (Hammer and Laghetti, 2005). In China, of almost 10,000 varieties of wheat in use in 1949, only about 1000 remained by the 1970s (Arunachalam, 1999). However, it is recognized that not all cases of loss are the result of the adoption of high-yielding varieties leading to either the abandonment of local varieties (Brush, 1995; Bellon, 1996; Perales *et al.*, 2003) or a reduction in locally available diversity (Lohar and Rana, 1998; Steele *et al.*, 2009) (see Thormann and Engels, 2015). For example, a long-term study on genetic erosion between 1976 and 2003 showed that although many local varieties of millet and sorghum in Niger had been replaced by improved ones, the overall diversity of pearl millet and sorghum varieties had not changed during this period, based on varietal names and DNA markers (Bezançon *et al.*, 2009). Ford-Lloyd *et al.* (2009) report stable levels of diversity based on the analysis of traditional rice varieties collected over decades across South and Southeast Asia and conserved in genebanks. Similarly, no loss of genetic diversity was found in a study that analysed samples of cultivated wheat (*Triticum aestivum* L.) collected over an interval of 40–50 years in four comparable geographical regions of Europe and Asia (Khlestkina *et al.*, 2004).

### 39.1.3 Genetic erosion in wild plant species

Genetic erosion has also been documented in wild plant species and crop wild relatives (CWR). For example, genetic erosion in wild rice has been documented by Kiang *et al.* (1977), Morishima and Oka (1995), Akimoto *et al.* (1999), Arunachalam (1999) and Gao (2003) in Asia; while Kiambi *et al.* (2005) has studied the threats to populations of African wild rice (*Oryza longistaminata* A. Chev. & Roehr., *Oryza barthii* A. Chev., *Oryza punctata* Kotschy ex Steud., *Oryza eichingeri* Peter and *Oryza brachyantha* A. Chev. & Roehr.) in eastern Kenya. In Latin America, several wild *Arachis* species were found to be threatened on the basis of their highly restricted distribution ranges and as a result of land-use pressures (Jarvis *et al.*, 2003). In these studies, invasive species, changes in land use, overgrazing and environmental changes caused by rapid population growth and new agricultural technologies, as well as economic and cultural change, were the major drivers.

### 39.1.4 Genetic erosion in *ex situ* collections

One of the main objectives of *ex situ* collections is to maintain the genetic integrity of their accessions. Genebank curators strive to prevent genetic erosion in their genebank collections through the proper management of their collections by adopting internationally agreed genebank standards (FAO, 2013a). However, in practice, there are many threats to the genetic integrity of *ex situ* collections. For example, Frankel (1975) reported in a survey of genetic resources that many collections had suffered genetic erosion due to hybridization, selection, genetic drift, unsuitable growing conditions or human error during propagation. In seed genebanks, regeneration is considered to be the most serious threat to genetic diversity (Khoury *et al.*, 2010). Studies on genetic change during regeneration showed that nearly 50% of the alleles discovered in original samples of rye accessions were no longer found in the most recent seed lots, while new alleles were detected in the most recently propagated subpopulations that had not been observed in the original seed lots (Chetobar *et al.*, 2003). Comparisons of oat and wheat seed samples that were conserved in different genebanks following collection confirmed that differences in genebank management and regeneration procedures affected diversity (Steiner *et al.*, 1997; Hirano *et al.*, 2009). In field genebanks, major threats include natural disasters, armed conflicts and economic instability. In Madagascar, for example, an entire field collection of coffee was devastated in a single cyclonic event, while the main coffee collection in Kianjavato has lost 46% of its originally collected materials since the 1960s, due mainly to non-adaptability of the species or accessions to the soil or climatic conditions at the site and lack of sufficient financial resources for the management of the field genebank (Dulloo *et al.*, 2009).

Despite the evidence for genetic erosion in crop genetic diversity on farm, *in situ* and *ex situ*, it is difficult to compare changes across space and time, because of a lack of consistent metrics at a global level. No monitoring mechanism is in place to measure the status and trends of genetic diversity at the global level (Dulloo *et al.*, 2010; Pereira *et al.*, 2013; Graudal *et al.*, 2014). This represents a major challenge in assessing the attainment of global targets established by governments at the global level, such as those of the Millennium Development Goals, the Sustainable Development Goals, the FAO State of the World reports on Plant Genetic Resources for Food and Agriculture, the Second Global Plan of Action for Plant Genetic Resources for Food and Agriculture (GPA2) and the Global Strategy on Plant Conservation and Strategic Plan for Bioversity 2011–2020 (particularly Aichi Target 13). Although biodiversity indicators have been developed by these global initiatives to monitor progress towards the achievement of their respective targets, indicators for monitoring the status and trends of crop genetic diversity *per se* are less developed. In this chapter, we outline the indicators used by some of these main initiatives and discuss their strengths and weaknesses.

## 39.2 Defining and Monitoring Genetic Diversity

A common feature of all living organisms is DNA, the basic units of which are the base pairs, which constitute the chromosomes of the organism. We know that no two individuals of a given taxon are alike, except for identical twins or clones, which have the same genetic make-up. The variations within a given trait between individuals of the same species are due to differences in the alleles present at the gene locus, and this element of heredity can be used as the metric par excellence for measuring genetic diversity within a species. Graudal *et al.* (2014) define genetic diversity or intraspecific variation as the total amount of genetic differences within species. It can be subdivided into inter- and intrapopulation variation (also called among and within population genetic diversity), and further into the diversity within an individual expressed by differences between alleles across chromosomes. Pereira *et al.* (2013) proposed allelic diversity as the essential biodiversity variable (EBV) for monitoring genetic diversity across species and organisms, which could form the basis of monitoring programmes worldwide. Metrics for measuring allelic diversity may include changes in allelic richness, allele frequency and levels of heterozygosity (Nei's index), or in the fixation index used to measure population differentiation. Recently, in a study on temporal indicators for genetic erosion, Hoban *et al.* (2014) showed that the number of alleles per genetic locus outperformed all other potential indicators of genetic diversity. Measuring these variables has now been made easier by rapid progress in molecular biology that has resulted in the development of a number of powerful new techniques for measuring genetic variation in plants and animals (Tittensor *et al.*, 2014). These measures are, however, expensive to measure or difficult to conduct on a large scale, and other proxy indicators for genetic diversity that focus on varietal diversity in landraces or ecogeographical diversity in CWR have been proposed (Brown, 2008; Jarvis *et al.*, 2008; Parra-Quijano *et al.*, 2008, 2012; Last *et al.*, 2014).

Crop genetic diversity is distributed across space and time. Thus, measuring change in status should take into account spatial distribution and evolutionary changes. Crop genetic diversity occurs in farmers' fields and wild habitats, and has been collected and conserved in *ex situ* collections such as genebanks. In developing indicators for monitoring genetic/varietal diversity, it is important to take account of the basic concepts of diversity in terms of richness, abundance and evenness (Eaton *et al.*, 2006; Brown, 2008). 'Richness' refers to the number of species or varieties regardless of their frequencies, while 'abundance' refers to the number or distribution of individuals of a given species or variety, and 'evenness' refers to the equality of the relative or proportional abundance of all species/varieties; that is, the degree of similarity of the frequencies of the different variants (Eaton *et al.*, 2006; Brown, 2008). In biodiversity assessments, area has been suggested as a surrogate indicator for diversity (Mac Arthur and Wilson, 1967). Brown and Brubaker (2002) have suggested that area planted under the landraces of a specific crop can serve as an indicator of genetic diversity for temporal and spatial comparative purposes, and Jarvis *et al.* (2008) have shown that, in fact, both richness and evenness are correlated significantly with farm area.

## 39.3 A Brief Review of the Development of Indicators for Crop Genetic Diversity

There are many efforts globally, led by FAO, the Organization for Economic Co-operation and Development (OECD) and the CBD Secretariat, to develop indicators for monitoring the status of PGRFA. In this respect, FAO has spearheaded the development of indicators, as mandated by the Commission on Genetic Resources for Food and Agriculture (CGRFA). The CGRFA deals with policy, sectorial and cross sectorial matters related to the conservation and utilization of genetic resources for food and agriculture, as well as providing guidance for the development and assessment of progress being made in the implementation of the global strategies for the management of plant, animal, forest, aquatic and microbial genetic resources. FAO has been reviewing the state of the art related to indicator development and use for genetic diversity, genetic erosion and genetic vulnerability, in order to facilitate the establishment of benchmark data and the elucidation of trends, as well as, ultimately, to allow for improved management of PGRFA (Collette, 2001).

At a joint workshop organized by FAO and the International Plant Genetic Resources Institute (now Bioversity International) in 2002, a list of 22 potential indicators for genetic diversity, genetic erosion and genetic vulnerability in several areas and applicable at different levels were identified to assess and monitor crop genetic diversity (GDEV, 2002). Some examples of the core indicators proposed include the number and kind of threatened and endangered species relevant to food and agriculture; the number and kind of CWR and wild food plants targeted for *in situ* conservation in National Environmental Action Plans; the number and kind of protected areas for *in situ* conservation; the number of *in situ* conservation sites and wild species conserved; the number of species and accessions preserved *ex situ* in medium- and long-term storage; the degree of genetic integrity of accessions preserved *ex situ*; and a list of major environmental constraints to *ex situ* conservation of PGRFA.

Brown (2008) provides a set of 22 indicators for genetic diversity and erosion both *in situ* and *ex situ* for cultivated and wild plant species (Table 39.1). He argues that the most practical ones are those based on the number of individuals or area occupied *in situ* and on the number of accessions and the number of genebanks *ex situ*. Table 39.1 provides a summary of these indicators.

The OECD proposed indicators for both crop and livestock breed diversity based on the number of varieties or breeds registered and certified for marketing, and the share of key crop or breed varieties in total marketed production for individual crops or breeds (OECD, 2001). The share of the key varieties refers to the dominant crops in the production system and is a measure of the

Table 39.1. Indicators of genetic diversity under four categories of PGRFA. Primary indicators in bold. (From Brown, 2008.)

| | *In situ* | *Ex situ* |
|---|---|---|
| Cultivated | • **Number and frequency of landraces, and proportion of the area planted that is growing them**<br>• Environmental amplitude of crop area<br>• Number of farmer selection criteria and evolution of farmer management<br>• Security of traditional knowledge | • **Number of crop species, subspecies or geographic categories adequately sampled in genebanks**<br>• Number of accessions held in the genebank<br>• Number of collections or genebanks<br>• Country distribution of seed genebanks<br>• Coverage in collection of crop diversity<br>• Backup duplication provisions<br>• Extent of usage and representation in core collections<br>• Collection health, accession viability<br>• Documentation and evaluation of collection |
| Wild | • **Number of species, subspecies or geographic subdivisions of taxa distributed in protected areas that cover the species environmental range**<br>• Abundance as population numbers and sizes, particularly of rare wild crop relatives<br>• Gene diversity, divergence and distribution | • **Number of wild species, subspecies or geographic subdivisions of taxa related to crops adequately sampled in the genebank**<br>• Coverage of species range<br>• Evolutionary relationships and taxonomic resolution<br>• Accession viability documentation and duplication<br>• Number and frequency of accessions used<br>• 'Pre-breeding' activities, including evaluation |

evenness of the diversity, meaning that the genetic diversity is larger if the share of the dominant variety is smaller. The number and share of dominant improved varieties can, however, be misleading, because an increase in the number of dominant varieties may not mean that diversity has increased because many new registered varieties are, in fact, only slightly modified versions of already popular improved varieties (Eaton *et al.*, 2006).

Wetterich (2003) proposed an expanded set of indicators to address the inadequacies of the OECD list, which in turn led Eaton *et al.* (2006) to draw on the OECD (2003) and Wetterich (2003) indicators to propose an extended list based on a pressure–state–response framework (which focuses on the cause–effect relationship by which the underlying and immediate pressures lead to changes and trends in the state of genetic diversity), as well as on a more restricted list of indicators (Table 39.2). Eaton *et al.* (2006) analysed the value of available indicators for the monitoring and evaluation of plant and animal genetic resources. They selected indicators for monitoring the loss of genetic diversity in wild and undomesticated species as well as in livestock and crop varieties. For wild species, they selected natural areas conserved and the number of key species as the main indicators. These two indicators are not suitable for domesticated agricultural species where the units considered are breeds and crop varieties and not the species itself. A fundamental difference between the two indicators is the unlimited level of gene flow between crop varieties as compared between wild species.

### 39.3.1 Indicators for monitoring the implementation of the FAO Global Plan of Action for Conservation and Sustainable Utilization of PGRFA

After the publication of the First State of the World Report on PGRFA and its accompanying First Global Plan of Action (GPA) for the Conservation and Sustainable Utilization of PGRFA adopted in Leipzig, Germany, in 1996, FAO developed a core set of 83 indicators and 68 complementary indicators for monitoring progress in the implementation of the 20 priority areas of the first GPA (FAO, 2002b). Subsequently, with the publication of the second GPA in 2012 (FAO, 2012), this list of indicators and

**Table 39.2.** Set of indicators for crop genetic diversity. (From Eaton *et al.*, 2006.)

| Extended list of indicators | Restricted list of indicators |
|---|---|
| • Average size of farms (in area in ha, kg crops produced, kg/ha)<br>• Number of species in use<br>• Crop species ratio/diversity index<br>• Number of registered key crop varieties (domesticated, non-domestic)<br>• Share of the three major crop varieties in seed production area/diversity index<br>• Number of crop endangered varieties<br>• Number of breeds conserved *ex situ*<br>• Number of accessions characterized<br>• Intensification and use of modern breeding strategies<br>• Percentage of seeds originating from own farm<br>• Share of the crop varieties and species adapted to landscapes/production environment important for biodiversity and/or characteristic for a region or country<br>• Area of low production/high biodiversity<br>• Number of breeders per crop<br>• Number of different breeding goals | • Area of varieties' characteristics for landscape/production environments important for biodiversity and characteristics for a region or country<br>• Area of low production/high biodiversity agriculture<br>• Percentage of seed originating on farm of three major (high-production) breeds<br>• Number of characteristics (low-production) varieties stored in genebank |

the associated reporting format were reviewed to ensure that a more effective system was in place to monitor the implementation of the 18 priority areas of the second GPA. A reduced set of 63 indicators was identified following a country-led participatory process, coordinated by FAO, including the Secretariats of the CGRFA and the ITPGRFA, and in collaboration with the Global Crop Diversity Trust and Bioversity International (FAO, 2013b). The core set covers the priority areas of the four main categories:

- *in situ* conservation and management (12 indicators);
- *ex situ* conservation (12 indicators);
- sustainable use (22 indicators); and
- building institutional and human capacities (20 indicators).

Some of the indicators developed cover more than one category. For example, the indicator, 'percentage of PGRFA threatened out of those surveyed/inventoried', used for priority 1 on surveying and inventorying PGRFA under the first category (*in situ* conservation and management), was also used to monitor progress on priority 16, namely 'developing and strengthening systems for monitoring and safeguarding genetic diversity and minimizing genetic erosion of PGRFA'. FAO also identified three targets within the planned or ongoing processes of monitoring the GPAs, as was requested by the CGRFA at its Thirteenth Regular Session in 2013 (FAO, 2013b), as follows:

- Target 1 – PGRFA Conservation: by 2020, an increasing proportion of the genetic diversity of cultivated plants and their wild relatives, as well as of wild food plant species, will be maintained *in situ*, on-farm and *ex situ* in a complementary manner.
- Target 2 – PGRFA Sustainable Use: by 2020, there will have been an increased use of plant genetic resources for food and agriculture to improve sustainable crop production intensification and livelihoods, while reducing the genetic vulnerability of crops and cropping systems.
- Target 3 – PGRFA Institutional and Human Capacities: by 2020, many more people will be aware of the values of plant genetic resources for food and agriculture, and institutional and human capacities will be strengthened to conserve and use them sustainably while minimizing genetic erosion and safeguarding their genetic diversity.

The core set of 63 indicators and targets were adopted at the FAO Fourteenth Regular Session of the CGRFA. At this session, the Commission requested that FAO elaborate higher-order composite indices (HCIs) for each of the PGRFA targets so as to assess progress towards the three PGRFA targets and to facilitate the comparison of performance across time, countries and regional areas. FAO (2014) developed a model for calculating HCIs, using the second GPA as the theoretical framework for the HCIs. Each of its 18 activities and 63 indicators contributes to one of the HCIs. The model depends on information provided by countries for each of the indicators in reporting format. In cases where there are uncertainties regarding data availability and data accuracy, or the need for contextualization, countries then provide an expert judgement on the level of achievement or implementation for each indicator. The HCIs are expressed through an ordered categorical scale from 1 to 9, defined on the basis of the expert judgement regarding country performance and progress on each indicator, where 1 represents the 'least degree of implementation or achievement' and 9 a 'full degree of implementation or achievement' (FAO, 2014). A reporting format for monitoring GPA implementation based on indicators will be used by national PGRFA stakeholders to collect information that will be made available under the FAO World Information and Early Warning System on PGRFA (WIEWS).

### 39.3.2 Indicators developed by the Convention on Biological Diversity

The Seventh Conference of Parties (COP) to the CBD adopted a strategic plan that included the 2010 biodiversity target for significant reduction in the current rate of biodiversity loss, at its meeting in April 2002 (Decision VII/26). It was endorsed by ministers responsible for CBD implementation during a Ministerial Roundtable discussion in April 2002 (Hague Ministerial Declaration), and was also endorsed by world leaders during the World Summit on Sustainable

Development in September 2002 (WSSD Plan of Implementation). The CBD Subsidiary Body on Scientific, Technical and Technological Advice (SBSTTA) recommended to the COP that a limited number of trial indicators be developed, tested and reviewed by SBSTTA prior to the eighth meeting of the COP in 2006. These global-level indicators should be selected to communicate effectively trends in the status of biodiversity and the impacts on ecosystem goods and services and human well-being during the present decade. An Ad hoc Technical Panel of Experts Group (AHTEG) on indicators was convened by SBSTTA to prepare a framework of 22 cross-disciplinary headline indicators, to measure progress towards the 2010 biodiversity target at a global level. A 2010 Biodiversity Indicators Partnership (www.bipindicators.net/) was established in 2007 with support from the Global Environment Facility (GEF) to lead the further development of indicators, bringing together 40 organizations worldwide to develop, strengthen, implement and communicate a suite of complementary indicators, which were agreed by parties to the CBD at COP 8 (Decision VIII/15) in 2006. The headline indicator on the genetic diversity of domesticated animals, cultivated plants and fish species of major socio-economic importance included two sub-indicators, namely:

- *Ex situ* crop collection – assessed as an Enrichment Index for measuring the increase in genetic diversity of cultivated plants and their crop wild relatives in *ex situ* crop collections.
- Genetic diversity of terrestrial domesticated animals – assessed in terms of the extinction risk faced by the world's livestock breeds.

However, by 2009, it was recognized that not all the indicators had been fully developed (Walpole *et al.*, 2009). Of the total, only nine were fully developed with well-established methodologies and global time series data. Others were underdeveloped or still in need of development (Walpole *et al.*, 2009). The incomplete development of the indicator framework and the absence of a global, harmonized observation system for delivering regular, timely data on biodiversity change (Pereira *et al.*, 2013) meant that the CBD 2010 target was not achieved. The tenth COP of the CBD held in Nagoya, Japan, in 2010 adopted a new Strategic Plan for Biodiversity for the decade 2011–2020 and 20 Aichi Targets. Target 13 is devoted to the genetic diversity of cultivated plants and farmed and domesticated animals, and of wild relatives, including other socio-economically as well as culturally valuable species. It calls for the genetic diversity of these species to be maintained, and strategies developed and implemented for minimizing genetic erosion and safeguarding their genetic diversity. To help monitor progress towards this target, the same headline indicators developed under the Biodiversity Indicators Partnership are being pursued. There is dire need to further develop indicators for the genetic diversity of cultivated plants and their crop wild relatives to monitor progress towards Target 13.

### 39.3.3 Other crop genetic diversity indicators

There are few indicators proposed that fully integrate the crop genetic diversity dimension of genetic resources for food and agriculture. In this respect, Bonneuil *et al.* (2012) have developed a new integrative index (HT*) for cultivated species, which not only takes account of varietal richness and spatial evenness, but also the between- and within-variety genetic diversity. They compared this index to existing indices using a comprehensive historical data set from a French territory dating back to 1878. The study revealed that more varieties (the varietal richness factor) can mean less diversity when their genetic structure is more similar (the effect of between-variety genetic diversity), or when more diverse landraces are replaced by many homogeneous lines (the effect of within-variety genetic diversity), or else when one or a few varieties become hegemonic in the landscape (the spatial evenness effect) (Bonneuil *et al.*, 2012). Furthermore, an increased evenness in variety distribution (varietal evenness) can also mean less diversity when varieties are genetically related.

More recently, Last *et al.* (2014), proposed five indicators for the estimation of genetic diversity of crop accession or breed diversity at farm level. These are 'crop species richness', 'crop cultivar diversity', 'type of crop accessions', 'livestock species richness' and 'breed diversity' (Table 39.3). These were developed and

Table 39.3. Key indicators for on-farm assessment of crop cultivar and livestock breed. (From Last *et al.*, 2014.)

| Indicator | Description |
|---|---|
| Crop species richness (CSR) | Total number of arable, vegetable, tree and forage crops expressed as crop species richness; that is, the total number of natural or cultivated populations of plant species for agricultural purpose per farm |
| Crop cultivar diversity (CCD) | Total number of accessions (commercial cultivars, landraces and 'unknown') divided by CSR (crop species richness) per farm<br>Sub-indicator: percentage of rare crop accessions per species as a supplement to CCD |
| Type of crop accessions (TCA) | Presence of landraces that are mostly heterogeneous populations of crop species locally adapted and valuable as genetic resources |
| Livestock species richness (LSR) | Total number of livestock species occurring per farm |
| Breed diversity (BD) | Total number of livestock breeds (breeds and crossbreeds) occurring per farm, divided by the livestock species richness per farm<br>Sub-indicator: percentage of rare breeds per species occurring per farm |

tested through a participatory approach involving stakeholders from Uganda and 12 European case studies. Additionally, they evaluated the potential role of such indicator values in developing strategies to conserve or increase crop cultivars and livestock breeds in agroecosystems.

### 39.3.4 Indicators of resilience in socioecological production landscapes

At a landscape level, Bergamini *et al.* (2013) proposed a 'holistic and inclusive' set of 20 ecological and sociocultural factors for socioecological production landscapes (SEPLs), developed over the course of a joint collaboration between Bioversity International and the United Nations University Institute of Advanced Studies (UNU-IAS) under the International Partnership for the Satoyama Initiative (IPSI). The indicators are classified into four major categories: ecosystem protection and maintenance of biodiversity; agricultural biodiversity; knowledge, learning and innovation; and social equity and infrastructure. Assessments for each indicator are made by assigning a Likert-scale type score (1–5). The trend in the indicator helps to depict an historical view of the landscape. Most of these indicators are qualitative in nature and, despite the use of a Likert-scale point, require subjective judgements, which lower the practicality and data generation of the indicators.

### 39.3.5 Indicators for agricultural biodiversity conservation services

Nguyen and Drucker (2013, unpublished results) reviewed the literature to identify the main groups of indicators that cover the public good benefits generated by agricultural biodiversity (agroecosystem resilience, maintenance of evolutionary processes/future option values and maintenance of traditional knowledge associated with specific genetic resources). Through the review of existing agricultural biodiversity (ABD) indicators, they provided recommendations for the development of a holistic indicator set for measuring the critical ecosystem services associated with agricultural biodiversity. They proposed a set of 12 indicators, with specific indicators being capable of simultaneously representing more than one ecosystem service category. Subsequently, Willaarts and Drucker (2015, unpublished report) have further adapted this indicator set to monitor ABD conservation incentive mechanism interventions, resulting in an expanded set of 26 indicators, which have now been applied for baseline purposes in a project in Ecuador researching Payments for Agrobiodiversity Conservation Services (PACS). These indicators cover the same categories as those of Nguyen and Drucker (2013) but, with a view to ensuring applicability to a wide range of ABD conservation interventions and ABD-relevant ecosystem services, they also incorporate indicators related to provisioning services (including food security) and seed system functionality, as well as soil and water quality.

## 39.4 Discussion

As described above, there have been many initiatives that have tried to establish indicators for monitoring diversity at various levels over the past couple of decades. However, there still remains no internationally agreed set of indicators that satisfactorily measure the state of crop genetic diversity. Most indicators focus on the drivers, pressures, responses and benefits, drawing on the DIPSR (driver–impact–pressure–state–response) framework, with very few focusing on the state of genetic diversity *per se*.

To be useful, indicators must be scientifically sound, technically realistic for monitoring and relevant for defining a policy baseline (Graudal *et al.*, 2014). A key discussion point on indicators for genetic diversity is how best to measure genetic diversity. Is it necessary to measure diversity directly at the genetic level, or can proxy variables be used instead? The more we can measure diversity directly at the DNA level, the more accurate will be our understanding of the status of genetic diversity. But it is argued that molecular techniques are still too expensive to be affordable and practicable (especially in developing countries), despite the declining costs of using molecular tools. Further, even if molecular characterization is possible, the statistical approaches needed to evaluate trends and provide indicators of genetic health are still less well developed (Levrel, 2007). The challenge remains of how to develop simple, inexpensive means to monitor genetic diversity at a global scale (Frankham, 2010). For this reason, proxy indicators of genetic diversity have been favoured over the direct measure of genetic diversity. Some authors have argued that the number of individuals of a species or variety is a good proxy for genetic diversity, because the number of individuals is directly related to genetic diversity (Eaton *et al.*, 2006), given that no two individuals are genetically identical (except in the case of twins or clones). This assumption is supported by Levrel (2007), who states that a good demographic health of a population reflects satisfactory genetic renewal and thus functions as a good indicator of genetic variation. While this may be true for wild species and cross-pollinating crops, for many inbreeding crops and vegetatively propagated crops, the number of individuals as an indicator of genetic variability can be very misleading. A combination of area planted or covered and number of varieties or species has also been suggested as an indicator for monitoring genetic diversity that captures both richness and evenness (Brown, 2008; Jarvis *et al.*, 2008). Balmford *et al.* (2003, 2005) proposed a more operational concept to measure the erosion of biodiversity. This concept is concerned with the variation in abundance within species. Balmford *et al.* (2003, 2005) argue that abundance is more sensitive to short-term dynamics and thus makes it possible to use indicator species. Some species are particularly informative, so monitoring their abundance can provide a suitable indicator for assessing the well-being of an ecosystem.

The set of 63 indicators proposed for monitoring the implementation of the Second GPA for PGRFA (FAO, 2014) are also proxies of genetic diversity and may not truly reflect the actual status on the ground. Further, the process of monitoring the implementation of the GPA is based on country reports on the set of indicators, which is often based on subjective assessment by country authorities, without scientific, evidence-based methods of assessment of the metrics for the indicators.

The number of indicators for monitoring plant genetic resources proposed by the various initiatives is proliferating, and this makes it hard to get common agreement on which indicators to use at a regional and global level. Often, indicators are developed with specific objectives and goals in mind, whether they are to monitor: the effectiveness of activities for the conservation and use of PGRFA (FAO, 2014); the state of genetic diversity and genetic erosion (Eaton *et al.* 2006; Brown, 2008; FAO, 2014); the contribution to ecosystem services (Nguyen and Drucker, 2013); the benefits to farmers (Bellon *et al.*, 2015); or the impacts of climate change (Holmes *et al.*, 2013). For example, in an assessment of the effectiveness of on-farm conservation projects of native crops in the high Andes of South America, Bellon *et al.* (2015) calculated crop diversity on farm using as an indicator factor analysis performed on the number of farmer varieties of all target crops planted by a household, as a means of capturing the structure of how the number of varieties was distributed among target crops per household and what benefits farmers derived from it, but it did not provide any indications of the on-farm genetic diversity.

One great challenge for monitoring crop genetic diversity is the availability of data over space and time. The availability and access of data, or the ease of measuring them in a cost-effective way, often determine the indicators that are used. For example, in monitoring the trends in genetic diversity for Aichi Target 13, an *ex situ* crop collection indicator is proposed because it is easier to know what has been collected and conserved in genebanks than what is existing on-farm and in the wild (Chenery *et al.*, 2013). While measuring the crop genetic diversity conserved in genebanks is relatively straightforward, it is a greater challenge to understand the extent and distribution of genetic diversity across a farm and *in situ* landscape. For *ex situ* collections, there exist global accession-level information systems, such as Genesys (www.genesys-pgr.org), which can provide useful information. At regional and national level, *ex situ* information systems are also fairly well developed.

However, information systems for *in situ* conservation and on-farm management of crop diversity are lacking. What is needed is a network of sites or observatories from across the continents and countries representing hotspots of crop diversity, similar to what has been proposed by Davies *et al.* (2012) on the establishment of an international genomic network, consisting of specific sites to act as genetic biodiversity observatories where special efforts are made to document and monitor genetic diversity within and among species in high spatiotemporal detail. One good example of monitoring genetic diversity *in situ* comes from managing the dynamic conservation of forest genetic diversity (Koskela *et al.*, 2014). Several international organizations, including Bioversity International, the International Center for Tropical Agriculture (CIAT), the International Potato Centre (CIP), the Global Biodiversity Information Facility (GBIF), supported by the Platform on Agrobiodiversity (PAR), Agropolis International and FAO, have proposed a case for developing collaboratively a global information system or systems for *in situ* conservation and on-farm management of PGRFA, complementary to existing *ex situ* information systems, as a contribution to implementation of Article 17 of the International Treaty on Plant Genetic Resources for Food and Agriculture (Anon., 2014).

While the debate on which indicators to use for monitoring crop genetic diversity continues, genetic erosion is continuing unabated. It is becoming more and more urgent to establish a global operational genetic monitoring system to be able to track changes in genetic diversity and inform policy makers so that conservation intervention can be taken to safeguard it.

## Acknowledgement

The authors would like to thank Arwen Bailey of Bioversity International for English editing the manuscript.

## References

Akimoto, M., Shimamoto, Y. and Morishima, H. (1999) The extinction of genetic resources of Asian wild rice, *Oryza rufipogon* Griff.: a case study in Thailand. *Genetic Resources and Crop Evolution* 46, 419–425.

Anon. (2014) *Global Information System for in situ Conservation and On-farm Management of Plant Genetic Resources for Food and Agriculture*. An input paper submitted to the International Treaty on Plant Genetic Resources for Food and Agriculture consultation on global information systems on plant genetic resources for food and agriculture. Bioversity International, International Center for Tropical Agriculture, International Potato Centre, and Global Biodiversity Information Facility. Available at: www.planttreaty.org/sister/default/files/cogis1Inf4a3.pdf (accessed 14 March 2016).

Arunachalam, V. (1999) Conservation, genetic erosion and early warning system: key issues. In: Serwinski, J. and Faberova, I. (eds) *Proceedings of the Technical Meeting on Methodology of the FAO World Information and Early Warning System on Plant Genetic Resources*, 21–23 June 1999. Research Institute of Crop Production, Prague, pp. 45–47.

Balmford, A., Green, R.E. and Jenkins M. (2003) Measuring the changing state of nature. *Trends in Ecology and Evolution* 18(7), 326–330.

Balmford, A., Crane, P., Dobson, A., Green, R.E. and Mace, G. (2005) The 2010 challenge: data availability, information needs and extraterrestrial insights. *Proceedings Transactions of the Royal Society B*, 360, 221–228.

Baur, E. (1914) Die Bedeutung der primitiven Kulturrassen und der wilden Verwandten unserer Kulturpflanzen fuer die Pflanzenzuechtung. *Jahrbuch Deutsche Landwirtschafliche Gesellschaft* 29, 104–110.

Bellon, M.R. (1996) The dynamics of crop infraspecific diversity: a conceptual framework at the farmer level. *Economic Botany* 50, 26–39.

Bellon, M.R. (2009) Do we need crop landraces for the future? Realizing the global option value of *in situ* conservation. In: Kontoleon, A., Pascual, U. and Smale, M. (eds) *Agrobiodiversity and Economic Development*. Routledge, London and New York, pp. 51–59.

Bellon, M.R., Gotor, E. and Caracciolo, F. (2015) Assessing the effectiveness of projects supporting on-farm conservation of native crops: evidence from the high Andes of South America. *World Development* 70, 172–176.

Bennett, E. (1968) *Record of the FAO/IBP Technical Conference on the Exploration, Utilization and Conservation of Plant Genetic Resources*, Rome, Italy, 18–26 September 1967. Food and Agriculture Organization of the United Nations (FAO), Rome.

Bergamini, N., Blasiak, R., Eyzaguirre, P., Ichikawa, K., Mijatovic, D., *et al.* (2013) *Indicators of Resilience in Socio-ecological Production Landscapes (SEPLs)*. United Nations University Institute of Advanced Studies (UNU-IAS), Japan.

Bezançon, G., Pham, J.-L., Deu, M., Vigouroux, Y., Sagnard, F., *et al.* (2009) Changes in the diversity and geographic distribution of cultivated millet (*Pennisetum glaucum* (L.) R. Br.) and sorghum (*Sorghum bicolor* (L.) Moench) varieties in Niger between 1976 and 2003. *Genetic Resources and Crop Evolution* 56(2), 223–236.

Bonneuil, C., Goffaux, R., Bonnin, I., Montalent, P., Hamon, C., *et al.* (2012) A new integrative indicator to assess crop genetic diversity. *Ecological Indicators* 23, 280–289.

Brown, A.H.D. (2008) *Indicators of Genetic Diversity, Genetic Erosion and Genetic Vulnerability for Plant Genetic Resources for Food and Agriculture*. Thematic Background Study, FAO, Rome.

Brown, A.H.D. and Brubaker, C.L. (2002) Indicators for sustainable management of plant genetic resources: How well are we doing? In: Engels, J.M.M., Ramanatha Rao, V., Brown, A.H.D. and Jackson, M.T. (eds) *Managing Plant Genetic Diversity*. CAB International, Wallingford, UK, pp. 249–262.

Brush, S.B. (1995) *In situ* conservation of landraces in centres of crop diversity. *Crop Science* 35, 346–354.

Brush, S.B. (2004) *Farmers' Bounty: Locating Crop Diversity in the Contemporary World*. Yale University Press, New Haven, Connecticut, and London.

CBD (2010a) *Strategic Plan for Biodiversity 2011–2020*. Secretariat of the Convention on Biological Diversity, Montreal, Canada.

CBD (2010b) *Global Strategy for Plant Conservation*. Secretariat of the Convention on Biological Diversity, Montreal, Canada.

Chaudhuri, S.K. (2005) Genetic erosion of agrobiodiversity in India and intellectual property rights: interplay and some key issues. *Patentmatics* 5(6), 1–10.

Chebotar, S., Röder, M.S., Korzun, V., Saal, B., Weber, W.E. and Börner, A. (2003) Molecular studies on genetic integrity of open-pollinating species rye (*Secale cereale* L.) after long-term genebank maintenance. *Theoretical and Applied Genetics* 107(8), 1469–1476.

Chenery, A., Plumpton, H., Brown, C., Walpole, M., Burfield, I., *et al.* (2013) *Aichi Targets Passport*. Biodiversity Indicators Partnership, UNEP-WCMC, Convention on Biological Diversity, Cambridge, UK, p. 90.

Collette, L. (2001) Indicators of agricultural genetic resources: FAO's contribution to monitoring agricultural biodiversity. Paper presented to the OECD Expert Meeting on Agri-Biodiversity Indicators, 5–8 November 2001, Zürich, Switzerland.

Davies, N., Meyer, C., Gilbert, J.A., Amaral-Zettler, L., Deck, J., *et al.* (2012) A call for an international network of genomic observatories (GOs). *GigaScience* 1, 5, doi:10.1186/2047-217X-1-5.

Dulloo, M.E. (2013) Global challenges for agricultural plant biodiversity and international collaboration. In: Normah, M.N., Chin, H.F. and Reed, B.M. (eds) *Conservation of Tropical Plant Species*. Springer, New York, Heidelberg, Dordrecht, London, pp. 491–510.

Dulloo, M.E., Ebert, A.W., Dussert, S., Gotor, E., Astorga, C., *et al.* (2009) Cost efficiency of cryopreservation as a long-term conservation method for coffee genetic resources. *Crop Science* 49, 2123–2138.

Dulloo, M.E., Hunter, D. and Borelli, T. (2010) *Ex situ* and *in situ* conservation of agricultural biodiversity: major advances and research needs. *Notulae Botanicae Horti Agrobotanici Cluj* 38(2) special issue, 123–135.

Dulloo, M.E., Hunter, D. and Leaman, D. (2014) Plant diversity in addressing food, nutrition and medicinal needs. In: Gurib-Fakim, A. (ed.) *Novel Plant Bioresources: Applications in Food, Medicine and Cosmetics.* John Wiley & Sons, Ltd, Chichester, UK, pp. 3–21.

Eaton, D., Windig, J., Hiemstra, S.J., van Veller, M., Trach, N.X., *et al.* (2006) *Indicators for Livestock and Crop Biodiversity*. CGN/DLO Foundation, Wageningen, the Netherlands.

Faith, D.P., Magallón, S., Hendry, A.P., Conti, E., Yahara, T., *et al.* (2010) Evosystem services: an evolutionary perspective on the links between biodiversity and human well-being. *Current Opinion in Environmental Sustainability* 2, 66–74.

FAO (Food and Agriculture Organization) (2002a) *The International Treaty on Plant Genetic Resources for Food and Agriculture*. FAO, Rome.

FAO (2002b) *Indicators and Reporting Format for Monitoring the Implementation of the Global Plan of Action for the Conservation and Sustainable Utilization of Plant Genetic Resources for Food and Agriculture*. Ninth Session of the Commission on Genetic Resources for Food and Agriculture Document CGRFA-9/02/Inf2. FAO, Rome.

FAO (2010) *The Second Report on the State of the World's Plant Genetic Resources for Food and Agriculture*. FAO, Rome.

FAO (2012) *The Second Global Plan of Action for the Conservation and Sustainable Utilization of Plant Genetic Resources for Food and Agriculture*. FAO, Rome.

FAO (2013a) *Genebank Standards for Plant Genetic Resources for Food and Agriculture*. FAO, Rome.

FAO (2013b) *Targets and Indicators for Plant Genetic Resources for Food and Agriculture*. Fourteenth Regular session of the Commission on Genetic Resources for Food and Agriculture. Document CGRFA-14/13/4.1 Rev.1. FAO, Rome.

FAO (2014) *Targets and Indicators for Plant Genetic Resources for Food and Agriculture*. Fifteenth Regular Session of the Commission on Genetic Resources for Food and Agriculture. Document CGRFA-15/15/4.1. FAO, Rome.

FAO (2015) *Coping with Climate Change – The Roles of Genetic Resources for Food and Agriculture.* FAO, Rome.

Ford-Lloyd, B.V., Brar, D., Khush, G.S., Jackson, M.T. and Virk, P.S. (2009) Genetic erosion over time of rice landrace agrobiodiversity. *Plant Genetic Resources: Characterization and Utilization* 7(2), 163–168.

Fowler, C. and Mooney, P. (1990) *Shattering Food, Politics, and the Loss of Genetic Diversity.* The University of Arizona Press, Tucson, Arizona.

Frankel, O.H. (1970) Genetic conservation in perspective. In: Frankel, O.H. and Bennett, E. (eds) *Genetic Resources in Plants – Their Exploration and Conservation*. IPB Handbook No 11. International Biological Programme, London, pp. 469–489.

Frankel, O.H. (1975) Genetic resources survey as a basis for exploration. In: Frankel, O.H. and Hawkes, J.G. (eds) *Crop Genetic Resources for Today and Tomorrow*. International Biological Programme 2. Cambridge University Press, Cambridge, UK, pp. 99–109.

Frankel, O.H. and Bennett, E. (1970) Genetic resources – introduction. In: Frankel, O.H. and Bennett, E. (eds) *Genetic Resources in Plants – Their Exploration and Conservation*. IPB Handbook No 11. International Biological Programme, London, pp. 7–18.

Frankham, R. (2010) Challenges and opportunities of genetic approaches to biological conservation. *Biological Conservation* 143, 1919–1927.

Gao, L.Z. (2003) The conservation of Chinese rice biodiversity: genetic erosion, ethnobotany and prospects. *Genetic Resources and Crop Evolution* 50, 17–32.

GDEV (2002) *Review and Development of Indicators for Genetic Diversity, Genetic Erosion and Genetic Vulnerability (GDEV)*. Summary report of a joint FAO/IPGRI workshop Rome, 11–14 September 2002. FAO, Rome.

Graudal, L., Aravanopoulos, F., Bennadji, Z., Changtragoon, S., Fady, B., *et al.* (2014) Global to local genetic diversity indicators of evolutionary potential in tree species within and outside forests. *Forest Ecology and Management* 333, 35–51.

Hajjar, R., Jarvis, D.I. and Gemmill-Herren, B. (2008) The utility of crop genetic diversity in maintaining ecosystem services. *Agriculture, Ecosystems and Environment* 123(4), 261–270, doi:10.1016/j.agee.2007.08.003.

Hammer, K. and Laghetti, G. (2005) Genetic erosion – examples from Italy. *Genetic Resources and Crop Evolution* 52, 629–634.

Hawkes, J.G. (1983) *The Diversity of Crop Plants*. Harvard University Press, Cambridge, Massachusetts.

Hirano, R., Jatoi, S.A., Kawase, M., Kikuchi, A. and Watanabe, N. (2009) Consequences of *ex situ* conservation on the genetic integrity of germplasm held at different gene banks: a case study of bread wheat collected in Pakistan. *Crop Science* 49, 2160–2166.

Hoban, S., Arntzen, J.A., Bruford, M.W., Godoy, J.A., Hoelzel, A.R., *et al.* (2014) Comparative evaluation of potential indicators and temporal sampling protocols for monitoring genetic erosion. *Evolutionary Applications* 7, 984–998.

Holmes, K.R., Nelson, T.A., Coops, N.C. and Wulder, M.A. (2013) Biodiversity indicators show climate change will alter vegetation in parks and protected areas. *Diversity* 5, 352–373.

Hutchinson, C.F. and Weiss, E. (1999) Remote sensing contribution to an early warning system for genetic erosion of agricultural crops. In: Serwinski, J. and Faberova, I. (eds) *Proceedings of the Technical Meeting on the Methodology of the FAO World Information and Early Warning System on Plant Genetic Resources,* held at the Research Institute of Crop Production, Prague, Czech Republic, 21–23 June 1999. FAO, Rome.

Jarvis, A., Ferguson, M.E., Williams, D.E., Guarino, L., Jones, P.J., *et al.* (2003) Biogeography of wild *Arachis*: assessing conservation status and setting future priorities. *Crop Science* 43, 1100–1108.

Jarvis, D.I., Brown, A.H.D., Hung Cuong, P., Collado-Panduro, L., Latournerie-Moreno, L., *et al.* (2008) A global perspective of the richness and evenness of traditional crop-variety diversity maintained by farming communities. *Proceedings of the National Academy of Sciences of the United States of America* 105, 5326–5331.

Khlestkina, E.K., Huang, X.Q., Quenum, F.J.-B., Chebotar, S., Roeder, M.S., *et al.* (2004) Genetic diversity in cultivated plants – loss or stability? *Theoretical and Applied Genetics* 108, 1466–1472.

Khoury, C., Laliberte, B. and Guarino, L. (2010) Trends in *ex situ* conservation of plant genetic resources: a review of global crop and regional conservation strategies. *Genetic Resources and Crop Evolution* 57, 625–639.

Kiambi, D., Ford-Lloyd, B., Jackson, M.T., Guarino, L., Maxted, N., *et al.* (2005) Collection of wild rice (*Oryza* L.) in east and southern Africa in response to genetic erosion. *Plant Genetic Resources Newsletter* 142, 10–20.

Kiang, Y.T., Antonovics, J. and Wu, L. (1977) The extinction of wild rice (*Oryza perennis formosana*) in Taiwan. *Journal of Asia Ecology* 1, 1–9.

Koskela, J., Lefevre, F., Schueler, S., Kraigher, H., Olrik, D.C., *et al.* (2014) Translating conservation genetics into management: Pan-European minimum requirements for dynamic conservation units of forest tree genetic diversity. *Biological Conservation* 157, 39–49.

Last, L., Arndorfer, M., Balázs, K., Dennis, P., Dyman, T., *et al.* (2014) Indicators for the on-farm assessment of crop cultivar and livestock breed diversity: a survey-based participatory approach. *Biodiversity and Conservation* 23, 3051–3071.

Levrel, H. (2007) *Selecting Indicators for Management of Biodiversity*. Les Cahiers de l'Institut Français de la Biodiversité, Paris, 93 pp.

Lohar, D.P. and Rana, R.B. (1998) The dichotomy of crop diversity management issues in subsistence and commercial hill farming systems in Nepal. In: Pratap, T. and Sthapit, B. (eds) *Managing Agrobiodiversity-Farmers' Changing Perspectives and Institutional Responses in the Hindu-Kush-Himalayan Region*. International Centre for Integrated Mountain Development (ICIMOD), Kathmandu, and International Plant Genetic Resources Institute, Rome, pp. 183–197.

Mac Arthur, R.H. and Wilson, E.O. (1967) *The Theory of Island Biogeography*. Princeton University Press, Princeton, New Jersey, USA.

Megersa, G. (2014) Genetic erosion of barley in North Shewa Zone of Oromiya Region, Ethiopia. *International Journal of Biodiversity and Conservation* 6(3), 280–289.

Millennium Ecosystem Assessment (2005) *Ecosystems and Human Well-being: Biodiversity Synthesis*. World Resources Institute, Washington, DC.

Morishima, H. and Oka, H.I. (1995) Genetic erosion in wild and cultivated rice species. *Rice Genetics Newsletter* 12, 168–170. National Institute of Genetics, Tsukuba, Japan.

OECD (2001) *Environmental Indicators for Agriculture*, Vol 3. Organization for Economic Co-operation and Development, Paris.

Parra-Quijano, M., Draper, D., Torres, E. and Iriondo, J.M. (2008) Ecogeographical representativeness in crop wild relative *ex situ* collections. In: Maxted, N., Ford-Lloyd, B.V., Kell, S.P., Iriondo, J.M., Dulloo, E. and Turok, J. (eds) *Crop Wild Relative Conservation and Use*. CAB International, Wallingford, UK, pp. 249–273.

Parra-Quijano, M., Iriondo, J.M., Frese, L. and Torres, E. (2012) Spatial and ecogeographical approaches for selecting genetic reserves in Europe. In: Maxted, N., Dulloo, M.E., Ford-Lloyd, B.V., Frese, L., Iriondo, J.M. and Pinheiro de Carvalho, M.A.A. (eds) *Agrobiodiversity Conservation: Securing the Diversity of Crop Wild Relatives and Landraces*. CAB International, Wallingford, UK, pp. 20–28.

Perales, H., Brush, S.B. and Qualset, C.O. (2003) Landraces of maize in Central Mexico: an altitudinal transect. *Economic Botany* 57, 7–20.
Pereira, H.M., Ferrier, S., Walters, M., Geller, G.N., Jongman, R.H.G., *et al.* (2013) Essential biodiversity variables. *Science* 339, 277–278. Supplementary material available at: www.sciencemag.org/cgi/content/full/339/6117/277/DC1 (accessed 9 August 2015).
Steele, K.A., Gyawali, S., Joshi, K.D., Shrestha, P., Sthapit, B.R., *et al.* (2009) Has the introduction of modern rice varieties changed rice genetic diversity in a high-altitude region of Nepal? *Field Crops Research* 113, 24–30.
Steiner, A.M., Ruckenbauer, P. and Goecke, E. (1997) Maintenance in genebanks, a case study: contaminations observed in the Nürnberg oats of 1831. *Genetic Resources and Crop Evolution* 44, 533–538.
Tapia, C.B. and Estrella, J.E. (2001) Genetic erosion quantification in ullucus (*Ullucus tuberosus* Caldas), oca (*Oxalis tuberosa* Mol.) and mashua (*Tropaeolum tuberosum* R. et P.) in agroecosystems of the provinces of Canar, Chimborazo and Tungurahua – Ecuador. In: *Proceedings of the International Symposium, Managing Biodiversity in Agricultural Ecosystems. Montreal, Canada, 8–10 September 2001.*
Thormann, I., and Engels, J.M.M. (2015) Genetic diversity and erosion – a global perspective. In: Ahuja, M.R. and Mohan Jain, S. (eds) *Genetic Diversity and Erosion in Plants: Indicators and Prevention*, Volume 1. Springer International Publishing, Basel, Switzerland, pp. 263–294.
Thrupp, L.A. (1998) *Cultivating Diversity: Agrobiodiversity and Food Security*. World Resources Institute, Washington, DC.
Tittensor, D.P., Walpole, M., Hill, S.L.L., Boyce, D.G., Britten G.L., *et al.* (2014) A mid-term analysis of progress toward international biodiversity targets. *Science* 346, 241–244.
Walpole, M., Almond, R.E.A., Besançon, C., Butchart, S.H.M., Campbell-Lendrum, D., *et al.* (2009) Tracking progress towards 2010 biodiversity target and beyond. *Science* 135, 1503–1504.
Wetterich, F. (2003) Biological diversity of livestock and crops: useful classification and appropriate agri-environmental indicators. In: OECD (eds) *Agriculture and Biodiversity: Developing Indicators for Policy Analysis*. Proceedings from an OECD Expert meeting Zurich, Switzerland, November 2001. Organization for Economic Co-operation and Development, Paris.
Wilkes, H.G. and Wilkes, K.K. (1972) The green revolution. *Environment* 14, 32–39.

# 40 Improved Utilization of Crop Diversity for Rationalized Breeding Using Data Interoperability

R. Finkers*
*Wageningen UR Plant Breeding, Wageningen, the Netherlands*

## 40.1 Introduction

A key societal challenge for the near future is to develop food crops with improved nutritional value, with higher yields while growing on more marginal soils and with increased ability to withstand more extreme climates, pests and diseases, which are the Food and Agriculture Organization's 2050 goals (FAO, 2015). Effective utilization of the available crop (including wild relative) diversity in plant breeding is key in addressing this challenge. Breeding nowadays is not simply crossing two parents and recurrent selection of the most promising offspring, but utilizes state-of-the art 'omics' technologies to phenotype and genotype progenies. Increasingly, the diversity available in accessions of crops and their wild relatives can be characterized further using the same technologies. Plant breeding has become a data-intensive science, and computational methodologies are required to bring these data together effectively. However, the data also need to be stored in an appropriate format to do so. In this chapter, I address why this is important without touching on the technical aspects and what the perspective is for plant breeding if we are able to do so.

## 40.2 Data and Databases in the Life Sciences

Why should we invest in proper data management and efforts to make data available to the wider research community? An often-heard comment is that an experiment is set up for a specific purpose and that other researchers cannot reuse the data as the results are only applicable in a specific context and environment, and knowing the metadata of the experiment. However, there is no meaningful information without data and, conversely, data cannot be generated nor valued without prior knowledge (Mons *et al.*, 2011). The harsh realization is that, as a scientific community, we are losing data at a rapid rate (Gibney and van Noorden, 2013; Vines *et al.*, 2013) while, for example, the amount of next-generation sequencing data is doubling rapidly (Cochrane *et al.*, 2011). The 2014 Nucleic Acids Research (NAR) database issues describe more than 180 resources in eight different categories, which is only a subset of the databases existing worldwide. So, even if the data and its metadata are available, it might be very hard to find the relevant data and even harder to combine data from different databases, while the need to do so is there and the semantic web technology to do so is available.

*E-mail: richard.finkers@wur.nl

## 40.3 Use-cases Interoperability in Plant Breeding

To illustrate the potential of data interoperability, using semantic web technology in plant breeding research, three examples will be given. The first example involves the identification of putative candidate genes for traits of interest from quantitative trait loci (QTL) regions; the second example illustrates how this concept could be extended to extract and integrate QTL information from the literature; while the third example outlines how genomics data could be utilized to obtain and test relevant allelic variants for these putative candidate genes for traits of interest.

### 40.3.1 Candidate genes for quantitative trait loci studies

QTL regions are regions on the genome statistically associated with a trait of interest, and molecular markers flanking such a region can be used in marker-assisted selection (MAS) strategies. However, QTL regions can easily contain up to 1000 genes. Classically, fine mapping and map-based cloning approaches were used to zoom into candidate genes of interest (Martin *et al.*, 1993; Ronen *et al.*, 2000; van Os *et al.*, 2006). However, this is a tedious process and can take up to several years to complete. Nowadays, with the availability of *de novo* genome assemblies for many crop species, this process could be speeded up by exploiting knowledge from previous experiments. Key in such an approach is a quality annotation of the *de novo* assembled genome and should contain elements, such as (predicted) gene models, regulatory elements, function categories, functional domains, etc., and experimental evidence, such as (tissue-specific) RNAseq expression data. Even more important, elements should be cross-referenced semantically to external resources, further describing knowledge about each element, such as literature, protein functions, metabolic reaction and product coexpression networks (Astola *et al.*, 2014). One can imagine that mining all this information, from different databases, by hand is a tedious and laborious process. However, many researchers will do this for genes which they might prioritize utilizing knowledge from previous experiments or literature reviews. If all data elements are described adhering the Semantic Web standards of triples (W3C, n.d.) or, alternatively, nanopublications (Groth *et al.*, 2010), this knowledge could be combined automatically and used to prioritize candidate genes of interest. The added advantage is that such an approach can explore all available information and is not limited to the number of articles that could be read by an individual scientist. The ability to integrate this information readily depends on the availability of proper vocabularies; for example, as defined in the crop ontology portal (Matteis *et al.*, 2013) or gene ontology (Ashburner *et al.*, 2000). The tool Marker2Sequences works according to the principle of aggregation of knowledge using the principles of sematic data integration (Chibon *et al.*, 2012). In this chapter, a use case is described showing the application of prioritization of candidate genes in QTL intervals. Marker2Sequence can be used for mining QTL regions in tomato (www.wageningenur.nl/en/show/Marker2sequence.htm); however, the principle is generic and can be used for any other crop for which a *de novo* assembled reference genome is available.

### 40.3.2 Candidate gene mining from the literature

Marker2Sequence can be used to prioritize candidates in a QTL region for traits of interest. The semantic data integration approach taken by Marker2Sequence should also be possible by re-using the results of previously published QTL studies. The concept summarized in Fig. 40.1 will help to reduce QTL intervals into a prioritized list of candidates, explaining the trait of interest. QTLs from the literature or own experimentation can be used to restrict the list of genes and regulators to those within the interval. The extended (crop) trait ontologies can be used to discover connections between these gene lists and knowledge stored implicitly and explicitly in the literature and databases. The outcome of this approach will be a prioritized list of candidate genes/gene regulators. Supposing that the literature and databases are available in an interoperable (nanopublication) format, we can

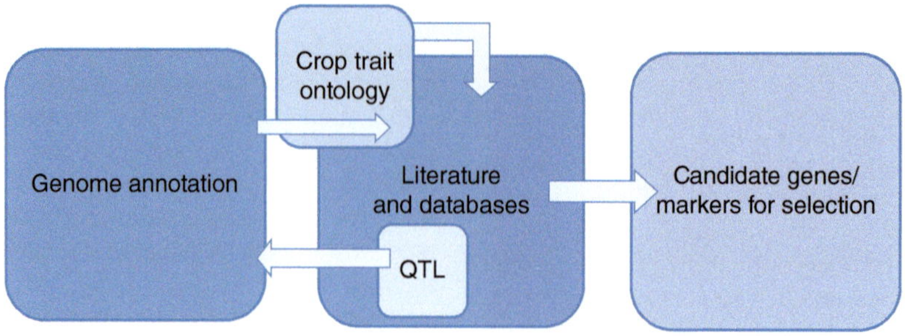


Fig. 40.1. Strategy to improve the identification of candidate genes for quantitative trait loci (QTL) regions of interest.

envisage that this information could be discovered. This would contribute to breeding research in two manners: (i) it will yield a resource of QTL regions that could be mined by researchers and used for visualization in genome browsers; and (ii) it will yield a resource that can be analysed computationally and which will result automatically in hypotheses for candidate genes for analysed traits. Plant breeders and researchers will then have the opportunity to mine these hypotheses and follow up novel leads, which should result in (i) insight into biological processes and (ii) alleles from accessions, contributing positively to the trait of interest. The extraction and comparison of QTLs from different experimental crosses with parents from different species from the literature and databases will not be trivial; however, we foresee that this concept will be technically feasible.

### 40.3.3 Studying allelic diversity for candidate genes of interest

Once we have a hypothesis for a gene(s) candidate for our trait of interest, the next step could be to utilize a reverse genetics strategy to study function and the impact of naturally occurring allelic variation for a gene. Such an approach will become increasingly feasible as results from re-sequencing studies are readily becoming available for many crops (e.g. Guo *et al.*, 2013; Huang *et al.*, 2013; Aflitos *et al.*, 2014) and seeds from these accessions can be obtained from genebanks. The allelic variation identified in resequencing collections will, to an increasing extent, be taken up by the genetic resources community and incorporated in the management of genebank collections, and will change the way genebank users can identify meaningful accessions from their collections. One now can assess the allelic variation of the candidates of interest, for example at the level of a gene-scale haplotype, by integrating variant data utilizing data interoperability strategies, and order a minimal set of accessions representing the available natural biodiversity. These genotypes can then be tested directly, or tested in a more uniform background after several rounds of backcrossing. Again, semantic data integration and software utilizing these resources play an important role (Finkers *et al.*, 2014).

### 40.3.4 Examples: concluding remarks

These three examples outline what can be achieved when data are available in an interoperable format and with the existence of proper tools to analyse such data effectively. A huge bonus of data interoperability and knowledge discovery pipelines is in the ability to combine different sets of data much faster, including all knowledge in contrast to the limited number of papers one can read, and good prioritization allows focusing on a biological question rather than mining through large piles of information by hand.

## 40.4 FAIR Data

There are both technical and societal considerations to take account of before data can effectively be made interoperable. To make full use of research data, the community needs to promote the growth of an open 'data commoning' culture (Sansone *et al.*, 2012). Data should be made FAIR (findable, accessible, interoperable and reusable; see datafairport.org). In short, the FAIR data guiding principles are: metadata of any data object should be described and uniquely identifiable via a persistent identifier (findable). Data should be accessible by machines and humans, upon appropriate authorization (accessible). Data objects utilize shared vocabularies and/or ontologies (interoperable), and data objects should be well described and rich, so that they can be linked or integrated (reusable) automatically (or with minimal human effort). Together with open initiatives, such as the Data FAIRport, commonly used data sets will be available in a FAIR format, and domain-specific data sets can be made FAIR in collaboration with these experts. In this way, technical obstacles will be greatly reduced.

## 40.5 Perspectives for Data Interoperability in Plant Breeding

In this chapter, I have described a few examples on how plant breeders can use data interoperability strategies to improve their research, and the importance of the genebank community in this process. Traditional plant breeding is changing rapidly by the increasing usage of 'omics' technologies (Langridge and Fleury, 2011; (Kusano and Fukushima, 2013). Resequencing projects allow the rapid identification of variation at the level of an individual gene (or gene haplotype) and study the effect of this gene-scale haplotype in diverse genetic backgrounds, while network analysis of large data sets provides insight in biological mechanisms (Acharjee *et al.*, 2011; Ligterink *et al.*, 2012). Making this information explicit and available in a FAIR format will add a novel layer to the principle of breeding by design (Peleman and Van Der Voort, 2003). Taking this one step further, the combined ability of alleles can be tested and the results can be stored as well. Finally, the development of decision support systems, in which this information is incorporated, will bring us one step closer to the *in silico* design of new crop varieties, specifically designed for a market segment and to be grown in a specified environment. The further development and incorporation of this technology by the breeding community will be instrumental in securing food safety and food security in the near future and in achieving the goals set by the FAO for 2050.

## References

Acharjee, A., Kloosterman, B., De Vos, R.C.H., Werij, J.S., Bachem, C.W.B., *et al.* (2011) Data integration and network reconstruction with ~omics data using Random Forest regression in potato. *Analytica Chimica Acta* 705(1–2), 56–63, doi:10.1016/j.aca.2011.03.050.

Aflitos, S., Schijlen, E., de Jong, H., de Ridder, D., Smit, S., *et al.* (2014) Exploring genetic variation in the tomato (*Solanum* section *Lycopersicon*) clade by whole-genome sequencing. *The Plant Journal: For Cell and Molecular Biology* doi:10.1111/tpj.12616.

Ashburner, M., Ball, C.A., Blake, J.A., Botstein, D., Butler, H., *et al.* (2000) Gene ontology: tool for the unification of biology. The Gene Ontology Consortium. *Nature Genetics* 25(1), 25–9, doi:10.1038/75556.

Astola, L., Stigter, H., van Dijk, A.D.J., van Daelen, R. and Molenaar, J. (2014) Inferring the gene network underlying the branching of tomato inflorescence. *PloS One* 9(4), e89689, doi:10.1371/journal.pone.0089689.

Chibon, P.-Y., Schoof, H., Visser, R.G.F. and Finkers, R. (2012) Marker2sequence, mine your QTL regions for candidate genes. *Bioinformatics (Oxford, England)* 28(14), 1921–1922, doi:10.1093/bioinformatics/bts298.

Cochrane, G., Karsch-Mizrachi, I. and Nakamura, Y. (2011) The International Nucleotide Sequence Database Collaboration. *Nucleic Acids Research* 39(Database issue) D15–8, doi:10.1093/nar/gkq1150.

FAO (Food and Agriculture Organization) (2015) How to Feed the World in 2050. Food and Agriculture Organization of the United Nations, Rome, Italy. Available at: www.fao.org/fileadmin/templates/wsfs/docs/expert_paper/How_to_Feed_the_World_in_2050.pdf (accessed 5 August 2015).

Finkers, R., Chibon, P., van Treuren, R., Visser, R. and van Hintum, T. (2014) Genebanks and genomics: how to interconnect data from both communities? *Plant Genetic Resources* 1–4, doi:10.1017/S1479262114000689.

Gibney, E. and van Noorden, R. (2013) Scientists losing data at a rapid rate. *Nature* doi:10.1038/nature.2013.14416.
Groth, P., Gibson, A. and Velterop, J. (2010) The anatomy of a nanopublication. *Information Services and Use* doi:10.3233/ISU-2010-0613.
Guo, S., Zhang, J., Sun, H., Salse, J., Lucas, W.J., *et al.* (2013) The draft genome of watermelon (*Citrullus lanatus*) and resequencing of 20 diverse accessions. *Nature Genetics* 45(1), 51–58, doi:10.1038/ng.2470.
Huang, X., Lu, T. and Han, B. (2013) Resequencing rice genomes: an emerging new era of rice genomics. *Trends in Genetics* 29(4), 225–232, doi:10.1016/j.tig.2012.12.001.
Kusano, M. and Fukushima, A. (2013) Current challenges and future potential of tomato breeding using omics approaches. *Breeding Science* 63(1), 31–41, doi:10.1270/jsbbs.63.31.
Langridge, P. and Fleury, D. (2011) Making the most of 'omics' for crop breeding. *Trends in Biotechnology* 29(1), 33–40. doi:10.1016/j.tibtech.2010.09.006
Ligterink, W., Joosen, R.V.L.L. and Hilhorst, H.W.M.M. (2012) Unravelling the complex trait of seed quality: using natural variation through a combination of physiology, genetics and -omics technologies. *Seed Science Research* 22(S1), S45–S52, doi:10.1017/S0960258511000328.
Martin, G., Brommonschenkel, S., Chunwongse, J., Frary, A., Ganal, M., *et al.* (1993) Map-based cloning of a protein kinase gene conferring disease resistance in tomato. *Science* 262(5138), 1432–1436, doi:10.1126/science.7902614.
Matteis, L., Chibon, P., Espinosa, H., Skofic, M., Finkers, R., *et al.* (2013) Crop ontology: vocabulary for crop-related concepts. In: Larmande, P., Arnaud, E., Mougenot, I., Jonquet, C., Libourel, T., *et al.* (eds) *Proceedings of the First International Workshop on Semantics for Biodiversity* (Vol 979). Montpellier, France, May 27, 2013. Available at: ceur-ws.org/Vol-979/WS_s4biodiv2013_paper_4.pdf (accessed 4 November 2014).
Mons, B., Van Haagen, H., Chichester, C., Hoen, P.-B. T. 't, Den Dunnen, J.T., *et al.* (2011) The value of data. *Nature Genetics* 43(4), 281–283, doi:10.1038/ng0411-281.
Peleman, J.D. and Van Der Voort, J.R. (2003) Breeding by design. *Trends in Plant Science* 8(7), 330–334, doi:10.1016/S1360-1385(03)00134-1.
Ronen, G., Carmel-Goren, L., Zamir, D. and Hirschberg, J. (2000) An alternative pathway to β-carotene formation in plant chromoplasts discovered by map-based cloning of beta and old-gold color mutations in tomato. *Proceedings of the National Academy of Sciences of the United States of America* 97(20), 11102–11107, doi:10.1073/pnas.190177497.
Sansone, S.-A., Rocca-Serra, P., Field, D., Maguire, E., Taylor, C., *et al.* (2012) Toward interoperable bioscience data. *Nature Genetics* 44(2), 121–126, doi:10.1038/ng.1054.
van Os, H., Andrzejewski, S., Bakker, E., Barrena, I., Bryan, G.J., *et al.* (2006) Construction of a 10,000-marker ultradense genetic recombination map of potato: providing a framework for accelerated gene isolation and a genomewide physical map. *Genetics* 173(2), 1075–1087, doi:10.1534/genetics.106.055871.
Vines, T.H., Albert, A.Y.K., Andrew, R.L., Débarre, F., Bock, D.G., *et al.* (2013) The availability of research data declines rapidly with article age. *Current Biology* 20, 94–97, doi:10.1016/j.cub.2013.11.014.
W3C (n.d.) SEMANTIC WEB. Available at: www.w3.org/standards/semanticweb/ (accessed 30 October 2013).

# 41 Implementation of a PGR Global Documentation System in Portugal

A.M. Barata,[1*] F. Rocha,[1] J. Oliveira,[2] J.M. Lima,[2] H. Nobrega,[3] M.Â.A. Pinheiro de Carvalho[3,4] and S. Dias[5]

[1]*Banco Português de Germoplasma Vegetal, Braga, Portugal;* [2]*Direcção Regional de Agricultura e Pescas do Norte (DRAPN), Braga, Portugal;* [3]*ISOPlexis Genebank, University of Madeira, Funchal, Portugal;* [4]*ICAAM, University of Évora, Évora, Portugal;* [5]*Bioversity International, Rome, Italy*

## 41.1 Introduction

The agricultural community needs ready access to information that enables the most efficient and economical choice of appropriate diversity to address climate change and food security challenges. While considerable amounts of diversity are stored in genebanks, information about them is fragmented and scattered, making it particularly difficult to access. It is often hard to find information about germplasm that can provide, for example, resistance to pests and diseases or other stresses or tolerances that reduce/increase productivity and yields (Arnaud *et al.*, 2010). For a long time, there have been efforts to gather, compile, aggregate and access information scattered around the world at the global level. However, there is equally a need to improve the documentation of genebank collections and strengthen staff capacity by investing more and providing technical support at the local level. This will help to improve access to locally available information and contribute to global-level information systems. These efforts specifically address the barriers faced by breeders, crop researchers and others in locating specific types of genetic diversity in the world's genebanks (Arnaud *et al.*, 2010).

Increasingly there is the need to apply new technologies widely, including the use of data standards. These facilitate the quantity and quality of data to be used for target purposes and by target users. However, it is difficult and expensive to apply correctly the advances in information technology, and for each genebank to develop and maintain its own documentation system. Until recently, many genebanks had no other option than to develop their own systems, either because of lack of long-term capacity (technology and staff) or because of the infeasibility and difficulty in adopting the existing tools developed by advanced genebanks.

Nowadays, with the rapidly evolving technology, operational differences, changing of purposes, specificity of domains and decreased budgets, increasingly the plant genetic resource (PGR) community will need to share software or use each other's code, in a community of developers and users. This is the opportunity to apply open-source documentation systems in the field of genebank documentation.

The global germplasm management information system currently under development (GRIN-Global – GG) specifically addresses information retrieval obstacles faced by breeders,

*Corresponding author; e-mail: ana.barata@iniav.pt

 

crop researchers and other users. The global information portal will use data standards and integrate data sets from existing information systems, or through a new scalable genebank information management system.

## 41.2 Why the GRIN-Global System?

The partnership between the Global Crop Diversity Trust, USDA/ARS National Plant Germplasm System and Bioversity International, for redesigning and software upgrades, will enable the new GRIN-Global system to be used by either small or large genebanks all over the world, making more information on plant genetic resources available to researchers worldwide. Many genebanks have computerized their information, but the database software and data models implemented may differ substantially between genebanks. As a result, these data sets often do not fully meet international quality standards for data exchange, which limits their value. Furthermore, because many data providers continue to use offline replication methods, the most up-to-date data are not readily available to researchers in a timely manner – a situation that is often exacerbated by the need for human interaction to replicate these offline data sets.

GRIN-Global aims at being a scalable genebank documentation system, one that will be suitable for use by any interested genebank in the world (Postman *et al.*, 2010; GRIN-Global, 2014a). With this aim, a shared open tool will solve the many issues for data documentation and management, due mainly to the high level of standardization that the system will set up. This will not only reduce costs and increase the efficiency of information in both quality and quantity but will also result in a very efficient information exchange, with the advantage of community benefit on the updating and sharing of developments across the world. This means that a country can apply the same tool and technology across existing collections, enabling for the first time an access and interoperability between documentation systems and giving a wider access to the existing material maintained across a country, which was only possible in situations where a country had a centralized system (Agrawal *et al.*, 2007). This opportunity goes beyond the information aspect, by having the possibility to share IT resources – staff and hardware.

Standardization and quality improvement allow genebanks to collaborate better, and thus improve efficiency, and to make better use of the limited financial resources available to them. In addition, it also helps genebanks to improve their service to users, to respond to the enormous challenges (e.g. climate change) ahead of us. So the logical step taken was to implement the same system in two of the genebanks operating in Portugal – Banco Português de Germoplasma Vegetal (BPGV) and Banco de Germoplasma ISOPlexis (BG ISOPlexis) – and provide access to information on the germplasm conserved in the genebanks and extended access to the material, thus contributing to a more complete and up-to-date national inventory. The main advantages of GG for genebank use are that:

- It is a user-friendly PGR information system that can promote access to, and the efficient and effective management and use of, PGRs worldwide.
- It is built on the well-proven GRIN expertise, which has been managing PGR information efficiently and effectively for more than 20 years.
- It is being developed in a joint effort with the Global Crop Diversity Trust, Bioversity International and the Agricultural Research Service of the USDA, and the system has been evaluated and tested with the international genebank community.
- It has a flexible system architecture that enables the customization of GG to meet local requirements.
- It includes a web publishing site for user searching and germplasm ordering.
- It can be implemented on a network computer system or on a stand-alone personal computer.
- Its scope and functionality results from the input of genebank managers throughout the world.
- It has the ability to drag and drop from GG views to spreadsheets or from spreadsheets directly to GG views.
- It has the ability to customize multiple screen views, forms and menus.

- It includes a 'Google-like' internal search engine.
- It is royalty free, with worldwide permission to use, to copy and/or modify.

## 41.3 The GRIN-Global Management System

The GG system is a new and scalable software version of the Germplasm Resource Information Network (GRIN), used by the USDA/ARS National Plant Germplasm System, that enables genebanks to store and manage information with a powerful, flexible, easy-to-use plant genetic resource (PGR) information management system (Postman *et al.*, 2010; GRIN-Global, 2014a; GRIN-Global, 2014b; USDA, 2014). The GG system has been built using the well-established, three-tier architecture design pattern consisting of a Presentation Tier, a Business Tier and a Data Tier (see Fig. 41.1).

The Data Tier is the database where all of the PGR data are permanently stored. According to the principle of database flexibility, the GG system supports free database systems – MySQL, Microsoft SQL Server and Oracle.

The Business Tier contains a full complement of web services (software modules that are accessible over the Internet via standard messaging protocols) that enable the entire database repository to be made accessible to any other computer system connected to the Internet. The web services implemented in the Business Tier provide data in XML format using the SOAP protocol to any computer system capable of processing this type of transaction.

The Presentation Tier can assume many different forms, with no compromise in data accessibility. Users access the data via the different GG applications: the Curator Tool (CT), the Search Tool (ST) and the Public Website (PW). The CT and the ST are closely integrated – the CT has a search button that invokes the ST; however, the ST can also run as a stand-alone application. Also, any third-party application or website wishing to retrieve data from the GG system can do so by connecting to the Business Tier web services in exactly the same way as the GG website or CT application (Cyr *et al.*, 2009).

### 41.3.1 Curator Tool

The Curator Tool (CT) is a Windows application that must be installed on the user's PC in order to connect to the remote GG server (Fig. 41.2).

The GG CT is used by curators and users to create and manage genebank data, for viewing and manipulating the data. In order for a PC to run the CT program, a copy of SQL Server Express must be installed on the user's PC when the

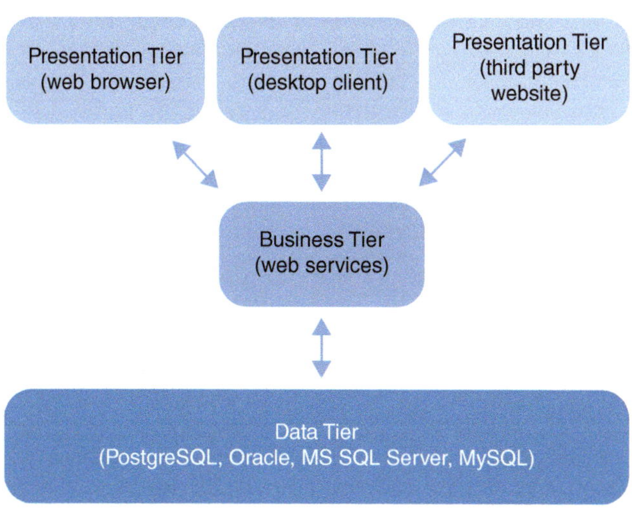


Fig. 41.1. GRIN-Global three-tier design.

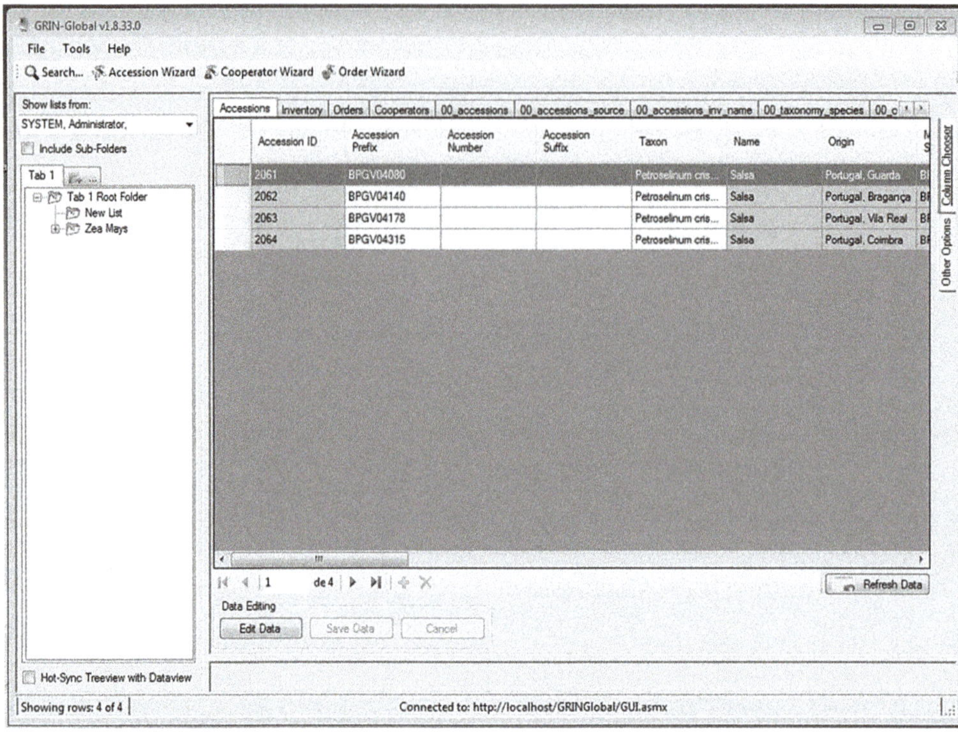

Fig. 41.2. Curator Tool.

CT is installed, regardless of what database product is used to run the main GG database hosted on the server. This SQL Server Express manages a set of lookup tables that are installed with the CT.

The GG CT user makes the access data through 'data views', which mask raw data descriptions and relationships. The CT contains many data views in which users display data without needing to know SQL or programming. A data view essentially retrieves data from tables via a programmed query. Fortunately, these data views have been created previously, and the users can display many data views and switch back and forth by clicking on their tabs (Grin-Global, 2014b).

### 41.3.2 Public Website

The Public Website (PW) is designed for users who will query the GG database for germplasm information or who will be ordering germplasm to meet specific research needs (Fig. 41.3). Besides researchers, breeders, scientists, curators and other internal genebank staff may also use the PW as a complement to the CT to look up information on Accessions, Orders, etc. No installation is required – the Public Website requires a standard browser and incorporates an easy-to-use interface for extracting plant genetic resource information from the database (Grin-Global, 2015a).

### 41.3.3 Admin Tool

The Admin Tool (AD) is a Windows application for administering GG and requires a direct database connection (Fig. 41.4). The AD for GG administrators handles diverse functions, including adding or modifying user accounts and their security permission, importing data and creating and editing data views (GRIN-Global, 2012). This tool is fundamental for completing two major steps when setting up GG for an organization, which is to:

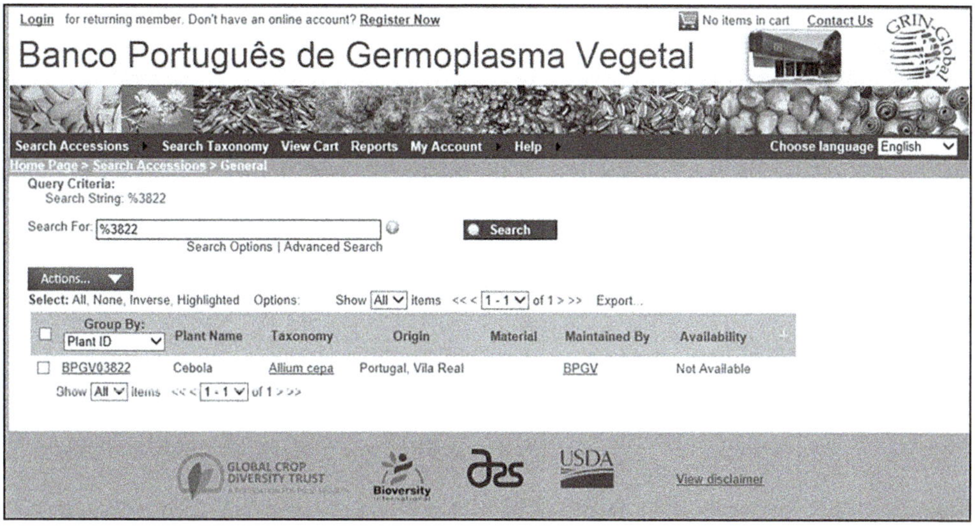


Fig. 41.3. Public Website.

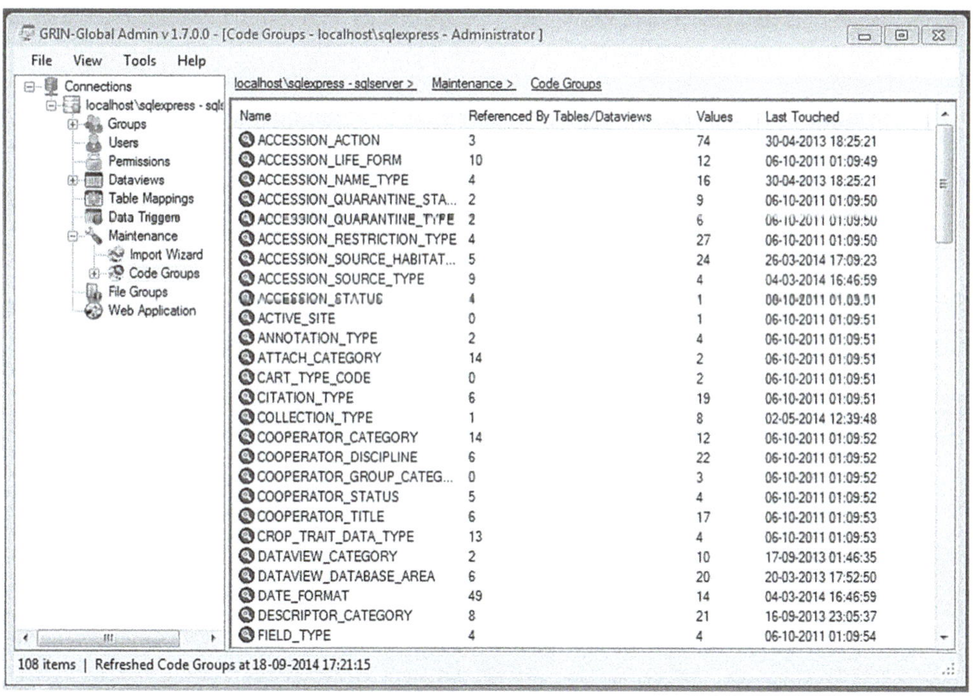


Fig. 41.4. Admin Tool.

- prepare and import any existing data that the organization had previously been maintaining in spreadsheets or other databases; and
- add users and grant them security privileges.

### 41.3.4 Import Wizard

The Import Wizard (IW) facilitates the bulk loading of records because it has been designed to copy data imported from a single, stored .csv file directly into multiple tables. The GG data are stored in many tables. This type of database is referred to as a relational database. Each table relates to other tables by key fields. A single spreadsheet, on the other hand, may be considered a database, but because the spreadsheet is not relational, it is sometimes referred to as a flat file database. Generally, relational databases are easier to use, reduce redundant data and consist of many tables that are used together to find the desired information quickly. In GG, the accession data are stored across more than 103 related tables (GRIN-Global, 2015b). Figure 41.5 illustrates the relationships between the primary GG tables.

GRIN-Global is accessible to all, is free and is without restriction. All source code (including the database schema) is in the public domain, it has been written using software development tools that are free and the data are supported in free database systems. The GG management system, because it is an improved and updated version of GRIN, has inherited all the knowledge and experience from its developers.

## 41.4 The GRIN-Global System Implementation Process in Portugal

The implementation process in Portugal was done in four phases (see Fig. 41.6) with several key steps:

**1.** System evaluation training (SET): the first step is to make a thorough evaluation of the genebank operations and the system's capacity to handle them or flexibility to adapt it. Training of personnel to become familiar with the tool is needed, so as to reduce the time for the staff to get acquainted with this complex and complete system.

**2.** Functionality analysis (FA): a second step is to understand fully the platform capacity and flexibility. Analysis of procedures in the genebanks – a comparison of both genebank's procedures was made to unravel the commonalities and the differences in procedures and operations in order to implement the system in both genebanks.

**3.** Implementation decision (ID): the third phase is to reach a decision on whether or not to implement the solution. To reach this decision, a matrix with two types of criteria was used: alignment with business objectives with 0.40 weight, and criteria with attractiveness for use with 0.60 weight. For each of these criteria, the subcriteria had also been weighted and each genebank scored them from 1 to 5 (1 low and 5 higher importance). Table 41.1 shows the two criteria and each genebank's scores. For the result analysis and communication of the outcome analysis, not only the matrix overall scores must be considered but also the pros and cons, technical considerations, level of coverage and costs to run it should be taken into account. The final step of this phase is the decision on the implementation of the solution, and at this point, having the matrix overall scores and the results analysis, there are three options:

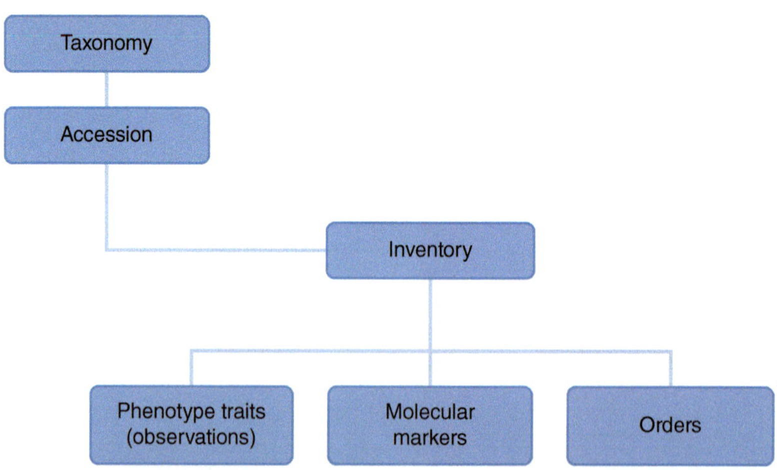


Fig. 41.5. GRIN-Global parent–child relationships.

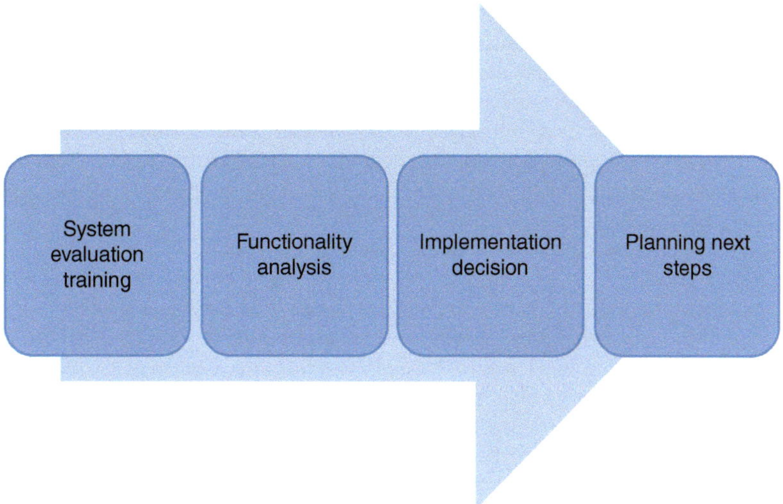


Fig. 41.6. Phases of implementation.

Table 41.1. Decision matrix.

| No | Criteria | Weight | BPGV | ISOPlex |
|---|---|---|---|---|
| 1.0 | Alignment with business objectives | 0.40 | | |
| 1.1 | Feasibility for adoption | 0.20 | 4 | 5 |
| 1.2 | How much the system covers the genebank operations | 0.40 | 5 | 5 |
| 1.3 | IT resources needed | 0.25 | 5 | 4 |
| 1.4 | Staff training | 0.15 | 3 | 3 |
| Alignment with business objectives: | | 1.00 | 4.5 | 4.45 |
| No | Criteria | Weight | BPGV | ISOPlex |
| 2.0 | Attractiveness for use | 0.60 | | |
| 2.1 | Be flexible for adoption covering up to 80% of operations | 0.40 | 5 | 4 |
| 2.2 | Having IT in-house capacity for long-term management | 0.45 | 4 | 2 |
| 2.3 | Decrease investment in the long term | 0.15 | 5 | 3 |
| Attractiveness for use: | | 1.00 | 4.55 | 2.95 |
| Overall score | | 1.00 | 4.53 | 3.55 |

- Yes, implementation without restrictions – when overall scores are above 3 and the results analysis is all positive.
- Yes, implementation with reservations – when overall scores are between 2.5 and 3, and the results analysis highlights items for consideration.
- No implementation – when overall scores are below 2.5, and the results analysis highlights major constraints.

**4.** Planning next steps (PNS): once the adoption decision has been made, planning can start. The last phase, after the decision has been made, is when the implementation planning begins to be drafted (for 41.3.1 and 41.3.2). Planning is divided in four main areas: (i) identification of functionalities needs – these are either functionalities that do not cover the remaining per cent of operations, or functionalities that are essential for serving the genebank mission, such as analysis, reporting, etc.; (ii) indexing of data – this activity is essential to set the new system data dictionary. This is to relay the information from your genebank into the system, which allows determining what is what. Note that most of the implementation time is taken up by this activity; (iii) load the data inventory – first start with the accession-level passport data for all collections maintained in the genebank, and subsequently plan for leading the remaining related data sets (such as characterization, evaluation, molecular data) in the system; and (iv) the public launching – a public launch should be planned to present and disseminate the genebank activities and website officially.

The implementation of the GRIN-Global system by the ISOPlexis Genebank, Madeira, and Banco Português de Germoplasma Vegetal (BPGV), located in Funchal and Braga, respectively (Fig. 41.7), began with a training session by Bioversity International in June 2011. The main objectives of the trainings were to show the tool and its potential for the genebanks' daily management of their collections, to provide an opportunity for the users to evaluate the tool, and to collect feedback for further improvement and to assess the feasibility of the possible adoption of GRIN-Global (Dias, 2011). With this training, the genebank staff became familiar with the system and its tools; installed the system at the genebank; and staff were trained in transferring the plant genetic resources and germplasm collection data from the existing internal documentation system to the GG system. These training sessions were the basis for covering the SET phase, which was followed by the FA phase that was done by each genebank. The next phase, the ID, was first done using the matrix (as per Table 41.1). The overall score weighted for

**Fig. 41.7.** Location of the genebanks.

BPGV was 4.53 and was 3.55 for ISOPlex. These scores immediately fell in the decision to implement without restriction for each genebank, which fully supported the results analysis made. Thus, the last phase was the planning to implement the solution – implementation of the GG system by each genebank. Each genebank has developed their own planning, which is described further in Sections 41.4.1 and 41.4.2 below, including descriptions of the lessons learned and the constraints encountered during the implementation process.

### 41.4.1 Implementation of GRIN-Global system in ISOPlexis Genebank

The Autonomous Region of Madeira is an Atlantic region of Portugal, located between latitude 33° 10′ and 32° 20′ N and longitude 16° 10′ and 17° 20′ W, approximately 630 km from North Africa, intersecting the former routes of the global exchange of genetic resources between Europe, South America, Africa and the Asiatic continents. The most recent incomplete inventories refer to the existence in the region of 111 crop and 432 crop wild relatives in use.

The ISOPlexis Genebank began its activities in 1996, with the main goal to inventory and survey, collect, conserve and evaluate crop genetic resources from Macaronesia. Since then, ISOPlexis has accumulated a considerable collection and its related documentation data. At present, information on 2805 ISOPs from 4169 accessions conserved belonging to 92 species are documented. The collection includes passport data, inventorying and agroecological data (ISOPex Teca), mission and management reports, morphological and agronomical information (ISOPex Fenogen), molecular data (ISOPex Fingerprinting) and biochemical and nutritional characterization data (ISOPex NutriValor). These data sets have been organized and stored in flat file databases, under the ISOPex acronym, but making their documentation, accessibility and analysis extremely difficult due to the absence of relational functions between the databases. The GRIN-Global being a management system using international agreed standards for plant genetic resources allows us to upload and document in a single database all the genebank information generated during genebank operational activities. To start this process, after the first staff training, the GRIN-Global database was installed on a curator stand-alone computer. The planning phase consisted of three main phases: one for loading passport data, the second to load all the related data (characterization, molecular, etc.) and the final one for the launching of the GG-ISOPLEX website. Currently, phase one has been achieved, with all the information relating to collection accessions having been uploaded into the system 'ISOPlexis GRIN-Global database'. The information included is basic passport data, such as crop species identification, accessions number and vernacular designations, sample donor, origin, seed quantity, etc. The next step will be the implementation of phase two, which will be dedicated to data organization and uploading of all the other ISOPex flat file databases (ISOPex Teca, ISOPex Fenogen, ISOPex Fingerprinting and ISOPex NutriValor); and the final phase will be to launch and make the 'GG ISOPLEX website' available online. For the first time we have been able to digitalize and manage the genebank's information, thus making it easier for genebank staff, researchers and the general public to access and further use the information and material available in the genebank.

#### *Lessons learned and constraints*

One of the lessons learned was the need to allocate dedicated staff to documentation/information activities. This is a steep learning curve and well-trained staff are critical for efficient implementation:

- There is a need to plan ahead, to include priority species/crop descriptors specifically for morpho-agronomic data of evaluated crop samples.
- Major constraints are related with database structure adjustment to the specificities of crop descriptors used by the ISOPlexis Genebank evaluation programme.
- ISOPlexis produces a battery of analytical work on several target species, generating agronomic, biochemical, nutritional and technological data that, once documented in the system, will make data analysis much easier.
- It will also enable better-informed decisions about collecting missions, plant material exchange and identification, as well as the

selection of the best target material for the use of stakeholders.
- The above steps will enable online publication of the data in the ISOPlexis Genebank website and guarantee the public access to this information.

### 41.4.2 Implementation in the Banco Português de Germoplasma Vegetal (BPGV)

Since 1977, BPGV has had a mandate to survey, collect, conserve, evaluate, document and distribute plant genetic resources in Portugal. The initial documentation system was based in FileMaker and Microsoft Access and actually holds data on 44,752 accessions from 173 species and respective families (Barata *et al.*, 2011). Passport data and evaluation data, as well as management information, are compiled on those databases. The GG being a management system will allow for the first time uploading of all the information generated in the genebank in one database. To start this new process, after the first training on the GRIN-Global database, the version available and its products were installed on a stand-alone computer.

The installation was followed by studying in detail the structure of the GG database. For this, the documentation aids such as the GRIN-Global Online Data Dictionary and Administrator Guide were used (GRIN-Global, 2012). The GG data are stored in a relational database composed of many tables, related by key fields, and it is essential to understand that the 'Accession identification', the 'Source data of the Accession' and the 'Inventory data associated with' are the most important tables. Once the GG Data Model was clearly understood, the next step was to focus on integrating existing information to meet the necessary programme requirements, with a mandatory assumption: never lose information considered relevant to the BPGV (e.g. using the 'note' field of the various tables). The migration strategy was tested with a set of 40 accessions of a single species. The total data were prepared using Microsoft Excel files, which allowed the development of a sustainable methodology to import all the data. Although data migration can be done using existing import wizards, it can also be done by drag and drop (or copy and paste) directly to the GG tables (when a specific wizard, such as Accession Source Map, does not exist). To help and guide the implementation process and the indexing of concepts and terms between GG and BPGV terminology, a revised 'GRIN-Global BPGV Data Dictionary' was produced.

For the data import, the first task was to create all the necessary information to meet the code group needs in the system. These were created considering the GG rules and matching the BPGV information needs. Examples of code groups used are: 'Collection Type' – codes for active and base collection – very important in BPGV management; 'Sites', 'Region' and 'Geography': 'Site BPGV' and other genebanks (e.g. backup sites); 'Taxonomy' (Family, Genus, and Species): the GG taxonomy was the starting point and was worked on further to meet the necessary collection needs; 'Cooperators': the cooperators were created at the expense of collection information ('Source Collected' and 'Source Offered').

Having completed the code group areas, the focus was on loading the remaining information, and for that the wizards were used, but splitting the import into two main groups: Group 1, Codes, Cooperators, Geography, etc. (Wizards 1–6); and Group 2, Accessions and Inventory (Wizards 9–11). For these data loads, Excel base files were designed to capture the information spread in the different sources into one file; followed by the conversion of the base files in the import wizard's format (creation of script files, in 'excel' to 'txt').

Quality tests were performed with the administrator, the curator tool and the website, using the test group of 40 accessions imported to the system, and results were compared with the original accession information held both in databases and on paper.

The website was configured according to BPGV needs and the network where the server was located. The user privileges, the address name (DNS) of the webserver and the aspect of the homepage of the GG website were defined to allow data searching through the Internet.

#### *Lessons learned and constraints*

Considering that, in accordance with the planned objectives, the BPGV information import process to the GG has been completed, we consider it important to register some of the difficulties encountered and some points to take into account when considering starting a similar process:

- Define cooperators and privileges: as mentioned previously, one of the stages was to define the genebank users (staff names and roles) and their respective privileges to handle data (add, delete, change and view). Only after this was the import process performed (using the registered users in GG) to fulfil properly the privileges defined for the columns 'owned by' and 'created by'.
- Attribution of UID to cooperators: the unique identifier (UID) is automatically defined in the GG cooperator table when each user is registered. Because of this, enormous difficulties were faced due to the fact that the name and address of a cooperator were not always recorded in the same manner (in the definition of the UID for the Accession, we adopted the existing logic in BPGV, making concatenation of data and using a single column for the UID (an example: accession prefix=BPGV01934, accession number and accession suffix with null values).
- Attribution of UID to Inventory: the processing of Inventory information (Sample Accession) was particularly complex in defining the UID of each record. The management logic of the cold chambers in BPGV shows that an Accession Inventory is recorded several times (different cold chambers, original material, multiplication/regeneration, etc.). To overcome this, the three columns provided by GG were used to identify each Inventory record (an example: inventory_number_part1=BPGV01934, inventory_number_part2=1 and inventory_number_part3=orig-1991).
- Validation issues–non-documented errors: even considering all the support documents and verifying all data sets, we still found some errors in the import process validated by the Wizard (incorrect dates, characters unforeseen, Code Groups missing, etc.).
- Versioning structure changes and indexing implications: while all this process of data import took place over a considerable period, at the same time GG had some updated versions affecting the infrastructure. Since there was no migration tool for version upgrade, it presented some difficulties in the preparation of the information (e.g. number of decimal places in quantity) and also in stabilizing the software versions (SQL Server, etc.).

## 41.5 Future Steps and Recommendations

### 41.5.1 Future steps

- Official launch of the GRIN-Global platform in the BPGV and ISOPlexis Genebank was in 2015.
- Document other data needs for further development.
- Be part of a community of practices for development and users.

### 41.5.2 Recommendations

These recommendations are not exhaustive but are the most important ones to facilitate the system implementation and to inform institutions of the long term commitment and support required to attain success:

- Follow the implementation process in its four steps: (i) SET (System Evaluation Training); (ii) FA (Functionality Analysis); (iii) ID (Implementation Decision); and (iv) PNS (Planning Next Steps).
- Ensure national and institutional recognition and long-term support.
- Genebank operations budget to include a budget line for documentation and information management activities.
- Proper training of genebank staff on the GG tools (Curator and Admin).
- Document your change processes and data index – make your own 'GRIN-Global [Name] Data Dictionary'.
- Establishment of a community of practices for GRIN-Global development and users.
- To facilitate upgrading, a priority must be to build a migration tool for version upgrade.
- Long-term support from USDA, TRUST and Bioversity International, for adoption and implementation.

## References


Agrawal, R.C., Behera, D. and Saxena, S. (2007) Genebank Information Management System (GBIMS). *Computers and Electronics in Agriculture* 59, 90–96.

Arnaud, E., Dias, S., Mackay, M., Cyr, P., Gardner, C., *et al.* (2010) A portal enabling worldwide access to information on conservation and use of biodiversity for food and agriculture. In: Maurer, L. and Tochtermann, K. (eds) *ICT-SECURE: Information and Communication Technologies for Biodiversity Conservation and Agriculture*. Shaker Verlag, Aachen, Germany, pp. 173–186.

Barata, A.M., Rocha, F., Reis, A. and Lopes, V.R. (2011) O Banco Português de Germoplasma Vegetal e a conservação dos recursos genéticos em Portugal. In: *Agrorrural. Contributos Científicos. Instituto Nacional de Recursos Biológicos*. I.P. e Empresa Nacional Casa da Moeda S.A., Depósito legal n° 335 133/11, Edição n° 1018454, Lisboa, Spain, pp. 964–974.

Cyr, P.D., Weaver, B.E., Mark, J., Millard, M.J., Gardner, C.A., *et al.* (2009) *GRIN-Global: An International Project to Develop a Global Plant Genebank and Information Management System.* Available at: www.ars.usda.gov/SP2UserFiles/Place/50301000/Posters/GRIN-Global_for_51st_Maize_Genetics_09.pdf (accessed 18 December 2014).

Dias, S. (2011) *EURISCO E-Bulletin, September 2011.* Available at: eurisco.ipk-gatersleben.de/apex/EURISCO_WEB.download_file?p_id=30 (accessed 18 September 2014).

GRIN-Global (2012) *Administrator Guide.* Available at: www.ars-grin.gov/npgs/gringlobal/docs/gg_admin_guide.pdf (accessed 18 September 2014).

GRIN-Global (2014a) *Main Page.* Available at: www.grin-global.org (accessed 18 September 2014).

GRIN-Global (2014b) Curator Tool User Guide Version 1.0. Available at: www.ars-grin.gov/npgs/gringlobal/docs/gg_curator_tool_user_guide.pdf (accessed 18 September 2014).

GRIN-Global (2015a) Public website. Available at: www.ars-grin.gov/npgs/gringlobal/docs/gg_public_website.pdf (accessed 9 July 2015).

GRIN-Global (2015b) Cookbook for Data Importing using the Admin Tool's Import Wizard. Available at: www.ars-grin.gov/npgs/gringlobal/docs/gg_cookbook.pdf (accessed 9 July 2015).

Postman, J., Hummer, K., Ayala-Silva, T., Bretting, P., Franko, T., *et al.* (2010) GRIN-Global: an international project to develop a global plant genebank information management system. *ISHS Acta Horticulturae* 859, 49–55.

USDA (2014) International Project to Develop a Global Plant Genebank and Information Management System, United States Department of Agriculture. Available at: ars.usda.gov/is/br/grin/index.htm (accessed 18 September 2014).

# 42 The GRIN Taxonomy Crop Wild Relative Inventory

J.H. Wiersema[1]* and B. León[2]

[1]US Department of Agriculture, Agricultural Research Service, National Germplasm Resources Laboratory, Beltsville, Maryland, USA; [2]Plant Resources Center, University of Texas at Austin, Austin, Texas, USA

## 42.1 Introduction

The recognized link between cultivated crops and wild related species is well established. The genetic diversity of the wild relatives of a crop provides the raw material for its genetic improvement to increase yield, provide disease or pest resistance, improve nutrition or processing quality of products, or reduce environmental effects from cultivation (Gepts, 2006). Faced with the ever-increasing threats of habitat loss or degradation and the uncertain future effects of climate change, the urgency to identify and conserve crop wild relatives (CWR) for the benefit of current and future agricultural crop production and protection has never been greater. The strategy of the *ex situ* conservation of plant genetic resources is crucial to agricultural research and development (Bömer, 2006), and the National Plant Germplasm System (NPGS) of the US Department of Agriculture's Agricultural Research Service has been an important participant in this activity since its inception in the mid-1900s (Shands, 1995). The NPGS is charged with acquiring and conserving the genetic diversity of all crop plants important to US agriculture, and over 565,000 accessions of some 14,700 species are already being preserved *ex situ* in NPGS.

To assist with the preservation of CWR, the Plant Exchange Office (PEO) of the NPGS's National Germplasm Resources Laboratory initiated in late 2008 a project to identify systematically the CWR of major and minor US crops. The Germplasm Resources Information Network (GRIN: npgsweb.ars-grin.gov) of the NPGS would then provide the means to link passport data on existing accessions with the accumulated taxonomic data on CWR from this project. These data, taken together, would allow us to evaluate gaps in the current NPGS holdings of CWR and set priorities for future germplasm conservation strategies.

Here, we provide a discussion of the concepts and terminology that underlie the CWR classification of this project. We also outline the procedures for compiling and evaluating this classification and discuss the public accessibility of the resulting data.

## 42.2 Classifying Crop Wild Relatives

In order to identify and evaluate the relative importance of CWR for various crops, it is necessary to establish a method of classifying the crop's relatives that prioritizes each species potential to donate genes for crop improvement.

*Corresponding author; e-mail: john.wiersema@ars.usda.gov

 

Maxted *et al.* (2006) have defined a crop wild relative as 'a wild plant taxon that has an indirect use derived from its relatively close genetic relationship to a crop', thereby making its preservation critical for current and future crop breeding. For decades, a classification proposed by Harlan and de Wet (1971) has offered a practical means of evaluating, for plant breeding purposes, the genetic resources available for crop improvement by defining the primary, secondary and tertiary genepools of a crop's relatives on the basis of their crossability with that crop (e.g. Gepts, 2006). While the genepool concept of Harlan and de Wet (1971) was a logical starting point for our classification of CWR, their genepool definitions can be applied to particular CWR only if their hybridization potential with a crop has been evaluated. However, crosses between many CWR and their related crop species have never been attempted, and crossability alone may not necessarily reflect genetic affinities. The limitation imposed by a lack of crossing data led Maxted *et al.* (2006) to develop the 'taxon group' concept, which classifies CWR into taxon groups based on their taxonomic relationship with the crop, with members of the same species, series or section, subgenus and genus being assigned to taxon groups one, two, three and four, respectively.

Although an acceptable infrageneric classification to apply the 'taxon group' concept does exist for some crop genera (e.g. *Cicer*: Davies *et al.*, 2007; *Hordeum*: Blattner, 2009), other genera lack such a classification (e.g. *Fragaria*: Potter *et al.*, 2000; Staudt, 2009), or the existing classification, traditionally based on morphology, is not supported by the most recent molecular phylogenetic studies (e.g. *Actinidia*: Li *et al.*, 2009; *Cucumis*: Ghebretinsae *et al.*, 2007; Renner *et al.*, 2007; Schaefer, 2007; *Daucus*: Spooner *et al.*, 2013; Arbizu *et al.*, 2014; *Helianthus*: Timme *et al.*, 2007; *Malus*: Robinson *et al.*, 2001, Forte *et al.*, 2002; *Saccharum*: Hodkinson *et al.*, 2002) and may remain in a state of flux, making the taxon group concept difficult to apply precisely. Also, for taxon groups with numerous members, such as the genus *Solanum*, a grouping of this size can become unworkable in practice (Knapp *et al.*, 2013). In such genera, we can make use of data from phylogenetic studies, which, even if not providing a fully resolvable classification at present, provide a more robust insight into potential genetic relationships among CWR and related crops than traditional plant classification alone. Other useful information on ploidy level, reproductive biology, natural hybridization/introgression, etc., can also offer clues to the genetic relatedness between individual CWR and their associated crop.

Therefore, despite the absence of actual crossing data, the degree of genetic relatedness of CWR to their crops, as might be reflected in the genepool status, can be estimated based on other information. The resulting CWR categorization thereby integrates Harlan and de Wet's (1971) genepool concept, for those CWR with available crossing data, with a taxon group concept (Maxted *et al.*, 2006) enhanced by data in addition to taxonomic classification, for other prospective CWR, into a single coherent classification. The resulting classification is thus descriptive for CWR of known crossability and predictive for CWR of unknown or uncertain hybridization potential. The hypothesized 'genepool' standing for CWR species not previously represented in crop hybridization studies can be directly evaluated by future researchers, and this classification upheld or refuted. Because we have defined our classes of crop/CWR genetic relatedness to include individuals lacking crossing data, we have refrained from labelling these groupings as 'genepools', preferring instead to designate them as primary, secondary and tertiary 'genetic relative' classes. Our treatment of the first pomological crops made it apparent that a fourth class of genetic relatives was necessary to account for CWR important for graft stock breeding and/or usage (Wiersema *et al.*, 2012), so the CWR classification was expanded to accommodate this group. These four classes are defined in Table 42.1.

These genetic relative classes provide the best approximation of the Harlan and de Wet genepools for a particular crop, based on a synthesis of all of the available evidence. Nevertheless, assigning a CWR to a particular genetic relative class using phylogenetic data, when crossing data are lacking, may overestimate the hybridization potential, that is, the genepool status, of that CWR. Taxa, or even individuals assigned to the same taxon, that are closely related in morphology or phylogeny may not necessarily be hybridized easily, due to differences in ploidy. This can be seen in our data for the crop lucerne,

Table 42.1. Definitions of four classes of genetic relatives of crops used by PEO to evaluate the relative importance of CWR. (From Wiersema *et al.*, 2012.)

| Genetic relative status | Definition |
|---|---|
| Primary | Taxa that cross readily with the crop (or can be predicted to do so based on their taxonomic relationships), yielding (or being expected to yield) fertile hybrids with good chromosome pairing, making gene transfer through hybridization simple. |
| Secondary | Taxa that will cross successfully with the crop (or can be predicted to do so based on their taxonomic relationships), but yield (or would be expected to yield) partially or mostly sterile hybrids with poor chromosome pairing, making gene transfer through hybridization difficult. |
| Tertiary | Taxa that can be crossed with the crop (or can be predicted to do so based on their taxonomic relationships), but hybrids are (or are expected to be) lethal or completely sterile. Special breeding techniques, some yet to be developed, are required for gene transfer. |
| Graft stock | Taxa used as rootstocks for grafting scions of a crop, or used as genetic resources in the breeding of such rootstocks. |

where tetraploids and diploids, which would otherwise be assigned to the same taxon, differ in their crossability with tetraploid lucerne. As this crop's data illustrate, such intra-taxon crossing differences can be accounted for in the CWR classification. In cotton, reproductive barriers apart from ploidy differences appear to be involved in inhibiting hybridization between related taxa (Pereira *et al.*, 2012). Nevertheless, owing to continuing advancement in the biotechnology of plant breeding, close relatives of a crop that may not now be readily crossed with it have the potential to become increasingly viable as potential gene donors, so an overestimation of their hybridization potential may become less relevant.

By the definition quoted above, which we have employed in our classification, CWR are 'wild' taxa. With few exceptions, we have therefore classified only taxa with a documented native or naturalized distribution among our CWR. In other words, other crops known only in cultivation cannot be considered as CWR of a given crop. However, for crop taxa that contain both wild and cultivated forms, the wild forms are treated as CWR for that crop or other related crops. Subcrops, such as broccoli, cauliflower, kohlrabi, etc., with essentially the same CWR within a crop grouping, such as cabbage/kale, are combined in a single classification. Crops such as groundnut (*Arachis hypogaea*), potato (*Solanum tuberosum*) or *Capsicum* peppers (*Capsicum* spp.) are treated similarly.

A CWR classification was developed in GRIN for each crop based on a thorough review of all pertinent taxonomic and phylogenetic literature, as well as reviews of the plant genetic, plant breeding, or other crop science literature. For each CWR, the resulting information was integrated to assign the appropriate genetic relative class. Genetic relative assignments are supported in GRIN by the literature or other references, and complete country/state occurrence data for both native and naturalized distributions have been gathered and documented in GRIN for all CWR. After the initial CWR genetic relative classification for a crop was developed, validation by specialists for that crop was solicited, usually with the assistance of NPGS crop curators or crop germplasm committees.

## 42.3 Current Progress of the Project

While the initial focus of our project was on the major agronomic crops such as wheat, maize, rice, sorghum, soybeans and lucerne, our attention gradually shifted to horticultural crops. Because most American agricultural production is in temperate regions, our initial concentration was also on temperate crops; however, several tropical crops have now been evaluated. To date, the CWR of 130 major and minor crops spanning 71 different genera have been classified. While those evaluated now include most major temperate food crops, as well as tropical ones such as cacao, cassava, coffee or papaya, little or no effort has yet been directed to forages or ornamentals, which are also important in plant breeding (Tay, 2007; Huyghe, 2010). Once the CWR classification of a crop is in place and the native/naturalized range of the CWR are documented in GRIN Taxonomy, comparisons of these data to GRIN accession passport data become possible. A tool to enable such comparisons has been developed for access by PEO and other NPGS personnel as an aid in identifying gaps in the NPGS collection and setting priorities for future germplasm acquisition. An additional tool to permit public access to the underlying CWR classification data has also been developed and is now available through GRIN Taxonomy (npgsweb.ars-grin.gov/gringlobal/taxon/taxonomysearchcwr.aspx). The web interface provides the opportunity to query these data in several ways: (i) by crops; (ii) by family or generic taxonomy; (iii) by genetic relative status; (iv) by country, region or continent of geographical distribution; or even (v) by NPGS germplasm repositories where crops are maintained. The resulting report enumerates all cultivated members (subcrops) of a particular crop, as well as all CWR for each genetic relative class selected. When geographical criteria are added to a query, the report evaluates only those selected CWR known to occur (based on GRIN distribution data) within the specified area. These reports thereby allow for the identification, for a given set of crops, of any genetic relative class or classes of CWR known to occur in a given country or region. Such data may assist agricultural workers in various countries in establishing conservation priorities for their CWR. Indeed, much of the data has already been made available to colleagues reporting on CWR globally (Vincent *et al.*, 2013) or nationally (Khoury *et al.*, 2013).

Integration and consolidation of the GRIN CWR data with those of the Harlan and de Wet Crop Wild Relative Inventory (www.cwrdiversity.org/checklist/) is planned in the future. We encourage the use of and, for those with specialized knowledge of a particular crop, the evaluation of these data. The external review of existing CWR classification data and the contribution of data on other crops are of crucial importance to this endeavour. We welcome and will acknowledge on the GRIN website any assistance from crop specialists in further developing and/or reviewing our CWR classification.

## 42.4 Conclusions

Procedures have been developed for the classification of CWR into primary, secondary, tertiary and graft stock groupings to express their degree of genetic relatedness to the associated crops. The CWR of a number of principal agronomic and horticultural crops mainly of interest to US agriculture, but of worldwide impact, have been evaluated. The data, when utilized in conjunction with GRIN country/state occurrence data for those CWR, have allowed the detection of gaps in the National Plant Germplasm System on a country-by-country basis and provided guidance for future germplasm conservation activities. Online public access to these CWR data is now available through GRIN Taxonomy, and this dynamic interface should enhance the usefulness of our CWR classification.

## Acknowledgements

Thanks are due to the EU Seventh Framework Programme Project, PGR Secure 'Characterization of biodiversity resources for wild crop relatives to improve crops by breeding' (Grant Agreement No 266394) for facilitating and supporting the participation of the senior author in the PGR Secure project and its closing symposium, and to Carole A. Ritchie and Gary R. Kinard for critically reviewing the manuscript.

## References

Arbizu, C., Ruess, H., Senalik, D., Simon, P.W. and Spooner, D.W. (2014) Phylogenomics of the carrot genus (*Daucus*, Apiaceae). *American Journal of Botany* 101, doi/10.3732/ajb.1400106.

Blattner, F.R. (2009) Progress in phylogenetic analysis and a new infrageneric classification of the barley genus *Hordeum* (Poaceae: Triticeae). *Breeding Science* 59, 471–480.

Bömer, A. (2006) Preservation of plant genetic resources in the biotechnology era. *Biotechnology Journal* 1, 1393–1404.

Davies, A.M.R., Maxted, N. and Van Der Maesen, L.J.G. (2007) A natural infrageneric classification for *Cicer* (Leguminosae, Cicereae). *Blumea* 52, 379–400.

Forte, A.V., Ignatov, A.N., Ponomarenko, V.V., Dorokhov, D.B. and Savelyev, N.I. (2002) Phylogeny of the *Malus* (apple tree) species, inferred from the morphological traits and molecular DNA analysis. *Russian Journal of Genetics* 38, 1150–1160.

Gepts, P. (2006) Plant genetic resources conservation and utilization: the accomplishments and future of a societal insurance policy. *Crop Science* 46, 2278–2292.

Ghebretinsae, A.G., Thulin, M. and Barber, J.C. (2007) Relationships of cucumbers and melons unraveled: molecular phylogenetics of *Cucumis* and related genera (Benincaseae, Cucurbitaceae). *American Journal of Botany* 94, 1256–1266.

Harlan, J.R. and de Wet, J.M.J. (1971) Toward a rational classification of cultivated plants. *Taxon* 20, 509–517.

Hodkinson, T.R., Chase, M.W., Lledó, D.M., Salamin, N. and Renvoize, S.A. (2002) Phylogenetics of *Miscanthus*, *Saccharum* and related genera (Saccharinae, Andropogoneae, Poaceae) based on DNA sequences from ITS nuclear ribosomal DNA and plastid trnL intron and trnL-F intergenic spacers. *Journal of Plant Research* 115, 381–392.

Huyghe, C. (2010) *Sustainable Use of Genetic Diversity in Forage and Turf Breeding*. Springer, Dordrecht, the Netherlands.

Khoury, C.K., Greene, S., Wiersema, J., Maxted, N., Jarvis, A., *et al.* (2013) An inventory of crop wild relatives of the United States. *Crop Science* 53, 1–13.

Knapp, S., Vorontsova, M.S. and Prohens, J. (2013) Wild relatives of the eggplant (*Solanum melongena* L.: Solanaceae): new understanding of species names in a complex group. *PloS ONE* 8(2), e57039.

Li, X.-W., Li, J.-Q. and Soejarto, D.D. (2009) Advances in the study of the systematics of *Actinidia* Lindley. *Frontiers of Biology in China* 4, 55–61.

Maxted, N., Ford-Lloyd, B.V., Jury, S., Kell, S. and Scholten, M. (2006) Towards a definition of a crop wild relative. *Biodiversity and Conservation* 15, 2373–2685.

Pereira, G.S., Sousa, R.L., Araújo, R.L., Hoffmann, L.V., Silva, E.F., *et al.* (2012) Selective fertilization in interspecific crosses of allotetraploid species of *Gossypium*. *Botany* 90, 159–166.

Potter, D., Luby, J.J. and Harrison, R.E. (2000) Phylogenetic relationships among species of *Fragaria* (Rosaceae) inferred from non-coding nuclear and chloroplast DNA sequences. *Systematic Botany* 25, 337–348.

Renner, S.S., Schaefer, H. and Kocyan, A. (2007) Phylogenetics of *Cucumis* (Cucurbitaceae): cucumber (*C. sativus*) belongs in an Asian/Australian clade far from melon (*C. melo*). *BMC Evolutionary Biology* 7, 58, doi:10.1186/1471-2148-7-58.

Robinson, J.P., Harris, S.A. and Juniper, B.E. (2001) Taxonomy of the genus *Malus* Mill (Rosaceae) with emphasis on the cultivated apple, *Malus domestica* Borkh. *Plant Systematics and Evolution* 226, 35–58.

Schaefer, H. (2007) *Cucumis* (Cucurbitaceae) must include *Cucumella*, *Dicoelospermum*, *Mukia*, *Myrmecosicyos*, and *Oreosyce*: a recircumscription based on nuclear and plastid DNA data. *Blumea* 52, 165–177.

Shands, H.L. (1995) The U.S. National Plant Germplasm System. *Canadian Journal of Plant Science* 75, 9–15.

Spooner, D.M., Rojas, P., Bonierbale, M., Mueller, L.A., Srivastav, M., *et al.* (2013) Molecular phylogeny of *Daucus*. *Systematic Botany* 38, 850–857.

Staudt, G. (2009) Strawberry biogeography, genetics and systematics. *Acta Horticulturae* 842, 71–84.

Tay, D. (2007) Herbaceous ornamental plant germplasm conservation and use: theoretical and practical treatments. In: Anderson, N.O. (ed.) *Flower Breeding and Genetics: Issues, Challenges and Opportunities for the 21st Century*. Springer, Dordrecht, the Netherlands, pp. 113–175.

Timme, R.E., Simpson, B.B. and Linder, C.R. (2007) High-resolution phylogeny for *Helianthus* (Asteraceae) using the 18S–26S ribosomal DNA external transcribed spacer. *American Journal of Botany* 94, 1837–1852.

Vincent, H., Wiersema, J., Kell, S., Fielder, H., Dobbie, S., *et al.* (2013) A prioritized crop wild relative inventory to help underpin global food security. *Biological Conservation* 167, 265–275.

Wiersema, J.H., León, B. and Garvey, E.J. (2012) Identifying wild relatives of subtropical and temperate fruit and nut crops. *Acta Horticulturae* 948, 285–288, doi:10.17660/ActaHortic.2012.948.33.

# Index